W9-ACU-882

Progress in Mathematics
Volume 92

Series Editors
J. Oesterlé
A. Weinstein

Operator Algebras, Unitary Representations, Enveloping Algebras, and Invariant Theory

Actes du colloque en l'honneur de Jacques Dixmier

Edited by
Alain Connes
Michel Duflo
Anthony Joseph
Rudolf Rentschler

1990

Birkhäuser
Boston • Basel • Berlin

Alain Connes
Institut des Hautes Etudes Scientifiques
91440 Bures-sur-Yvette
France

Michel Duflo
Université Paris 7
75251 Paris Cedex 05
France

Anthony Joseph
Université Pierre et Marie Curie
Paris, France, and
Department of Mathematics
The Weizmann Institute of Science
Rehovot 76100
Israel

Rudolf Rentschler
Université Pierre et Marie Curie
Laboratoire de Mathématiques Fondementales
75230 Paris Cedex 05
France

Library of Congress Cataloging-in-Publication Data
Operator algebras, unitary representations, enveloping algebras, and
invariant theory : actes du colloque en l'honneur de Jacques Dixmier
/ editors, Alain Connes . . . [et al.].
p. cm. – (Progress in mathematics ; v. 92)
"Articles contributed . . . at the colloquium celebrating the sixty-fifth birthday of Professor Jacques Dixmier held during 22–26 May 1989 at the Institute Henri Poincaré in Paris" – Pref.
Includes bibliographical references.
ISBN 0-8176-3489-4. – ISBN 3-7643-3489-4
1. Lie groups – Congresses. 2. Lie algebras – Congresses.
3. Operator algebras – Congresses. 4. Invariants – Congresses.
I. Dixmier, Jacques. II. Connes, Alain. III. Series: Progress in mathematics (Boston, Mass.) ; vol. 92.
QA387.064 1990
512'.55 – dc20 90-49716

Printed on acid-free paper.

© 1990 Birkhäuser Boston.
All rights reserved. This work may not be translated or copied in whole or in part without the written permission of the publisher (Birkhäuser Boston, 675 Massachusetts Avenue, Cambridge, Massachusetts 02139, USA), except for brief excerpts in connection with reviews or scholarly analysis. Use in connection with any form of information storage and retrieval, electronic adaptation, computer software, or by similar or dissimilar methodology now known or hereafter developed is forbidden.
Permission to photocopy for internal or personal use, or the internal or personal use of specific clients, is granted by Birkhäuser, Boston for libraries registered with the Copyright Clearance Center (CCC), provided that the base fee of $0.00 per copy, plus $0.20 per page is paid directly to CCC, 21 Congress St., Salem, MA 01970, USA. Special requests should be addressed directly to Birkhäuser Boston, 675 Massachusetts Avenue, Cambridge, Massachusetts 02139, USA.

Camera-ready copy provided by the editors.
Printed and bound by Edwards Brothers, Inc., Ann Arbor, MI.
Printed in the United States of America.

9 8 7 6 5 4 3 2 1

ISBN 0-8176-3489-4
ISBN 3-7643-3489-4

C stack
SCIMON

Jacques Dixmier, 1990

Allocution de Jacques Dixmier

le 26 Mai 1989

Mesdames, Messieurs,

Récemment, j'ai assisté au colloque organisé pour les 60 ans du professeur Springer. Dans son allocution, Springer s'est demandé : qu'est-ce que le nombre 60 a donc de particulier ? Immédiatement, Jacques Tits a remarqué que 60 est le nombre d'éléments du premier groupe simple. Aussi, quand j'ai appris l'existence du présent colloque, j'ai feuilleté mon catalogue des groupes simples : pas trace d'un groupe simple à 65 éléments! J'ai interrogé mon ordinateur : 65 est-il premier ? Non, m'a-t-il répondu. J'allais abandonner, quand j'ai lu quelque part que le nombre d'électrons de l'univers est 2^{65}. Ce rapprochement entre mon âge et une constante cosmique m'a paru éblouissant, et justifiant un colloque d'une certaine ampleur.

J'ai donc 65 ans, aujourd'hui même. J'en suis un peu surpris, comme tous ceux, sans doute, qui font la même expérience. Voici quelques souvenirs personnels.

J'ai eu au lycée, de la classe de 5ème à la classe de mathématiques spéciales, d'excellents professeurs de mathématiques : MM. Schirmer, Bellivier, Chrétien, Vasseur, Maillard. C'est à M. Schirmer, mon professeur de la 5ème à la 3ème, que je dois ma vocation mathématique. C'était un homme chaleureux. Il est mort assez jeune. Je doute qu'il m'entende ; et, s'il m'entend, il faut espérer que cela lui est indifférent. Néanmoins, je tenais à lui rendre hommage aujourd'hui. A l'Université, j'ai été influencé par Gaston Julia, qui m'a orienté vers l'étude des espaces hilbertiens, et par les leçons de Henri Cartan à l'école Normale. Nous étions, étudiants à l'école Normale, tout disposés à travailler, et même à travailler beaucoup, mais dans une direction peut-être un peu routinière ; les leçons décapantes de Cartan secouaient cette routine ; nous ne suivions pas toujours, mais l'effet global était considérable. Par exemple, Cartan nous a fait plusieurs cours sur les groupes de Lie ; vous imaginez comme il était difficile d'enseigner à des débutants, en 1943, des résultats substantiels sur les groupes de Lie ; Cartan y réussissait. Pour vous prouver mon antiquité, je mentionne que j'ai aussi suivi à l'école Normale des conférences d'Elie Cartan et des conférences de Louis de Broglie. Après la rédaction de ma thèse, j'ai lu les mémoires de von Neumann et de Murray-von Neumann ; et, là encore, j'ai eu de la chance, car ces mémoires sont rédigés de manière exceptionnellement claire. Ça amusera sans doute les chercheurs en algèbres d'opérateurs de savoir que, vers 1950, nous n'étions guère plus d'une demi-douzaine de chercheurs dans le monde entier, dont 2 en France, sur ce sujet. Je ne compte pas Murray ni von Neumann qui, autant que je sache, ne travaillaient plus alors dans cette direction. Il est vrai que le mémoire de von Neumann sur la théorie de la réduction est paru en 1949, mais il était rédigé depuis 1939. J'ai dit : 2 chercheurs en France. Il s'agissait de Roger Godement et moi ; et c'est le moment de signaler que Godement m'a beaucoup influencé ; entre autres choses, il m'a initié aux représentations des groupes localement compacts. Notre bible à l'époque était le livre d'André Weil sur l'intégration dans les groupes topologiques. Un peu plus tard, un mémoire de Kaplansky a joué pour moi un grand rôle ; c'est après sa lecture que les C*-algèbres sont devenues pour moi des objets familiers. Puisque je parle des influences qui m'ont marqué, je dois citer Bourbaki. Chacun devine ce que peuvent apporter des conversations quotidiennes avec Cartan, Chevalley, Dieudonné, Schwartz, Weil, etc. D'autre part, et pour moi ce n'est pas un détail, Bourbaki m'a appris à rédiger. Ecrire un texte, et puis le voir épluché, ligne à ligne, pendant des heures, amicalement et férocement, voilà une expérience irremplaçable, et je regrette que ce genre de discipline soit si peu répandu. Bien des gens vont sourire, en pensant que le mode de rédaction de Bourbaki, hum ..... Soit ! De toutes façons, la perfection en matière de rédaction est inaccessible. Mais au moins, il y a quantité d'erreurs élémentaires, d'erreurs parfois grossières, qu'il faut apprendre à éviter ; eh bien, Bourbaki m'a appris à en éviter pas mal. Et à cet égard, certains des critiques les plus virulents de Bourbaki feraient bien, à mon sens, de relire dans les coins leurs propres mémoires. Comme influence encore que j'ai subie, mais cette fois inconsciemment, il y a eu le tournant des mathématiques vers des tendances plus concrètes. Quand j'étudiais les algèbres d'opérateurs, les exemples étaient rares et difficiles à décortiquer complètement. Quand je suis passé sérieusement aux groupes de Lie, j'ai constaté avec surprise qu'on pouvait faire des expériences, quantité d'expériences ; dresser des plans complets d'expériences. A partir de là, j'ai écrit au brouillon des milliers de pages de calcul. Des milliers, certainement ; sans doute des dizaines de milliers. Attention : je n'affirme pas que c'est un exemple à suivre ; je n'oserais pas donner un conseil à ce sujet.

J'ai enseigné à l'Université de 23 ans à 60 ans. J'ai donc été témoin de bien des changements ; inutile d'insister, c'est une constatation banale ; disons seulement que, dans les années 50, l'Université permettait à ses professeurs de faire de la recherche sans acrobaties ; dans les années 80, les obstacles à la recherche, sous forme de commissions, conseils, assemblées générales, scrutins, examens partiels, etc ... , sont devenus difficilement supportables. Je n'en ai apprécié que mieux le paradis pour mathématiciens que constitue l'I.H.E.S.

J'ai eu d'excellents élèves de thèse. Notre travail en commun a été pour moi très enrichissant, et je les en remercie aujourd'hui.

Ce colloque est un succès. J'ai été souvent surpris cette semaine, en constatant que des sujets avec lesquels j'ai perdu contact se sont renouvelés en quelques années. Même pour ceux d'entre vous qui sont plongés dans ces sujets, je pense que le colloque a apporté beaucoup d'informations. Cette vitalité scientifique est, pour quelqu'un qui vieillit, une source bien utile d'optimisme. Nous entendons tous les jours des critiques acerbes de la science, souvent justifiées. Il y a des réponses traditionnelles : beaucoup d'entre nous seraient morts à l'heure actuelle sans les progrès de la médecine ; etc . Je pense à d'autres réponses. L'humanité sera confrontée un jour ou l'autre à des dangers (climatologiques, biologiques, cosmiques ...) auxquels bien d'autres espèces ont succombé. Alors, de deux choses l'une. Ou bien on freine le développement de la science ; alors ces grandes fluctuations, inévitables, auxquelles j'ai fait allusion seront catastrophiques, en fait mortelles. Ou bien on accepte le développement scientifique, et l'on peut au moins espérer que la science maîtrisera ces dangers, et maîtrisera de plus les dangers que son développement même suscitera. Il est peut-être ridicule, je m'en rends compte, de s'inquiéter de l'avenir lointain ; mais je n'arrive pas, je l'avoue, à me désintéresser du résultat de milliards d'années d'évolution. Bref, je crois naïvement à la science, et je suis, de ce point de vue, un homme du 19ème siècle. Du 19ème, et même du 18ème, car voici ce qu'écrivait notre collègue Condorcet : "A force de temps et d'efforts, l'homme a pu enrichir son esprit de vérités nouvelles, perfectionner son intelligence, étendre ses facultés, apprendre à les mieux employer, et pour son bien-être, et pour la félicité commune". Soit dit en passant, Condorcet a écrit ces lignes quelques semaines avant de mourir en prison. Je parlais de ma naïveté. Je ne suis pas, personne ne peut plus être, hélas, aussi naïf que l'était Condorcet.

En votre nom à tous, je remercie les organismes qui ont aidé à la préparation du colloque : le C.N.R.S., les ministères des affaires étrangères, de la recherche et de la technologie, de l'éducation nationale ; l'université Paris 6, les U.E.R. 47 et 48. Je remercie Madame Brohan qui a bien voulu ajouter à ses tâches habituelles le secrétariat du colloque. Je remercie les conférenciers : ils m'ont fait un immense honneur en parlant ici. Je remercie le comité d'organisation et le comité scientifique. Je sais la somme de temps et d'efforts que demande la préparation d'un tel colloque. Je rappelle les noms des organisateurs : Nicole Berline, Alain Connes, Michel Duflo, Alain Guichardet, Anthony Joseph, Colette Moeglin, Rudolf Rentschler, Jean-Marie Schwartz, Michèle Vergne. J'espère assister, plus jeune que jamais, aux futurs colloques qui marqueront leurs 65 ans.

Preface

The present volume is a collection of twenty-two research articles contributed by the speakers at the colloquium celebrating the sixty-fifth birthday of Professor Jacques Dixmier held during 22-26 May 1989 at the Institute Henri Poincaré in Paris. The guiding principle of the conference was to bring together specialists in the four great areas of research which have occupied Dixmier during his long and fruitful scientific career. These were the following. One, operator algebras in the C^* and von Neumann sense. In these, Dixmier founded his first school or research. Two, the theory of unitary Lie group representations which he notably advanced in the nilpotent case, a work which was to inspire the orbit method. Three, enveloping algebras of Lie algebras in which he instigated the study of the primitive spectrum as a natural algebraic alternative to the unitary dual. These led to the formation of two intermingling schools of research at Paris 6 and Paris 7. Four, invariant theory to which he was drawn by the desire to understand the relationship between the geometry of group actions and representation theory in which he has a continued and very active interest.

The colloquium was followed with great enthusiasm. Most of the speakers had been in close contact with Dixmier and his scientific work for a long period. Together with younger colleagues they have contributed original works of great quality which afford a fitting tribute to Dixmier's exemplary career and in whose contents the reader may glean many significant developments in current research. Most of the articles are based on the talks of the speakers; whereas three articles (De Concini-Kac, Kazhdan, Littelmann-Procesi) are different contributions.

In the theory of operator algebras a central theme has been the classification of factors defined in the classical work of F. Murray and J. von Neumann. The article of E. Størmer classifies equivalence classes of normal states of a von Neumann algebra M through the centre Z of a certain continuous crossed product of M over $\mathbb{R}$. He gives an injective map from equivalence classes to a positive cone in Z_* and describes its image for type II and type III factors. The paper of M. Takesaki provides a proof of a result announced by A. Connes concerning the structure of the automorphism group of an AFD factor of type III. The contribution of D. Voiculescu is in developing probabilistic methods to study type II_1 factors of free groups.

The natural interdependence of Lie group representation theory and of enveloping algebras, which formed the second and third subjects of Dixmier's research, is reflected in the contributed papers which span across analytic, geometric and algebraic methods. Therefore we have made no attempt to draw even the finest dividing line between them.

The work of L. Pukanszky formed a bridge between C^*-algebras and the representation theory of Lie groups. It was a lifetime contribution to the classification of the primitive spectrum of the C^*-algebra of a Lie group. In his paper Pukanszky gives a historical survey on the character theory developed from orbital integration as prescribed by the orbit correspondence. The article of A.A. Kirillov details a number of interesting, new and important constructions which may prove crucial to the orbit method for the infinite dimensional Lie groups arising from diffeomorphisms of varieties. For example he describes the symplectic and invariant complex structures on co-adjoint orbits for the Bott-Virasoro group.

B. Kostant examines the minimal unitary representation of $SO(4,4)$. One key aspect is that the unitary structure is derived from the triality automorphism and a second aspect is a relation with differential geometry involving the vanishing of scalar curvature on 6 manifolds. Another interest arises from space-time geometry and the fulfillment of Einstein's equations. D. Kazhdan constructs a unitary representation associated to a minimal unipotent orbit for a simple quasi-split group of type D_4 defined over a degree 3 extension E of p-adic field F. An important point is that restriction gives an explicit construction of the correspondence between unitary representations of $SL_3(F)$ and the characters of the group of norm 1 elements of E.

Two results were reported that were developed from the theory of finite dimensional Lie groups through a study of infinite dimensional flag varieties. Thus M. Kashiwara and T. Tanisaki give a proof of the Kazhdan-Lusztig conjectures as extended to symmetrizable Kac-Moody Lie algebras. Again V. Lakshmibai describes a standard monomial theory for the Kac-Moody Lie group $\widehat{SL_n}$. This gives a characteristic free basis for modules arising as global sections of line bundles on the flag variety of this group.

The contribution of C. Procesi is joint work with P. Littelmann. It describes in some detail the G equivariant cohomology ring $H^*_{G\times G}(X, \mathbb{Q})$ where G is a reductive algebraic group and X is what they construct and call a wonderful compactification of G.

W. Rossmann computes, up to a slight restriction, the canonical measure on a real nilpotent coadjoint orbit through a polynomial on a Cartan subalgebra which transforms according to the appropriate Springer representation. This extends the work of Barbasch-Vogan developed in the complex case, though the method of Rossmann based on an integration theory of equivariant forms is more intrinsic. R. Brylinski studies the degeneration of line bundles on a regular semisimple orbit to these on the regular nilpotent orbit N° in the dual of the Lie algebra of a complex semisimple Lie group G. This recovers a graded refinement of Kostant's classical result concerning the G module structure of the ring of regular functions on the closure N of N°. It is further used to describe the ideal of definition of $N \setminus N^{\circ}$.

D. Barbasch computes a character formula for unipotent representation arising from complex exceptional Lie groups when the corresponding left cell in the Weyl group is not multiplicity-free. In earlier work D.A. Vogan has introduced the concepts of Dixmier algebras and orbit data with an aim of extending the bijectivity property of the Dixmier map from the solvable case. In the work reported here he studies the behaviour of these objects under induction procedures. This aims at a sheet-like classification and a reduction of their conjectured interrelation to the rigid case. Open questions are whether distinct sheets of orbit data are disjoint and if induction of Dixmier algebras is only dependent on the Levi factor. W. McGovern constructs notably three Dixmier algebras which are common bedfellows of just three violations of a natural rule connecting Weyl group representations associated to a special nilpotent orbit and to it Langlands' dual. One of these is realized as a primitive quotient arising from a Lie algebra of type D_4 containing as its S_3 fixed subalgebra a primitive quotient in type G_2.

J. Bernstein constructs a trace functor on the category of $U(\mathfrak{g})$ modules (for $\mathfrak{g}$ complex semisimple). This gives a simple proof of W. Soergel's crucial result describing $End\ P$ as the cohomology algebra on the flag variety. Here P is the projective cover in the $\mathcal{O}$ category of the Verma module of highest weight $\mathcal{O}$.

The contribution of J.-E. Björk is joint work with E.K. Ekström. It describes a Gabber filtration $\mathcal{G}$ on a finitely generated module M over a filtered Auslander Gorenstein ring A. This has the great virtue that purity passes to graded modules. Thus in particular if M is simple and grA is commutative, then $Ann\ gr_{\mathcal{G}}\ M$ has no embedded primes. In his paper T. Levasseur considers the action of a reductive algebraic group G on an affine variety V. He gives a sufficient condition for a differential operator on $V//G$ to be the image of an invariant differential operator on V and having the same order. This recovers both an earlier result of his concerning the surjectivity of the restriction map $\mathcal{D}(V)^G \to \mathcal{D}(V//G)$ when G is finite and has no pseudo reflections and a surjectivity result of Levasseur-Stafford concerning enveloping algebras.

V. Kac and C. de Concini study the quantum group associated to a semisimple Lie algebra. In particular they show that the "minimal" specialization at a root of unity is finite dimensional over its centre and by this means are able to describe its finite dimensional modules.

Some of the last topics interleave with those of the remaining section of this volume devoted to invariant theory. In this the contribution of M. Brion is joint work with C. Procesi. It concerns the action of a torus T on the projective variety $P(V)$ associated to T module V. Through data involving the T weights of V, they define stable and semistable points and apply these considerations to the asymptotic behaviour of the dimension of T weight spaces of graded $S(V)$ modules. Such questions relate to identifying associated varieties of highest weight modules.

V.L. Popov considers G modules V for which a G orbit in V being closed implies finiteness of the stabilizer of one point. He lists all such occurrences when G is a connected simple algebraic group and V a faithful G module for which $V^G = 0$. H. Kraft shows that any algebraic $\mathbb{C}^*$ action on $\mathbb{C}^3$ is linearizable whenever $\mathbb{C}^3/\mathbb{C}^*$ is smooth. It is not known if this still holds without the smoothness hypothesis; while work of G. Schwarz reported at this meeting (but not published in this volume) gives non-linearizable actions of certain algebraic groups on $\mathbb{C}^n$.

The conference was supported by the Ministries of Education, of Foreign Affairs and of Research and Technology. Further support came from the C.N.R.S., the University of P. and M. Curie and several research groups within that university.

The editors would like to thank the Minister of Research and Technology, Monsieur Hubert Curien for his speech in honour of Professor Dixmier and Madame Janine Brohan for her skillful and devoted handling of secretarial and organisational matters.

Paris, July 17th 1990
The Editors

Contents

Allocution de Jacques Dixmier vii
Preface xi

CHAPTER I: $\mathbb{C}^*$-ALGEBRAS

E. Størmer 1
Equivalence of Normal States of von Neumann Algebras

M. Takesaki 19
The Structure of the Automorphism Group of an AFD Factor

D. Voiculescu 45
Circular and Semicircular Systems and Free Product Factors

CHAPTER II: LIE GROUPS AND LIE ALGEBRAS

L. Pukanszky 63
On the Characters of Connected Lie Groups

A. A. Kirillov 73
La géométrie des moments pour les groupes de difféomorphismes

B. Kostant 85
The Vanishing of Scalar Curvature and the Minimal Representation of $SO(4,4)$

D. Kazhdan 125
The Minimal Representation of D_4

M. Kashiwara and T. Tanisaki 159
Kazhdan-Lusztig Conjecture for Symmetrizable Kac-Moody Lie Algebra. II. Intersection Cohomologies of Schubert Varieties

V. Lakshmibai 197
Standard Monomial Theory for $\widehat{SL}_n$

P. Littelmann and C. Procesi 219
Equivariant Cohomology of Wonderful Compactifications

W. Rossmann 263
Nilpotent Orbital Integrals in a Real Semisimple Lie Algebra and Representations of Weyl Groups

R.K. Brylinski 289
Twisted Ideals of the Nullcone

D. Barbasch 317
Representations with Maximal Primitive Ideal

D.A. Vogan, Jr. 333
Dixmier Algebras, Sheets, and Representation Theory

W.M. McGovern 397
Dixmier Algebras and the Orbit Method

J. Bernstein 417
Trace in Categories

J.-E. Björk and E.K. Ekström 425
Filtered Auslander-Gorenstein Rings

T. Levasseur 449
Relèvements d'opérateurs différentiels sur les anneaux d'invariants

C. De Concini and V.G. Kac 471
Representations of Quantum Groups at Roots of 1

CHAPTER III: INVARIANT THEORY

M. Brion and C. Procesi 509
Action d'un tore dans une variété projective

V.L. Popov 541
When are the Stabilizers of all Nonzero Semisimple Points Finite?

H. Kraft 561
C^*-Actions On Affine Space

I. ℂ*-ALGEBRAS

Equivalence of normal states of von Neumann algebras

E. STØRMER

Dedicated to Jacques Dixmier on the occasion of his 65th birthday

1. Introduction

Let M be a von Neumann algebra with predual M_* with positive cone M_*^+. For each $\varphi \in M_*^+$ let $[\varphi]$ be the norm closure of its orbit under the action of inner automorphisms by $\varphi \to u\varphi u^* = \varphi \circ \text{Ad}u$. From the norm identity

$$||\varphi - u\psi u^*|| = ||u^*\varphi u - \psi||$$

it is clear that we have an equivalence relation on M_*^+ by $\varphi \sim \psi$ if $\psi \in [\varphi]$. The quotient space $M_*^+/\sim$ is a metric space with metric

$$\varphi([\varphi],[\psi]) = \inf\{||\varphi' - \psi'|| : \varphi' \sim \varphi, \psi' \sim \psi\}.$$

The first result on $M_*^+/\sim$ is due to Powers [11], who showed that for factors of type I_n, $n \in \mathbb{N}, M_*^+/\sim$ is isometric to the set of decreasing nonnegative functions on the set $\{1, 2, \ldots, n\}$ with L^1-norm and counting measure. The map was as follows: Let Tr be the usual trace and $\varphi \in M_*^+$ be defined by $\varphi(x) = Tr(hx)$. Then h has spectrum $\lambda_1 \geq \lambda_2 \geq \cdots \geq \lambda_n \geq 0$, and the function corresponding to φ is $f_\varphi(j) = \lambda_j$.

Later on, Connes and the author [4] showed that for factors of type III_1 with separable predual any two states are equivalent, from which it follows that $[\varphi] \to \varphi(1)$ defines an isometry of $M_*^+/\sim$ onto $\mathbb{R}_+ \cup \{0\}$. In her thesis [1] Bion-Nadal characterized $M_*^+/\sim$ for factors of types III with separable predual by exhibiting a bijection of $M_*^+/\sim$ onto a certain class of Borel measures on a point set realization of the flow of weights. In particular, for factors of type $\mathrm{III}_\lambda, 0 < \lambda < 1, M_*^+/\sim$ was mapped onto the set of Borel measures on the unit circle, taking states onto probability measures. The next step towards our understanding of the quotient space was taken by Connes, Haagerup and the author in [3], where the diameter of the quotient state space in $M_*^+/\sim$ was shown to be 2 for factors of types II and I_∞ and $2\,\frac{1-\sqrt{\lambda}}{1+\sqrt{\lambda}}$ for factors of type III_λ with separable predual, $\lambda \in [0,1]$.

In the present article I shall describe the completion of the program as put forward by Haagerup and the author in [8]. In the semifinite case one can find a characterization similar to that of Powers. For factors of type III_λ, $0 < \lambda < 1$, with separable predual this is also possible, namely we show $M_*^+/\sim$ is isometric to the set of nonnegative decreasing functions on $\mathbb{R}_+$ continuous from the right satisfying $f(a) = \lambda f(\lambda a)$, $a \in \mathbb{R}_+$, considered as a subset of $L^1(\lambda, 1)$. For factors of type III_1 the characterization is trivial, as remarked above. However, for factors of type III_0 the above type of characterization seems impossible as pointed out by Bion-Nadal [1], who showed that it is necessary to consider the "flow of weights". We shall follow this approach and reformulate the results in terms of the continuous crossed product $M \times_{\sigma^\omega} \mathbb{R}$ with respect to the modular automorphism of a faithful normal semifinite weight ω on M. We then obtain for general von Neumann algebras an isometry of $M_*^+/\sim$ into the positive cone of the predual of the center of $M \times_{\sigma^\omega} \mathbb{R}$ and then characterize the range of this map when M has no direct summand of type I. The type I part does not have a nice characterization in this formulation, but can be described by a global version of Powers' theorem [11], so we shall ignore type I algebras in our discussion. For details I refer to the original paper [8].

In the last section of this article we shall discuss applications of the theorem to the set of normal states $S_n(M)$ and in particular the quotient space $S_n(M)/\sim$.

2. The equivalence relation on M_*^+

As described in the introduction, if M is a von Neumann algebra and $\varphi, \psi \in M_*^+$ then φ and ψ are called equivalent, written $\varphi \sim \psi$, if ψ belongs to the norm closure $[\varphi]$ of the set of functionals $u\varphi u^*$, $u \in U(M)$ - the unitary group in M. The quotient space $M_*^+/\sim$ is a complete metric space with metric

$$d([\varphi],[\psi]) = \inf\{||\varphi' - \psi'||, \varphi' \sim \varphi, \psi' \sim \psi\}.$$

While the definition of the equivalence relation $\sim$ is given in terms of unitary operators it is often preferable to work with partial isometries instead. We next state a result to this effect. As usual $\mathrm{supp}(\varphi)$ denotes the support projection of φ in M when $\varphi \in M_*^+$.

Theorem 2.1. *Let M be a von Neumann algebra and $\varphi, \psi \in M_*^+$. Then the following are equivalent.*

(i) $\varphi \sim \psi$.

(ii) *For every $\varepsilon > 0$ there exists a partial isometry $u \in M$ such that $u^*u = \mathrm{supp}(\varphi), uu^* = \mathrm{supp}(\psi)$, and $||u\varphi u^* - \psi|| < \varepsilon$.*

3. The first characterization of $M_*^+/\sim$ for semifinite and III_λ-factors

We first consider the case when M is a semifinite factor with separable predual. Let τ_0 be a faithful normal trace on M normalized so that $\tau_0(e) = 1$ if M is of type I and e a minimal projection in M, and $\tau_0(1) = 1$ if M is of type II_1. Let $P_f(M)$ denote the set of finite projections in M, and let

$$J = \{\tau_0(p) : p \in P_f(M)\}.$$

Then it is not hard to show there is a map $p : J \to P_f(M)$ such that

(i) $s, t \in J,\ s \leq t \Rightarrow p(s) \leq p(t)$.

(ii) $\tau_0(p(t)) = t,\ t \in J$.

We next define the functional representation of M_*^+ which will yield our first characterization of $M_*^+/\sim$. χ_E denotes as usual the characteristic function of a set E.

Lemma 3.1. *Let M be a semifinite factor as above. For $\varphi \in M_*^+$ define a nonnegative function f_φ on $\mathbb{R}_+$ by*

$$f_\varphi(a) = \tau_0(\chi_{(a,\infty)}(\frac{d\varphi}{d\tau_0})), \quad a > 0,$$

where $\frac{d\varphi}{d\tau_0}$ is the Radon-Nikodym derivative of φ with respect to τ_0. Then we have for $\varphi, \psi \in M_*^+$.

(a) $\varphi(1) = \int_0^\infty f_\varphi(a)\, da$.

(b) *If* $\varphi \leq \psi$ *then* $f_\varphi \leq f_\psi$.

(c) $d([\varphi],[\psi]) \geq \int_0^\infty |f_\varphi(a) = f_\psi(a)|\ da$.

The proof of this result gives a good description of some of the techniques in a simple case. It goes as follows. From the identity

$$t = \int_0^\infty \chi_{(a,\infty)}(t)\ da, \quad t > 0,$$

we have

$$\frac{d\varphi}{d\tau_0} = \int_0^\infty \chi_{(a,\infty)}\left(\frac{d\varphi}{d\tau_0}\right) da,$$

whence

$$\varphi(1) = \tau_0\left(\frac{d\varphi}{d\tau_0}\right) = \int_0^\infty \tau_0\left(\chi_{(a,\infty)}\left(\frac{d\varphi}{d\tau_0}\right)\right)\ da = \int_0^\infty f_\varphi(a)\ da,$$

proving (a).

It is well known that if h, k are positive self-adjoint operators affiliated with a von Neumann algebra N and $h \leq k$ (in the obvious sense of unbounded operators) then

$$\chi_{(a,\infty)}(h) \precsim \chi_{(a,\infty)}(k), \quad a > 0,$$

as projections in N. Thus if $\varphi \leq \psi$ then $\frac{d\varphi}{d\tau_0} \leq \frac{d\varphi}{d\tau_0}$, whence

$$f_\varphi(a) = \tau_0\left(\chi_{(a,\infty)}\left(\frac{d\varphi}{d\tau_0}\right)\right) \leq \tau_0\left(\chi_{(a,\infty)}\left(\frac{d\psi}{d\tau_0}\right)\right) = f_\psi(a), \qquad a > 0$$

proving (b).

If $\varphi, \psi \in M_*^+$ and $\varphi - \psi = (\varphi - \psi)^+ + (\varphi - \psi)^-$ is the orthogonal decomposition of $\varphi - \psi$ as a difference of two positive functionals, we define

$$\varphi \vee \psi = \varphi + (\varphi - \psi)^- = \psi + (\varphi - \psi)^+.$$

Then it is easy to show

$$\begin{aligned} ||\varphi - \psi|| &= 2\varphi \vee \psi(1) - \varphi(1) - \psi(1) \\ &= \inf\{2\omega(1) - \varphi(1) - \psi(1) : \omega \in M_*^+, \omega \geq \varphi, \omega \geq \psi\}. \end{aligned}$$

Thus to complete the proof we have by (a) and (b)

$$\begin{aligned}||\varphi - \psi|| &= \int_0^\infty (2f_{\varphi\vee\psi}(a) - f_\varphi(a) - f_\psi(a))\, da \\ &\geq \int_0^\infty (2f_\varphi(a) \vee f_\psi(a) - f_\varphi(a) - f_\psi(a))\, da \\ &= \int_0^\infty |f_\varphi(a) - f_\psi(a)|\, da.\end{aligned}$$

Since clearly $f_{u\varphi u^*} = f_\varphi$ for $u \in U(M)$, (c) follows. Q.E.D.

From the observation $f_{u\varphi u^*} = f_\varphi$ in the above proof it is immediate that if $\varphi \sim \psi$ then $f_\varphi = f_\psi$. The converse also true, hence $\varphi \sim \psi$ if and only if $f_\varphi = f_\psi$.

The map $\varphi \to f_\varphi$ is a surjection onto the set of functions $f : R_+ \to J$ in $L^1(\mathbf{R}_+)$ which are decreasing and continuous from the right. Indeed by spectral theory we can for given f find a positive self-adjoint operator h affiliated with M such that $\chi_{(a,\infty)}(h) = p(f(a))$, $a > 0$, where p is the function $J \to P_f(M)$ described at the beginning of the section. Thus if $\varphi(x) = \tau_0(hx), x \in M$, then by conditions (i) and (ii) for p,

$$f_\varphi(a) = \tau_0(\chi_{(a,\infty)}(h)) = \tau_0(p(f(a))) = f(a),$$

proving surjectivity.

Using the properties of p once more we obtain:

Theorem 3.2. *The map $\varphi \to f_\varphi$, $\varphi \in M_*^+$, determines a bijective map of $M_*^+/\sim$ onto the set of decreasing L^1-functions $f : \mathbf{R}_+ \to J$, which are continuous from the right. Furthermore, if $\varphi, \psi \in M_*^+$ then*

$$d([\varphi],[\psi]) = \int_0^\infty |f_\varphi(a) - f_\psi(a)|\, da.$$

The above theorem is different from the original characterization of Powers for I_n-factors and its generalizations to semifinite factors in that the functions f_φ maps $\mathbf{R}_+$ into J and not the other way around. To accomplish the original characterization we use the distribution function f_φ^* of f_φ defined by

$$f_\varphi^*(a) = \int_0^\infty \chi_{(a,\infty)}(f_\varphi(s))\, ds, \quad a \in J.$$

Then f_φ^* is a nonnegative decreasing function on J continuous from the right, and the map $f_\varphi \to f_\varphi^*$ is an L^1-isometry. Then $\varphi \to f_\varphi^*$ is the known characterization of $M_*^+/\sim$.

We next consider the case when M is a factor of type III_λ, $0 < \lambda < 1$, with separable predual. Then there are a factor P of type II_∞ with trace tr and an automorphism α of P for which $\mathrm{tr} \circ \alpha = \lambda\, \mathrm{tr}$, and $M = P \times_\alpha \mathbf{Z}$. Let ω be the dual weight of tr on M. Then its modular automorphism σ^ω satisfies $\sigma^\omega_{t_0} = id$, where $t_0 = -\frac{2\pi}{\log \lambda}$. Put

$$N_0 = M \times_{\sigma^\omega} (\mathbf{R}/t_0\mathbf{Z}).$$

Then N_0 is generated by operators $\pi_0(x)$, $x \in M$, and $\lambda_0(t)$, $t \in \mathbf{R}$, defined by

$$\left.\begin{aligned}(\pi_0(x)\xi)(s) &= \sigma^\omega_{-s}(x)\xi(s)\\ (\lambda_0(t)\xi)(s) &= \xi(s-t)\end{aligned}\right\} \xi \in L^2(\mathbf{R}/t_0\mathbf{Z}, H).$$

Furthermore N_0 is a factor of type II_∞, and each normal semifinite weight ψ on M has a canonical dual weight $\bar{\psi}$ on N_0. N_0 also has a canonical trace τ_0 scaling the dual action of α by λ. If we put

$$f_\varphi(a) = \tau_0\left(\chi_{(a,\infty)}\left(\frac{d\bar{\varphi}}{d\tau_0}\right)\right), \quad a > 0,$$

then we have a nonnegative decreasing function on $\mathbf{R}_+$ which is continuous from the right and satisfies

(i) $\lambda f_\varphi(\lambda a) = f_\varphi(a)$, $\quad a > 0$,

(ii) $\varphi(1) = \int_\lambda^1 f_\varphi(a)\, da$.

Modifying the arguments from the semifinite case we then show $f_\varphi = f_\psi$ if and only if $\varphi \sim \psi$, and obtain the following characterization of $M_*^+/\sim$.

Theorem 3.3. *Let M be a factor of type* III_λ, $0 < \lambda < 1$, *with separable predual, then*

(i) *the map $[\varphi] \to f_\varphi$ is a bijection of $M_*^+/\sim$ onto the set of decreasing functions from $\mathbf{R}_+$ into $\mathbf{R}_+$ which are continuous from the right and satisfy*

$$\lambda f(\lambda a) = f(a), \quad a > 0.$$

(ii) *For* $\varphi, \psi \in M_*^+$ *we have*

$$d([\varphi],[\psi]) = \int_\lambda^1 |f_\varphi(a) - f_\psi(a)|\, da.$$

4. The continuous crossed product

Let M be a von Neumann algebra acting on a Hilbert space H. Let ω be a faithful normal semifinite weight on M with modular group σ^ω. Then the crossed product $N = M \times_{\sigma^\omega} \mathbf{R}$ is defined as follows. Let π and λ be the representations of M and $\mathbf{R}$, respectively, on the Hilbert space $L^2(\mathbf{R}, H)$:

$$\begin{aligned}(\pi(x)\xi)(s) &= \sigma^\omega_{-s}(x)\xi(s), \quad && \xi \in L^2(\mathbf{R}, H)\\ (\lambda(t)\xi)(s) &= \xi(s-t), && s \in \mathbf{R}.\end{aligned}$$

Then λ is a unitary representation of $\mathbf{R}$ on $L^2(\mathbf{R}, H)$ such that

$$\lambda(t)\pi(x)\lambda(t)^* = \pi(\sigma_t^\omega(x)), \quad x \in M,\ t \in \mathbf{R}.$$

N is the von Neumann algebra generated by $\pi(x)$, $x \in M$, and $\lambda(t)$, $t \in \mathbf{R}$. It should be remarked at this point that the way N is defined it depends on ω. But as pointed out by Takesaki [15] both N and the concepts defined below are up to isomorphism independent of the particular choice of ω.

We define the dual automorphism group $(\theta_s)_{s\in\mathbf{R}}$ of σ^ω on N by the properties

$$\begin{aligned}\theta_s(\pi(x)) &= \pi(x), \quad && x \in M,\ s \in \mathbf{R}\\ \theta_s(\lambda(t)) &= e^{-ist}\lambda(t), && s, t \in \mathbf{R}.\end{aligned}$$

Then $\pi(M)$ is the fixed point algebra of θ in N. By [7] there is a faithful normal semifinite operator valued weight T from N on $\pi(M)$ given by

$$T(y) = \int_{-\infty}^{\infty} \theta_s(y)\, ds, \quad y \in N^+,$$

where ds denotes the Lebesgue measure on $\mathbf{R}$. Then for any normal semifinite weight on φ on M its dual weight $\tilde{\varphi}$ on N is given by

$$\tilde{\varphi} = \varphi \circ \pi^{-1} \circ T.$$

By [15] there is a positive self-adjoint operator affiliated with N such that $\lambda(t) = h^{it}$, and the weight τ defined by

$$\tau(y) = \tilde{\omega}(h^{-1}y)$$

is a faithful normal semifinite trace on N such that

$$\tau \circ \theta_s = e^{-s}\tau, \quad s \in \mathbf{R}.$$

We call τ the canonical trace on N.

If φ is a normal semifinite weight on M we put $h_\varphi = \frac{d\tilde{\varphi}}{dt}$. Then h_φ is a positive self-adjoint operator affiliated with N such that $\tilde{\varphi}(y) = \tau(h_\varphi y)$, $y \in N^+$. Put

$$e_\varphi = \chi_{(1,\infty)}(h_\varphi)\,, \quad \varphi \in M_*^+.$$

Then it is not hard to show $\tau(e_\varphi) = \varphi(1)$.

Definition 4.1. If $\varphi \in M_*^+$ we denote by $\hat{\varphi} \in Z(N)_*^+$, $Z(N)$ being the center of N, the functional defined by

$$\hat{\varphi}(z) = \tau(e_\varphi z), \quad z \in Z(N).$$

Then by the above $\hat{\varphi}(1) = \varphi(1)$. Furthermore from the definition of $\hat{\varphi}$ it is immediate that

$$h_{u\varphi u^*} = \pi(u)h_\varphi\pi(u)^*, \quad u \in U(M),$$

hence $e_{u\varphi u^*} = \pi(u)e_\varphi\pi(u)^*$, and therefore $(u\varphi u^*)^\wedge = \hat{\varphi}$. It follows from our next result that $\hat{\varphi} = \hat{\psi}$ whenever $\varphi \sim \psi$.

Lemma 4.2. *The map $\varphi \to \hat{\varphi}$ of M_*^+ into $Z(N)_*^+$ preserves order, and for $\varphi, \psi \in M_*^+$ we have*

$$||\hat{\varphi} - \hat{\psi}|| \leq d([\varphi],[\psi]).$$

Proof. From the arguments in the proof of Lemma 3.1 it follows that if $\varphi, \psi \in M_*^+$, $\varphi \leq \psi$, then $e_\varphi \precsim e_\psi$, so that

$$\hat{\varphi}(z) = \tau(e_\varphi z) \leq \tau(e_\psi z) = \hat{\psi}(z), \quad z \in Z(N)^+.$$

Thus the map $\varphi \to \hat{\varphi}$ preserves order. Also from the proof of Lemma 3.1 since $(\varphi \vee \psi)^\wedge \geq \hat{\varphi}, \hat{\psi}$ we have

$$\begin{aligned} \|\hat{\varphi} - \hat{\psi}\| &\leq 2(\varphi \vee \psi)^\wedge(1) - \hat{\varphi}(1) - \hat{\varphi}(1) \\ &= 2(\varphi \vee \psi)(1) - \varphi(1) - \psi(1) \\ &= \|\varphi - \psi\|. \end{aligned}$$

In particular, if $u \in U(M)$ then

$$\|\hat{\varphi} - \hat{\psi}\| = \|(u\varphi u^*)^\wedge - \hat{\psi}\| \leq \|u\varphi u^* - \psi\|,$$

whence $\|\hat{\varphi} - \hat{\psi}\| \leq d([\varphi], [\psi])$. Q.E.D.

It follows that the map $\varphi \to \hat{\varphi}$ induces a contraction $[\varphi] \to \hat{\varphi}$ of $M_*^+/\sim$ into $Z(N)_*^+$. Our next result points out the key property of the functionals in the range of this map.

Lemma 4.3. *For all $\varphi \in M_*^+$, $\hat{\varphi} \circ \theta_s \geq e^{-s}\hat{\varphi}$ for $s > 0$.*

Proof. Since $\theta_s(h_\varphi) = e^{-s}h_\varphi$ it follows that

$$\theta_{-s}(e_\varphi) = \chi_{(e^{-s},\infty)}(h_\varphi) \geq e_\varphi \quad \text{for} \quad s \geq 0.$$

Hence for $z \in Z(N)^+$ and $s > 0$

$$\hat{\varphi} \circ \theta_s(z) = \tau(e_\varphi \theta_s(z)) = e^{-s}\tau(\theta_{-s}(e_\varphi)z) \geq e^{-s}\tau(e_\varphi z) = e^{-s}\hat{\varphi}(z). \text{ Q.E.D.}$$

The last two lemmas describe the easy part of our main result, which we now state.

Main Theorem (4.4). *Let M be a von Neumann algebra. Then with notation as above we have:*

(i) *The map $[\varphi] \to \hat{\varphi}$ is an isometry of $M_*^+/\sim$ into $Z(N)_*^+$.*

(ii) *If M is properly infinite with no direct summand of type* I *then the range of the map $[\varphi] \to \hat{\varphi}$ is the cone*

$$P(M) = \{\chi \in Z(N)_*^+ : \chi \circ \theta_s \geq e^{-s}\chi, \ s \in \mathbf{R}_+\}.$$

(iii) *If M is of type* II_1 *then the range of the map* $[\varphi] \to \hat{\varphi}$ *is the set* $\{\chi \in P(M) : \chi \leq \tau \mid Z(N)\}$.
Furthermore in case (ii) $Z(N)_*$ *is the closed linear span of* $P(M)$.

We remark that the description of the range in the II_1-case may look perplexing. However, one should keep in mind that if $\varphi \in M_*^+$ and $\lambda > 0$ then we do not have equality between $(\lambda\varphi)^\wedge$ and $\lambda\hat{\varphi}$.

To obtain the theorem from our characterizations in Theorem 3.2 and 3.3 for semifinite and III_λ-factors with separable preduals, we have to give quite detailed descriptions of the crossed product $N = M \times_{\sigma^\omega} \mathbf{R}$ in those cases, and then to write $\hat{\varphi}$ in terms of the function f_φ. For example, in the semifinite case N is generated by the operators $\pi(x) = x \otimes 1 \in M \otimes L^\infty(\mathbf{R})$, and since $\sigma_t^{\tau_0} = id$, $t \in \mathbf{R}$, $\lambda(t) = 1 \otimes m(e^{i\gamma t})$, where $m(f)$ denotes the multiplication operator $(m(f)g)(\gamma) = f(\gamma)g(\gamma)$, on $L^2(\mathbf{R})$, $f \in L^\infty(\mathbf{R})$. Thus $N = M \otimes L^\infty(\mathbf{R})$, and we show that when $Z(N)$ is identifed with $L^\infty(\mathbf{R})$,

$$\hat{\varphi}(z) = \int_{-\infty}^{\infty} z(\gamma) f_\varphi(e^{-\gamma}) e^{-\gamma} d_\gamma, \quad z \in Z(N).$$

For factors of type III_λ, $0 < \lambda < 1$, the situation is more complicated. Let

$$N_0 = M \times_{\sigma^\omega} (\mathbf{R}/t_0\mathbf{Z})$$

as described in section 3 with $t_0 = -\frac{2\pi}{\log\lambda}$, and generators $\pi_0(x)$, $x \in M$, $\lambda_0(t)$, $t \in \mathbf{R}$. Then there is a spatial isomorphism

$$\Phi : N \to N_0 \otimes L^\infty(0, -\log\lambda)$$

such that

$$\Phi(\pi(x)) = \pi_0(x) \otimes 1,$$
$$\Phi(\lambda(t)) = \lambda_0(x) \otimes m(e^{it}).$$

Using this result we can compute the basic ingredients associated with N from those associated with N_0, and thus translate Theorem 3.3 into the formulation of the Main Theorem. We omit the rather technical details.

5. A martingale theorem

The strategy for proving the Main Theorem is first to prove it for factors with separable preduals, then to use direct integral theory to get it for all von Neumann algebras with separable preduals. Having that we then extend it to σ-finite (= countably decomposable) von Neumann algebras, and then finally to general von Neumann algebras. In the cases of III_0-factors and σ-finite von Neumann algebras the proof will consist of extending the theorem from subalgebras for which the theorem is proved by use of the following martingale theorem, which we state without proof.

Theorem 5.1. *Suppose M is a von Neumann algebra such that $M = (\cup_{\alpha\in A} M_\alpha)^-$ is the σ-weak closure of an increasing family of von Neumann algebras for which there is for each $\alpha \in A$ a faithful normal conditional expectation $E_\alpha : M \to M_\alpha$ of M onto M_α satisfying $E_\alpha E_\beta = E_\alpha$ whenever $M_\alpha \subset M_\beta$, and*

$$\lim_\alpha ||\varphi \circ E_\alpha - \varphi|| = 0 \quad \textit{for all} \quad \varphi \in M_*.$$

Suppose the conclusion of the Main Theorem holds for each M_α. Then

(i) *Part (i) of the Main Theorem holds for M.*

(ii) *If each M_α is properly infinite with no direct summand of type I, then part (ii) of the Main Theorem holds for M.*

(iii) *If each M_α and M are of type* II_1 *part (iii) of the Main Theorem holds for M.*

6. Von Neumann algebras with separable preduals

Let M be a von Neumann algebra with separable predual, which we assume acts on a separable Hilbert space H. Then (see [9, 14.2.1] or [6, Ch. II]) there are a locally compact complete separable metric space Z, a positive measure ν on Z with support Z, a ν-measurable field $\zeta \to H(\zeta)$ of nonzero Hilbert spaces, a ν-measurable field $\zeta \to M(\zeta)$ of factors on $H(\zeta)$, and an isomorphism of H on $\int_Z^\oplus H(\zeta) d\nu(\zeta)$ carrying M onto $\int_Z^\oplus M(\zeta) d\nu(\zeta)$.

Assume we have the above situation, except that we do not assume the $M(\zeta)$ to be factors. It follows from work of Lance [10] and Sutherland [12,13] that with $N = M \times_{\sigma^\omega} \mathbf{R}$, $Z(N)$ its center, θ_s the dual automorphism group, τ the canonical trace on N, then all these objects are direct integrals of the corresponding ones for $M(\zeta)$. The same is also true for $\tilde{\varphi}$, e_φ, $\hat{\varphi}$ with $\varphi \in M_*^+$. Thus it is not surprising that we have

Theorem 6.1. *Let M be a von Neumann algebra with separable predual and direct integral decomposition*

$$M = \int_Z^\oplus M(\zeta)\, d\nu(\zeta).$$

Suppose the Main Theorem holds for each $M(\zeta)$. Then it holds for M.

It was shown by Connes [2] that a III_0-factor with separable predual is the closure of the union of an increasing sequence of von Neumann subalgebras of type-II_∞. Refining the construction slightly we can obtain a sequence such that the assumptions in the martingale theorem (5.1) hold. Since the Main Theorem holds for factors of type II_∞ with separable predual, and each von Neumann algebra of type II_∞ with separable predual is a direct integral of factors of type II_∞, the Main Theorem holds for all von Neumann algebras of type II_∞ with separable preduals by Theorem 6.1. Thus by the martingale theorem it holds for all factors of type III_0 with separable preduals. If M is a factor of type III_1 with M_* separable then by [4] $M_*^+/\sim \cong \mathbf{R}_+$, and by [15] $N = M \times_{\sigma^\omega} \mathbf{R}$ is a factor, hence $Z(N) \cong \mathbf{C}$. Thus the theorem is true in that case too, and thus for all factors with separable preduals. Using Theorem 6.1 once more we conclude

Theorem 6.2. *The Main Theorem holds for all von Neumann algebras with separable preduals.*

7. The general case

Let M be a σ-finite von Neumann algebra and ω a faithful normal state on M. Note that if $P \subset M$ is a countably generated von Neumann algebra then the σ^ω-invariant von Neumann algebra

$$\left(\bigcup_{t\in\mathbf{R}} \sigma_t^\omega(P)\right)'' = \left(\bigcup_{t\in\mathbf{Q}} \sigma_t^\omega(P)\right)''$$

is also countably generated. We may thus write M as an increasing union $M = (\cup_{\alpha\in A} M_\alpha)^-$, where A is an index set such that $\alpha \le \beta$ implies $M_\alpha \subset M_\beta$, and each M_α is countably generated and σ^ω-invariant. By [14] there are faithful normal conditional expectations $E_\alpha : M \to M_\alpha$ of M onto M_α, $\alpha \in A$, such that $\alpha \le \beta$ implies $E_\alpha E_\beta = E_\alpha$, and $\lim_\alpha \|\varphi \circ E_\alpha - \varphi\| = 0$ for all $\varphi \in M_*$. Thus the first part of the Main Theorem holds for M by the martingale theorem (5.1) and Theorem 6.2.

To show parts (ii) and (iii) we must be more careful because in the martingale theorem it is important that the M_α either are all of type II_1 if M is, or properly infinite with no type I direct summand when M has this property. We therefore have to show we can choose the M_α's so this is true.

Theorem 7.1. *Let M be a σ-finite von Neumann algebra of type III. Then there exists a countably generated von Neumann subalgebra $N \subset M$*

such that any von Neumann subalgebra P of M with $N \subset P \subset M$ is of type III.

A similar result holds for M of types II_1 and II_∞. The proof of Theorem 7.1 relies on the following characterization of type III von Neumann algebras in terms of their normal states.

Theorem 7.2. *Let M be a von Neumann algebra with a faithful normal state of φ. Then M is of type III if and only if φ belongs to the convex set*

$$\mathrm{conv}\left\{ \tfrac{1}{2}(v\varphi v^* + w\varphi w^*) : v, w \in M,\ v^*v = w^*w = 1,\ vv^* + ww^* = 1 \right\}^- .$$

To show that 7.2 implies 7.1 let φ be as in 7.2 and choose sequences (v_n) and (w_n) of isometries in M with $v_n v_n^* + w_n w_n^* = 1$ and such that

$$\varphi \in \mathrm{conv}\left\{ \tfrac{1}{2}(v_n \varphi v_n^* + w_n \varphi w_n^*) \right\}^- .$$

Let N be the von Neumann algebra generated by all the v_n's and w_n's. If P is a von Neumann algebra with $N \subset P \subset M$ put $\varphi' = \varphi \mid p$. Then φ' is a faithful normal state on P, and

$$\varphi' \in \mathrm{conv}\left\{ \tfrac{1}{2}(v_n \varphi' v_n^* + w_n \varphi' w_n^*) \right\}^- ,$$

whence P is of type III by Theorem 7.2.

To show the Main Theorem in the general case we first show by straightforward arguments that if $M = \bigoplus_{i \in A} M_i$ is a direct sum of von Neumann algebras for which the Main Theorem holds then it holds for M. If M is of type II_1 we can assume each M_i is σ-finite, and so it holds for M. If M is properly infinite we may write each M_i in the form $P_i \otimes B(H_i)$ with $B(H_i)$ the bounded operators on a Hilbert space H_i and P_i a σ-finite von Neumann algebra. Thus we have reduced to the case $M = P \otimes B(H)$ with $P\sigma$-finite and properly infinite. For such M there are natural isometries π of $M_*^+/\sim$ onto P_*^+/N and δ of $\{\hat{\psi} : \psi \in M_*^+\}$ onto $\{\hat{\varphi} : \varphi \in P_*^+\}$ such that the following diagram is commutative.

$$(*) \qquad \begin{array}{ccc} M_*^+/\sim & \xrightarrow{\pi} & P_*^+/\sim \\ \downarrow & & \downarrow \\ \{\hat{\psi} : \psi \in M_*^+\} & \xrightarrow{\delta} & \{\hat{\varphi} : \varphi \in P_*^+\} \end{array}$$

where the vertical maps are the canonical ones $[\varphi] \to \hat{\varphi}$. This then completes the proof of the Main Theorem.

8. Applications

If M is a von Neumann algebra we denote by $S_n(M)$ the set of normal states of M. Since $M_*^+/\sim$ is a complete metric space the quotient space

$$S_n(M)/\sim = \{[\varphi] \in M_*^+/\sim : \varphi \in S_n(M)\}$$

is the same and with diameter bounded by 2. In [3] the diameter of $S_n(M)/\sim$ was computed. It is trivially 2 when M is not a factor because $d([\varphi],[\psi]) = 2$ when φ and ψ are centrally disjoint states. When M is a semifinite factor the diameter is easily computed and is 2 unless M is of type I_n, $n < \infty$. When M is of type III with separable predual the Main Theorem yields a new proof of the diameter formula.

Theorem 8.1 [3]. *If M is a factor of type* III_λ, $0 \le \lambda \le 1$, *with separable predual then the diameter*

$$\mathrm{diam}(S_n(M)/\sim) = 2\frac{1-\lambda^{1/2}}{1+\lambda^{1/2}}.$$

By the Main Theorem the diameter is the same as the diameter of the convex set

$$P_1(M) = \{\chi \in S_n(Z(N)) : \chi \circ \theta_s \ge e^{-s}\chi, s > 0\}.$$

Hence the problem is reduced to the study of the flow $(\theta_s)_{s\in\mathbb{R}}$ on the abelian von Neumann algebra $Z(N)$. One reduces the study to the flow (θ_s) on the abelian C^*-subalgebra A of $Z(N)$ consisting of those $z \in Z(N)$ for which $s \to \theta_s(z)$ is norm continuous. The adjoint flow (α_s) on the dual $\hat{A}$ is defined by $\alpha_s(\delta) = \delta \circ \theta_s$, and we consider the three cases when

(i) $\hat{A}$ has more than one α-orbit. This corresponds to the III_0-case.

(ii) $A \cong C(\mathbb{R}/t_0\mathbb{Z})$ for some t_0, and α_s corresponds to translation by s on $\mathbb{R}/t,\mathbb{R})$. This corresponds to the III_λ-case.

(iii) $A = \mathbb{C}$, which corresponds to the III_1-case.

The diameter of $P_1(M)$, or rather the corresponding set $P_1(A)$ defined by A, is 2 in case (i),

$$2\frac{1-\exp(-t_0/2)}{1+\exp(-t_0/2)}$$

in case (ii) and 0 in case (iii). This then yields the theorem.

The set $P_1(M)$ can be viewed as the normal state space collapsed under the equivalence relation $\sim$. In our next theorem part (i) is an extension of [1, Thm. 3.4].

Theorem 8.2. *Let M be a properly infinite von Neumann algebra with no type I direct summand. Then $P_1(M)$ is a norm closed Choquet simplex. We have*

(i) *If M is a factor of type III_λ, $0<\lambda<1$, then $P_1(M)$ is a compact Bauer simplex naturally isomorphic to the set of probability measures on $\mathbf{R}/t_0\mathbf{Z}$ where $t_0=-\log\lambda$. In particular $S_n(M)/\sim$ is compact.*

(ii) *If M is a factor of type III_0 the simplex $P_1(M)$ has no extreme points, so in particular $P_1(M)$, and hence $S_n(N)/\sim$, is noncompact.*

Recall that the centralizer M_φ of a state φ on M is the set of $x\in M$ such that $\varphi(xy-yx)=0$ for all $y\in M$, or equivalently, if φ is faithful and normal M_φ is the fixed point algebra of the modular automorphism group σ^φ. The following is an extension of a result in [4] and shows in particular that in the properly infinite case without type I summands the equivalence classes $[\varphi]$ are very large.

Theorem 8.3. *Let M be a von Neumann algebra with no direct summand of type I. Suppose φ is a faithful normal state on M. Then there is a faithful normal state $\psi\sim\varphi$ whose centralizer is of type II_1.*

The proof shows some of the flexibility of the equivalence relation $\sim$, also indicated by the diagram (*) at the end of section 7. Namely, write M in the theorem in the form $M=P\otimes M_2(\mathbf{C})$. Let ω be a faithful normal state (or semifinite weight) on P, and let τ_0 be the tracial state on $M_2(\mathbf{C})$. Then there is an isometry

$$\gamma^*: Z(P\times_{\sigma^\omega}\mathbf{R})_*\to Z(M\times_{\sigma^{\omega\otimes t_0}}\mathbf{R})_*$$

such that

$$\gamma^*(\hat\rho)=(\rho\otimes\tau_0)^\wedge,\quad \rho\in P_*^+.$$

(In the diagram (*) δ is in this case the inverse of γ^* restricted to the set $\{\hat\psi:\psi\in M_*^+\}$). If φ is a faithful normal state on M then by the diagram there is $\varphi_0 \in S_n(P)$ such that $\gamma^*(\hat{\varphi}_0) = \hat{\varphi}$. Since also $\gamma^*(\hat{\varphi}_0) = (\varphi_0 \otimes \tau_0)^\wedge$ we have $\varphi \sim \varphi_0 \otimes \tau_0$, where $\varphi_0 \otimes \tau_0$ has $M_2(\mathbf{C})$ in its centralizer. This is the first step in an induction process, which eventually yields a state $\psi \sim \varphi$ containing the hyperfinite II_1-factor in its centralizer.

There is another set of applications of the Main Theorem, namely to automorphisms of a von Neumann algebra M. Each $\alpha \in \mathrm{Aut}(M)$ has a canonical extension to $\tilde{\alpha} \in \mathrm{Aut}(N)$, $N = M \times_{\sigma^\omega} \mathbf{R}$, whose restriction to $Z(N)$ is mod (α) in the sense of Connes and Takesaki [5]. Dualizing we get isometries of $M_*^+/\sim$ and $Z(N)_*^+$ reflecting properties of α. Since this theory has been covered in other survey articles I'll not discuss it here. I just want to point out that the isometry $\bar{\alpha}_*$ of $M_*^+/\sim$ is given by $\bar{\alpha}_*([\varphi]) = [\varphi \circ \alpha^{-1}]$. From the Main Theorem and the identification of the pair $(Z(N), \theta\,|_{Z(N)})$ with the flow of weights it follows that $\mathrm{mod}(\alpha) = id$ if and only if $\bar{\alpha}_* = id$. Thus the kernel of mod consists of exactly those α for which there is for each $\varepsilon > 0$ and $\varphi \in M_*^+$ a unitary $u \in U(M)$ such that

$$\|\varphi \circ \alpha^{-1} - u\varphi u^*\| < \varepsilon,$$

[8, Thm. 12.4]. This characterization of ker(mod) should be compared with the one in Takesaki's article of ker(mod) when M is an injective factor.

REFERENCES

[1] J. Bion-Nadal, *Espace des états normaux d'une facteur de type* III_λ, $0 < \lambda < 1$, *et d'une factuer de type* III_0, Canad. J. Math **36** (1984), 830-882.

[2] A. Connes, *Une classification des facteurs de type* III, Ann. Ec. Norm. Sup. **6** (1973), 133-252.

[3] A. Connes, U. Haagerup and E. Størmer, *Diameters of state spaces of factors of type* III. Lecture Notes in Math. **1132** (1985), 91-116, Springer-Verlag.

[4] A. Connes and E. Størmer, *Homogeneity of the state space of factors of type* III_1, J. Funct. Anal. **28** (1978), 187-196.

[5] A. Connes and M. Takesaki, *The flow of weights of factors of type* III, Tôhoku Math. J. **29** (1977), 473-575.

[6] J. Dixmier, *Les algébres d'opérateurs dans l'espace hilbertien,* Gauthier-Villars, Paris 1969.

[7] U. Haagerup, *On the dual weights for crossed products of von Neumann algebras*, II, Math. Scand. **43** (1978), 119-140.

[8] U. Haagerup and E. Størmer, *Equivalence of normal states on von Neumann algebras and the flow of weights*, Advances in Math. To appear.

[9] R. V. Kadison and J. R. Ringrose, *Fundamentals of the theory of operator algebras,* Vol. II, Academic Press, Orlando 1986.

[10] C. Lance, *Direct integrals of left Hilbert algebras*, Math. Ann. **216** (1975), 11-28.

[11] R. T. Powers, *Representation of uniformly hyperfinite algebras and their associated von Neumann rings*, Ann. Math. **86** (1967), 138-171.

[12] C. Sutherland, *Crossed products, direct integrals and Connes classifiction of type* III *factors*, Math. Scand. **40** (1977), 209-214.

[13] C. Sutherland, *Cartan subalgebras, transverse measures and non-type* I *Plancherel formulae*, J. Funct. Anal. **60** (1985), 281-308.

[14] M. Takesaki, *Conditional expectation in von Neumann algebra*, J. Funct. Anal. **9** (1972), 306-321.

[15] M. Takesaki, *Duality for crossed products and the structure of von Neumann algebras of type* III, Acta. Math. **131** (1973), 249-310.

Received November 3, 1989

University of Oslo
Department of Mathematics
PO Box 1053 Blindern
0316 Oslo 3, Norway

and

Department of Mathematics
University of California,
Los Angeles, CA 90024

The Structure of the Automorphism Group of an AFD Factor

M. TAKESAKI

Dedicated to Professor Jacques Dixmier on his 65th birthday

0. Introduction

In his seminal work [4,6], A. Connes determined the outer conjugacy class of an automorphism α of an AFD factor $\mathcal{R}_0$ of type II_1 by means of the associated invariants: the outer period $p_0(\alpha)$ and the obstruction $\mathrm{Ob}(\alpha)$, and showed the simplicity of the group: $\mathrm{Out}(\mathcal{R}_0) = \mathrm{Aut}(\mathcal{R}_0)/\mathrm{Int}(\mathcal{R}_0)$. For AFD factor $\mathcal{R}_{0.1}$ of type II_∞, he showed that the module, $\mathrm{mod}(\alpha)$, the outer period $p_0(\alpha)$ and the obstruction $\mathrm{Ob}(\alpha)$ together determine the outer conjugacy class of α. He further announced in [5] the outer conjugacy classification of $\alpha \in \mathrm{Aut}(\mathcal{R})$ for an AFD factor $\mathcal{R}$ of type III. Despite its importance, the proof has not yet appeared. The author observed that some of the experts were able to prove partial results, but not the full account. Thus, the author decided to provide the proof for the entire results of Connes. He kindly permitted the author to present this proof. On the other hand, he mentioned to the author that his own proof is quite different from the present one. According to him, his proof does not require to split the arguments into the cases of type III_0, type III_λ and type III_1, while our arguments depends heavily on the type of the base factor.

Let $\mathcal{R}$ be an AFD factor of type III. First, we determine the kernel of the module as the closure of $\mathrm{Int}(\mathcal{R})$. Next, we prove that $\alpha \in \mathrm{Aut}\,(\mathcal{R})$ acts trivially on the strongly central sequences if and only if α is of the form $\alpha = \mathrm{Ad}(u)\cdot\bar{\sigma}_c^\varphi$, where $\bar{\sigma}_c^\varphi$ is the extended modular automorphism corresponding to a one cocycle c on the flow of weight and φ a dominant weight on $\mathcal{R}$.

On the course of the proof, we prove that $\mathcal{R}_\omega$ is a factor of type II_1 if $\mathcal{R}$ is an AFD factor of type III_λ, $\lambda \neq 0$.

The results of this paper are taken from the forthcoming joint paper with Kawahigashi and Sutherland [17].

1. The Automorphism Group of an AFD Factor of Type III

A bounded sequence $\{x_n\}$ in a factor $\mathcal{R}$ is said to be *strongly central* if $\|[x_n, \psi]\| \to 0$ as $n \to \infty$ for every $\psi \in \mathcal{R}_*$. For a free ultra filter ω on $\mathbf{N}$, the sequence $\{x_n\}$ is called *strongly ω-central* if $\lim_{n\to\omega} \|[x_n, \psi]\| = 0$, $\psi \in \mathcal{R}_*$. Let $\mathcal{C}(\omega)$ be the set of all strongly ω-central sequences, and $\mathcal{I}(\omega)$ denote the set of all ω-null sequences, where $\{x_n\}$ is called ω-null if $\{x_n\}$ converges to zero in the σ-strong$*$ topology as $n \to \omega$. It is then known by Connes [3] that (i) $\mathcal{C}(\omega)$ is a $\mathcal{C}^*$-algebra under the pointwise algebraic operation and the sup norm; (ii) $\mathcal{I}(\omega)$ is a closed ideal of $\mathcal{C}(\omega)$ such that the quotient $\mathcal{C}^*$-algebra $\mathcal{R}_\omega = \mathcal{C}(\omega)/\mathcal{I}(\omega)$ is a von Neumann algebra with a faithful normal trace τ; (iii) The trace τ is given by

$$\tau(X) = \lim_{n\to\omega} x_n, \quad X = \{x_n\},$$

where the convergence is taken in the σ-weak topology of $\mathcal{R}$ and the limit is a scalar multiple of the identity due to the strong centrality of $\{x_n\}$.

A strongly central sequence $\{x_n\}$ is called *strongly hypercentral* if $[x_n, y_n] = x_n y_n - y_n x_n$ converges to zero in the σ-strong* topology for every strongly central sequence $\{y_n\}$. A strongly ω-hypercentral sequence is also defined similarly.

Proposition 1.1. *In an AFD factor $\mathcal{R}$ of* type III_λ, $\lambda \neq 0$, *all strongly hypercentral sequences are equivalent to trivial ones. Therefore $\mathcal{R}_\omega$ is a factor of type* II_1 *for any free ultrafilter ω.*

Proof. By the uniqueness of AFD factors of type III_λ with fixed $\lambda \neq 0$, Connes [7,8] and Haagerup [13], $\mathcal{R}$ can be identified with an infinite tensor product of matrix algebras. Thus, there exists an increasing sequence $\{M_n\}$ of finite factors of type I and a faithful normal state φ such that $\mathcal{R} = (\cup M_n)''$ and each M_n is globally invariant under $\{\sigma_t^\varphi\}$. Each M_n is generated by a finite group G_n of unitaries. Set $\alpha_g(x) = gxg^{-1}$, $g \in G_n$, and

$$\mathcal{E}_n(x) = \int_{G_n} \alpha_g(x) dg, \qquad x \in \mathcal{R},$$

where the integral means the summation relative to the normalized Haar measure, i.e. $\frac{1}{|G_n|}\sum_{g\in G_n}$. Then $\mathcal{E}_n$ is a projection of norm one from $\mathcal{R}$ onto M_n^c such that $\|x - \mathcal{E}_n(x)\|_\varphi^\# \leqq \sup_{g\in G_n} \|x - \alpha_g(x)\|_\varphi^\#$. Hence if

$\{x_k\}$ is strongly central, then there exists a subsequence $\{x_{k_n}\}$ such that $\|x_{k_n} - \mathcal{E}_n(x_{k_n})\|_\varphi^\# \to 0$.

Suppose that $\{x_k\}$ is strongly central and not equivalent to a trivial sequence. With $\{k_n\}$ as above, set $y_n = \mathcal{E}_n(x_{k_n}), n \in \mathbf{N}$. If we have chosen $\{x_k\}$ so that $\liminf \|x_k - \varphi(x_k)\|_\varphi^\# = \alpha > 0$, which is possible by passing to a subsequence, we get $\liminf \|y_n - \varphi(y_n)\|_\varphi^\# \geq \alpha > 0$. Replace y_n by $y_n - \varphi(y_n)$ to obtain a strongly central sequence $\{y_n\}$ with $\lim_{n\to\infty} y_n \neq 0$; $\varphi(y_n) = 0$ and $y_n \in M_n^c$. Since M_n^c is an AFD factor of type III_λ, there exists a unitary $u_n \in M_n^c$ such that

$$\|y_n - u_n y_n u_n^*\|_\varphi^\# \geq \frac{1}{2}\|y_n\|_\varphi^\#; \quad \|[u_n, \varphi\,|_{M^c}]\| < \frac{1}{4^n},$$

because 0 is in the σ-weak convex closure of

$$\left\{u y_n u^* : u \in \mathcal{U}(M_n^c)\|[u, \varphi|_{M_n^c}]\| < \frac{1}{4^n}\right\}$$

by Haagerup [13] for the type III_1 case, and for the type III_λ, $0 < \lambda < 1$, case, 0 is in the σ-weak convex closure of $\{u y_n u^* : u \in \mathcal{U}(M_n^c),\ [u, \varphi|_{M_n^c}] = 0\}$ since φ can be chosen as the standard product state of the Powers factor. Since $\varphi = \varphi|_{M_n} \otimes \varphi|_{M_n^c}$, we have $\|[u_n, \varphi]\| < \frac{1}{4^n}$. If ξ_φ is the representing vector of φ in the natural cone in a standard form, we have

$$\|[u_n, \xi_\varphi]\| \leqq \|[u_n, \varphi]\|^{1/2} < 1/2^n.$$

We claim that $\{u_n\}$ is strongly central. Since $\{u_n\}$ commutes with φ asymptotically, we have only to show that $\{u_n\}$ is central. Given $\varepsilon > 0$, and $a \in \mathcal{R}$, choose $a_0 \in M_k$ such that $\|a - a_0\|_\varphi^\# < \varepsilon$ and $\|a_0\| \leqq \|a\|$. We then have

$$\begin{aligned}\|[a, u_n]\xi_\varphi\| &\leqq \|[a - a_0, u_n]\xi_\varphi\| + \|[a_0, u_n]\xi_\varphi\| \\ &\leqq \|(a - a_0)\xi_\varphi u_n\| + \|(a - a_0)[u_n, \xi_\varphi]\| + \|u_n(a - a_0)\xi_\varphi\| + \|[a_0, u_n]\xi_\varphi\| \\ &\leqq 2\|a - a_0\|_\varphi^\# + 2\|a\|\,\|[u_n, \xi_\varphi]\| + \|[a_0, u_n]\xi_\varphi\| \leqq 2\varepsilon + \|a\|/2^{n-1}\end{aligned}$$

for $n \geq k$. Hence $\lim\|[a, u_n]\xi_\varphi\| = 0$. Similarly we have $\lim\|[a, u_n]^*\xi_\varphi\| = 0$. Thus $\{u_n\}$ is central. But $\{u_n\}$ does not commute with $\{y_n\}$ asymptotically. Therefore, $\{x_n\}$ is not strongly hypercentral. Q.E.D.

Before going into the discussion of the structure of $\mathrm{Aut}(\mathcal{R})$ for an AFD factor $\mathcal{R}$ of type III, let us determine the structure of $\mathrm{Aut}(\mathcal{R}_{0,1})$ where $\mathcal{R}_{0,1} = \mathcal{R}_0 \bar{\otimes} \mathcal{L}(\ell^2(\mathbf{Z}))$ with $\mathcal{R}_0$ the AFD factor of type II_1.

Theorem 1.2. *Let $\mathcal{R}_{0,1}$ be an AFD factor of type* II_∞. *Then we have a split short exact sequence:*

$$1 \to \overline{\mathrm{Int}}(\mathcal{R}_{0,1}) \to \mathrm{Aut}(\mathcal{R}_{0,1}) \overset{\mathrm{mod}}{\to} \mathbf{R}_+^* \to 1. \tag{1}$$

Proof. Let $\mathcal{R}$ be an AFD factor of type III_1, and $\mathcal{R} = \mathcal{R}_{0,1} \rtimes_\theta \mathbf{R}$ be its continuous decomposition. Since $\mathrm{mod}(\theta_s) = e^s$, $s \in \mathbf{R}$, the map:$a \in \mathbf{R}_+^* \to \theta_{\log(a)} \in \mathrm{Aut}(\mathcal{R}_{0,1})$ gives the right inverse of the module map of $\mathcal{R}_{0,1}$.

On the other hand, it is well known that $\alpha \in \overline{\mathrm{Int}}(\mathcal{R}_{0,1})$ if and only if α preserves the trace of $\mathcal{R}_{0,1}$ i.e. $\mathrm{mod}(\alpha) = 1$. Q.E.D.

For a factor $\mathcal{M}$, we denote by $\mathrm{Cnt}(\mathcal{M})$ the group of all those automorphisms α of $\mathcal{M}$ such that $\{\alpha(x_n) - x_n\}$ converges σ^*-strongly to zero for every strongly central sequence $\{x_n\}$.

We now state the main theorem announced by A. Connes [5]:

Theorem 1.3. *Let $\mathcal{R}$ be an AFD factor of type* III. *Then we have the following:*

(i) $\mathrm{Ker}(\mathrm{mod}) = \overline{\mathrm{Int}}(\mathcal{R})$;

(ii) $\mathrm{Cnt}(\mathcal{R}) = \mathrm{Int}(\mathcal{R}) \circ \mathrm{Aut}(\mathcal{R}/\mathcal{R}_\varphi)$

for a dominant weight φ where $\mathrm{Aut}(\mathcal{R}/\mathcal{R}_\varphi)$ *means the group of all those automorphisms of $\mathcal{R}$ which fixes $\mathcal{R}_\varphi$ elementwise. Therefore, each $\alpha \in$* $\mathrm{Cnt}(\mathcal{R})$ *is of the form $\alpha = \mathrm{Ad}(u) \circ \overline{\sigma}_c^\varphi$ for some $u \in \mathcal{U}(\mathcal{R})$ and a one cocycle $c \in Z_\theta^1(\mathbf{R}, \mathcal{U}(\mathcal{F}(\mathcal{R})))$, where $\overline{\sigma}_c^\varphi$ is the extended modular automorphism corresponding to a cocycle c of the flow of weights* $\mathcal{F}(\mathcal{R})$ of $\mathcal{R}$.

We shall prove this result according to the type of $\mathcal{R}$, i.e. the type III_0 case, the type III_λ, $0 < \lambda < 1$, case and the type III_1 case.

It was proven in [11] that $\overline{\mathrm{Int}}(\mathcal{M})$ is contained in $\mathrm{Ker}(\mathrm{mod})$ for an arbitrary factor $\mathcal{M}$ of type III. We now handle the case of type III_0.

Lemma 1.4. *Let $\mathcal{M}$ be an AFD factor of type* III_0 *and $\alpha \in$* $\mathrm{Aut}(\mathcal{M})$. *If* $\mathrm{mod}(\alpha) =$ *id, then there exist a faithful lacunary weight ψ on $\mathcal{M}$ with infinite multiplicity and $u \in \mathcal{U}(\mathcal{M})$, the group of unitaries of $\mathcal{M}$, with the properties:*

(i) *In the discrete decomposition $\mathcal{M} = \mathcal{M}_\psi \rtimes_\theta \mathbf{Z}$, we have* $\mathrm{Ad}(u) \circ \alpha|_{\mathcal{C}_\psi} =$ *id, where $\mathcal{C}_\psi$ means the center of $\mathcal{M}_\psi$;*

(ii) $\psi \circ \mathrm{Ad}(u) \circ \alpha|_{\mathcal{M}_\psi} = \psi|_{\mathcal{M}_\psi}$;

(iii) $\mathrm{Ad}(u) \circ \alpha(U) = U$, *where U is the unitary of $\mathcal{M}$ corresponding to the crossed product of* (i).

Proof. The existence of ψ and u satisfying (i) and (ii) follows from a result of Connes-Takesaki, [11; Theorem IV.1.10, p.555]. Replacing α by $\mathrm{Ad}(u) \circ \alpha$, we compute for $x \in \mathcal{C}_\psi$:

$$\alpha(U)U^*x = \alpha(U)\theta^{-1}(x)U^* = \alpha(U\theta^{-1}(x))U^* = \alpha(xU)U^* = x\alpha(U)U^*.$$

By [11, Corollary I.2.10], $\alpha(U)U^*$ belongs to $\mathcal{M}_\psi$. Then the stability of θ, [11, Theorem III.5.1] implies the existence of $v \in \mathcal{U}(\mathcal{M}_\psi)$ with

$$\alpha(U)U^* = v^*\theta(v) = v^*UvU^*.$$

Therefore $\mathrm{Ad}(v) \circ \alpha$ fixes U. Q.E.D.

Lemma 1.5. *Let $\mathcal{M}$, ψ and θ be as in the last lemma, and set $\mathcal{N} = \mathcal{M}_\psi$. Consider a free ultrafilter ω on $\mathbf{N}$. Then for any $n \in \mathbf{N}$ and any countable subset $\{x_j : j \in \mathbf{N}\}$ of $\mathcal{N}_\omega$, there exists a partition $\{F_k : 1 \leq k \leq n\}$ of unity in $\mathcal{N}_\omega$ such that*

(i) *each F_j commutes with $\{x_k\}$;*

(ii) *$\theta_\omega(F_k) = F_{k+1}$, $k = 1, 2, \cdots, n$, where $F_{n+1} = F_1$*

Proof. Since the action of θ on the center $\mathcal{C}$ of $\mathcal{N}$ is free and ergodic, the proof of Lemma 2.1.4 of [4] based on the usual Rohlin Lemma applies. Since $\mathcal{C}^\omega$ is contained in the center $\mathcal{Z}(\mathcal{N}_\omega)$, we complete the proof. Q.E.D.

Lemma 1.6. *With $\mathcal{M}$, $\mathcal{N}$, θ and ω as in the previous lemma, θ_ω is stable on $\mathcal{N}_\omega$ in the sense that*

$$\mathcal{U}(\mathcal{N}_\omega) = \{v\theta_\omega(v^*) : v \in \mathcal{U}(\mathcal{N}_\omega)\}.$$

Proof. The proof of [4; Theorem 2.1.3] works with our Lemma 1.5.

Proof of Theorem 1.3. (i) for AFD factors $\mathcal{M}$ of type III_0: Suppose $\alpha \in \mathrm{Ker}(\mathrm{mod})$. Applying Lemma 1.4 to α, we replace α by $\mathrm{Ad}(\mathcal{U}) \circ \alpha$. With $\mathcal{N} = \mathcal{M}_\psi$ as in Lemma 1.5, ψ is invariant under α and $\mathcal{N}$ is isomorphic to $L^\infty(X, \mu)\overline{\otimes}\mathcal{R}_{0,1}$, where $\{X, \mu\}$ is a non-atomic standard measure space and $\mathcal{R}_{0,1}$ is the AFD factor of type II_∞, i.e. $\mathcal{R}_{0,1} = \mathcal{R}_0\overline{\otimes}\mathcal{L}(\ell^2)$ with $\mathcal{R}_0$ the AFD II_1-factor. It follows from [; Corollary 6] that $\alpha|_\mathcal{N} \in \overline{\mathrm{Int}}(\mathcal{N})$, so that there exists a sequence $\{u_n\}$ in $\mathcal{U}(\mathcal{N})$ such that $\alpha|_\mathcal{N} = \lim_{n\to\infty} \mathrm{Ad}(u_n)$. Since θ and α commute, $\{u_n\theta(u_n^*)\}$ is strongly central, so that there exists a strongly ω-central sequence $\{v_n\}$ in $\mathcal{U}(\mathcal{N})$ such that

$$\pi_\omega(\{u_n\theta(u_n^*)\}) = \pi_\omega(\{v_n\theta(v_n^*)\})$$

by Lemma 1.6, where π_ω means the canonical map from the C*-algebra of strongly ω-central sequences onto $\mathcal{N}_\omega$. Replacing $\{u_n\}$ by $\{v_n^*U_n\}$ and choosing a subsequence, we conclude that $\alpha|_\mathcal{N} = \lim_{n\to\infty} \mathrm{Ad}(u_n)$ and $\{u_n - \theta(u_n)\}$ converges to zero σ^*-strongly, and $u_n \in \mathcal{U}(\mathcal{N})$.

We now prove that $\alpha = \lim_{n\to\infty} \mathrm{Ad}(u_n)$. It suffices to prove that $\|\varphi \circ \alpha - u_n^* \varphi u_n\| \to 0$ and $\|\varphi \circ \alpha^{-1} - u_n \varphi u_n^*\| \to 0$ for each φ in a total subset of $\mathcal{M}_*$.

Let $\mathcal{E}_0$ be the normal conditional expectation of $\mathcal{M}$ onto $\mathcal{N}$ relative to ψ. Set $\mathcal{E}_k(x) = \mathcal{E}_0(xU^{-k})U^k$, $k \in \mathbf{Z}$. Then $\{\varphi \circ \mathcal{E}_k : \varphi \in \mathcal{M}_*, k \in \mathbf{Z}\}$ is total in $\mathcal{M}_*$. Fix $x \in \mathcal{M}$ and $k \in \mathbf{Z}$. With $z_k = \mathcal{E}_0(xU^{-k}) \in \mathcal{N}$, we have

$$\begin{aligned}
&|\langle x, \varphi \circ \mathcal{E}_k \circ \alpha - u_n^*(\varphi \circ \mathcal{E}_k)u_n\rangle| = |\langle \alpha(x) - u_n x u_n^*, \varphi \circ \mathcal{E}_k\rangle| \\
&\quad = |\langle \alpha(z_k)U^k - u_n z_k U^k u_n^*, \varphi\rangle| = |\langle \alpha(z_k)U^k - u_n z_k \theta^k(u_n^*)U^k, \varphi\rangle| \\
&\quad \leqq |\langle \alpha(z_k) - u_n z_k u_n^*, U^k\varphi\rangle| + |\langle u_n z_k u_n^*, (1 - u_n\theta^k(u_n^*))U^k\varphi\rangle| \\
&\quad \leqq \|(U^k\varphi)|_{\mathcal{N}} \circ (\alpha - \mathrm{Ad}(u_n))\| \, \|z_k\| + \|((1 - u_n\theta^k(u_n^*))U^k\varphi))|_{\mathcal{N}}\| \, \|z_k\| \\
&\quad \leqq \|x\| \, \|\psi_k \circ (\alpha - \mathrm{Ad}(u_n))\| + \|(1 - u_n\theta^k(u_n^*))\psi_k\| \, \|x\|,
\end{aligned}$$

where $\psi_k = (U^k\varphi)|_{\mathcal{N}}$. This estimate shows the convergence of $\|\varphi \circ [\alpha - \mathrm{Ad}(u_n)]\|$. The other convergence follows similarly. Q.E.D.

Lemma 1.7. *Let $\mathcal{M}$ be a factor of type* III_0 *and ψ a lacunary weight with infinite multiplicity. For each $\alpha \in \mathrm{Aut}(\mathcal{M})$, there exists a projection $e \in \mathcal{C}_\psi$ and a partial isometry $u \in \mathcal{M}$ such that*

(i) $u^*u = \alpha(e)$ *and* $uu^* = e$*;*

(ii) $u\alpha(\mathcal{M}_{\psi,e})u^* = \mathcal{M}_{\psi,e}$.

Proof. Choose $\delta > 0$ so small that $[-\delta, \delta] \cap \mathrm{Sp}(\sigma^\psi) = \{0\}$. Let $\overline{\psi} = \psi \circ \alpha^{-1} \oplus \psi$ be the balanced weight on $\mathcal{M} \otimes M_2(\mathbf{C})$. Choose a nonzero element $\overline{x} = x \otimes e_{12} \in \mathcal{M}_{\overline{\psi}}[c - \delta/3,\ c + \delta/3]$, the Arveson spectral subspace for $\sigma^{\overline{\psi}}$ for some $c \in \mathbf{R}$. For every $a \in \mathcal{M}_\psi$, we have

$$xax^* \otimes e_{11} = \overline{x}(a \otimes e_{22})\overline{x}^* \in \mathcal{M}_{\psi\circ\alpha^{-1}}[-2\delta/3, 2\delta/3] \otimes e_{11}\,.$$

But the choice of δ forces $\mathcal{M}_{\psi\circ\alpha^{-1}}[-2\delta/3,\ 2\delta/3] = \mathcal{M}_{\psi\circ\alpha^{-1}}$, so that we conclude

$$x\mathcal{M}_\psi x^* \subset \mathcal{M}_{\psi\circ\alpha^{-1}}\ ;\ x^*\mathcal{M}_{\psi\circ\alpha^{-1}}x \subset \mathcal{M}_\psi.$$

Let $x = uh$ be the polar decomposition. Then we have $h \in \mathcal{M}_\psi$, $u^*u = f \in \mathcal{M}_\psi$ and $uu^* = g \in \mathcal{M}_{\psi\circ\alpha^{-1}}$, and that

$$u\mathcal{M}_{\psi,f}u^* = \mathcal{M}_{\psi\circ\alpha^{-1},g}; \quad u^*\mathcal{M}_{\psi\circ\alpha^{-1},g}u = \mathcal{M}_{\psi,f}\ .$$

Since we may replace x by $x \otimes 1$ and ψ by $\psi \otimes \mathrm{Tr}$ on $\mathcal{M}\overline{\otimes}\mathcal{L}(\ell^2)$, we may assume that f (resp. g) is properly infinite in $\mathcal{M}_\psi$ (resp. $\mathcal{M}_{\psi\circ\alpha^{-1}}$). Let $\overline{f}$ (resp $\overline{g}$) be the central support of f in $\mathcal{M}_\psi$ (resp. g in $\mathcal{M}_{\psi\circ\alpha^{-1}}$). Then

$f \sim \overline{f}$ in $\mathcal{M}_\psi$ and $g \sim \overline{g}$ in $\mathcal{M}_{\psi\circ\alpha^{-1}}$. There exists a partial isometry $v \in \mathcal{M}_\psi$ (resp. $w \in \mathcal{M}_{\psi\circ\alpha^{-1}}$) such that

$$vv^* = f,\ v^*v = \overline{f};\ w^*w = g,\ ww^* = \overline{g}.$$

Let $U = wuv$. Then we have $U^*U = \overline{f}$, $UU^* = \overline{g}$ and $U(\mathcal{M}_{\psi,\overline{f}})U^* = \mathcal{M}_{\psi\circ\alpha^{-1},\overline{g}}$. Since $\mathcal{M}_{\psi\circ\alpha^{-1}} = \alpha(\mathcal{M}_\psi)$, $\overline{g}$ is of the form $\overline{g} = \alpha(e)$, $e \in \operatorname{Proj}(\mathcal{C}_\psi)$. We now compare e and $\overline{f}$ in $\mathcal{C}_\psi$ under the Hopf equivalence given by the ergodic automorphism θ on $\mathcal{C}_\psi$, where θ is the automorphism of Lemma 1.4.(i). Let $p \simeq q$ denote the Hopf equivalence of $p, q \in \operatorname{Proj}(\mathcal{C}_\psi)$. Decompose $e = e_1 + e_2 + \cdots + e_n + e_{n+1}$ in such a way that

$$e_i \simeq \overline{f},\ 1 \le i \le n, \text{ and } e_{n+1} \simeq \overline{f}' \le \overline{f},\ \overline{f}' \in \operatorname{Proj}(\mathcal{C}_\psi).$$

Since $\mathcal{M}_\psi$ is properly infinite, there exists a partition:

$$\overline{f} = f_1 + f_2 + \cdots + f_n + f_{n+1}$$

such that $\overline{f} \sim f_i$, $1 \le i \le n$, and $f_{n+1} \sim \overline{f}'$ in $\mathcal{M}_\psi$. The Hopf equivalence $e_i \simeq \overline{f}$ implies the existence of partial isometries v_i such that $v_i^* v_i = e_i$, $v_i v_i^* = \overline{f}$ and $v_i \mathcal{M}_\psi v_i^* \subset \mathcal{M}_\psi$. Putting these things together, we obtain a partial isometry V such that

$$V^*V = e, \quad VV^* = \overline{f}, \quad V\mathcal{M}_\psi V^* = \mathcal{M}_{\psi,\overline{f}}.$$

Thus we come to the situation that

$$UV\mathcal{M}_{\psi,e}V^*U^* = \mathcal{M}_{\psi\circ\alpha^{-1},\overline{f}} = \alpha(\mathcal{M}_{\psi,e}).$$

This completes the proof. Q.E.D.

Lemma 1.8. *If $\mathcal{M}$ is a factor of type* III_0 *and $\alpha \in \operatorname{Aut}(\mathcal{M})$, then there exists a lacunary weight $\overline{\psi}$ with infinite multiplicity and $V \in \mathcal{U}(\mathcal{M})$ such that*

$$\operatorname{Ad}(V) \circ \alpha(\mathcal{M}_{\overline{\psi}}) = \mathcal{M}_{\overline{\psi}}.$$

Proof. By the last lemma, there exists a partial isometry u and a lacunary weight ψ with infinite multiplicity such that

$$uu^* = e \in \operatorname{Proj}(\mathcal{C}_\psi),\ u^*u = \alpha(e);\ u\alpha(\mathcal{M}_{\psi,e})u^* = \mathcal{M}_{\psi,e}.$$

Let v be an isometry in $\mathcal{M}_\psi$ with $e = vv^*$. Set

$$\overline{\psi}(x) = \psi(vxv^*), x \in \mathcal{M}, V = v^*u\alpha(v) \in \mathcal{U}(\mathcal{M}).$$

Then we have $\mathrm{Ad}(V) \circ \alpha(\mathcal{M}_{\overline{\psi}}) = \mathcal{M}_{\overline{\psi}}$. Q.E.D.

Lemma 1.9. *Let $\mathcal{M} = \mathcal{N} \rtimes_\theta \mathbf{Z}$ be the discrete decomposition of a strongly stable factor $\mathcal{M}$ of type* III_λ, $0 \leq \lambda < 1$. *The set of all strongly central sequence of $\mathcal{M}$ is given, within the equivalence, as the set of all those strongly central sequences $\{x_n\}$ in $\mathcal{N}$ such that $\{\theta(x_n) - x_n\}$ converges to zero σ^*-strongly.*

Proof. Let ψ be a faithful normal state on $\mathcal{M}$. Let σ be the action of $\mathbf{T} = \hat{\mathbf{Z}}$ on $\mathcal{M}$ dual to θ. Then the conditional expectation $\mathcal{E}_0$ of $\mathcal{M}$ onto $\mathcal{N}$ is given by

$$\mathcal{E}_0(x) = \int_{\mathbf{T}} \sigma_t(x)dt, \ x \in \mathcal{M}.$$

Since σ_t is an extended modular automorphism of the lacunary weight $\overline{\psi}$ associated with the discrete decomposition $\mathcal{M} = \mathcal{N} \rtimes_\theta \mathbf{Z}$, i.e. $\overline{\psi} = \hat{\tau}$ dual to the trace τ on $\mathcal{N}$, and the cocycle corresponding to σ_t is invariant under $\overline{\mathrm{Int}}(\mathcal{M})$, see [11; Theorem IV.2.6], $\mathcal{E}(\sigma_t)$, the image of σ_t in $\mathrm{Out}\mathcal{M} = \mathrm{Aut}(\mathcal{M})/\mathrm{Int}(\mathcal{M})$, belongs to the center of $\mathrm{Out}(\mathcal{M})$. Therefore it follows from [3; Corollary 2.3.2] that σ_t is centrally trivial, i.e. $\lim_{n\to\infty} \|\sigma_t(x_n) - x_n\|_\psi^\# = 0$ for every strongly central sequence $\{x_n\}$. The Lebesgue dominated convergence theorem entails that $\|\mathcal{E}_0(x_n) - x_n\|_\psi^\# \to 0$, so that $\{x_n\} \sim \{\mathcal{E}_0(x_n)\}$. Since $[x_n, U] \to 0$ σ^*-strongly, where U means of course the unitary corresponding to the crossed product decomposition, we obtain $\{\theta(\mathcal{E}_0(x_n))\} \sim \{x_n\}$.

Conversely, suppose that $\{y_n\}$ is a strongly central sequence in $\mathcal{N}$ such that $\|\theta(y_n) - y_n\|_\psi^\# \to 0$. We want to prove that $\|[y_n, \psi]\| \to 0$ for every $\varphi \in \mathcal{M}_*$. By the boundedness of $\{y_n\}$, it suffices to prove that $\|[y_n, \psi]\| \to 0$ for a total subset of ψ in $\mathcal{M}_*$. Let $\mathcal{E}_k$ be as in the proof of Lemma 1.6. As before, we have only to prove that $\|[y_n, \varphi \circ \mathcal{E}_k]\| \to 0$. With $z_k = \mathcal{E}_0(xU^{-k}) \in \mathcal{N}$ as before, we compute

$$\begin{aligned}
|\langle x, [y_n, \varphi \circ \mathcal{E}_k]\rangle| &= |\langle xy_n - y_n x, \varphi \circ \mathcal{E}_k\rangle| = |\langle z_k U^k y_n - y_n z_k U^k, \varphi\rangle| \\
&\leqq |\langle (z_k \theta^k(y_n) - z_k y_n)U^k, \varphi\rangle| + |\langle z_k y_n - y_n z_k, U^k \varphi\rangle| \\
&\leqq \|z_k\| \, \|[(\theta^k(y_n) - y_n)U^k \varphi]|_{\mathcal{N}}\| + \|[y_n, U^k \varphi]|_{\mathcal{N}}\| \, \|z_k\|
\end{aligned}$$

which converges to zero uniformly in x with $\|x\| \leq 1$. Therefore, $\{y_n\}$ is strongly central. Q.E.D.

Lemma 1.10. *Let $\mathcal{M}$ be an AFD factor of type* III_0 *and* $\alpha \in \mathrm{Cnt}(\mathcal{M})$. *Then there exists a lacunary weight ψ with infinite multiplicity on $\mathcal{M}$ and $u \in \mathcal{U}(\mathcal{M})$ such that*

(i) $\mathrm{Ad}(u) \circ \alpha(\mathcal{M}_\psi) = \mathcal{M}_\psi$;

(ii) $\mathrm{Ad}(u) \circ \alpha|_{\mathcal{C}_\psi} = id$;

(iii) $\mathrm{Ad}(u) \circ \alpha(U) = U$, *where U is the unitary corresponding to the decomposition $\mathcal{M} = \mathcal{M}_\psi \rtimes_\theta \mathbf{Z}$.*

Proof. By Lemma 1.8, we may assume that $\alpha(\mathcal{M}_\psi) = \mathcal{M}_\psi$ for some lacunary weight ψ with infinite multiplicity. Fix a free ultrafilter ω on $\mathbf{N}$. If $\{x_n\}$ is a bounded sequence in $\mathcal{C}_\psi$ such that $\lim_{n\to\omega}(\theta(x_n) - x_n) = 0$ in the σ^*-strong convergence, then $\{x_n\}$ is strongly ω-central by the previous lemma, so that $\lim_{n\to\omega}(\alpha(x_n) - x_n) = 0$ in the σ^*-strong topology. This means that $(\alpha|_{\mathcal{C}_\psi})_\omega = id$ on $(\mathcal{C}_\psi)_\omega$. Since $\alpha \in \mathrm{Aut}(\mathcal{M})$, $\alpha|_{\mathcal{C}_\psi}$ belongs to the normalizer $N[\theta|_{\mathcal{C}_\psi}]$ of the full group of $\theta|_{\mathcal{C}_\psi}$. Thus Lemma 2.4 in [9] implies $\alpha|_{\mathcal{C}_\psi} \in [\theta|_{\mathcal{C}_\psi}]$. Thus there exists $u \in \mathcal{M}$ such that $\mathrm{Ad}(u) \circ \alpha|_{\mathcal{C}_\psi} = id$, and $\mathrm{Ad}(u) \circ \alpha(\mathcal{M}_\psi) = \mathcal{M}_\psi$. The exactly same arguments of Lemma 1.4. (iii) applied to $\mathrm{Ad}(u) \circ \alpha$ shows that we can further perturb $\mathrm{Ad}(u) \circ \alpha$ by $\mathrm{Int}(\mathcal{M}_\psi)$ so that the resulted automorphism leaves U fixed. Q.E.D.

Lemma 1.11. *In the context of Lemma* 1.10, *α can be perturbed by* $\mathrm{Int}(\mathcal{M}_\psi)$ *into the form $\alpha_0 \otimes id$ on $\mathcal{M}_\psi = \mathcal{R}_{0,1} \bar{\otimes} L^\infty(X, \mu)$.*

Proof. By Lemma 1.10, we may view θ and α as an action of $\mathbf{Z}^2$ on $\mathcal{M}_\psi$. Then by Theorems 1.2 and 3.1 in [19], we can perturb α to the desired form. Q.E.D.

Lemma 1.12. *If T is a non-singular ergodic transformation on a probability measure space $\{X, \mu\}$, then for any $n \in \mathbf{N}$, there exists a measurable subset E of X such that*

(i) *$E, T^{-1}E, \dots, T^{-n+1}E$ are mutually disjoint;*

(ii) $\mu\left(\bigcup_{j=0}^{n-1} T^{-j}E\right) \geq 1 - 1/n$;

(iii) $\mu(E) < 1/n$, $\mu(T^{-n}E) \leq 2/n$.

Proof. Choose a measurable subset A_0 of X such that $\mu(T^{-j}A_0) < 1/2n^2$ for $j = 0, 1, \dots, 2(n-1)$, and set

$$A_m = \{x \in X : T^m x \in A_0,\ T^j x \notin A_0,\ 0 \leq j \leq m-1\}.$$

We then have $X = \bigcup_{m=0}^\infty A_m$. Set $F = \bigcup_{k=1}^\infty A_{kn}$. As in the usual proof of the Rohlin lemma, we conclude that $F, T^{-1}F, \cdots, T^{-(n-1)}F$ are mutually disjoint and $\bigcup_{j=0}^{n-1} T^{-j}F \supset X - \bigcup_{j=0}^{n-1} T^{-j}A_0$. Therefore, we have

$\mu(T^{-j_0}F) \leq 1/n$ for some j_0, $0 \leq j_0 \leq n-1$. We then set $E = T^{-j_0}F$. It follows that $E, T^{-1}E, \dots, T^{-(n-1)}E$ are mutually disjoint. Since we have

$$\text{(a)}\quad T^{-n}F \subset F \cup A_0 \cup \cdots \cup A_{n-1}; \quad \text{(b)}\quad T^{n}F \subset F \cup A_0,$$

we get

$$\begin{aligned} &\left(\bigcup_{k=0}^{n-1} T^{-k}E\right) \cup \left(\bigcup_{\ell=0}^{j_0-1} T^{-(n+\ell)}A_0\right) \\ &\quad = \left(\bigcup_{k=0}^{j_0-1} T^{-k}E\right) \cup \left(\bigcup_{\ell=0}^{j_0-1} T^{-(n+\ell)}(F \cup A_0)\right) \supset \bigcup_{k=0}^{j_0} T^{-k}F, \end{aligned}$$

so that

$$1 - \frac{1}{2n} \leqq \mu\left(\bigcup_{j=0}^{n-1} T^{-j}F\right) \leqq \mu\left(\bigcup_{j=0}^{n-1} T^{-j}E\right) + n \cdot \frac{1}{2n^2}.$$

Hence (ii) follows. By (a), we have

$$\begin{aligned} \mu(T^{-n}E) &= \mu(T^{-j_0}T^{-n}F) \\ &\leqq \mu(T^{-j_0}F) + \mu(T^{-j_0}A_0) + \cdots + \mu(T^{-j_0}A_{n-1}) \\ &\leqq \frac{1}{n} + n \cdot \frac{1}{2n^2} \leq \frac{2}{n}, \end{aligned}$$

which proves (iii). Q.E.D.

Lemma 1.13. *Let $\mathcal{M}$ be a strongly stable factor and $\alpha \in \mathrm{Aut}(\mathcal{M})$. If $\alpha \notin Cnt(\mathcal{M})$, then for any free ultrafilter ω, there exists $\gamma \in \mathbf{C}, |\gamma| = 1,\ \gamma \neq 1$, and $u \in \mathcal{U}(\mathcal{M}_\omega)$ such that $\alpha_\omega(u) = \gamma\mu$..*

Proof. Since $p_a(\alpha) \neq 1$, α is outer conjugate to $\alpha \otimes s_p$ for some $p > 1$ in $\mathcal{M} \cong \mathcal{M}\overline{\otimes}\mathcal{R}_0$, where $\mathcal{R}_0$ is the AFD factor of type II_1 and s_p denotes the action of $\mathbf{Z}/p\mathbf{Z}$ on $\mathcal{R}_0$ constructed by Connes [6]. Thus, we may assume that α is of the form $\alpha \otimes s_p$. Then we can construct a central sequence $\{u_n\}$ of unitaries in $\mathcal{R}_0$ such that $s_p(u_n) = \gamma u_n$, $\gamma = \exp(2\pi i/p)$. The sequence $\{1 \otimes u_n\}$ in $\mathcal{M}\overline{\otimes}\mathcal{R}_0 \cong \mathcal{M}$ is strongly central, and we set this sequence to be u in $\mathcal{M}_\omega$. Q.E.D.

Lemma 1.14. *The automorphism α_0 of $\mathcal{R}_{0,1}$ in Lemma 1.11 belongs to* $\mathrm{Cnt}(\mathcal{R}_{0,1})$.

Proof. We keep the notations of Lemma 1.10. Since θ is ergodic on $\mathcal{C}_\psi$, we can apply Lemma 1.12 to $\{\mathcal{C}_\psi, \theta\}$ to find a projection $e_n \in \mathcal{C}_\psi$ such that

(i) $\{\theta^j(e_n) : 0 \leqq j \leqq n-1\}$ are orthogonal;

(ii) $\mu\left(\sum_{j=0}^{n-1} \theta^j(e_n)\right) \geqq 1 - \frac{1}{n}$;

(iii) $\mu(\theta^n(e_n)) \leqq \frac{2}{n}$, $\mu(e_n) \leqq \frac{1}{n}$.

Suppose $\alpha_0 \notin \mathrm{Cnt}(\mathcal{R}_{0,1})$, and choose a strongly central sequence $\{u_m\}$ of unitaries in $\mathcal{R}_{0,1}$ such that $\alpha_0(u_m) - \gamma u_m \to 0$ σ^*-strongly as $m \to \infty$ for some $\gamma \in \mathbf{C}$, $|\gamma| = 1$, $\gamma \neq 1$. Choose a normal state φ on $\mathcal{R}_{0,1}$ and a dense sequence $\{\varphi_n\}$ in $(\mathcal{R}_{0,1}\overline{\otimes}\mathcal{C}_\psi)_*$. For each n, the sequence $\{u_m \otimes e_n\}_{m=1}^\infty$ is strongly central in $\mathcal{M}_\psi = \mathcal{R}_{0,1}\overline{\otimes}\mathcal{C}_\psi$. Thus there exists an integer $m = m(n)$ such that

(a) $\|(\alpha_0(u_m) - \gamma u_m) \otimes e_n\|^\#_{(\varphi\otimes\mu)\circ\theta^j} \leqq \frac{1}{n^2}$, $0 \leqq j \leqq n-1$;

(b) $\|\,[u_m \otimes e_n, \varphi_k \circ \theta^{-j}]\,\| \leqq \frac{1}{n^2}$, $0 \leqq j \leqq n-1$, $k = 1, 2, \cdots, n$.

Set $x_n = \sum_{j=0}^{n-1} \theta^j(u_{m(n)} \otimes e_n)$. We show that $\{x_n\}$ is strongly central in $\mathcal{M}$. First, we see $\|x\| = 1$. For each $k < n$, we have

$$\|\,[x_n, \varphi_k]\,\| \leqq \sum_{j=0}^{n-1} \|\,[\theta^j(u_{m(n)} \otimes e_n), \varphi_k]\,\| \leqq \frac{1}{n} \to 0$$

as $n \to \infty$ by (b). We also have

$$\begin{aligned}\|\theta(x_n) - x_n\|^\#_{\varphi\otimes\mu} &= \|\theta^n(u_{m(n)} \otimes e_n) - u_{m(n)} \otimes e_n\|^\#_{\varphi\otimes\mu} \\ &\leqq \mu(\theta^n(e_n))^{1/2} + \mu(e_n)^{1/2} \leqq 2\left(\frac{2}{n}\right)^{1/2} \to 0\end{aligned}$$

as $n \to \infty$. Hence $\{x_n\}$ is strongly central in $\mathcal{M}$ by Lemma 1.9. On the other hand,

$$\begin{aligned}\|(\alpha_0 \otimes id)(x_n) - \gamma x_n\|^\#_{\varphi\otimes\mu} &= \|\sum_{j=0}^{n-1} \theta^j((\alpha_0(u_{m(n)}) - \gamma u_{m(n)}) \otimes e_n)\|^\#_{\varphi\otimes\mu} \\ &\leqq \sum_{j=0}^{n-1} \|(\alpha_0(u_{m(n)}) - \gamma u_{m(n)}) \otimes e_n\|^\#_{(\varphi\otimes\mu)\circ\theta^j} \\ &\leqq \frac{1}{n} \to 0\end{aligned}$$

as $n \to \infty$ by (a). Since $\gamma \neq 1$, $\alpha_0 \otimes id$ does not belong to $\mathrm{Cnt}(\mathcal{M})$. Q.E.D.

Proof of Theorem 1.3. (ii) **for AFD** III_0**-Factors:** We keep the previous notations. Suppose $\alpha \in \mathrm{Cnt}(\mathcal{M})$ and that $\mathcal{M}$ is an AFD factor of type III_0. By [4; Lemma 5] and the last lemma, $\alpha|_{\mathcal{M}_\psi}$ is inner. So after inner perturbation, we get $\alpha|_{\mathcal{M}_\psi} = id$. Then α must be an extended modular automorphism by the relative commutant theorem in [11]. Q.E.D.

We now move on to the case AFD factors of type III_λ, $0 < \lambda < 1$.

Proof of Theorem 1.3 for AFD Factors of type III_λ: (i) By Connes-Takesaki, [11; p. 554], $\overline{\mathrm{Int}}(\mathcal{M}) \subset \mathrm{Ker}\,(\mathrm{mod})$. Suppose that $\mathrm{mod}(\alpha) = 1$, $\alpha \in \mathrm{Aut}(\mathcal{M})$, where $\mathcal{M}$ is an AFD factor of type III_λ, $0 < \lambda < 1$. Let φ be a generalized trace on $\mathcal{M}$. Then we have $\varphi \circ \alpha \circ \mathrm{Ad}(u) = \varphi$ for some $u \in \mathcal{U}(\mathcal{M})$. Replacing α by $\alpha \circ \mathrm{Ad}(u)$, we assume $\varphi \circ \alpha = \varphi$, so that α and $\{\sigma_t^\varphi\}$ commute. Hence $\alpha(\mathcal{M}_\varphi) = \mathcal{M}_\varphi$. Let $\mathcal{M} = \mathcal{N} \rtimes_\theta \mathbf{Z}$ be the discrete decomposition with $\mathcal{N} = \mathcal{M}_\varphi$, and U be the corresponding unitary of $\mathcal{M}$, i.e. $\sigma_t^\varphi(U) = \lambda^{it}U$. It then follows that $\mathcal{N} \cong \mathcal{R}_{0,1}$ and $\alpha(U)U^* = v \in \mathcal{N}$. By the stability of θ, there exists $w \in \mathcal{U}(\mathcal{N})$ with $v = w^*\theta(w)$, which means that $\mathrm{Ad}(w) \circ \alpha(U) = U$. Replace α by $\mathrm{Ad}(w) \circ \alpha$, so that $\varphi \circ \alpha = \varphi$ and $\alpha(U) = U$. Since $\varphi|_\mathcal{N} = \tau$, a trace on $\mathcal{N}$, we have $\mathrm{mod}(\alpha|_\mathcal{N}) = 1$, so that $\alpha_0 = \alpha|_\mathcal{N}$ is approximately inner by [4; Cor. 6]. Let $\{u_n\}$ be a sequence in $\mathcal{U}(\mathcal{N})$ such that $\alpha_0 = \lim_{n\to\infty} \mathrm{Ad}(u_n)$. Since θ and α_0 commute, we have also $\alpha_0 = \lim_{n\to\infty} \mathrm{Ad}(\theta(u_n))$. Hence $\{u_n^*\theta(u_n)\}$ is strongly central in $\mathcal{N}$. Hence by the stability of θ_ω, [4; Theorem 2.1.3], there exists a strongly central sequence $\{v_n\}$ in $\mathcal{U}(\mathcal{N})$ such that $\{u_n^*\theta(u_n)\} \sim \{v_n^*\theta(v_n)\}$. Hence $\{u_n v_n^* - \theta(u_n v_n^*)\}$ converges to zero σ^*-strongly and we have

$$\alpha_0 = \lim_{n\to\infty} \mathrm{Ad}(u_n) \lim_{n\to\infty} \mathrm{Ad}(v_n^*) = \lim_{n\to\infty} \mathrm{Ad}(u_n v_n^*).$$

An argument similar to the type III_0 case shows that $\lim_{n\to\infty} \mathrm{Ad}(u_n v_n^*) = \alpha$ in $\mathrm{Aut}(\mathcal{M})$. Hence α belongs to $\overline{\mathrm{Int}}(\mathcal{M})$.

(ii) We keep the above notations. By [3; Prop. 2.3], we know $\sigma^\varphi(\mathbf{R}) \cdot \mathrm{Int}(\mathcal{M}) \subset \mathrm{Cnt}(\mathcal{M})$. Suppose $\alpha \in \mathrm{Cnt}(\mathcal{M})$. We first prove that $\mathrm{mod}(\alpha) = 1$. Suppose $\mathrm{mod}(\alpha) \neq 1$. Then there exists μ, $\lambda < \mu < 1$, such that $\varphi\circ\alpha \sim \mu\varphi$, i.e. $F_{-\log\mu} = \mathrm{mod}\,(\alpha)$. Thus $\varphi \circ \alpha \circ \mathrm{Ad}(u) = \mu\varphi$ for some $u \in \mathcal{U}(\mathcal{M})$. Replacing α by $\alpha \circ \mathrm{Ad}(u)$, we assume $\varphi \circ \alpha = \mu\varphi$, $\lambda < \mu < 1$. It then follows that α and $\{\alpha_t^\varphi\}$ commute and $v = \alpha(U)U^* \in \mathcal{N}$. As in (i), we have $v = w^*\theta(w)$ for some $w \in \mathcal{U}(\mathcal{N})$ and $\mathrm{Ad}(w) \circ \alpha(U) = U$. Replacing α by $\mathrm{Ad}(w)\circ\alpha$, we can assume that α is the canonical extension of $\alpha_0 = \alpha|_\mathcal{N}$, i.e. $\alpha(U) = U$. Since $\varphi\circ\alpha = \mu\varphi$, $\tau\circ\alpha_0 = \mu\tau$. Furthermore, α_0 and $\theta = \mathrm{Ad}(v)|_\mathcal{N}$ commute. Since $\mathrm{mod}(\alpha_0) \notin (\log\lambda)\mathbf{Z}$, α_0 and θ gives rise to a free action β of $\mathbf{Z}^2$ on $\mathcal{N}$, hence on $\mathcal{N}_\omega$ for a free ultrafilter ω on $\mathbf{N}$, which means that $\Gamma(\beta) = \mathbf{T}^2 = (\mathbf{Z}^2)^\wedge$. Hence for any $\gamma \in \mathbf{T}$, we have a strongly central sequence $\{x_n\}$ in $\mathcal{N}$ such that $\{\alpha_0(x_n)\} \sim \{\gamma x_n\}$ and $\{\theta(x_n)\} \sim \{x_n\}$. By Lemma 1.9, $\{x_n\}$ is strongly central in $\mathcal{M}$ and $\{\alpha(x_n)\} \sim \{\gamma x_n\}\psi = \psi\{x_n\}$.

Hence α does not belong to $\mathrm{Cnt}(\mathcal{M})$. Thus we have proved $\mathrm{mod}(\alpha) = 1$ for every $\alpha \in \mathrm{Cnt}(\mathcal{M})$.

We continue the discussion with $\alpha \in \mathrm{Cnt}(\mathcal{M})$. We know that α may be assumed to leave φ and U invariant. We claim that $\alpha_0 = \alpha|_{\mathcal{N}}$ is inner. Suppose $\alpha_0 \notin \mathrm{Int}(\mathcal{N})$. Let $\beta_{m,n} = \theta^m \alpha_0^n$ for $(m,n) \in \mathbf{Z}^2$. By the assumption, $\{0\} \times \mathbf{Z} \not\subset \beta^{-1}(\mathrm{Int}(\mathcal{N}) = H \subset \mathbf{Z}^2$. Thus, $\Gamma(\beta^\omega)$ does not annihilate $\{0\} \times \mathbf{Z}$, where β^ω is the action of $\mathbf{Z}^2$ on $\mathcal{N}_\omega$ given by β for a free ultrafilter ω on $\mathbf{N}$. This means that $\alpha_{0,\omega}$ is not trivial on $(\mathcal{N}_\omega)^\theta$, which implies that $\alpha \notin \mathrm{Cnt}(\mathcal{M})$ by Lemma 1.9. Thus $\alpha_0 = \mathrm{Ad}(u)$ for some $u \in \mathcal{U}(\mathcal{N})$. Since θ and α_0 commute, we have $\theta(u) = \lambda^{is} u$ for some $s \in \mathbf{R}$, so that $\mathrm{Ad}(u)(U) = \lambda^{-is} U$. Hence $\mathrm{Ad}(U^*) \circ \alpha$ is trivial on $\mathcal{N}$ and $\mathrm{Ad}(U^*) \circ \alpha(U) = \lambda^{is} U$. Therefore, we conclude that $\mathrm{Ad}(U^*) \circ \alpha = \sigma_s^\varphi$; and therefore $\alpha = \mathrm{Ad}(u) \circ \sigma_s^\varphi$. Q.E.D.

Lemma 1.15. *Fix $0 < \lambda < 1$. If $\{e_{i,j}(k) : 1 \le i,j \le 2\}$ and $\{f_{i,j}(k) : 1 \le i,j \le 2\}$ are respectively mutually commuting sequences of 2×2-matrix units in a separable factor $\mathcal{M}$ such that for every $\psi \in \mathcal{M}_*$*

$$\lim_{k\to\infty} \|\psi e_{i,j}(k) - \lambda^{i-j} e_{i,j}(k)\psi\| = 0;$$
$$\lim_{k\to\infty} \|\psi f_{i,j}(k) - \lambda^{i-j} f_{i,j}(k)\psi\| = 0,$$

Then there exists $\sigma \in \overline{\mathrm{Int}}(\mathcal{M})$ and a increasing sequence $\{k_n : n \in \mathbf{N}\}$ in $\mathbf{N}$ such that

$$\sigma(e_{i,j}(k_n)) = f_{i,j}(k_n), \quad n \in \mathbf{N},\ i,j = 1,2.$$

Proof. Let $\{\psi_\nu\}$ be a dense sequence in the space $\mathfrak{S}_*$ of a normal states on $\mathcal{M}$. Passing to subsequences, we assume that

$$\sum_{k=1}^{\infty} \|\psi_\nu e_{i,j}(k) - \lambda^{i-j} e_{i,j}(k)\psi_\nu\| < +\infty;$$
$$\sum_{k=1}^{\infty} \|\psi_\nu f_{i,j}(k) - \lambda^{i-j} e_{i,j}(k)\psi_\nu\| < +\infty.$$

for $i,j = 1,2$ and $\nu \in \mathbf{N}$, so that the subfactor $\mathcal{P}$ (resp. $\mathcal{Q}$) generated by $\{e_{i,j}(k) : k \in \mathbf{N}, i,j = 1,2\}$ (resp. $\{f_{i,j}(k) : k \in \mathbf{N},\ i,j = 1,2\}$) decomposes $\mathcal{M}$ into a tensor product: $\mathcal{M} = \mathcal{P}\overline{\otimes}\mathcal{P}^c$ (resp. $\mathcal{M} = \mathcal{Q}\overline{\otimes}\mathcal{Q}^c$) by a result of Araki, [1; Theorem 1.3]. Let ω be a free ultrafilter on $\mathbf{N}$. Since every strongly central sequence of $\mathcal{P}$ (resp. $\mathcal{Q}$) is strongly central in $\mathcal{M}$, $\mathcal{P}_\omega$ and $\mathcal{Q}_\omega$ are both von Neumann subalgebras of $\mathcal{M}_\omega$. Since $\mathcal{P}$ and $\mathcal{Q}$ are both an AFD factors of type III_λ, $\mathcal{P}_\omega$ and $\mathcal{Q}_\omega$ are both subfactors of $\mathcal{M}_\omega$ by Proposition 1.1. Thus all tracial states on $\mathcal{M}_\omega$ take the same values on $\mathcal{P}_\omega$

(resp $\mathcal{Q}_\omega$). This means that to prove the equivalence of the projections E and F represented respectively by $\{e_{11}(k)\}$ and $\{f_{11}(k)\}$ we have only to prove that E and F take the same trace value. Let φ be a faithful normal state on $\mathcal{M}$. We then have

$$\begin{aligned}\tau_\omega(E) &= \lim_{k\to\omega}\varphi(e_{11}(k)) = \lim_{k\to\omega}\varphi(e_{12}(k)e_{21}(k))\\ &= \lim_{k\to\omega}\lambda^{-1}\varphi(e_{21}(k)e_{12}(k))\\ &= \frac{1}{\lambda}\tau_\omega(1-E),\end{aligned}$$

so that $\tau_\omega(E) = 1/(1+\lambda)$. Similarly, $\tau_\omega(F) = 1/(1+\lambda)$. Hence E and F are equivalent in $\mathcal{M}_\omega$.

By induction, we construct sequences $\{k_n\}$ in $\mathbf{N}$ and $\{u_n\}$ in $\mathcal{U}(\mathcal{M})$ such that

$$\begin{aligned}&\text{(a)}\quad [u_n, f_{i,j}(k_\nu)] = 0, \quad \nu = 1,2,\cdots,n-1;\\ &\text{(b)}\quad \text{with } v_n = u_n u_{n-1}\cdots u_1, \quad v_n e_{i,j}(k_\nu)v_n^* = f_{i,j}(k_\nu),\ 1\le\nu\le n;\\ &\text{(c)}\quad \|\psi_\nu\circ \mathrm{Ad}(v_n) - \psi_\nu\circ\mathrm{Ad}(v_{n-1})\| < 1/2^n,\\ &\qquad\ \|\psi_\nu\circ \mathrm{Ad}(v_n^*) - \psi_\nu\circ\mathrm{Ad}(v_{n-1}^*)\| < 1/2^n,\ \nu = 1,2,\cdots,n.\end{aligned}$$

Suppose that $\{k_\nu\}$ and $\{u_\nu\}$ have been constructed for $\nu = 1,2,\cdots,n-1$. Let

$$\mathcal{N} = \{f_{i,j}(k_\nu) : \nu = 1,2,\cdots,n-1,\ i,j = 1,2\}^c.$$

Since $v_{n-1}e_{i,j}(k_\nu)v_{n-1}^* = f_{i,j}(k_\nu), 1\le\nu\le n-1$, we have $v_{n-1}e_{i,j}(k)v_{n-1}^* \in \mathcal{N}$ for $k > k_{n-1}$. Let E and F be the projections of $\mathcal{N}_\omega$ corresponding respectively $\{v_{n-1}e_{1,1}(k)v_{n-1}^*;\ k > k_{n-1}\}$ and $\{f_{11}(k) : k > k_{n-1}\}$. As seen above, we have $E \sim F$ in $\mathcal{N}_\omega$, so that there exists a strongly central sequence $\{w_k\}$, passing to a subsequence if necessary, such that

$$\begin{aligned}w_k^* w_k &= v_{n-1}e_{11}(k)v_{n-1}^*\ ,\ k > k_{n-1};\\ w_k w_k^* &= f_{11}(k).\end{aligned}$$

Set

$$x_k = \sum_{j=1}^{2} f_{j,1}(k)w_k v_{n-1}e_{1,j}v_{n-1}^*.$$

Then $\{x_k\} \subset \mathcal{U}(\mathcal{N})$ is strongly central. If k is sufficiently large, then $u_n = x_k$ satisfies the above (a), (b) and (c).

By (c), $\{\mathrm{Ad}(v_n)\}$ is a Cauchy sequence in $\mathrm{Aut}(\mathcal{M})$. With $\sigma = \lim_{n\to\infty} \mathrm{Ad}(v_n) \in \overline{\mathrm{Int}}(\mathcal{M})$, we have

$$\sigma(e_{i,j}(k_n)) = f_{i,j}(k_n).$$

Q.E.D.

Corollary 1.16. *Let $\mathcal{M}$ be a separable factor. Let $\mathcal{P}$ and $\mathcal{Q}$ be AFD subfactors of type* III_λ, $0 < \lambda < 1$. *If $\mathcal{M} = \mathcal{P} \vee \mathcal{P}^c$ and $\mathcal{M} = \mathcal{Q} \vee \mathcal{Q}^c$ are both tensor product factorizations such that $\mathcal{P}^c \cong \mathcal{Q}^c \cong \mathcal{M}$, then there exist $\sigma \in \overline{\mathrm{Int}}(\mathcal{M})$ such that $\sigma(\mathcal{P}) = \mathcal{Q}$.*

Proof. By the last lemma, there exists $\sigma_1 \in \overline{\mathrm{Int}}(\mathcal{M})$ and tensor product factorizations $\mathcal{P} = \mathcal{P}_1 \vee \mathcal{P}_2$ and $\mathcal{Q} = \mathcal{Q}_1 \vee \mathcal{Q}_2$ such that $\sigma_1(\mathcal{P}_1) = \mathcal{Q}_1$. By the assumption: $\mathcal{P}^c \cong \mathcal{Q}^c \cong \mathcal{M}$, we have tensor product factorizations: $\mathcal{P}^c = \mathcal{P}_3 \vee \mathcal{P}_4$ and $\mathcal{Q}^c = \mathcal{Q}_3 \vee \mathcal{Q}_4$ such that $\mathcal{P}_3 \cong \mathcal{Q}_3 \cong \mathcal{P}$ and $\mathcal{P}_4 \cong \mathcal{Q}_4 \cong \mathcal{M}$. Then there exist $\sigma_2 \in \overline{\mathrm{Int}}(\mathcal{P}\vee\mathcal{P}_3)$ and $\sigma_3 \in \overline{\mathrm{Int}}(\mathcal{Q}\vee\mathcal{Q}_3)$ such that $\sigma_2(\mathcal{P}) = \mathcal{P}_1$ and $\sigma_3(\mathcal{Q}_1) = \mathcal{Q}$. Clearly, we can view σ_2 and σ_3 as elements of $\overline{\mathrm{Int}}(\mathcal{M})$, and we have $\sigma = \sigma_3 \circ \sigma_1 \circ \sigma_2 \in \overline{\mathrm{Int}}(\mathcal{M})$, which carries $\mathcal{P}$ onto $\mathcal{Q}$. Q.E.D.

Proof of Theorem 1.3. (i) *for AFD factors of type* III_1: Let $\mathcal{M}$ be an AFD factor of type III_1 and $\alpha \in \mathrm{Aut}(\mathcal{M})$. Fix λ, $0 < \lambda < 1$. Since $\mathcal{M}$ is strongly λ-stable, it admits a sequence $\{e_{i,j}(k) : 1 \le i,j \le 2,\ k \in \mathbf{N}\}$ of mutually commuting 2×2-matrix units such that

$$\lim \|\psi e_{i,j}(k) - \lambda^{i-j} e_{ij}(k)\psi\| = 0, \quad \psi \in \mathcal{M}_*.$$

Applying Lemma 1.15 to $\{e_{i,j}(k)\}$ and $\{f_{i,j}(k)\}$ with $f_{i,j}(k) = \alpha(e_{i,j}(k))$, we find $\sigma \in \overline{\mathrm{Int}}(\mathcal{M})$ and a sequence $\{k_n\} \subset \mathbf{N}$ such that $\alpha(e_{i,j}(k_n)) = \sigma(e_{i,j}(k_n))$. To prove $\alpha \in \overline{\mathrm{Int}}(\mathcal{M})$, we may replace α by $\sigma^{-1}\alpha$ since $\overline{\mathrm{Int}}(\mathcal{M})$ is a subgroup. Thus, we come to the situation: $\alpha(e_{i,j}(k_n)) = e_{i,j}(k_n)$. Considering a subsequence if necessary, we may assume that $\{e_{i,j}(k_n)\}$ generates a subfactor $\mathcal{P}$ of type III_λ which factorizes $\mathcal{M}$ tensorially in such a way that $\mathcal{M} \cong \mathcal{P}^c$. With the decomposition: $\mathcal{M} = \mathcal{P}\bar{\otimes}\mathcal{P}^c$, α is of the form: $\alpha = id_{\mathcal{P}} \otimes \alpha|_{\mathcal{P}^c}$. Repeating the same arguments for $\{\mathcal{P}^c, \alpha|_{\mathcal{P}^c}\}$ with $0 < \mu < 1$ such that $\log\lambda / \log\mu \notin \mathbf{Q}$, we obtain a further tensorial factorization, and an $\overline{\mathrm{Int}}(\mathcal{M})$-perturbation α so that

$$\mathcal{M} = \mathcal{P}\bar{\otimes}\mathcal{Q} \otimes \mathcal{N}, \quad \mathcal{N} \cong \mathcal{M}:$$
$$\alpha = id_{\mathcal{P}\bar{\otimes}\mathcal{Q}} \otimes \alpha|_{\mathcal{N}}.$$

We know, however, that $\mathcal{P}\bar{\otimes}\mathcal{Q} \cong \mathcal{M}$, thanks to the uniqueness of AFD factors of type III_1. Therefore, α is, up to module $\overline{\mathrm{Int}}(\mathcal{M})$, of the form:

$\alpha \sim \alpha_1 \otimes id$ relative to a tensor product factorization $\mathcal{M} = \mathcal{M}_1 \overline{\otimes} \mathcal{N}$ where $\mathcal{M}_1 \cong \mathcal{N} \cong \mathcal{M}$. Let $\mathcal{M}_n$ be a replica of $\mathcal{M}$ for $n \in \mathbf{N}$ and write $\mathcal{M} = \prod_{n \in \mathbf{N}}^{\otimes} \{\mathcal{M}_n, \varphi_n\}$ with $\varphi_n = \varphi$ for a fixed faithful normal state φ on $\mathcal{M}$. Since $\mathcal{M}_n$ can be written as an infinite tensor product of 2×2-matrix algebras and φ can be taken to be a product state, the automorphism $\sigma_n \in \mathcal{M}$ which exchanges $\mathcal{M}_1$ and $\mathcal{M}_n$ and leaves the other components fixed is approximately inner, cf [12; Lemma 2.1], we may assume that α is of the form: $\alpha = \alpha_1 \otimes id$, where $\alpha_1 \in \mathrm{Aut}(\mathcal{M}_1)$ and id acts on $\prod_{n \geq 2}^{\otimes} \{\mathcal{M}_n, \varphi_n\}$. Since $\overline{\mathrm{Int}}(\mathcal{M})$ is a normal subgroup of $\mathrm{Aut}(\mathcal{M})$, $\alpha\sigma_n\alpha^{-1}\sigma_n^{-1}$ belongs to $\overline{\mathrm{Int}}(\mathcal{M})$. But we have

$$\alpha\sigma_n\alpha^{-1}\sigma_n^{-1} = \alpha_1 \otimes id \otimes \alpha_1^{-1} \otimes id,$$

where α_1^{-1} appears on the n-th component. Therefore, it remains only to prove that for any $\alpha_1 \in \mathrm{Aut}(\mathcal{M}_1)$ there exists a sequence $\{u_n\}$ in $\mathcal{U}(\mathcal{M})$ such that

$$\lim_{n \to \infty} \sigma_n((\mathrm{Ad}(u_n) \circ \alpha_1) \otimes id)\sigma_n^{-1} = id,$$

because this will show that

$$\alpha_1 \otimes id = \lim_{n \to \infty} (\alpha_1 \otimes id)\sigma_n(\alpha_1^{-1} \otimes id)\sigma_n^{-1}\sigma_n(\mathrm{Ad}(u_n^*) \otimes id)\sigma_n^{-1}$$

is approximately inner.

By the density of the orbit of φ under $\mathrm{Int}(\mathcal{M}_1)$, [10; Theorem 4], there exists a sequence $\{u_n\}$ in $\mathcal{U}(\mathcal{M}_1)$ such that $\|\varphi \circ \mathrm{Ad}(u_n) \circ \alpha_1 - \varphi\| < 1/2^n$. In the space $\mathfrak{S}_*$ of normal states on $\mathcal{M}$, the set of states of the form: $\psi \otimes \prod_{n=N+1}^{\infty}{}^{\otimes} \varphi_n$ with ψ a normal state on $\prod_{k=1}^{N}{}^{\otimes} \mathcal{M}_k$, is dense. Then we have, for $n > N$,

$$\left\| \psi \otimes \prod_{k=N+1}^{\infty}{}^{\otimes} \varphi_k - (\psi \otimes \prod_{k=N+1}^{\infty}{}^{\otimes} \varphi_k) \circ \sigma_n((\mathrm{Ad}(u_n) \circ \alpha_1) \otimes id)\sigma_n^{-1} \right\|$$
$$= \|\varphi - \varphi \circ \mathrm{Ad}(u_n) \circ \alpha_1\| < 1/2^n \to 0$$

as $n \to \infty$. This implies that

$$\lim_{n \to \infty} \sigma_n \circ (\mathrm{Ad}(u_n) \circ \alpha_1 \otimes id)\sigma_n^{-1} = id.$$

Q.E.D.

Lemma 1.17. *Let $\mathcal{M}$ be an AFD factor of type* III_1 *and* $\alpha \in \mathrm{Cnt}(\mathcal{M})$. *For any λ, $0 < \lambda < 1$, there exists $a \in \mathcal{U}(\mathcal{M})$ and a tensor product factorization: $\mathcal{M} = \mathcal{P}_1 \overline{\otimes} \mathcal{P}_2$ with the following properties:*

(i) $\mathrm{Ad}(a) \circ \alpha = \alpha_1 \otimes \alpha_2$ *relative to* $\mathcal{P}_1 \overline{\otimes} \mathcal{P}_2$;

(ii) $\mathcal{P}_1$ *is an AFD factor of type* III_λ*; therefore* α_1 *is of the form:* $\mathrm{Ad}(u) \circ \sigma_{T_1}^{\varphi_1}$ *with* $u \in \mathcal{U}(\mathcal{P}_1)$, $T_1 \in \mathbf{R}$, *and* φ_1 *a faithful normal state on* $\mathcal{P}_1$;
(iii) $\mathcal{P}_2 \cong \mathcal{M}$.

Proof. By assumption, there exists a sequence $\{e_{i,j}(k) : 1 \le i,j \le 2,\ k \in \mathbf{N}\}$ of mutually commuting 2×2-matrix units such that

$$\lim_{k\to\infty} \|\psi e_{i,j}(k) - \lambda^{-j} e_{i,j}(k)\psi\| = 0, \psi \in \mathcal{M}_*.$$

Passing to a subsequence, we may assume that $\{e_{i,j}(k)\}$ generates an AFD subfactor $\mathcal{P}_1$ of type III_λ such that $\mathcal{M} = \mathcal{P}_1 \overline{\otimes} \mathcal{P}_1^c \cong \mathcal{M}$. Since $\alpha \in \mathrm{Cnt}(\mathcal{M})$, $\{e_{11}(k)\}$ and $\{e_{22}(k)\}$ are both strongly central, we have

$$\lim_{k\to\infty} \|\alpha(e_{ii}(k)) - e_{ii}(k)\|_\varphi^\# = 0, \ i = 1,2.$$

We claim that there exists $\gamma \in \mathbf{C}$, $|\gamma| = 1$, such that $\lim_{k\to\infty} \|\alpha(e_{12}(k)) - \gamma e_{12}(k)\|_\varphi^\# = 0$. First, observe that $\{\alpha(e_{12}(k))e_{21}(k)\}$ is strongly central. With a free ultrafilter ω on $\mathbf{N}$, set

$$E = \pi_\omega(\{e_{11}(k)\}, \quad U = \pi_\omega(\{\alpha(e_{12}(k))e_{21}(k)\}).$$

where π_ω is the map in the proof of Lemma 1.6. Since $\mathcal{M}_\omega$ is a factor, to show that $U = \gamma E$ for some $\gamma \in \mathbf{C}$, $|\gamma| = 1$, it suffices to prove that U is in the center of $\mathcal{M}_{\omega,E}$. Let X be an element of $\mathcal{M}_{\omega,E}$ and represent X by a sequence $\{x(k)\}$ such that $x(k) = e_{11}(k)x(k)e_{11}(k)$. Set $y(k) = e_{21}(k)x(k)e_{12}(k)$. Then $\{y(k)\}$ is strongly ω-central. We now compute:

$$\begin{aligned} XU &= \pi_\omega(\{x(k)\alpha(e_{12}(k))e_{21}(k)\}) \\ &= \pi_\omega(\{\alpha(x(k)e_{12}(k))e_{21}(k)\}) \quad \text{since} \quad \alpha \in \mathrm{Cnt}(\mathcal{M}), \\ &= \pi_\omega(\{\alpha(e_{12}(k)y(k))e_{21}(k)\}) \\ &= \pi_\omega(\{\alpha(e_{12}(k))y(k)e_{21}(k)\}) \\ &= \pi_\omega(\{\alpha(e_{12}(k))e_{21}(k)x(k)\}) = UX\ . \end{aligned}$$

which shows that U and X commute. Therefore $U = \gamma E$ for some $\gamma \in \mathbf{C}$. Since U is a partial isometry, we have $|\gamma| = 1$. Thus, we obtain a subsequence $\{k_n\}$ such that

$$\lim_{n\to\infty} \|\alpha(e_{12}(k_n)) - \gamma e_{12}(k_n)\|_\varphi^\# = 0\ .$$

Passing to a subsequence, we get a sequence $\{e_{i,j}(k)\}$ of mutually commuting 2×2-matrix units such that for any $\psi \in \mathcal{M}_*$

$$\lim_{k\to\infty} \|\psi e_{i,j}(k) - \lambda^{i-j} e_{ij}(k)\psi\| = 0;$$
$$\lim_{k\to\infty} \|\alpha(e_{12}(k)) - \gamma e_{12}(k)\|_\varphi^\# = 0;$$
$$\lim_{k\to\infty} \|\alpha(e_{21}(k) - \overline{\gamma} e_{21}(k)\|_\varphi^\# = 0;$$
$$\lim_{k\to\infty} \|\alpha(e_{ii}(k)) - e_{ii}(k)\|_\varphi^\# = 0, \ i = 1,2.$$

A slight modification of Lemma 1.15 yields a sequence $\{k_n\}$ in $\mathbf{N}$ and $\{u_n\}$ in $\mathcal{U}(\mathcal{M})$ such that

(a) $[u_n, e_{i,j}(k_\nu)] = 0, \ 1 \le \nu \le n-1;$

(b) with $v_n = u_n u_{n-1} \cdots u_1, \quad v_n \alpha(e_{i,j}(k_\nu))v_n^* = \gamma^{j-i} e_{i,j}(k_\nu), \ 1 \le \nu \le n;$

(c) $\|u_n - 1\|_\varphi^\# < 1/2^n$.

Condition (c) guarantees the convergence $v = \lim_{n\to\infty} v_n \in \mathcal{U}(\mathcal{M})$ and $\mathrm{Ad}(v) \circ \alpha e_{ij}(k_n) = \gamma^{j-i} e_{ij}(k_n)$.

Now, $\mathrm{Ad}(v) \circ \alpha$ leaves the von Neumann subalgebra $\mathcal{P}_1$ generated by $\{e_{i,j}(k_n) : 1 \le i,j \le n, \ n \in \mathbf{N}\}$ globally invariant. If we choose a subsequence from $\{e_{i,j}(k_n)\}$ further denoted by $\{e_{ij}(k_n)\}$ again, then $\mathcal{P}_1$ factorizes $\mathcal{M}$ and $\mathcal{P}_1^c \cong \mathcal{M}$. We know also that $\mathcal{P}_1$ is an AFD factor of type III_λ. Now $\mathrm{Ad}(v) \circ \alpha$ is of the form: $\alpha_1 \otimes \alpha_2$ relative to the factorization $\mathcal{M} = \mathcal{P}_1 \overline{\otimes} \mathcal{P}_1^c$. Furthermore, if φ_1 is the product state of $\mathcal{P}_1$ such that

$$\varphi_1(e_{11}(k_n)) = 1/1+\lambda, \quad \varphi_1(e_{22}(k_n)) = \lambda/(1+\lambda).$$

Then we have $\alpha_1 = \sigma_{T_1}^{\varphi_1}$ for $T_1 \in \mathbf{R}$ such that $\gamma = \lambda^{-iT_1}$. Q.E.D.

Lemma 1.18. *Suppose $\mathcal{P}$ and $\mathcal{Q}$ are AFD factors of type* III_λ *and type* III_μ *respectively, $0 < \lambda, \mu < 1$, and that $\log\lambda/\log\mu \notin \mathbb{Q}$. Let φ and ψ be faithful normal states on $\mathcal{P}$ and $\mathcal{Q}$ respectively. If $\sigma_{T_1}^\varphi \otimes id \in \mathrm{Cnt}(\mathcal{P}\overline{\otimes}\mathcal{Q})$, then there exists $T \in \mathbf{R}$ such that*

$$\sigma_{T_1}^\varphi \otimes id \sim \sigma_T^\varphi \otimes \sigma_T^\psi.$$

Proof. We may assume $T_1 \log\lambda \notin 2\pi\mathbf{Z}$. Otherwise $\sigma_{T_1}^\varphi \in \mathrm{Int}(\mathcal{P})$ and there is nothing to prove. If $\log\mu = 2\pi n \log\lambda/(T_1\log\lambda + 2\pi m)$ for some $m, n \in \mathbf{Z}$, then we set

$$T = \frac{2\pi n}{\log\mu} = \frac{T_1\log\lambda + 2\pi m}{\log\lambda} \ .$$

It then follows that $\sigma^{\varphi}_{T_1} \equiv \sigma^{\varphi}_{T}$ mod(Int($\mathcal{P}$)) and $\sigma^{\psi}_{T} \in$ Int($\mathcal{Q}$). Thus, we now assume that $\log \mu \neq (2\pi n \log \lambda)/(T_1 \log \lambda + 2\pi m)$ for any $m, n \in \mathbf{Z}$.

Consider the quotient group $\mathbf{T} = \mathbf{R}/(2\pi\mathbf{Z})$, and define a subgroup:

$$A = \left\{(2\pi k \frac{\log \lambda}{\log \mu}, kT_1 \log \lambda) : \ k \in \mathbf{Z}\right\}$$

of the Cartesian product $\mathbf{T}^2$. Let $B = \overline{A}$. We claim that $\{0\} \times \mathbf{T} \cap B \neq \{0\}$. Suppose $\{0\} \times \mathbf{T} \cap B = \{0\}$. Since the projection of A to the first coordinate is a dense subgroup of $\mathbf{T}$ by the irrationality of $\log\lambda/\log\mu$, the projection of B to the first coordinate covers the entire $\mathbf{T}$ since B is compact. Hence the assumption that $B \cap \{0\} \times \mathbf{T} = \{0\}$ means that B is the group of a continuous homomorphism of $\mathbf{T}$ into $\mathbf{T}$, so that there exists $n \in \mathbf{Z}$ such that $B = \{(\alpha, n\alpha) : \alpha \in \mathbf{T}\}$. In particular, we have

$$T_1 \log \lambda = 2\pi n \frac{\log \lambda}{\log \mu} - 2\pi m$$

for some $m \in \mathbf{Z}$, which means however that $\log \mu = (2\pi n \log \lambda)/(T_1 \log \lambda + 2\pi m)$, the case we have excluded. Therefore, there must exist $x \in B$, $x \neq 0$, such that $(0, x) \in B$. Since $B = \overline{A}$, there exist two sequences $\{k(n)\}$ and $\{\ell(n)\}$ of integers such that

$$\lim_{n\to\infty} \left[2\pi k(n)\frac{\log \lambda}{\log \mu} - 2\pi \ell(n)\right] = 0 \quad \text{in} \quad \mathbf{R};$$
$$\lim_{n\to\infty} k(n) T_1 \log \lambda = x \quad \text{in} \quad \mathbf{T}.$$

We may take both $k(n)$ and $\ell(n)$ in $\mathbf{N}$. Note that the above two convergence mean that

$$\lambda^{k(n)}\mu^{-\ell(n)} \to 1 \quad \text{and} \quad e^{iT_1 k(n)} \to e^{ix} \neq 1.$$

Let $\mathcal{P}_0$ and $\mathcal{Q}_0$ be AFD factors of type III$_\lambda$ and III$_\mu$ respectively. Choose faithful normal states φ_0 on $\mathcal{P}_0$ and ψ_0 on $\mathcal{Q}_0$ such that their modular automorphism groups σ^{φ_0} and σ^{ψ_0} have respectively the period $2\pi/\log\lambda$ and $2\pi/\log\mu$. It follows that the centralizers $\mathcal{P}_{0,\varphi_0}$ and $\mathcal{Q}_{0,\psi_0}$ have both trivial relative commutants. Suppose k, $\ell \in \mathbf{N}$ are given. Then there exist isometries $u_1 \in \mathcal{P}_0$ and $v_1 \in \mathcal{Q}_0$ such that

$$\varphi_0 u_1 = \lambda^k u_1 \varphi_0 \qquad u_1^* u_1 = 1, \ u_1 u_1^* = e_1 \in \mathcal{P}_{0,\varphi_0};$$
$$\psi_0 v_1 = \mu^\ell v_1 \psi_0, \qquad v_1^* v_1 = 1, \ v_1 v_1^* = f_1 \in \mathcal{Q}_{0,\psi_0}.$$

Here the projections $e_1 \in \mathcal{P}_{0,\varphi_0}$ and $f_1 \in \mathcal{Q}_{0,\psi_0}$ can be arbitrary as long as the conditions $\varphi_0(e_1) = \lambda^k$ and $\psi_0(f_1) = \mu^\ell$ are satisfied. Applying the same argument to $\mathcal{P}_{0,1-e_1}$ and $\mathcal{Q}_{0,1-f_1}$, and repeating it inductively, we obtain sequences of partial isometries $\{u_n\} \subset \mathcal{P}_0$ and $\{v_n\} \subset \mathcal{Q}_0$ such that with $e_n = u_n^* u_n$ and $f_n = v_n^* v_n$,

(a) $\{e_n\}$ and $\{f_n\}$ are both orthogonal respectively;

(b) $\varphi_0 u_n = \lambda^k u_n \varphi_0, \quad u_n u_n^* = 1 - \sum_{j=1}^{n-1} e_j,$

$\psi_0 v_n = \mu^\ell v_n \psi_0, \quad v_n v_n^* = 1 - \sum_{j=1}^{n-1} f_j.$

Set $w = \sum_{n=1}^\infty u_n \otimes v_n^* \in \mathcal{P}\overline{\otimes}\mathcal{Q}$. Then we have

$$(\varphi_0 \otimes \psi_0) w = \lambda^k \mu^{-\ell} w (\varphi_0 \otimes \psi_0)$$
$$(\sigma_{T_1}^{\varphi_0} \otimes id)(w) = \lambda^{ikT_1} w.$$

Since $\varphi_0(e_n) = \lambda^k (1-\lambda^k)^{n-1}$ and $\psi_0(f_n) = \mu^\ell (1-\mu^\ell)^{n-1}$, we have

$$\begin{aligned}(\|w\|^\#_{\varphi_0\otimes\psi_0})^2 &= \frac{1}{2}(\varphi_0 \otimes \psi_0)\left(\sum_{n=1}^\infty (u_n^* u_n \otimes v_n v_n^* + u_n u_n^* \otimes v_n^* v_n)\right)\\ &= \frac{1}{2}\sum_{n=1}^\infty \left((1-\lambda^k)^{n-1}\mu^\ell(1-\mu^\ell)^{n-1} + \lambda^k(1-\lambda^k)^{n-1}(1-\mu^\ell)^{n-1}\right)\\ &= \frac{1}{2}\cdot\frac{\lambda^k+\mu^\ell}{\lambda^k+\mu^\ell-\lambda^k\mu^\ell} \geqq \frac{1}{2}.\end{aligned}$$

With $\mathcal{P}_n = \mathcal{P}_0$, $\mathcal{Q}_n = \mathcal{Q}_0, \varphi_n = \varphi_0$ and $\psi_n = \psi_0$, we set

$$\{\mathcal{P}, \varphi\} = \prod_{n=1}^{\infty}{}^{\otimes} \{\mathcal{P}_n, \varphi_n\}, \quad \{\mathcal{Q}, \psi\} = \prod_{n=1}^{\infty}{}^{\otimes} \{\mathcal{Q}_n, \psi_n\}\ .$$

For the sequences $k(n)$ and $\ell(n)$ obtained above, we apply the above construction to get $w(n)$ in the n-th factor $\mathcal{P}_n\overline{\otimes}\mathcal{Q}_n \subset \mathcal{P}\overline{\otimes}\mathcal{Q}$. By construction, we have

$$(\varphi \otimes \psi) w(n) = \lambda^{k(n)} \mu^{-\ell(n)} w(n) (\varphi \otimes \psi)\ ,$$

and $\lambda^{k(n)}\mu^{-\ell(n)} \to 1$, $w(n) \in \mathcal{P}_n\overline{\otimes}\mathcal{Q}_n$ and $\|w(n)\|^\#_{\varphi\otimes\psi} \geqq \frac{1}{2}$, so that $\{w(n)\}$ is a non-zero strongly central sequence in $\mathcal{P}\overline{\otimes}\mathcal{Q}$. But $(\sigma_{T_1}^\varphi \otimes id)(w(n)) = \lambda^{ik(n)T_1} w(n)$ and $\lambda^{ik(n)T_1} \to e^{ix} \neq 1$, which means that $\sigma_{T_1}^\varphi \otimes id \notin \mathrm{Cnt}(\mathcal{P}\overline{\otimes}\mathcal{Q})$. Q.E.D.

Proof of Theorem 1.3. (ii) for AFD Factors of Type III_1: Let $\mathcal{M}$ be an AFD factor of type III_1 and $\alpha \in \mathrm{Cnt}(\mathcal{M})$. By Lemma 1.17, there exists an AFD subfactor $\mathcal{P}_1$ of type III_λ, $0 < \lambda < 1$, such that $\mathcal{M} =$

$\mathcal{P}_1\overline{\otimes}\mathcal{P}_1^c, \mathcal{P}_1^c \cong \mathcal{M}$ and $\alpha \equiv \alpha_1 \otimes \alpha'$ mod(Int($\mathcal{M}$)). Since $\alpha_1 \otimes \alpha' \in \mathrm{Cnt}(\mathcal{M})$, we have $\alpha_1 \in \mathrm{Cnt}(\mathcal{P}_1)$ and $\alpha' \in \mathrm{Cnt}(\mathcal{P}_1^c)$. Applying the same arguments to α', we obtain a factorization of $\{\mathcal{P}_1^c, \alpha'\}$:

$\mathcal{P}_1^c = \mathcal{P}_2\overline{\otimes}\mathcal{Q}$, $\mathcal{Q} \cong \mathcal{M}$, $\alpha' \equiv \alpha_2 \otimes \beta \mod(\mathrm{Int}(\mathcal{P}_2^c))$, where $\mathcal{P}_2$ is an AFD subfactor of type III_μ with $\log\lambda/\log\mu \notin \mathbb{Q}$. Thus, we obtain a factorization:

$$\mathcal{M} = \mathcal{P}_1\overline{\otimes}\mathcal{P}_2\overline{\otimes}\mathcal{Q}, \quad \alpha \sim \alpha_1 \otimes \alpha_2 \otimes \beta.$$

Since $\alpha_1 \in \mathrm{Cnt}(\mathcal{P}_1)$ and $\alpha_2 \in \mathrm{Cnt}(\mathcal{P}_2)$, we have

$$\alpha_1 \sim \sigma_{T_1}^{\varphi_1} \qquad \text{and} \quad \alpha_2 \sim \alpha_{T_2}^{\varphi_2},$$

where φ_1 and φ_2 are respectively faithful normal states on $\mathcal{P}_1$ and $\mathcal{P}_2$. Since $\alpha_1 \otimes \alpha_2 \in \mathrm{Cnt}(\mathcal{P}_1\overline{\otimes}\mathcal{P}_2)$, $\sigma_{T_1}^{\varphi_1} \otimes \sigma_{T_2}^{\varphi_2}$ must be in $\mathrm{Cnt}(\mathcal{P}_1\overline{\otimes}\mathcal{P}_2)$. Since $\sigma_{-T_2}^{\varphi_1} \otimes \sigma_{-T_2}^{\varphi_2} = \sigma_{-T_2}^{\varphi_1\otimes\varphi_2} \in \mathrm{Cnt}(\mathcal{P}_1\overline{\otimes}\mathcal{P}_2)$, $\sigma_{T_1-T_2}^{\varphi_1\otimes\varphi_2}$ belongs to $\mathrm{Cnt}(\mathcal{P}_1\overline{\otimes}\mathcal{P}_2)$, which means that we may assume $T_1 = T_2$, say T, by Lemma 1.18. Thus we obtain a decomposition $\alpha \sim \sigma_T^{\varphi_1} \otimes \sigma_T^{\varphi_2} \otimes \beta$. Since $\mathcal{M} \cong \mathcal{P}_1\overline{\otimes}\mathcal{P}_2$, with $\varphi = \varphi_1 \otimes \varphi_2$ we come to the situation that $\mathcal{M} = \mathcal{P}\overline{\otimes}\mathcal{Q}$, $\alpha \sim \sigma_T^\varphi \otimes \beta$ and $\mathcal{M} \cong \mathcal{P} \cong \mathcal{Q}$. Now, let σ be the flip of $\mathcal{P}\overline{\otimes}\mathcal{Q}$ after identifying $\mathcal{P}$ and $\mathcal{Q}$, i.e, $\sigma(x \otimes y) = y \otimes x$. Since $\sigma \in \overline{\mathrm{Int}}(\mathcal{M})$, and $\mathrm{Cnt}(\mathcal{M})$ and $\overline{\mathrm{Int}}(\mathcal{M})$ commute module $\mathrm{Int}(\mathcal{M})$, we have

$$\sigma_T^\varphi\beta^{-1} \otimes \beta\sigma_{-T}^\varphi = (\sigma_T^\varphi \otimes \beta)\sigma(\sigma_T^\varphi \otimes \beta)^{-1}\sigma \in \mathrm{Int}(\mathcal{M}),$$

which means that $\sigma_T^\varphi \sim \beta$. Thus, we conclude

$$\alpha \sim \sigma_T^\varphi \otimes \beta \sim \sigma_T^\varphi \otimes \sigma_T^\varphi = \sigma_T^{\varphi\otimes\varphi}.$$

Q.E.D.

2. Outer Conjugacy of Automorphisms of AFD Factors of Type III

Let $\mathcal{M}$ be an AFD factor of type III with flow of weights $\mathcal{F}$. Let φ be a dominant weight on $\mathcal{M}$ and $\mathcal{M} = \mathcal{N} \rtimes_\theta \mathbf{R}$ be the associated continuous decomposition with $\mathcal{N} = \mathcal{M}_\varphi$. Let $\{u(s) : s \in \mathbf{R}\}$ be the corresponding one parameter unitary group in $\mathcal{M}$. Let α be an automorphism of $\mathcal{M}$. The module, mod(α), of α is clearly an outer conjugacy invariant of α. The asymptotic outer period $p_a(\alpha)$ is, by definition, the period of α in the quotient group $\mathrm{Aut}(\mathcal{M})/\mathrm{Cnt}(\mathcal{M})$. It is also an outer conjugacy invariant of α. We know that an inner perturbation of α leaves φ and $\{u(s)\}$ invariant, so we assume that $\varphi \circ \alpha = \varphi$ and $\alpha(u(s)) = u(s)$, $s \in \mathbf{R}$. Let $p =$

$p_a(\alpha)$. Then $\alpha^p = \mathrm{Ad}(u)\circ\sigma_c^\varphi$ by Theorem 1.3, where $u \in U(\mathcal{M})$ and σ_c^φ is the extended modular authomorphism corresponding to $c \in \mathcal{Z}^1(\mathfrak{F})$. Since $\varphi\circ\alpha = \varphi$, u must belong to $\mathcal{M}_\varphi = \mathcal{N}$. Furthermore, in the choice of u and c, we have exactly the ambiguity of $\mathcal{U}(\mathcal{C})$, where $\mathcal{C} = \mathcal{C}_\varphi(= \mathfrak{F})$. Hence the cohomology class $\nu(\alpha) = [c] \in H^1(\mathfrak{F})$ should be a candidate for an outer conjugacy invariant of α. In fact, it will be shown that $\nu(\alpha)$ is an outer conjugacy invariant. Also, we have

$$\alpha^p = \alpha\circ\alpha^p\circ\alpha^{-1} = \mathrm{Ad}(\alpha(u))\circ\alpha\circ\sigma_c^\varphi\circ\alpha^{-1} = \mathrm{Ad}(\alpha(u))\circ\sigma^\varphi_{\mathrm{mod}(\alpha(c))}.$$

Hence $\alpha(u)u^* = \gamma \in \mathcal{U}(\mathcal{C})$, i.e. $\alpha(u) = \gamma u$. Furthermore, we have

$$\begin{aligned} u &= \mathrm{Ad}(u)\circ\sigma_c^\varphi(u) = \alpha^p(u) \\ &= \alpha^{p-1}(\gamma)\alpha^{p-2}(\gamma)\cdots\alpha(\gamma)\gamma u\ , \end{aligned}$$

so that

$$\alpha^{p-1}(\gamma)\alpha^{p-2}(\gamma)\cdots\alpha(\gamma)\gamma = 1\ .$$

which means that γ gives rise to a cocycle over $\mathbf{Z}_p = \mathbf{Z}/p\mathbf{Z}$ relative to $\{\mathcal{U}(\mathcal{C}), \mathrm{mod}(\alpha)\}$ defined by $\gamma_k = \alpha^{k-1}(\gamma)\cdots\alpha(\gamma)\gamma$, $k \in \mathbf{Z}$. Here we note that when the flow of weights $\mathfrak{F}$ and $\{\mathcal{C}, \mathbf{Z}, \theta\}$ is identified, $\mathrm{mod}(\alpha)$ is precisely the restriction of α to $\mathcal{C}$ and $\mathrm{mod}(\alpha)$ has the period p. Let $\mathrm{Ob}_m(\alpha)$ be the cohomology class of $\{\gamma_k\}$ in $H^1_{\mathrm{mod}(\alpha)}(\mathbf{Z}_p, \mathcal{U}(\mathfrak{F}))$. Now, we state the main result of this section:

Theorem 2.1. *Let $\mathcal{M}$ be an AFD factor of type* III *with flow of weights $\mathfrak{F}$, and $\alpha \in \mathrm{Aut}(\mathcal{M})$. The quartet $\{p_a(\alpha), \mathrm{mod}(\alpha), \nu(\alpha), \mathrm{Ob}_m(\alpha)\}$ is a complete invariant for the outer conjugacy of α in the sense that*

(i) $p_a(\sigma\circ\mathrm{Ad}(w)\circ\alpha\circ\sigma^{-1}) = p_a(\alpha)$;
(ii) $\mathrm{mod}(\sigma\circ\mathrm{Ad}(w)\circ\alpha\circ\sigma^{-1}) = \mathrm{mod}(\sigma)\circ\mathrm{mod}(\alpha)\circ\mathrm{mod}(\sigma)^{-1}$.
(iii) $\nu(\sigma\circ\mathrm{Ad}(w)\circ\alpha\circ\sigma^{-1}) = \mathrm{mod}(\sigma)(\nu(\alpha))$;
(iv) $\mathrm{Ob}_m(\sigma\circ\mathrm{Ad}(w)\circ\alpha\circ\sigma^{-1}) = \mathrm{mod}(\sigma)(\mathrm{Ob}_m(\alpha))$.

Proof. If $\mathcal{M}$ is of type III_λ, $0 \le \lambda < 1$, then our assertion follows from a more general result of Sutherland-Takesaki, [19; Theorem 5.9]. Thus, we present here a proof in the case of type III_1. So, we assume that $\mathcal{M}$ is an AFD factor of type III_1. In this case, the flow of weights becomes trivial; thus the module, $\mathrm{mod}(\alpha)$, disappears; $\nu(\alpha)$ is a real number such that $\alpha \sim \mathrm{Ad}(u)\circ\sigma^\varphi_{\nu(\alpha)}$; finally $\mathrm{Ob}_m(\alpha)$ is a $p_a(\alpha)$-th root of unity.

Suppose now $p_a(\alpha) = p_a(\beta) = p, \nu(\alpha) = \nu(\beta) = T$ and $\mathrm{Ob}_m(\alpha) = \mathrm{Ob}_m(\beta) = \gamma$ for $\alpha, \beta \in \mathrm{Aut}(\mathcal{M})$. From the discussion in the last section, we may assume that α and β admit an invariant dominant weight φ and the associated crossed product decomposition of $\mathcal{M} = \mathcal{N}\rtimes_\theta\mathbf{R}$ with $\mathcal{N} = \mathcal{M}_\varphi$ such that $\alpha(u(s)) = \beta(u(s)) = u(s)$, $s \in \mathbf{R}$, where $\{u(s)\}$ is the one parameter unitary group in $\mathcal{M}$ associated with the crossed product. If $p = 0$, then α and β are outer conjugate by the fundament result of Connes, [4; Theorem 2] since $\mathrm{Aut}(\mathcal{M}) = \overline{\mathrm{Int}}(\mathcal{M})$ by Theorem 1.3. Suppose $p \geqq 2$, $\alpha^p = \mathrm{Ad}(u)\circ\sigma_T^\varphi$ and $\beta^p = \mathrm{Ad}(v)\circ\sigma_T^\varphi$ with $u, v \in \mathcal{U}(\mathcal{N})$. Set $\tilde{\alpha} = \alpha\circ\sigma^\varphi_{-T/p}$ and $\tilde{\beta} = \beta\circ\sigma^\varphi_{-T/p}$. We then have

$$p = p_0(\tilde{\alpha}) = p_a(\tilde{\alpha}) = p_0(\tilde{\beta}) = \tilde{p}_a(\tilde{\beta});$$
$$\tilde{\alpha}^p = \mathrm{Ad}(u), \quad \tilde{\beta}^p = \mathrm{Ad}(v)$$

and furthermore

$$\tilde{\alpha}(u) = \gamma u; \tilde{\beta}(u) = \gamma v$$

Therefore, $\tilde{\alpha}$ and $\tilde{\beta}$ are outer conjugate by an earlier result of Connes, [4], see also Ocneanu [18]. Hence there exist $\theta \in \mathrm{Aut}(\mathcal{M})$ and $w \in \mathcal{U}(\mathcal{M})$ such that

$$\mathrm{Ad}(w) \circ \tilde{\alpha} = \theta^{-1} \circ \tilde{\beta} \circ \theta.$$

Therefore, we get

$$\begin{aligned}
\mathrm{Ad}(w) \circ \alpha \circ \sigma^\varphi_{-T/P} &= \theta^{-1} \circ \beta \circ \sigma^\varphi_{-T/p} \circ \theta \\
&= \theta^{-1} \circ \beta \circ \theta \circ \sigma^{\varphi\circ\theta}_{-T/p} \\
&= \theta^{-1} \circ \beta \circ \theta \circ \mathrm{Ad}((D\varphi\circ\theta : D\varphi)_{-T/p}) \circ \sigma^\varphi_{-T/p},
\end{aligned}$$

so that

$$\alpha \equiv \theta^{-1} \circ \beta \circ \theta \qquad \mathrm{mod} \quad \mathrm{Int}(\mathcal{M}).$$

This completes the proof. Q.E.D.

Definition 2.2. The invariants $\mathrm{Ob}_m(\alpha)$ and $\nu(\alpha)$ are respectively called the *modular obstruction* and the *modular invariant* of α.

The modular obstruction $\mathrm{Ob}_m(\alpha)$ and the modular invariant of $\alpha \in \mathrm{Aut}(\mathcal{M})$ are not independent. To find the relation, let $[\gamma] = \mathrm{Ob}_m(\alpha) \in H^1_{\mathrm{mod}(\alpha)}(\mathbf{Z}_p\mathcal{U}(\mathcal{F})$ and $\nu(\alpha) = [c] \in H^1(\mathcal{F}$ be given by

$$\alpha^p = \mathrm{Ad}(u) \circ \sigma^\varphi_c, \quad \alpha(u) = \gamma u .$$

Assume that α is so adjusted that i) φ is an α-invariant dominant weight and ii) $\alpha(u)(s)) = u(s)$, $s \in \mathbf{R}$, for the one parameter unitary group $\{u(s)\}$ corresponding to the decomposition : $\mathcal{M} = \mathcal{M}_\varphi \rtimes_\theta \mathbf{Z}$. We then have

$$\begin{aligned}
\mathrm{Ad}(u) \circ \sigma^\varphi_c &= \alpha^p = \alpha \circ \alpha^p \circ \alpha^{-1} \\
&= \mathrm{Ad}(\alpha(u)) \circ \sigma^\varphi_{\mathrm{mod}(\alpha(c)} \\
&= \mathrm{Ad}(\gamma u) \circ \sigma^\varphi_{\mathrm{mod}(\alpha)(c)}
\end{aligned}$$

Hence we have

$$\mathrm{Ad}(\gamma) = \sigma^{\varphi}_{c^*\mathrm{mod}(\alpha)(c)}\,,$$

which means that

$$\gamma\theta_s(\gamma^*) = c_s^* \ \mathrm{mod}(\alpha)(c_s), \ s \in \mathbf{R}\,. \tag{*}$$

Therefore, the cohomology class $\mathrm{Ob}_m(\alpha) = [\gamma] \in H_{\mathrm{mod}(\alpha)}(\mathbf{Z}_p, \mathcal{U}(\mathcal{F}))$ is invariant under the action of the flow $\{\theta_s\}$ of weights. Hence we have

$$\mathrm{Ob}_m(\alpha) \in H^1_{\mathrm{mod}(\alpha)}(\mathbf{Z}_p, \mathcal{U}(\mathcal{F}))^{\mathbf{R}}.$$

Furthermore, the identity (*) motivates us to define maps δ_1 and δ_2 respectively from $H^1_{\mathrm{mod}(\alpha)}(\mathbf{Z}_p, \mathcal{U}(\mathcal{F}))^{\mathbf{R}}$ and $H^1(\mathcal{F})$ into $H^1(\mathbf{R}, B_{\mathrm{mod}(\alpha)}(\mathbf{Z}_p, \mathcal{U}(\mathcal{F})))$ as follows: $\delta_1([\gamma])$ is the cohomology class of the one cocycle, $s \in \mathbf{R} \mapsto \gamma\theta_s(\gamma^*) \in B^1_{\mathrm{mod}(\alpha)}(\mathbf{Z}_p, \mathcal{U}(\mathcal{F}))$, and $\delta_2([c])$ is the cohomology class of the cocycle, $s \in \mathbf{R} \mapsto c_s^* \ \mathrm{mod}(\alpha)(c_s) \in B^1_{\mathrm{mod}(\alpha)}(\mathbf{Z}_p, \mathcal{U}(\mathcal{F}))$. The identity (*) means that

$$\delta_1(\mathrm{Ob}_m(\alpha)) = \delta_2(\nu(\alpha)).$$

Now, we conclude the following:

Theorem 2.3. *Let $\mathcal{M}$ be an AFD factor of type* III *with flow of weights $\mathcal{F}$.*

(i) *The following sequence is a split short exact sequence:*

$$1 \to \overline{\mathrm{Int}}(\mathcal{M}) \to \mathrm{Aut}(\mathcal{M}) \overset{\mathrm{mod}}{\to} \mathrm{Aut}(\mathcal{F}) \to 1.$$

(ii) *If* $\mathrm{mod}(\alpha)$, $\alpha \in \mathrm{Aut}(\mathcal{M})$, *is a periodic, then* $p_a(\alpha) = 0$. *Therefore,* $\mathrm{mod}(\alpha)$ *is a complete invariant for the outer conjugacy of* α.

(iii) *Let* $\bar{\alpha} \in \mathrm{Aut}(\mathcal{F})$ *be periodic and* $p > 1$ *divides the period of* $\bar{\alpha}$. *Then* $([\gamma], [c]) \in H^1_{\bar{\alpha}}\mathbf{Z}_p, \mathcal{U}(\mathcal{F}))^{\mathbf{R}} \times H^1(\mathcal{F})$ *appears as* $\mathrm{Ob}_m(\alpha), \nu(\alpha))$ *for some* $\alpha \in \mathrm{Aut}(\mathcal{M})$ *with* $\mathrm{mod}(\alpha) = \bar{\alpha}$ *if and only if*

$$\delta_1([\gamma]) = \delta_2([c]).$$

Proof. (i) This follows from the existence of a functor from the conjugacy category of ergodic flows into the orbit equivalence category of ergodic AF equivalence relations shown by R. Wang [21].

(ii) This is an immediate consequence of Theorem 2.1.

(iii) The necessity of the condition was already established before the Theorem. Suppose that $[\gamma] \in H^1_{\bar{\alpha}}(\mathbf{Z}_p, \mathfrak{U}(\mathcal{F}))^{\mathbf{R}}$ and $[c] \in H^1(\mathcal{F})$ satisfy the equation. If $\mathcal{M}$ is of type III_λ, $\lambda \neq 1$, then the existence of $\alpha \in \mathrm{Aut}(\mathcal{M})$ with $(\mathrm{mod}(\alpha), \mathrm{Ob}_m(\alpha), \nu(\alpha)) = (\bar{\alpha}, [\gamma], [c])$ follows from a result of Sutherland-Takesaki, [20]. Thus, we assume that $\mathcal{M}$ is of type III_1. In this case, the situation is so simple, that $[\gamma]$ is a p-th root of unity and $[c]$ is a real number. Thus, let γ be a complex number with $\gamma^p = 1$ and $T \in \mathbf{R}$. Set $\alpha = \sigma^{\varphi}_{T/p} \otimes s^{\gamma}_p$ on $\mathcal{M} \bar{\otimes} \mathcal{R}_0$, where $\mathcal{R}_0$ is the AFD-factor of type II_1 and s^{γ}_p is the periodic automorphism of $\mathcal{R}_0$ with $\gamma = \mathrm{Ob}(s^{\gamma}_p)$ constructed by Connes [6]. It is then clear that $\mathrm{Ob}_m(\alpha) = \gamma$ and $\nu(\alpha) = T$. Q.E.D.

If $\mathcal{M}$ is an AFD factor of type III, then $\mathrm{Aut}(\mathcal{M})$ admits three distinguished normal subgroups:

$$\mathrm{Int}(\mathcal{M}), \quad \mathrm{Cnt}(\mathcal{M}), \quad \overline{\mathrm{Int}}(\mathcal{M})$$

in the increasing order. The quotient groups are identified as follows:

$$\mathrm{Cnt}(\mathcal{M})/\mathrm{Int}(\mathcal{M}) \cong H^1(\mathcal{F});$$
$$\overline{\mathrm{Int}}(\mathcal{M})/\mathrm{Cnt}(\mathcal{M}) \cong \mathrm{Out}(\mathcal{R}_0)$$

provided that $\mathcal{M}$ is of type III_λ, $\lambda \neq 0$, with $\mathcal{R}_0$ the AFD factor of type II_1;

$$\mathrm{Aut}(\mathcal{M})/\overline{\mathrm{Int}}(\mathcal{M}) \cong \mathrm{Mod}(\mathcal{M}) = \mathrm{Aut}(\mathcal{F}).$$

If $\mathcal{M}$ is of type III_λ, $\lambda \neq 0$, then we have

$$H^1(\mathcal{F}) \cong \mathbf{R}/(-2\pi/\log\lambda)\mathbf{Z}$$
$$\mathrm{Aut}(\mathcal{F}) \cong \mathbf{R}/(\log\lambda)\mathbf{Z}.$$

REFERENCES

[1] H. Araki, *Asymptotic ratio set and property* L'_λ, Publ. RIMS. Kyoto Univ. **6** (1970/71), 443-460.

[2] A. Connes, *Une classification des facteurs de type* III, Ann. Sci. École Norm. Sup. **6** (1973), 133-252.

[3] A. Connes, *Almost periodic states and factors of type* III_1, J. Funct. Anal. **16** (1974), 415-445.

[4] A. Connes, *Outer conjugacy classes of automorphisms of factors*, Ann. Sci. École Norm. Sup. **8** (1975), 383-419.

[5] A. Connes, *On the classification of von Neumann algebras and their automorphisms*, Symposia Math. XX (1976), 435-478.

[6] A. Connes, *Periodic automorphisms of the hyperfinite factor of type* II_1, Acta Sci. Math. **39** (1977), 39-66.
[7] A. Cones, *Classification of injective factors, Cases* II_1, II_∞, III_λ, $\lambda \neq 1$, Ann. Math. **104** (1976), 73-115.
[8] A. Connes, *Factors of type III_1, property L'_λ and closure of inner automorphisms*, J. Operator Theory **14** (1985), 189-211.
[9] A. Connes & W. Krieger, *Measure space automorphism groups, the normalizer of their full groups, and approximate finiteness*, J. Func. Anal. **24** (1977), 336-352.
[10] A. Connes & E. Størmer, *Homogenuity of the state spaces of factors of type* III_1, J. Funct. Anal. **28** (1978), 187-196.
[11] A. Connes & M. Takesaki, *The flow of weights on factors of type* III, Tohoku Math. J. **29** (1977), 473-555.
[12] A. Connes & E. J. Woods, *A construction of approximately finite dimensional non-ITPFI factors,* Canad. Math. Bull. **23** (1980), 227-230.
[13] U. Haagerup, *Connes bicentralizer problem and uniqueness of the injective factor of type* III_1, Acta Math. **158** (1987), 95-147.
[14] U. Haagerup & E. Størmer, *Pointwise inner automorphisms of von Neumann algebras* with an appendix by C. Sutherland, preprint 1988.
[15] V.F.R. Jones, *Actions of finite groups on the hyperfinite type II_1 factor*, Mem. Amer. Math. Soc. **237** (1980).
[16] V.F.R. Jones & M. Takesaki, *Actions of compact abelian groups on semifinite injective factors*, Acta Math. **153** (1984), 213-258.
[17] Y. Kawahigashi, C.E. Sutherland and M. Takesaki, *The structure of the automorphism group of an injective factor and the cocycle conjugacy of discrete abelian group actions*, to appear.
[18] A. Ocneaunu, *Actions of discrete amenable groups on factors.* Lecture Notes in Math. No. 1138, Springer, Berlin, 1985.
[19] C.E. Sutherland & M. Takesaki, *Actions of discrete amenable groups and groupoids on von Neumann algebras*, Publ. RIMS Kyoto Univ. **21** (1985), 1087-1120.
[20] C.E. Sutherland & M. Takesaki, *Actions of discrete amenable groups on injective factors of type* III_λ, $\lambda \neq 1$, Pacific J. Math. **137** (1989), 405-444.
[21] S.Y.R. Wong, *On the dictionary between ergodic transformations, Krieger factors and ergodic flows*, Thesis (1986), University of New South Wales.

Received November 3, 1989

Department of Mathematics
UCLA
Los Angeles, CA 90024

Circular and Semicircular Systems and Free Product Factors

DAN VOICULESCU[1]

Dedicated to Prof. Jacques Dixmier on the occasion of his 65th birthday

Many basic questions about the type II_1 factors of free groups, and more generally of free products of cyclic groups, are still unanswered. On the other hand the free group factors, like the hyperfinite factor, arise naturally from a functor taking Hilbert spaces and contractions to C^*-algebras and completely positive maps. This functor is the analogue of Gaussian measure on Hilbert space in a kind of non-commutative probability theory ([**1**], [**2**], [**3**], [**4**], [**5**]) in which free products are given a treatment similar to tensor products, i.e. to independence. One of the aims of the present paper is to apply these probabilistic results to free group factors.

To be more specific, here are two results we obtained in this way. By $F(\mathbb{N})$ we denote the free group with generators indexed by the natural numbers.

1°. *If G is an at most countable free product of cyclic groups, then the* II_1 *factors of $F(\mathbb{N})$ and of $G * F(\mathbb{N})$ are isomorphic.*

2°. *The fundamental group of the* II_1 *factor of $F(\mathbb{N})$ contains the positive rational numbers.*

The circular and semicircular systems, we study, are analogues of families of independent complex and respectively real Gaussian random variables.

[1] Research supported in part by a grant from the National Science Foundation.

They provide convenient systems of generators for free group factors. Actually, the main source for our results is the recent realization ([**5**]) that Gaussian random $N \times N$ matrices in the large N limit, behave like objects in the non-commutative probabilistic context for free products. Also, conversely, the results we obtain here, have consequences for random matrices: the components of the polar decomposition of a Gaussian random matrix are asymptotically free (see Remark 2.7).

The semicircular systems are called so because of the semicircle distribution of its elements. The semicircle law for the eigenvalues of random matrices was discovered by E. Wigner ([**6**], [**7**]). In [**1**] we found that the semicircle distribution is the analogue for free products of the Gaussian distribution. This coincidence was the first clue for the connection between random matrices and free products.

This paper has three sections.

Section 1 is a large section of preliminaries, designed to make the paper self-contained.

Section 2 is devoted to results on circular and semicircular systems. Among these we mention Propositon 2.6 concerning the polar decomposition of a circular element.

Section 3 contains applications to free group factors.

1. Preliminaries

This section is devoted to preliminaries. We recall definitions and facts from ([**1**], [**4**], [**5**]) and present some further definitions and general remarks which will be used in the next sections.

A "*non-commutative probability space*" (A, φ) consists of a unital algebra A over $\mathbb{C}$ equipped with a *state* $\varphi : A \to \mathbb{C}$, i.e. a linear functional φ such that $\varphi(1) = 1$. If A is a $*$-algebra and $\varphi(x^*) = \overline{\varphi(x)}$ we call (A, φ) a $*$-*probability space*. Similarly, a $*$-probability space (A, φ) where A is a C^*-algebra and φ is positive, will be called a C^*-*probability space*. Further, a C^*-probability space (A, φ) with A a W^*-algebra and φ a normal state, will be called a W^*-*probability space*.

If (A, φ) is a non-commutative probability space, elements of A will often be referred to as *non-commutative random variables* (or simply random variables).

1.1. DEFINITION: Let (A, φ) be a non-commutative probability space. A family of subalgebras $1 \in A_\iota \subset A$ $(\iota \in I)$ is called a *free family of subalgebras* if $\varphi(a_1 \ldots a_n) = 0$ whenever $a_j \in A_{\iota(j)}$, with $\iota(j) \neq \iota(j+1)$ $(1 \leq j \leq n-1)$ and $\varphi(a_j) = 0$ $(1 \leq j \leq n)$. A *family of subsets* $(\Omega_\iota)_{\iota \in I}$ *is* *called free* if the subalgebras A_ι generated by $\{1\} \cup \Omega_\iota$ form a free family of subalgebras. A *family of random variables* $(f_\iota)_{\iota\in I}$ *in* A *is called free* if the family of subsets $(\{f_\iota\})_{\iota\in I}$ is free. If (A,φ) is a $*$-probability space, *a family of subsets* $(\Omega_\iota)_{\iota\in I}$ *is called* $*$*-free* if the family of subsets $(\Omega_\iota \cup \Omega_\iota^*)_{\iota\in I}$ is free. A *family of random variables* $(f_\iota)_{\iota\in I}$ *is called* $*$*-free* if the family of subsets $(\{f_\iota\})_{\iota\in I}$ is $*$-free.

1.2. REMARK: If $(\Omega_\iota)_{\iota\in I}$ is a $*$-free family of subsets in a C^*-probability space (respectively in a W^*-probability space) then the C^*-subalgebras (respectively the W^*-subalgebras) generated by $\{1\} \cup \Omega_\iota$ $(\iota \in I)$ form a free family of subalgebras.

1.3. REMARK: Let (A,φ) be a non-commutative probability space.

(i) If $(A_\iota)_{\iota\in I}$ is a free family of subalgebras in (A,φ) and B is the subalgebra generated by $\bigcup_{\iota\in I} A_\iota$, then $\varphi \mid B$ is completely determined by the $\varphi \mid A_\iota$ $(\iota \in I)$.

(ii) If $(\Omega_\iota)_{\iota\in I}$ is a free family of subsets in (A,φ) and if $(I_j)_{j\in J}$ is a partition of I and if $\Sigma_j = \bigcup_{\iota\in I_j} \Omega_\iota$, then $(\Sigma_j)_{j\in J}$ is also a free family of subsets.

(iii) If $(A_\iota)_{\iota\in I}$ is a free family of subalgebras in (A,φ) and if $(\omega_{\iota k})_{k\in K_\iota}$ is a free family of subsets in $(A_\iota, \varphi \mid A_\iota)$, then $(\omega_{\iota k})_{(\iota,k)\in K}$ where $K = \coprod_{\iota\in I} \{\iota\} \times K_\iota$, is a free family of subsets in (A,φ).

1.4. EXAMPLE: Let G be a discrete group which is the free product $\ast_{\iota\in I} G_\iota$ of its subgroups G_ι, and let $(L(G),\tau)$ be the von Neumann algebra generated by the left regular representation of G in $\ell^2(G)$ and τ the canonical trace state, i.e. $\tau(T) = \langle T\xi, \xi\rangle$ where $\xi(g) = \delta_{g,e}$. Let further $L(G_\iota)$ be identified with the corresponding subalgebra in $L(G)$. Then $(L(G_\iota))_{\iota\in I}$ is a free family of subalgebras in $(L(G),\tau)$.

1.5. Given non-commutative probability spaces (A_k,φ_k) $(k \in J)$ there is a state φ on the free product algebra $A = \ast_{k\in J} A_k$ (amalgamation over $\mathbb{C}1$ is assumed) such that $\varphi \mid A_k = \varphi_k$ (the A_k being canonically identified with subalgebras of A) and such that the $(A_k)_{k\in J}$ form a free family of subalgebras in (A,φ). The state φ is uniquely determined by these conditions and is called the *free product of the states* φ_k, denoted $\ast_{k\in J} \varphi_k$.

If the (A_k,φ_k) are W^*-probability spaces such that the A_k are finite W^*-algebras and the φ_k are faithful trace states, then φ is a trace state on the $*$-algebra A and the GNS-construction applied to (A,φ) yields a von Neumann algebra $\tilde{A}$ with a normal faithful trace state $\tilde{\varphi}$ which will be called the *reduced W^*-free product* of the (A_k, φ_k). Each A_k identifies with a von Neumann subalgebra of $\tilde{A}$ so that $\varphi_k = \tilde{\varphi} \mid A_k$ and the A_k form a free family of subalgebras. If the A_k's are II_1 factors then A is a II_1 factor.

If $(G_\iota)_{\iota\in I}$ is a family of discrete groups, then the reduced W^*-free product of the $(L(G_\iota), \tau_\iota)$ is $(L(G), \tau)$ where $G = \underset{\iota\in I}{*} G_\iota$.

1.6. DEFINITION: If $(f_\iota)_{\iota\in I}$ is a family of random variables in (A, φ), let $\mathbf{C}\langle\{X_\iota \mid \iota \in I\}\rangle$ be the free algebra with unit over $\mathbf{C}$ and generators X_ι $(\iota \in I)$ and let $h : \mathbf{C}\langle\{X_\iota \mid \iota \in I\}\rangle \to A$ be the homomorphism such that $h(X_\iota) = f_\iota$ $(\iota \in I)$. The *joint distribution* of the $(f_\iota)_{\iota\in I}$ is the functional $\mu : \mathbf{C}\langle\{X_\iota \mid \iota \in I\}\rangle \to \mathbf{C}$ defined by $\mu = \varphi \circ h$. The *moments* of $(f_\iota)_{\iota\in I}$ are the numbers $\mu(Y)$ where $Y = X_{\iota_1} X_{\iota_2} \ldots X_{\iota_k}$ is a monomial. If $(f_\iota)_{\iota\in I}$ is a family of random variables in a $*$-probability space the distribution of $(f_\iota)_{\iota\in I} \cup (f_\iota^*)_{\iota\in I}$ will be called the *$*$-distribution of* $(f_\iota)_{\iota\in I}$.

1.7. DEFINITION: If $(f_{\iota,n})_{\iota\in I}$ are random variables in (A_n, φ_n) and μ_n is their joint distribution, then μ_n is called the *limit distribution* of these families as $n \to \infty$, if $\lim_{n\to\infty} \mu_n(a) = \mu(a)$ for every $a \in \mathbf{C}\langle\{X_\iota \mid \iota \in I\}\rangle$. If $(f_{\iota,n})_{\iota\in I}$ are families of random variables in (A_n, φ_n) and if $I = \bigcup_{s\in S} I_s$ is a partition of I, then the family of subsets $((f_{\iota,n})_{\iota\in I_s})_{s\in S}$ is called *asymptotically free as $n \to \infty$* if the distributions μ_n of $(f_{\iota,n})_{\iota\in I}$ converge to a limit distribution μ and if the family of subsets $((X_\iota)_{\iota\in I_s})_{s\in S}$ is free in $(\mathbf{C}\langle\{X_\iota \mid \iota \in I\}\rangle, \mu)$.

1.8. REMARK: Let (A, φ), (B, ψ) be two C^*-probability spaces (respectively W^*-probability spaces) such that the GNS representations associated with φ and ψ are faithful. Let further $(f_\iota)_{\iota\in I} \subset A$ and $(g_\iota)_{\iota\in I} \subset B$ be families of random variables which generate A and B as C^*-algebras (respectively as W^*-algebras). If the $*$-distributions of $(f_\iota)_{\iota\in I}$ and $(g_\iota)_{\iota\in I}$ are equal then there is an isomorphism of C^*-algebras (respectively W^*-algebras) $\gamma : A \to B$ such that $\varphi = \psi \circ \gamma$ and $\gamma(f_\iota) = g_\iota$ for $\iota \in I$.

1.9. DEFINITION: A family of random variables $(f_\iota)_{\iota\in I}$ in a $*$-probability space (A, φ) is called a *semicircular family* if it is a free family, $f_\iota = f_\iota^*$ for all $\iota \in I$ and the distribution of each f_ι is given by the semicircle law

$$\varphi(f_\iota^k) = \frac{2}{\pi} \int_{-1}^{1} t^k (1 - t^2)^{1/2}\, dt.$$

A family of random variables $(g_\iota)_{\iota\in I}$ is called *circular* if the family $(x_\iota)_{\iota\in I} \cup (y_\iota)_{\iota\in I}$ where $x_\iota = 2^{-1/2}(g_\iota + g_\iota^*)$, $y_\iota = -i2^{-1/2}(g_\iota - g_\iota^*)$, is a semicircular family.

1.10. REMARK: If $(f_\iota)_{\iota\in I}$ is a semicircular system in the $*$-probability space (A,φ) and B is the subalgebra generated by the f_ι's, then $\varphi \mid B$ is completely determined and is a trace. If (A,φ) is a C^*-probability space or a W^*-probability space such that the GNS representation associated with φ is faithful, then the C^*-algebra or respectively the W^*-algebra generated by the $(f_\iota)_{\iota\in I}$ is completely determined. In the W^*-algebra case this W^*-algebra is isomorphic to a type II_1 factor of a free group on card I generators and φ is the canonical trace. This follows from 1.3(i), the fact that a free product of trace states is a trace state and 1.8.

1.11. REMARK: The analogue of the Gaussian functor constructed in [**1**] provides a canonical model for a semicircular family. We recall only part of the construction. Let H be a Hilbert space, $TH = \mathbb{C}1 \oplus \bigoplus_{n\geq 1} H^{\otimes n}$ the full Fock space and if $h \in H$ let $\ell(h)\xi = h\otimes\xi$ for $\xi \in TH$. If $(e_\iota)_{\iota\in I}$ is an orthonormal basis in H, let A be the C^*-algebra generated by the $s(e_\iota) = 1/2(\ell(e_\iota)+\ell(e_\iota)^*)$, $\iota \in I$ and let $\varphi(a) = \langle a1, 1\rangle$. Then φ is a faithful trace state and $(s(e_\iota))_{\iota\in I}$ is a semicircular family. The von Neumann algebra generated by A is isomorphic to the free group factor on card I generators and $\langle\cdot 1, 1\rangle$ is its normal trace state. Moreover, as an algebra of operators on TH it is in standard form.

We conclude this section with some facts from [**5**] concerning the connection between Gaussian random matrices and free random variables.

1.12. The natural framework for random matrices is the following. $(\Sigma, d\sigma)$ is a standard non-atomic measure space with a probability measure $d\sigma$ and $L = \bigcap_{p\geq 1} L^p(\Sigma)$ is a $*$-algebra endowed with the state $E : L \to \mathbb{C}$ given by $Ef = \int_\Sigma f(s)d\sigma(s)$. Let further $\mathfrak{M}_n$ be the complex $n\times n$ matrices endowed with the normalized trace $\tau_n : \mathfrak{M}_n \to \mathbb{C}$ (i.e. $\tau_n = 1/n\ Tr$). We shall denote by e_{ij} or $e(i,j;n)$ when n needs to be emphasized, the $n\times n$ matrix with zero entries except for the (i,j)-entry which is 1. The algebra of random $n\times n$ matrices is the algebra $M_n = \mathfrak{M}_n(L) = \mathfrak{M}_n \otimes L$ and is equipped with a trace state $\varphi_n : M_n \to \mathbb{C}$ given by $\varphi_n = \tau_n \otimes E$. $\mathfrak{M}_n$ can be naturally identified with $\mathfrak{M}_n \otimes 1 \subset M_n$, i.e. with the constant matrices in M_n. By $\Delta_n \subset \mathfrak{M}_n$ we shall denote the constant diagonal matrices.

1.13. THEOREM ([5]). *Let* $Y(s,n) = \sum_{1\leq i,j\leq n} a(i,j;n,s)e(i,j;n) \in M_n$ *be random matrices,* $s \in \mathbb{N}$. *Assume that* $a(i,j;n,s) = \overline{a(j,i;n,s)}$ *and that*

$$\{\mathrm{Re}\, a(i,j;n,s) | 1 \leq i \leq j \leq n,\ s \in \mathbb{N}\} \cup \{\mathrm{Im}\, a(i,j;n,s) | 1 \leq i < j \leq n, s \in \mathbb{N}\}$$

are independent Gaussian random variables such that

$$E(a(i,j;n,s)) = 0 \quad \text{for} \quad 1 \le i,j \le n,$$

$$E((\operatorname{Re} a(i,j;n,s))^2) = E((\operatorname{Im} a(i,j;n,s))^2) = (2n)^{-1} \quad \text{for} \quad 1 \le i < j \le n,$$

$$E((a(k,k;n,s)^2) = n^{-1} \quad \text{for} \quad 1 \le k \le n.$$

Let further $(D(j,n))_{j\in\mathbf{N}}$ *be elements in* Δ_n *such that* $\sup_{n\in\mathbf{N}} \|D(j,n)\| < \infty$ *for* $j \in \mathbf{N}$ *and* $(D(j,n))_{j\in\mathbf{N}}$ *has a limit distribution as* $n \to \infty$. *Then the family of subsets of random variables*

$$\{(\{Y(s,n)\}_{s\in\mathbf{N}}, \{D(j,n) \mid j \in \mathbf{N}\}\}$$

is asymptotically free as $n \to \infty$. *Moreover, the limit distribution of the family* $(1/2\, Y(s,n))_{s\in\mathbf{N}}$ *as* $n \to \infty$ *is that of a semicircular family.*

1.14. THEOREM ([5]). *Let* $Y(s,n) = \sum a(i,j;n,s)e(i,j;n,s) \in M_n$ *and* $Z(s,n) = \sum b(i,j;n,s)e(i,j;n) \in M_n$ *be random matrices,* $s \in \mathbf{N}$. *Assume that* $\operatorname{Im} a(i,j;n,s) = 0$, $\operatorname{Re} b(i,j;n,s) = 0$, $a(i,j;n,s) = a(j,i;n,s)$, $b(i,j;n,s) = -b(j,i;n,s)$ *and that* $\{a(i,j;n,s) \mid 1 \le i \le j \le n,\ s \in \mathbf{N}\} \cup \{ib(p,q;n,s) \mid 1 \le p < q \le n,\ s \in \mathbf{N}\}$ *are independent Gaussian random variables such that*

$$E(a(i,j;n,s)) = E(b(i,j;n,s)) = 0 \quad \text{for} \quad 1 \le i,j \le n$$

$$E((a(i,j;n,s))^2) = -E((b(p,q;n,s))^2) = n^{-1} \quad \text{for} \quad 1 \le i,j,p,q \le n$$

and $p \neq q$. *Let further* $(D(j,h))_{j\in\mathbf{N}}$ *be elements in* Δ_n *such that*

$$\sup_{n\in\mathbf{N}} \|D(j,n)\| < \infty$$

for $j \in \mathbf{N}$ *and* $(D(j,n))_{j\in\mathbf{N}}$ *has a limit distribution as* $n \to \infty$. *Then the family of sets of random variables*

$$\{(\{Y(s,n)\})_{s\in\mathbf{N}}, (\{Z(s,n)\})_{s\in\mathbf{N}}, \{D(j,n) \mid j \in \mathbf{N}\}\}$$

is asymptotically free as $n \to \infty$. *Moreover, the limit distribution of* $\{1/2\, Y(s,n) \mid s \in \mathbf{N}\} \cup \{1/2\, Z(s,n) \mid s \in \mathbf{N}\}$ *is that of a semicircular family.*

Note that Theorem 1.14 actually implies Theorem 1.13.

2. Circular and Semicircular Systems

This section deals with properties of circular and semicircular systems.

2.1. PROPOSITION. *Let (A, φ) be a C^*-probability space such that φ is a faithful trace and let $(f_\iota)_{\iota \in I}$ be a semicircular system in (A, φ). Let further $\mathcal{F}$ be the closed real linear span of the $(f_\iota)_{\iota \in I}$ in A. Then the following hold:*

a) *$\mathcal{F}$ is a real Hilbert space with orthonormal basis $(f_\iota)_{\iota \in I}$.*
b) *An orthonormal system in $\mathcal{F}$ is a semicircular system.*

PROOF: Since φ is a faithful trace, the GNS representation of the C^*-subalgebra B of A generated by the $(f_\iota)_{\iota \in I}$. Hence we may apply 1.10 to $(B, \varphi \mid B)$ and to the semicircular system in 1.11. After this identification a) and b) become obvious. □

2.2. PROPOSITION. *Let (A, φ) be a C^*-probability space such that φ is a faithful trace and let $(g_j)_{j \in J}$ be a circular system in (A, φ). Let further $\mathcal{G}$ be the closed complex linear span of the $(g_j)_{j \in J}$ in A. Then the following hold:*

a) *$\mathcal{G}$ is a complex Hilbert space with orthonormal basis $(g_j)_{j \in J}$.*
b) *An orthonormal system in $\mathcal{G}$ is a circular system.*

PROOF: Using the preceding proposition, let us show that the propositon follows if we know that $\|g\| = 1$ if g is circular in (A, φ). Let $x_j = 2^{-1/2}(g_j + g_j^*)$ and $y_j = -i2^{-1/2}(g_j - g_j^*)$ so that $\{x_j \mid j \in J\} \cup \{y_j \mid j \in J\}$ is semicircular. We will show first that $\|\sum_{j \in J} \lambda_j g_j\| = (\sum_{j \in J} |\lambda_j|^2)^{1/2}$ and $\sum_{j \in J} \lambda_j g_j$ is circular. Note first that it is sufficient to prove this for finite J. Next note that the family $(e^{i\theta_j} g_j)_{j \in J}$ is $*$-free and that if g is circular then $e^{i\theta} g$ is also circular. Indeed, if $2^{-1/2}(g + g^*) = x$, $-i2^{-1/2}(g - g^*) = y$ then $\{x, y\}$ is circular and hence $\{(\cos\theta)x - (\sin\theta)y,\ (\cos\theta)y + (\sin\theta)x\}$ is also semicircular, since it is an orthonormal system in the Hilbert space with basis $\{x, y\}$. Hence $e^{i\theta} g$ is circular. Thus it will be sufficient to prove our assertion in case $\lambda_j \geq 0$ $(j \in J)$. Indeed if $\sum \lambda_j g_j = 2^{-1/2}(a + ib)$ with $a = s^*$, $b = b^*$ then $x = \sum \lambda_j x_j$, $y = \sum \lambda_j y_j$ and $\{\mu a, \mu b\}$, where $\mu = (\sum \lambda_j^2)^{-1/2}$, is semicircular in view of the preceding proposition, so that $\mu \sum \lambda_j g_j$ is circular, which implies our assertion.

The fact we just proved clearly implies a). To prove b) note that it is sufficient to consider the case where the orthonormal system is a basis.

Let $U : \mathcal{G} \to \mathcal{G}$ be the unitary mapping $(g_j)_{j\in J}$ to the given basis. It will be sufficient to note that there is an orthogonal transformation V of the real Hilbert space H spanned by $\{x_j \mid j \in J\} \cup \{y_j \mid j \in J\}$ such that on the complexification $H + iH \supset \mathcal{G}$ the complexification of V extends U. Indeed the orthogonal transformation V of H extends to a trace-preserving automorphism of the C^*-algebra spanned by H and hence maps circular system to a circular system.

To conclude the proof let g be circular. In view of 1.10 and 1.11 we may assume we are in the context of 1.11. Thus we may assume $g = 2^{-3/2}(\ell(e_1) + \ell(e_1)^* + i\ell(e_2) + i\ell(e_2)^*)$. Let $h_1 = 2^{-1/2}(e_1 + ie_2)$, $h_1 = 2^{-1/2}(e_1 - ie_2)$, which is an orthonormal system so that $2g = \ell(h_1) + \ell(h_2)^*$. Clearly $||g|| \leq 1$. Conversely, let

$$\xi_n = n^{-1/2}(h_2 \otimes h_1 + h_2 \otimes h_1 \otimes h_2 \otimes h_1 + \cdots + h_2 \otimes h_1 \otimes \cdots \otimes h_2 \otimes h_1)$$

so that $||\xi_n|| = 1$, and $||2g\xi_n - 2\xi_n|| \to 0$ showing that $||g|| \geq 1$. □

2.3. Proposition. *Let (A, φ) be a C^*-probability space and let $1 \in D \subset A$ be a commutative $*$-subalgebra and let $(f(s))_{s\in S} \subset A$ be a semicircular system. Assume D and $\{f(s) \mid s \in S\}$ form a free pair of sets and suppose $p \in D$ is a self-adjoint idempotent such that $\varphi(p) \neq 0$. Then $(\varphi(p)^{-1/2}pf(s)p)_{s\in S}$ is a semicircular system in $(pAp, \varphi(p)^{-1}\varphi \mid pAp)$ and the pair of sets $\{\varphi(p)^{-1}pf(s)p \mid s \in S\}$ and pDp is free.*

Proof: It is clearly sufficient to prove the proposition in case S is finite and D is generated as an algebra by a finite set $\{p, d_1, \ldots, d_m\}$. Passing now to the context of 1.13 let $Y(s,n)$ $(s \in S)$ be independent self-adjoint Gaussian matrices as in 1.13 and let $\{p(n), d_1(n), \ldots, d_m(n)\} \subset \Delta_n$ be such that $p(n) = p(n)^* = p(n)^2$ and the joint distribution of $\{p(n), d_1(n), \ldots, d_m(n)\}$ converges to the joint distribution of $\{p, d_1, \ldots, d_m\}$. In particular $p(n)$ is a diagonal matrix with entries 0 and 1 and if k_n is the number of 1's then $\lim_{n\to\infty} k_n n^{-1} = \varphi(p)$. We may clearly assume that the non-zero entries of $p(n)$ procede the zero entries. Then, leaving aside the zeros which border the non-zero $k_n \times k_n$ matrix, the matrices $(nk_n^{-1})^{1/2}p(n)Y(s,n)p(n)$ are random $k_n \times k_n$ matrices of the same type as $Y(s,n)$ and $p(n)d_j(n)$ correspond to diagonal matrices in Δ_{k_n}. Therefore, since $k_n \to \infty$ as $n \to \infty$, 1.13 applies to these matrices. Hence the limit distribution of $((nk_n^{-1})^{1/2}p(n)Y(s,n)p(n))_{s\in S}$ is that of a semicircular family and the sets $\{(nk_n^{-1})^{1/2}p(n)Y(s,n)p(n) \mid s \in \mathbb{N}\}$ and $\{p(n)d_j(n) \mid j = 1, \ldots, m\}$ are asymptotically free. Since the limit distribution of

$$((nk_n^{-1})^{1/2}p(n)Y(s,n)p(n))_{s\in S} \cup (p(n)d_j(n))_{1\leq j\leq m}$$

is the distribution of

$$(\varphi(p))^{-1/2} pf(s)p)_{s \in S} \cup (pd_j)_{1 \leq j \leq m}$$

in $(pAp, \varphi(p)^{-1}\varphi \mid pAp)$ the assertion follows. □

We pass now to the study of the polar decomposition of a circular random variable. This requires some preparation.

2.4. LEMMA. *Let (A, φ) be a non-commutative probability space. Let $u, v \in A$ be invertible elements and let $1 \in B \subset A$ be a subalgebra such that the following conditions hold:*

a) *$(\{u, u^{-1}\}, \{v, v^{-1}\} \cup B)$ is a free pair of sets in (A, φ).*
b) *$\varphi(u) = \varphi(u^{-1}) = \varphi(v) = \varphi(v^{-1}) = 0$*
c) *$\varphi(vb) = \varphi(bv^{-1}) = \varphi(vbv^{-1}) = 0$ whenever $b \in B$ and $\varphi(b) = 0$.*

Then $(\{uv, (uv)^{-1}\}, B)$ is a free pair of sets in (A, φ).

PROOF: Let $t = uv$. In view of a) and b) we have that $\varphi(t^k) = 0$ if $k \neq 0$. Thus we must show that

$$\varphi(b_0 t^{k_1} b_1 t^{k_2} b_2, \ldots, t^{k_n} b_n) = 0$$

where $k_j \neq 0$ for $1 \leq j \leq n$ and $\varphi(b_p) = 0$ for $1 \leq p \leq n-1$. Clearly we may assume that either $\varphi(b_0) = 0$ or $b_0 = 1$ and similarly $\varphi(b_n) = 0$ or $b_n = 1$. Note that the monomial $b_0 t^{k_1} b_1 t^{k_2} b_2, \ldots, t^{k_n} b_n$ is actually a product of terms of the form u, u^{-1} alternating with terms of one of the following types $v, v^{-1}b_j, vb_jv^{-1}, vb_j, b_jv^{-1}$ (for those j such that $\varphi(b_j) = 0$). Since all these elements have expectation equal to zero, we conclude in view of a) that the expectation of the product is zero. □

2.5. COROLLARY. *Let (A, φ) be a non-commutative probability space and suppose φ is a trace. Let α be an automorphism of A such that $\varphi \circ \alpha = \varphi$ and let $1 \in B \subset A$ be a subalgebra such that $\alpha(b) = b$ for $b \in B$. Let further $u, v \in A$ be invertible elements such that the following conditions are satisfied:*

a) *$(\{u, u^{-1}\},\ B \cup \{v, v^{-1}\})$ is a free pair of sets in (A, φ).*
b) *$\varphi(u) = \varphi(u^{-1}) = 0$*
c) *$\alpha(v) = cv$ where $c \in \mathbb{C}$, $c \neq 1$.*

Then $(\{uv, (uv)^{-1}\}, B)$ is a free pair of sets in (A, φ).

PROOF: If $b \in B$, we have

$$\varphi(vb) = \varphi(\alpha(vb)) = c\varphi(vb)$$

and hence $\varphi(vb) = 0$. Similarly $\varphi(bv^{-1}) = 0$ because

$$\varphi(bv^{-1}) = \varphi(\alpha(bv^{-1})) = c^{-1}\varphi(bv^{-1}).$$

Also, $\varphi(vbv^{-1}) = \varphi(b)$ since φ is a trace. Thus, the conditions in Lemma 2.4 are satisfied and the desired conclusion follows. □

2.6. PROPOSITION. *Let (M, τ) be a W^*-probability space such that τ is a faithful trace state. Let $x \in M$ be a circular element and let $x = vb$ be its polar decomposition. Then we have:*

a) *The pair (v, b) is $*$-free in (M, τ).*
b) *v is unitary and $\tau(v^k) = 0$ if $k \in \mathbb{Z}\backslash\{0\}$*
c) *$\tau(b^k) = \frac{4}{\pi} \int_0^1 t^k(1-t^2)^{1/2}dt$ if $k \geq 0$.*

PROOF: a) It will be useful to pass to a larger W^*-probability space. Consider the W^*-probability space $(L(\mathbb{Z}), \lambda)$, where λ is the canonical trace, and let $(A, \varphi) = (L(\mathbb{Z}), \lambda) * (M, \tau)$. We shall identify M and $L(\mathbb{Z})$ with W^*-subalgebras in A. Let $u \in L(\mathbb{Z}) \subset A$ be the unitary corresponding to translation by $1 \in \mathbb{Z}$. There is also no loss of generality to assume that M is generated as a W^*-algebra by x. Then there is an unique automorphism α of (A, φ) such that $\alpha(u) = u$ and $\alpha(x) = -x$. Let $B \subset A$ be the $*$-algebra generated by $b = (x^*x)^{1/2}$. It is easily seen that $(A, \varphi), \alpha, u, v, B$ satisfy the assumptions of Corollary 2.5 (the invertibility of v follows by examining the circular element $1/2(\ell(h_1) + \ell(h_2)^*)$ considered in the proof of Proposition 2.2). It follows that the pair (uv, b) is $*$-free.

Let us show that in order to conclude the proof of a) it suffices to prove that x and ux have the same $*$-distribution. Indeed, this implies the existence of an injective weakly continuous homomorphism $\mu : M \to A$ such that $\varphi \circ \mu = \tau$ and $\mu(x) = ux$. This, in turn, would imply that $\mu(v) = uv$ and $\mu(b) = b$ and the $*$-freeness of (v, b) would follow from that of (uv, b).

To show that x and ux have the same $*$-distribution we use our results on random matrices (Theorem 1.13). Let $x_n = 2^{-3/2}(Y(1,n)+iY(2,n)) \in M_n$ and let $u_n = \sum_{1 \leq j \leq n} \exp(2\pi ij/n)e(j,j;n) \in \Delta_n$. It follows from Theorem 1.13 that the $*$-distribution of (x_n, u_n) converges to the $*$-distribution of (x, u). Thus, the equality of the $*$-distribution of x and ux will follow if we show that the $*$-distributions of x_n and u_nx_n. The random matrix x_n is characterized by the fact that the $2n^2$ random variables which are the real and imaginary parts of its entries are Gaussian, independent, have first moments equal to 0 and second moments $(8n)^{-1}$. It is clear that u_nx_n is a matrix with the same properties and hence that the $*$-distributions of x_n and u_nx_n are equal.

b) As we already mentioned in the proof of a) the unitarity of v follows by examining a concrete circular element $1/2(\ell(h_1) + \ell(h_2)^*)$. The fact that $\tau(v^k) = 0$ if $k \in \mathbf{Z}\setminus\{0\}$ is a consequence of the fact that $e^{2\pi i\theta}x$ is also circular and hence $\tau(v^k) = \tau((e^{2\pi i\theta}v)^k)$.

c) Here again we use the concrete circular element $x = 1/2(\ell(h_1) + \ell(h_2)^*)$. In 3.5 of [5] we have shown that the moments of x^*x are the same as those of the square of a semicircular element, i.e.

$$\tau((x^*x)^k) = \frac{2}{\pi}\int_{-1}^{1} t^{2k}(1-t^2)^{1/2}dt.$$

This gives

$$\tau(b^{2k}) = \frac{4}{\pi}\int_{0}^{1} t^{2k}(1-t^2)^{1/2}dt$$

and in view of the Stone–Weierstrass theorem this equality extends also to the odd moments. □

2.7. Remark: It is not hard to derive from Proposition 2.6a) the following fact about the polar decomposition of random matrices. If $X(n) = \sum_{1\le i,j\le n} a(i,j;n)e(i,j;n) \in M_n$ is such that the $2n^2$ random variables $\mathrm{Re}\, a(i,j;n)$, $\mathrm{Im}\, a(i,j;n)$ are independent, Gaussian and have first moments equal zero and second moments equal n^{-1}, let $u_n b_n = X(n)$ be the polar decomposition, then the pair (u_n, b_n) is asymptotically $*$-free. Similarly, consider a real matrix, i.e. $\mathrm{Im}\, a(i,j;n) = 0$ and the n^2 random variables $a(i,j;n)$ are independent, Gaussian and have their first two mements equal to 0 and respectively n^{-1}. Then again the two components of the polar decomposition are asymptotically $*$-free as $n \to \infty$.

We pass to another kind of results which can be derived from the results on random matrices.

2.8. Proposition. *Let (A, φ) be a C^*-probability space and let $D \subset A$ be a commutative $*$-subalgebra. Let*

$$\Omega = \{h(p,q;s) \mid 1 \le p \le q \le N,\ s \in S\} \cup \{f(p,q;s) \mid 1 \le p < q \le N,\ s \in S\}$$

be a semicircular family such that the pair of sets (Ω, D) is free. Further, let $f(p,p;s) = 0$ and if $p > q$ let $h(p,q;s) = h(q,p;s)$, $f(p,q;s) = -f(q,p;s)$ and consider

$$k(s) = N^{-1/2} \sum_{1\le p,q\le N} (2^{-1/2} + \delta_{p,q}(1-2^{-1/2}))(h(p,q;s) + if(p,q;s)) \otimes e_{pq}$$

$$d(t) = \sum_{1\le p\le N} d(p,t) \otimes e_{pp}$$

in $(A \otimes \mathfrak{M}_N,\ \varphi \otimes \tau_N)$ *where* $\{d(p,t) \mid 1 \le p \le N,\ t \in T\} \subset D$. *Then* $(k(s))_{s\in S}$ *is a semicircular family and the pair of sets* $(\{k(s) \mid s \in S\}, \{d(t) \mid t \in T\})$ *is free.*

PROOF: The proof is along the same lines as the other proofs based on 1.13. The $*$-distribution of (Ω, D) is approximated by that of matrices, the matrices for Ω being Gaussian random matrices like in 1.13 and the ones for D constant diagonal matrices. Then the matrices built out of the $n \times n$ approximants, are $Nn \times Nn$ matrices to which 1.13 applies yielding the desired conclusion. □

3. Factors

This section contains some applications of circular and semicircular systems to factors.

3.1. LEMMA. *Let* A *be a* $*$*-algebra with unit and let* $(w_{ij})_{1\le i,j\le n}$ *be a system of matrix units in* A *with* $\Sigma\, w_{ii} = 1$. *Let further* $\Omega \subset A$ *be a set which generates* A *as a* $*$*-algebra. Then the set* $\bigcup_{1\le j,k\le n} w_{1j}\Omega w_{k1}$ *generates* $w_{11}Aw_{11}$ *as a unital* $*$*-algebra. The same assertion holds for* $*$*-algebra replaced by* C^**-algebra or* W^**-algebra.*

PROOF: Define $\Theta(\Sigma) = \bigcup_{1\le j,k\le n} w_{1j}\Omega w_{k1}$ for any set $\Sigma \subset A$. It is easily seen that

$$\Theta(\Sigma^*) = (\Theta(\Sigma))^*$$

and

$$\Theta(\Sigma_1)\Theta(\Sigma_2) \supset \Theta(\Sigma_1\Sigma_2).$$

Hence if Σ is the semigroup generated by $\Omega \cup \Omega^*$ and if S is the semigroup generated by $\Theta(\Omega \cup \Omega^*)$ we have $S \supset \Theta(\Sigma)$.

Hence the $*$-algebra with unit generated in $w_{11}Aw_{11}$ by $\Theta(\Omega)$ contains $\Theta(S) + \mathbf{C}w_{11}$ which spans $\Theta(A) = w_{11}Aw_{11}$.

The assertions for C^*-algebras and W^*-algebras are immediate corollaries. □

If $\mathcal{M}$ is a II_1-factor and $\alpha \in \mathbf{R}_{>0}$ we denote by $\mathcal{M}_\alpha$ a II_1-factor isomorphic to $e(\mathcal{M} \otimes B(H))e$ where $e \in \mathcal{M} \otimes B(H)$ is a self-adjoint idempotent in the II_∞-factor with trace α. If $\alpha \le 1$ we may define $\mathcal{M}_\alpha$ to be isomorphic to $e\mathcal{M}e$ with $e \in \mathcal{M}$, $e = e^* = e^2$ and trace of e equal α.

Besides writing a free group as a free product, it will also be convenient to denote by $F(S)$ the free group with generators indexed by a set S. We shall also sometimes write $F(|S|)$ where $|S|$ is the cardinal of S.

3.2. DEFINITION: A family of unitary elements $(u_\iota)_{\iota\in I}$ in a $*$-probability space is said to have *a free $*$-distribution* if $\varphi(\pi(g)) = \delta_{g,e}$ for $g \in F(I)$, with $\pi : F(I) \to U(A)$ denoting the homomorphism such that $\pi(g_\iota) = u_\iota$ $(\iota \in I)$.

3.3. THEOREM. *Let $N \in \mathbf{N}$ and let S be a set with at least 2 elements. We have*

a) $(L(F(S) * \mathbf{Z}/N\mathbf{Z}))_{1/N} \simeq L(F(|S|N^2 - N + 1)$
b) $(L(F(S)))_{1/N} \simeq L(F(|S|N^2 - N^2 + 1))$

PROOF: a) We shall realize $L(F(S) * \mathbf{Z}/N\mathbf{Z})$ as a W^*-algebra of $A \otimes \mathfrak{M}_N$ where (A, φ) is a W^*-probability space with φ a faithful trace and containing a sufficiently large semicircular system. We shall use to this end Proposition 2.8. Let $\{f(p;s) \mid 1 \le p \le N,\ s \in S\} = \omega_1 \subset A$ be a semicircular system and let $\{g(p,q;s) \mid 1 \le p < q \le N,\ s \in S\} = \omega_2 \subset A$ be a circular system such that the pair of sets (ω_1, ω_2) is $*$-free in (A, φ). Then $L(F(S) * \mathbf{Z}/N\mathbf{Z})$ can be identified with the W^*-algebra $\mathcal{X}$ generated by

$$k(s) = \sum_{1\le p\le N} f(p;s)\otimes e_{pp} + \sum_{1\le p<q\le N} (g(p,q;s)\otimes e_{p,q} + g(p,q;s)^*\otimes e_{qp}) \quad (s\in S)$$

and

$$d = \sum_{1\le j\le N} \exp(2\pi ij/N)1\otimes e_{jj}.$$

It is easily seen that this set of generators is equivalent to

$$\begin{array}{ll} f(p;s)\otimes e_{pp} & 1\le p\le N,\ s\in S \\ g(p,q;s)\otimes e_{pq} & 1\le p<q\le N,\ s\in S. \end{array}$$

If $g(p,q;s) = v(p,q;s)b(p,q;s)$ is the polar decomposition, then we get another set of generators

$$\begin{array}{ll} f(p;s)\otimes e_{pp} & 1\le p\le N,\ s\in S \\ v(p,q;s)\otimes e_{pq} & 1\le p<q\le N,\ s\in S \\ b(p,q;s)\otimes e_{qq} & 1\le p<q\le N,\ s\in S. \end{array}$$

Since the distribution of $f(p;s)$ and $b(p,q;s)$ are measures without atoms (see Proposition 2.6) it follows that there are unitary elements $F(p;s)$ and $B(p,q;s)$ in A, such that they generate the same W^*-subalgebras as $f(p;s)$ and respectively $b(p,q;s)$ and $\varphi((F(p;s))^k) = \varphi((B(p,q;s)^k) = 0$ if $h \in \mathbf{Z}\backslash\{0\}$. Thus there is also the following set of generators of $\mathcal{X}$:

$$\begin{array}{ll} F(p;s)\otimes e_{pp} & 1\le p\le N,\ s\in S \\ v(p,q;s)\otimes e_{pq} & 1\le p<q\le N,\ s\in S \\ B(p,q;s)\otimes e_{qq} & 1\le p<q\le N,\ s\in S. \end{array}$$

Note that in view of Proposition 2.6 the family of unitary elements

$$\Gamma = \{F(p;s)\mid 1\le p\le N,\ s\in S\}\cup\{v(p,q;s)\mid 1\le p<q\le N,\ s\in S\}\cup$$
$$\cup\{B(p,q;s)\mid 1\le p<q\le N,\ s\in S\}$$

has a free $*$-distribution.

We have $(L(F(S) * \mathbf{Z}/N\mathbf{Z}))_{1/N} \simeq (1\otimes e_{11})\mathcal{X}(1\otimes e_{11})$. Let $w_{11} = 1\otimes e_{11}$ and $w_{1p} = v(1,p;\sigma)\otimes e_{1p}$ (here $2\le p\le N$ and $\sigma\in S$ is a fixed element) and consider the system of matrix units $w_{pq} = w_{1p}^* w_{1q}$ in $\mathcal{X}$. In view of Lemma 3.1 we have that $(1\otimes e_{11})\mathcal{X}(1\otimes e_{11}) = \mathcal{Y}\otimes e_{11}$ where $\mathcal{Y}$ is generated by

$$\begin{aligned} \gamma = {} & \{v(1,p;\sigma)v(p,q;s)v(1,q;\sigma)^{-1}\mid 2\le p<q\le N,\ s\in S\backslash\{0\}\}\cup \\ & \cup\{v(1,q;s)v(1,q;\sigma)^{-1}\mid 1<q\le N,\ s\in S\backslash\{0\}\}\cup \\ & \cup\{v(1,p;\sigma)v(p,q;\sigma)v(1,q;\sigma)^{-1}\mid 2\le p<q\le N\}\cup \\ & \cup\{v(1,q;\sigma)B(p,q;s)v(1,q;\sigma)^{-1}\mid 1\le p<q\le N,\ s\in S\}\cup \\ & \cup\{v(1,p;\sigma)F(p,s)v(1,p;\sigma)^{-1}\mid 2\le p\le N,\ s\in S\}\cup \\ & \cup\{F(1;s)\mid s\in S\}. \end{aligned}$$

It is easily seen that in $F(\Gamma)$ the set γ is a free set of generators for a subgroup. Hence since Γ has a free $*$-distribution it follows that $\mathcal{Y}\simeq L(F(\gamma))$. Counting elements, we have $|\gamma| = N^2|S| - N + 1$, which gives the desired conclusion.

b) The proof of b) is a slight modification of the proof of a). Thus $L(F(S))$ can be identified with the W^*-algebra $\mathcal{R}$ generated by

$$k(s) = \sum_{1\le p\le N} f(p;s)\otimes e_{pp} + \sum_{1\le p<q\le N} (g(p,q;s)\otimes e_{pq} + g(p,q;s)^*\otimes e_{qp})\ \ (s\in S')$$

and

$$h = \sum_{1 \le j \le N} ja \otimes e_{jj}$$

where $S' = S \backslash \{\sigma_1\}$, $f(p;s)$ and $g(p,q;s)$ are as in the proof of a) and a is a unitary element with a free $*$-distribution and such that $(\{a\}, \{\omega_1 \cup \omega_2\})$ is a free pair.

An equivalent set of generators is

$$\begin{array}{ll} f(p;s) \otimes e_{pp} & 1 \le p \le N,\ s \in S' \\ g(p,q;s) \otimes e_{pq} & 1 \le p < q \le N,\ s \in S' \\ a \otimes e_{pp} & 1 \le p \le N. \end{array}$$

We introduce again the polar decomposition $g(p,q;s) = v(p,q;s)b(p,q;s)$ and the unitary elements $B(p,q;s)$, $F(p,s)$. Thus we get another set of generators

$$\begin{array}{ll} F(p;s) \otimes e_{pp} & 1 \le p \le N,\ s \in S' \\ v(p,q;s) \otimes e_{pq} & 1 \le p < q \le N,\ s \in S' \\ B(p,q;s) \otimes e_{qq} & 1 \le p < q \le N,\ s \in S' \\ a \otimes e_{pp} & 1 \le p \le N. \end{array}$$

The family of unitary elements $F(p;s)$, $v(p,q;s)$, $B(p,q;s)$, a has a free $*$-distribution. Let further $\sigma_2 \in S'$ and $S'' = S' \backslash \{\sigma_2\}$ and let $w_{11} = 1 \otimes e_{11}$, $w_{1p} = v(1,p;\sigma_2) \otimes e_{1p}$ $(2 \le p \le N)$, $w_{pq} = w_{1p}^* w_{1q}$. Then $(1 \otimes e_{11})\mathcal{R}(1 \otimes e_{11}) = \mathcal{P} \otimes e_{11}$ where $\mathcal{P}$ is generated by

$$\begin{aligned} \chi = &\{v(1,p;\sigma_2)av(1,p;\sigma_2)^{-1} \mid 2 \le p \le N\} \cup \\ &\cup \{a\} \cup \\ &\cup \{v(1,p;\sigma_2)v(p,q;s)v(1,q;\sigma_2)^{-1} \mid 2 \le p < q \le N,\ s \in S''\} \cup \\ &\cup \{v(1,q;s)v(1,q;\sigma_2)^{-1} \mid 1 < q \le N,\ s \in S''\} \cup \\ &\cup \{v(1,p;\sigma_2)v(p,q;\sigma_2)v(1,q;\sigma_2)^{-1} \mid 2 \le p < q \le N\} \cup \\ &\cup \{v(1,q;\sigma_2)B(p,q;s)v(1,q;\sigma_2)^{-1} \mid 1 \le p < q \le N,\ s \in S'\} \cup \\ &\cup \{v(1,p;\sigma_2)F(p,s)v(1,p;\sigma_2)^{-1} \mid 2 \le p \le N,\ s \in S'\} \cup \\ &\cup \{F(1;s) \mid s \in S'\}. \end{aligned}$$

Again, it is easy to check that these elements form a set of free generators for a subgroup of the free group generated by the $F(p;s)$, $v(p,q;s)$, $B(p,q;s)$ and a. The assertion follows by counting the elements of χ. □

We have as an immediate consequence of Theorem 3.3b) the following corollary.

3.4. COROLLARY. *The fundamental group of $L(F(\mathbb{N}))$ contains the multiplicative gorup of positive rational numbers.*

3.5. COROLLARY. *Let $G = \underset{\iota\in I}{*} \mathbb{Z}/n_\iota\mathbb{Z}$, where I is an at most countable set and $n_\iota \geq 2$ for all $\iota \in I$. Then we have*

$$L(F(\mathbb{N})) \simeq L(G * F(\mathbb{N})).$$

PROOF: It follows from Theorem 3.3a) and b) that:

$$(L(F(\mathbb{N}) * \mathbb{Z}/m\mathbb{Z}))_{1/m} \simeq L(F(\mathbb{N})) \simeq (L(F(\mathbb{N})))_{1/m}.$$

This in turn implies

$$L(F(\mathbb{N}) * \mathbb{Z}/m\mathbb{Z}) \simeq L(F(\mathbb{N})).$$

Taking free products of II_1-factors we have:

$$L(F(\mathbb{N})) \simeq (L(F(\mathbb{N})))^{*I} \simeq \underset{\iota\in I}{*} L(F(\mathbb{N}) * \mathbb{Z}/n_\iota\mathbb{Z}) \simeq L(F(\mathbb{N}) * G).$$

□

REFERENCES

1. Voiculescu, D., *Symmetries of some reduced free product C^*-algebras*, in "Operator Algebras and Their Connections with Topology and Ergodic Theory", Lecture Notes in Math., Vol. II, **32**, Springer–Verlag (1985), 556–588.
2. ——————, *Additon of certain non-commuting random variables*, J. Functional Analysis, Vol. 66, No. 3 (1986), 323–346.
3. ——————, *Multiplication of certain non-commuting random variables*, J. Operator Theory **18** (1987), 223–235.
4. ——————, *Non-commutative random variables and spectral problems in free product C^*-algebras*, Preprint, Berkeley (1988), to appear in Rocky Mountain Journal.
5. ——————, *Limit laws for random matrices and free products*, Preprint, IHES (1989).
6. Wigner, E., *Characteristic vectors of bordered matrices with infinite dimensions*, Ann. of Math. **62** (1955), 548–564.
7. —————, *On the distribution of the roots of certain symmetric matrices*, Ann. of Math. **67** (1958), 325–327.

Received October 27, 1989

Department of Mathematics
University of California, Berkeley
Berkeley, CA 94720

II. Lie Groups and Lie Algebras

On the Characters of Connected Lie Groups

L. PUKANSZKY

To J. Dixmier on his 65th birthday with admiration and gratitude

The objective of this talk is to sketch certain aspects of the development of the unitary representations of general (that is not necessarily semi-simple) Lie groups during the last decades, with particular emphasis on the inspiration due to the work of J. Dixmier in this field. Because of the limitation on my time, the point of view taken will be necessarily selective, and the focus will be on phenomena appearing, when no assumption on the type of the group in question is made. Much of Dixmier's direct or indirect contributions to the subject belong to earlier phases of his work. Therefore for their proper appreciation, it is indispensable that we should start by taking a brief look at the status of these questions as they unfolded in the late forties.

Subsequent to the success of the theory of locally compact abelian and of compact groups, the classical exposition of which to the present day has remained the book of André Weil ([34]), important results of a different nature came to light. Here I mean (1) Unitary representations of classical and even already of general semi-simple groups ([14], [20]). From our present point of view the most important outgrowth of this line of investigation was the extension of the notion of the group character. Given an irreducible unitary representation T of such a group, if f is a smooth function of compact support, the operator $T(f) = \int_G f(s)T(s).ds$, where ds is the element of the here necessarily biinvariant Haar measure, turned out to be of trace class, and the linear form $f \mapsto \mathrm{Tr}\,(T(f))$ a distribution (cf. [21]). It is the latter, which we take as the character of T. As a brief parenthetical comment let us note here that, upon leaving the semi-simple domain, the situation changes drastically. If the connected Lie group G is of two dimensions, there is still some choice of f as above, such that $T(f)$ is a nonzero compact operator. In five dimensions, however, for some G and T, $T(f)$ can be compact only, if it is zero. (2) Just as importantly from our present point of view there emerged, in the form of the theory of positive definite functions, due to Gelfand-Raikov and Godement ([13],

[15]), a general framework, promising rapid clarification of the basic questions. It was also during this period, that J. v. Neumann published his Reduction Theory ([23]), providing new proofs and important insights. Some of these were of a disturbing nature for the prospects of a general theory along the classical lines. Thus while it became possible to represent a unitary representation as direct integral of irreducibles, it also turned out that this decomposition, as soon as the group under consideration fails to be of type one, (that is, if it admits unitary representations generating continuous factors), is, in general, badly non-unique. I recall, incidentally, that connected Lie groups fail to be of type one already in five dimensions. Uniqueness features could be expected only from the central disintegration into factor representations. In other words, one appeared to be forced to abandon the classical idea of a one-to-one correspondence between representations generating a ring of operators with a trivial center (factor representations) modulo quasi-equivalence, and irreducible representations. Just as importantly, as noted by Godement (cf. [18], in particular 8, p.76 and also [8]), closer inspection of concrete examples revealed that one had to distinguish between factor representations. The real interest belongs only to those which admit a property somewhat analogous to what I have just quoted as being valid in the semi-simple case: when integrated against by a suitably chosen Radon measure, they give rise to nonzero operators, which are of trace class in the sense of the, possibly continuous, factor they generate. These we call normal representations. In particular, for a factor representation to claim interest in this sense, it is neither necessary, nor sufficient, that it should be a multiple of an irreducible representation. As it was proved later by Glimm, all irreducible representations are normal, if and only if the group in question is of type one. Granted the said property, one can proceed, somewhat similarly as in the semi-simple case, to consider as generalization of the character, the linear form $f \mapsto \mathrm{Tr}\,(T(f))$, where Tr( ) is to be construed in the sense of the factor generated by T and where, of course, the domain of f still requires clarification. All this being so, the first basic question to be faced was the existence of characters, or that of normal representations. Taking one's cue from the compact case it is plausible, via Reduction Theory, that at least a separating family of such representations can be obtained, as soon as one can produce on the operator ring generated by the left regular representation, the left ring, what we call in our current terminology a faithful, normal, semifinite trace, such that sufficiently many convolution operators be of trace class. If so, however, in particular the left ring itself must be semifinite, that is a central continuous direct sum of factors of type I or II.

This leads us to the next basic question: Is the left ring semifinite? It was recognized by R. Godement and I. E. Segal that the answer is yes, if the group is unimodular (cf. [16], [33]). On the other hand, as pointed out by Godement, upon deleting this stipulation the conclusion may fail in the category of all separable locally compact groups; in fact, the left ring can be a purely infinite or type III factor. The technical reason for which the unimodular case presents a simpler situation, is that here the left ring can be obtained via a unitary algebra. This is defined as an associative algebra R (say) over the complex field with an involution $x \to x^s$; R is also a unitary space such that (i) $(x^s, x^s) = (x, x)$, (ii) $(xy, z) = (y, x^s z)$, (iii) For each x fixed, $y \to xy$ is continuous, and (iv) The set of linear combinations of elements xy is dense. An important example is obtained by taking a locally compact unimodular group G by considering the totality $L(G)$ of continuous functions of compact support with the usual inner product with respect to some Haar measure, with convolution as multiplication and $f^s(x) \equiv \overline{f(x^{-1})}$ as involution. Given a unitary algebra R we can form the Hilbert space which is its completion (H, say). The axioms then imply the existence, for each x in R, of two bounded operators U_x and V_x, which on $R \subset H$ reduce to multiplication on the left and right resp. by x. Let R^g and R^d be the operator rings, called the left and right rings of R, generated by (U_x) and (V_x); they are commutants of each other. If R belongs to G as above, R^g is simply the left ring of G. This being so the decisive observation is that R^g carries canonically a trace F with respect to which, for all $x \in R$, U_x is of Hilbert-Schmidt class such that $F(U_x U_x^*) = (x, x)$. Suppose now, that G is not unimodular and let dx be the element of a right-invariant Haar measure, such that $d(sx) = r(s)dx$. To comply with $(x, x) = (x^s, x^s)$, we are forced to change the involution on $L(G)$ to $f^s(x) \equiv f(x^{-1})/(r(x))^{\frac{1}{2}}$. But then one needs an automorphism j defined by $f^j(x) \equiv (r(x))^{\frac{1}{2}} f(x)$, since in this case $\overline{f(x^{-1})} \equiv f^{sj}(x)$ and thus we have for any three $x, y, z \in L(G) : (xy, z) = (y, x^{sj} z)$. This is one of the first observations leading to Dixmier's theory of quasi-unitary algebras, published in 1952 ([3]). This is not only a masterful analysis of the reasons for the failure of the unitary algebra approach in the general case, and an outline of the remedial steps needed, but with its wealth of genial insights turned out subsequently fundamental for the theory of operator rings as well as for representation theory. Much of what I shall say in my further account of the theory of the regular representation owes its inspiration to this paper.

Formalizing the previous hints, a quasi-unitary algebra R is again an associative algebra over the complex field, which is also a unitary space, on which there are given an involution $x \mapsto x^s$ and a positive automorphism $x \mapsto x^j$, such that we have (i) $(x, x) = (x^s, x^s)$, (ii) $(xy, z) = (y, x^{sj} z)$, (iii) For each x fixed, $y \mapsto xy$ is continuous, and (iv) The set of linear combinations of elements $xy + (xy)^j$ is dense. One forms, as in the unitary case, the left and right rings; they are again commutants of each other.

On the other hand, they not only did not have to be semifinite but, as pointed out by Dixmier, all known examples of operator rings were $*$-isomorphic to the left ring of some quasi-unitary algebra. This being so, the problem to develop criteria under which R^g is semifinite posed itself. Dixmier's answer, as completed by me, is as follows (cf. [3], V-VI, pp.289-298 and [24]). The axioms imply the existence of a positive self-adjoint operator J on the completion of R, which continues the map $x \mapsto x^j$. Let S be the continuation of $x \mapsto x^s$. Then R^g is semifinite if and only if there is an operator M, which is positive, self-adjoint and non-singular, commutes with operators of R^g such that if we set $M' = SMS$, J is a minimal closed extension of $M'M^{-1}$. If such an M exists, we can arrange a change of R, while retaining R^g, such that there exists a trace F on R^g satisfying $F(U_x U_x^*) = ||Mx||^2$ for a dense subset in R. If R is unitary, we have $J = I$ and we can choose $M = I$. In the context of the time, a striking indication of the depth of Dixmier's ideas was that all this could be turned into a new proof for the existence of type III factors, which was completely different from J. v. Neumann's original proof.

The exploitation of these results for the investigation of the regular representation of Lie groups depended much on an extension of our knowledge of the representation theory of particular classes of such groups. In the non-semi-simple domain this study was initiated and advanced to a substantial degree by Dixmier through his papers on algebraic groups, nilpotent groups and with various contributions to more general solvable Lie groups (cf. in particular [4], [5], and [6]). His work served as a basis for further refinements, the best known of these being the unitary representation theory of connected nilpotent Lie groups due to Kirillow ([22]). The central statement of this is the character formula, itself the analogue of the Weyl character formula for compact semi-simple groups (cf. 2, p.258 in [26]).

Since now it represents the simplest instance of a class of results I want to speak about later on, it is essential that I recall here briefly its statement. Let G be a connected and simply connected nilpotent Lie group with the Lie algebra $\mathfrak{g}$. Then the exponential map establishes an analytic isomorphism between $\mathfrak{g}$ and the underlying manifold of G. Let T be an irreducible unitary representation of G and f a smooth function with compact support. It was shown by Dixmier that, just as in the semisimple case, the integral $T(f)$ of f with respect to T is of trace class, and the pullback, via the exponential map, of the linear form $f \mapsto \mathrm{Tr}\,(T(f))$ a tempered distribution on $\mathfrak{g}$ (cf. [5], Corollaire 1, p.78). Kirillov's character formula provides for it the following analytic expression. There is an orbit O of the coadjoint representation of G such that $\mathrm{Tr}(T(f)) = \int_O (\widehat{f \circ \exp}).dv$, where dv is the element of an invariant measure on O, and $\hat{}$ is the Fourier transform of some function on $\mathfrak{g}$. The integral on the right converges absolutely. The main assertion of Kirillov's theory, as proposed by him, is the existence of a certain bijection between the underlying set of the unitary dual of G and the orbit space $\mathfrak{g}^*/G$. The character formula can be used to define such a bijection, which coincides with that originally exhibited by Kirillov using induced representations (cf. [25]).

In my own work on the structure of the regular representation my principal guide was the theorem of Dixmier quoted before, supplemented by Kirillov's theory. I came to recognize first, that in many instances of connected solvable Lie groups, one can associate with their orbits on $\mathfrak{g}^*$ normal representations, such that the analogue of Kirillov's formula holds true (cf. [27], pp. 459-460). Here, again, it is understood, that the trace Tr( ) is to be taken in the sense of the factor generated by our normal representation which, in general, is not of type one. Also, the support of the test function f must lie in an open set, on which log is defined. Still, the mechanism to recover $f(e)$ from the various $\mathrm{Tr}(T(f))$'s remained the same as in the nilpotent case, and thus it became possible to centrally disintegrate the regular representation into normal representations. In other words, in these cases of connected solvable groups the left ring had to be semifinite.

An investigation along different lines in the general case soon led to the operator M of Dixmier I spoke about before. If G is simply connected solvable, then M is quotient of two differential operators with many special properties (cf. [27], Chapter IV, in particular Theorem 4, p.593). The solvable case having thus been settled, Dixmier turned to the same problem for general connected Lie groups (cf. [10]). His approach inspired, with much of its proof, the following statement, which not only settled the problem, but turned out to be pivotal for the character theory of connected Lie groups as well, about which I shall say more presently. For this reason, I take the liberty of quoting it in full, in spite of its technical nature (cf. [28], Theorem, p.379). Suppose that $\mathfrak{g}$ and $\mathfrak{h}$ are algebraic Lie algebras of endomorphisms of a finite-dimensional real vector space, and that $\mathfrak{h}$ is an ideal of $\mathfrak{g}$. Let G be a connected and simply connected Lie group with the Lie algebra $\mathfrak{g}$, and H the connected subgroup belonging to $\mathfrak{h}$. Then the natural action of G on the dual of H is countably separated. Here I recall that, by an earlier result of Dixmier, connected Lie groups, locally isomorphic to real algebraic groups, are of type one (cf. [4], Théorème 1, p. 326).

The above proof of the semifiniteness of the left ring in the solvable case implies, in a routine manner, the existence of a separating family of normal representations: For any group element, different from the unity, there is a normal representation assuming a value, different from the unit operator. Even this, however, leaves open the question of a proper definition of the completeness, for some connected Lie group, of a set of normal representations. A persuasive way to proceed is as follows. In the case of a finite group there is a canonical bijection between the set of prime ideals of the group algebra and quasi-equivalence classes of factor representations. Let us take now a separable C^*-algebra A. We know from Dixmier that the notion of a closed prime ideal is synonymous with that of a primitive ideal, or kernel of an irreducible representation (cf. [7], Corollaire 1, p.100). On the other hand, again by Dixmier, kernels of factor representations, too, are primitive ideals (cf. *loc cit.* Corollaire 3). In this fashion, we have a canonical map from the set of quasi-equivalence classes of normal representations, denoted by $\widehat{A}_{\mathrm{norm}}$, into the set of all primitive ideals Prim(A). To the latter we can always guarantee a size on the basis of general principles. On the other hand, it can very well happen, that there is no nontrivial normal representation; in other words, $\widehat{A}_{\mathrm{norm}}$ can be trivial (cf. [2]). However, the canonical map $\widehat{A}_{\mathrm{norm}} \to$ Prim(A) is bijective in important special cases, such as when A is of type one (cf. [9], §9). Thus it is permitted to ask, if this remains so, if A is replaced by the group C^*-algebra of a connected Lie group, and this indeed turned out to be the case. The proof of this took more time, and was arrived at in essentially three stages.

(1) The dual of a connected and simply connected solvable Lie group of type one was described by L. Auslander and B. Kostant by aid of the coadjoint representation (cf. [1]). In the nilpotent case this reduces to the theory of Kirillov. To get hold of the general solvable case, first I introduced the notion of the generalized orbit of a group of the said kind (cf. [27], Chapter II, p.512 and also [31], 3, p.48). To this end one starts by showing the existence, for arbitrary connected Lie group G with the Lie algebra $\mathfrak{g}$, of the finest G-invariant equivalence relation R on $\mathfrak{g}^*$, such that $\mathfrak{g}^*/R$ is a T_O-space (cf. [32], Lemma 1, p.817).

Any locally closed orbit of the coadjoint representation is an R-orbit; such are the R-orbits of type one solvable groups. A generalized orbit is a certain principal bundle over an R-orbit, the structure group being some closed subgroup of the multitorus, which is the dual of the fundamental group of an arbitrary transitive G-orbit contained in it. It is a G-space, and the group action defines a foliation, the leaves of which, as G-spaces, are isomorphic to coadjoint orbits.

There presented itself a compelling way to associate a factor representation to each generalized orbit, such as to extend the algorithm of Auslander and Kostant; these I shall quote as distinguished representations for a while. As it turned out, these suffice for a central disintegration of the regular representation (cf. [27], Chapter III, p. 544, in particular Theorem 3, p. 565).

(2) As noted earlier, the kernel of a factor representation is a closed prime ideal, or primitive ideal. The next question was if, just as in the case of type one groups, the map, assigning to each generalized orbit the kernel of the corresponding distinguished factor representation, would yield a bijection between the set of all generalized orbits and the set of all primitive ideals Prim(G) of the group C^*-algebra? I showed that the answer is affirmative (cf. [29], Theorem 1, p.114). Here the proof depended much on an extension of existing techniques for induced representations of C^*-algebras.

(3) So far nothing has been said about normal representations. Let now G be an arbitrary connected Lie group. Here the principal question is if the canonical map from $\widehat{G}_{\text{norm}}$ into Prim(G) provides a bijection. A different problem was the relation, in the solvable case, between the distinguished factor representations, parametrized by the generalized orbits, and the normal representations. The final result turned out to be optimal: (a) In the general case we have a bijection $\widehat{G}_{\text{norm}} \to \text{Prim}(G)$ (cf. [34], Theorem 1, p.119); (b) If G is solvable, the distinguished factor representations coincide with the normal representations modulo quasi-equivalence, and thus we are left with a threefold bijection: (Generalized orbits) $\leftrightarrow \text{Prim}(G) \leftrightarrow \widehat{G}_{\text{norm}}$.

BIBLIOGRAPHY

[1] L. Auslander, B. Kostant, *Polarization and unitary representations of solvable groups*, Invent. Math. **14** (1971), 255-354.

[2] J. Cuntz, *Simple C^*-algebras generated by isometries*, Comm. Math. Phys. **57** (1977), 173-185.

[3] J. Dixmier, *Algèbres quasi-unitaires*, Comment. Math. Helv. **26** (1952), 275-322.

[4] J. Dixmier, *Sur les représentations unitaires des groupes de Lie algébriques*, Ann. Inst. Fourier **7** (1957), 315-318.

[5] J. Dixmier, *Sur les représentations unitaires des groupes de Lie nilpotents*, V., Bull. Soc. Math. France **87** (1959), 65-79.

[6] J. Dixmier, *Sur les représentations unitaires des groupes de Lie résolubles*, Math. Jour. Okayama University **11** (1962), 1-18.

[7] J. Dixmier, *Sur les C^*-algèbres*, Bull. Soc. Math. France **88** (1960), 95-112.

[8] J. Dixmier, *Traces sur les C^*-algèbres*, Ann. Inst. Fourier **13** (1963), 219-262.

[9] J. Dixmier, *Les C^*-algèbres et leurs représentations*, Gauthier-Villars, Paris, 1964.

[10] J. Dixmier, *Sur la représentation régulière d'un groupe localement compact connexe*, Ann. Sci. Ecole Norm. Sup. **2** (1959), 423-436.

[11] M. Duflo, *Caractères des groupes et des algèbres de Lie résolubles*, Ann. Sci. Ecole Norm. Sup **3** (1970), 23-74.

[12] M. Duflo, M. Rais, *Sur l'analyse harmonique sur des groupes de Lie résolubles*, Ann. Sci. Ecole Norm. Sup. **9** (1976), 107-144.

[13] I. M. Gelfand, D. Raikov, *Irreducible unitary representations of locally compact groups*, Mat. Sb. **13** (1943), 301-316. (Russian).

[14] I. M. Gelfand, M. A. Naimark, *Unitary representations of classical groups*, Moscow, 1951. (Russian).

[15] R. Godement, *Les fonctions de type positif et la théorie des groupes*, Trans. Amer. Math. Soc. **63** (1948), 1-84.

[16] R. Godement, *Mémoire sur la théorie des caractères dans les groupes localement compacts unimodulaires*, Jour. de Math. Pures et Appl. **30** (1951), 1-110.

[17] R. Godement, *Théorie des caractères, I. Algèbres unitaires*, Ann. of Math. **59** (1954), 47-62.

[18] R. Godement, *Théorie des caractères, II. Définitions et propriétés génerales des caractères*, Journal **59** (1954), 63-85.

[19] A. Guichardet, *Caractères des algèbres de Banach involutives*, Ann. Inst. Fourier **13** (1962), 1-81.

[20] Harish-Chandra, *Representations of semi-simple Lie groups*, Proc. Nat. Acad. Sci. **37** (1951), 366-369.

[21] Harish-Chandra, *Representations of semi-simple Lie groups III*, Trans. Am. Math. Soc. **76** (1954), 234-253.

[22] A. A. Kirillov, *Unitary representations of nilpotent Lie groups*, Usp. Mat. Nauk **17** (1962), 57-110. (Russian).

[23] J. v. Neumann, *On rings of operators. Reduction Theory*, Ann. of Math. **50** (1949), 401-485.

[24] L. Pukanszky, *On the theory of quasi-unitary algebras*, Acta Sci. Math. Szeged **16** (1955), 105-121.

[25] L. Pukanszky, *Leçons sur les représentations des groupes*, Dunod, Paris, 1966.

[26] L. Pukanszky, *On the characters and the Plancherel formula of nilpotent groups*, Jour. of Funct. Anal. **1** (1967), 255-280.

[27] L. Pukanszky, *Unitary representations of solvable Lie groups*, Ann. Sci. Ecole Norm. Sup. **4** (1972), 457-608.

[28] L. Pukanszky, *Action of algebraic groups of automorphisms on the dual of a class of type I groups*, Ann. Sci. Ecole Norm. Sup. **5** (1972), 379-396.

[29] L. Pukanszky, *The primitive ideal space of solvable Lie groups*, Invent. Math. **22** (1973), 74-118.

[30] L. Pukanszky, *Characters of connected Lie groups*, Acta Math. **133** (1974), 81-137.

[31] L. Pukanszky, *Symplectic structure in generalized orbits of solvable Lie groups*, Jour. Reine Angew. Math. **347** (1984), 33-68.

[32] L. Pukanszky, *Quantization and Hamiltonian G-foliations*, Trans. Am. Math. Soc. **295** (1986), 811-847.

[33] I. E. Segal, *An extension of Plancherel's formula to separable unimodular groups*, Ann. of Math. **52** (1950), 272-292.

[34] A. Weil, *L'intégration dans les groupes topologiques et ses applications*, Hermann, Paris, 1953.

Received May 26, 1989

Department of Mathematics
University of Pennsylvania
Philadelphia, PA, 19104

La géométrie des moments pour les groupes de difféomorphismes

A. A. KIRILLOV

Dédié à Jacques Dixmier pour son 65ème anniversaire

1. Introduction

La méthode des orbites dans la théorie des représentations pose le problème de la classification et de l'étude des orbites d'un groupe de Lie G dans l'espace dual $\mathfrak{g}^*$ d'algèbre de Lie $\mathfrak{g}$ de ce groupe par rapport à l'action coadjointe. Toutes ces orbites sont des G-espaces symplectiques homogènes. Cette affirmation reste vraie aussi pour les groupes de Lie de dimension infinie. De cette façon on obtient beaucoup d'exemples intéressants de variétés homogènes symplectiques de dimension infinie. L'étude de ces variétés et de leur rôle dans la théorie des représentations a été commencée dans [1], [2].

Nous ne considérons ici qu'une classe spéciale de groupes de Lie de dimension infinie : les groupes des difféomorphismes des variétés lisses et les sous-groupes qui préservent certaines structures géométriques sur ces variétés.

Les exemples les plus importants sont :

a) Le groupe $Diff_c(M)$ des difféomorphismes à support compact d'une variété lisse M (c'est-à-dire des difféomorphismes qui ne diffèrent de l'application identique que sur un sous-ensemble compact de M).

Parfois il est utile de considérer au lieu de ce groupe, le groupe $\widetilde{Diff}_c^o(M)$ qui est le revêtement universel de la composante connexe de l'unité du groupe $Diff_c(M)$. Les deux groupes ont la même algèbre de Lie $Vect_c(M)$ composée des champs de vecteurs lisses sur M à support compact.

b) Soit M une variété symplectique avec la forme de structure ω. Nous posons

$$Diff_c(M,\omega) = \{\phi \in Diff_c(M) \mid \phi^*\omega = \omega\}$$

L'algèbre de Lie de ce groupe consiste en les champs de vecteurs hamiltoniens ξ satisfaisant la condition :

$$L_\xi \omega = 0$$

où L_ξ est la dérivation de Lie le long de ξ. Cette condition est équivalente à la condition :

$$d\iota_\xi(\omega) = 0$$

où ι_ξ est la multiplication intérieure par ξ. Si la forme $\iota_\xi \omega$ est non seulement fermée mais exacte, nous dirons que ξ est strictement hamiltonien et noterons $f_\xi \in C^\infty(M)$ une fonction génératrice pour ξ c'est-à-dire une fonction telle que :

$$\iota_\xi \omega = df_\xi.$$

On sait ([1], [3]) que l'ensemble des champs strictement hamiltoniens est un idéal dans $Vect_c(M)$ qui, pour M compact, coïncide avec l'algèbre dérivée de cette algèbre de Lie. L'algèbre de Lie abélienne correspondante est isomorphe à $H^1(M, \mathbb{R})$.

En particulier si M est compacte et simplement connexe, l'algèbre de Lie $Vect_c(M, \omega)$ s'identifie avec l'algèbre de Poisson $P_o(M, \omega)$ des fonctions lisses sur M de moyenne nulle.

c) Soit M une variété de contact avec la forme de structure α. Nous posons

$$Diff_c(M, \omega) = \{\phi \in Diff_c(M) \mid \phi^* \alpha \wedge \alpha = 0\}$$

Plus précisément α n'est définie que localement et à un facteur fonctionnel près. Mais il est utile de supposer que l'on peut choisir une forme α définie globalement sur M. Dans ce cas l'algèbre de Lie

$$Vect_c(M, \alpha) = \{\xi \in Vect_c(M) \mid L_\xi \alpha \wedge \alpha = 0\}$$

s'identifie avec l'espace $C_c^\infty(M)$ des fonctions lisses sur M à support compact :

$$\xi \to f_\xi := \alpha(\xi). \tag{1}$$

En effet, la forme α a la propriété $\alpha \wedge (d\alpha)^k \neq 0$, où $2k + 1 = dim\, M$. Donc, $ker\alpha \cap ker\, d\alpha = 0$ et un champ vectoriel ξ sur M est déterminé uniquement par $\iota_\xi \alpha$ et $\iota_\xi d\alpha$. Soit ϵ le *champ canonique*, défini par

$$\iota_\epsilon \alpha = 1\,,\ \iota_\epsilon d\alpha = 0.$$

Il est facile de vérifier que pour le champ $\xi \in Vect_c(M,\alpha)$ on a

$$L_\xi \alpha = \epsilon f_\xi \alpha \text{ et } \iota_\xi d\alpha = \epsilon f_\xi - df_\xi. \tag{2}$$

L'opération de commutation dans $Vect_c(M,\alpha)$ se transforme en termes de fonctions génératrices dans le crochet de contact :

$$[f_\xi, f_\eta] = 1/2(\xi f_\eta - \eta f_\xi + f_\xi \epsilon f_\eta - f_\eta \epsilon f_\xi). \tag{3}$$

d) Soit v une forme volume sur une variété M. Posons

$$Diff_c(M,v) = \{\phi \in Diff_c(M) \mid \phi^* v = v\}$$

et

$$Vect_c(M,v) = \{\xi \in Vect_c(M) \mid L_\xi v = 0\}.$$

Les exemples *a*) – *d*) sont les quatre types de "groupes de Lie infinis simples" découverts par E. Cartan. Certains de ces groupes possèdent des extensions centrales non triviales, donc ils admettent des représentations projectives non linéarisables.

2. Espace des moments $\mathfrak{g}^*$

Dans tous les exemples précédents l'espace $\mathfrak{g}$ était un espace de sections lisses d'un fibré vectoriel *naturel* E sur une variété M (c'est-à-dire un fibré avec action du groupe $Diff_c(M)$). Nous supposons que M est compacte et orientée. Soit $E^* = E' \otimes V$ où E' est le fibré dual de E et V est le fibré des formes de volume sur M. Alors le dual topologique $\mathfrak{g}^*$ s'identifie avec l'espace des sections généralisées (sections-distributions) du fibré E^*. Pour notre but il est essentiel de remarquer que l'action coadjointe du groupe G en question coïncide avec l'action du groupe $Diff_c(M)$ (ou son sous-groupe G) sur $\Gamma(E^*, M)$ comme sur l'espace des sections d'un fibré naturel.

Nous introduisons le sous-espace $\mathfrak{g}^*_{rég}$ formé des sections lisses de E^*. Les éléments de ce sous-espace s'appellent moments réguliers de G.

Dans l'exemple *a*), $E = TM, E^* = T^*M \otimes \Lambda^n T^*M, n = dim\, M$. La valeur d'un élément $\alpha \otimes v \in \mathfrak{g}^*_{rég}$ sur un élément $\xi \in \mathfrak{g}$ est donnée par

$$< \alpha \otimes v, \xi >= \int_M \alpha(\xi) v. \tag{4}$$

Le problème de la classification des G-orbites dans $\mathfrak{g}^*_{rég}$ a été étudié dans [4], [5]. Le rôle principal ici est joué par la notion de moment *générique* (ou *de Morse*).

Dans l'exemple *b*), en supposant M simplement connexe, on peut considérer E, E' et E^* comme des fibrés triviaux. Notamment, chaque champ hamiltonien ξ sur M est strictement hamiltonien et se donne par la fonction génératrice f_ξ modulo constante. De plus on peut trivialiser V en choisissant $\omega^{dim M/2}$ comme forme de volume sur M. Donc, l'espace $\mathfrak{g}^*$ dans ce cas n'est que l'espace $D_o(M)$ des distributions sur M qui annulent les constantes. La classification des G-orbites dans $\mathfrak{g}^*_{rég}$ pour ce cas est assez facile pour le cas le plus simple où M est la sphère de dimension 2 avec la structure symplectique standard. Il y a deux types d'invariants de $f \in \mathfrak{g}^*_{rég} \cong C^\infty_o(S^2)$: l'espace topologique dont les points sont les composantes connexes des lignes de niveaux de la fonction f (*"l'arbre des composantes"*) $\Gamma(f)$ et la fonction Φ_f réelle sur $\Gamma(f)$ qui donne la surface de la partie de S^2 bornée par la composante donnée. Si f n'a qu'un seul maximum M et un seul minimum m sur M, son arbre des composants est un segment $[m, M]$ et un système complet d'invariants de f comme moment régulier de G est formé par les intégrales

$$I_n = \int_M f^n \cdot \omega,$$

ou par la seule fonction croissante Φ_f sur $[m, M]$ pour laquelle on a

$$I_n = \int_m^M t^n d\Phi_f(t).$$

3. La notion de dimension fonctionnelle

Toutes les algèbres considérées ci-dessus sont de dimension infinie. On peut néanmoins les distinguer par leur *"dimension fonctionnelle"*. Plus précisément nous dirons que l'ensemble E a la *dimension fonctionnelle* (k, ℓ) et écrirons

$$Dim f E = (k, \ell)$$

si on peut le réaliser comme l'espace de toutes les applications lisses d'une variété de dimension k dans une variété de dimension ℓ. Il est commode d'écrire $\ell.\infty^k$ au lieu de (k,ℓ) parce que c'est en accord avec la loi de multiplication des dimensions fonctionnelles. Si E_i est "modelé" sur l'espace vectoriel topologique V_i avec $Dimf V_i = \ell_i \, . \, \infty^{k_i}$, et si E est modelé sur $V = V_1 \hat{\otimes} V_2$ (produit tensoriel complété) alors $Dimf E = \ell_1 \ell_2 \, . \, \infty^{k_1 k_2}$. Les algèbres $a) - d)$ de 1 ont les dimensions fonctionnelles suivantes :

$$\begin{aligned} Dimf\; Vect_c(M) &= n.\infty^n, & n &= dim\, M \\ Dimf\; Vect_c(M,\omega) &= \infty^{2n}, & 2n &= dim\, M \\ Dimf\; Vect_c(M,\alpha) &= \infty^{2n+1}, & 2n+1 &= dim\, M \\ Dimf\; Vect_c(M,v) &= (n-1) & \cdot\infty^n, n &= dim\, M \end{aligned}$$

Mentionnons pour comparaison que pour l'algèbre $\hat{\mathfrak{g}}$ de Kac-Moody associée à une algèbre semi-simple $\mathfrak{g}$ de dimension n on a

$$Dimf\, \hat{\mathfrak{g}} = n \cdot \infty.$$

Un problème très intéressant et provocatif est de définir une notion de dimension fonctionnelle pour une représentation linéaire d'un groupe de Lie, de telle façon que pour la représentation unitaire irréductible T associée à une orbite $\Omega \subset \mathfrak{g}^*$ on ait l'égalité

$$Dimf\; T = 1/2 dim\, \Omega. \tag{5}$$

Comme exemple d'une réponse particulière à ce problème on peut citer :

Théorème (Dixmier-Kirillov). *Soit G un groupe de Lie nilpotent et T la représentation unitaire irréductible associée à une orbite $\Omega \subset \mathfrak{g}^*$. Alors*

$$T_*(U(\mathfrak{g})) \cong A_m, \; m = 1/2 \cdot dim\, \Omega, \tag{6}$$

où A_m est l'algèbre de Weyl (isomorphe à l'algèbre des opérateurs différentiels à coefficients polynomiaux sur $\mathbb{R}^m$).

Une approche à ce problème consiste en l'étude de la structure algébrique des algèbres de Lie de champs de vecteurs. Il y a deux caractéristiques algébriques de ces algèbres qui sont liées avec la dimension fonctionnelle.

1) *La croissance* de l'algèbre de Lie $\mathfrak{g}$, c'est-à-dire le comportement asymptotique des dimensions des composantes homogènes de l'algèbre de Lie graduée engendrée par deux (ou plus généralement par un nombre fini)

éléments génériques de $\mathfrak{g}$. Par exemple (voir [8]), pour $\mathfrak{g} = Vect_c(M)$ cette dimension s'accroît avec n comme

$$exp(c \cdot n^{m/m+1}),\ m = dim\, M \tag{7}$$

Pour $M = \mathbb{R}$ on a le résultat beaucoup plus précis :

$$dim\mathfrak{a}^{k,\ell} = p_k(k+\ell-1) + p_\ell(k+\ell-1) - p(k+\ell-1) \tag{8}$$

où $\mathfrak{a} = \oplus \mathfrak{a}^{k,\ell}$ est l'algèbre bigraduée engendrée par deux champs vectoriels génériques sur $\mathbb{R}, p(n)$ est le nombre de partitions de n et $p_k(n)$ est le nombre de partitions de n en $\leq k$ parties.

2) *La réserve des identités qui sont vérifiées* dans l'algèbre de Lie $\mathfrak{g}$. Soit par exemple

$$T_k(x_1 x_2, \dots x_k\ ; y) = Alt(adx_1 \cdot adx_2 \dots adx_k)y \tag{9}$$

où Alt signifie la somme alternée de toutes les permutations des x_i. C'est un élément non nul de l'algèbre de Lie libre à $k+1$ générateurs. On a le résultat suivant :

Théorème ([9]). *L'identité T_k est vérifiée*
a) dans l'algèbre de Lie $Vect_c(M)$ pour $k \geq (n+1)^2, n = dim\ M$,
b) dans l'algèbre de Lie $Vect_c(M,\omega)$ pour $k \geq n(2n+5), 2n = dim\ M$,
c) dans l'algèbre de Lie $Vect_c(M,\alpha)$ pour $k \geq 2n^2+5n+5, 2n+1 = dim\ M$.

La description complète des identités dans les algèbres de Lie de champs de vecteurs est un problème très intéressant et non résolu, même dans le cas le plus simple $\mathfrak{g} = Vect_c(\mathbb{R})$ où on conjecture que toutes les identités sont conséquences de T_4 (voir [10], [11]).

4. Le cas $M = S^1$

Ce cas est intéressant tout d'abord parce qu'il est lié avec des travaux récents dans la physique mathématique. Il s'agit du rôle qui est joué par le groupe $Diff\, S^1$ dans la théorie de cordes et dans certains modèles de la mécanique statistique quantique. (Voir, par exemple, [17]).

Le problème de la classification des orbites coadjointes pour le groupe de Virasoro-Bott (une extension centrale du groupe $Diff_+ S^1$ des difféomorphismes du cercle préservant l'orientation) est lié avec plusieurs domaines des mathématiques. Il est équivalent au moins à trois autres problèmes de classification :

a) des équations de Hill (ou de Sturm-Liouville ou de Schrödinger avec un potentiel périodique)

$$2cu'' + pu = 0 \ ;$$

b) des structures projectives sur S^1 ;

c) des classes de conjugaison dans le revêtement universel du groupe $SL(2, \mathbf{R})$.

La réponse a été obtenue indépendamment en [12], [13], [14], [15] et [16]. En gros, il y a un ensemble de dimension 2 d'orbites, et la partie ouverte de cet ensemble est formée par les orbites passant par les points

$$p_{h,c} = (h \cdot dt^2, c) \in \mathfrak{g}^*_{rég}.$$

Si $2h/c \neq n^2, n \in \mathbb{Z}$, le stabilisateur de $p_{h,c}$ dans $Diff_+ S^1$ est égal à $Rot\, S^1$. (Remarquons que le centre d'un groupe G agit trivialement dans $\mathfrak{g}^*$, donc les orbites coadjointes du groupe de Virasoro-Bott sont en effet des orbites de $Diff_+ S^1$). Presque toutes les orbites considérées sont isomorphes comme espaces homogènes à $M = Diff_+ S^1 / Rot\, S^1$. Mais, comme variétés symplectiques homogènes, elles sont différentes. Autrement dit, il y a sur M une famille à deux paramètres de structures symplectiques invariantes.

L'espace tangent à M au point initial s'identifie avec l'espace $C^\infty_o(S^1)$ des fonctions réelles lisses de moyenne nulle sur S^1. Les formes symplectiques, dans cette réalisation, s'écrivent

$$< v_1, v_2 > = \int_o^{2\pi} v_1(t)\{2hv_2\,'(t) + cv_2'''(t)\}dt. \tag{10}$$

De plus, on a sur M une structure complexe invariante. En effet, l'existence d'une structure presque complexe invariante est évidente : elle est donnée par l'opérateur J (transformation de Hilbert) :

$$J(v) = \frac{1}{2\pi} \int_o^{2\pi} \frac{v(s) - v(t)}{tg(s-t)/2} dt. \tag{11}$$

Cette structure satisfait à la condition d'intégrabilité, mais dans la situation de dimension infinie il n'y a pas de théorème général qui garantisse l'existence de structure complexe. Néanmoins dans ce cas la structure complexe a été construite ([5], [18]). Il se trouve que la variété M est isomorphe, comme variété complexe, à l'ensemble S des fonctions f holomorphes et univalentes sur le disque unité $D_+ = \{z \in \mathbb{C} \mid |z| \leq 1\}$ du plan complexe normalisées par les conditions

$$f(0) = 0, f'(0) = 1. \tag{12}$$

La correspondance infinitésimale entre M et S est très simple : à une fonction $v \in C_o^\infty(S^1)$ correspond la fonction holomorphe ϕ sur D telle que

$$v(t) = Im\, \phi\, (e^{it}). \tag{13}$$

La correspondance globale est beaucoup plus délicate (voir [5], [18]).

Les structures symplectiques et complexes sur M se réunissent dans la famille des structures Kälhériennes. En termes de coordonnées naturelles sur S (c'est-à-dire les coefficients c_k de la série de Taylor pour $f \in S$) les formes Kählériennes au point initial $f \equiv 0$ s'écrivent :

$$w_{h,c} = \sum_{k \geq 1} (2hk + ck) \cdot dc_k \overline{dc_k}. \tag{14}$$

On peut calculer explicitement le tenseur de courbure riemannienne et le tenseur de Ricci ([19], [20]). Il est intéressant que ce dernier ne dépende pas de h et c (justement comme dans le cas de dimension finie - voir [21]) et coïncide avec $w_{h,c}$ pour $h = 1/12$, $c = -26/12$. Or, si $c/h = -26$, notre variété est d'Einstein !

Un problème très intéressant est de généraliser à ce cas la formule intégrale pour les caractères des représentations irréductibles. On a le dilemme : ou bien cette intégrale se réduit à l'intégrale gaussienne à l'aide d'un changement de variable approprié (donc, on obtient de cette façon la paramétrisation explicite de la variété S des fonctions univalentes et l'expression commode pour le potentiel Kählérien des formes $w_{h,c}$), ou nous aurons le premier exemple d'intégrale fonctionnelle calculable de type non Gaussien. Je tiens à supposer que la troisème possibilité - que la formule universelle pour le caractère ne soit pas vraie dans ce cas - n'a pas lieu.

5. Le cas $dim M = 2$

Soit M une surface orientée compacte de dimension 2 avec une forme symplectique ω. L'algèbre de Lie $\mathfrak{g} = Vect\,(M,\omega)$ contient l'idéal $[\mathfrak{g}, \mathfrak{g}]$ qui est isomorphe à l'algèbre de Poisson $P_o(M,\omega)$ (voir numéro 2) et on a la suite exacte :

$$0 \to P_o(M,\omega) \to \mathfrak{g} \xrightarrow{w} H_1(M,\mathbf{R}) \to 0. \tag{15}$$

où w est l'application qui envoit un champ $\xi \in \mathfrak{g}$ dans son nombre d'enroulement $w(\xi)$ dans le sens [22].

Il est commode de fixer un plongement $\iota : H_1(M,\mathbf{R}) \hookrightarrow Vect(M,\omega)$ tel que $w.\iota = id$. Alors, comme espace vectoriel, $\mathfrak{g} = P_o(M,\omega) \oplus \iota(H_1(M,\mathbf{R}))$: $\xi = s - grad\,(f) + \iota(b), b \in H_1(M,\mathbf{R})$. D'autre part on peut construire une deuxième suite exacte de la forme

$$0 \to H^1(M,\mathbf{R}) \to \tilde{\mathfrak{g}} \to P_o(M,\omega) \to 0. \tag{16}$$

Elle est définie par le 2-cocyle sur $P_o(M,\omega)$ à valeurs dans $H^1(M,\mathbf{R}) = (H_1(M,\mathbf{R}))^*$ donné par

$$c(f,g) = \pi(fdg) \tag{17}$$

où π est la projection de $\Omega^1(M)$ sur $H^1(M,\mathbf{R})$ duale de l'injection ι :

$$< \pi(fdg), b >=< fdg, \iota(b) >= \int_M f.\iota(b)g.\omega.$$

Il se trouve qu'on peut réunir les deux suite (15) et (16) dans le complexe

$$0 \to H^1(M,\mathbf{R}) \to \hat{\mathfrak{g}} \to H_1(M,\mathbf{R}) \to 0,$$

dont les cohomologies sont respectivement $0, P_o(M,\omega), 0$.

De plus, l'algèbre $\hat{\mathfrak{g}}$ possède une forme symétrique invariante : cette forme se réduit sur $P_o(M,\omega)$ au produit scalaire ordinaire $(f,g) = \int fg\omega$ et sur $H_1(M,\mathbf{R}) \oplus H^1(M,\mathbf{R})$ à la dualité canonique. Grâce à cette propriété de $\hat{\mathfrak{g}}$, l'étude des orbites coadjointes régulières pour le groupe correspondant est équivalente à l'étude des orbites adjointes. Cette étude est très intéressante de plusieurs points de vue. Par exemple, on obtient de cette façon une structure Kählérienne invariante à gauche sur le groupe G qui est le groupe dérivé de $Diff(\mathbf{T}^2,\omega)$.

REFERENCES

[1] A.A. Kirillov, *Unitary representations of the group of diffeomorphisms and of some of its subgroups*, Preprint 82 of the Keldysh Institut of Applied Mathematics, USSR Academy of Sciences 1974, (translated into English in Selecta Math. Sov.,**1** :4, (1981), 351–372).

[2] A.A. Kirillov, *Infinite dimensional Lie groups, their representations, orbits, invariants,* Proc. Internat. Congr. Math. (Helsinki, 1978). Acad. Sci. Fenn., Helsinki, **2** (1980), 705–708.

[3] V.I. Arnold,*One-dimensional cohomologies of Lie algebras of nondivergent vector fields and on the rotation numbers of dynamical systems,* Funct. Anal. Appl., **1**:4, 1967.

[4] A.A. Kirillov, *The orbit method and representations of infinite-dimensional Lie groups,* Amer. Math. Soc. Transl. (2) **137**, 1987.

[5] A.A. Kirillov, M.I. Golenishcheva-Kutuzova, *La géométrie des moments pour les groupes de difféomorphismes*, preprint du Keldysh Institut de Mathématiques appliquées, Moscou, **101** (1986), (en russe).

[6] M.I. Golenishcheva-Kutuzova, *La classification locale des moments pour les groupes de difféomorphismes*, Uspekhi Math. nauk., **3**, 181–182 (en russe).

[7] A.A. Kirillov, M.L. Kontsevich, *La croissance de l'algèbre de Lie engendrée par deux champs vectoriels sur la droite*, Vestnik MGU, ser. mat., **4** (1983), 15–20. (En russe).

[8] A.I. Molev, *Sur la structure algébrique de l'algèbre de Lie des champs de vecteurs sur la droite*, Mat. Sbornik, **134**: 1 (1987), 82–92 (en russe).

[9] A.A. Kirillov, *Sur les identités dans l'algèbre de Lie des champs de vecteurs*, Vestnik MGU, ser. mat., **2** (1989), 11–13. (en russe).

[10] A.A. Kirillov, V. Yu Ovsienko, O.D. Udalova, *Sur les identités dans l'algèbre de Lie des champs de vecteurs sur la droite*, preprint 135 of the Keldysh Institut of Applied Mathematics, USSR Academy of Sciences 1984; (translated into English in Selecta Math. Sov., 1989) (en russe).

[11] Yu. P. Razmyslov, *L'identité des algèbres et leur représentations*, Moscou, "Nauka", 1989 (en russe).

[12] N.H. Kuiper, *Locally projective spaces of dimension one*, Michigan Math. J., **2** (1954), 95–97.

[13] V.F. Lazutkin, T.F. Pankratova, *Normal forms and versal deformations for Hill's equation*, Funkt. Anal., **9**:4 (1975), 41–48; English transl. Funct. Anal. Appl., **9** (1975).

[14] G. Segal, *Unitary representations of some infinite dimensional groups*, Commun. Math. Phys., **80**:3, (1981), 301–342.

[15] A.A. Kirillov, *Infinite dimensional Lie groups : their orbits, invariants and representations. The geometry of moments, Twistor Geometry and Nonlinear Systems* (Lectures, Fourth Bulgarien Summer School, Primorsko, 1980), Lecture notes in Math., **970** (1982), Springer-Verlag, 101–123.

[16] E. Witten, *Coadjoint orbits of the Virasoro group*, Commun. Math. Phys. **114**:1 (1988), 1–53.

[17] A. Belavin, A. Polyakov, A. Zamolodchikov, *Infinite dimensional symmetries in two-dimensional quantum field theory*, Nucl. Phys. B **241** (1984), 333–380.

[18] A.A. Kirillov, *A Kähler structure on K-orbits of the group of diffeomorphisms of a circle*, Funkt. Anal., **21**:2 (1987), 42–45; English transl. FAA, **21** (1987).

[19] A.A. Kirillov, D.V. Yuriev, *Kähler geometry of the infinite dimensional homogeneous manifold* $M = Diff_{+}S^1/Rot\,S^1$, Funkt. Anal. **20**:4 (1986), 79–80 et **21**:4 (1987), 34–46 ; English. transl. FAA, **20** (1986) **21** (1987).

[20] M.J. Bowick, S.G. Rageev, *String theory as the Kähler geometry of loop space*, Phys. Rev. Letters (1987), **58**:6, 535–538.

[21] D.V. Alekseevskii, A.M. Perelomov, *Invariant Kähler - Einstein metrics on compact homogeneous spaces*, Funkt. Anal. **20**:3 (1986), 1–16; English transl. FAA, **20** (1986).

[22] I.M. Gelfand, I.I. Piatetski-Shapiro, *On a theorem of Poincaré*, Dokl. Akad. Nauk SSSR, 1959. (Voir aussi I.M. Gelfand, Collected Papers, **1**, Springer, 1987).

Received January 12, 1990

Chair of Functional Analysis
Department of Mechanics and Mathematics
Moscow University
Moscow, 117234 USSR

The Vanishing of Scalar Curvature and the Minimal Representation of SO(4,4)

BERTRAM KOSTANT*

Dedicated to Jacques Dixmier
on the occasion of his 65th birthday

1. Introduction

1.1 *Summary of contents.* The much-studied metaplectic representation π' is a "minimal" representation of the metaplectic group $Mp(2n, \mathbf{R})$. The group $Mp(2n, \mathbf{R})$ is a split group of type C_n. On the other hand the group $SO(n, n)$ is a split group of type D_n and, analogous to π', there is a "minimal" representation π of $SO(n, n)$. For the case of $Mp(8, \mathbf{R})$ the representation π' has well-known strong connections with physical theories — particularly with regard to models for the states of elementary particles of mass zero. We have been motivated to study the analogous representation π of SO(4,4) not only for connections with physical theory but as a route for the introduction of the exceptional Lie groups into physical theory.

Taking G to be the simply-connected covering group of the identity component of SO(4,4) we have constructed π in [Ko], established its unitarity (Theorems 6.25 and 6.28 in [Ko]) and exhibited the action (Theorem 7.4 in [Ko]) of a triality automorphism on K-types of π (roughly speaking it converts the product of traces of operators to the trace of their products).

A central theme of the present paper is the establishment of a connection between π (where G is as above) and geometry. In particular to establish a connection between π and a phenomenon peculiar only to 6-dimensional manifolds — namely the conformal invariance of the vanishing scalar curvature. The stage is set for this in that the group G operates as a conformal group on a conformal (with signature $+++---$) real projective variety X which has $S^3 \times S^3$ as a 2-fold covering manifold.

There are 4 main sections in the paper. The following is a brief description of their contents.

In §2 we recall the construction of a space H of smooth vectors for

* Supported in part by NSF Grant No. DMS-8703278

π — where n is arbitrary. The conformal structure on X is introduced thereby focusing attention on the case of SO(4,4).

In §3 we prove that the annihilator J in the enveloping algebra U of the Lie algebra of G is the Joseph ideal. The argument is novel in that the proof arises from the very specialized nature of the K-types. In fact in the case of SO(4,4) (*i.e.* $n = 4$) it is shown that J is generated by its intersection with the center of the centralizer U^K of K in U. For the case where $n > 4$ the argument succeeds only after a certain modification. We thank David Vogan for this rescue of the argument in this case –*i.e* where $n > 4$.

In §4 for the case where $n = 4$ we give a new (analytic and geometric) proof of the unitarity of π which is much more interesting and informative than the algebraic proof given in [Ko]. The idea is a joint one of V. Guillemin, E. Grinberg and myself. In this section I work out the details and come up with an explicit formula for the unitary structure — fortunately agreeing with the one in [Ko]. The space X is the space of isotropic lines in $\mathbf{R}^8$ relative to a G-invariant bilinear form of signature $+ + + + - - - -$. Let Y be one of the two components of the space of isotropic 4-planes in $\mathbf{R}^8$. Then classically there exists an automorphism τ of G — corresponding to one of the exceptional symmetries of the Dynkin diagram of D_4 — which induces a diffeomorphism $\tau_Y : Y \longrightarrow X$. The unitary structure for π arises from a composite of 3 maps: (1) a Radon transform R involving X and Y, (2) a map of line bundles induced by τ_Y and, (3) the operator $\pi(\tau)$.

In §5 (again for the case where $n = 4$) we relate π to the vanishing of scalar curvature. Assume that M is any manifold where dim $M > 1$ and that g and g' are conformally equivalent pseudo-Riemannian metrics on M. Thus $g' = fg$ for a positive function f on M. It seems to have been overlooked by geometers (at least by the ones we have spoken to) that if dim $M = 6$ (and only in that case) is it always true that if Scal $g = 0$ then Scal $g' = 0$ if and only if f is harmonic with respect to g. (Here Scal denotes scalar curvature). This fact then leads to the result (0) below. Let $\mathcal{M}$ denote the G-stable cone of all metrics in the conformal class of X. Then H and $\mathcal{M}$ can be regarded as sets of sections of the same line bundle. Put $H_+ = H \cap \mathcal{M}$. Then not only is the set $\{g \in \mathcal{M} | \text{ Scal } g = 0\}$$G$-stable but in fact

$$H_+ = \{g \in \mathcal{M} | \text{ Scal } \; g = 0\} \tag{0}$$

It seems very far-out and very speculative but I find the equality (0) suggestive of the following: Let $g \in H_+$ and let $M \subset X$ be a 4-dimensional manifold such that $h = g|M$ is Lorentzian. Then regarding (M,h) as a model for space-time the Einstein tensor on M should arise from normal data of M in X. In "philosophic" terms *matter fields on space-time M should arise from an embedding $M \longrightarrow X$ where X is pseudo-Riemannian with respect to a metric $g \in H_+$.* In fact in §5 we state (without proof) a result along these lines due to Witold Biedrzycki in the case where M is the (naturally embedded) compactified Minkowski space. Let $g \in \mathcal{M}$ and let $h = g|M$. Let R_{ij} be the Ricci tensor on M with respect to h and let $R = \text{Scal } h$. Then Biedrzycki has defined an "energy-momentum" tensor T in terms of the second fundamental form and the sectional curvature of the normal bundle to M in X and proved that Einstein's equations

$$R_{ij} - R/2 \cdot h_{ij} = 8\pi T_{ij}$$

are satisfied if and only if Scal g vanishes along M. In particular these equations are satisfied for all G transforms of g if and only if $g \in H_+$.

1.2 *Related information and acknowledgments.* Coincidentally we note here that D. Kazhdan (see [Ka]), following another agenda, has written a paper entitled "On the Minimal Representation of D_4 which is also to appear in the proceedings of this conference. Kazhdan is looking at π in the case of an arbitrary local field.

We wish to thank W. Biedrzycki, V. Guillemin, S. Sternberg, D. Vogan for conversations and cited contributions on matters in this paper.

2. The minimal G-module H and Ker Δ_X

2.1 *Representations and nilpotent orbits.* Let G be a connected non-compact semisimple Lie group with finite center and let $\mathfrak{g}$ be its Lie algebra. We will identify $\mathfrak{g}$ with its dual space, using the Killing form. Complexifications will be denoted by subscript $\mathbb{C}$. Orbits in $\mathfrak{g}$ are with respect to Ad $\mathfrak{g}$ (or, equivalently, with respect to G) and in $\mathfrak{g}_{\mathbb{C}}$ with respect to Ad $\mathfrak{g}_{\mathbb{C}}$. Let $U = U(\mathfrak{g})$ and $S = S(\mathfrak{g})$ be, respectively, the universal enveloping algebra and symmetric algebra of $\mathfrak{g}$ over $\mathbb{C}$. We may regard S as the ring of polynomial functions on $\mathfrak{g}_{\mathbb{C}}$. Also if $\{U_n\}$ denotes the usual filtration of U and Gr U denotes the corresponding graded algebra then (PBW theorem) we may regard Gr $U = S$.

Let $\mathcal{H}$ be a Hilbert space and let $B(\mathcal{H})$ be the algebra of bounded operators on $\mathcal{H}$. Assume that

$$\pi : G \longrightarrow B(\mathcal{H})$$

is a Hilbert space representation of G. Let $\mathcal{H}^\infty$ be the space of smooth vectors in $\mathcal{H}$ so that π induces a homomorphism $\pi^\infty : U \longrightarrow \text{End } \mathcal{H}^\infty$. Thus if the ideal $J = \text{Ker } \pi^\infty$ one has an exact sequence

$$J \longrightarrow U \longrightarrow \text{End } \mathcal{H}^\infty \tag{1}$$

Let $\{J_n\}$ be the filtration of J defined by putting $J_n = J \cap U_n$ so that $I = \text{Gr } J$ is an Ad G invariant ideal in S. Let Spec $I \subset \mathfrak{g}_{\mathbf{C}}$ be the zero spectrum of the ideal I. The following is one of the most important results in representation theory. It may be attributed to the work of Barbasch, Borho, Brylinski, Ginzburg, Joseph, and Vogan.

Theorem 1. *One has*

$$\text{Spec } I = \overline{Z} \tag{2}$$

where Z is a nilpotent orbit in $\mathfrak{g}_{\mathbf{C}}$ and $\overline{Z}$ is its closure.

By (2) the Gelfand-Kirillov dimension d of U/J may be given as the even integer

$$d = \dim_{\mathbf{C}} Z \tag{3}$$

On the other hand other than 0 the smallest possible value of $\dim_{\mathbf{C}} Z$ for a nilpotent orbit Z in $\mathfrak{g}_{\mathbf{C}}$ is uniquely achieved when Z is the orbit Z_o of the highest root vector in $g_{\mathbf{C}}$. Let $d_o = \dim_{\mathbf{C}} Z_o$. In case Z_o is defined over $\mathbf{R} - for$ example when $\mathfrak{g}$ is split — we will put

$$O = \mathfrak{g} \cap Z_o \tag{4}$$

It is easy to see, using the finite dimensional representational theory of real semisimple groups that either O is connected — in which case it is a single orbit — or it is a union of at most 2 orbits O_1 and O_2 where $O_1 = -O_2$.

2.2 *The G-module* $\Gamma^s(X)$. Now if $\mathfrak{g}_{\mathbf{C}}$ is not of type A_l then A. Joseph has defined a completely prime ideal J_o in U — which one refers to as the Joseph ideal — such that Spec Gr $J_o = Z_o$. Perhaps the most familiar example of the case where $J = J_o$ (using the notation of (1)) is when G is the double cover of the real symplectic group $Sp(2n, \mathbf{R})$ and π is either one of the two irreducible components (even or odd) of the metaplectic representation. Correspondingly $\mathcal{H}$ may be taken to be either the space of even functions or odd functions in $L_2(\mathbf{R}^n)$. Also in this case $d = 2n$, O is defined over $\mathbf{R}$ and O is a union of two orbits O_1 and O_2. (Vogan understands the two orbits as corresponding to the two irreducible components of the metaplectic representation.) Either O_1 or O_2 may be identified with $\mathbf{R}^{2n}$ - {origin} with respect to the natural transitive action of $Sp(2n, \mathbf{R})$

Now let G be a covering of the identity component $SO(p+1, q+1)_e$ of $SO(p+1, q+1)$ where $p \geq 1$ and $q \geq 1$. Since G is simple we also have to assume that $p + q > 2$. One then has that

$$d_o = 2(p + q - 1) \tag{5}$$

Furthermore O is always defined over $\mathbf{R}$ and O is a single G orbit whenever $p \geq 2$ and $q \geq 2$.

It is naively suggestive from (5) that in order to construct π so that $J = J_o$ one should look for a $p+q-1$ dimensional space on which G acts and to consider the action of G on line bundles over the space. One is quickly disabused of this idea when one notes in general that G does not operate on a space of such low dimension (in particular O does not have an invariant real polarization). However G does operate on a compact manifold X of dimension $p+q$. To define X let B be a non-zero symmetric bilinear form of signature $(p+1, q+1)$ on $\mathbf{R}^{p+q+2}$ which is invariant under $SO(p+1, q+1)$ and hence under G. For $u, v \in \mathbf{R}^{p+q+2}$ we will write $(u, v) = B(u, v)$. Let

$$C = \{v \in \mathbf{R}^{p+q+2} | (v, v) = 0\} \tag{6}$$

so that $C^* = C - \{\text{origin}\}$ is a G-homogeneous space of dimension $p+q+1$. Let $\mathbf{R}^*$ be the multiplicative group $\mathbf{R} - \{0\}$ which we can regard as operating freely by scalar multilplication on C^*. One puts $X = C^*/\mathbf{R}^*$ so that X is the space of all isotropic lines in $\mathbf{R}^{p+q+2}$. Clearly X is a $p+q$ dimensional G-homogeneous space and one has the structure of a principal $\mathbf{R}^*$ bundle

$$\begin{array}{ccc} \mathbf{R}^* & \longrightarrow & C^* \\ & & \downarrow{\scriptstyle \alpha} \\ & & X \end{array} \tag{7}$$

Now let r_i, $i = 1, \ldots, p+q+2$, be the natural linear coordinate system in $\mathbf{R}^{p+q+2}$. Let $\epsilon_i = 1$ for $1 \leq i \leq p+1$ and equal to -1 for $p+2 \leq i \leq p+q+2$. We may assume that the quadratic form $Q(v) = (v, v)$ is given by

$$Q = \sum_{i=1}^{p+q+2} \epsilon_i r_i^2 \tag{8}$$

Let $E \subset \mathbf{R}^{p+q+2}$ (resp. F) be the subspace defined by the equations $r_i = 0$ for $i > p+1$ (resp. $\leq p+1$) so that B is positive definite in E and negative definite in F and

$$\mathbf{R}^{p+q+2} = E \oplus F \tag{9}$$

and if $|v| = |(v, v)|^{1/2}$ for $v \in \mathbf{R}^{p+q+2}$ one has

$$C = \{e + f | e \in E,\ f \in F,\ |e| = |f|\} \tag{10}$$

It follows then that if relative to $|\,|$ one puts S_E (respectively S_F) equal to the unit sphere in E (resp. F) then

$$S_E \cong S^p \text{ and } S_F \cong S^q \tag{11}$$

and one has

$$S_E \times S_F \subset C^*.$$

Also X is compact and in fact the projection α (see (7)) induces a double covering

$$S_E \times S_F \longrightarrow X \tag{12}$$

Now recalling (7) for any $s \in \mathbf{C}$ let $L^s(X)$ be the line bundle over X which is associated to the character $t \longrightarrow |t|^{-s}$ of $\mathbf{R}^*$ and let $\Gamma^s(X)$ be the space of smooth sections of $L^s(X)$. We may then identify $\Gamma^s(X)$ with the space given by

$$\Gamma^s(X) = \{f \in C^\infty(C^*) | f(tv) = |t|^s f(v) \text{ for all } v \in C^*\} \tag{13}$$

Now one notes that $\Gamma^{-(p+q)}(X)$ may be identified with the space of all smooth volume elements on X (i.e with the space of differential forms of maximal degree on X). Indeed C^* is easily seen to admit a non-trivial G-invariant $p+q+1$ form ω. (see *e.g.* the argument of §4 in [Ko]). The space $\Gamma^s(X)$ is a G-module (and by differentiation a U-module) with respect to a representation π_s where if $f \in \Gamma^s(X)$, $g \in G$ and $v \in C^*$ then $(\pi_s(g)f)(v) = f(g^{-1} \cdot v)$.

2.3 *The conformal structure on X.* One knows that the indefinite metric on $\mathbf{R}^{p+q+2}$ given by B defines the structure of a conformal manifold on X and the group G operates as a group of conformal transformations. That is, B defines a set $\mathcal{M}$ of smooth pseudo-Riemannian metrics γ on X of signature (p,q) so that if $\gamma \in \mathcal{M}$ then $\gamma' \in \mathcal{M}$ if and only if γ' is of the form $\gamma' = f\gamma$ where $f \in C^\infty(X)$ and $f > 0$. In case $g \in G$ then $g \cdot \gamma = f_g \gamma$ where $f_g > 0$. We recall the appropriate definitions. If $v \in C^*$ let $v^\perp \subset \mathbf{R}^{p+q+2}$ be the orthogonal subspace to $\mathbf{R}v$ relative to B. Thus $\mathbf{R}v \subset v^\perp$ and B induces a non-singular bilinear form on $v^\perp/\mathbf{R}v$ with signature (p,q). The pairing of $w, z \in v^\perp/\mathbf{R}v$ will be denoted by (w,z). If $x \in X$ we will use the notation $v \in x$ to mean that $v \in C^*$ and $x = \mathbf{R}^* v$. Using the bundle projection α one sees that the tangent space $T_x(X)$ to X at x may be given by

$$T_x(X) = \text{Hom}\,(\mathbf{R}v, v^\perp/\mathbf{R}v) \text{ where } v \in x \tag{14}$$

The conformal class $\mathcal{M}$ is defined so that if γ is a smooth pseudo-Riemannian metric on X then $\gamma \in \mathcal{M}$ if and only if for any $x \in X$ and $\xi, \eta \in \text{Hom}\,(\mathbf{R}v, v^\perp/\mathbf{R}v)$, using the notation of (14), one has

$$\gamma(\xi,\eta) = \lambda(\xi(v), \eta(v)) \tag{15}$$

for some $\lambda > 0$.

We now observe that $\mathcal{M}$ may be regarded as a cone in $\Gamma^{-2}(X)$. Let $\Gamma^{-2}(X)_+ = \{f \in \Gamma^{-2}(X) | f > 0\}$ so that $\Gamma^{-2}(X)_+$ is a G-invariant cone in $\Gamma^{-2}(X)$.

Proposition 2. *If $f \in \Gamma^{-2}(X)_+$ then we may define an element $\gamma_f \in \mathcal{M}$ by the relation*

$$\gamma_f(\xi,\eta) = f(v)(\xi(v),\eta(v)) \tag{16}$$

for any $x \in X$ and $\xi, \eta \in T_x(X)$ using the notation of (14). Furthermore the map

$$\Gamma^{-2}(X)_+ \longrightarrow \mathcal{M},\ f \longrightarrow \gamma_f \tag{17}$$

is a G-isomorphism.

The proof is obvious and left to the reader. The point to be noted of course is that the right side of (16) is independent of the choice $v \in x$.

2.4 *The G-invariant differential operator Δ_X.* For generic values of s the representation π_s completes to an infinitesimally irreducible Hilbert space representation where in the notation of (3) one has $d = 2(p+q)$. To get down to $d_o = 2(p+q-1)$ one might look for a submodule of $\Gamma^s(X)$ which is defined as the space of solutions of an intertwining differential operator. We shall consider this for the case where G is split. Namely the case where $p = q$ so that (11) and (12) yields a double covering

$$S^p \times S^p \longrightarrow X \tag{18}$$

and

$$d_o = 2(2p-1) \tag{19}$$

Now dim $X = 2p$ and if $\Omega^{2p}(X)$ denotes the space of all smooth differential forms of maximal degree then one knows that there is a G-isomorphism

$$\Gamma^{-2p}(X) \longrightarrow \Omega^{2p}(X),\ h \longrightarrow \mu(h) \tag{20}$$

To define $\mu(h)$ one notes that there is a G-invariant differential form ω_{2p+1} of degree $2p+1$ on C^* and that ω_{2p+1} (since dim $\mathbb{R}^{2p+2}$ is even) is invariant under the map $\zeta : C^* \longrightarrow C^*$ where $\zeta(v) = -v$. One may obtain ω_{2p+1} by "dividing" the Lebesgue volume form on $\mathbb{R}^{2p+2}$ by dQ and then restricting to C^*. It follows that if

$$\mathcal{E} = \sum_{i=1}^{2p+2} r_i \partial / \partial r_i \tag{21}$$

is the Euler vector field on $\mathbb{R}^{2p+2}$ then $\mathcal{E}$ is tangent to C^* and if θ denotes Lie differentiation and ι denotes interior product one has (since $\theta(\mathcal{E})dQ =$

$2 \cdot dQ)\theta(\mathcal{E})\omega_{2p+1} = 2p \cdot \omega_{2p+1}$. Thus if $\omega = \iota(\mathcal{E})\omega_{2p+1}$ then ω is a G-invariant 2p form on C^* and one has

$$\iota(\mathcal{E})\omega = 0 \text{ and } \theta(\mathcal{E})\omega = 2p \cdot \omega \tag{22}$$

But then if $h \in \Gamma^{-2p}(X)$ it follows that $h\omega$ is invariant under $\zeta, \iota(\mathcal{E})$ and $\theta(\mathcal{E})$. Thus recalling (7) it follows that $h\omega$ descends to a differential 2p form $\mu(h)$ on X. One notes that X is orientable and we may choose an orientation so that $\mu(f)$ is positive for a positive function f. Thus

$$\int_X \mu(h) < \infty$$

is defined for any $h \in \Gamma^{-2p}(X)$ and

$$\int_X \mu(h) = \int_X \mu(a \cdot h)$$

for any $a \in G$. But then for any $s \in \mathbf{C}$ one has a G-invariant pairing

$$\Gamma^s(X) \times \Gamma^{-s-2p}(X) \longrightarrow \mathbf{C}$$

defined by putting

$$< f, h >= \int_X \mu(fh) \tag{23}$$

for $f \in \Gamma^s(X)$ and $h \in \Gamma^{-s-2p}(X)$.

Now let

$$\Delta = \sum_{i=1}^{2p+2} \epsilon_i(\partial/\partial_i)^2 \tag{23.5}$$

so that Δ can be regarded as an ultrahyperbolic operator defined on $C^\infty(U)$ for open $U \subset \mathbf{R}^{2p+2}$. If U is stable under the action of $\mathbf{R}^*$ then both $\mathcal{E}$ and Q (regarded as multiplication operator) are defined on $C^\infty(U)$. If I denotes the identity operator one has the commutation relation

$$[\Delta, Q] = 4(\mathcal{E} + (p+1)I) \tag{24}$$

Now assume that $0 \notin U$ and that $\varphi \in C^\infty(U)$ is such that for any $u \in U$ and $t \in \mathbf{R}^*$

$$\varphi(tu) = |t|^s \varphi(u) \tag{25}$$

It follows easily then that

$$\Delta\varphi(tu) = |t|^{s-2} \Delta\varphi(u) \tag{26}$$

On the other hand if $C^* \subset U$ and φ vanishes on C^* we can write $\varphi = Q\psi$ where

$$\psi(tu) = |t|^{s-2}\psi(u) \tag{27}$$

This is clear since $dQ|C^*$ vanishes nowhere. One notes that (27) implies

$$\mathcal{E}\ \psi = (s-2)\psi \tag{28}$$

But then (24)

$$\begin{aligned}\Delta\varphi = \Delta Q\psi &= Q\Delta\psi + 4(\mathcal{E} + (p+1)I)\psi \\ &= Q\Delta\psi + 4((s-2) + (p+1))\psi\end{aligned} \tag{29}$$

Now one would like to use the relation (26) to induce a G-invariant operator

$$\Delta_X : \Gamma^s(X) \longrightarrow \Gamma^{s-2}(X) \tag{30}$$

The strategy might be as follows: Choose $f \in \Gamma^s(X)$. Let U as above be a neighborhood of C^* and extend f to an element $\varphi \in C^\infty(U)$ satisfying (25). Apply Δ and restrict $\Delta\varphi$ to C^*. In fact there is no difficulty as far as the extension is concerned. For example, recalling (9), one might choose

$$U = \{w \in \mathbf{R}^{2p+2} | w = u + v,\ u \in E,\ v \in F, \text{ for } \ u \neq 0 \neq v\} \tag{31}$$

and in the notation of (31) define

$$\varphi(w) = (|u| \cdot |v|)^{s/2} f(u/|u| + v/|v|) \tag{32}$$

The problem of course is that $\Delta\varphi|C^*$ depends on the choice of the extension φ (One notes that the U above or any other suitable choice is not G-invariant). This however would be no problem if $\Delta\varphi|C^*$ vanishes whenever $\varphi|C^*$ vanishes. But by (29) this is clearly the case if

$$s = 1 - p \tag{33}$$

The argument above and the following conclusion were taught to us by V. Guillemin.

Proposition 3. *By extension of functions the ultrahyperbolic operator* $\Delta = \sum_{i=1}^{2p+2} \epsilon_i(\partial/\partial_i)^2$ *induces a G-invariant intertwining operator*

$$\Delta_X : \Gamma^{1-p}(X) \longrightarrow \Gamma^{-1-p}(X) \tag{34}$$

2.5 *The G-module* H. We may choose a maximal compact subgroup $K \subset G$ such that the decomposition (9) of $\mathbf{R}^{2p+2}$ is stable under K. One then notes that the product of spheres $S_E \times S_F$ in $\mathbf{R}^{2p+2}$ is a homogeneous space for K. Let ν be the unique K-invariant measure on $S_E \times S_F$ of total volume 1. Now the map (20) is clearly unique up to a scalar (*i.e.* it depends upon a choice of Lebesgue measure in $\mathbf{R}^{2p+2}$). Note that since (20) is, in particular a $K - map$, it may be normalized so that

$$\int_X \mu(h) = \int_{S_E \times S_F} h\ \nu \tag{35}$$

for any $h \in \Gamma^{-2p}(X)$.

Now let Δ_E (respectively, Δ_F) be the Laplace-Beltrami operator on E with respect to $B|E$ (respectively, with respect *to* $-B|F$). By the obvious extension, using the direct sum (9), we may regard Δ_E and Δ_F as differential operators on $\mathbf{R}^{2p+2}$ so that in particular

$$\Delta = \Delta_E - \Delta_F \tag{36}$$

Now Δ_E, as an operator on E, upon restriction to S_E, decomposes, using the classical polar decomposition, into a sum of its radial and its tangential (to S_E) components. Let Δ_{S_E} be the tangential component. By the obvious extension we may regard Δ_{S_E} as a differential operator on $S_E \times S_F$. One defines Δ_{S_F} similarly.

If f is a function on a subset of $\mathbf{R}^{2p+2}$ which contains $S_E \times S_F$ let $\tilde{f}$ denote its restriction to $S_E \times S_F$. Recalling Proposition 3 one has

Proposition 4. *Let $f \in \Gamma^{-p+1}(X)$ then*

$$(\widetilde{\Delta_X f}) = (\Delta_{S_E} - \Delta_{S_F})\tilde{f} \tag{37}$$

Proof. We may define $\Delta_X f$ by applying Δ to the extension φ of f given by (32) and then restrict to C^*. But clearly φ is an eigenvector for the radial part of Δ_E and of Δ_F. But in fact since

$$\dim\ E = \dim\ F \tag{38}$$

the eigenvalues are the same and hence the radial parts cancel out in the subtraction (36). This immediately yields (38). QED

Now the covering map

$$G \longrightarrow SO(p+1, p+1)_e \tag{39}$$

restricted to K yields a covering map

$$K \longrightarrow SO(p+1) \times SO(p+1) \tag{40}$$

For $k \in \mathbf{Z}_+$ let γ'_k be the natural irreducible representation of $SO(p+1)$ on the space of homogeneous harmonic polynomials of degree k on $\mathbf{R}^{p+1}$. Then $\gamma_k \times \gamma_k$ defines an irreducible representation of $SO(p+1) \times SO(p+1)$ and by composition with (40) an irreducible representation ζ_k of K.

Now let $H \subset \Gamma^{-p+1}(X)$ be the kernel of Δ_X and let H' be the cokernel so that one has an exact sequence of G-modules

$$0 \longrightarrow H \longrightarrow \Gamma^{-p+1}(X) \xrightarrow{\Delta_X} \Gamma^{-p-1}(X) \longrightarrow H' \longrightarrow 0 \qquad (41)$$

We can now be sure that H (and also H') is not trivial.

Theorem 5. *Let H_K be the space of K-finite elements in H. Then one has the direct sum*

$$H_K = \sum_{k=0}^{\infty} H_k \qquad (42)$$

where H_k is an irreducible K-module which transforms according to the representation ζ_k.

Proof. Now as one knows the eigenspaces W_k of Δ_{S_E} may be parameterized by $k \in \mathbf{Z}_+$ where $W_k(E)$ is the restriction to S_E of the space of homogeneous Δ_E harmonic polynomials functions on $\mathbf{R}^{p+1}$ of degree k. One defines $W_k(F)$ similarly. But now since dim $E =$ dim F one has that

$$W_j(E) \otimes W_k(F) \subset \text{Ker}\ (\Delta_{S_E} - \Delta_{S_F}) \text{ if and only if } j = k. \qquad (43)$$

One the other hand one easily sees that there exist $Z_k \subset \Gamma^{-p+1}(X)$ such that

$$Z_k|(S_E \times S_F) = W_k(E) \otimes W_k(F) \qquad (44)$$

The result then follows from (37). QED

3. Irreducibility, the Joseph Ideal and the Annihilator of *H*

3.1 *The decomposition of* $S^2(\mathfrak{g})$. Now H_K is clearly a $(\mathfrak{g}, K)$ Harish-Chandra module. We wish to prove that H_K is irreducible and its annihilator J in U is the Joseph ideal J_o. Since this ideal is defined for the case where $\mathfrak{g}$ is not of type A_ℓ we will assume that $p \geq 3$.

Now let $\mathfrak{a}$ be a Cartan subalgebra of $\mathfrak{g}$ so that dim $\mathfrak{a} = p+1$. Then as one knows the $\mathfrak{a}$-weights of the representation of $\mathfrak{g}$ on $\mathbf{R}^{2p+2}$ are of the form $\pm\lambda_i$, $i = 1, \ldots, p+1$, where the λ_i are the elements of a (orthonormal — with respect to a suitable multiple of the co-Killing form) basis of the dual space $\mathfrak{a}'$ to $\mathfrak{a}$. In addition the set $R = \{\pm\lambda_i \pm \lambda_j \text{ where } i \neq j\}$ is the set of roots for the pair $(\mathfrak{g}, a)$. We may choose a positive system $R_+ \subset R$ so that

$$R_+ = \{\lambda_i \pm \lambda_j | \text{ for } i < j\}$$

In that case the highest root (*i.e.* the highest weight of the adjoint representation) is $\psi = \lambda_1 + \lambda_2$.

Now consider the decomposition of the tensor product $\mathfrak{g} \otimes \mathfrak{g}$ of the adjoint representation with itself into a sum of irreducible $\mathfrak{g}$-modules. One knows that the highest weight of any irreducible component Y in $\mathfrak{g} \otimes \mathfrak{g}$ is necessarily of the form $\psi + \varphi$ where $\varphi \in R$ or $\varphi = 0$. If $\varphi \in R$ then its multiplicity is one and hence the multiplicity of Y is 1. Furthermore since $\psi + \varphi$ is necessarily dominant the only possibilities for $\psi + \varphi$ where $\varphi \in R$ is $\psi + \varphi = (1)\, 2\lambda_1 + 2\lambda_2$, (2) $2\lambda_1$, (3) 0, (4) $\lambda_1 + \lambda_2 + \lambda_3 + \lambda_4$, in case $p = 3$ also (4′) $\lambda_1 + \lambda_2 + \lambda_3 - \lambda_4$, (5) $2\lambda_1 + \lambda_2 + \lambda_3$,. But now as a $\mathfrak{g}$-module one has

$$\mathfrak{g} \otimes \mathfrak{g} = S^2(\mathfrak{g}) \oplus \wedge^2 \mathfrak{g}$$

However by (4.4.4), p.156 and Theorem 8, p.158 in [Ko 2] one has a decomposition $\wedge^2 \mathfrak{g} = E_1 \oplus E_2$ where E_1 is equivalent to the adjoint representation and the highest weight of E_2 is $2\lambda_1 + \lambda_2 + \lambda_3$(*i.e.* case (5) above). On the other hand since $S^3(\mathfrak{g})$ contains no non-trivial $\mathfrak{g}$-invariant it follows that 2 is not an exponent of $\mathfrak{g}$ and hence the adjoint representation (case where $\varphi = 0$) does not occur in $S^2(\mathfrak{g})$. See *e.g.* Theorem 17, p.396 in [Ko 1]. Thus the only possibilities for highest weights (and then with multiplicity 1) in $S^2(\mathfrak{g})$ are (1), (2), (3), (4), and if $p = 3$, (4′). This is all we need to prove Theorems 8 and 9. However it is the case that the representations with these highest weights do in fact occur in $S^2(\mathfrak{g})$. This is established by applying Weyl's dimension formula and noting that the sum of the dimensions of these representations is exactly dim $S^2(\mathfrak{g})$. Thus one has

Proposition 6. *If* $p = 3$ *then there are 5 irreducible components in* $S^2(\mathfrak{g})$ *each occurring with multiplicity 1 and the corresponding highest weights are* $(1)\, 2\lambda_1 + 2\lambda_2$, $(2)\, 2\lambda_1$, $(3)\, 0$, $(4) \lambda_1 + \lambda_2 + \lambda_3 + \lambda_4$, *and* $(4') \lambda_1 + \lambda_2 + \lambda_3 - \lambda_4$.

If $p > 3$ *then there are* 4 *irreducible components in* $S^2(g)$ *each occurring with multiplicity* 1 *and the corresponding highest weights are* $(1) 2\lambda_1 + 2\lambda_2$, $(2) 2\lambda_1$, $(3) 0$, $(4) \lambda_1 + \lambda_2 + \lambda_3 + \lambda_4$.

3.2 Cent U^K *and generators of the Joseph ideal.* Using characterizations of the Joseph ideal in [J] and a result of mine determining the ideal of polynomial functions which vanish on the (any semisimple) G-orbit of the highest weight vector in any finite dimensional irreducible G-module, Devra Garfinkle in [*G*] gave a very useful characterization of the Joseph ideal. To apply this characterization to the case at hand let $Q_2 \subset S^2(\mathfrak{g})$

be the Cartan product of $\mathfrak{g}$ with itself so that Q_2 is the irreducible component of $S^2(\mathfrak{g})$ whose highest weight is twice the highest root. That is Q_2 is the component whose highest weight is $2\lambda_1 + 2\lambda_2$. Let R_2 be the unique $\mathfrak{g}$-stable complement of Q_2 in $S^2(\mathfrak{g})$. Her result implies

Theorem 7. *Assume that J is any ideal in U of infinite codimension (so that Spec Gr J does not reduce to the origin). Let $I = Gr\ J$. Then J is the Joseph ideal J_o if and only if $I_2 = R_2$.*

The argument used in the proof of the next theorem to determine the annihilator J in $U of$ a Harish-Chandra module M we believe is novel. However it probably has a very limited application. Let k be the Lie algebra of K and let U^K be the centralizer of K in U. Clearly

$$\text{Cent } U(\mathfrak{k}) \cdot \text{Cent } U(\mathfrak{g}) \subset \text{Cent } U^K \tag{44.1}$$

(In fact it is likely that one has equality in (44.1)). Briefly stated, in the case at hand, the K-types occurring in M are of such a restricted nature that not only is (Cent $U(\mathfrak{k})) \cap J$ non-trivial but in fact (Cent $U^K) \cap J$ generates J.

Theorem 8. *Assume in Theorem 5 that $p = 3$ so that G is the simply-connected covering group of the identity component of $SO(4,4)$. Then H_K is an irreducible Harish-Chandra module and the annihilator J in U of H_K is the Joseph ideal J_o. Furthermore J_o is the ideal in U generated by (Cent $U^K) \cap J_o$.*

Proof. Let u be the Lie algebra of $SU(2)$. If $p = 3$ then clearly one has a decomposition

$$\mathfrak{k} = \mathfrak{u}_1 \oplus \mathfrak{u}_2 \oplus \mathfrak{u}_3 \oplus \mathfrak{u}_4$$

where $\mathfrak{u}_i \cong \mathfrak{u}$ for $i = 1, \ldots, 4$. The irreducible representations of $\mathfrak{u}$ are of course parameterized by integers $n \in \mathbb{Z}_+$ where $n+1$ is the dimension of the corresponding irreducible module. Thus we may parameterize the irreducible representations of $\mathfrak{k}$ by a 4-tuple $(k, \ell, m, n) \in \mathbb{Z}_+^4$. But recalling the definition of ζ_k it is straightforward to see that

$$\zeta_k \text{ corresponds to } (k,k,k,k) \tag{44.2}$$

Now for $i = 1, \ldots, 4$ let $c_i \in \text{Cent } U(\mathfrak{u}_i)$ be the Casimir element. Obviously $c_i - c_j \in \text{Cent } U(\mathfrak{k})$. But by (44.2) one clearly has

$$c_i - c_j \in J$$

where $J \subset U$ is the annihilator of H_K. On the other hand clearly

$$X = G/P$$

for a parabolic subgroup P and $\Gamma^{-2}(X)$ is the space of smooth elements of an induced (to G) one-dimensional representation of P. Thus if c is the Casimir element of $\mathfrak{g}$ then there exists a constant c_o such that $c - c_o \in J$. Thus

$$c - c_o,\ c_i - c_j \in (\text{Cent } U^K) \cap J_2. \tag{44.3}$$

Since the ζ_k are all distinct the image of $U(\mathfrak{k})$ in U/J is infinite dimensional and hence

$$\dim\ U/J = \infty \tag{44.4}$$

But (44.4) implies that, in the notation of Theorem 7, that

$$I_2 \subset R_2 \tag{44.5}$$

Indeed otherwise one would have $e_\psi^2 \in I_2$ where ψ is the highest root clearly implying that Spec I reduces to the origin. But this implication easily contradicts (44.4).

Now one notes that not only are there 4 independent elements on the left side of (44.3) but in fact one must have

$$\dim\ I_2^K \geq 4$$

(44.6)
On the other hand if M is any finite dimensional irreducible G module then one knows that dim $M^K \leq 1$. Thus (44.6) implies that I_2 has at least 4 irreducible G-modules. But then by (44.5) and Proposition 6 one has

$$I_2 = R_2$$

(Note that we do not need the full strength of Proposition 6 but only that there are at most 5 irreducible $\mathfrak{g}$-components in $S^2(\mathfrak{g})$ which our argument preceding Proposition 6 implies). But recalling (44.4) one then has $J = J_o$ by Theorem 7. But by (44.3), Theorem 7, and the argument just given one also has that J_o is generated by (Cent $U^K) \cap J_2$.

But now if

$$\mathfrak{g} = \mathfrak{k} + \mathfrak{p}$$

is the Cartan decomposition of $\mathfrak{g}$ one notes that as a $\mathfrak{k}$-module

$$\mathfrak{p} \text{ corresponds to } \zeta_1$$

But also one readily has that

$$\zeta_{k+1} \text{ is the Cartan product of } \zeta_1 \text{ and } \zeta_k$$

It follows then from Theorem 5 that the subset of $\mathbf{Z}_+$ which parameterizes the K-types in any irreducible $\mathfrak{g}$-subquotient of H_K is necessarily "connected". But then if H_K was not $\mathfrak{g}$-irreducible it must contain a non-trivial finite dimensional $\mathfrak{g}$-subquotient. But the Joseph ideal is a maximal ideal and hence cannot be in the kernel of a finite dimensional U-module. This implies that H_K is $\mathfrak{g}$-irreducible. QED

3.3 *The case where* $p > 3$. The case in Theorem 8 where $p = 3$ is the case which really interests us in this paper — primarily because of a connection with geometry which we will take up in §5. The argument in the proof specifically uses the condition that $p = 3$ because with that condition one has 4 linearly independent elements on the left side of (44.3) whereas $S^2(\mathfrak{g})$ has 5 irreducible components for the action of $\mathfrak{g}$. In case $p > 3$ then there would be only 2 linearly independent elements on the left side of (44.3) whereas $S^2(\mathfrak{g})$ has 4 irreducible components for the action of $\mathfrak{g}$. It was therefore my initial impression that the argument of Theorem 8 would fail to imply that the annihilator J of H_K is the Joseph ideal J_o – *in* case $p > 3$. However when I discussed this with David Vogan he pointed out to me that my argument still works since the irreducible component corresponding to the highest weight $\lambda_1 + \lambda_2 + \lambda_3 + \lambda_4$ (See Proposition 6) automatically annihilates $\Gamma^{-2}(X)$ and this irreducible component is non-spherical (for $p > 3$).

It is not clear whether the final statement in Theorem 8 is true for $p > 3$. Generalizing the first statement of Theorem 8 one has

Theorem 9. *Assume in Theorem 5 that* $p \geq 3$. *Then* H_K *is an irreducible Harish-Chandra module and the annihilator* J *in* U *of* H_K *is equal to the Joseph ideal* J_o.

Proof. By Theorem 8 we may assume that $p > 3$. But then one has

$$\mathfrak{k} = \mathfrak{u}_1 \oplus \mathfrak{u}_2$$

where $\mathfrak{u}_i$, $i = 1, 2$ are the simple components of $\mathfrak{k}$. Let c_i be the Casimir element of $\mathfrak{u}_i$ and let c be the Casimir element of $\mathfrak{g}$. Then from the definition of ζ_k and, as argued in the proof of Theorem 8, there exists a constant c_o such that

$$c - c_o,\ c_1 - c_2 \in (\text{Cent } U^K) \cap J_2. \tag{44.7}$$

Now let $M \subset U$ be an irreducible ad $\mathfrak{g}$ stable subspace which as a $\mathfrak{g}$ module has $\lambda_1 + \lambda_2 + \lambda_3 + \lambda_4$ as highest weight. One notes that as $\mathfrak{g}$ modules

$$M \cong \wedge^4 \mathbf{R}^{2p+2}.$$

But then it is a simple matter to verify (since $p > 3$) that M is a non-spherical representation of $\mathfrak{g}$. (One may also verify this by noting that the

Cartan-Helgason condition is not satisfied for $\lambda_1 + \lambda_2 + \lambda_3 + \lambda_4$). On the other hand we assert that

$$M \subset J \tag{44.8}$$

To prove (44.8) let $\mathfrak{p}$ here denote the Lie algebra of the parabolic subgroup P such that $X = G/P$. Then if $\mathfrak{r}$ is the semisimple part of the Levi factor of $\mathfrak{p}$ one notes that $\mathfrak{r}$ is isomorphic to the Lie algebra of $SO(p,p)$ and one easily has

$$M \cong \wedge^2\mathbf{R}^{2p} \oplus \wedge^3\mathbf{R}^{2p} \oplus \wedge^3\mathbf{R}^{2p} \oplus \wedge^4\mathbf{R}^{2p} \tag{44.9}$$

as $\mathfrak{r}$ modules.

But now $\Gamma^{-p+1}(X)$ is the space of sections of a line bundle on X. Let L be the left ideal in U which annihilates the dual fiber at the origin of X. As one knows, to prove (44.8) it suffices to prove that

$$\tilde{M} \subset L$$

where $\tilde{M}$ is the image of M under the anti-involution $u \longrightarrow \tilde{u}$ of U such that $\tilde{x} = -x$ for $x \in \mathfrak{g}$. But since M is clearly a self-contragredient $\mathfrak{g}$-module it suffices to prove that

$$M \subset L$$

But $\mathfrak{r}$ clearly annihilates the dual fiber at the origin of X so that $\mathfrak{r} \subset L$. Furthermore U/L as a $\mathfrak{r}$ module is easily seen to be equivalent to the natural action of $\mathfrak{r}$ on the symmetric algebra $S(\mathbf{R}^{2p})$ and from the usual decomposition involving harmonic polynomials one notes from (44.9) that $\mathrm{Hom}_{\mathfrak{r}}(M, U/L) = 0$. But the map $M \longrightarrow U/L$ defined by projection is clearly a $\mathfrak{r} -$ *map* and hence must reduce to 0 so that one has $M \subset L$ and hence (44.8).

To prove the theorem, as argued in the proof of Theorem 8 it suffices to show that $I_2 = R_2$. But, as in that proof, one has $I_2 \subset R_2$. But since the elements on the left side of (44.7) are linearly independent one has $I_2 = R_2$, by Proposition 6, the argument in the proof of Theorem 8, and (44.8) (recalling that M is non-spherical). QED

4. Unitarity, a Radon Transform, and Triality

4.1 *The pairing of* $\Gamma^{-2}(X)$ *and* $\Gamma^{-4}(X)$. Now we recall from §2.3 that X has a natural conformal structure and the set $\mathcal{M}$ of smooth metrics in the conformal class corresponds to the cone $\Gamma^{-2}(X)_+ \subset \Gamma^{-2}(X)$. Of course for the case at hand any $g \in \mathcal{M}$ has signature (p,p). On the other hand, $\Gamma^{-p+1}(X)$, by Theorem 5, admits a proper G-submodule $H(= Ker\ \Delta_X)$.

Thus from the point of view of geometry we are motivated to consider the case where $-p+1=-2$. That is we assume that

$$p = 3 \tag{45}$$

so that (41) becomes the exact sequence

$$0 \longrightarrow H \longrightarrow \Gamma^{-2}(X) \xrightarrow{\Delta_X} \Gamma^{-4}(X) \longrightarrow H' \longrightarrow 0 \tag{45.5}$$

Furthermore X is 6-dimensional and the double covering (18) becomes a double covering

$$S^3 \times S^3 \longrightarrow X \tag{46}$$

Remark 10. The map (46) is readily seen to be an identification of antipodal points so that one has a diffeomorphism

$$X \cong SO(4) \tag{47}$$

We will pursue the geometric tie-ins of the present case ($p=3$) in §5. For the present we will deal with the unitary aspects of the representation of G (we recall that G is the covering group of the identity component of $SO(4,4)$) defined by H_K.

In [Ko] we established that H_K does define a unitary representation of G and in fact we computed (1) the inner product on each of the K-types (See Theorem 6.28, p.106 in [Ko]). We also (2) explicitly determined the action of the triality automorphism on the various K-types (See Theorem 7.4, p. 107 in [Ko]). The proof of (1) in [Ko] depended on some algebraic facts from general representation theory. We wish to present here another (analytic and geometric) proof of the unitarity which I think is much more interesting. It is intrinsic to the construction of the representation and it seems to be peculiarly related to this particular representation of this particular group. The idea behind it involves a composition of a Radon type transform together with an isomorphism induced by triality. The idea is a joint effort of V. Guillemin, E. Grinberg and myself. Here I will work out the details and verify that the inner product is the same as the one given in [Ko].

Now for the case at hand one has

$$K \cong SU(2) \times SU(2) \times SU(2) \times SU(2) \tag{48}$$

and

$$S_E \times S_F \cong S^3 \times S^3$$

Remark 11. Note that if we pick a point in $S_E \times S_F$ we may identify $S_E \times S_F$ with $SU(2) \times SU(2)$ in such a fashion that the action of K, as described by (48), is such that the first two factors operate by left and right multiplication on S_E and the last two factors operate by left and right multiplication on S_F.

Now recall that ν is the unique K-invariant measure of total volume 1 on $S_E \times S_F$. We recall (see (23)) that there is a G-invariant pairing of $\Gamma^{-2}(X)$ and $\Gamma^{-4}(X)$ which by (35) may be given by

$$< f, h >= \int_{S_E \times S_F} fh\ \nu \tag{49}$$

for $f \in \Gamma^{-2}(X)$ and $h \in \Gamma^{-4}(X)$.

Now recalling (45.5) one has

Proposition 12. *Let* $f_i \in \Gamma^{-2}(X)$, $i = 1, 2$, *then*

$$< \Delta_X f_1, f_2 >=< f_1, \Delta_X f_2 > \tag{50}$$

Proof. Using the notation of (37) it suffices by (49) to show that

$$\int_{S_E \times S_F} ((\Delta_{S_E} - \Delta_{S_F})\tilde{f}_1)\tilde{f}_2\nu = \int_{S_E \times S_F} \tilde{f}_1((\Delta_{S_E} - \Delta_{S_F})\tilde{f}_2)\nu$$

But this follows since the Laplacians Δ_{S_E} and Δ_{S_F} are clearly symmetric operators with respect to "Haar measure" on the "groups" S_E and S_F respectively. QED

4.2. *The manifold Y of isotropic* 4-*planes*. It follows from (45.5) and Proposition 12 that the G-invariant pairing of $\Gamma^{-2}(X)$ and $\Gamma^{-4}(X)$ induces a G-invariant pairing of H and H'. But then a G-invariant sesquilinear form on H would arise from a G-invariant conjugate linear map from H to H'. It is slightly more awkward to deal with but, instead, our pre-Hilbert structure on H_K will arise from a G- invariant conjugate linear map

$$Q : H' \longrightarrow H \tag{51}$$

which we will now procede to define.

Recalling (9) let $s : E \longrightarrow F$ be the isomorphism defined by the condition that $r_{i+4}(s(e)) = r_i(e)$ for $e \in E$, and $i = 1, \ldots, 4$ and let

$$V = \{e + s(e) | e \in E\} \tag{52}$$

Then clearly V is an isotropic 4-plane in $\mathbf{R}^8$ with respect to the bilinear form B.

As one knows, classically, the set of all isotropic 4-planes in $\mathbf{R}^8$ decomposes into 2 orbits under the action of G. (For a discussion of this see § 2 in [Ko].) As in [Ko] let Y denote that orbit which contains V. It is easy to see that Y is a closed subvariety of the Grassmannian of all 4-planes in $\mathbf{R}^8$. If $P_2 = \{g \in G | gV = V\}$ then as G-homogeneous spaces

$$G/P_2 \cong Y \tag{53}$$

For each $W \in Y$ there exists a unique map, which in fact is an isometry with respect to the norm $| \, |$,

$$s_W : E \longrightarrow F$$

such that

$$W = \{e + s_W(e) | e \in E\} \tag{54}$$

One notes that

$$s_V = s \tag{54.5}$$

Also note that Y is a principal homogeneous space with respect to the group $SO(E)$ by the relation

$$s_{a \cdot W}(e) = s_W(a^{-1}e) \tag{55}$$

for all $e \in E$. In particular then one has a diffeomorphism

$$Y \cong SO(4). \tag{56}$$

Remark 13. It is clear that $P_2 \cap K$ is a maximal compact subgroup of P_2. Since this group preserves an orientation of E it follows easily that P_2 preserves an orientation of V. Consequently once an orientation of V is fixed then the action of G confers an orientation on any $W \in Y$.

4.3 *The diffeomorphism* $\tau_Y : Y \longrightarrow X$. Let $v_1 \in V$ be defined by the conditions $r_1(v_1) = r_5(v_1) = 1$ and $r_i(v_1) = 0$ for the remaining values of i. Let $P_1 = \{g \in G | gv_1 \in \mathbf{R}v_1\}$ so that as homogeneous spaces

$$G/P_1 \cong X \tag{57}$$

Now corresponding to one of the "extraordinary symmetries" of the Dynkin diagram of D_4 we have constructed, in [Ko], an involutory automorphism τ of G such that $\tau(P_1) = P_2$. In particular τ induces a diffeomorphism $\tau_X : X \longrightarrow Y$ such that if $x \in X$ and $g \in G$ then $\tau_X(g \cdot x) =$

$\tau(g) \cdot \tau_X(x)$. See Theorem 2.6, p. 73 in [Ko]. Since τ is involutory one also has

$$\tau(P_2) = P_1 \tag{58}$$

and hence τ also induces a diffeomorphism

$$\tau_Y : Y \longrightarrow X \tag{59}$$

such that if $y \in Y$ and $g \in G$ then

$$\tau_Y(g \cdot y) = \tau(g) \cdot \tau_Y(y) \tag{60}$$

Now we may choose E, F, (and hence V), and the Cartan subalgebra $\mathfrak{a}$ of §3.1 , so that it agrees with the choices of §2 in [Ko]. But then τ stabilizes $\mathfrak{a}$ and hence operates in $\mathfrak{a}'$. Furthermore we may assume the λ_i of §3.1 here are the same as the λ_i of §1.1 in [Ko] so that the λ_i are the weights of $\mathfrak{a}$ in V. Also λ_1 is the highest weight of $\mathbf{R}^8$ as a $\mathfrak{g}$-module (recall that $\mathfrak{g}$ is split). In addition v_1 here is the same as v_1 in [Ko] so that v_1 is a highest weight vector in $\mathbf{R}^8$.

4.4 *The τ-twisted isomorphism $\hat{\tau}_Y : L^{-2}(Y) \longrightarrow L^{-2}(X)$ of line bundles.* Now by §1.1 in [Ko] the simple roots α_i, $1 = 1, \ldots, 4$, are chosen so that $\alpha_1 = \lambda_1 - \lambda_2$, $\alpha_2 = \lambda_3 + \lambda_4$, $\alpha_3 = \lambda_3 - \lambda_4$, and $\alpha_4 = \lambda_2 - \lambda_3$. The involution τ has the property that τ interchanges α_1 and α_2 but leaves α_3 and α_4 fixed. It follows that τ interchanges the highest weights of the fundamental representations of G corresponding to α_1 and α_2. But then by Proposition 6.1 in [Ko] one has for twice these weights

$$\tau(2\lambda_1) = \lambda_1 + \lambda_2 + \lambda_3 + \lambda_4 \tag{61}$$

and

$$\tau(\lambda_1 + \lambda_2 + \lambda_3 + \lambda_4) = 2\lambda_1 \tag{62}$$

On the other hand note that the action of P_1 on $\mathbf{R}v_1$ is defined by a real character of P_1 which is naturally parameterized by λ_1 (in the sense that the restriction of this character to exp a corresponds to the weight λ_1). The square of this character is then parameterized by $2\lambda_1$ and is clearly realized by the natural action of P_1 on $\mathbf{R}\ v_1 \otimes v_1$. By complexification we note the obvious identification

$$L^{-2}(X) = G \times_{P_1} \mathbf{C}\ v_1 \otimes v_1 \tag{63}$$

using the notation of §1.2 so that if $x_1 = \alpha(v_1)$ is the element of X which corresponds to P_1 in (57) then

$$L^{-2}(X)_{x_1} = \mathbf{C}\ v_1 \otimes v_1 \tag{64}$$

On the other hand, in a similar fashion, the character of P_2 defined by its action on $\wedge^4 V$ is clearly parameterized by $\lambda_1+\lambda_2+\lambda_3+\lambda_4$. Thus the line bundle $L^{-2}(Y)$ over Y corresponding to this weight is defined by

$$L^{-2}(Y) = G \times_{P_2} \mathbf{C} \wedge^4 V \tag{65}$$

and hence the fiber at $W \in Y$ for any W is just the line

$$L^{-2}(Y)_W = \mathbf{C} \wedge^4 W \tag{66}$$

Now having defined $v_1 \in V$ let $v_2, v_3, v_4 \in V$ be such that the v_i are a weight basis corresponding to the weights λ_i so that

$$\mathbf{C}\; v_1 \wedge v_2 \wedge v_3 \wedge v_4 = \wedge^4 V \tag{67}$$

Now by (58) and (62), for any positive constant κ (to be fixed later) the diffeomorphism (59) lifts to an isomorphism of line bundles

$$\hat{\tau}_Y : L^{-2}(Y) \longrightarrow L^{-2}(X) \tag{68}$$

such that $\hat{\tau}_Y(L^{-2}(Y)_V) = L^{-2}(X)_{x_1}$ where

$$\hat{\tau}_Y(v_1 \wedge v_2 \wedge v_3 \wedge v_4) = \kappa\; v_1 \otimes v_1 \tag{69}$$

and for any $g \in G$ and $\ell \in L^{-2}(Y)$

$$\hat{\tau}_Y(g \cdot \ell) = \tau(g) \cdot \hat{\tau}_Y(\ell) \tag{70}$$

One notes that both $L^{-2}(X)$ and $L^{-2}(Y)$, as complex line bundles are in fact defined over $\mathbf{R}$ and that the corresponding real line bundles have G-invariant orientations. For $L^{-2}(X)$ we choose that orientation which at x_1 contains $v_1 \otimes v_1$. For $L^{-2}(Y)$ we choose the orientation which at V contains $v_1 \wedge v_2 \wedge v_3 \wedge v_4$. Having done so note then each $W \in Y$ has been given an orientation. See Remark 13. Furthermore $\hat{\tau}_Y$ clearly induces an orientation preserving map for the corresponding real line bundles.

4.5. *An explicit computation of* $\hat{\tau}_Y$. It will be necessary for us to be much more specific about the map $\hat{\tau}_Y$. We will be able to this since K is transitive on both X and Y for both $L^{-2}(X)$ and $L^{-2}(Y)$ and there exists non-trivial K-invariant sections, σ_X and σ_Y. Since τ stabilizes K (see Proposition 3.1 in [Ko]) this implies that $\hat{\tau}_Y$ carries σ_Y into a constant multiple of σ_X. One may define σ_X so that for any $x \in X$

$$\sigma_X(x) = v \otimes v \text{ where } v \in S_E \times S_F \text{ is such that } \alpha(v) = x \tag{71}$$

One notes of course that although v in (71) is not unique σ_X is, however, uniquely defined.

To define σ_Y let

$$\pi_E : \mathbf{R}^8 \longrightarrow E \tag{72}$$

be the projection defined relative to the decompositon (9). Now the bilinear form $B|E$ is positive definite and by (54) for any $W \in Y$ the restriction $\pi_E|W$ is a bijection onto E. Let B_W be the positive definite bilinear form on W so that $\pi_E|W$ is an isometry with $B|E$. It is clear that if $a \in K$ then

$$a \cdot B_W = B_{a \cdot W} \tag{73}$$

One may then define σ_Y so that $\sigma_Y(W)$, for any $W \in Y$, is the unique element of $\wedge^4 W$ which is positively oriented and has unit length with respect to B_W. Clearly we may normalize the v_i so that

$$\sigma_Y(V) = v_1 \wedge v_2 \wedge v_3 \wedge v_4 \tag{74}$$

Now in §3.4 of [Ko] we, in effect, identified K with $SU(2) \times SU(2) \times SU(2) \times SU(2)$ and defined diffeomorphisms

$$h_E : SU(2) \longrightarrow S_E \text{ and } h_F : SU(2) \longrightarrow S_F \tag{75}$$

with the following properties: (1) if 1 here denotes the identity of SU(2) then

$$h_E(1) = e_1 \text{ and } h_F(1) = f_1 \tag{76}$$

where $e_1 \in E$ and $f_1 \in F$ are such that

$$v_1 = e_1 + f_1 \tag{77}$$

(2) one has a two-sided action of $SU(2)$ on both S_E and S_F so that for $a, b, c, d, x, y \in SU(2)$

$$h_E(axb^{-1}) = ah_E(x)b^{-1} \text{ and } h_F(cyd^{-1}) = ah_F(y)d^{-1} \tag{78}$$

(3) recalling (54.5), for any $a, b \in SU(2)$ and $e \in S_E$

$$s_V(aeb^{-1}) = as_V(e)b^{-1} \tag{79}$$

(4) the action of K on $S_E \times S_F$ is given by

$$(a, b, c, d) \cdot (e + f) = aeb^{-1} + cfd^{-1} \tag{80}$$

where $a, b, c, d \in SU(2)$, $e \in S_E$ and $f \in S_F$.

Now by Theorem 3.6 in [Ko] the automorphism τ stabilizes K and in fact for any $a,b,c,d \in SU(2)$ one has

$$\tau(a,b,c,d) = (a,c,b,d) \tag{81}$$

Proposition 14. *Let $x \in X$ and let $v \in S_E \times S_F$ be such that $x = \alpha(v)$ so that $\sigma_X(x) = v \otimes v$. Then there exists uniquely determined $a,b \in SU(2)$ such that*

$$v = ae_1 + bf_1 \tag{82}$$

Moreover (recalling (69))

$$\hat{\tau}_Y(\sigma_Y(W)) = \kappa\ \sigma_X(x) \tag{83}$$

where, recalling (52),

$$W = \{e + a^{-1}s(e)b | e \in E\} \tag{84}$$

Proof. The uniqueness of $a,b \in SU(2)$ satisfying (82) is obvious. But it then follows from (80) that

$$v = (a,1,b,1)\cdot v_1 \tag{85}$$

and hence $v \otimes v = (a,1,b,1)\cdot v_1 \otimes v_1$ or $\sigma_X(x) = (a,1,b,1)\cdot v_1 \otimes v_1$. But then by (69),(70) and (81)

$$\begin{aligned}\hat{\tau}_Y^{-1}(\sigma_X(x)) &= \kappa^{-1}\cdot(a,b,1,1)\cdot\sigma_Y(V)\\ &= \kappa^{-1}\cdot\sigma_Y((a,b,1,1)\cdot V)\end{aligned} \tag{86}$$

But $V = \{e + s(e) | e \in E\}$ so that if $W = (a,b,1,1)\cdot V$ then $W = \{aeb^{-1} + s(e) | e \in E\}$. But if $e' = aeb^{-1}$ then $aeb^{-1} + s(e) = e' + s(a^{-1}e'b)$. But $s(a^{-1}e'b) = a^{-1}s(e')b$ by (79). This clearly proves the proposition. QED

4.6 *The Radon Transform R.* Now let $f \in \Gamma^{-4}(X)$ so that f is a function of C^* satisfying

$$f(tv) = t^{-4}f(v)$$

for any $t \in \mathbf{R}^*$ and $v \in C^*$. Let $W \in Y$ and let $\mu \in \wedge^4 W'$ so that we can regard μ as a translation invariant volume form on the oriented vector space W. Now the Euler field $\mathcal{E}$ (see (21)) on $\mathbf{R}^8$ is of course tangent to

W and hence the interior product $\iota(\mathcal{E})\mu$ is a 3-form on W. Furthermore if $\theta(\mathcal{E})$, as in (22), denotes Lie differentiation with respect to $\mathcal{E}$ then clearly $\theta(\mathcal{E})\mu = 4\mu$ so that if $\beta(f, W, \mu) = f|W \cdot \iota(\mathcal{E})\mu$ one has

$$\theta(\mathcal{E})\beta(f, W, \mu) = \iota(\mathcal{E})\beta(f, W, \mu) = 0 \tag{87}$$

Since $\beta(f, W, \mu)$ is also clearly invariant under the involution $v \longrightarrow -v$ of W it follows that, with respect to the principal bundle,

$$\begin{array}{ccc} \mathbf{R}^* & \longrightarrow & W \\ & & \downarrow \\ & & P(W) \end{array} \tag{88}$$

where $P(W)$ is the projective 3-space defined by W, the orientation on W induces an orientation on $P(W)$ and $\beta(f, W, \mu)$ descends to a 3-form $\beta'(f, W, \mu)$ on $P(W)$. One therefore obtains a linear functional $\ell_f(W)$ on $\wedge^4 W'$ by putting

$$\ell_f(W)(\mu) = \int_{P(W)} \beta'(f, W, \mu) \tag{89}$$

Remark 15. For clarification we note that the orientation on $P(W)$ is such that (89) is positive for positive functions f, when μ defines the orientation on W.

But the dual space to $\wedge^4 W'$ is $\wedge^4 W$ so that we can regard

$$\ell_f(W) \in L^{-2}(Y)_W \tag{90}$$

In fact if $\Gamma^{-2}(Y)$ denotes the space of smooth sections of $L^{-2}(Y)$ then it is easy to see that $\ell_f \in \Gamma^{-2}(Y)$ where ℓ_f is the section $W \longrightarrow \ell_f(W)$. In fact we have defined a Radon Transform

$$R : \Gamma^{-4}(X) \longrightarrow \Gamma^{-2}(Y) \tag{91}$$

where we put $Rf = \ell_f$ and we note that R is manifestly a G-map.

Now as in § 5.1 of [Ko] for any $n \in \mathbf{Z}_+$ let

$$\rho_n : SU(2) \longrightarrow \text{Aut } Z_n \tag{92}$$

be an irreducible representation of $SU(2)$ of dimension $n+1$. By taking the 4-fold tensor product one has for any $(k, \ell, m, n) \in \mathbf{Z}_+^4$ an irreducible representation $\rho_{(k,\ell,m,n)}$ of K on a vector space of dimension $(k+1)(\ell+1)$ $(m+1)(n+1)$ and any irreducible representation of K is equivalent to $\rho_{(k,\ell,m,n)}$ for a unique $(k, \ell, m, n) \in \mathbf{Z}_+^4$.

If M is a K-module let M_K denote the subspace of all K-finite elements in M and for any $(k,\ell,m,n) \in \mathbb{Z}_+^4$ let $M_{(k,\ell,m,n)}$ denote the $\rho_{(k,\ell,m,n)}$ isotypic subspace of M_K. Now any element $f \in \Gamma^{-2}(X)$, for any $s \in \mathbb{C}$, is determined by its restriction $\varphi = f|(S_E \times S_F)$ and subject only to the condition that $\varphi(v) = \varphi(-v)$ for $v \in S_E \times S_F$ the function $\varphi \in C^\infty(S_E \times S_F)$ is otherwise arbitrary. It follows then from (80) that

$$\Gamma^s(X)_K = \sum_{n,m} \Gamma^s(X)_{(n,n,m,m)} \tag{93}$$

where the sum is over $n, m \in \mathbb{Z}_+$ such that $n \equiv m \bmod 2$. Furthermore all isotypic components $\Gamma^s(X)_{(n,n,m,m)}$ are K-irreducible. On the other hand applying the K-finite functor to (45.5) yields the exact sequence

$$0 \longrightarrow H_K \longrightarrow \Gamma^{-2}(X)_K \xrightarrow{\Delta_X} \Gamma^{-4}(X)_K \longrightarrow H'_K \longrightarrow 0 \tag{94}$$

But then by Theorem 5 one has

$$H_K = \sum_{n=0}^{\infty} H_{(n,n,n,n)} \tag{95}$$

where each isotypic component $H_{(n,n,n,n)}$ is irreducible.
This immediately yields the following description of H'_K and the image of Δ_X.

Proposition 16. *One has*

$$\Delta_X(\Gamma^{-2}(X)_K) = \sum_{n \neq m} \Gamma^{-4}(X)_{(n,n,m,m)} \tag{96}$$

where the sum is over the same set as in (93) *except that now there is, in addition, the condition that* $n \neq m$. *Furthermore*

$$H'_K = \sum_{n=0}^{\infty} H'_{(n,n,n,n)} \tag{97}$$

where each isotypic component $H'_{(n,n,n,n)}$ *is irreducible.*

Remark 17. Using standard arguments concerning the fall off in the "size" of the Fourier components on $S_E \times S_F$ of elements in $\Gamma^s(X)$ versus the growth in the eigenvalues of $\Delta_{S_E} - \Delta_{S_F}$ on $S_E \times S_F$ one notes that H is the space of all $f \in \Gamma^{-2}(X)$ such that all of the Fourier components of f are in $\Gamma^{-2}(X)_{(n,n,n,n)}$ for some n. Furthermore $\Delta_X(\Gamma^{-2}(X))$ is the space of all $f \in \Gamma^{-4}(X)$ such that all of the Fourier components of f are in $\Gamma^{-4}(X)_{(n,n,m,m)}$ for some n, m where $n \neq m$ and $n \equiv m \bmod 2$. It follows then that H' is canonically isomorphic to the space of all $f \in \Gamma^{-4}(X)$ such that all of the Fourier components of f are in $\Gamma^{-4}(X)_{(n,n,n,n)}$ for some n.

4.7 *The image of* Δ_X *and the kernel of* R. Our next objective is to show that the Radon transform R vanishes on the image of Δ_X and hence induces a G-map

$$R : H' \longrightarrow \Gamma^{-2}(Y) \tag{98}$$

To do this we need a more explicit computation of the element $\ell_f(W) \in \wedge^4 W$ for any $f \in \Gamma^{-4}(X)$ and $W \in Y$.

Note first that there exists an orientation on E such that the diffeomorphism

$$\pi_E | W : W \longrightarrow E \tag{99}$$

is an orientation preserving map for any $W \in Y$. But the orientation on E induces an orientation on S_E so that if μ is an oriented non-trivial translation invariant 4-form on E then the integral of $\nu = \iota(\mathcal{E})\mu$ on S_E is positive. Let μ_E be that positive multiple of μ such that

$$1/2 \cdot \int_{S_E} \nu_E = 1 \tag{100}$$

where $\nu_E = \iota(\mathcal{E})\mu_E$.

Now for any $W \in Y$ let μ_W be the 4-form on W which corresponds to μ_W by the map (99). Let $\nu_W = \iota(\mathcal{E})\mu_W$ and let S_W be the unit sphere in W with respect to B_W. One notes that the orientation in W induces an orientation on S_W so that the map

$$S_W \longrightarrow P(W) \tag{101}$$

induced by the bundle projection (88) is an orientation preserving double covering. Now identify μ_W with its corresponding element in $\wedge^2 W'$. Then for any $f \in \Gamma^{-4}(X)$ one clearly has, by definition of $\ell_f(W)$,

$$\ell_f(W)(\mu_W) = 1/2 \cdot \int_{S_W} f \cdot \nu_W \tag{102}$$

Furthermore the map (99) induces a diffeomorphism $S_W \longrightarrow S_E$ of oriented manifolds and a correspondence of ν_W and ν_E. Recalling (54) the equation (102) clearly establishes

Lemma 18. *For any $f \in \Gamma^{-4}(X)$ and $W \in Y$ one has*

$$\ell_f(W)(\mu_W) = 1/2 \cdot \int_{S_E} f(e + s_W(e)) \cdot \nu_E \tag{103}$$

Now K operates on S_E and there is an associated action of K on $C^\infty(S_E)$. On the other hand, by (100), the map

$$SU(2) \longrightarrow S_E,\ a \longrightarrow ae_1 \tag{104}$$

is a diffeomorphism which puts Haar measure da on $SU(2)$ in correspondence with the measure $\hat{\nu}_E$ on S_E defined by the 3-form $1/2 \cdot \nu_E$.

Recall (92). Let $n \in \mathbb{Z}_+$. For any $A \in \operatorname{End} Z_n$ let $\varphi_A \in C^\infty(S_E)$ be the function on S_E defined by putting

$$\varphi_A(ae_1) = (n+1) \cdot \operatorname{tr} A\rho_n(a^{-1}) \tag{105}$$

One knows, by the Peter-Weyl theorem as applied to $SU(2)$, that

$$\varphi_A \in C^\infty(S_E)_{(n,n,1,1)} \tag{106}$$

and if $a, b, c, d \in SU(2)$ one has

$$(a, b, c, d) \cdot \varphi_A = \varphi_{\rho_n(a)A\rho_n(b^{-1})} \tag{107}$$

Furthermore, since ρ_n is self-contragredient, there exists a (complex linear) transpose operation, A $\longrightarrow A^t$, on End Z_n, such that

$$\varphi_{A^t}(ae_1) = \varphi_A(a^{-1}e_1) \tag{108}$$

On the other hand, from general theory, there exists a (conjugate linear) Hermitian adjoint operation, $A \longrightarrow A^*$, on End Z_n, such that

$$\varphi_{A^*} = \bar{\varphi}_{A^t} \tag{109}$$

Again from the Peter-Weyl theory one has for $A \in \operatorname{End} Z_n$ and $B \in \operatorname{End} Z_m$

$$\int_{S_E} \varphi_A \varphi_B \hat{\nu}_E = \begin{cases} 0 & \text{if } n \neq m \\ n+1 \operatorname{tr} AB^t & \text{if } n = m \end{cases} \tag{110}$$

Recall (77). Let $n, m \in \mathbb{Z}_+$ be such that $n \equiv m \bmod 2$ and for $A \in \operatorname{End} Z_n$ and $B \in \operatorname{End} Z_m$ let $\xi(A, B) \in \Gamma^{-4}(X)$ be such that its restriction to $S_E \times S_F$ is given by

$$\xi(A,B)(ae_1 + bf_1) = (n+1)(m+1) \cdot \operatorname{tr} A\rho_n(a^{-1}) \cdot \operatorname{tr} B\rho_m(b^{-1}) \tag{111}$$

for any $a, b \in SU(2)$. By the Peter-Weyl Theorem one has that

$$\xi(A,B) \in \Gamma^{-4}(X)_{(n,n,m,m)} \tag{112}$$

and that furthermore $\Gamma^{-4}(X)_{(n,n,m,m)}$ is spanned by $\xi(A,B)$ for all $A \in \text{End } Z_n$ and $B \in \text{End } Z_m$.

Theorem 19. *The Radon Transform R (see (91)) vanishes on the image of Δ_X so that it induces a G-map*

$$R : H' \longrightarrow \Gamma^{-2}(Y) \tag{113}$$

Proof. By taking Fourier components it suffices by (96) and Remark 17 to show that $Rf = 0$ for any $f \in \Gamma^{-4}(X)_{(n,n,m,m)}$ where $n \neq m$ but $n \equiv m \bmod 2$. In fact for such values of n and m it suffices to show that if $A \in \text{End } Z_n$ and $B \in \text{End } Z_m$ then $R\xi(A,B) = 0$.

Let $W \in Y$. Now by (54) and (80) there clearly exists $c, d \in SU(2)$ so that $s_W(e) = cs(e)d$ for all $e \in S_E$. On the other hand for $e \in E$ we can uniquely write $e = ae_1$ for $a \in SU(2)$. But $s(ae_1) = as(e_1)$, by (79) and $s(e_1) = f_1$ so that $s_W(e) = caf_1d$. But $f_1d = df_1$ by (76) and (78) so that $s_W(e) = cadf_1$ and hence

$$e + s_W(e) = ae_1 + cadf_1 \tag{114}$$

and hence by (111), if $\delta = (n+1)(m+1)$

$$\xi(A,B)(e + s_W(e)) = \delta \cdot \text{tr } A\rho_n(a^{-1}) \cdot \text{tr } B\rho_m((cad)^{-1})$$

and hence if $C = \rho_m(c^{-1})B\rho_m(d^{-1})$ then

$$\xi(A,B)(e + s_W(e)) = \delta \cdot \text{tr } A\rho_n(a^{-1}) \cdot \text{tr } C\rho_m(a^{-1})$$

But then by (100) and (103)

$$\ell_{\xi(A,B)}(\mu_W) = \delta \cdot \int_{SU(2)} \text{tr } A\rho_n(a^{-1}) \cdot \text{tr } C\rho_m(a^{-1}) \tag{115}$$

where da is Haar measure on $SU(2)$. But (115) vanishes by the Peter-Weyl theorem

4.8 *The composite map $\tilde{r}_Y \circ R$.* As a corollary of the proof we note that if $A, B \in \text{End } Z_n$ then

$$\xi(A,B) \in \Gamma^{-4}(X)_{(n,n,n,n)} \tag{116}$$

and by (110) and (115) one has, for any $W \in Y$,

$$\ell_{\xi(A,B)}(\mu_W) = (n+1) \cdot \operatorname{tr} A\rho_n(d)B^t\rho_n(c) \tag{117}$$

(since $C^t = \rho_n(d)B^t\rho_n(c)$ by *e.g.* (108)) where $c, d \in SU(2)$ are such that

$$W = \{e + cs(e)d | e \in E\} \tag{118}$$

Now the isomorphism $\hat{\tau}_Y$ (see (68)) of line bundles induces an isomorphism

$$\tilde{\tau}_Y : \Gamma^{-2}(Y) \longrightarrow \Gamma^{-2}(X)$$

of vector spaces which, by (70), as G-modules, satisfies,

$$\tilde{\tau}_Y(g \cdot \sigma) = \tau(g) \cdot \tilde{\tau}_Y(\sigma) \tag{119}$$

for any $g \in G$ and $\sigma \in \Gamma^{-2}(Y)$. But now if $\psi_1 \in \Gamma^{-2}(X)$ is that function which is the constant 1 on $S_E \times S_F$ then $\psi_1 = \sigma_X$ in the notation of (83) and (83) implies that

$$\tilde{\tau}_Y \sigma_V = \kappa \cdot \psi_1 \tag{120}$$

We now fix the constant κ so that

$$< v_1 \wedge v_2 \wedge v_3 \wedge v_4, \mu_V >= \kappa \tag{121}$$

Now consider the composite $(\tilde{\tau}_Y \circ R) : H' \longrightarrow \Gamma^{-2}(X)$ of the Radon transform R with the map $\tilde{\tau}_Y : \Gamma^{-2}(Y) \longrightarrow \Gamma^{-2}(X)$ induced by "triality". Since the representation $\rho_{(n,n,n,n)}$ of K is stable under the automorphism of K induced by τ it follows that the Fourier components of any f in the image of $\tilde{\tau}_Y \circ R$ lies in $\Gamma^{-2}(X)_{(n,n,n,n)}$ so that one has

$$(\tilde{\tau}_Y \circ R) : H' \longrightarrow H \tag{122}$$

and that for $h \in H'$ and $g \in G$ one has

$$(\tilde{\tau}_Y \circ R)(g \cdot h) = \tau(g) \cdot (\tilde{\tau}_Y \circ R)(h) \tag{123}$$

Lemma 20. *Let $v \in S_E \times S_F$ so that there exist $c, d \in SU(2)$ such that $v = c^{-1}e_1 + df_1$. Let $n \in \mathbf{Z}_+$ and let $A, B \in \operatorname{End} Z_n$. Let $\xi(A,B)$ be as in* (111) *so that by* (112) *we can regard $\xi(A,B)$ as an element of $H'_{(n,n,n,n)}$. Then*

$$\tilde{\tau}_Y \circ R)(\xi(A,B))(v) = (n+1) \cdot \operatorname{tr} A\rho_n(d)B^t\rho_n(c) \tag{124}$$

Proof. Let $\alpha(v) = x \in X$ and let $W \in Y$ be given by (118). Then

$\hat{\tau}_Y(\kappa^{-1}\cdot\sigma_Y(W)) = \sigma_X(x)$ by Proposition 14. *But* $<\kappa^{-1}\cdot\sigma_Y(W), \mu_W>= 1$ by (121). Thus $R\xi(A,B)(W) = (n+1)\cdot \text{tr } A\rho_n(d)B^t\rho_n(c)\cdot\kappa^{-1}\cdot\sigma_Y(W)$ by (117) and hence

$$\hat{\tau}_Y(R\xi(A,B)(W)) = (n+1)\cdot \text{tr } A\rho_n(d)B^t\rho_n(c)\cdot\sigma_X(x) \tag{125}$$

But (125) implies (124) since $\sigma_X = f_1$. QED

4.9 *The operator* $Q : H' \longrightarrow H$. Now (see § 7 in [Ko] noting that H in [Ko] is H_K here) we defined an involutory operator $\pi(\tau)$ on H_K in [Ko] such that $H_{(n,n,n,n)}$ is stable under $\pi(\tau)$ and (see (154) in [Ko])

$$\pi(\tau)\pi^\infty(u)\pi(\tau)^{-1} = \pi^\infty(\tau\cdot u) \tag{126}$$

on H_K for any $u\in U$. This implies that $\pi(\tau)\pi(a)\pi(\tau)^{-1} = \pi(\tau\cdot a)$ for any $a\in K$. It follows then that $\pi(\tau)$ is unitary on $H_{(n,n,n,n)}$ with respect to any K-invariant Hermitian structure. By an obvious continuity argument (upon restriction to $S_E\times S_F$) this implies that $\pi(\tau)$ extends to an operator on H and that in fact

$$\pi(\tau)\pi(g)\pi(\tau)^{-1} = \pi(\tau\cdot g) \tag{127}$$

on H for any $g\in G$. One obtains (127) from (126) on analytic elements of H and then on all of H by continuity.

Now by (123) and (127) the triple composite $\pi(\tau)\circ\tilde{\tau}_Y\circ R$ mapping H' to H is a G-map. But it is linear and not conjugate linear. However since G leaves the real elements of H stable one obtains a conjugate linear G-map

$$Q : H' \longrightarrow H \tag{128}$$

by defining for $h\in H'$

$$Qh = \pi(\tau)(\text{conj}((\tilde{\tau}_Y\circ R)h)) \tag{129}$$

Let $n\in \mathbf{Z}_+$ and let $A, B\in \text{End } Z_n$. Let $\eta(A,B)\in H_{(n,n,n,n)}$ be such that $\eta(A,B) = \xi(A,B)$ on $S_E\times S_F$. That is for $a, b\in SU(2)$

$$\eta(A,B)(ae_1 + bf_1) = (n+1)^2\cdot \text{tr } A\rho_n(a^{-1})\cdot \text{tr } B\rho_n(b^{-1}) \tag{130}$$

We have determined in [Ko] an explicit formula for the action of $\pi(\tau)$ on $H_{(n,n,n,n)}$. The formula essentially says that upon applying $\pi(\tau)$ to $\eta(A,B)$ the product of traces gets replaced by the trace of the product — almost. In fact (155) in [Ko] asserts that

$$(\pi(\tau)\eta(A,B))(ae_1 + bf_1) = (n+1)^2\cdot \text{tr } A^t\rho_n(a)B\rho_n(b^{-1}) \tag{131}$$

Remark 21. The last statement in the proof of (155) in [Ko] is somewhat unclear. In order to prove (155) we first observe that since both sides of (155) are bilinear in A and B there an analogous statement asserting that two linear maps of End $Z_n \otimes$ End Z_n are the same. On the other hand, using (81) here, one easily sees that the set of all decomposable elements $A \otimes B$ satisfying (155) is stable under the action of K. By the irreducibility of K on End $Z_n \otimes$ End Z_n it suffices therefore to show that (155) is true for a single non-trivial pair (A, B). But the argument in the proof of (155) in [Ko] certainly exhibits such a pair.

4.10 *An explicit formula for the G-invariant unitary structure.* We wish to explicitly compute Q on $H'_{(n,n,n,n)}$.

If $A \in$ End Z_n then one readily has $A^{t^*} = A^{*t}$. Let $\bar{A} = A^{t^*}$. One of course has tr $A^t =$ tr A and tr $A^* =$ conj tr A.

The following lemma is the main result as far as computing the unitary structure for the representation π of G.

Lemma 22. *Let $n \in \mathbb{Z}_+$ and let $A, B \in$ End Z_n. Then regarding $\xi(A,B)$ as an element of $H'_{(n,n,n,n)}$ one has*

$$Q\xi(A,B) = (n+1)^{-1} \cdot \eta(\bar{A}, \bar{B}) \tag{132}$$

Proof. Let $c, d \in SU(2)$. Then recalling (124) one has

$$(\text{conj}(\tilde{\gamma}_Y \circ R)(\xi(A,B)))(c^{-1}e_1 + df_1) = (n+1) \cdot \text{conj}(\text{tr } A\rho_n(d)B^t\rho_n(c))$$

But then upon Hermitian adjoints one has

$$\begin{aligned}(\text{conj}(\tilde{\gamma}_Y \circ R)(\xi(A,B)))(c^{-1}e_1 + df_1) &= (n+1) \cdot \text{tr } \rho_n(c^{-1})\bar{B}\rho_n(d^{-1})A^* \\ &= (n+1) \cdot \text{tr } A^*\rho_n(c^{-1})\bar{B}\rho_n(d^{-1}) \\ &= (n+1)^{-1} \cdot (\pi(\tau)\eta(\bar{A},\bar{B}))(c^{-1}e_1 + df_1)\end{aligned}$$

by (131). Thus

$$(\text{ conj}(\tilde{\gamma}_Y \circ R)(\xi(A,B))) = (n+1)^{-1} \cdot \pi(\tau)\eta(\bar{A},\bar{B})$$

The lemma then follows since $\pi(\tau)$ is involutory. QED

For any $h \in \Gamma^{-4}(X)$ let $h_o \in \Gamma^{-2}(X)$ be that element such that $h = h_o$ on $S_E \times S_F$. Then as a generalization of Lemma 22 one has

Lemma 23. *Let $n \in \mathbf{Z}_+$ and let $h \in \Gamma^{-4}(X)_{(n,n,n,n)}$. Then regarding h as an element of $H'_{(n,n,n,n)}$ one has*

$$Qh = (n+1)^{-1} \cdot \bar{h}_o \tag{133}$$

Proof. If $a \in SU(2)$ and $A \in \text{End}\, Z_n$ one has tr $\bar{A}\rho_n(a^{-1}) =$ tr $\rho_n(a)A^*$ since tr $B =$ tr B^t for any $B \in \text{End}\, Z_n$. But tr $\rho_n(a)A^* =$ conj tr $A\rho_n(a^{-1})$ since tr $B =$ conj tr B^*. Thus recalling (132), one has $\eta(\bar{A}, \bar{B}) =$ conj $\eta(A, B)$. B it $\eta(A, B) = \xi(A, B)_o$. Thus (133) follows by linearity. QED

It follows immediately from Lemma 23 that the map $Q : H' \longrightarrow H$ is bijective and hence there is an inverse $Q^{-1} : H \longrightarrow H'$.

Theorem 24. *One defines a G-invariant pre-Hilbert space structure $\{f, g\}$ on H by putting, for $f, g \in H$,*

$$\{f, g\} = < f, Q^{-1}g > \tag{134}$$

so that π extends to an irreducible unitary representation

$$\pi : G \longrightarrow U(\mathcal{H}) \tag{135}$$

where $\mathcal{H}$ is the Hilbert space completion of H. On K-types one explicitly has $\{f, g\} = 0$ if $f \in H_{(n,n,n,n)}$ and $g \in H_{(m,m,m,m)}$ where $n \neq m$. However if $n = m$ then

$$\{f, g\} = (n+1) \cdot \int_{S_E \times S_F} f\bar{g} \cdot \nu \tag{136}$$

where ν is the unique K-invariant measure on $S_E \times S_F$ of total measure 1.

Proof . For $f, g \in H$ let $\{f, g\}$ be the sesquilinear form defined by (134). It is clearly G-invariant. On the other hand if $h = Q^{-1}g$ then

$$< f, h >= \int_{S_E \times S_F} fh \cdot \nu$$

by (49). The result then follows from (133) and some obvious continuity properties of Q. QED

Remark 25. It is satisfying to note that the Hilbert space structure on H_K arrived at here from the Radon transform R and the maps $\hat{\tau}_Y$ and $\pi(\tau)$ (arising from "triality) agree with that given by Theorem 6.28 in [Ko]. The notation in the latter reference for the inner product is $< f|g >$.

5. The Vanishing of Scalar Curvature, Einstein's Equations and π

5.1 *The vanishing of scalar curvature and a remarkable property of 6-manifolds.* We return to the conformal structure $\mathcal{M}$ on X. We recall (see Proposition 1) that the cone $\Gamma^2(X)_+ \subset \Gamma^{-2}(X)$ bijectively corresponds to the cone $\mathcal{M}$ of metrics on X by the map $f \longrightarrow \gamma_f$. See (16) and (17). But now since $p = 3$ we have a G-submodule H of $\Gamma^{-2}(X)$ and consequently if we put $H_+ = H \cap \Gamma^{-2}(X)_+$ we get a G-stable subcone $\mathcal{M}_o$ of metrics on X defined by putting

$$\mathcal{M}_o = \{\gamma_f | f \in H_+\} \tag{137}$$

One of the key points of this paper is that we can characterize the metrics in $\mathcal{M}_o$ purely in terms of the machinery of Riemannian geometry.

Assume that M is any C^∞ pseudo-Riemannian manifold. Let g be the metric tensor. Now let g' be any C^∞ metric tensor on M which is conformally equivalent to g so that there exists a smooth positive function f on M such that

$$g' = fg \tag{138}$$

One refers to f as the conformal factor. If ξ, η are vector fields on M let (ξ, η) be the function on M defined by taking the inner product of ξ and η with respect to g. Since f is positive we may write $f = e^{2\sigma}$. Let ζ_σ be the vector field on M defined so that for any vector field ξ on M one has $\xi\sigma = (\xi, \zeta_\sigma)$. Let Δ be the Laplace-Beltrami operator on M with respect to g. It is straightforward to see that

$$\Delta f = (2\Delta\sigma - 4(\zeta_\sigma, \zeta_\sigma))f \tag{139}$$

Now let Scal g be the scalar curvature of M with respect to g and let Scal g' be defined similarly. Then if dim $M = n$ one readily computes that

$$\text{Scal } g' = e^{-2\sigma}(\text{Scal } g + 2(n-1)\Delta\sigma - (n-1)(n-2)(\zeta_\sigma, \zeta_\sigma)) \tag{140}$$

See *e.g* [Go], (3.9.4), p.115.

Multiplying both sides of (140) by $e^{4\sigma} = f^2$ one has, by (139),

$$f^2\text{Scal } \ g' = (n-1)\Delta f + f(\text{Scal } \ g - (n-1)(n-6)(\zeta_\sigma, \zeta_\sigma)) \tag{141}$$

But (141) clearly implies the following property of 6-dimensional manifolds.

Theorem 26. *Assume M is a 6-dimensional manifold. Let g and g' be pseudo-Riemannian metric tensors on M such that $g' = fg$ where f is a positive function on M — so that g and g' are conformally equivalent. If* Scal g *and* Scal g' *denote the scalar curvatures of M with respect to g*

and g' respectively and Scal $g = 0$ *one has* Scal $g' = 0$ *if and only if the conformal factor f is harmonic with respect to g. That is, in the notation above, if and only if $\Delta f = 0$.*

Remark 27. Note that (141) also implies, conversely, that if dim $M > 1$ and the statement of Theorem 26 is true for all conformally equivalent metrics then one necessarily has dim $M = 6$.

5.2 *A view of the metrics in the conformal class $\mathcal{M}$.* Now with the assumption (45) let $u, y \in \mathbf{R}^8$ be such that $(u,u) = (y,y) = 0$ and $(u,y) = 1/2$. Let Z be the orthocomplement $\mathbf{R}u + \mathbf{R}y$ in $\mathbf{R}^8$ so that Z has a basis e_i, $i = 1, \ldots, 6$, where $(e_i, e_j) = 0$ if $i \neq j$ and $(e_i, e_i) = 1$ if $i \leq 3$ and $(e_i, e_i) = -1$ if $i > 3$. Now any $v \in \mathbf{R}^8$ may be uniquely written

$$v = au - by + z \tag{142}$$

for $a = a(v), b = b(v) \in \mathbf{R}$ and $z = z(v) \in Z$. Let U_o be the open set in $\mathbf{R}^8$ defined by putting $U_o = \{v \in \mathbf{R}^8 | a(v) \neq 0\}$. Let $C_o^* = U_o \cap C^*$ so that C_o^* is clearly, using the notation (142) the open subcone of C^* defined by

$$C_o^* = \{v \in \mathbf{R}^8 | ab = (z,z),\, a \neq 0\} \tag{143}$$

Recalling (7) let $X_o = \alpha(C_o^*)$ so that X_o is open in X. Let $\beta : U_o \longrightarrow C_o^*$ be the map defined by putting

$$\beta(v) = au - (z,z)/a \cdot y + z \tag{144}$$

using the notation of (142) and put $\delta = \alpha \circ \beta$. Then we note that there exists a unique coordinate system u_i, $i = 1, \ldots, 6$, on X_o such that if $\tilde{u}_i = u_i \circ \delta$ then for any $v \in U_o$ one has, using the notation of (142)

$$z(v) = a \cdot \sum_{i=1}^{6} \tilde{u}_i(v) e_i \tag{145}$$

Now let $Z_o = \{v \in C^* | a(v) = 1\}$ so that $\alpha | Z_o$ is a diffeomorphism mapping Z_o onto X_o. Let $\alpha_o : X_o \longrightarrow Z_o$ be the inverse of $\alpha | Z_o$.

Recalling the notation of Proposition 2 we can be very explicit about the restriction to X_o of the metric tensor γ_f defined by any $f \in \Gamma^{-2}(X)_+$.

Proposition 28. *Let $f \in \Gamma^{-2}(X)_+$. Let f_o be the function on X_o defined by putting $f_o = f \circ \alpha_o$. Then on X_o one has*

$$\gamma_f = f_o \cdot (\sum_{i=1}^{3} du_i^2 - \sum_{j=4}^{6} du_j^2) \tag{146}$$

Proof. Let $x \in X_o$ and put $v = \alpha_o(x) \in Z_o$. Then using the notation of (142),(145) and Proposition 2 we may find $\xi_i \in T_x(X)$, $i = 1, \ldots, 6$, such that, mod $\mathbf{R}v$,

$$\xi_i(v) = -2\rho_i \tilde{u}_i(v)y + e_i \in v^\perp \tag{147}$$

where $\rho_i = 1$ if $i \leq 3$ and $\rho_i = -1$ for $i > 3$. But then by (16)

$$\gamma_f(\xi_i, \xi_j) = f_o(x)\rho_i\delta_{ij} \tag{148}$$

On the other hand if we identify $\mathbf{R}^8$ with the tangent space $T_v(\mathbf{R}^8)$ then clearly $e_i' = -2\rho_i\tilde{u}_i(v)y + e_i$ is tangent to Z_o at v and

$$\delta_*(e_i') = \xi_i \tag{149}$$

But since in (145) $a = 1$ on Z_o one clearly has, by (145), $e_i'\tilde{u}_j = \delta_{ij}$. Thus

$$\xi_i = \partial/\partial u_i \tag{150}$$

But this together with (148) implies (146). QED

5.3 *The cone H_+ and the G-invariance of the vanishing of scalar curvature.* Now let D_u, D_y and for $i = 1, \ldots, 6$, let D_i be the translation invariant vector fields on $\mathbf{R}^8$ corresponding respectively to u, y, and to e_i. Then using the notation of (147) and (23.5) we note that

$$\Delta = \sum_{i=1}^{6} \rho_i D_i^2 + 4D_u D_y$$

Now let $f \in \Gamma^{-2}(X)$. By continuity it is obvious that if $\Delta_X f = h \in \Gamma^{-4}(X)$ then $h = 0$ if and only if $h|X_o = 0$. Let f' be the function on U_o defined by putting $f' = f \circ \beta$. Clearly f' is an extension of $f|C_o^*$ and $f'(rw) = r^{-2}f'(w)$ for $r \in \mathbf{R}^*$ and $w \in U_o$. It follows from the argument of §1.4 that $h|X_o$ is given by $\Delta f'|C_o^*$. But by the definition of β one has

$$D_y f' = 0 \tag{151}$$

Thus

$$\Delta f' = \sum_{i=1}^{6} \rho_i D_i^2 f' \tag{152}$$

Now let $\Delta_o = \sum_{i=1}^{6} \rho_i(\partial/\partial u_i)^2$ so that $-\Delta_o$ is the Laplace-Beltrami operator on X_o with respect to the metric $\sum_{i=1}^{6} \rho_i du_i^2$.

Lemma 29. *Let $f \in \Gamma^{-2}(X)$ and let f_o be the function on X_o defined as in* (146)*. Then $\Delta_X f = 0$ if and only if $\Delta_o f_o = 0$. That is if and only if f_o is harmonic with respect to the metric $g_o = \sum_{i=1}^{6} \rho_i du_i^2$.*

Proof. Let ∂_i be the vector field on Z_o such that $\alpha_* \partial_i = \partial/\partial u_i$. But then by (149) and (150) it follows that upon replacing v in (147) by any point v' in Z_o that at v' one has $\partial_i = D_i + cD_y$ for some scalar c, depending on v'. Thus $D_i f' = \partial_i f$ on Z_o by (151). But also $D_y D_i f' = 0$. Hence $D_i^2 f' = \partial_i^2 f$. By (152) this implies

$$\Delta f' | Z_o = \sum_{i=1}^{6} \rho_i \partial_i^2 f \tag{153}$$

But $\Delta_X f = 0$ if and only if the left side of (153) $= 0$. But by (153) one has $\Delta f' \circ \alpha_o = \Delta_o f_o$. QED

We can now characterize geometrically the set of metrics $\mathcal{M}_o$ which are in the G-submodule. The condition is just the vanishing of scalar curvature.

Theorem 30. *Let $\mathcal{M}$ be the set of pseudo-Riemannian metrics in the G-invariant natural conformal class (signature $+++---$) on the 6-dimensional projective variety X— so that (see* Proposition 1*) any such metric is uniquely of the form γ_f for $f \in \Gamma^{-2}(X)_+$. Let $H \subset \Gamma^{-2}(X)$ be the G-submodule defined as in (45.5) and let $\mathcal{M}_o = \{\gamma_f | f \in H_+\}$ where $H_+ = H \cap \Gamma^{-2}(X)_+$. Then*

$$\mathcal{M}_o = \{\gamma_f \in \mathcal{M} |\ Scal\ \ \gamma_f = 0\} \tag{154}$$

In particular the set of all metrics in the conformal class $\mathcal{M}$ which have vanishing scalar curvature is stable under the conformal group G.

Proof. Clearly, in the notation of Lemma 29, the metric g_o on X_o has vanishing scalar curvature. Indeed the curvature tensor for g_o vanishes. But now if $f \in \Gamma^{-2}(X)_+$ and Scal $\gamma_f = 0$ then by Proposition 28 and Theorem 26 one has $\Delta_o f_o = 0$. But this implies $\Delta_X f = 0$ by Lemma 29. That is $f \in H$. Conversely if $f \in H$ then $\Delta_o f_o = 0$ by Lemma 29. But then Scal $\gamma_f = 0$ by Proposition 28 and Theorem 26. QED

5.4 *The vanishing of scalar curvature on X and Einstein's equations on compactified Minkowski space.* The existence of a special family of metrics on the 6-dimensional manifold X raises the question as to whether there is any application in physics for these metrics. Two related structures which do have strong connections with physical theory come to mind.

(1) The group $SO(4,4)$ contains $SO(2)\times$ the "conformal group" $SO(2,4)$ as a maximal subgroup and correspondingly (2) X contains compactified Minkowski space as a submanifold.

In this section we will present a result from the doctoral thesis, [B], of Witold Biedrzycki which relates Einstein's field equations in general relativity with the metrics in $\mathcal{M}_o$.

Assume that M is a 4-dimensional manifold and assume that h is a Lorentzian metric tensor on M (*i.e.* a pseudo-Riemannian metric of signature $+---$). Let $R = \text{Scal } h$ be the scalar curvature on M with respect to h. Then using classical coordinate notation for tensor fields one has a space-time model in the sense of general relativity in case Einstein's field equations

$$R_{ij} - R/2 \cdot h_{ij} = 8\pi T_{ij} \tag{155}$$

are satisfied where R_{ij} is the Ricci tensor for the metric h and T_{ij} is the energy-momentum tensor for certain matter fields. There is of course a vast literature on these equations. One rather delicate point is how T_{ij} arises in each of the various examples of matter fields (*e.g.* an electromagnetic field, a charged scalar field, a perfect fluid, ...). See *e.g.* $[H-E]$, Chapter 3, for a discussion of the variational machinery which gives rise to the energy-momentum tensor. A mathematical elaboration of some of the ideas in [H-E] can be found in [Ko 3].

One of the simplest instances of a matter field which gives rise to an energy-momentum tensor is the case of an electromagnetic field or more simply a closed 2-form ω on M. But now if M happened to be a submanifold of our 6-dimensional manifold X and, say, $h = g|M$ where $g \in \mathcal{M} = \Gamma^{-2}(X)_+$ then besides the metric tensor M also inherits a closed form ω from the embedding

$$M \longrightarrow X \tag{156}$$

Indeed if N is the normal 2-plane bundle of M in X then $g|N$ defines the structure of a complex line bundle L on M and that furthermore g defines a connection in L and hence ω would arise from the curvature of L.

In the case of a perfect fluid scalar functions on M enter into the construction of T_{ij}. On the other hand an embedding $M \longrightarrow X$ gives rise to scalar functions on M in a number of ways. The sectional curvature of N with respect to g is one such way. Since in fact there exists special metrics g on X — namely those in $\mathcal{M}_o$(*i.e.* those in the Hilbert space of our representation π) – *the* following daring but far-out possibility suggested itself to us; namely, *matter fields on space-time M should arise from an embedding $M \longrightarrow X$ where the metric g on X is an element in $\mathcal{M}_o$. In particular one has* Scal $g = 0$. For this to be the case one would have to construct T_{ij} from g so that (155) is satisfied in case $h = g|M$. In a special case having to do with compactified Minkowski space Witold Biedrzycki has constructed such a tensor T_{ij}.

Biedrzycki's tensor T_{ij} is made up of the sectional curvature of N together with components of the second fundamental form associated to the embedding (156).

Let $g \in \mathcal{M}$ and assume M is a 4-dimensional submanifold of X such that $h = g|M$ is a Lorentzian metric on M. Let R^i_{jkl} be the curvature tensor on X with respect to g. If v is a tangent vector on X let ∇_v denote covariant differentiation by v with respect (Levi-Civita connection) to g. Thus if ∂_i are the local vector fields on X given by the coordinate system and $\nabla_i = \nabla_{\partial_i}$ then with the usual summation notation

$$(\nabla_k \nabla_l - \nabla_l \nabla_k)\partial_j = R^i_{jkl} \cdot \partial_i$$

With the usual notation for raising and lowering indices the sectional curvature $K(\Pi)$ on 2-dimensional subspaces Π of tangent spaces of X where $g|\Pi$ is definite is such that if $g|\Pi_{ij}$ is definite on the planes Π_{ij} spanned by ∂_i and ∂_j, $i \neq j$, then

$$K(\Pi_{ij}) = R_{ijij}/(g_{ii}g_{jj} - g_{ij}^2).$$

Let K be the scalar function on M defined by putting for any $p \in M$

$$K(p) = K(N_p) \tag{157}$$

A field S of tangent space endomorphisms is a tensor field of type S^i_j. The curvature tensor can be regarded as a map $S \longrightarrow R(S)$ from the space of such tensor fields to the space of covariant tensor fields U of type U_{kl}(*i.e.* fields $U(\ ,\ )$ of tangent space bilinear forms where $U(\partial_k, \partial_l) = U_{kl}$). The map in coordinate notation is given by

$$R(S)_{kl} = R^j_{k\,li} S^i_j \tag{158}$$

It is easy to see that if S is symmetric with respect to g then $R(S)$ is a field of symmetric bilinear forms. Note also that if S is the field of identity operators I then $R(I)$ is the Ricci tensor.

Next one defines a distinguished cross-section H of the normal bundle N as follows. Let $p \in M$ and let $x \in N_p$. One defines a symmetric operator C_x on $T_p(M)$, using inner products, by the relation

$$(C_x u, v) = (x, \nabla^X_u \eta) \tag{159}$$

where $u, v \in T_p(M)$ and where η is any vector field on X whose value at p is v. Then there exists a unique vector $H_p \in N_p$ such that

$$(x, H_p) = \text{tr } C_x \tag{160}$$

One then defines a scalar function L on M by putting

$$L(p) = (H_p, H_p) \tag{161}$$

Then Biedrzycki's tensor T on M as a symmetric bilinear form on $T_p(M)$ at any $p \in M$ is given by

$$1/8\pi \cdot T(u,v) = R(P_p)(u,v) + (K(p) - 3/16 \cdot L(p)) \cdot (u,v) \tag{162}$$

where $u, v \in T_p(M)$ and $P_p : T_p(X) \longrightarrow N_p$ is the orthogonal projection.

Now recalling (see (8)) the coordinate system r_i on $\mathbf{R}^8$ let

$$G_1 = \{g \in G | g \cdot r_1 = r_1,\ g \cdot r_2 = r_2\}$$

so that the covering map (39) induces a covering

$$G_1 \longrightarrow SO(2,4)_e \tag{163}$$

of the "conformal group". Furthermore with regard to the action of G_1 on X one notes there is exactly one closed orbit M, that M is 4-dimensional and the covering map (46) induces a double covering

$$S^1 \times S^3 \longrightarrow M \tag{164}$$

Moreover M may be identified with the usual compactification of Minkowski space and if $h = g|M$ where $g \in \mathcal{M}$ then h is a Lorentzian metric in the conformal class on M for which $SO(2,4)$ is the proper conformal group. For this choice of M one has

Theorem 31. [Witold Biedrzycki]. *Let $g \in \mathcal{M}$ and let h be the Lorentzian metric on compactified Minkowski space $M \subset X$ given by the restriction $h = g|M$. Let T be the tensor field on M defined by* (162). *Then Einstein's equations*

$$R_{ij} - R/2 \cdot h_{ij} = 8\pi T_{ij} \tag{165}$$

are satisfied on M if and only if the scalar curvature Scal g *vanishes on M. In particular one has* (165) *if $g \in \mathcal{M}_o$. In fact the Einstein equation* (165) *is satified for all G transforms of g if and only if $g \in \mathcal{M}_o$.*

The proof of Theorem 31 will be published by Biedrzycki elsewhere.

REFERENCES

[B] Biedrzycki, W., MIT Doctoral Thesis in preparation.

[G] Garfinkle, D., *A New Construction of the Joseph Ideal*, MIT Doctoral Thesis, 1982.

[Go] Goldberg, S., *Curvature and Homology*, Academic Press, New York and London, 1962.

[J] Joseph, A., *The Minimal Orbit in a Simple Lie Algebra and its Associated Maximal Ideal*, Ann. Scient. Ec. Norm. Sup. (4) **9** (1976), 1–30.

[H-E] Hawking, S. W., and Ellis, G. F. R., *The Large Scale Structure of Space-Time*, Cambridge Monographs on Math. Phys., Cambridge Univ. Press, 1973.

[Ka] Kazhdan, D., *The Minimal Representation of D_4*, Submitted to the Proceedings of the Colloque en l'Honneur de Jacques Dixmier, mai 1989.

[Ko 1] Kostant, B. *Lie Group Representations on Polynomial Rings*, Am. J. Math., **85** (1963), 327–404.

[Ko 2] Kostant, B. *Eigenvalues of a Laplacian and Commutative Lie Subalgebras*, Topology, **13** (1965), 147–159.

[Ko 3] Kostant, B., *A Course in the Mathematics of General Relativity*, Ark Publications Inc., Newton, 1988.

[Ko] Kostant, B., *The Principle of Triality and a Distinguished Representation of $SO(4,4)$*, Differential Geometrical Methods in Theoretical Physics, edited by K. Bleuler and M. Werner, Series C: Math. and Phy. Sci. **250** Kluwer Acad. Pub. (1988), 65–109.

Received January 25, 1990

Department of Mathematics
MIT
Cambridge, MA 02139

The Minimal Representation of D_4

DAVID KAZHDAN[1]

Dedicated to Jacques Dixmier on his 65th birthday

Introduction

In this paper we study the minimal representation of quasi-split groups of type D_4 over a local field, F. We give an explicit realization of such a representation and show that it can be used to prove an explicit realization of the lifting from multiplicative characters of a cubic extension E of F to irreducible representations of $\mathrm{GL}_3(F)$.

I want to express my gratitude to Joseph Bernstein for a number of very useful and clarifying discussions and to the referee of the paper who did a marvelous job.

1. Let $\underline{G}$ be an adjoint semisimple quasisplit group over a local field F of characteristic zero, $\underline{B} = \underline{T}\,\underline{U}$ a Borel subgroup of $\underline{G}$, $X(\underline{T}) \stackrel{\text{def}}{=} \mathrm{Hom}(\underline{T}, \underline{G}_m)$ the lattice of weights, $\Phi \subset X(\underline{T})$ the subset of roots and Φ^+ the positive roots and $\Sigma \subset \Phi^+$ the subset of simple roots. We will write $G, B, \dots$, instead of $\underline{G}(F), \underline{B}(F), \dots$. For simplicity we assume that Φ is a reduced root system – that is, for any $\alpha \in \Phi$, $2\alpha \notin \Phi$. Then for any $\alpha \in \Phi^+$ there exists a finite extension F_α of F and an algebraic group homomorphism $j_\alpha : SL_2(F_\alpha) \to G$ such that $j_\alpha\begin{pmatrix} c & 0 \\ 0 & c^{-1} \end{pmatrix} \subset T \; \forall c \in F_\alpha^*$ and j_α defines an isomorphism of the subgroup $\left\{\begin{pmatrix} 1 & * \\ 0 & 1 \end{pmatrix}\right\}$ of $SL_2(F_\alpha)$ with the subgroup U_α of U where

$$U_\alpha = \{u \in U | ln(tut^{-1}) = \alpha(t) ln\ u \quad \text{for all } t \in T).$$

The homomorphism j_α is uniquely defined up to a conjugation by an element $t \in T$. We fix j_α for all $\alpha \in \Sigma$ and for any $a \in F$, $\alpha \in \Sigma$ define $E_\alpha(a) \stackrel{\text{def}}{=} j_\alpha\begin{pmatrix} 1 & a \\ 0 & 1 \end{pmatrix}$, $E_{-\alpha}(a) \stackrel{\text{def}}{=} j_\alpha\begin{pmatrix} 1 & 0 \\ a & 1 \end{pmatrix}$, and $s_\alpha(a) \stackrel{\text{def}}{=} j_\alpha\begin{pmatrix} 0 & 1 \\ -1 & 0 \end{pmatrix}$. As is well known (see [T1]), $\{s_\alpha\}$ $\alpha \in \Sigma_0$ satisfy the braid relations: for any $\alpha, \beta \in \Sigma$, $\alpha \neq \beta$, we have

$$(*_{\alpha,\beta}) \qquad \underbrace{s_\alpha s_\beta \dots}_{n_{\alpha,\beta}} = \underbrace{s_\beta s_\alpha \dots}_{n_{\alpha,\beta}}$$

[1] Partially supported by an NSF Grant.

where

$$n_{\alpha,\beta} = 2 \quad \text{if } \langle\alpha,\beta\rangle = 0$$
$$= 3 \quad \text{if } \langle\alpha,\beta\rangle = -1$$
$$= 4 \quad \text{if } \langle\alpha,\beta\rangle = -2$$
$$= 6 \quad \text{if } \langle\alpha,\beta\rangle = -3.$$

For any $\alpha \in \Sigma$, we denote by $P_\alpha = L_\alpha \cdot U_\alpha$ the minimal parabolic subgroup corresponding to α and define $B_\alpha \stackrel{\text{def}}{=} T \cdot U_\alpha \subset B$. It is clear that s_α normalizes B_α for all $\alpha \in \Sigma$.

Let $\pi : G \to Aut\ W$ be an irreducible unitary representation. We say that π is small if for any $\alpha \in \Sigma$ the restriction τ_α of π on B_α is irreducible.

LEMMA 1. *Let π_1, π_2 be small representations of G such that the restrictions of π_1 and π_2 on B are equivalent. Then π_1 and π_2 are equivalent.*

PROOF: Since the restriction of π_1, π_2 on B are equivalent we may assume that there exists a unitary representation $\tau : B \to Aut\ W$ and small representations $\pi_1, \pi_2 : G \to Aut\ W$ such that for any $b \subset B$, $\pi_1(b) = \pi_2(b) = \tau(b)$. We want to show that $\pi_1 = \pi_2$. Fix $\alpha \in \Sigma$ and consider operator $A_i \stackrel{\text{def}}{=} \pi_i(s_\alpha) \in Aut\ W$, $i = 1,2$. Since s_α normalizes B_α we have $A_i\tau(b)A_i^{-1} = \tau(s_\alpha b s_\alpha^{-1})$ for all $b \in B_\alpha$, $i = 1,2$. Since representations π_1, π_2 are small, the restriction of τ on B_α is irreducible and we conclude that $A_1 = cA_2$ for $c \in \mathbb{C}^*$. Since $s_\alpha^2 \in T \subset B_\alpha$ we have $A_1^2 = A_2^2 = \tau(s_\alpha^2)$. Therefore $c = \pm 1$. To prove that $c = 1$ we use the equality $\left(\begin{pmatrix} 0 & 1 \\ 1 & 0 \end{pmatrix}\begin{pmatrix} 1 & 1 \\ 0 & 1 \end{pmatrix}\right)^3 = Id$. Therefore $(A_1\tau(E_\alpha(1)))^3 = (A_2\tau(E_\alpha(1)))^3 = Id$. Therefore $c = 1$. So we see that for any $\alpha \in \Sigma$, $\pi_1(s_\alpha) = \pi_2(s_\alpha)$. Since B and $\{s_\alpha\}$, $\alpha \in \Sigma$ generate G we see that $\pi_1 \equiv \pi_2$. Lemma 1 is proved.

We can reformulate Lemma 1 in the following way: given an irreducible unitary representation of B such that the restriction of τ on B_α is irreducible for all $\alpha \in \Sigma$ there exists at most one way to extend τ to a representation of G on W. It is natural then to ask: given an irreducible unitary representation τ of B when is it possible to extend it to a representation of G. For any representation τ of B we denote by $\tau_\alpha, \tilde{\tau}_\alpha$ the representations of B_α given by $\tau_\alpha(b) = \tau(b), \tilde{\tau}_\alpha(b) = \tau(s_\alpha b s_\alpha^{-1})$ for $b \in B_\alpha$.

LEMMA 2. *Let $\tau : B \to Aut\ W$ be a unitary representation such that*

a) the representation $\tau_\alpha, \tilde{\tau}_\alpha : B_\alpha \to Aut\ W$ are irreducible and equivalent.

b) For any $\alpha \in \Sigma$ there exists $A_\alpha \in Aut\ W$ such that $A_\alpha\tau_\alpha(b)A_\alpha^{-1} = \tilde{\tau}_\alpha(b)$, $b \in B_\alpha$, $A_\alpha^2 = \tau(W_\alpha^2)$, and for any $a \in F^*$,

$$A_\alpha\tau(E_\alpha(a))A_\alpha = \tau(j_\alpha\left\{\begin{pmatrix} -a^{-1} & 0 \\ 0 & -a \end{pmatrix} E_\alpha(-a)\right\})A_\alpha\tau(E_\alpha(-a^{-1})).$$

c) For any $\alpha, \beta \in \Sigma, \alpha \neq \beta$ we have

$$(*_{\alpha,\beta}) \qquad \underbrace{A_\alpha A_\beta \ldots}_{n_{\alpha,\beta}} = \underbrace{A_\beta A_\alpha \ldots}_{n_{\alpha,\beta}}$$

Then there exists a representation $\pi : G \to Aut\ W$ such that $\pi(b) = \tau(b)$ for $b \in B \subset G$.

PROOF: For any field E let $H \subset SL_2(E)$ the subgroup of diagonal matrices, N be the normalizer of H in $SL_2(E)$ and $Q \supset H$ be the subgroup of upper-triangular matrices. Let $\widehat{S}$ be the free product of Q and N with H as an amalgamated subgroup. We have a natural map $\widehat{S} \to SL_2(E)$. As is well known (see [JL], p. 7), the kernel of this map is generated as a normal subgroup by elements $(su_a s)^{-1} h_a u_{-a} s u_{-a^{-1}}$, $a \in F^*$, where $s = \begin{pmatrix} 0 & 1 \\ -1 & 0 \end{pmatrix} \in H$ and $u_a = \begin{pmatrix} 1 & a \\ 0 & 1 \end{pmatrix}, h_a = \begin{pmatrix} -a^{-1} & 0 \\ 0 & -a \end{pmatrix} \in Q$. Therefore there exists a representation $\sigma_\alpha : SL_2(F_\alpha) \to Aut\ W$ such that $\sigma_\alpha \begin{pmatrix} 1 & t \\ 0 & 0 \end{pmatrix} = \tau(E_\alpha(t))$, $\sigma_\alpha \begin{pmatrix} 0 & 1 \\ -1 & 0 \end{pmatrix} = A_\alpha$. It is clear that we can extend σ_α to a representation $\sigma_\alpha : P_\alpha \to Aut\ W$. Now Lemma 2 follows from Proposition 13.3 in [T2], which says that G is a quotient of the free product $\widehat{G}$ of groups P_α, amalgamated by B by a normal subgroup in $\widehat{G}$ generated by relations $(*)_{\alpha,\beta}$, $\alpha, \beta \in \Sigma$, $\alpha \neq \beta$.

How to construct representations τ of B satisfying the conditions of Lemma 2?

We do not have a general answer but we can propose a "quasiclassical" picture. Let $\mathcal{L}_G, \mathcal{L}_B, \ldots$ be Lie algebras of $G, B, \ldots$ and $\mathcal{L}'_G, \mathcal{L}'_B, \ldots$ be the dual spaces. The imbeddings $B_\alpha \hookrightarrow G$ induce projections $p_\alpha : \mathcal{L}'_G \to \mathcal{L}'_{B_\alpha}$. Given a coadjoint G-orbit $\Omega \subset \mathcal{L}'_G$ we say that Ω is small if there exists a B_α-orbit $\Omega_\alpha \subset \mathcal{L}'_{B_\alpha}$ such that $p_\alpha(\Omega)$ lies in the closure of Ω_α and the restriction of p_α on $\Omega \cap p_\alpha^{-1}(\Omega_\alpha)$ is an isomorphism $\Omega \cap p_\alpha^{-1}(\Omega_\alpha) \xrightarrow{\sim} \Omega_\alpha$. Let q, q_α be the projections $q : \mathcal{L}'_G \to \mathcal{L}'_U$, $q_\alpha : \mathcal{L}'_{B_\alpha} \to \mathcal{L}'_U$. Let $\bar{\Omega} \subset \mathcal{L}'_U$ be any U-orbit in the image $q_\alpha(\Omega_\alpha) \subset \mathcal{L}'_U$. It is clear $q(\Omega) \supset Ad(T)\overline{\Omega}$ and $Ad(T)\overline{\Omega}$ is open and dense in $q(\Omega)$. Let $T_{\overline{\Omega}} \subset T$ be the normalizer of $\overline{\Omega}$ in T and $\rho_{\overline{\Omega}} : U \to Aut\ R$ be the unitary irreducible representation of U corresponding to $\overline{\Omega}$ by Kirillov's theory. As is well known (see [H]), we can extend $\rho_{\overline{\Omega}}$ to a representation $\widetilde{\rho}_{\overline{\Omega}} : T_{\overline{\Omega}} \cdot U \longrightarrow Aut\ R$ which is unique up to a multiplication by a character of $T_{\overline{\Omega}}$. Let $\widetilde{\tau}$ be the unitary induced representation of B, $\widetilde{\tau} = Ind^B_{T_{\overline{\Omega}}U} \widetilde{\rho}_{\overline{\Omega}}$.

CONJECTURE: All representations τ of B satisfying the conditions of Lemma 2 can be constructed in this way.

The case when G is a split simply laced group and Ω is the smallest nonzero coadjoint orbit is analyzed in a joint work with G. Savin. In this paper we consider the case when $G = G^E$ is a quasisplit group of type D_4 corresponding to an arbitrary cubic extension E of F.

Let F be a field of characteristic zero. From now on we restrict our attention to the case when $\underline{G}$ is a quasisplit adjoint simple F-group of type D_4. Since the group of outer automorphisms of split adjoint group $\underline{G}^0$ of type D_4 is isomorphic to the symmetric group S_3 the adjoint quasisplit groups $\underline{G}$ of type D_4 are parametrized by commutative semisimple 3-dimensional F-algebras E. To describe this correspondence we fix a Borel subgroup $\underline{B}^0 \subset \underline{G}^0$ a maximal torus $\underline{T}^0 \subset \underline{B}^0$ and imbedding $S_3 \hookrightarrow Aut\ \underline{G}^0$ such that S_3 preserves $\underline{B}^0, \underline{T}^0$ and $Aut\ \underline{G}^0 = S_3 \ltimes \underline{G}^0$ Let $\overline{F}$ be the algebraic closure of F.

For any 3-dimensional commutative semisimple algebra E over F there exists an algebraic isomorphism $E \otimes \overline{F} \xrightarrow{\sim} \overline{F}^3$. Since $Aut(\overline{F}^3) \approx S_3$ we see that E defines a homomorphism $\tau_E : \mathcal{G} \to S_3$ uniquely up to a conjugation where $\mathcal{G} \overset{\text{def}}{=} Gal(\overline{F} : F)$. Since we have fixed an imbedding $S_3 \to Aut\ \underline{G}^0$ we can consider τ_E as a homomorphism $\tau_E : \mathcal{G} \to Aut\ \underline{G}^0$. We consider τ_E as an element of $H^1(\overline{F}/F, Aut\ \underline{G}^0)$ and denote by $\underline{G}^E$ the form of $\underline{G}^0$ coresponding to τ_E (see [S], III.5).

It is easy to see that the map $E \to \underline{G}^E$ is well defined and provides a parametrization of adjoint quasisplit F-groups of type D_4. We denote by $\underline{B}^E, \underline{T}^E$ the Borel subgroup and a maximal torus of $\underline{G}^E$ such that $\underline{B}^E(\overline{F}) = \underline{B}^0(\overline{F})$ and $\underline{T}^E(\overline{F}) = \underline{T}^0(\overline{F})$.

To describe the group G^E of F-points of $\underline{G}^E$ we consider the action of $\mathcal{G}$ on $\underline{G}^0(\overline{F})$ given by $\gamma \circ g \overset{\text{def}}{=} \tau_E(\gamma)(g^\gamma)$ where $g \to g^\gamma$ is defined by the action of $\mathcal{G}$ on $\overline{F}$. It is easy to see that $G^E = \{g \in \underline{G}^0(\overline{F}) | \gamma \circ g = g \text{ for all } \gamma \in \mathcal{G}\}$. The Dynkin diagram of $\underline{G}^0$ has the form

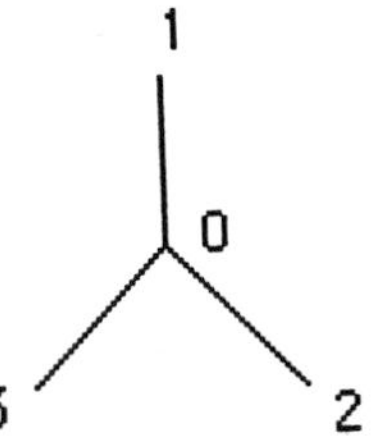

We denote by α_i, $1 \leq i \leq 3$, β_0 the corresponding simple roots $\alpha_i, \beta_0 : \underline{T}^0 \to \underline{G}_m$ and define $\alpha \overset{\text{def}}{=} \alpha_1 + \alpha_2 + \alpha_3$. We consider β_0, α as characters of B^0 trivial on the maximal uniponent radical U^0 of B^0. Since the characters β_0, α are $\mathcal{G}$-invariant they define characters $\beta_0, \alpha : \underline{B}^E \longrightarrow \underline{G}_m$. It is clear that $\beta_0, \gamma_0 \overset{\text{def}}{=} \beta_0 + \alpha$ and $\omega \overset{\text{def}}{=} \gamma_0 + \beta$ are roots of (G^E, B^E, T^E).

Let $\underline{\widetilde{P}}^0, \underline{P}^0 \subset \underline{G}^0$ be the parabolic subgroups containing $\underline{B}^0$ and corresponding to subsets $\{\beta_0\}$ and $\{\alpha_1, \alpha_2, \alpha_3\}$ of the set of simple roots of $(\underline{G}^0, \underline{B}^0, \underline{T}^0)$ correspondingly. Since those subsets are S_3-invariant $\underline{\widetilde{P}}^0, \underline{P}^0$

then define parabolic subgroups $\underline{\widetilde{P}}^E$, $\underline{P}^E$ of $\underline{G}^E$.

Let $\underline{P}^E = \underline{L}^E \ltimes \underline{H}$, $\underline{\widetilde{P}}^E = \underline{\widetilde{L}}^E \ltimes \underline{\widetilde{H}}$ be the Levi decompositions such that $\underline{T}^E = \underline{L}^E \cap \underline{\widetilde{L}}^E$, $\underline{L}^0 \subset \underline{L}^E$, $\underline{\widetilde{L}}^0 \subset \underline{\widetilde{L}}^E$ be the commutator subgroups. It is easy to see that there exist isomorphisms $j : SL_2(E) \xrightarrow{\sim} L^0, \tilde{j} : SL_2 \xrightarrow{\sim} \widetilde{L}^0$ which map diagonal subgroups inside T^E and upper triangular subgroups $C \subset SL_2(E)$, $\widetilde{C} \subset SL_2$ inside B^E. We define $s \stackrel{\text{def}}{=} j\begin{pmatrix} 0 & 1 \\ -1 & 0 \end{pmatrix} \subset P^E$, $\tilde{s} \stackrel{\text{def}}{=} \tilde{j}\begin{pmatrix} 0 & 1 \\ -1 & 0 \end{pmatrix} \subset \widetilde{P}^E$. As is well known $(\tilde{s}s)^3 = (s\tilde{s})^3$.

We will need the following description of the group G^E. Let $\widehat{G}^E$ be the free product of P^E and $\widetilde{P}^E$ with the amalgamated subgroup B^E. We have a natural surjection $\epsilon : \widehat{G}^E \longrightarrow G^E$.

LEMMA 3. ker ϵ *is generated as a normal subgroup by the element* $(s\tilde{s})^3(\tilde{s}s)^{-3}$.

PROOF: Follows from [T2], 13.1 and the standard description of the Weyl group of G^E in terms of generators and relations ([T2], 2.14).

For any (i,j), $1 \leq i \neq j \leq 3$ we denote by E_i, the 3×3 matrix with 1 in the place (i,j) and 0 in all other places and for any $a \in F$ define $e_{ij}(a) = Id + aE_{ij} \in SL_3(F)$.

LEMMA 4. a) *There exists a group imbedding* $i : \underline{SL}_3 \hookrightarrow \underline{G}^E$ *such that*

$$i(e_{12}(a)) = E_{\beta_0}(a), \quad i(e_{23}(a)) = E_{\gamma_0}(a), \quad i(e_{13}(a)) = E_{\omega}(a)$$

$$i(e_{21}(a)) = E_{-\beta_0}(a), \quad i(e_{32}(a)) = E_{-\gamma_0}(a), \quad i(e_{31}(a)) = E_{-\omega}(a)$$

where $E_{\pm\beta_0}, E_{\pm\gamma_0}, E_{\pm\omega}$ *are root subgroups in* $\underline{G}^E$.

b) *The centralizer of* $i(\underline{SL}_3)$ *in* $\underline{G}^E$ *is isomorphic to the kernel* $\underline{E}'$ *of the norm map* $N : \underline{E}^* \longrightarrow \underline{F}^*$, *where* $\underline{E}^* \stackrel{\text{def}}{=} R_{E/F}(\underline{G}_m)$.

PROOF: a) As follows from the Corollary 1 to Theorem 7 in [St] there exists an algebraic homomorphism $i : \underline{SL}_3 \to \underline{G}^E$ satisfying the condition of Lemma 2. For any root ν of $\underline{G}^E$ we denote by $\check{h}_\nu$ the corresponding coroot $\check{h}_\nu : \underline{G}_m \to \underline{H} \subset \underline{G}^E$. The center of $\underline{SL}_3$ is isomorphic to the group μ_3 of cubic roots of unity and for any $\epsilon \in \mu_3 \subset \underline{SL}_3$ we have $i(\epsilon) = \check{h}_\beta(\epsilon)\check{h}_{\gamma_0}(\epsilon^{-1})$. To show that $i : \underline{SL}_3 \to \underline{G}^E$ is an imbedding it is sufficient to check that $Ad(i(\epsilon)) \neq Id$ for $\epsilon \in \mu_3 - \{1\}$. But for any i, $1 \leq i \leq 3$, $a \in F$ we have $i(\epsilon)E_\alpha(a)i(\epsilon)^{-1} = E_\alpha(\epsilon^2 a)$. Therefore $i : \underline{SL}_3 \to \underline{G}^E$ is an imbedding.

b) For any root ν of $\underline{G}^E$ we can find $\mu \in (\pm\beta_0, \pm\gamma_0)$ such that $\nu + \mu$ is also a root. Therefore the centralizer Z of $i(SL_3)$ in $\underline{G}^E$ lies in the Cartan subgroup H of $\underline{G}^E$. Moreover $\underline{Z} = \{h \in H | \beta_0(h) = \gamma_0(h) = e\}$. Let $\check{h}$: $\underline{G}_m \to \underline{H}$ be the coroots over $\bar{F}$ corresponding to simple roots α_i, $i = 1, 2, 3$,

$\check{h}_\alpha : \underline{E}^* \to \underline{H}$ be the map given by $\check{h}_\alpha(a_1, a_2, a_3) \overset{\text{def}}{=} \check{h}_1(a_1)\check{h}_2(a_2)\check{h}_3(a_3)$. It is clear that h_α commutes with the action of the Galois group $\mathcal{G}$. Therefore, $\check{h}_\alpha$ defines a morphism of algebraic f-groups $\check{h}_\alpha : \underline{E}^* \to \underline{H}$. It is clear that $\check{h}_\alpha$ defines an isomorphism of the algebraic group $\check{h}_\alpha : \underline{E}' \xrightarrow{\sim} \underline{Z}$. Lemma 4 is proved.

We will consider $\underline{SL_3}$ and $\underline{E}'$ as subgroups of $\underline{G}^E$. It is easy to see that $\underline{SL_3} \cap \underline{E} = \mu_3$ – the group of cubic roots of unity, and i extends to a group monomorphism $i : (\underline{SL_3} \times \underline{E})/\mu_3 \longrightarrow \underline{G}^E$, where we imbed μ_3 in $SL_3 \times E$, $\epsilon \to (\epsilon\, Id, \epsilon^{-1})$.

Let $\underline{\widehat{Y}} \subset \underline{GL_3} \times \underline{E}^*$ be the subgroup of elements (g, e) such that $\det g \cdot N(e) = 1$. We can imbed $\underline{G_m}$ into $\underline{\widehat{Y}}$ by $\lambda \xrightarrow{\zeta} (\lambda Id, \lambda^{-1})$. It is clear that the quotient $\underline{Y}$ of $\underline{\widehat{Y}}$ by the image of ζ is isomorphic to $(\underline{SL_3} \times \underline{E}')/\mu_3$. Therefore we can consider i as an imbedding of $\underline{Y}$ into $\underline{G}^E$. Let $Y, \widehat{Y}$ be the groups of F-points of $\underline{Y}$ and $\underline{\widehat{Y}}$, and $G_3^E \subset GL_3(F)$ be the image of $\widehat{Y}$ under the natural projection $\widehat{Y} \to GL_3(F)$.

Assusme that E, F are a local fields, and $\pi : Y \to AutW$ is a smooth representation (see [BZ]). For any character $\chi : E^* \to C^*$ of E^* denote by $W_\chi \subset W$ the subspace of vectors $w \in W$ such that $\pi(1 \times e)w = \chi(e)w$ for all $e \in E'$ and define the representation $\widetilde{\pi}_\chi : G_3^E \to AutW_\chi$ by the formula $\widetilde{\pi}_\chi(g) = \chi^{-1}(e)\pi(g, e)W_\chi$ for any $e \in E^*$ such that $\det g \cdot N(e) = 1$. It is clear that the operator $\widetilde{\pi}_\chi(g)$ does not depend on a choice of $e \in E^*$ and the map $g \to \widetilde{\pi}_\chi(g)$ is a representation of G_3^E. We denote by π_χ the induced representation of $GL_3(F)$.

We can give an alternative construction of the representation π_χ which is applicable in the case when E is any semisimple 3-dimensional F-algebra. We consider π as a representation of $\widehat{Y}$ and denote by $\widetilde{\pi} : GL_3(F) \times E^* \to Aut\widetilde{W}$ the induced representation $(\widetilde{\pi}, \widetilde{W}) = cnd_Y^{GL_3(F)\times E^*}(\pi, W)$. For any character χ of E^* we denote by W_χ the quotient of $\widetilde{W}$ by the subspace spend by vectors of the form $\widetilde{\pi}(1 \times e)w - \chi(e)w, w \in \widetilde{W}, e \in E^*$. We denote by π_χ the natural action of the group on W_χ.

Now we can formula the main results of the paper. Let $\underline{\Omega} \subset \mathcal{L}^1_{G^E}$ be the minimal nonzero coadjoint orbit of $\underline{G}^E$. It is easy tosee that the centralizer of any element of $\underline{\Omega}$ in $\underline{G}^E$ is unipotent and therefore $\Omega(\overset{\text{def}}{=}\underline{\Omega}(F))$ is a G^E-orbit. Then Ω defines a unitary, irreducible representation τ of B uniquely up to a multiplication by a character.

THEOREM A. *There exists a unique unitary irreducible representation $\pi_2 : G^E \longrightarrow AutW_2$ of G^E such that the restriction of π_2 on B is isomorphic to a representation τ corresponding to Ω.*

Moreover we give an explicit construction of (π_2, W_2). Let $W \subset W_2$ be the subspace of smooth vectors and π be the restriction of π_2 on W.

THEOREM B. *For any character χ of E^* the representation π_χ of $GL_3(F)$ on W_χ is nonempty and irreducible and it corresponds to χ by the local lifting (see [J-PS-S]).*

We can interpret Theorem B by saying that $\pi : G^E \to AutW$ provides an explicit realization of the lifting : characters $E^* \Longrightarrow$ representations of $GL_3(F)$ in the same way in which the metaplectic representation of $Sp(4,F)$ provides the lifting : characters $K^* \Longrightarrow$ representations of $GL_2(F)$ for any quadratic extension K of F.

We end this introduction with the description of the place of π in the Langlands picture.

The Levi component $\widetilde{\underline{L}}^E$ can be described as the quotient of the group $\underline{GL}_2 \times \underline{E}^*$ by the subgroup G_m which is imbedded in it by $\lambda \to (\lambda Id, \lambda^{-1})$. By the Hilbert 90 theorem we see that $\tilde{L}^E = GL_2(E) \times E^*/F^*$.

The group S_3 has unique irreducible representation $\mu : S_3 \to GL_2(\mathbf{C})$. Combining it with the morphism $\mu_E : \mathcal{G} \to S_3$ we obtain a 2-dimensional representation μ_E of the Galois group $\mathcal{G}$. By ([JL]) it corresponds to an irreducible unitary representation ρ_E^0 of $GL_2(F)$ such that $\pi_E^0(\lambda Id) = \epsilon_E(\lambda)Id$ where ϵ_E is the quadratic character of F^* corresponding to the character $\det \mu_E : \mathcal{G} \to (\pm 1) \subset \mathbf{C}^*$.

We denote by ρ_E the representation of the group $GL_2(F) \times E^*$ given by $\rho_E(g,e) \overset{\text{def}}{=} \rho_E^0(g)\epsilon_E(N(e))$, where $N : E^* \to F^*$ is the norm map. It is clear that for all $\lambda \in F^*$, $\rho_E(\lambda Id, \lambda^{-1}) = Id$ and therefore we can consider ρ_E as a representation of $\tilde{L}^E$.

Let $\theta : \tilde{L}^E \to F^*$ be the character given by $\theta(g,e) \overset{\text{def}}{=} (\det g)^3 N(e)^2$ and Π^E be the representation of G^E unitary induced from $\tilde{P}^E$ by the representation $\rho_E \otimes |\theta|$. As follows from the Langlands theory Π^E has unique irreducible quotient which we denote by $\overline{\Pi}^E$.

THEOREM C. *The representation π of G^E is the equivalent to the representation of G^E on $\overline{\Pi}^E$.*

If E is not a field we can give a more elementary description of the group G^E and its representation π. In this case $E = F \oplus K$ where K is a semisimple 2-dimensional F algebra. The norm $N_K : K \to F$ can be considered as a quadratic form on the 2-dimensional F-space K. Let φ_K be the direct sum of N_K and a six-dimensional split form and let SO_K be the proper orthogonal group of φ_K. It is easy to see that for $E = F \oplus K$, G^E is isomorphic to the quotient of SO_K by the center.

We have an imbedding $SO_K \times SL_2(F) \to Sp(16,F)$ and the representation π cooincides with the action of SO_K on the space $SL_2(F)$-invariant functionals on the space of the metaplectic representation of $Sp(16,F)$.

In the case when $F = \mathbf{R}$ the representation π was discussed in [Ko] in detail.

2. In this section we give an explicit description of the parabolic subgroup $P^E \subset G^E$. We denote by N, T the norm and trace maps $E \to F$. We denote by B_E the bilinear form $(e_1, e_2) \to T(e_1 e_2)$ on E and by $\delta_{E/F} \in F^*/F^{*2}$ the discriminant of the bilinear form B_E. For any $\alpha \in F^*$ we denote by α_a the homomorphism $\alpha_a : \mathcal{G} \to \{\pm 1\}$ given by $\alpha_a(\gamma)\sqrt{a} = (\sqrt{a})^\gamma$ for all $\gamma \in \mathcal{G}$. We will interpret α_a as an element of $H^1(\mathcal{G}, \{\pm 1\})$. It is clear that α_a depends only o the image of a in F^*/F^{*2}. For any $a, b \in F^*/F^{*2}$ we define $(a, b) \stackrel{\text{def}}{=} \alpha_a \cdot \alpha_b \in H^2(\mathcal{G}, \{\pm 1\})$.

The algebra $\overline{E} \stackrel{\text{def}}{=} E \otimes_F \overline{F}$ is isomorphic to the direct sum $\overline{F}^3$ of three copies of $\overline{F}$. The permutation group S_3 acts naturally as the group of automorphisms of $\overline{F}^3$ over $\overline{F}$ and an isomorphism $i : \overline{E} \simeq \overline{F}^3$ is unique up to a composition with an element in $Aut_{\overline{F}}\overline{F}^3 = S_3$. In particular, we have a homomorphism $\tilde{i}_E : \mathcal{G} \to S_3$ which is well defined up to a conjugation. Consider the map $\theta^{(3)} : \overline{F}^3 \to \overline{F}^3$ given by $\theta^{(3)}(\alpha_1, \alpha_2, \alpha_3) = (\alpha_2\alpha_3, \alpha_1\alpha_3, \alpha_1\alpha_2)$ and define $\overline{\theta} : \overline{E} \to \overline{E}$ by $\overline{\theta} = i^{-1} \circ \theta^{(3)} \circ i$. It is clear that $\overline{\theta}$ is a quadratic map from $\overline{E}$ into itself (that is, a linear map from $sym^2_{\overline{F}}\overline{E}$ into $\overline{E}$) which does not depend on a choice of an isomorphism $i : \overline{E} \simeq \overline{F}^3$ and commutes with the action of $\mathcal{G}$. Therefore the restriction of $\overline{\theta}$ on E defines a quadratic map $\theta : E \to E$. It is clear that $\theta(e) = e^{-1}N(e)$ for all invertible $e \in E$.

Let $\Lambda \stackrel{\text{def}}{=} F \oplus E$, for any $t \in E$ we denote by Q_t the quadratics form on Λ given by $Q_t(x_0, x) \stackrel{\text{def}}{=} N(t)x_0^2 + x_0 T(\theta(t)x) + T(t\theta(x))$. Let $d_t \in F^*/F^{*2}$, $S_t \in H^2(\mathcal{G}, \{\pm 1\})$ be the discriminant and the Hasse invariant of Q_t (see [0], §58).

LEMMA 1. *(a) For all $x_0 \in F, x, t \in E$ we have $x_0 Q_t(x_0, x) = N(x + x_0 t) - N(x)$.*

(b) For all $t \in E^*$ the form Q_t is nondegenerate and $d_t = d_1$.

(c) $S_t = S_1(\delta_{E/F}, N(t))$ for $t \in E^*$.

PROOF: Let $\overline{\Lambda} = \Lambda_F \otimes_F \overline{F}$ and $\overline{Q}_t : \overline{\Lambda} \to \overline{F}$ be quadratic forms obtained from Q_t by the extension of scalars. Using an isomorphism $i : \overline{E} \simeq \overline{F}^3$ we can consider $\overline{Q}_t$ as a quadratic form on $\overline{F}^4$. It is clear that for $i(t) = (t_1, t_2, t_3)$ we have $x_0\overline{Q}_t(x_0, x_1, x_2, x_3) = (x_1 + x_0 t_1)(x_2 + x_0 t_2)(x_3 + x_0 t_3) - x_1 x_2 x_3$ for all $(x_1, x_2, x_3) \in \overline{F}^3$, $x_0 \in \overline{F}$. The part (a) is proved. To prove (b) we compute the discriminant Δ_E of Q_t in terms of the standard basis in $\overline{F}^4$. We find $\Delta_t = N(t)^2$ and therefore Q_t is not degenerate if $N(t) \neq 0$. Since the identification $id \oplus i : \overline{\Lambda} \simeq \overline{F}^4$ does not depend on t we see that $d_t = d_1$ for all $t \in E^*$. Part (b) is proved.

We first prove (c) in the case when $t \in F^* \subset E^*$. Let $E^0 = \langle e \in E, Te =$

$0\rangle$. We have $E = F \oplus E^0$ and $\Lambda = F \oplus F \oplus E^0$ and $Q_t(x_0, y_0, e_0) = (P_t(x_0, y_0) + tP(e^0), x_0, y_0 \subset F, e^0 \in E^0$ where $P_t(x_0, y_0) = t^3x_0^2 + t^2x_0y_0 + ty_0^2$ and $P : E^0 \to F$ is the quadratic form $P(e^0) = T\theta(e^0)$. It is clear that the quadratic form P_t is equivalent to the form $t\widetilde{P}$ where $\widetilde{P}(x_0, y_0) = x_0^2 + x_0y_0 + y_0^2$. So Q_t is equivalent to the form $t(\widetilde{P} \oplus P)$. Since $dim\, \Lambda = 4$ it follows from the definition of S_t that $S_t = S_1(t, d_1)$ where $d_1 = d(\widetilde{P}) \cdot d(P) = 3d(P)$. To compute $d(P)$ we observe that we have $T\theta(e^0) = -\frac{1}{2}T((e^0))^2$ for all $e^0 \in E^0$. Really let $(\alpha_1, \alpha_2, \alpha_3) = i(e^0) \in \overline{F}^3$. Then $i(\theta(e^0)) = (\alpha_2\alpha_3, \alpha_1\alpha_3, \alpha_1\alpha_2)$ and $T\theta(e^0) = \alpha_1\alpha_2 + \alpha_1\alpha_3 + \alpha_2\alpha_3 = -\frac{1}{2}(\alpha_1^2 + \alpha_2^2 + \alpha_3^3)$ $+\frac{1}{2}(\alpha_1 + \alpha_2 + \alpha_3)^2 = -\frac{1}{2}T(e^0)^2$ since $Te^0 = 0$.

Now it is clear that $d_1 = \delta_{E/F}$ and (c) is proved in the case when $t \in F^* \subset E^*$.

To prove c) for the general $t \in E^*$ we will use the following result.

Claim. *If $t = uv^2\tilde{t}$ where $u, v, \tilde{t} \in E^*$, $N(u) = 1$ then $S_t = S_{\tilde{t}}$.*

Proof of the Claim: The map $(x_0, x) \to (x_0', x')$ given by $x_0' = N(v)x_0$, $x' = \theta(v)v^{-1}u^{-1}x$ defines an equivalence of quadratic forms Q_t and $Q_{\tilde{t}}$. The Claim is proved.

Since any $t \in E^*$ can be written in the form $t = uv^2\tilde{t}$ where $u = N(t)^{-1}t^3, v = t^{-1}$ and $\tilde{t} = N(t) \in F^*$, we have $S_t = S_{N(t)}$. Lemma 1 is proved.

Let $\Lambda' \stackrel{\text{def}}{=} Hom(\Lambda, F)$ be the dual space to $\Lambda, V \stackrel{\text{def}}{=} \Lambda \oplus \Lambda'$. We consider V as a symplectic vector space over F where the symplectic form $[\ ,\ ] : V \times V \to F$ is defined by

$$[(\lambda_1, \lambda_1'), (\lambda_2, \lambda_2')] \stackrel{\text{def}}{=} \lambda_2'(\lambda_1) - \lambda_1'(\lambda_2), \lambda_1, \lambda_2 \in \Lambda, \lambda_1, \lambda_1' \in \Lambda'.$$

We will write elements $v \in V$ as $v = (x_0, x, x', x_0')$, $x_0 \in F$, $x \in E$, $x' \in E' \stackrel{\text{def}}{=} Hom_F(E, F)$, $x_0' \in Hom_F(F, F) = F$. We denote by $\beta : \Lambda \to \Lambda'$ the linear map such that $\beta(x_0, x)(y_0, y) = -x_0y_0 - T(xy)$ for all $x_0, y_0 \in F, x, y \in E$. For any $t \in E$ we denote by $\alpha_t : \Lambda \to \Lambda$ the map $\alpha_t(x_0, x) = (x_0, x + x_0t)$, by $\alpha_t' : \Lambda' \to \Lambda'$ the dual map and by $\ell_t : V \to V$ the map given by $\ell_t(\lambda, \lambda') = (\alpha_t(\lambda), (\alpha_t')^{-1}(\lambda' + \widehat{Q}_t(\lambda))$ for $(\lambda, \lambda') \in V$, where $\widehat{Q}_t : \Lambda \to \Lambda'$ the symmetric morphism such that $(\widehat{Q}_t(\lambda))(\lambda) = 2Q_t(\lambda)$ for all $\lambda \in \Lambda$. For any $e \in E^*$ we define $r_e : V \to V$ by $r_e(x_0, x, y, y_0) = (N(e)x_0, \theta(e)x, ey, y_0)$, $x_0, y_0 \in F, x, y \in E$. Let $\underline{GSp}(V)$ be the algebraic f-group of linear transformation $g : V \to V$ such that there exists $\omega(g) \in F^*$ such that $[v_1g, v_2g] = \omega(g)[v_1, v_2]$ for all $v_1, v_2 \in V$. Let $\underline{E}^*, \underline{GL}_2(E)$ as before be algebraic F-groups obtained from $\underline{G}_m$ and $\underline{GL}_2$ by the restriction of scalars from E to F, $\underline{N}$ be the norm map $\underline{N} : \underline{E}^* \longrightarrow \underline{G}_m$ and $\underline{E}' \stackrel{\text{def}}{=} \ker \underline{N}$.

LEMMA 2. *There exists a group homomorphism* $\underline{r} : \underline{GL}_2(E) \to \underline{GSp}(V)$ *such that*

$$r\begin{pmatrix} 1 & t \\ 0 & 1 \end{pmatrix} = \ell_t,\ t \in E, \quad r\begin{pmatrix} e & 0 \\ 0 & 1 \end{pmatrix} = r_e,\ e \in E^*$$

and

$$r\begin{pmatrix} 0 & 1 \\ -1 & 0 \end{pmatrix} = \begin{pmatrix} 0 & \beta \\ -\beta^{-1} & 0 \end{pmatrix}.$$

PROOF: To prove the lemma we may assume that $F = \overline{F}$ and fix an isomorphism $E = F^3$. Let $\sigma_1, \sigma_2, \sigma_3 : \underline{GL}_2(E) \to \underline{GL}_2$ be the natural projections. We consider σ_i as two-dimensional representations of $\underline{GL}_2(E)$ and define $\sigma : \underline{GL}_2(E) \to Aut\ \tilde{V}$ to be the tensor product $r = \sigma_1 \otimes \sigma_2 \otimes \sigma_3$. We choose bases (e_i, f_i), $1 \leq i \leq 3$ in the spaces of representations σ_i and define an isomorphism $A : V \to \tilde{V}$ by $A((x_0, x), (y, y_0)') = x_0 e_1 \otimes e_2 \otimes e_3 + x_1 f_1 \otimes e_2 \otimes e_3 + x_2 e_1 \otimes f_2 \otimes e_3 + x_3 e_1 \otimes e_2 \otimes f_3 + y_1 e_1 \otimes f_2 \otimes f_3 + y_2 f_1 \otimes e_2 \otimes f_3 + y_3 f_1 \otimes f_2 \otimes e_3 + y_0 f_1 \otimes f_2 \otimes f_3$ where for all $x_0, y_0 \in F, x = (x_1, x_2, x_3)$, $y = (y_1, y_2, y_3) \in E$ where we consider $(y, y_0)'$ as an element in Λ' given by $(y, y_0)'(x_0, x) \stackrel{\text{def}}{=} \sum_{i=0}^{3} x_i y_i - x_0 y_0$. It is clear that $\ell_t = A^{-1} \cdot \sigma\begin{pmatrix} 1 & t \\ 0 & 1 \end{pmatrix} A, r_e = A^{-1} \cdot \sigma\begin{pmatrix} e & 0 \\ 0 & 1 \end{pmatrix} A$ and $\begin{pmatrix} 0 & \beta \\ -\beta^{-1} & 0 \end{pmatrix} = A^{-1} \sigma \begin{pmatrix} 0 & 1 \\ -1 & 0 \end{pmatrix} A$ for all $t \in E, e \in E^*$. So we can define the group homomorphism $\underline{r} : \underline{GL}_2(E) \to \underline{GSp}(V)$ by $r = A^{-1} \sigma \cdot A$. Lemma 2 is proved.

Let $\overline{V} \stackrel{\text{def}}{=} V \otimes_F \overline{F}$.

We denote by $\overline{\Omega} \subset \tilde{V}$ the subset of nonzero decomposable vectors. Using the isomorphism $A : \overline{V} \xrightarrow{\sim} \tilde{V}$ we can consider $\overline{\Omega}$ as a subvariety of $\overline{V} - \{0\}$. It is clear that $\overline{\Omega}$ is $\mathcal{G}$-invariant. Therefore there exists an F-subvariety $\underline{\Omega} \subset \underline{V} - \{0\}$ such that $\overline{\Omega} = \underline{\Omega}(\overline{F})$. Let $q : \Omega \to \Lambda$ be the restriction of the projection $V \to \Lambda$ on Ω.

LEMMA 3.

a) *dim* $\Omega = 4$

b) $\underline{\Omega}$ *is an* $\underline{L}$*-orbit*

c) Ω *is an* L^E*-orbit*

d) *For any* $v \in V - \Omega$ *dim* $\underline{L}v > 4$

e) $\Omega \subset V - \{0\}, V = F \oplus E \oplus E \oplus F$ *can be defined by the equations* $\Omega = \{(x_0, x, y, y_0) \in V - \{0\} | xy = x_0 y_0, \theta(x) = x_0 y, \theta(y) = y_0 x$

f) $q : \Omega \to \Lambda$ *is birational and for any* $(x_0, x) \in \Lambda = F \oplus E, x_0 \neq 0$, $q^{-1}(x_0, x) = (x_0, x, x_0^{-1}\theta(x), x_0^{-2} N(x))$.

PROOF: a), b), d), e), f) are obvious. c) follows easily from b) and e).

COROLLARY. *Assume that E is a field.*
a) If $(x_0, x, y, y_0) \in \Omega$ is such that $y = 0$, then $x = 0$.
b) There is no point $(x_0, x, y, y_0) \in \Omega$ such that $x_0 = y_0 = 0$.

PROOF: Clear.

Let $\underline{M} \subset \underline{GL}_2(E)$ be the image of the imbedding $\underline{E}' \hookrightarrow \underline{GL}_2(E)$ given by $\lambda \to \lambda Id$ and $\underline{L}^E$ be the quotient group $\underline{L}^E \stackrel{\text{def}}{=} \underline{GL}_2(E)/\underline{M}$. It is clear that $\underline{r}$ defines an imbedding $\underline{L}^E \to \underline{GSp}(V)$. The morphism $\underline{\omega} : \underline{GSp}(V) \to \underline{G}_m$ defines the character $\underline{\omega} : \underline{L}^E \to \underline{G}_m$. The intersection $\underline{M}^0 \stackrel{\text{def}}{=} \underline{M} \cap \underline{SL}_2(E)$. $\underline{M}^0$ is a finite F-subgroup in the center of $\underline{SL}_2(E)$. We denote by $\underline{L}_0^E$ be the quotient group $\underline{L}_0^E \stackrel{\text{def}}{=} \underline{SL}_2(E)/\underline{M}^0$.

Let $\underline{p} : \underline{GL}_2(E) \longrightarrow \underline{L}^E$ be the natural projection. The imbedding $\underline{G}_m \to \underline{GL}_2$ given by $\lambda \to \begin{pmatrix} \lambda & 0 \\ 0 & 1 \end{pmatrix}$ defines an imbedding $\underline{E}^* \to \underline{GL}_2(E)$ and therefore map $\underline{E}^* \xrightarrow{\kappa} \underline{L}^E$. It is clear that κ is an imbedding and it induces the imbedding $\underline{E}' \overset{\kappa}{\hookrightarrow} \underline{L}_0^E$.

Consider the morphism $\underline{\tilde{h}} : \underline{G}_m \to \underline{GS}_p(V)$ given by

$$\tilde{h}(a)(x_0, x, y, y_0) \stackrel{\text{def}}{=} (a^2 x_0, ax, y, a^{-1} y_0).$$

It is easy to see that $\underline{\tilde{h}}(\underline{G}_m) \subset \underline{L}^E$ and $\underline{\omega} \circ \underline{\tilde{h}} = \underline{Id}$.

As before, we denote by H the group $\underline{H}(F)$ of F points of an algebraic F-group $\underline{H}$. For example, $G_m = F^*$.

LEMMA 4. *a) The group L^E is generated by its normal subgroup L_0^E and the image $\tilde{h}(F^*)$ of $\tilde{h}$.*
b) The map $\omega : L^E \to F^*$ is surjective.
c) The image $p(GL_2(E))$ in L^E consists of $\ell \in L^E$ such that $\omega(\ell) \subset N(E^*)$.
d) L_0^E is generated by its normal subgroup $p(SL_2(E))$ and the image of E' in L_0^E.

PROOF: a) and b) follow immediately from the equalities $\omega \circ \tilde{h} = Id$ and $L_0^E = \ker \omega$.

c) Let $\mathcal{G} = Gal(\overline{F} \cdot F)$. It is clear that $p(GL_2(E))$ is a normal subgroup in L^E and we have an imbedding of the quotient $L^E/p(GL_2(E))$ into $H'(\mathcal{G}, \underline{M}(\bar{F})) = H'(\mathcal{G}, \underline{E}'(\bar{F}))$. From the exact sequence $0 \longrightarrow \underline{E}' \longrightarrow \underline{E}^* \xrightarrow{N} \underline{G}_m \longrightarrow 0$ we conclude that $H'(\mathcal{G}, \underline{E}'(\bar{F})) = F^*/N(E^*)$. Therefore, we have an imbedding $L^E/p(GL_2(E)) \hookrightarrow F^*/N(E^*)$. On the other hand, for any $g \in GL_2(E)$ we have $\omega(p(g)) = N(\det g) \in N(E^*)$ and ω defines a morphism $\bar{\omega} : L^E/p(GL_2(E) \longrightarrow F^*/N(E^*)$. As follows from b) $\bar{\omega}$ is surjective. Therefore it is an isomorphism. c) is proved.

d) As follows from c) for any $\ell \in L^0_E$ we can find $g \in GL_2(E)$ such that $\ell = p(g)$. Let $\lambda = \det g \in E^*$. It is clear that $\lambda \in E'$ and $s \stackrel{\text{def}}{=} \begin{pmatrix} \lambda & 0 \\ 0 & 1 \end{pmatrix}^{-1} g \in SL_2(E)$. Then $g = \kappa(\lambda)p(s)$. Lemma 4 is proved.

We can identify L^E_0 with the kernel of the restriction of ω on L^E and L^E is the semidirect product of L^E_0 and Δ ($\approx G_m$), where Δ is the image of $\widetilde{h}$.

Let $\underline{H}$ be the Heisenberg group associated to V. That is, H is isomorphic to $\underline{V} \oplus \underline{G_a}$ as an algebraic variety and the multiplication on $\underline{H}$ is given by $(v_1; a_1) \cdot (v_2; a_2) = (v_1 + v_2; a_1 + a_2 + \frac{1}{2}[v_1, v_2])$. It is clear that the map $a \longrightarrow z_a \stackrel{\text{def}}{=} (0; a)$ defines an isomorphism of $\underline{G_a}$ with the center $\underline{Z}$ of $\underline{H}$. The maps $\lambda \to ((\lambda, 0); 0)$, $\lambda' \to ((0, \lambda'); 0)$ identify Λ, Λ' with commutative subgroups of H and the product defines an isomorphism $\underline{\Lambda} \times \underline{\Lambda}' \times \underline{Z} \to \underline{H}$ of algebraic varieties. The group $\underline{GSp}(V)$ acts naturally on $\underline{H}$. In particular, we have the action of $\underline{L}^E$ on $\underline{H}$, $h \to h^\ell$, where for $h = (v; a)$, $h^\ell \stackrel{\text{def}}{=} (v^\ell; \omega(\ell)a)$.

Let $\underline{P}^E \subset \underline{G}^E$ be the parabolic subgroup as in §2.

PROPOSITION 1. *$\underline{P}^E$ is isomorphic to the semidirect product $\underline{L}^E \ltimes \underline{H}$.*

Let $\overline{P} = P^E(\overline{F})$, $\overline{L} = L^E(\overline{F})$, $\overline{H} = H(\overline{F}), \overline{V} = V(\overline{F})$. The group $\mathcal{G}$ acts naturally on $\overline{P}, \overline{L}$ and $\overline{H}$. It is sufficient to construct an algebraic isomorphism $\overline{L} \ltimes \overline{H} \xrightarrow{\sim} \overline{P}$ compatible with the action of $\mathcal{G}$.

Let $\beta_0, \alpha_1, \alpha_2, \alpha_3$ be the simple roots on G^0, $\gamma_0 \stackrel{\text{def}}{=} \beta_0 + \alpha_1 + \alpha_2 + \alpha_3$. Then γ_0 is a root and $\langle \beta_0, \alpha_i, \beta_i, \gamma_i, \gamma_0, \omega \rangle$ is the set of positive roots where $\beta_i = \beta_0 + \alpha_i$, $\gamma_i = \gamma_0 - \alpha_i$, $\omega = \gamma_0 + \beta$, $1 \leq i \leq 3$. For any root of G^0 we denote by the same letter the root subspace in the Lie algebra of G^0. The Lie algebra of the nilpotent radical of $\overline{P}$ is spanned by $\beta_a, \gamma_a, \omega, 0 \leq a \leq 3$, and is isomorphic to the Heisenberg Lie algebra. $\kappa = V \oplus F \cdot \omega$ where V is the 8-dimensional symplectic F-space spanned by $\beta_\alpha, \gamma_\alpha$, $0 \leq a \leq 3$ and the symplectic form $\langle\ ,\ \rangle$ and V is given by $[v_1, v_2] = \langle v_1, v_2 \rangle \omega$. Let $\Lambda, \Lambda' \subset V$ be subspaces spanned by β_a and γ_a, respectively, $0 \leq a \leq 3$. It is easy to see that Λ, Λ' are maximal isotropic $\mathcal{G}$-invariant subspaces of V. The Levi subgroup of $\overline{P}$ is generated by the Cartan subgroup of $\overline{G}$ and one-parameter subgroups corresponding to roots $\pm\alpha_i$, $1 \leq i \leq 3$. So this Levi subgroupis isomorphic to $\overline{L}$.

It is easy to see that $\overline{P}$ is isomorphic to the semidirect product of $\overline{L}$ and $\overline{H}$ and this isomorphism is compatible with the action of $\mathcal{G}$. We leave to the reader to check the action of the Levi component of $\overline{P}$ on its unipotent radical coincides with the action of L on H constructed earlier. Proposition 1 is proved.

3. In this section we consider the case when F is a local field of characteristic zero. Then $H^2(\mathcal{G},\{\pm 1\}) = \{\pm 1\}$ if $F \neq \mathbf{C}$ and $H^2(\mathcal{G},\{\pm 1\}) = \{1\}$ if $F = \mathbf{C}$. Therefore, for local fields we can consider symbols $\{\alpha,\beta\}, \alpha,\beta \in F^*$ as elements on $\{\pm 1\} \subset \mathbf{C}^*$. We fix a non-trivial additive character $\psi : F \to \mathbf{C}^*$. Let P^0 be the semidirect product $P^0 \stackrel{\text{def}}{=} SL_2(E) \ltimes H$.

LEMMA 1. *(a) There exists unique (up to an equivalence) irreducible unitary representation $\sigma_{2,\psi} : H \to Aut\ W_2^0$ such that $\sigma_{2,\psi}(z_a) = \psi(a)Id$ for all $a \in F$.*

(b) There exists unique extension of $\sigma_{2,\psi}$ to a representation $\sigma_{2,\psi} : P^0 \to Aut\ W_2^0$.

PROOF: We consider only the case when E is a field. Other cases are even simpler.

(a) is well known (see [W]).

As follows from [W] there exists a double covering $\widetilde{Sp} \xrightarrow{r} Sp(V)$ and a representation $\tilde{\sigma} : \widetilde{Sp} \to Aut\ W_2^0$ such that for any $\tilde{g} \in \widetilde{Sp}$, $h \in H$ we have $\tilde{\sigma}(\tilde{g}^{-1})\sigma_{2,\psi}(h)\tilde{\sigma}(\tilde{g}) = \sigma_{2,\psi}(h^{r(\tilde{g})})$. The restriction of r to $SL_2(E) \hookrightarrow Sp(V)$ defines a double covering on $SL_2(E)$ or, in other words, an element $\alpha \in H^2(SL_2(E), \mathbf{Z}/2\mathbf{Z})$. Since $[E : F] = 3(\equiv 1(mod\ 2))$ it follows from [M], Th. 14) that $\alpha = f^*\beta$ where f^* is the transfer homomorphism and $\beta \in H^2(SL_2(F), \mathbf{Z}/2\mathbf{Z})$ is the restriction of α on $SL_2(F)$. In other words, β corresponds to the restriction of r on the image of $SL_2(F)$. It is easy to see that $\beta = 0$. Therefore $\alpha = 0$ and there exists a group homomorphism $\tilde{i} : SL_2(E) \longrightarrow \widetilde{Sp}$ such that $i = r \circ \tilde{i}$. Now we can extend $\sigma_{2,\psi}$ to $SL_2(E)$ by $\sigma_{2,\psi} = \tilde{\sigma} \circ \tilde{i}$. This proves b).

To give an explicit realization of $\sigma_{2,\psi}$ we will use the notion of Fourier transform. let Λ be a finite-dimensional F-space, Λ' - the dual space $d\lambda$ a Haar measure on Λ. Let $W^0(\Lambda)$ be the space of Schwartz-Bruhat function[1] on Λ, $W'(\Lambda)$ the dual space of distributions and $W_2^0(\Lambda) = L^2(\Lambda, d\lambda)$ (see [W]). The Haar measure $d\lambda$ defines the imbeddings $W^0(\Lambda) \subset W_2^0(\Lambda) \subset W'(\Lambda)$. Let $\mathcal{F} : W^0(\Lambda) \to W^0(\Lambda')$ be the Fourier transform given by $\mathcal{F}(f)(\lambda') \stackrel{\text{def}}{=} \int_\Lambda \psi(\lambda'(\lambda))f(\lambda)d\lambda$. As is well known, $\mathcal{F}$ extends to continuous isomorphisms $\mathcal{F} : W_2^0(\Lambda) \xrightarrow{\sim} W_2^0(\Lambda')$ and $\mathcal{F} : W'(\Lambda) \xrightarrow{\sim} W'(\Lambda')$ and we can choose a Haar measure $d\lambda$ in such a way that $\mathcal{F}$ is unitary.

For any nondegenerate quadratic for Q on Λ there exists a symmetric invertible linear map $\hat{Q} : \Lambda \to \Lambda'$ such that $Q(\lambda) = \frac{1}{2}\hat{Q}(\lambda)(\lambda)$. We denote by Q^{-1} the quadratic form on Λ' given by $Q^{-1}(\lambda') = \frac{1}{2}\lambda'(\hat{Q}^{-1}(\lambda'))$.

One may consider the locally constant function $\psi(Q)$ on Λ as a distribution and define its Fourier transform $\mathcal{F}(\psi(Q) \in W'(\Lambda')$. As is well known,

[1] In the case when F is a local nonarchimedean field and $\underline{X}$ is an F-variety by Schwartz-Bruhat functions on X we understand locally constant functions on X with compact support.

([W], Th. 2 and Prop. 3) $\mathcal{F}(\psi(Q)) = \gamma_Q c_Q \cdot \psi(Q^{-1})$ where $c_Q \in \mathbf{R}^+$, $\gamma_Q \in \mathbf{C}^*$, $|\gamma_Q| = 1$ and c_Q and γ_Q depend only on the image of Q in the Witt group of F.

LEMMA 2. *(a) There exists a unitary representation*

$$\sigma_{2,\psi} : H \longrightarrow Aut\ W_2^0$$

such that

$$\sigma_{2,\psi}(\lambda_0)f(\lambda) = f(\lambda + \lambda_0)$$
$$(\sigma_{2,\psi}(\lambda'_0)f)(\lambda) = \psi(\lambda'_0)(\lambda))f(\lambda)$$
$$(\sigma_{2,\psi}(z_a)f)(\lambda) = \psi(a)f(\lambda)$$

for all $f \in W_2^0$, $\lambda, \lambda_0 \in \Lambda, \lambda'_0 \in \Lambda', a \in F$.

(b) The representation $\sigma_{2,\psi}$ is irreducible.

(c) There exists unique extension of $\sigma_{2,\psi}$ to a representation $\sigma_{2,\psi} : P^0 \longrightarrow Aut\ W_2^0$ and for all $f \in W_2^0$, $t \in F$, we have

$$\sigma_{2,\psi}\begin{pmatrix} 1 & t \\ 0 & 1 \end{pmatrix} f(x_0, x) = \psi(Q_t(x_0, x))f(x_0, x + x_0 t)$$

$$\sigma_{2,\psi}\begin{pmatrix} 0 & 1 \\ -1 & 0 \end{pmatrix} f(x_0, x) = \zeta\mathcal{F}(f)\beta(x_0, x) \quad |\zeta| = 1.$$

(d) The subspace $W^0 \subset W_2^0$ is $\sigma_{2,\psi}$ invariant.

PROOF: Lemma 2 follows immediately from Lemma 1 and [W], n. 12 and n. 13. We denote by σ_ψ^0 the restriction of $\sigma_{2,\psi}$ on $W^0 \subset W_2^0$.

LEMMA 3. *(a)*

$$\sigma_\psi^0(h_e)f(x_0, x) = (N(e), \delta_{E/F})\|N(e)\|^{-1}f(N(e)^{-1}x_0, N(e)^{-1}e^2x)$$

where $h_e \stackrel{\text{def}}{=} \begin{pmatrix} e & 0 \\ 0 & e^{-1} \end{pmatrix}$, $e \in E^*$.

(b) $\zeta = \gamma_{Q_1}$.

PROOF: It follows from Lemma 1 and [W] n. 13 that there exists a character $\chi : E^* \to C^*$ such that $|\chi| = 1$ and

$$\sigma_\psi^0(h_e)f(x_0, x) = \chi(e)\|N(e)\|^{-1}f(N(e)^{-1}x_0,\ N(e)^{-1}e^2x).$$

Let $w_e \stackrel{\text{def}}{=} \begin{pmatrix} 0 & e \\ -e^{-1} & 0 \end{pmatrix}$. We have

$$w_e = \begin{pmatrix} 1 & e \\ 0 & 1 \end{pmatrix}\begin{pmatrix} 1 & 0 \\ -e^{-1} & 1 \end{pmatrix}\begin{pmatrix} 1 & e \\ 0 & 1 \end{pmatrix}$$
$$= \begin{pmatrix} 1 & e \\ 0 & 1 \end{pmatrix}\begin{pmatrix} 0 & 1 \\ -1 & 0 \end{pmatrix}^{-1}\begin{pmatrix} 1 & e^{-1} \\ 0 & 1 \end{pmatrix}\begin{pmatrix} 0 & 1 \\ -1 & 0 \end{pmatrix}\begin{pmatrix} 1 & e \\ 0 & 1 \end{pmatrix}.$$

The representation σ^0 defines the representation σ'_ψ of $SL_2(E)$ on the space W' of distributions (that is, linear functionals on W^0). Let $\delta \in W'$ be given by $\delta(f) = f(0)$. We have

$$\begin{aligned}\sigma'_\psi(w_e)(\delta) &= \sigma'_\psi\begin{pmatrix}1 & e\\ 0 & 1\end{pmatrix}\mathcal{F}^{-1}\sigma'_\psi\begin{pmatrix}1 & -e^{-1}\\ 0 & 1\end{pmatrix}\mathcal{F}\sigma'_\psi\begin{pmatrix}1 & e\\ 0 & 1\end{pmatrix}\delta\\ &= \sigma'_\psi\begin{pmatrix}1 & e\\ 0 & 1\end{pmatrix}\mathcal{F}^{-1}\psi(Q_{e^{-1}})\end{aligned}$$

where $\psi(Q_t) \stackrel{\text{def}}{=} \psi(Q_t(x_0, x))$ is considered as a distribution on Λ. It follows from [W], Theorem 2, that $\sigma'(w_e)(\delta)$ is a locally constant function on Λ' and $\sigma'(w_e)(\delta)(0) = \gamma(Q_{-e^{-1}}) \cdot c$ where c is a positive number. The part (b) is proved. To prove (a) we remark that $h_e w_1 = w_e$. Therefore we find $\chi(e) = \gamma(Q_{-e^{-1}})\gamma(Q_{-1})^{-1}$. Lemma 3 follows from Lemma 3.1. Theorem 63.23 in [0] and Propositions 3 and 4 in [W].

Corollary. $\zeta^2 = (-1, \delta_{E|F})$.

Let P^E be the semidirect product $P^E = L^E \ltimes H$. We extend $\omega : L^E \to G_m$ to the character $\omega : P^E \to G_m$ trivial on H and define $P^0 = \ker\, \omega$. It is clear that P^0 is the group of F-points of a connected algebraic group $\underline{P}^0$.

Since $\underline{L}^E$ is isomorphic to the semidirect product $\underline{L}^E = \underline{\widetilde{h}}(\underline{G}_m) \times \underline{SL}_2(E)$ we have $\underline{P}^E = \underline{\widetilde{h}}(\underline{G}_m) \times \underline{P}^0$ and we can identify P^E/P^0 with F^*. Let dy be an additive Haar measure on $F \supset F^*$.

We denote $({}^0\sigma, {}^0W)$ the subspace of the induced representation $ind_{P^0}^{P^E} W^0$ which consists of smooth functions $f : P^E \to W^0$ which satisfy two conditions

a) $f(p^0 p) = \sigma^0_\psi(p^0)f(p), \quad p^0 \in P^0, p \in P^E$.

b) The function $h^*(f)$ on F^* given by $h^*(f)(y) = f(\widetilde{h}(y))$ belongs to the space of Schwartz-Bruhat functions on F^*.

As follows from Lemma 3 the equivalence class of $({}^0\sigma, {}^0W)$ does not depend on a choice of ψ.

We can realize ${}^0W$ as a subspace of $W_2 \stackrel{\text{def}}{=} L^2(F \times F \times E, dydx_0dx)$ which consists of Schwartz-Bruhat functions on $F^* \times F \times E$.

It will be convenient to conjugate this representation with an operator $\phi : f(y, x_0, x) \to f(y, yx_0, yx)$. Then we obtain the following realization of the representation $({}^0\sigma, {}^0W)$.

Lemma 4. *The representation* ${}^0\sigma : P \longrightarrow Aut\ {}^0W_0$ *is given by the formulas*

$$({}^0\sigma(z_a)f)(y,x_0,x) = \psi(ay)f(y,x_0,x)$$
$$({}^0\sigma(t_0',t')f)(y,x_0x) = \psi(t_0'x_0 + t'(x))f(y,x_0,x)$$
$$({}^0\sigma(t_0,t)f)(y,x_0,x) = f(y,x_0+yt_0,x+yt)$$
$$({}^0\sigma\begin{pmatrix}1 & t\\ 0 & 1\end{pmatrix}f)(y,x_0,x) = \psi(y^{-1}Q_t(x_0,x))f(y,x_0,x+x_0t)$$
$$({}^0\sigma\begin{pmatrix}0 & 1\\ -1 & 0\end{pmatrix}f)(y,x_0x) = \gamma_{Q_1}(y,\delta_{E/F})|y|^{-2}$$
$$\int_{F\times E} f(y,\tilde{x}_0,\tilde{x})\psi(y^{-1}(x_0\tilde{x}_0 + T(x\tilde{x}))d\tilde{x}_0 d\tilde{x}$$
$$(\sigma(\tilde{h}(c))f)(y,x_0,x) = f(cy,c^{-1}x_0,x)$$
$${}^0\sigma(h(e)f)(y,x_0,x) = (N(e),\delta_{E/F})|N(e)|f(y,N(e)x_0,N(e)e^{-2}x)$$
$$({}^0\sigma(h(u))f)(y,x_0,x) = f(y,x_0,u^{-1}x)$$

for all $t_0, a \in f,\ c \in F^*,\ t \in E,\ u \in E',\ e \in E^*,\ t_0' \in Hom_F(F,F) = F,\ t' \in Hom_F(E,F)$.

PROOF: Clear.

Let $W_2 \overset{\text{def}}{=} L^2(F\times F\times E, dydx_0dx)$. We denote by ${}^0\sigma_2 : P \to Aut\ W_2$ the unitary representation of P given by the same formulas.

Define operators $\tilde{A}, R_a \in Aut\ W_2,\ a \in F$ by formulas

$$(\tilde{A}f)(y,x_0,x) \overset{\text{def}}{=} \psi\left(-\frac{N(x)}{x_0y}\right) f(-x_0,y,x)$$
$$(R_af)(y,x_0,x) \overset{\text{def}}{=} \psi\left(\frac{aN(x)}{y(y-ax_0}\right) f(y-ax_0,x_0,x).$$

Let $\epsilon : \widehat{G}^E \to G^E$ be as in §1. Since F is a local field, we may consider $\widehat{G}^E$ as a topological group. Let $\tilde{s} \subset \tilde{L}^0 \subset G^E$ be as in §1, $\tilde{C} = \overset{\text{def}}{=} \tilde{L}^0 \cap B$. Then $\tilde{L}^0 = SL_2(F)$, $\tilde{C} = \left\{\begin{pmatrix} a & b\\ 0 & a^{-1}\end{pmatrix}\right\} \subset \tilde{L}^0$.

PROPOSITION 1. *a) The representation* $\sigma_2 : P \to Aut\ W_2$ *can be extended to a representation* $\hat{\pi} : \widehat{G}^E \to Aut\ W_2$ *such that* $\hat{\pi}(\tilde{s}) = \tilde{A}$, $\hat{\pi}\left(\tilde{j}\begin{pmatrix}1 & 0\\ a & 1\end{pmatrix}\right) = R_a$.

b) Let $\hat{\pi}'$ *be a representation of* $\widehat{G}$ *on* W_2 *such that* $\hat{\pi}'|_P = \sigma_2$. *Then* $\hat{\pi}' = \hat{\pi}$.

PROOF: a) As well known the group $\tilde{L}^0$ is the quotient of the free product of groups $\{\tilde{s}^Z\}$ and $\tilde{C}$ by the relations

(a) $\tilde{s}\begin{pmatrix} c & 0\\ 0 & c^{-1}\end{pmatrix} = \begin{pmatrix} c^{-1} & 0\\ 0 & c\end{pmatrix}\tilde{s}.$

(b) $\widetilde{s}^2 = \begin{pmatrix} -1 & 0 \\ 0 & -1 \end{pmatrix}$ and

(c) $\begin{pmatrix} 1 & c \\ 0 & 1 \end{pmatrix}\widetilde{s}\begin{pmatrix} 1 & c^{-1} \\ 0 & 1 \end{pmatrix}\widetilde{s}^{-1}\begin{pmatrix} 1 & c \\ 0 & 1 \end{pmatrix} = \begin{pmatrix} c & 0 \\ 0 & c^{-1} \end{pmatrix}\widetilde{s},\ c \in F^*$.

It is easy to see that $\widetilde{A}\sigma_2(\widetilde{h}(c)) = \sigma_2(\widetilde{h}((c^{-1}))\widetilde{A}, \widetilde{A}^2 = \sigma_2(\widetilde{h}(-1))$ and $\sigma_2(c,0)\widetilde{A}\sigma_2(c^{-1},0)\widetilde{A}^{-1}\sigma_2(c,0) = \sigma_2(\widetilde{h}(c))\widetilde{A}$. Therefore there exists unique representation $\widehat{\pi}_{|\widetilde{L}^0} : \widetilde{L}^0 \to Aut\ W_2$ such that $\widehat{\pi}|_{L^0 \cap B} = \widehat{\sigma}_2|_{L^0 \cap B}$ and $\widehat{\pi}(\widetilde{s})) = \widetilde{A}$. It is easy to see that $\widehat{\pi}(\widetilde{j}\begin{pmatrix} 1 & 0 \\ a & 1 \end{pmatrix}) = R_a$ and $\widehat{\pi}|_{\widetilde{L}^0}$ extends uniquely to a representation $\widehat{\pi}|_{\widetilde{L}}$ of $\widetilde{L}$ on W_2 such that $\widehat{\pi}|_{\widetilde{L}}(t) = \sigma(t)$ for all $t \in T^E$.

For any $\widetilde{h} \in \widetilde{H}$ we have $\sigma_2(\widetilde{s}\widetilde{h}\widetilde{s}^{-1}) = \widetilde{A}\sigma_2(\widetilde{h})\widetilde{A}^{-1}$. Since $\widetilde{P}^E = \widetilde{L}^E \ltimes \widetilde{H}$ we see that $\widetilde{\sigma}_{\widetilde{L}}$ extends uniquely to a representation $\widehat{\pi}|_{\widetilde{P}} : \widetilde{P}^e \to Aut\ W_2$ such that $\widehat{\pi}|_{\widetilde{P}}(b) = \sigma_2(b)$ for all $B \in B^E$. Therefore there exists a representation $\widehat{\sigma} : \widehat{G}^E \to Aut\ W_2$ such that $\widehat{\pi}(\widehat{S}) = A$ and $\widehat{\pi}(\widetilde{j}\begin{pmatrix} 1 & 0 \\ a & 1 \end{pmatrix}) = R_a$.

b) Let $\widehat{\pi}' : \widehat{G}^E \to Aut\ W_2$ be a representation such that $\widehat{\pi}'|_P = \sigma$ and $\widetilde{A} \stackrel{\text{def}}{=} \widehat{\pi}'(\widetilde{s})$. The subgroup $\widetilde{B} \stackrel{\text{def}}{=} T\widetilde{H} \subset B^E$ is normalized by $\widetilde{S}$. Therefore $\widetilde{A}$ satisfies the conditions. $\widetilde{A}\sigma_2(\widetilde{b})\widetilde{A}^{-1} = \sigma_2(\widetilde{s}\widetilde{b}\widetilde{s}^{-1})$ for all $\widetilde{b} \in \widetilde{B}$. This implies the existence of a function α on F^* such that $(\widetilde{A}f)(y,x_0,x) = \alpha((yx_0)\psi\left(-\frac{N(x)}{x_0y}\right) f(-x_0,y,x)$ and $\alpha(N(t)t_0) = \alpha((t_0)$ for all $t_0 \in F^*$, $t \in E^*$. The relation (c) in $\widetilde{L}^0$ implies that for any $x_0, y, t \in F$ we have $\alpha((x_0 - ty)y)\alpha(t^{-1}x_0(x_0 - ty)) = \alpha(x_0y)$. Since the image $Im\ N$ of the norm map $N : E^* \to F^*$ is an open subgroup in F^* we have $1 - t \in Im\ N$. If $|t|\mathbb{L}1$, take $x_0 = y = 1$. We find $\alpha(1)\alpha(t^{-1}) = \alpha(1)$ for all sufficiently small t. This implies $\alpha \equiv 1$. Proposition 1 is proved.

We can formulate the main results of this paper more precisely.

THEOREM 1. *The representation $\overline{\Pi}^E$ of G^E is unitarizable and $\widehat{\pi}$ is equivalent to the representation $\overline{\Pi}_2^E \circ \epsilon$ where $\overline{\Pi}_2^E$ is the unitary completion of $\overline{\Pi}^E$.*

REMARK: We will denote $\overline{\Pi}^E$ by (π, W) and its unitary completion by (π_2, W_2). It is clear now that Theorem 1 implies Theorems A and C.

Let $A \subset Aut\ W$ be the operator given by $A \stackrel{\text{def}}{=} {}^0\sigma\left(\begin{pmatrix} 0 & 1 \\ -1 & 0 \end{pmatrix}\right)$.

COROLLARY 1. $(A\widetilde{A})^3 = (\widetilde{A}A)^3$.

REMARK: It would be interesting to find a direct proof of Corollary 1. As follows from Theorem 1.2, Corollary 1 implies Theorem A.

PROPOSITION 2. *Theorem C is true in the case when $E = F^3$.*

PROOF: In this case the representation ρ_E of the Levi subgroup $\widetilde{L}^E$ is the representation unitary induced from the trivial character of its Borel subgroup. Therefore the representation $\overline{\Pi}^E$ is a quotient of a representation of the principal series of G^E and Proposition 2 follows immediately from Theorem 4 in [KS].

4. In this section F will be a global field of characteristic zero, Σ the set of places of F, $\Sigma' \subset \Sigma$ the subset of nonarchimedean places, $\mathbb{A}$ the adele ring of F, $E \supset F$ a cubic extension, A_E the adele ring of E. For any $\nu \in \Sigma$ we denote by F_ν the completion of F at ν. Then $E_\nu \overset{\text{def}}{=} E \otimes_F F_\nu$ is a separable 3-dimensional
F_ν-algebra. If $\nu \in \Sigma'$ we denote by $\mathcal{O}_\nu \subset F_\nu$ the ring of integers in F_ν. We fix a nontrivial additive character ψ on $\mathbb{A}$ such that $\psi|_F \equiv 1$. For any $\nu \in \Sigma$ the restriction of ψ on F_ν defines a nontrivial additive character ψ_ν on F_ν. Let $H_\nu = H(F_\nu)$, $P^0_\nu = P^0(F_\nu)$, $P_\nu = P(F_\nu)$ and $H_{\mathbb{A}}, P^0_{\mathbb{A}}, P^E_{\mathbb{A}}$ be the corresponding adelic groups. The isomorphism $P^E = G_m \ltimes P^0$ defines an isomorphism $P^E_{\mathbb{A}} = I_{\mathbb{A}} \ltimes P^0_{\mathbb{A}}$ where $I_{\mathbb{A}} \subset \mathbb{A}$ is the group of ideles. For any $\nu \in \Sigma$ we have a representation $\sigma^0_\nu : P^0_\nu \longrightarrow Aut\ W^0_\nu$ and for almost all $\nu \in \Sigma'$ we have $dim(W^0_\nu)^{P^0_{\mathcal{O}_\nu}} = 1$ where $P^0_{\mathcal{O}_\nu} \subset P^0_\nu$ is the subgroup of $\mathcal{O}_\nu$-points. Let $\sigma^0_{\mathbb{A}} : P^0_{\mathbb{A}} \longrightarrow Aut\ W_{\mathbb{A}}$ be the restricted tensor product of $\sigma^0_\nu, \nu \in \Sigma$.

Let W^0_ψ be the space of smooth functions f on $H_{\mathbb{A}}$ such that $f(\gamma z_a h) = \psi(a) f(h)$ for all $\gamma \in H_F$, $h \in H_{\mathbb{A}}$, $a \in F$ and $|f|$ has compact support on $H_{\mathbb{A}}/H_F Z_{\mathbb{A}}$. We define an L^2-norm $\|\ \|_{W^0_\psi}$ on W^0_ψ in a natural way. The group $H_{\mathbb{A}}$ acts on W^0_ψ by right shifts

$$(\sigma^0_\psi(h_0)f)(h) = f(hh_0),\ f \in W^0_\psi,\ h_0, h \in H_{\mathbb{A}}.$$

LEMMA 1. *The restriction of $\sigma^0_{\mathbb{A}}$ on $H_{\mathbb{A}}$ is equivalent to σ^0_ψ.*

PROOF: Follows from [W], §18.

We denote by $\alpha : W^0_\psi \overset{\sim}{\longrightarrow} W^0_{\mathbb{A}}$ an intertwining operator. α is unique up to a multiplication by a scalar. We extend σ^0_ψ to a representation $\sigma^0_\psi : P^0_{\mathbb{A}} \to Aut\ W^0_\psi$ by $\sigma^0_\psi \overset{\text{def}}{=} \alpha^{-1} \circ \sigma^0_{\mathbb{A}} \circ \alpha$. It is clear that the action of the subgroup $SL_2(E)$ of $SL_2(\mathbb{A}_E))$ on $H_{\mathbb{A}}$ preserves H_F and therefore defines a natural action of $SL_2(E)$ on $W^0_\psi : g : f \to f^g$ where $f^g(h) = f(h^{g^{-1}})$.

LEMMA 2. *For any $g \in SL_2(E)$, $f \in W^0_\psi$ we have*

$$\sigma^0_\psi(g)f = f^g.$$

PROOF: For any $g \in SL_2(E)$ consider the map $B : W^0_\psi \to W^0_\psi$ given by $f \to Bf \overset{\text{def}}{=} \sigma^0_\psi(g^{-1})f^g$. It is clear that B commutes with the action of

$H_{\mathbb{A}}$. Since the representation σ_ψ^0 is irreducible we have $B = \lambda_g \cdot 1$ for some $\lambda_g = \mathbb{C}^*$. It is clear that the map $g \longrightarrow \lambda_g$ defines a group homomorphism $SL_2(F) \longrightarrow \mathbb{C}^*$. Since the group $SL_2(\underline{F})$ is perfect we have $\lambda \equiv 1$. Lemma 2 is proved.

Let W_ψ be the space of functions $f : P_{\mathbb{A}}^E \to W_\psi^0$ such that a) $f(p^0p) = \sigma_\psi^0(p^0)f(p)$, $p^0 \in P_{\mathbb{A}}^0$, $p \in P_{\mathbb{A}}^E$, b) there exists a smooth function $\widetilde{f} : \mathbb{A} \to W_\psi^0$ such that $\widetilde{f}(\ell) = f(\widetilde{h}(\ell))$, for all $\ell \in I_{\mathbb{A}} \subset \mathbb{A}$ and c) $|\widetilde{f}||_{W_\psi} \in L^2(\mathbb{A}, dx_{\mathbb{A}})$ where $dx_{\mathbb{A}}$ is a Haar measure on $\mathbb{A}$. $P_{\mathbb{A}}$ acts on W_ψ by right shifts.

Let $\pi : P_{\mathbb{A}} \to Aut\ D$ be a representation and D' be the dual space to D. For any linear functional $d' \in D'$ we define a map $\varphi_{d'} : D \to \mathbb{C}(P_{\mathbb{A}})$ by $\varphi_{d'}(d)(p) \overset{\text{def}}{=} d'(\pi(p)d)$,
$d \in D, p \in P_{\mathbb{A}}$. Let $D'_F = \{d' \in D' | d'(\pi(\gamma)d) = d'(d)\ \forall d \in D,\ \gamma \in P_F\}$. For any $d' \in D'_F$ we define $d'_\psi \in D'$ by $d'_\psi(d) \overset{\text{def}}{=} \int_{\mathbb{A}/F} \psi(-a)d'(\pi(z_a)d)da$.

Lemma 3. *Let $d' \in D'_F$ be such that* ker $\varphi_{d'} = \{0\}$. *Then for any $d \in$ ker $\varphi_{d'_\psi}$ we have $\pi(z_a)d = d$ for all $a \in \mathbb{A}$.*

Proof: Fix $d \in$ ker $\varphi_{d'_\psi}$, and any character $\widetilde{\psi}$ of $\mathbb{A}/F$ define $d_{\widetilde{\psi}} \overset{\text{def}}{=} \int_{\mathbb{A}/F} \widetilde{\psi}(-a)\pi(z_a)\ da$. We claim that for any $\widetilde{\psi} \neq Id$ we have $d_{\widetilde{\psi}} = 0$. Really there exists $x \in F^*$ such that $\widetilde{\psi}(a) = \psi(xa)$ $\forall a \in \mathbb{A}$. By Lemma 3.3 we can find $\gamma \in P_F$ such that $\gamma^{-1}z_a\gamma = z_{x^{-1}a}$ $\forall a \in \mathbb{A}$. We have $\varphi_{d'}(d_{\widetilde{\psi}})(p) = \varphi_{d'_\psi}(d)(\gamma p)$ for all $p \in P_A$. So $d_{\widetilde{\psi}} \in$ ker $\varphi_{d'} = \{0\}$. Since $d_{\widetilde{\psi}} = 0$ for all nontrivial characters $\widetilde{\psi}$ of $\mathbb{A}/F$ we have $\pi(z_a)d = d$ $\forall a \in \mathbb{A}$. Lemma 3 is proved.

Suppose that for each $\nu \in \Sigma$ we have an irreducible unitary representation $\pi_\nu : G_\nu \to Aut\ D_{2,\nu}$ such that $dim\ D_{2,\nu}^{G_{\mathcal{O}_\nu}} = 1$ for almost all $\nu \in \Sigma'$. We choose $d_\nu^0 \in D_{2,\nu}^{G_{\mathcal{O}_\nu}} - \{0\}$ for such $\nu \in \Sigma'$. Let $D_\nu \subset D_{2,\nu}$ be the subspace of smooth vectors and $(\pi, D_{\mathbb{A}})$ be the restricted tensor product of (π_ν, D_ν). That is, the space $D_{\mathbb{A}}$ is the space of finite linear combinations of vectors of the form $d = \otimes_{\nu\in\Sigma} d_\nu, d_\nu \in D_\nu, d_\nu = d_\nu^0$ for almost all $\nu \in \Sigma'$. The group $G_{\mathbb{A}}$ acts naturally on $D_{\mathbb{A}}$. Let Σ_0 be the set of places $\nu \in \Sigma$ such that E splits at ν.

Proposition 1. *Let $\pi : G_{\mathbb{A}} \to Aut\ D_{\mathbb{A}}$, $\pi = \otimes_{\nu\in\Sigma}\pi_\nu$ be the tensor product of irreducible representations of G_ν. Suppose that $D'_F \not\subset \{0\}$ and that for some $\nu_0 \in \Sigma_0$ the restriction of $(\pi_\nu, D_{2,\nu})$ to P_ν is equivalent to $\sigma_{2,\nu}$. Then the restriction of (π_ν, D_{2,ν_0}) to P_{ν_0} is equivalent to $\sigma_{2,\nu}$ for all $\nu \in \Sigma$.*

PROOF: Fix $d' \in D'_F, d' \neq 0$. Let $\varphi_{d'} : D \to \mathbb{C}(P_{\mathbb{A}})$ be the map as in Lemma 1.11.

LEMMA 4. *$\varphi_{d'}$ is an embedding.*

PROOF OF LEMMA 4: Let $\phi_{d'} : D \to \mathbb{C}(G_{\mathbb{A}})$ be the map defined by $\phi_{d'}(d)(g) \overset{\text{def}}{=} d'(\pi(g)d)$, $d \in D, g \in G_{\mathbb{A}}$. It is clear that $\varphi_{d'}$ intertwines the actions of $G_{\mathbb{A}}$ on D and on $\mathbb{C}(G_{\mathbb{A}})$. Therefore ker $\varphi_{d'}$ is a $G_{\mathbb{A}}$-invariant subspace of D. Since $D = \otimes_\nu D_\nu$ and the representations $\pi_\nu : G_\nu \to Aut\ D_\nu$ are irreducible, we see that ker $\phi_{d'} = 0$. Take $d \in \ker_{\varphi_{d'}}$ and define $f_d \overset{\text{def}}{=} \phi_{d'}(d) \in \mathbb{C}(G_{\mathbb{A}})$. It is clear from the definition that $f_d(\gamma g) = f_d(g)$ for all $\gamma \in G_F$, $g \in G_{\mathbb{A}}$. Since $\phi_{d'} = 0$ we have $f_d|_{P_{\mathbb{A}}} = 0$. Therefore $f_d|_{G_F P_{\mathbb{A}}} = 0$.

The subset $G_F P_{\mathbb{A}} \subset G_{\mathbb{A}}$ is dense and we conclude that $f_d \equiv 0$. Since $\phi_{d'}$ is an imbedding, we conclude that $d = 0$. Lemma 4 is proved.

Define a linear functional d'_ψ on D as in Lemma 3, and write φ instead of $\varphi_{d'_\psi} : D \to \mathbb{C}(P_{\mathbb{A}})$. For any $p \in P_{\mathbb{A}}$, $d \in D$ defines a function $w_d(p)$ on $H_{\mathbb{A}}$ by $w_d(p)(h) = \varphi(d)(hp)$. It is clear that $w_d(p) \in W^0_\psi$.

LEMMA 5. *For any $d \in D$, $p^0 \in P^0_{\mathbb{A}}$, $p \in P_{\mathbb{A}}$ we have $w_d(p^0 p) = \sigma^0_\psi(p^0) w_d(p)$.*

PROOF: Fix $d \in D$, $p \in P_{\mathbb{A}}$ and define a function $\Phi : P^0_{\mathbb{A}} \to W_\psi$ by $\Phi(p^0) \overset{\text{def}}{=} \sigma^0_\psi(p^0)^{-1} w_d(p^0 p)$. By the assumptions of Proposition 1 we have $\Phi(p^0 p^0_v) = \Phi(p^0)$ for any $p^0_{v_0} \subset P_{\mathbb{A}}$. On the other hand, it follows from Lemma 2 that $\Phi(\gamma p) = \Phi(p)$ for all $\gamma \in P_F$. Since the set $P^0_E \cdot P^0_{\nu_0}$ is dense in $P^0_{\mathbb{A}}$ we concluded that $\varphi \equiv$ constant. Lemma 5 is proved. It follows from Lemmas 3 and 4 that φ is an imbedding. Now Proposition 1 follows from Lemma 1.

Now we can prove Theorem 3.1. Let $\check{F}$ be a local field and $\check{E} \supset \check{F}$, $dim_F \check{E} = 3$ be a commutative semisimple $\check{F}$-algebra. We can find a global field F, a non-Galois cubic extension E and a place $\nu_0 \in \Sigma$ such that $\check{F} = F_{\nu_0}$, $\check{E} = E_{\nu_0}$.

Let $\mathcal{G}$ be the Galois group $\mathcal{G} = Gal(\overline{F} : F)$ and $\tau_E : \mathcal{G} \to S_3$ be the morphism corresponding to the cubic extension E over F. Since E is not normal, τ_E is surjective. Let $\mu_E : \mathcal{G} \to GL_2(\mathbb{C})$ be the representation as in §1. We can associate with μ_E a cuspidal automorphic representation ρ_E of the group $\widetilde{L}^E_{\mathbb{A}}$ in a subspace $M^E \subset \mathbb{C}(\widetilde{L}_{\mathbb{A}}/\widetilde{L}_F)$. For any good function f on $G^E_{\mathbb{A}}/\tilde{\rho}_F \cdot \tilde{U}_{\mathbb{A}}$ with values in M^E we denote by $E_f(g,s)$ the corresponding Eisenstein series of the group G^E (see [L]).

It is easy to see that the corresponding L-funciton $L(\rho_E, s)$ has a pole of the first order at $s = 1$. Therefore (see [L]) functions $E_f(g,s)$ have a first order pole at $s = 1$. We define $\varphi_f \overset{\text{def}}{=} \lim_{s\to 1}(s-1)E_f(g,s)$. As follows from [L] ϕ_f are automorphic forms from $G^E_{\mathbb{A}}/G^E_F$ they span a nonzero $G_{\mathbb{A}}$-invariant subspace $W_{\mathbb{A}}$ of $L^2(G^E_F \backslash G^E_{\mathbb{A}})$ and the representation $\overline{\Pi}_{\mathbb{A}}$ of $G^E_{\mathbb{A}}$ on $W_{\mathbb{A}}$ is irreducible. Moreover it follows from arguments in [J] §2.4 that $\overline{\Pi}_{\mathbb{A}}$ is equivalent to the restricted tensor product $\bigotimes_{\nu\in\Sigma}\overline{\Pi}_\nu$ where representations $\overline{\Pi}^{E_\nu}_\nu$ of G^{E_ν} where defined in §1.

As follows from Proposition 4.3 the representation $\overline{\Pi}_{\mathbb{A}}$ satisfies assumptions of Proposition 1. Therefore for all $v \in \Sigma$ the restriction of $\overline{\Pi}_\nu$ on P_ν is equivalent to the representation $\overline{\Pi}_{2,\nu}$. In particular, we have a representation $\sigma_{\nu_0} : G^{\check{E}} \to Aut\ W_{\nu_0}$ such that the restriction of σ_{ν_0} on $P^{\check{E}}$ is equivalent to σ_2. As follows from Proposition 3.1, $\widehat{\sigma} = \sigma_{\nu_0} \circ \nu$. Theorem 3.1 is proved.

5. In this section we fix a local nonarchimedean field F of characteristic zero, a field $E \supset F$, $dim_F E = 3$, and will write G, P instead of G^E, P^E. The restriction of (π, W) (see §3) to P we denote by (σ, W) or simply σ. The goal of this section is to get some information about σ.

For any algebraic F-group H we denote by $\mathcal{A}(H)$ the category of smooth representations of H (see [BZ] 2.1) and by $\mathcal{E}(H)$ the set of equivalence classes of algebraically irreducible smooth representations of H.

Let U be a unipotent F-group, $\theta : U \to \mathbf{C}^*$ a character of U. For any smooth representation $\tau : U \to Aut\ R$ we denote by $R(U,\theta) \subset R$ the subspace generated by vectors $\langle\tau(u)r - \theta(u)r\rangle$, $u \in U$, $r \in R$ and by $R_{U,\theta}$ the quotient space $R_{U,\theta} \stackrel{\text{def}}{=} R/R(U,\theta)$. We will write R_U instead of $R_{U,Id}$.

LEMMA 1. *For unipotent group U and a character θ of U the functor $R \to R_{U,\theta}$ from $\mathcal{A}(U)$ to the category of vector spaces is exact.*

PROOF: See ([BZ] 2.30).

We start with some general results on representation of the group P. For simplicity, we will write $\mathcal{A}$ instead of $\mathcal{A}(P)$ and define subcategories $\mathcal{A}_0, \mathcal{A}_1, \mathcal{A}_2$ of $\mathcal{A}$ as follows.

$\mathcal{A}_0 = \{(\pi, R) \in \mathcal{A} | R_Z = \{0\}\}$

$\mathcal{A}_1 = \{(\pi, R) \in \mathcal{A} | \pi_Z = Id$ and $R_H = \{0\}\}$ and

$\mathcal{A}_2 = \{(\pi, R) \in \mathcal{A} | \pi_{|H} = Id\}$ where $\pi_{|H}, \pi_{|Z}$ are the restrictions of π on H and Z.

For any representation $(\pi, R) \subset \mathcal{A}$ we define subspaces R_0, R_1 of R by $R_0 \stackrel{\text{def}}{=} \ker(R \to R_Z)$ and $R_1 \stackrel{\text{def}}{=} \ker(R \to R_H)$ where $R \to R_Z$, $R \to R_H$ are natural projections. It is clear that R_0, R_1 are P-invariant subspaces of R and $R_0 \subset R_1$.

Let $(\tau_0, R_0) \in \mathcal{A}$ be the restriction of τ on R_0, $(\tau_1, \overline{R}_1) \in \mathcal{A}$ the natural action of P on $\overline{R}_1 = R_1/R_0$ and $(\tau_2, \overline{R}_2) \in \mathcal{A}$ be the action of P on $\overline{R}_2 \stackrel{\text{def}}{=} R/R_1$.

LEMMA 2. *$(\tau_0, R_0) \in \mathcal{A}_0$, $(\tau_1, \overline{R}_1) \in \mathcal{A}_1$ and $(\tau_1, \overline{R}_2) \in \mathcal{A}_2$.*

PROOF: Follows from Lemma 1.

For any $v \in V$ we define by θ_v the character of V defined by

$$\theta_v(\tilde{v}) \stackrel{\text{def}}{=} \psi(\langle v, \tilde{v}\rangle).$$

Since V is a quotient of H we can and will consider θ_v as a character of H.

Let $\Omega \subset V - \{0\}$ be a subvariety as in Lemma 3.4. We fix $v_0 \in \Omega$ and denote by S_0 the stabilizer of v_0 in L and by θ_0 the character θ_{v_0}. It is clear that S_0 is isomorphic to the semidirect product $S_0 = E^* \ltimes E$, where E^* is the multiplicative group of E. For any character χ of E^* we extend it to a character χ on S_0 trivial on E and extend θ_0 to a character θ_0^χ of the semidirect product $S_0 \ltimes H$ by $\theta_0^\chi(s,h) \stackrel{\text{def}}{=} \theta_0(h)\chi(s)$. We denote by $\rho_0^\chi \in \mathcal{A}_1(P)$ the unitary induced representation $\rho_0^\chi \stackrel{\text{def}}{=} ind(P, S_0 \ltimes H, \theta_0^\chi)$.

For any representation $(\tau, R) \in \mathcal{A}$ the action of S_0 on R preserves $R(H, \theta_0)$ and therefore defines the action of S_0 on $R_{H,\theta_0} : (s, \overline{r}) \to s\overline{r}$, $s \in S_0$, $\overline{r} \in R_{H,\theta_0}$. We denote by τ_{θ_0} the representation of S_0 on R_{H,θ_0} given by $\tau_{\theta_0}(s)\overline{r} = |s|^{1/2} s\overline{r}$, where for $s = (e^*, e) \in S_0$, $|s| \stackrel{\text{def}}{=} |e^*|$. For any representation $(\pi, R) \in \mathcal{A}$ we define
$X(\tau) = \{v \in V - \{0\} | R_{H,\theta_v} \neq \{0\}\}$.

LEMMA 3. *a) $X(\tau)$ is an L-invariant subset of $V - \{0\}$.*

b) Let $(\tau, R) \in \mathcal{A}$ be such that $X(\tau) = \Omega$, $\dim R_{H,\theta_0} = 1$. Then $\tau = \rho_0^\chi$, where χ is the restriction of θ_0 on $E^* \subset S_0$.

PROOF: a) is clear. b) follows from the Mackey theory (see [BZ]§5).

Let σ be the restriction of π on P, $0 \subset W_0 \subset W_1 \subset W$ be the filtration as in Lemma 2, ${}^0W \stackrel{\text{def}}{=} S(F^* \times F \times E) \subset W$ as in the end of §3.

We denote by σ_1, σ_2 the representations of P on $\overline{W}_1 \stackrel{\text{def}}{=} W_1/W_0$ and $\overline{W} \stackrel{\text{def}}{=} W/W_1$.

PROPOSITION 1. *a) $W_0 = {}^0W$.*

b) $\sigma_1 \simeq \rho_0^{\chi_E}$ where χ_E is the character of E^* given by $\chi_E(c) = (N(c), \delta_{E/F})$, $c \in E^*$.

c) If $E \subset F$ is not normal then $\overline{W} = \{0\}$.

d) If $E \subset F$ is normal, then $N(E^*)$ is a subgroup of index 3 in F^*. Let $\mathcal{E}$ be a nontrivial character on $F^*/N(E^*)$. Then $\overline{W} = \mathbf{C}_{\mathcal{E}} \oplus \mathbf{C}_{\mathcal{E}^2}$ where $\mathbf{C}_{\mathcal{E}}$ is a 1-dimensional space on which P acts by the character $p \to |\omega(p)|\mathcal{E}(\omega(p))$ for all $p \in P$.

PROOF: a) It is clear that ${}^0W$ is a P-invariant subspace of $W_0 \subset W$. To prove a) we have to show that Z acts trivially on $W/{}^0W$. In other words, we have to show that for any $f \in W$ and $y \in F$ we have $\sigma(E_\omega(y))f - f \in {}^0W$. But this follows immediately from formulas for $\sigma_{|H}$.

b) Let $X(\sigma_1)$ be the subset of $V = F \oplus E \oplus E \oplus F$ as in Lemma 3, $V_1 \subset V$ be the affine subspace $N = (x_0, x, \tilde{x}, 1)$, $x_0 \in F, x, \tilde{x} \in E$ and $X \stackrel{\text{def}}{=} X(\sigma_1) \cap V_1$.

LEMMA 4. *a)* $X = \{(N(x), \theta(x), x, 1)\}$, $x \in E$ *and*

b) for any $v \in X$, *dim* $W_{H,\theta_v} = 1$ *and* $\chi_{\theta_v}(c) = (N(c), \delta_{E/F})$ *for all* $c \in E^*$.

PROOF: a) Let $\check{H} \stackrel{\text{def}}{=} \tilde{s}H\tilde{s}^{-1}$. For any $v \in N$ we denote by $\check{\theta}_v$ the character of $\check{H}$ given by $\check{\theta}_v(\check{h}) \stackrel{\text{def}}{=} \theta_v(\tilde{s}^{-1}\check{h}\tilde{s})$. Let $U \stackrel{\text{def}}{=} \Lambda' Z \subset H$. It is clear that $\tilde{s}U\tilde{s}^{-1} = U$.

For any $x \in E$ we denote by χ_x the character of U given by $\chi_x(\tilde{x}, \tilde{x}_0, a) \stackrel{\text{def}}{=} \psi(T(x\tilde{x}) - \tilde{x}_0)$ and by $\check{\chi}_x$ the character of U given by $\hat{\chi}_x(u) = \chi_x(\tilde{s}u\tilde{s}^{-1})$. As follows from Proposition 1 and formulas for $\sigma|_U$ we have *dim* $W_{1,U,\check{\chi}_x} = 1$ for all $x \in E$ and the natural projection $W \to W_{U,\hat{\chi}_x} = \mathbf{C}$ is given by the formula $f \to f(-1, 0, x)$ for $f = f(y, x_0, x) \in W$. Therefore for any $v \in N \subset V$ we have $W_{\check{H},\check{\theta}_v} \neq 0$ if and only if $v = (N(x), \theta(x), x, 1)$ for some $x \in E$ and in this case *dim* $W_{\check{H},\check{\theta}_v} = 1$. Therefore $X = \{N(x), \theta(x), x, 1\}$ and *dim* $W_{H,\theta_v} = 1$ for all $v \in X$. It is easy to check that $\pi_{\theta_v}(c) = (N(c), \delta_{E/F})$ for $v \in X$. Lemma 4 is proved.

Now we can prove Proposition 1b). As follows from Lemma 1 we have $X(\sigma_1) = X(\sigma)$. Therefore $X(\sigma_1)$ is an L-invariant subset of $V - \{0\}$ and by Lemma 4, $X(\sigma_1) \cap N = \Omega \cap N$. So $X(\sigma_1) \supset \Omega$ and $X(\sigma_1) - \Omega$ is an L-invariant subset of V such that $(X(\sigma_1) - \Omega) \cap N = \emptyset$. It is easy to see that this implies $X(\sigma_1) = \Omega$. Proposition 1b) follows from Lemmas 1, 2, 3, and 4 and formulas for σ.

c) We have $W_H = W/W_1$. (In other words, the representation of L on W/W_1 is the value of Jacquet functor $\Gamma^L_G(W)$.) As follows from Theorem 3.1, π is a subquotient of a representation Π induced from a cuspidal residual representation of $\widetilde{P} = \widetilde{L} \cdot \widetilde{U}$. Therefore $\Gamma^L_G(W) = \{0\}$.

d) The group L acts naturally on $\bar{W}$ and it follows from Theorem 3.1 that $\bar{W}$ has exactly two 1-dimensional quotients which are isomorphic to $\mathbf{C}_{\mathcal{E}}$ and $\mathbf{C}_{\mathcal{E}^2}$. The following result implies the validity of Proposition 1d).

Let ψ_E be the additive character of E given by $\psi_E(e) = \psi(Tr_{E/F}(e))$. We can identify E with the root subgroup E_α of G^E. We extend ψ_E from E_α to the unipotent radical U_B of B in such a way that $\psi_E|_U \equiv 1$.

LEMMA 5. $W_{U_B,\psi_E} = \{0\}$.

PROOF: Let $\widetilde{U_B} \stackrel{\text{def}}{=} \tilde{s}U_B\tilde{s}k^{-1} \subset G^E$ and $\tilde{\psi}$ be the character of $\widetilde{U_B}$ given by $\tilde{\psi}(n) = \psi_E(\tilde{s}^{-1}n\tilde{s})$. It is sufficient to show that $W_{\widetilde{U_B},\widetilde{\psi}_E} = \{0\}$. But this follows immediately from Proposition 1b) and Corollary to Lemma 2.3.

Proposition 1 is proved.

Let $\underline{U_3}$ be the subgroup of $\underline{G}^E$ generated by root subgroups $\underline{E}_{\beta_0}$, $\underline{E}_{\gamma_0}$, and $\underline{E}_\omega$. If $\widehat{\psi} : U_3 \to \mathbf{C}^*$ is a character of U_3 we say that $\widehat{\psi}$ is nondegenerate if the restriction on $\widehat{\pi}$ on the subgroups E_{β_0} and E_{γ_0} is nontrivial. Let $\widehat{\psi}$ be a nondegenerate character of U_3. Then there exists an element $\alpha_\psi \in F^*$ such that for any $a \in F$, $\widehat{\psi}(E_{\beta_0}(a)) = \widehat{\psi}(s\ E_{\beta_0}(\alpha_\psi \alpha))s^{-1})$ where the element $s \in L^E$ is as in Lemma 1.3. It is easy to see that the image $\bar{\alpha}_\psi$ of α_ψ in the group $F^*/N(E^*)$ does not depend on a choice of the element s.

Let E' be the kernel of the norm map $N : E^* \to F^*$ and χ be a character of E^*. Let i be the imbedding $i : (SL_3(F) \times E')/\mu_3 \to G^E$ as in §1. We will identify U_3 with the maximal unipotent subgroup of $SL_3(F)$. Let $\widetilde{W}_\chi \subset W$ be the subspace of $w \in W$ such that $\pi(i(1 \times e'))w = \chi'(e')w$ for all $e' \in E'$. The natural action of $SL_3(F)$ on W_χ extends (see §1) to a representation $\tilde{\pi}_\chi$ of the group G_3^E on $\widetilde{W}_\chi$.

LEMMA 6. *Let $\widehat{\psi}$ be a nondegenerate character of U_3. We have*

$$dim(\widetilde{W}_\chi)_{U_3,\widehat{\psi}} = \begin{cases} 1 & \text{if } \bar{\alpha}_\psi = 1 \\ 0 & \text{otherwise.} \end{cases}$$

PROOF: Follows from Proposition 1b) and Lemma 2.3.

PROPOSITION 2. *Let (π_χ, W_χ) be the representation of $GL_3(F)$ induced from the representation $(\tilde{\pi}_\chi, \widetilde{W}_\chi)$ of G_3^E. Then π_χ is an irreducible representation.*

PROOF: As follows from Lemma 6 for any nondegenerate character $\widehat{\psi}$ of U_3 we have $dim(W_\chi)_{U_3,\psi} = 1$. Therefore there exists an irreducible nondegenerate (see [BZ]) subrepresentation $W_\chi^0 \subset W_\chi$ such that the quotient $\bar{W}_\chi$ is degenerate. As follows from Proposition 1 b) and Corollary to Lemma 2.4 that $(\bar{W}_\chi)_{U_3,Id} = \mathbf{C}_\mathcal{E} \oplus \mathbf{C}_{\mathcal{E}^2}$ if $E \supset F$ is normal. This implies immediately that $\bar{W}_\chi = \{0\}$. Proposition 2 is proved.

Assume now that E is a normal extension of F and $\mathcal{E} : F^*/N(E^*) \to \mathbf{C}^*$ is a nontrivial character. Consider the function φ on $F^* \times E$ given by $\varphi_\mathcal{E}(x_0, x) = \frac{\mathcal{E}(x_0}{|x_0|}\psi\left(\frac{N(x)}{x_0}\right)$. We can easily extend $\varphi_\mathcal{E}$ to a distribution on $F \times E$. As in §3 we can define the Fourier transform $\mathcal{F}$ on the space of distributions on $F \times E$.

PROPOSITION 3. *$\mathcal{F}(\varphi_\mathcal{E}) = \gamma\varphi_\mathcal{E}$ for some $\gamma \in \mathbf{C}^*$.*

PROOF: As follows from Proposition 1d) there exists a functional λ^+ on W such that $\lambda^+(pw) = \mathcal{E}(w(p))|w(p)|\lambda^+(w)$ for all $w \in W, p \in P$.

Let $\underline{P}^- = \underline{LU}^- \subset \underline{G}^E$ be the parabolic opposite to $\underline{P}$. The subgroups $\underline{P}$ and $\underline{P}^-$ of $\underline{G}^E$ are conjugate. Therefore, there exists a nonzero functional

$\lambda \in W'$ such that $\pi'(h^-)\lambda = \lambda$ and $\pi'(\ell)\lambda = \mathcal{E}^{-1}(\omega(\ell))|\omega(\ell)|^{-1}\lambda$ for $\ell \in L, u^- \in U^-$.

Let ${}^0W \subset W_2$ be the subspace of locally constant functions on $F^* \times F \times E$ with compact support. As follows from formulas for π_2 we have ${}^0W \subset W$.

LEMMA 7. *Let ${}^0\lambda$ be the restriction of λ on ${}^0W$. Then ${}^0\lambda \neq 0$.*

PROOF: Suppose that $\lambda\big|_{{}^0W} = 0$. Then we could consider λ as a functional on $W/{}^0W$. In this case, as follows from Proposition 1a), λ is Z-invariant. But P^- and Z generate G. Therefore λ is a G-eigenfunctional. Since W is a nontrivial irreducible representation of G we have a contradiction. Therefore ${}^0\lambda \neq 0$. Lemma 7 is proved.

Since L preserves ${}^0W$, it acts on ${}^0W'$ and ${}^0\lambda$ is an L-eigenfunctional. Consider the subspace $\check{W} = S(F^* \times F^* \times E) \subset {}^0W = S(F^* \times F \times E)$ and denote by $\check{\lambda}$ the restriction of ${}^0\lambda$ on $\check{W}$. Since ${}^0\lambda$ is an L-eigenfunctional we see that ${}^0\lambda$ is invariant under the Fourier transform A on ${}^0W$. This implies that $\check{\lambda} \neq 0$.

The subgroup $B_L \stackrel{\text{def}}{=} B \cap L$ of P^- preserves $\check{W}$. The condition $\pi'(b)\lambda = |\omega(b)|^{-1}\lambda\ \mathcal{E}(\omega(b))^{-1}$ for all $b \in B_L$ shows the existence of a function $\mathcal{H}$ on $F^*/N(E^*)$ such that $\check{\lambda}(f) = \int(\lambda_{\mathcal{H}} \cdot f)(y, x_0, x)dydx_0dx$ for all $f \in \check{W}$ where $\lambda_{\mathcal{H}}$ is a locally constant function on $F^* \times F^* \times E$ given by $\lambda_{\mathcal{H}}(y, x_0, x) = \mathcal{H}(yx_0)\frac{(x_0, \delta_{E/F})}{|x_0|}\psi\left(\frac{N(x)}{yx_0}\right)$.

Let $f \in \check{W}$ be a function such that for any $x_0 \in F^*, x \in E$ we have $f(x_0, x_0, x) = 0$. Consider the function $\tilde{f}$ on $F^* \times F^* \times E$ given by $\tilde{f}(y, x_0, x) = \psi\left(\frac{N(x)}{y(y-x_0)}\right) f(y - x_0, x_0, x)$. We have $\tilde{f} \in S(F^* \times F^* \times E) = \check{W}$. Since λ is an $E_{-\beta_0}(1)$-invariant functional we have $\check{\lambda}(f) = \check{\lambda}(\tilde{f})$ for any $f \in \check{W}$ such that $f(x_0, x_0, x) = 0$ for all $x_0 \in F^*, x \in E$. Therefore, for any $x_0, y \in F^*, x_0 \neq y_0$ we have $\mathcal{H}(y(y - x_0)) = \mathcal{H}(yx_0)$. As in the proof of Proposition 3.1b) this implies $\mathcal{H} =$ constant.

Now it is clear that ${}^0\lambda$ coincides with the functional $\varphi_{\mathcal{E}}$. Proposition 3 is proved.

6. In this section we consider the case when F is a local nonarchimedean field and E is a direct sum $E = F \oplus K$ where K is a semisimple 2-dimensional F-algebra. We denote by $z \to \bar{z}$ the nontrivial involution of K. It is clear that for any $z \in K$, $N(1, z) = z\bar{z}$ and $T(0, z) = z \otimes \bar{z}$. Since $E = F \oplus K$ we have $L_0 = SL_2(F) \times SL_2(K)$ and we can consider α_1, γ_1 as roots of G. Let $E_{\pm\alpha_1}, E_{\pm\gamma_1} : G_a \hookrightarrow G$ be the corresponding one-parametric subgroups.

LEMMA 1. *a) There exists an algebraic group homomorphism $\check{i} : SL_3(F) \to$*

G such that

$$\check{i}(e_{12}(a)) = E_{\alpha_1}(a),\ \check{i}(e_{21}(a)) = E_{-\alpha_1}(a),\ \check{i}(e_{23}(a)) = E_{\gamma_1}(a),\ \check{i}(e_{32}(a)) = E_{-\gamma_1}(a), \qquad a \in F.$$

Moreover, we can extend $\check{i}$ *to a group homomorphism* $\check{i} : Sl_3(F) \times K^* \to G$ *such that for all* $u \in K^*$

$$\check{i}(u) = j\left(\begin{pmatrix} u & 0 \\ 0 & u^{-1} \end{pmatrix}\right) \cdot h^2_{\beta_0}(N(u)),$$

where $h_{\beta_0} : G_m \to T$ *is the coroot corresponding to* β_0.

b) Let $i : SL_3(F) \times E' \to G$ *be the morphism as in the end of* §2. *Then the morphisms* $i, \check{i} : SL_3(F) \times E' \to G$ *are conjugate under an action of an element from Aut* G.

PROOF: Clear.

Our immediate goal is to describe the unitary representation of $SL_3(F) \times E'$ on W_2 obtained by the composition of σ_2 and i. Remember that we have $E' = K^*$. Let $\overline{R}$ be the subspace of Schwartz-Bruhat functions on $F^* \times K$.

LEMMA 2. *There exists a unitary representation* $\tau : GL_2(F) \times K^* \to Aut\,\overline{R}$ *such that*

$$(\overline{\tau}\begin{pmatrix} 1 & a \\ 0 & 1 \end{pmatrix}\varphi)(y,z) = \psi(y^{-1}a\, N(z))\varphi(y,z)$$

$$(\overline{\tau}\begin{pmatrix} 0 & 1 \\ -1 & 0 \end{pmatrix}\varphi)(y,z) = \gamma \int_{\tilde{y} \in K} \psi(y^{-1}T(\tilde{z}\ \bar{z}))f(y,\tilde{z})d\tilde{z}$$

$$(\overline{\tau}\begin{pmatrix} c & 0 \\ 0 & c^{-1} \end{pmatrix}\varphi)(y,z) = \epsilon_K(c)|c|f(y,cz)$$

$$(\overline{\tau}\begin{pmatrix} c & 0 \\ 0 & 1 \end{pmatrix}\varphi)(y,z) = \epsilon_K(c)|c|^{3/2}f(cy,cz)$$

$$(\overline{\tau}(u)\varphi)(y,z) = |N(u)|f(N(u)y,uz),$$

$a \in F, c \in F^*, u \in K^*$ *where* $\gamma \in \mathbb{C}^*$ *is as in [JL]* §1, *and* $\epsilon_K : F^* \to \mathbb{C}(\pm 1) \subset \mathbb{C}^*$ *is the character such that* $\ker\, \epsilon_K = Im(N : K^* \to F^*)$.

PROOF: See [JL] §1.

Let $Q = M\overline{U} \subset SL_3(F)$ be the standard parabolic subgroup $Q = \begin{pmatrix} m & * \\ 0 & \det m^{-1} \end{pmatrix}$, $m \in GL_2(F)$. We extend τ to a representation of $Q \times K^*$ trivial on the unipotent radical of Q and denote by (τ_2, R_2) the representation of $SL_3(F) \times E'$ unitary induced from $Q \times E'$ the representation $\overline{\tau}$ to $SL_3(F) \times E'$.

PROPOSITION 1. *The representation $\sigma_2 \circ i$ of $SL_3(F) \times E'$ is equivalent to the representation τ_2.*

PROOF: Since morphism $i, \check{i} : SL_3(F) \times E' \to G$ are conjugate, it is sufficient to show that $\sigma_2 \circ \check{i}$ is equivalent to τ_2. Let $\|\ \| : GL_2(F) \to \mathbb{C}^*$ be the map given by $\|m\| = |\det\ m|^{3/2}$. We extend it to a character $|\ | : Q \to \mathbb{C}^*$. By the definition R_2 consists of functions $\varphi : SL_3(F) \to L_2(F \times K)$ such that $f(q\ g) = \overline{\tau}(q)\|q\|f(g)$ for all $q \in Q, g \in G$. Therefore we can consider R_2 as a subspace of the space of functions on $SL_3(F) \times F^* \times K$.

We construct a map $\varphi \to \varphi_f$ from W into the space of functions on $SL_3(F) \times F^* \times K$ by the formula

$$\varphi_f(g, y, z) \stackrel{\text{def}}{=} (\sigma(g)f)(y, 0, 0, z).$$

It is easy to see that for any $f \in {}^0W \subset W \subset W_2$, $\varphi_f \subset R_2$ and $\|\varphi_f\|_{R_2} = \|f\|_{W_2}$. Let $\varphi_2 : W_2 \to R_2$ be the closure of φ. It is clear that φ_2 is an isomorphism of Hilbert spaces. Proposition 1 is proved.

Let $\overline{\pi} : GL_2(F) \times E' \to Aut\ \overline{W}$ be the image of Jacquet r_G^M in respect to the unipotent radical $\overline{U}$ of Q applied to $(\sigma \circ i, W)$. For any character χ of K^* we can define representation $(\overline{\pi}_\chi, \overline{W}_\chi)$ and $(\overline{\tau}_\chi, \overline{R}_\chi)$ of $M = GL_2(F)$ as quotient of $\overline{W}$ and $\overline{R}$ by subspaces spanned by vectors $\{\overline{\pi}(u)\overline{w} - \chi(u)\overline{w}\}$, $u\ inE', \overline{w} \in \overline{W}$ and vectors $\overline{\tau}(u)\overline{r} - \chi(u)\overline{r}$, $u \in K^*, \overline{r} \in \overline{R}$ correspondingly.

If K is a field, we say that a character χ of K^* is regular if $\chi_{|K'} \not\equiv 1$ where $K' = \{z \in K^* | z\overline{z} = 1\}$. If $K = F^2$ we say that a character $\chi = (\chi_1, \chi_2)$ of K^* is regular if $\chi_i\chi_j^{-1} \neq |\ |^\epsilon$, $1 \leq i, j \leq 3$ where $\epsilon = \pm 1$ or 0 and $\chi_3 : F^* \to \mathbb{C}^*$ is given by $\chi_3 = (\chi_1\chi_2)^{-1}$.

LEMMA 3. *a) If K is a field and χ is a regular unitary character of K^* then the representation $\overline{\tau}_\chi$ and $\overline{\pi}_\chi$ are equivalent.*

b) If $K = F^2$ and $\chi = (\chi_1, \chi_2)$ is a regular character of K^*, then

$$\overline{\pi_\chi} = \overline{\tau}_{(\chi_1, \chi_2)} \oplus \overline{\tau}_{(\chi_1, \chi_3)} \oplus \overline{\tau}_{(\chi_2, \chi_3)}.$$

PROOF: We consider only the case a), so we assume that K is a field. Since the image $i(\overline{U})$ is a normal subgroup of $\tilde{P}$ the group $\tilde{P} \supset M$ acts naturally on $\overline{W}$. As follows from Proposition 6.1 there exists an imbedding of $S(F^* \times K)$ in $\overline{W}$ as a B_{β_0}-invariant subspace and the subgroup $i(K') \subset G$ acts trivially on the quotient space. Since $\tilde{s} = i\begin{pmatrix} 0 & 1 \\ -1 & 0 \end{pmatrix}$ normalizers B_{β_0} we can find explicitly the action of $\tilde{s}$ on $\overline{W}(\chi)$ for any regular character χ of K^*. Since $i(M)$ is generated by $\tilde{s}$ and $i(M) \cap B_{\beta_0}$, Lemma 3a) is proved. We leave the case b) for the reader.

Lemma 3 is proved.

PROPOSITION 2. *Theorem B is true in the case when $E = F \oplus K$.*

PROOF: In our case the norm map $N : E^* \to F^*$ is surjective and $G_3^E = GL_3(F)$. On the other hand, we have $E^* = (F^* \times \{1\}) \times E'$. Therefore for any character θ of F^* we can extend χ^1 to a character χ_θ of E^* defining χ_θ to be θ on $F^* \times \{1\}$. So to prove Theorem 4.3 in our case we have to construct for every $\theta : F^* \to \mathbf{C}^*$ the extension of $\pi_{\chi'}$ to a representation π_χ of $GL_3(F)$ and to show that π_χ corresponds to χ_θ.

LEMMA 4. $(\pi_{\chi'}, W_{\chi'}) \simeq ind_Q^{SL_3(F)}(\overline{\tau}_\chi, \overline{R}_\chi)$.

PROOF: As follows from Proposition 1 the representation of $SL_3(F)$ on R_2 does not have supercuspidal subquotients. Therefore the representation $\pi_{\chi'}$ of $SL_3(F)$ on W_χ also does not have supercuspidal subquotients and $\pi_{\chi'}$ is completely determined by its image under the Jacquet functor $r_G^M(\pi_{\chi'})$. Since the functor r_G^M is exact, we have $r_G^M(\pi_{\chi'}) = \overline{\pi}_{\chi'}$. By Lemma 3 and [BZ] we have $r_G^M(\pi_{\chi'}) = r_G^M(ind_Q^G \overline{\tau}_{\chi'})$. Lemma 4 is proved.

Let θ be a character of F^*. We extend the representation $\overline{\tau}_{\chi'} : M \to Aut\ \overline{R}_\chi$ to a representation $\overline{\tau}_\chi : \tilde{M} \to Aut\ \overline{R}_\chi$ where $\tilde{M} \stackrel{\text{def}}{=} M \times F^*$ and $\overline{\tau}_\chi(\{1\} \times c) = \theta(c) Id$, $c \in F^*$. We consider $\tilde{M}$ as a Levi subgroup of a parabolic $\tilde{Q} \subset GL_3(F)$, $\tilde{Q} = \overline{M U}$ and define $\pi_\chi = ind_{\tilde{Q}}^{GL_3(F)} \overline{\tau}_\chi$.

As follows from Lemma 4, $\pi_\chi = \pi_\chi^E$. Proposition 2 is proved.

6. In this section we consider the case when F is a local nonarchimedean field and E is a direct sum $E = F \oplus K$ where K is a semisimple 2-dimensional F-algebra. We denote by $z \to \overline{z}$ the nontrivial involution of K. It is clear that for any $z \in K$, $N(1, z) = z\overline{z}$ and $T(0, z) = z \otimes \overline{z}$. Since $E = F \oplus K$ we have $L_0 = SL_2(F) \times SL_2(K)$ and we can consider α_1, γ_1 as roots of G. Let $E_{\pm\alpha_1}, E_{\pm\gamma_1} : G_a \hookrightarrow G$ be the corresponding one-parametric subgroups.

LEMMA 1. a) *There exists an algebraic group homomorphism $\check{i} : SL_3(F) \to G$ such that*

$$\begin{aligned} \check{i}(e_{12}(a)) &= E_{\alpha_1}(a),\ \check{i}(e_{21}(a)) \\ &= E_{-\alpha_1}(a),\ \check{i}(e_{23}(a)) \\ &= E_{\gamma_1}(a),\ \check{i}(e_{32}(a)) \\ &= E_{-\gamma_1}(a), \qquad a \in F. \end{aligned}$$

Moreover, we can extend $\check{i}$ to a group homomorphism $\check{i} : Sl_3(F) \times K^* \to G$ such that for all $u \in K^*$

$$\check{i}(u) = j\left(\begin{pmatrix} u & 0 \\ 0 & u^{-1} \end{pmatrix}\right) \cdot h_{\beta_0}^2(N(u)),$$

where $h_{\beta_0} : G_m \to T$ is the coroot corresponding to β_0.

b) Let $i : SL_3(F) \times E' \to G$ be the morphism as in the end of §2. Then the morphisms $i, \check{i} : SL_3(F) \times E' \to G$ are conjugate under an action of an element from Aut G.

PROOF: Clear.

Our immediate goal is to describe the unitary representation of $SL_3(F) \times E'$ on W_2 obtained by the composition of σ_2 and i. Remember that we have $E' = K^*$. Let $\overline{R}$ be the subspace of Schwartz-Bruhat functions on $F^* \times K$.

LEMMA 2. *There exists a unitary representation $\tau : GL_2(F) \times K^* \to Aut\, \overline{R}$ such that*

$$(\overline{\tau}\begin{pmatrix} 1 & a \\ 0 & 1 \end{pmatrix}\varphi)(y, z) = \psi(y^{-1}a\ N(z))\varphi(y, z)$$

$$(\overline{\tau}\begin{pmatrix} 0 & 1 \\ -1 & 0 \end{pmatrix}\varphi)(y, z) = \gamma \int_{\tilde{y} \in K} \psi(y^{-1}T(\tilde{z}\ \bar{z}))f(y, \tilde{z})d\tilde{z}$$

$$(\overline{\tau}\begin{pmatrix} c & 0 \\ 0 & c^{-1} \end{pmatrix}\varphi)(y, z) = \epsilon_K(c)|c|f(y, cz)$$

$$(\overline{\tau}\begin{pmatrix} c & 0 \\ 0 & 1 \end{pmatrix}\varphi)(y, z) = \epsilon_K(c)|c|^{3/2}f(cy, cz)$$

$$(\overline{\tau}(u)\varphi)(y, z) = |N(u)|f(N(u)y, uz),$$

$a \in F, c \in F^*, u \in K^*$ where $\gamma \in \mathbb{C}^*$ is as in [JL] §1, and $\epsilon_K : F^* \to \mathbb{C}(\pm 1) \subset \mathbb{C}^*$ is the character such that $\ker \epsilon_K = Im(N : K^* \to F^*)$.

PROOF: See [JL] §1.

Let $Q = M\overline{U} \subset SL_3(F)$ be the standard parabolic subgroup $Q = \begin{pmatrix} m & * \\ 0 & \det m^{-1} \end{pmatrix}$, $m \in GL_2(F)$. We extend τ to a representation of $Q \times K^*$ trivial on the unipotent radical of Q and denote by (τ_2, R_2) the representation of $SL_3(F) \times E'$ unitary induced from $Q \times E'$ the representation $\overline{\tau}$ to $SL_3(F) \times E'$.

PROPOSITION 1. *The representation $\sigma_2 \circ i$ of $SL_3(F) \times E'$ is equivalent to the representation τ_2.*

PROOF: Since morphism $i, \check{i} : SL_3(F) \times E' \to G$ are conjugate, it is sufficient to show that $\sigma_2 \circ \check{i}$ is equivalent to τ_2. Let $\|\ \| : GL_2(F) \to \mathbb{C}^*$ be the map given by $\|m\| = |\det m|^{3/2}$. We extend it to a character $|\ | : Q \to \mathbb{C}^*$. By the definition R_2 consists of functions $\varphi : SL_3(F) \to L_2(F \times K)$ such that $f(q\ g) = \overline{\tau}(q)\|q\|f(g)$ for all $q \in Q, g \in G$. Therefore we can consider R_2 as a subspace of the space of functions on $SL_3(F) \times F^* \times K$.

We construct a map $\varphi \to \varphi_f$ from W into the space of functions on $SL_3(F) \times F^* \times K$ by the formula

$$\varphi_f(g, y, z) \stackrel{\text{def}}{=} (\sigma(g)f)(y, 0, 0, z).$$

It is easy to see that for any $f \in {}^0W \subset W \subset W_2$, $\varphi_f \subset R_2$ and $||\varphi_f||_{R_2} = ||f||_{W_2}$. Let $\varphi_2 : W_2 \to R_2$ be the closure of φ. It is clear that φ_2 is an isomorphism of Hilbert spaces. Proposition 1 is proved.

Let $\overline{\pi} : GL_2(F) \times E' \to Aut\, \overline{W}$ be the image of Jacquet r_G^M in respect to the unipotent radical $\overline{U}$ of Q applied to $(\sigma \circ i, W)$. For any character χ of K^* we can define representation $(\overline{\pi}_\chi, \overline{W}_\chi)$ and $(\overline{\tau}_\chi, \overline{R}_\chi)$ of $M = GL_2(F)$ as quotient of $\overline{W}$ and $\overline{R}$ by subspaces spanned by vectors $\{\overline{\pi}(u)\overline{w} - \chi(u)\overline{w}\}$, $u\ inE', \overline{w} \in \overline{W}$ and vectors $\overline{\tau}(u)\overline{r} - \chi(u)\overline{r}$, $u \in K^*, \overline{r} \in \overline{R}$ correspondingly.

If K is a field, we say that a character χ of K^* is regular if $\chi_{|K'} \not\equiv 1$ where $K' = \{z \in K^* | z\overline{z} = 1\}$. If $K = F^2$ we say that a character $\chi = (\chi_1, \chi_2)$ of K^* is regular if $\chi_i\chi_j^{-1} \neq |\ |^\epsilon$, $1 \leq i,j \leq 3$ where $\epsilon = \pm 1$ or 0 and $\chi_3 : F^* \to \mathbf{C}^*$ is given by $\chi_3 = (\chi_1\chi_2)^{-1}$.

LEMMA 3. *a) If K is a field and χ is a regular unitary character of K^* then the representation $\overline{\tau}_\chi$ and $\overline{\pi}_\chi$ are equivalent.*

b) If $K = F^2$ and $\chi = (\chi_1, \chi_2)$ is a regular character of K^*, then

$$\overline{\pi_\chi} = \overline{\tau}_{(\chi_1,\chi_2)} \oplus \overline{\tau}_{(\chi_1,\chi_3)} \oplus \overline{\tau}_{(\chi_2,\chi_3)}.$$

PROOF: We consider only the case a), so we assume that K is a field. Since the image $i(\overline{U})$ is a normal subgroup of $\tilde{P}$ the group $\tilde{P} \supset M$ acts naturally on $\overline{W}$. As follows from Proposition 6.1 there exists an imbedding of $S(F^* \times K)$ in $\overline{W}$ as a B_{β_0}-invariant subspace and the subgroup $i(K') \subset G$ acts trivially on the quotient space. Since $\tilde{s} = i\begin{pmatrix} 0 & 1 \\ -1 & 0 \end{pmatrix}$ normalizers B_{β_0} we can find explicitly the action of $\tilde{s}$ on $\overline{W}(\chi)$ for any regular character χ of K^*. Since $i(M)$ is generated by $\tilde{s}$ and $i(M) \cap B_{\beta_0}$, Lemma 3a) is proved. We leave the case b) for the reader.

Lemma 3 is proved.

PROPOSITION 2. *Theorem B is true in the case when $E = F \oplus K$.*

PROOF: In our case the norm map $N : E^* \to F^*$ is surjective and $G_3^E = GL_3(F)$. On the other hand, we have $E^* = (F^* \times \{1\}) \times E'$. Therefore for any character θ of F^* we can extend χ^1 to a character χ_θ of E^* defining χ_θ to be θ on $F^* \times \{1\}$. So to prove Theorem 4.3 in our case we have to construct for every $\theta : F^* \to \mathbf{C}^*$ the extension of $\pi_{\chi'}$ to a representation π_χ of $GL_3(F)$ and to show that π_χ corresponds to χ_θ.

LEMMA 4. $(\pi_{\chi'}, W_{\chi'}) \simeq ind_Q^{SL_3(F)}(\overline{\tau}_\chi, \overline{R}_\chi)$.

PROOF: As follows from Proposition 1 the representation of $SL_3(F)$ on R_2 does not have supercuspidal subquotients. Therefore the representation $\pi_{\chi'}$ of $SL_3(F)$ on W_χ also does not have supercuspidal subquotients and $\pi_{\chi'}$

is completely determined by its image under the Jacquet functor $r_G^M(\pi_{\chi'})$. Since the functor r_G^M is exact, we have $r_G^M(\pi_{\chi'}) = \overline{\pi}_{\chi'}$. By Lemma 3 and [BZ] we have $r_G^M(\pi_{\chi'}) = r_G^M(ind_Q^G \overline{\tau}_{\chi'})$. Lemma 4 is proved.

Let θ be a character of F^*. We extend the representation $\overline{\tau}_{\chi'} : M \to Aut\ \overline{R}_\chi$ to a representation $\overline{\tau}_\chi : \tilde{M} \to Aut\ \overline{R}_\chi$ where $\tilde{M} \stackrel{\text{def}}{=} M \times F^*$ and $\overline{\tau}_\chi(\{1\} \times c) = \theta(c)Id,\ c \in F^*$. We consider $\tilde{M}$ as a Levi subgroup of a parabolic $\tilde{Q} \subset GL_3(F)$, $\tilde{Q} = \overline{M}U$ and define $\pi_\chi = ind_{\tilde{Q}}^{GL_3(F)} \overline{\tau}_\chi$.

As follows from Lemma 4, $\pi_\chi = \pi_\chi^E$. Proposition 2 is proved.

7. In §6 we have proved Theorem B in the case when E is not a field. In this section we consider the case when $\tilde{E} \supset \tilde{F}$ are local fields, $dim_F \tilde{E} = 3$, and $\tilde{\chi}$ is a multiplicative character of $\tilde{E}^*$. We fix a global field $E \supset F$ and a place $\nu_0 \in \Sigma$ such that $F_{\nu_0} = \tilde{F}, E_{\nu_0} = \tilde{E}$ and the extension $E \supset F$ is not normal.

Let $\sigma_{\mathbb{A}} : G_{\mathbb{A}}^E \to Aut\ W_{\mathbb{A}}$ be the representation as in the end of §5 and $\sigma'_{\mathbb{A}} : G_{\mathbb{A}}^E \to Aut\ W'_{\mathbb{A}}$ be the dual representation. We have $\sigma_{\mathbb{A}} = \bigotimes_{\nu \in \Sigma} \sigma_\nu$, where $\sigma_\nu : G_{E_\nu}^{E_\nu}(F_\nu) \to Aut\ W_\nu$ are representations as in Theorem 3.1. Since $\sigma_{\mathbb{A}}$ is an automorphic representation, there exists a nonzero functional $\gamma \in W'$ such that $\sigma'(\gamma)w' = w'$ for all $\gamma \in G^E(F)$. Define $w'_{\tilde{\chi}} \in W'_{\mathbb{A}}$ by $w'_{\tilde{\chi}} = \int_{E^1_{\nu_0}} \tilde{\chi}(u)\sigma'(i(u))w'du$ where as before i is an imbedding $E' \hookrightarrow G^E$.

LEMMA 1. $w'_{\tilde{\chi}} \neq 0$.

PROOF: The same as for Proposition 1 in [K].

For any character χ' of $E'_{\mathbb{A}}/E'$ we define a quotient space W_χ of $W_{\mathbb{A}}$ as before, $W_{\chi'} = \bigotimes_{\nu \in \Sigma} (W_\nu)_{\chi_\nu}$. As follows from Lemma 1 there exists a character χ' of $E'_{\mathbb{A}}/E'$ such that $\chi'_{|E'_{\nu_0}} = \tilde{\chi}_{|E'_{\nu_0}}$ and $w'_\chi \neq 0$ where

$$w'_{\chi'} \stackrel{\text{def}}{=} \int_{u \in E'_{\mathbb{A}}/E'} \chi(u)\sigma'(i(u))w'du.$$

The group $SL_3(\mathbb{A})$ acts naturally on the space $W_{\chi'}$. We choose an extension χ on χ' to a unitary character on $E^*_{\mathbb{A}}/E^*$. Then, as in §4 we can define a representation $\pi_\chi^E : G_3^{E/\mathbb{A}} \to Aut\ W_\chi$ where $G_3^{E_{\mathbb{A}}} \stackrel{\text{def}}{=} \{g \in GL_3(\mathbb{A}) |\ \det\ g \subset Im\ N : E^*_{\mathbb{A}} \to \mathbb{A}^*\}$. In other words, $G_3^{E_{\mathbb{A}}} = \{g = (g_\nu) \in GL_3(\mathbb{A}) | g_\nu \in G_3^{E_\nu}\ \forall \nu \in \Sigma\}$.

Let π_χ be the induced representation of the group $GL_3(\mathbb{A})$ and $\Sigma_0 \subset \Sigma$ be the subset of places ν such that E_ν is not a field.

PROPOSITION 1. *a) π_χ is an automorphic representation.*

b) $\pi_\chi = \otimes_{\nu\in\Sigma}\pi_{\chi_\nu}$ where π_{χ_ν} is a representation of $GL_3(F_\nu)$ as in Theorem 4.3.

c) For all $\nu \in \Sigma_0$, π_{χ_ν} corresponds to the character χ_ν of E_ν^*.

PROOF: a) Since the extension $E \supset F$ is not normal, the norm map N induces a surjection of the idele class group of E to the one of F. Therefore we have $GL_3(F)G_3^{E_{\mathbb{A}}} = GL_3(\mathbb{A})$ and

$$G_3^{E_{\mathbb{A}}} \cap GL_3(F) \setminus G_3^{E_{\mathbb{A}}} \xrightarrow{\sim} GL_3(F) \setminus GL_3(\mathbb{A}).$$

Consider the map $\varphi_\chi : W_\chi \to \mathbb{C}(GL_3(F)\setminus GL_3(\mathbb{A}))$ given by $\varphi_\chi(w)(g) = w'_\chi(\pi_\chi^E(g)w)$ for $g \in G_3^{E_{\mathbb{A}}}$. By Proposition 5.2 we know that W_χ is irreducible. Since $\varphi_\chi \not\equiv 0$ it is an imbedding. Therefore π_χ is automorphic. a) is proved.

b) Follows from the definition of $W_{\chi'}$.

c) Follows from Proposition 6.2.

Proposition 1 is proved.

Now we can prove Theorem B.

Consider the representation π_χ from Proposition 1. Since the density of Σ_0 is equal to $2/3 > 1/2$ it follows from [G] and Proposition 1 that $\pi_{\chi_{\nu_0}}$ corresponds to χ_{ν_0}. Since $\chi_{\nu_0}|_{\tilde{E}'} = \tilde{\chi}|_{\tilde{E}'}$, therefore there exists a character θ of $\tilde{F}^*$ such that $\tilde{\chi} = \chi_{\nu_0}\cdot\theta$. It is clear then that $\pi_{\tilde{\chi}} = \pi_{\chi_{\nu_0}} \otimes \theta(\det)$. Since $\pi_{\chi_{\nu_0}}$ corresponds to χ_{ν_0}. We see that $\pi_{\tilde{\chi}}$ corresponds to $\tilde{\chi}$. Theorem B is proved.

8. In the conclusion we discuss some open problems.

1) Let F be a local field, V be a symplectic vector space over F, $L \subset Sp(V)$ be a semisimple subgroup such that the double covering $\widetilde{Sp}(V) \to Sp(V)$ splits over L, $\sigma : L \to Aut\ W$ be the restriction of the metaplectic representation on L. Suppose that there exists an L-orbit $\Omega \subset V - \{0\}$ which is a closed Lagrangian subvariety. Could we define $\varphi_\Omega \in (W')^L$? Proposition 5.3 shows how to define φ_Ω in the case when $L = SL_2(E)$, $V = \Lambda \oplus \Lambda'$.

2) For any $n \in \mathbb{Z}^+$ let $\Lambda_n = F^n \oplus F^{n+2}$, $\lambda = (y_1, \dots, y_{n-2}, x_1, \dots, x_n)$. Let φ be a distribution given on $(F^*)^{n-2} \times F^n$ by $\varphi(y_1, \dots, y_n, x_1, \dots, x_{n+2}) = \frac{1}{|y_1\cdots y_n|}\psi\left(\frac{x_1\cdots x_{n+2}}{y_1\cdots ,y_n}\right)$. It is easy to see that $\mathcal{F}(\varphi) = \varphi$ where $\mathcal{F}$ is the Fourier transform.

Let E, K be cyclic extensions of F of degree $n+2$ and n. $\Lambda = K \oplus E$. Could we find a multiplicative character $\mathcal{E}$ on $K^* \times E^*$ such that the distribution
$\varphi_{\mathcal{E}}(k,e) \overset{\text{def}}{=} \mathcal{E}(k,e)\cdot\psi\left(\frac{N_E(e)}{N_K(k)}\right)$ is invariant under the Fourier transform?

It is true in the case $n = 0$ by A. Weil's theorem and in the case $n = 1$ by Proposition 5.3.

3) In the case when F is a finite field we can consider $\psi\left(\frac{N_E(e)}{N_K(k)}\right)$ as a locally free sheaf of rank 1 on $K^* \times E$. Let φ be the perverse extension of this sheaf to $K \times E$. It is easy to see that $\mathcal{F}(\varphi) = \varphi$. Is it possible to use the technique of perverse sheaves to prove that (*) in the case where E, K are unramified extensions of F?

4) We have proven Theorem 3.1 using the global arguments. Is it possible to prove (*) if we assume that it is known in the unramified case?

5) It will be very interesting to compute a character t_σ of the representation σ. One could hope that t_σ is given by a nice explicit formula and the restriction of t_σ on $SL_3(F) \times E'$ generalizes the formulas in [GGPs]2.5 from SL_2 to SL_3.

REFERENCES

[BZ] J. Bernstein, A. Zelvinsky, *Representation of the group $GL(n, F)$, where F is a local nonarchimedean field,* Uspekhi Mat. Nauk, **31** No. 3 (1976), 5-70.

[G] G.T. Gilbert, *Multiplicity theorems for $GL(2)$ and $GL(3)$*, Contemp. Math. **53**, Amer. Math. Soc., Providence, R.I. (1986), 201-206.

[GGP] I.M. Gelfand, M.I. Graev and I.I. Pyatetskii-Shapiro, *Representation Theory and Automorphic Forms*, Philadelphia, W.B. Saunders Co., 1969.

[H] R. Howe, *The Fourier transform on nilpotent locally compact groups I.* Pacific J. Math. **73** (1977), 307-327.

[J] H. Jacquet, *On the residual spectrum of GL_n*, Springer-Verlag **1041**, 1984.

[JL] H. Jacquet and R. Langlands, *Automorphic forms on $GL(2)$*, Springer-Verlag **114**, 1970.

[J-PS-S] H. Jacquet, I.I. Piatetskii-Shapiro et J. Shalika, *Relèvement cubique non normal*, C.R. Acad. Sci., Paris, **292** (1981), 567-571.

[K] D. Kazhdan, *Some applications of the Weil representation.* Journal d'Analyse Math., **32** (1977), 235-248.

[KS] D. Kazhdan and G. Savin, *On the smallest representation of simply placed groups over local fields*, preprint.

[Ko] B. Kostant, *The principle of triality and a distinguished unitary representation of $SO(4, 4)$*, Diff. Geom. Methods in Theoretical Physics, Edited by K. Bleuler and M. Werner, Series C: Math. and Phy. Sci. Vol. 250, Kluwer Academic Pub. (1988), 65–109.

[L] R. Langlands, *On the functional equation satisfied by Eisenstein series.* Springer Lecture Notes in Math. **544** (1976).
[M] J. Milnor, *Introduction to Algebraic K-theory*, Princeton Univ. Press 1971.
[O] O.T. O'Meara, *Introduction to quadratic forms*, Berlin, Springer, 1963.
[S] J.-P. Serre, *Cohomologie Galoisienne*, Collège de France, Paris, 1963.
[St] R. Steinberg, *Lectures on Chevalley groups*, Yale, 1967.
[T1] J. Tits, *Normalisateurs de Tores.* J. of Algebra **4** (1966), 96-116.
[T] J. Tits, *Buildings of Special Type and Finite BN-pair*, Springer-Verlag **386**, 1974.
[W] A. Weil, *Sur certains groupes d'opérateurs unitaires*, Acta Math. **111**, 143-211.

Received January 5, 1990

Department of Mathematics
Harvard University
Cambridge, MA 02138

Kazhdan-Lusztig Conjecture for Symmetrizable Kac-Moody Lie Algebra. II Intersection Cohomologies of Schubert Varieties

MASAKI KASHIWARA
TOSHIYUKI TANISAKI

Dedicated to Professor Jacques Dixmier on his sixty-fifth birthday

0. Introduction

0.0. This article is a continuation of Kashiwara [**K3**]. We shall complete the proof of a generalization of the Kazhdan-Lusztig conjecture to the case of symmetrizable Kac-Moody Lie algebras.

0.1. The original Kazhdan-Lusztig conjecture [**KL1**] describes the characters of irreducible highest weight modules of finite-dimensional semisimple Lie algebras in terms of certain combinatorially defined polynomials, called Kazhdan-Lusztig polynomials. It was simultaneously solved by two parties, Beilinson-Bernstein and Brylinski-Kashiwara, by similar methods ([**BB**], [**BK**]). The proof consists of the following two parts.

(i) The algebraic part — the correspondence between $\mathcal{D}$-modules on the flag variety and representations of the semisimple Lie algebra.
(ii) The topological part — the description of geometry of Schubert varieties in terms of the Kazhdan-Lusztig polynomials.

Note that the topological part had been already established by Kazhdan and Lusztig themselves ([**KL2**]).

0.2. Our proof of the generalization of the Kazhdan-Lusztig conjecture in the symmetrizable Kac-Moody Lie algebra case is similar to that in the finite-dimensional case mentioned above. The algebraic part has already appeared in [**K3**] and this paper is devoted to the topological part. The proof is again similar to the finite-dimensional case except two points.

The first point is that we use the theory of mixed Hodge modules of M. Saito [**S**] instead of the Weil sheaves. Note that mixed Hodge modules and Weil sheaves are already employed by several authors in order to relate the Hecke-Iwahori algebra of the Weyl group with the geometry of Schubert varieties ([**LV**], [**Sp**], [**T**]).

The second point is that we interpret the inverse Kazhdan-Lusztig polynomials as the coefficients of certain elements of the dual of the Hecke-Iwahori algebra. The appearance of the dual of the Hecke-Iwahori algebra is natural because the open Schubert cell corresponds to the identity element of the Weyl group, contrary to the finite-dimensional case in which the open Schubert cell corresponds to the longest element.

0.3. We shall state our results more precisely. Let $\mathfrak{g}$ be a symmetrizable Kac-Moody Lie algebra, $\mathfrak{h}$ the Cartan subalgebra and W the Weyl group (see [**K'**]). For $\lambda \in \mathfrak{h}^*$ let $M(\lambda)$ (resp. $L(\lambda)$) be the Verma module (resp. irreducible module) with highest weight λ. For $w \in W$ we define a new action of W on $\mathfrak{h}^*$ by $w \circ \lambda = w(\lambda+\rho)-\rho$, where ρ is an element of $\mathfrak{h}^*$ such that $\langle \rho, h_i \rangle = 1$ for any simple coroot $h_i \in \mathfrak{h}$. For $w, z \in W$ let $P_{w,z}(q)$ be the Kazhdan-Lusztig polynomial and $Q_{w,z}(q)$ the inverse Kazhdan-Lusztig polynomial ([**KL1**], [**KL2**]). They are defined through a combinatorics in the Hecke-Iwahori algebra of the Weyl group, and are related by

$$\sum_{w \in W} (-1)^{\ell(w)-\ell(y)} Q_{y,w} P_{w,z} = \delta_{y,z}. \tag{0.3.1}$$

Our main result is the following.

THEOREM. *For a dominant integral weight $\lambda \in \mathfrak{h}^*$ we have*

$$\operatorname{ch} L(w \circ \lambda) = \sum_{z \in W} (-1)^{\ell(z)-\ell(w)} Q_{w,z}(1) \operatorname{ch} M(z \circ \lambda),$$

or equivalently

$$\operatorname{ch} M(w \circ \lambda) = \sum_{z \in W} P_{w,z}(1) \operatorname{ch} L(z \circ \lambda).$$

Here ch *denotes the character and $\ell(w)$ is the length of w.*

0.4. Let X be the flag variety of $\mathfrak{g}$ constructed in [**K2**] and let X_w be the Scubert cell corresponding to $w \in W$. Note that X_w is a finite-codimensional locally closed subvariety of the infinite-dimensional variety X.

By the algebraic part [**K3**] $\mathfrak{g}$-modules correspond to holonomic $\mathcal{D}_X$-modules. Hence by taking the solutions of holonomic $\mathcal{D}_X$-modules, we obtain a correspondence between $\mathfrak{g}$-modules and perverse sheaves on X. Since $M(w \circ \lambda)$ and its dual $M^*(w \circ \lambda)$ have the same characters and since the perverse sheaf corresponding to the highest weight module $L(w \circ \lambda)$ (resp. $M^*(w \circ \lambda)$) is ${}^\pi\mathbf{C}_{X_w}[-\ell(w)]$ (resp. $\mathbf{C}_{X_w}[-\ell(w)]$), the proof of the theorem is reduced to

$$(0.4.1) \qquad [{}^\pi\mathbf{C}_{X_w}[-\ell(w)]] = \sum_{z \in W} (-1)^{\ell(z)-\ell(w)} Q_{w,z}(1)[\mathbf{C}_{X_z}[-\ell(z)]]$$

(in the Grothendieck group of perverse sheaves). We shall prove it for any (not necessarily symmetrizable) Kac-Moody Lie algebra in §6 by using Hodge modules.

0.5. We finally remark that the Kazhdan-Lusztig conjecture for symmetrizable Kac-Moody Lie algebras is explicitly stated in Deodhar-Gabber-Kac [**DGK**]. We also note that we have received the following short note announcing the similar result: L. Cassian, Formule de multiplicité de Kazhdan-Lusztig dans le cas de Kac-Moody, preprint.

1. Infinite-dimensional schemes

1.0. In this section we shall briefly discuss infinite-dimensional schemes.

1.1. A scheme X is called *coherent* if the structure ring $\mathcal{O}_X$ is coherent. A scheme X over $\mathbf{C}$ is said to be *of countable type* if the $\mathbf{C}$-algebra $\mathcal{O}_X(U)$ is generated by a countable number of elements for any affine open subset U of X (cf. [**K2**]). A morphism $f: X \to Y$ of schemes is called *weakly smooth* if $\Omega^1_{X/Y}$ is a flat $\mathcal{O}_X$-module, where $\Omega^1_{X/Y}$ is the sheaf of relative differentials.

1.2. We say that a $\mathbf{C}$-scheme X satisfies (S) if $X \simeq \varprojlim_{n \in \mathbf{N}} S_n$ for some projective system $\{S_n\}_{n \in \mathbf{N}}$ of $\mathbf{C}$-schemes satisfying the following conditions:

(1.2.1) S_n is quasi-compact and smooth over $\mathbf{C}$ for any n.
(1.2.2) The morphism $p_{nm}: S_m \to S_n$ is smooth and affine for $m \geqq n$.

In particular, X is quasi-compact.

Remark that by [**EGA IV**, Proposition (8.13.1)], the pro-object "$\varprojlim$"S_n is uniquely determined in the category of $\mathbf{C}$-schemes of finite type. More precisely, we have

$$(1.2.3) \qquad \varinjlim_n \operatorname{Hom}(S_n, Y) \xrightarrow{\sim} \operatorname{Hom}(X, Y)$$

for any **C**-scheme Y locally of finite type.

Note that the projection $p_n : X \to S_n$ is flat and and we have

$$\Omega^1_X \simeq \varinjlim_n (p_n)^* \Omega^1_{S_n}, \tag{1.2.4}$$

where $\Omega^1_X = \Omega^1_{X/\mathbf{C}}$. Thus we obtain

(1.2.5) Ω^1_X is locally a direct sum of locally free $\mathcal{O}_X$-modules of finite rank.

We see from the following lemma that, if X is separated, we may assume that S_n is also separated for any n.

LEMMA 1.2.1. *Let X be an affine (resp. separated) scheme such that $X \simeq \varprojlim_n S_n$, where $\{S_n\}_{n \in \mathbf{N}}$ is a projective system of schemes satisfying the following conditions:*

(1.2.6) *S_n is quasi-compact and quasi-separated for any n.*
(1.2.7) *$p_{nm} : S_m \to S_n$ is affine for $m \geqq n$.*

Then S_n is also affine (resp. separated) for $n \gg 0$.

PROOF: Let $p_n : X \to S_n$ be the projection.

(1) Assume that X is affine. We see from the assumptions that there exist an affine open covering $S_0 = \cup_{i \in I} U_i$ and $f_i \in \Gamma(X; \mathcal{O}_X)$ $(i \in I)$ such that $p_0^{-1}(U_i) \supset X_{f_i}$ and $X = \cup_{i \in I} X_{f_i}$, where I is a finite index set and $X_{f_i} = X \setminus \operatorname{Supp}(\mathcal{O}_X / \mathcal{O}_X f_i)$. Setting $A = \Gamma(X; \mathcal{O}_X)$ and $A_n = \Gamma(S_n; \mathcal{O}_{S_n})$, we have $A = \varinjlim_n A_n$ by [**EGA IV**, Theorem(8.5.2)], and hence there exists some n satisfying $f_i \in A_n$ $(i \in I)$. Thus we may assume that $f_i \in A_0$ from the beginning. It is easily seen from the assumptions that $(S_n)_{f_i} \subset p_{0n}^{-1} U_i$ and $A_n = \sum_{i \in I} A_n f_i$ for $n \gg 0$. Then $(S_n)_{f_i}$ is affine, and hence $S_n \to \operatorname{Spec}(A_n)$ is an affine morphism.

(2) Assume that X is separated. In order to prove that S_n is separated for $n \gg 0$, it is enough to show that, for any affine open subsets U and V of S_0, $p_{0n}^{-1}(U \cap V) \to p_{0n}^{-1}(U) \times p_{0n}^{-1}(V)$ is a closed embedding for $n \gg 0$. Since $p_0^{-1}(U \cap V)$ is affine, $p_{0n}^{-1}(U \cap V)$ is affine for $n \gg 0$ by (1), and hence we may assume from the beginning that $U \cap V$ is affine. Since $\mathcal{O}_{S_0}(U \cap V)$ is of finite type over $\mathcal{O}_{S_0}(U)$, $\mathcal{O}_{S_0}(U \cap V)$ is generated by finitely many elements a_i over $\mathcal{O}_{S_0}(U)$. Since $p_0^{-1}(U \cap V) \to p_0^{-1}(U) \times p_0^{-1}(V)$ is a closed embedding, $(p_0)^* a_i$ is contained in the image of $\mathcal{O}_X(p_0^{-1}(U)) \otimes \mathcal{O}_X(p_0^{-1}(V)) \to \mathcal{O}_X(p_0^{-1}(U \cap V))$. Thus $(p_{0n})^* a_i$ is contained in the image of $\mathcal{O}_{S_n}(p_{0n}^{-1}(U)) \otimes \mathcal{O}_{S_n}(p_{0n}^{-1}(V)) \to \mathcal{O}_{S_n}(p_{0n}^{-1}(U \cap V))$ for $n \gg 0$. Therefore $\mathcal{O}_{S_n}(p_{0n}^{-1}(U)) \otimes \mathcal{O}_{S_n}(p_{0n}^{-1}(V)) \to \mathcal{O}_{S_n}(p_{0n}^{-1}(U \cap V))$ is surjective. □

1.3. Let (L) (resp. (LA)) denote the category of quasi-compact smooth **C**-schemes and smooth (resp. smooth affine) morphisms.

PROPOSITION 1.3.1. *Let X be a* **C***-scheme satisfying* (S). *Then* "$\varprojlim$"S_n *as a pro-object in* (LA) *does not depend on the choice of the projective system* $\{S_n\}_{n\in\mathbf{N}}$ *as in* §1.2.

PROOF: It is enough to show that, for any quasi-compact smooth **C**-scheme Y, the natural map

(1.3.1)
$$\varinjlim_n \mathrm{Hom}_{(\mathrm{L})}(S_n, Y) \to \{ f \in \mathrm{Hom}(X,Y)\,;\, (f^*\Omega^1_Y)(x) \to \Omega^1_X(x) \text{ is injective for any } x \in X \}$$

is bijective. Here, for an $\mathcal{O}_X$-module $\mathcal{F}$ and $x \in X$, $\mathcal{F}(x)$ denotes $\mathcal{F}_x/\mathfrak{m}_x\mathcal{F}_x$, where $\mathfrak{m}_x$ is the maximal ideal of $\mathcal{O}_{X,x}$. In fact, by Lemma 1.2.1, we then have

(1.3.2)
$$\varinjlim_n \mathrm{Hom}_{(\mathrm{LA})}(S_n, Y) \xrightarrow{\sim} \{ f \in \mathrm{Hom}(X,Y)\,;\, f \text{ is affine and } (f^*\Omega^1_Y)(x) \to \Omega^1_X(x) \text{ is injective for any } x \in X \}.$$

The injectivity of (1.3.1) follows from (1.2.3). Let $f\colon X \to Y$ be a **C**-morphism such that $(f^*\Omega^1_Y)(x) \to \Omega^1_X(x)$ is injective for any $x \in X$. Then f splits into the composition of $p_n\colon X \to S_n$ and $\tilde{f}\colon S_n \to Y$ for some n. Since $(\tilde{f}^*\Omega^1_Y)(p_n(x)) \to (f^*\Omega^1_Y)(x)$ is injective for any $x \in X$, $(\tilde{f}^*\Omega^1_Y)(s) \to \Omega^1_{S_n}(s)$ is also injective for any $s \in p_n(X)$. Hence there exists an open neighborhood Ω of $p_n(X)$ such that $(\tilde{f}^*\Omega^1_Y)(s) \to \Omega^1_{S_n}(s)$ is injective for any $s \in \Omega$. Now [**EGA IV**, Proposition (1.9.2)] guarantees that there exists $m \geqq n$ such that $p_{nm}^{-1}(\Omega) = S_m$, and hence $((\tilde{f} \circ p_{nm})^*\Omega^1_Y)(s) \to \Omega^1_{S_m}(s)$ is injective for any $s \in S_m$. This means that $\tilde{f} \circ p_{nm}$ is smooth. □

LEMMA 1.3.2. *Let $f\colon X \to Y$ be a morphism of* **C***-schemes satisfying* (S). *Then the following conditions are equivalent.*

(i) *f is weakly smooth (i.e. $\Omega^1_{X/Y}$ is flat).*

(ii) *For any $x \in X$, $(f^*\Omega^1_Y)(x) \to \Omega^1_X(x)$ is injective.*

(iii) *There exist projective systems $\{X_n\}$, $\{Y_n\}$ satisfying* (1.2.1), (1.2.2) *and a morphism $\{f_n\}\colon \{X_n\} \to \{Y_n\}$ of projective systems such that $X \simeq \varprojlim_n X_n$, $Y \simeq \varprojlim_n Y_n$, $f = \varprojlim_n f_n$, and f_n is smooth for any n.*

PROOF: (i)⇒(ii) is evident. (iii)⇒(i) follows from the fact that $\Omega^1_{X/Y}$ is the inductive limit of the flat $\mathcal{O}_X$-modules $(p_n)^*\Omega^1_{X_n/Y_n}$, where $p_n : X \to X_n$ is the projection. Assume (ii). By (1.2.3), there exist $\{X_n\}$, $\{Y_n\}$ and $\{f_n\}$ such that $X \simeq \varprojlim_n X_n$, $Y \simeq \varprojlim_n Y_n$, $f = \varprojlim_n f_n$. Then we see from the bijectivity of (1.3.1) that, for any n, there exists some $m \geqq n$ such that the composition $X_m \to X_n \to Y_n$ is smooth. This implies (iii). □

1.4. A **C**-scheme X is called *pro-smooth* if it is covered by open subsets satisfying (S).

LEMMA 1.4.1 (CF. [**K2**]). *A pro-smooth* **C***-scheme is coherent and of countable type.*

LEMMA 1.4.2. *Let $f: X \to Y$ be a smooth morphism of* **C***-schemes. If Y satisfies* (S)*, so does X.*

PROOF: Let $Y \simeq \varprojlim_n S_n$, where $\{S_n\}$ is as in §1.2. By [**EGA IV**, Theorem (8.8.2)] there exist some n and a morphism $f_n : X_n \to S_n$ satisfying $f \simeq f_n \times_{S_n} Y$. Then, by [**EGA IV**, Proposition (17.7.8)], $f_n \times_{S_n} S_m$ is smooth for $m \gg 0$. □

COROLLARY 1.4.3. *A* **C***-scheme smooth over a pro-smooth* **C***-scheme is also pro-smooth.*

1.5. We give several examples of pro-smooth **C**-schemes.

(a) $\mathbf{A}^\infty = \mathrm{Spec}(\mathbf{C}[X_n\,;\, n = 1, 2, \ldots])$ (cf. [**K2**]). Denoting by $p_n : \mathbf{A}^\infty \to \mathbf{A}^n$ the projection given by $(X_1, \ldots, X_n)$, we have $\mathbf{A}^\infty \simeq \varprojlim_n \mathbf{A}^n$.

(b) $\mathbf{P}^\infty$ (cf. [**K2**]).

(c) Let E be a countable subset of **C**, and let A be the **C**-subalgebra of the rational function field $\mathbf{C}(x)$ generated by x and $\{\,(x-a)^{-1}\,;\, a \in E\,\}$. Then $X = \mathrm{Spec}(A)$ is a pro-smooth **C**-scheme and we have $X(\mathbf{C}) \simeq \mathbf{C} - E$.

(d) Let A be the **C**-algebra which is generated by the elements e_n ($n \in \mathbf{Z}$) satisfying the fundamental relations $e_n e_m = \delta_{n,m} e_n$. Let x_n ($n \in \mathbf{Z}$) and ξ be the points of $X = \mathrm{Spec}(A)$ given by the prime ideals $A(1 - e_n)$ ($n \in \mathbf{Z}$) and $\sum_{n \in \mathbf{Z}} A e_n$ respectively. Then X is a pro-smooth **C**-scheme consisting of x_n ($n \in \mathbf{Z}$) and ξ. The underlying topological space is homeomorphic to the one-point compactification of **Z** with discrete topology, and the structure sheaf $\mathcal{O}_X$ is isomorphic to the sheaf of locally constant **C**-valued functions.

1.6. A **C**-scheme X is called *essentially smooth* if it is covered by open subsets U, each of which is either smooth over **C** or isomorphic to $W \times \mathbf{A}^{\infty}$ for a smooth **C**-scheme W. An essentially smooth **C**-scheme is obviously pro-smooth.

PROPOSITION 1.6.1. *If W is a **C**-scheme of finite type such that $W \times \mathbf{A}^{\infty}$ is pro-smooth, then W is smooth.*

PROOF: We may assume that $W \times \mathbf{A}^{\infty}$ satisfies (S). Hence we have $W \times \mathbf{A}^{\infty} \simeq \varprojlim S_n$ for some $\{S_n\}$ satisfying (1.2.1) and (1.2.2). Then there exist n and m such that the morphism $p_{0m} : S_m \to S_0$ splits into $S_m \to W \times \mathbf{A}^n \to S_0$. Hence $W \times \mathbf{A}^n \to S_0$ is smooth at the image of S_m. Therefore $W \times \mathbf{A}^n$ is smooth and hence so is W. □

PROPOSITION 1.6.2. *Let X and Y be **C**-schemes and let $f: Y \to X$ be a morphism of finite presentation. Assume $X \simeq W \times \mathbf{A}^{\infty}$ for a **C**-scheme W of finite type. Then there exist some n and a **C**-morphism $f': U \to W \times \mathbf{A}^n$ of finite type satisfying $f = f' \times \mathbf{A}^{\infty}$ (Note that we have $\mathbf{A}^{\infty} \simeq \mathbf{A}^n \times \mathbf{A}^{\infty}$).*

PROOF: Since $X \simeq \varprojlim_n W \times \mathbf{A}^n$, there exist some n and a **C**-scheme U of finite presentation over $W \times \mathbf{A}^n$ such that $Y \simeq X \times_{W \times \mathbf{A}^n} U$ by [**EGA IV**, Theorem (8.8.2)]. Then $f': U \to W \times \mathbf{A}^n$ satisfies the desired condition. □

COROLLARY 1.6.3. *A **C**-scheme smooth over an essentially smooth **C**-scheme is also essentially smooth.*

LEMMA 1.6.4. *Any essentially smooth **C**-scheme is a disjoint union of open irreducible subsets.*

PROOF: Let X be an essentially smooth **C**-scheme. Since X is covered by open irreducible subsets, it is enough to show that, if U is an open irreducible subset of X, then $\overline{U}$ is also an open subset of X. Let $x \in \overline{U}$ and let W be an irreducible open subset of X containing x. Since $W \cap U \neq \emptyset$, we have $\overline{W} = \overline{W \cap U} = \overline{U}$. This shows that $\overline{U}$ is a neighborhood of x. □

1.7. We shall recall the definition of $\mathcal{D}_X$ and admissible $\mathcal{D}_X$-modules for a pro-smooth **C**-scheme X.

For a morphism $f\colon X \to Y$ of pro-smooth $\mathbf{C}$-schemes we set

(1.7.1)
$$F_n(\mathcal{D}_{X\to Y}) = 0 \quad (n < 0),$$
(1.7.2)
$$F_n(\mathcal{D}_{X\to Y}) = \{ P \in \mathcal{H}om_{\mathbf{C}}(f^{-1}\mathcal{O}_Y, \mathcal{O}_X);\ [P, a] \in F_{n-1}(\mathcal{D}_{X\to Y}) \text{ for any } a \in \mathcal{O}_Y \} \quad (n \geqq 0),$$
(1.7.3)
$$F(\mathcal{D}_{X\to Y}) = \cup_n F_n(\mathcal{D}_{X\to Y}) \subset \mathcal{H}om_{\mathbf{C}}(f^{-1}\mathcal{O}_Y, \mathcal{O}_X).$$

We have

$$F_0(\mathcal{D}_{X\to Y}) \simeq \mathcal{O}_X, \tag{1.7.4}$$
$$F_1(\mathcal{D}_{X\to Y}) \simeq \mathcal{O}_X \oplus \Theta_{X\to Y}, \tag{1.7.5}$$

where $\Theta_{X\to Y} := \mathcal{H}om_{f^{-1}\mathcal{O}_Y}(f^{-1}\Omega_Y^1, \mathcal{O}_X)$ is the sheaf of derivations.

Let I_{Δ_Y} denote the defining ideal of the diagonal Δ_Y in $Y \times_{\mathbf{C}} Y$ and set $\mathcal{O}_{\Delta_Y(n)} = \mathcal{O}_{Y\times_{\mathbf{C}} Y}/(I_{\Delta_Y})^{n+1}$. Then $\mathcal{O}_{\Delta_Y(n)}$ is locally a direct sum of locally free $\mathcal{O}_Y$-modules of finite rank with respect to the $\mathcal{O}_Y$-module structure induced by the first projection. Then we have

$$F_n(\mathcal{D}_{X\to Y}) = \mathcal{H}om_{f^{-1}\mathcal{O}_Y}(f^{-1}\mathcal{O}_{\Delta_Y(n)}, \mathcal{O}_X),$$

and hence $F_n(\mathcal{D}_{X\to Y})$ has a structure of a sheaf of linear topological spaces induced from the pseudo-discrete topology of $\mathcal{O}_{\Delta_Y(n)}$ (cf. [**EGA 0**, 3.8]). More concretely, for an affine open subset U of X and an affine open subset V of Y such that $U \subset f^{-1}(V)$,

$$\{ P \in F_n(\mathcal{D}_{X\to Y})(U);\ P(f_i) = 0 \quad (i \in I) \}$$

form a neighborhood system of 0 in $\Gamma(U; F_n(\mathcal{D}_{X\to Y}))$, where $\{f_i\}_{i\in I}$ ranges over finite subsets of $\mathcal{O}_Y(V)$.

If $g\colon Y \to Z$ is also a morphism of pro-smooth $\mathbf{C}$-schemes, we can define the composition

$$\mathcal{D}_{X\to Y} \otimes f^{-1}\mathcal{D}_{Y\to Z} \to \mathcal{D}_{X\to Z}. \tag{1.7.6}$$

In particular, $D_X := D_{X\overset{\mathrm{id}}{\to}X}$ is a sheaf of rings and $\mathcal{D}_{X\to Y}$ is a $(\mathcal{D}_X, f^{-1}\mathcal{D}_Y)$-bimodule.

If Y is a smooth **C**-scheme, we have

$$\mathcal{D}_{X\to Y} \simeq \mathcal{O}_X \otimes_{f^{-1}\mathcal{O}_Y} f^{-1}\mathcal{D}_Y. \tag{1.7.7}$$

Definition 1.7.1. Let X be a pro-smooth **C**-scheme. A $\mathcal{D}_X$-module $\mathfrak{M}$ is called *admissible* if it satisfies the following conditions:

(1.7.8) For any affine open subset U of X and any $s \in \Gamma(U;\mathfrak{M})$, there exists a finitely generated subalgebra A of $\Gamma(U;\mathcal{O}_X)$ such that $Ps = 0$ for any $P \in \Gamma(U;\mathcal{D}_X)$ satisfying $P(A) = 0$.

(1.7.9) $\mathfrak{M}$ is quasi-coherent as an $\mathcal{O}_X$-module.

The condition (1.7.8) is equivalent to saying that $\mathcal{D}_X$ acts continuously on $\mathfrak{M}$ with the pseudo-discrete topology.

1.8. Let $f\colon X \to Y$ be a morphism of pro-smooth **C**-schemes. Then for any admissible $\mathcal{D}_Y$-module $\mathfrak{N}$, $f^*\mathfrak{N} = \mathcal{O}_X \otimes_{f^{-1}\mathcal{O}_Y} f^{-1}\mathfrak{N}$ has a structure of $\mathcal{D}_X$-module. Moreover $f^*\mathfrak{N}$ is admissible (cf. §1.9).

1.9. Let X be a **C**-scheme satisfying (S). Let $\{S_n\}_{n\in\mathbf{N}}$ be a projective system as in §1.2 and let $p_n : X \to S_n$ be the projection. Then we have

(1.9.1)
$$F_k(\mathcal{D}_{X\to S_n}) \simeq \varinjlim_m p_m^{-1} F_k(\mathcal{D}_{S_m\to S_n}) = \mathcal{O}_X \otimes_{p_n^{-1}\mathcal{O}_{S_n}} p_n^{-1} F_k(\mathcal{D}_{S_n}),$$

$$F_k(\mathcal{D}_X) \simeq \varprojlim_n F_k(\mathcal{D}_{X\to S_n}). \tag{1.9.2}$$

If $\mathfrak{M}$ is an admissible $\mathcal{D}_X$-module locally of finite type, then there exist some n and a coherent $\mathcal{D}_{S_n}$-module $\mathfrak{N}$ such that

$$\mathfrak{M} \simeq (p_n)^*\mathfrak{N} = \mathcal{O}_X \otimes_{p_n^{-1}\mathcal{O}_{S_n}} p_n^{-1}\mathfrak{N} \simeq \mathcal{D}_{X\to S_n} \otimes_{p_n^{-1}\mathcal{D}_{S_n}} p_n^{-1}\mathfrak{N}. \tag{1.9.3}$$

Conversely, for a quasi-coherent $\mathcal{D}_{S_n}$-module $\mathfrak{N}$, the $\mathcal{D}_X$-module $(p_n)^*\mathfrak{N}$ is an admissible $\mathcal{D}_X$-module.

1.10. Let X be a pro-smooth **C**-scheme. A $\mathcal{D}_X$-module $\mathfrak{M}$ is called *holonomic* (resp. *regular holonomic*) if it satisfies the following condition:

(1.10.1) For any point $x \in X$, there exist a morphism $f\colon U \to Y$ from an open neighborhood U of x to a smooth **C**-scheme Y and a holonomic (resp. regular holonomic) $\mathcal{D}_Y$-module $\mathfrak{N}$ such that $\mathfrak{M}|U$ is isomorphic to $f^*\mathfrak{N}$.

Let $X \simeq \varprojlim_n S_n$, where $\{S_n\}_{n\in\mathbf{N}}$ is a projective system as in §1.2 and let $\mathfrak{N}$ be a coherent $\mathcal{D}_{S_0}$-module. It is seen that, if $(p_0)^*\mathfrak{N}$ is a holonomic (resp. regular holonomic) $\mathcal{D}_X$-module, then $(p_{0n})^*\mathfrak{N}$ is a holonomic (resp. regular holonomic) $\mathcal{D}_{S_n}$-module for any n.

2. The analytic structure on C-schemes

2.0. In this section, ringed spaces, schemes and their morphisms are all over **C**.

2.1. Let X be an affine **C**-scheme. We define a local ringed space X_{an} as follows. The underlying set of X_{an} is the set $X(\mathbf{C})$ of the **C**-valued points of X. The topology on X_{an} is the weakest one such that, for any $f \in \mathcal{O}_X(X)$, $f(\mathbf{C})\colon X(\mathbf{C}) \to \mathbf{C}$ is continuous with respect to the Euclidian topology on **C**. We define the sheaf of rings $\mathcal{O}_{X_{\mathrm{an}}}$ on $X(\mathbf{C})$ by

$$\mathcal{O}_{X_{\mathrm{an}}} = \varinjlim_{f} f(\mathbf{C})^{-1}\mathcal{O}_{S_{\mathrm{an}}}, \tag{2.1.1}$$

where $f\colon X \to S$ ranges over morphisms with schemes S of finite type as targets. Here S_{an} denotes the complex analytic space associated to S.

2.2. More generally, let X be a **C**-scheme. We endow with $X(\mathbf{C})$ the weakest topology such that, for any affine open subset U of X, any open subset of U_{an} is open in $X(\mathbf{C})$. Let X_{an} denote this topological space. We define the sheaf of rings $\mathcal{O}_{X_{\mathrm{an}}}$ by $\mathcal{O}_{X_{\mathrm{an}}}|U_{\mathrm{an}} \simeq \mathcal{O}_{U_{\mathrm{an}}}$ for any affine open subset U of X.

Then we can check easily the following.

LEMMA 2.2.1. (i) *The ringed space* $(X_{\mathrm{an}}, \mathcal{O}_{X_{\mathrm{an}}})$ *is well-defined.*

(ii) *The correspondence* $X \mapsto (X_{\mathrm{an}}, \mathcal{O}_{X_{\mathrm{an}}})$ *is functorial.*

(iii) $(X \times Y)_{\mathrm{an}} = X_{\mathrm{an}} \times Y_{\mathrm{an}}$ *as a topological space.*

(iv) *If* $X \to Y$ *is an open (resp. closed) embedding, so is* $X_{\mathrm{an}} \to Y_{\mathrm{an}}$.

(v) *Let* $\{X_n\}_{n\in\mathbf{N}}$ *be a projective system of* **C**-*schemes such that* $X_m \to X_n$ *is affine for* $m \geqq n$. *Then we have* $(\varprojlim X_n)_{\mathrm{an}} \simeq \varprojlim (X_n)_{\mathrm{an}}$ *as a ringed space.*

(vi) *If* X *is separated, then* X_{an} *is Hausdorff.*

(vii) *If* X *is quasi-compact and of countable type, then* X_{an} *has a countable base of open subsets.*

2.3. There exists a natural morphism of ringed spaces

$$\iota = \iota_X : X_{\mathrm{an}} \to X \tag{2.3.1}$$

functorial in X, and we have a natural $\iota_X^{-1}\mathcal{D}_X$-module structure on $\mathcal{O}_{X_{\mathrm{an}}}$.

2.4. A *quasi-compact stratification* of a **C**-scheme X is a locally finite family $\{X_\alpha\}$ of locally closed subsets of X such that

(2.4.1) $X = \sqcup X_\alpha$ as a set,
(2.4.2) $\overline{X}_\alpha \cap X_\beta \neq \emptyset$ implies $\overline{X}_\alpha \supset X_\beta$,
(2.4.3) The inclusion $X_\alpha \hookrightarrow X$ is a quasi-compact morphism.

2.5. Let X be a coherent $\mathbf{C}$-scheme and let k be a field.

Definition 2.5.1 A sheaf F of k-vector spaces on X_{an} is called *weakly constructible* if there exists a quasi-compact stratification $X = \sqcup X_\alpha$ such that $F|(X_\alpha)_{\mathrm{an}}$ is locally constant. If moreover F_x is finite-dimensional for any $x \in X_{\mathrm{an}}$, we call F *constructible.*

Let $D(X_{\mathrm{an}}; k)$ be the derived category of the category of sheaves of k-vector spaces on X_{an}. An object K of $D(X_{\mathrm{an}}; k)$ is called *constructible* (resp. *weakly constructible*) if it satisfies the following conditions.

(2.5.1) $H^n(K)$ is constructible (resp. weakly constructible) for any n.
(2.5.2) For any quasi-compact open subset Ω of X, $H^n(K)|\Omega_{\mathrm{an}} = 0$ except for finitely many n.

We denote the full subcategory of $D(X_{\mathrm{an}}; k)$ consisting of constructible (resp. weakly constructible) objects by $D_c(X; k)$ (resp. $D_{w.c.}(X; k)$).

PROPOSITION 2.5.2. *Let W be a smooth $\mathbf{C}$-scheme. Set $X = W \times \mathbf{A}^\infty$, $X_n = W \times \mathbf{A}^n$ and let $p_n : X \to X_n$ be the projection. Then for any $K \in \mathrm{Ob}(D_c(X; k))$, there exist some n and $K_n \in \mathrm{Ob}(D_c(X_n; k))$ satisfying $(p_n)_{\mathrm{an}}^{-1} K_n \simeq K$.*

PROOF: We can take a finite coherent stratification $X = \sqcup X_\alpha$ such that $H^j(K)|X_\alpha$ is locally constant for any j and α. Then there exist some n and a stratification $X_n = \sqcup \tilde{X}_\alpha$ such that $X_\alpha = p_n^{-1} \tilde{X}_\alpha$. Let $i: X_n \to X_n \times \mathbf{A}^\infty$ be the embedding by the origin $\in \mathbf{A}^\infty$. Since $X \simeq X_n \times \mathbf{A}^\infty$ and $(\mathbf{A}^\infty)_{\mathrm{an}}$ is contractible to the origin, we have $K \simeq (p_n)_{\mathrm{an}}^{-1} K_n$ with $K_n = (i_{\mathrm{an}})^{-1} K$. □

PROPOSITION 2.5.3. *Let $X = \mathbf{A}^\infty \times W$, where W is a smooth $\mathbf{C}$-scheme, and let $p: X \to W$ be the projection. Then for a cohomologically bounded object K of $D(W_{\mathrm{an}}; k)$, we have*

$$\mathbf{R}\mathcal{H}om((p_{\mathrm{an}})^{-1} K, k_{X_{\mathrm{an}}}) \simeq (p_{\mathrm{an}})^{-1} \mathbf{R}\mathcal{H}om(K, k_{W_{\mathrm{an}}}).$$

PROOF: We shall show that the functorial morphism

$$(p_{\mathrm{an}})^{-1} \mathbf{R}\mathcal{H}om(K, k_{W_{\mathrm{an}}}) \to \mathbf{R}\mathcal{H}om((p_{\mathrm{an}})^{-1} K, k_{X_{\mathrm{an}}})$$

is an isomorphism. In order to see this, it suffices to show that

$$\mathbf{R}\Gamma((p_{\text{an}})^{-1}V;(p_{\text{an}})^{-1}\mathbf{R}\mathcal{H}om(K,k_{W_{\text{an}}}))\to\mathbf{R}\Gamma((p_{\text{an}})^{-1}V;\\ \mathbf{R}\mathcal{H}om((p_{\text{an}})^{-1}K,k_{X_{\text{an}}}))$$

is an isomorphism for any open subset V of W_{an} (Observe that $U\times(\mathbf{A}^{\infty})_{\text{an}}\times V$ form a base of open subsets of $(\mathbf{A}^{\infty}\times W)_{\text{an}}$, where $\mathbf{A}^{\infty}\simeq\mathbf{A}^{n}\times\mathbf{A}^{\infty}$, U is an open subset of $(\mathbf{A}^{n})_{\text{an}}$ and V is an open subset of W_{an}). This follows from the following lemma.

LEMMA 2.5.4. *Let X, Y and S be topological spaces and let $p_X: X\to S$ and $p_Y: Y\to S$ be continuous maps. Let $p: X\times[0,1]\to X$ be the projection, and let $h: X\times[0,1]\to Y$ be a continuous map satisfying $p_Y\circ h=p_X\circ p$. Define $i_\nu: X\to X\times[0,1]$ ($\nu=0,1$) by $i_\nu(x)=(x,\nu)$, and set $f_\nu=h\circ i_\nu$. Let K (resp. F) be a cohomologically bounded (resp. lower bounded) object in the derived category of the category of sheaves of k-vector spaces on S and let $f_\nu^\sharp$ be the composition of*

$$\mathbf{R}(p_Y)_*\mathbf{R}\mathcal{H}om(p_Y^{-1}K,p_Y^{-1}F)\to\mathbf{R}(p_Y)_*\mathbf{R}(f_\nu)_*\mathbf{R}\mathcal{H}om(f_\nu^{-1}p_Y^{-1}K,f_\nu^{-1}p_Y^{-1}F)\\ \xrightarrow{\sim}\mathbf{R}(p_X)_*\mathbf{R}\mathcal{H}om(p_X^{-1}K,p_X^{-1}F).$$

Then we have $f_0^\sharp=f_1^\sharp$.

PROOF: Set $Z=X\times[0,1]$ and $p_Z=p_X\circ p$. Then $f_\nu^\sharp$ is obtained by

$$\begin{aligned}\mathbf{R}(p_Y)_*\mathbf{R}\mathcal{H}om(p_Y^{-1}K,p_Y^{-1}F)&\to\mathbf{R}(p_Y)_*\mathbf{R}h_*\mathbf{R}\mathcal{H}om(h^{-1}p_Y^{-1}K,h^{-1}p_Y^{-1}F)\\ &\simeq\mathbf{R}(p_Z)_*\mathbf{R}\mathcal{H}om(p_Z^{-1}K,p_Z^{-1}F)\\ &\to\mathbf{R}(p_Z)_*\mathbf{R}(i_\nu)_*\mathbf{R}\mathcal{H}om(i_\nu^{-1}p_Z^{-1}K,i_\nu^{-1}p_Z^{-1}F)\\ &\simeq\mathbf{R}(p_X)_*\mathbf{R}\mathcal{H}om(p_X^{-1}K,p_X^{-1}F).\end{aligned}$$

Set $\tilde{K}=p_X^{-1}K$ and $\tilde{F}=p_X^{-1}F$. Since $\mathbf{R}(p_Z)_*=\mathbf{R}(p_X)_*\mathbf{R}p_*$, it is enough to show that the morphism

$$\begin{aligned}i_\nu^\sharp:\mathbf{R}p_*\mathbf{R}\mathcal{H}om(p^{-1}\tilde{K},p^{-1}\tilde{F})&\to\mathbf{R}p_*\mathbf{R}(i_\nu)_*\mathbf{R}\mathcal{H}om(i_\nu^{-1}p^{-1}\tilde{K},i_\nu^{-1}p^{-1}\tilde{F})\\ &\simeq\mathbf{R}\mathcal{H}om(\tilde{K},\tilde{F})\end{aligned}$$

does not depend on ν. Since

$$\begin{aligned}p^*:\mathbf{R}\mathcal{H}om(\tilde{K},\tilde{F})&\to\mathbf{R}p_*\mathbf{R}\mathcal{H}om(p^{-1}\tilde{K},p^{-1}\tilde{F})\\ &\simeq\mathbf{R}\mathcal{H}om(\tilde{K},\mathbf{R}p_*p^{-1}\tilde{F})\end{aligned}$$

is an isomorphism and $i_\nu^\sharp\circ p^*=\mathrm{id}$, we obtain the desired result. □

For a quasi-compact separated essentially smooth $\mathbf{C}$-scheme X we set

(2.5.3) $$\mathbf{D}_X(K)=\mathbf{R}\mathcal{H}om(K,k_{X_{\text{an}}})$$

for $K\in D_c(X;k)$.

COROLLARY 2.5.5. *Let X be a quasi-compact separated essentially smooth* $\mathbf{C}$*-scheme. Then*

(i) $\mathbf{D}_X$ *preserves* $D_c(X;k)$.

(ii) $\mathbf{D}_X \circ \mathbf{D}_X \simeq \mathrm{id}$.

2.6. Let X be a quasi-compact separated essentially smooth $\mathbf{C}$-scheme. Define full subcategories ${}^pD_c^{\leqq 0}(X;k)$ and ${}^pD_c^{\geqq 0}(X;k)$ of $D_c(X;k)$ by

(2.6.1) K belongs to ${}^pD_c^{\leqq 0}(X;k)$ if and only if $\operatorname{codim}\operatorname{Supp} H^n(K) \geqq n$ for any n.

(2.6.2) K belongs to ${}^pD_c^{\geqq 0}(X;k)$ if and only if $\mathbf{D}_X(K)$ belongs to ${}^pD_c^{\leqq 0}(X;k)$.

The following theorem is similarly proven as in the finite-dimensional case (see [**BBD**], [**KS**]), and we omit the proof.

THEOREM 2.6.1. (i) $({}^pD_c^{\leqq 0}(X;k), {}^pD_c^{\geqq 0}(X;k))$ *is a t-structure of* $D_c(X;k)$.

(ii) *For* $K_1 \in \mathrm{Ob}({}^pD_c^{\leqq 0}(X;k))$ *and* $K_2 \in \mathrm{Ob}({}^pD_c^{\geqq 0}(X;k))$, *we have*

$$H^n(\mathbf{R}\mathcal{H}om(K_1, K_2)) = 0 \ (n < 0).$$

(iii) $\mathrm{Perv}(X;k) = {}^pD_c^{\leqq 0}(X;k) \cap {}^pD_c^{\geqq 0}(X;k)$ *is a stack, i.e.*

(a) *For* $K_1, K_2 \in \mathrm{Ob}(\mathrm{Perv}(X;k))$, $U \mapsto \mathrm{Hom}(K_1|U_{\mathrm{an}}, K_2|U_{\mathrm{an}})$ *is a sheaf on* X.

(b) *Let* $X = \cup_j U_j$ *be an open covering. Assume that we are given objects* K_j *of* $\mathrm{Perv}(U_j;k)$ *and isomorphisms* $f_{ij}\colon K_j|(U_i)_{\mathrm{an}} \cap (U_j)_{\mathrm{an}} \to K_i|(U_i)_{\mathrm{an}} \cap (U_j)_{\mathrm{an}}$ *such that* $f_{ij} \circ f_{jk} = f_{ik}$ *on* $(U_i)_{\mathrm{an}} \cap (U_j)_{\mathrm{an}} \cap (U_k)_{\mathrm{an}}$. *Then there exist* $K \in \mathrm{Ob}(\mathrm{Perv}(X;k))$ *and isomorphisms* $f_i\colon K|(U_i)_{\mathrm{an}} \to K_i$ *such that* $f_{ij} \circ f_j = f_i$ *on* $(U_i)_{\mathrm{an}} \cap (U_j)_{\mathrm{an}}$.

We call an object of $\mathrm{Perv}(X;k)$ a *perverse sheaf*. When X is smooth, this definition coincides with the one in [**BBD**] up to shift.

PROPOSITION 2.6.2. *Let X be a separated essentially smooth* $\mathbf{C}$*-scheme such that* $X \simeq \varprojlim_n X_n$ *for some projective system* $\{X_n\}$ *satisfying* (1.2.1) *and* (1.2.2). *Then we have* $\mathrm{Perv}(X;k) \simeq \varinjlim_n \mathrm{Perv}(X_n;k)$; *i.e. the following two properties hold.*

(2.6.3) *For* $M_1, M_2 \in \mathrm{Perv}(X_n;k)$ *we have* $\varinjlim_m \mathrm{Hom}((p_{nm})^*M_1, (p_{nm})^*M_2) \simeq \mathrm{Hom}((p_n)^*M_1, (p_n)^*M_2)$.

(2.6.4) *For any* $M \in \mathrm{Perv}(X;k)$ *there exist some* n *and* $M_n \in \mathrm{Perv}(X_n;k)$ *such that* $M \simeq (p_n)^*M_n$.

Here, $p_{nm}: X_m \to X_n$ and $p_n: X \to X_n$ are the projections.

2.7. Let X be a **C**-scheme satisfying (S) and let $\{S_n\}_{n\in\mathbf{N}}$ be a projective system as in §1.2. We denote by $p_{nm}: S_m \to S_n$ $(m \geqq n)$ and $p_n: X \to S_n$ the projections. Let $\mathfrak{B}^{(p,q)}_{(S_n)_{\mathrm{an}}}$ be the sheaf of (p,q)-forms on $(S_n)_{\mathrm{an}}$ with hyperfunction coefficients. Then we have natural homomorphisms

$$(p_{nm})^{-1}_{\mathrm{an}}\mathfrak{B}^{(p,q)}_{(S_n)_{\mathrm{an}}} \to \mathfrak{B}^{(p,q)}_{(S_m)_{\mathrm{an}}}, \tag{2.7.1}$$

$$\iota_X^{-1}p_m^{-1}\mathcal{D}_{S_m\to S_n} \times (p_n)^{-1}_{\mathrm{an}}\mathfrak{B}^{(0,p)}_{(S_n)_{\mathrm{an}}} \to (p_m)^{-1}_{\mathrm{an}}\mathfrak{B}^{(0,p)}_{(S_m)_{\mathrm{an}}}. \tag{2.7.2}$$

By (2.7.1) we obtain a sheaf $\mathfrak{B}^{(p,q)}_{X_{\mathrm{an}}} = \varinjlim (p_n)^{-1}_{\mathrm{an}}\mathfrak{B}^{(p,q)}_{(S_n)_{\mathrm{an}}}$ on X_{an}, and this does not depend on the choice of $\{S_n\}_{n\in\mathbf{N}}$ by Proposition 1.3.1. Taking the inductive limit in (2.7.2) with respect to m, we obtain

$$\iota_X^{-1}\mathcal{D}_{X\to S_n} \times (p_n)^{-1}_{\mathrm{an}}\mathfrak{B}^{(0,p)}_{(S_n)_{\mathrm{an}}} \to \mathfrak{B}^{(0,p)}_{X_{\mathrm{an}}}. \tag{2.7.3}$$

Taking again the limit in (2.7.3) with respect to n, we obtain

$$\iota_X^{-1}\mathcal{D}_X \times \mathfrak{B}^{(0,p)}_{X_{\mathrm{an}}} \to \mathfrak{B}^{(0,p)}_{X_{\mathrm{an}}}, \tag{2.7.4}$$

and this defines a structure of an $\iota_X^{-1}\mathcal{D}_X$-module on $\mathfrak{B}^{(0,p)}_{X_{\mathrm{an}}}$. We have also

$$\begin{aligned}&\varinjlim_m (p_m)^{-1}_{\mathrm{an}} \mathcal{H}om_{\iota_{S_m}^{-1}\mathcal{D}_{S_m}}(\iota_{S_m}^{-1}\mathcal{D}_{S_m\to S_n}, \mathfrak{B}^{(0,p)}_{(S_m)_{\mathrm{an}}}) \\ &\simeq \mathcal{H}om_{\iota_X^{-1}\mathcal{D}_X}(\iota_X^{-1}\mathcal{D}_{X\to S_n}, \mathfrak{B}^{(0,p)}_{X_{\mathrm{an}}}).\end{aligned} \tag{2.7.5}$$

2.8. More generally, let X be a pro-smooth **C**-scheme. We can patch the sheaves $\mathfrak{B}^{(p,q)}_{U_{\mathrm{an}}}$ for affine open subschemes U of X satisfying (S), and obtain a sheaf $\mathfrak{B}^{(p,q)}_{X_{\mathrm{an}}}$ on X_{an} such that

$$\mathfrak{B}^{(p,q)}_{X_{\mathrm{an}}}|U_{\mathrm{an}} \simeq \mathfrak{B}^{(p,q)}_{U_{\mathrm{an}}}. \tag{2.8.1}$$

We can define the derivatives

$$\partial\colon \mathfrak{B}^{(p,q)}_{X_{\mathrm{an}}} \to \mathfrak{B}^{(p+1,q)}_{X_{\mathrm{an}}}, \tag{2.8.2}$$

$$\overline{\partial}\colon \mathfrak{B}^{(p,q)}_{X_{\mathrm{an}}} \to \mathfrak{B}^{(p,q+1)}_{X_{\mathrm{an}}}, \tag{2.8.3}$$

and we have the exact sequence:

$$0 \to \mathcal{O}_{X_{\mathrm{an}}} \to \mathfrak{B}_{X_{\mathrm{an}}}^{(0,0)} \xrightarrow{\bar\partial} \mathfrak{B}_{X_{\mathrm{an}}}^{(0,1)} \xrightarrow{\bar\partial} \cdots \tag{2.8.4}$$

of $\iota_X^{-1}\mathcal{D}_X$-modules. The Dolbeault complex:

$$\mathfrak{B}_{X_{\mathrm{an}}}^{(0,0)} \xrightarrow{\bar\partial} \mathfrak{B}_{X_{\mathrm{an}}}^{(0,1)} \xrightarrow{\bar\partial} \cdots$$

is denoted by $\mathfrak{B}^{\cdot}_{X_{\mathrm{an}}}$.

2.9. Let X be a pro-smooth $\mathbf{C}$-scheme. For a holonomic $\mathcal{D}_X$-module $\mathfrak{M}$, we set

$$\mathrm{Sol}(\mathfrak{M}) = \mathcal{H}om_{\mathcal{D}_X}(\mathfrak{M}, \mathfrak{B}^{\cdot}_{X_{\mathrm{an}}}) \left(= \mathcal{H}om_{\iota_X^{-1}\mathcal{D}_X}(\iota_X^{-1}\mathfrak{M}, \mathfrak{B}^{\cdot}_{X_{\mathrm{an}}})\right), \tag{2.9.1}$$

and regard this as an object of $D(X_{\mathrm{an}}; \mathbf{C})$. When X is smooth, we have

$$\mathrm{Sol}(\mathfrak{M}) = \mathbf{R}\mathcal{H}om_{\mathcal{D}_X}(\mathfrak{M}, \mathcal{O}_{X_{\mathrm{an}}})$$

by [**K1**].

Let U be an open subset of X satisfying (S) and let $\{S_n\}$ be a projective system of $\mathbf{C}$-schemes satisfying (1.2.1), (1.2.2) such that $U \simeq \varprojlim S_n$. Then we have $\mathfrak{M}|U \simeq (p_n)^*\mathfrak{N}$ for some n and some holonomic $\mathcal{D}_{S_n}$-module $\mathfrak{N}$, where $p_n : U \to S_n$ is the projection. By (2.7.5) we have

(2.9.2)

$$\begin{aligned}
&\mathcal{H}om_{\mathcal{D}_U}(\mathfrak{M}, \mathfrak{B}_{U_{\mathrm{an}}}^{(0,p)}) \\
\simeq &\mathcal{H}om_{\mathcal{D}_U}(\mathcal{D}_{U\to S_n} \otimes_{\mathcal{D}_{S_n}} \mathfrak{N}, \mathfrak{B}_{U_{\mathrm{an}}}^{(0,p)}) \\
\simeq &\mathcal{H}om_{\mathcal{D}_{S_n}}(\mathfrak{N}, \mathcal{H}om_{\mathcal{D}_U}(\mathcal{D}_{U\to S_n}, \mathfrak{B}_{U_{\mathrm{an}}}^{(0,p)})) \\
\simeq &\varinjlim_m \mathcal{H}om_{\mathcal{D}_{S_n}}(\mathfrak{N}, (p_m)_{\mathrm{an}}^{-1}\mathcal{H}om_{\mathcal{D}_{S_m}}(\mathcal{D}_{S_m\to S_n}, \mathfrak{B}_{(S_m)_{\mathrm{an}}}^{(0,p)})) \\
\simeq &\varinjlim_m (p_m)_{\mathrm{an}}^{-1}\mathcal{H}om_{\mathcal{D}_{S_m}}((p_{nm})^*\mathfrak{N}, \mathfrak{B}_{(S_m)_{\mathrm{an}}}^{(0,p)})).
\end{aligned}$$

On the other hand we have

$$\begin{aligned}
&H^q(\mathcal{H}om_{\mathcal{D}_{S_m}}((p_{nm})^*\mathfrak{N}, \mathfrak{B}^{\cdot}_{(S_m)_{\mathrm{an}}})) \\
\simeq &\mathcal{E}xt^q_{\mathcal{D}_{S_m}}((p_{nm})^*\mathfrak{N}, \mathcal{O}_{(S_m)_{\mathrm{an}}}) \\
\simeq &(p_{nm})_{\mathrm{an}}^{-1}\mathcal{E}xt^q_{\mathcal{D}_{S_n}}(\mathfrak{N}, \mathcal{O}_{(S_n)_{\mathrm{an}}}).
\end{aligned} \tag{2.9.3}$$

Thus

(2.9.4) $$H^q(\mathrm{Sol}(\mathfrak{M})|U_{\mathrm{an}}) \simeq H^q((p_n)_{\mathrm{an}}^{-1}\,\mathrm{Sol}(\mathfrak{N})),$$

and we finally obtain

(2.9.5) $$\mathrm{Sol}(\mathfrak{M})|U_{\mathrm{an}} \simeq (p_n)_{\mathrm{an}}^{-1}\,\mathrm{Sol}(\mathfrak{N}) \quad \text{in} \quad D^b(U_{\mathrm{an}};\mathbf{C}).$$

This shows in particular

LEMMA 2.9.1. *Let X be a quasi-compact separated essentially smooth* **C**-*scheme. If $\mathfrak{M}$ is a holonomic $\mathcal{D}_X$-module, then* Sol($\mathfrak{M}$) *is a perverse sheaf, and* Sol *is a contravariant exact functor from the category of holonomic $\mathcal{D}_X$-modules to* Perv(X).

3. Mixed Hodge modules on essentially smooth C-schemes

3.0. We shall study mixed Hodge modules on essentially smooth **C**-schemes. In this section all schemes are over **C** and assumed to be *quasi-compact and separated.*

3.1. In [S], M. Saito constructed mixed Hodge modules on finite-dimensional manifolds. In his formulation, the weights behave well under direct images, but not under inverse images. Since we treat infinite-dimensional manifolds, we have to modify his definition so that the weights behave well under inverse images.

3.2. Let X be a quasi-compact essentially smooth **C**-scheme. Let $\tilde{M}FW(X)$ be the category consisting of $M = (\mathfrak{M}, F, K, W, \iota)$, where

(3.2.1) $\mathfrak{M}$ is a regular holonomic $\mathcal{D}_X$-module,

(3.2.2) F is a filtration of $\mathfrak{M}$ by coherent $\mathcal{O}_X$-submodules which is compatible with $(\mathcal{D}_X, F)$,

(3.2.3) $W(\mathfrak{M})$ is a finite filtration of $\mathfrak{M}$ by regular holonomic $\mathcal{D}_X$-modules,

(3.2.4) K is an object of Perv$(X;\mathbf{Q})$,

(3.2.5) $W(K)$ is a finite filtration of K in Perv$(X;\mathbf{Q})$,

(3.2.6) ι is an isomorphism $\mathbf{C}_X \otimes_{\mathbf{Q}_X} K \xrightarrow{\sim} \mathrm{Sol}(\mathfrak{M})$ in $D_c(X;\mathbf{C})$, compatible with W; i.e. ι induces a commutative diagram

$$\begin{array}{ccc}
\mathbf{C}_X \otimes_{\mathbf{Q}_X} W_k(K) & \xrightarrow{\sim} & \mathrm{Sol}(\mathfrak{M}/W_{-k-1}(\mathfrak{M})) \\
\downarrow & & \downarrow \\
\mathbf{C}_X \otimes_{\mathbf{Q}_X} K & \xrightarrow{\sim} & \mathrm{Sol}(\mathfrak{M}).
\end{array}$$

We define morphisms of $\tilde{M}FW(X)$ so that $M \mapsto K \in \mathrm{Perv}(X;\mathbf{Q})$ is a covariant functor and $M \mapsto \mathfrak{M}$ is a contravariant functor.

Sometimes, ι in $(\mathfrak{M}, F, K, W, \iota)$ will be omitted.

3.3. Let X be a smooth $\mathbf{C}$-scheme and let $MHM(X)$ be the category of mixed Hodge modules on X defined in Saito [**S**]. We define a contravariant functor

$$\varphi_X : MHM(X) \to \tilde{M}FW(X) \tag{3.3.1}$$

as follows. Let $M = (\mathfrak{M}, F, K, W)$ be an object of $MHM(X)$ and let $\mathbf{D}_X(M) = (\mathfrak{M}^*, F, K^*, W)$ be the dual of M (cf. [**S**]). Then we define $\varphi_X(M) = (\mathfrak{N}, F, \tilde{K}, W)$ by

$$\mathfrak{N} = \mathfrak{M}^* \otimes_{\mathcal{O}_X} (\Omega_X^{\dim X})^{\otimes -1}, \tag{3.3.2}$$

$$F_p(\mathfrak{N}) = F_p(\mathfrak{M}^*) \otimes_{\mathcal{O}_X} (\Omega_X^{\dim X})^{\otimes -1}, \tag{3.3.3}$$

$$W_k(\mathfrak{N}) = W_{k+\dim X}(\mathfrak{M}^*) \otimes_{\mathcal{O}_X} (\Omega_X^{\dim X})^{\otimes -1}, \tag{3.3.4}$$

$$\tilde{K} = K[-\dim X], \tag{3.3.5}$$

$$W_k(\tilde{K}) = W_{k+\dim X}(K)[-\dim X]. \tag{3.3.6}$$

Note that $\mathfrak{N}$ is a left $\mathcal{D}_X$-module since $\mathfrak{M}$ and $\mathfrak{M}^*$ are right $\mathcal{D}_X$-modules.

Let $\tilde{M}HM(X)$ be the image of φ_X. It is a full subcategory of $\tilde{M}FW(X)$, isomorphic to $MHM(X)$.

We define

$$\varphi_X : D^b(MHM(X)) \xrightarrow{\sim} D^b(\tilde{M}HM(X)) \tag{3.3.7}$$

by $M^{\cdot} \mapsto \varphi_X(M^{\cdot})[\dim X]$. Hence φ_X is compatible with

$$i_X : D^b(MHM(X)) \to D_c(X;\mathbf{Q})$$

and $i_X : D^b(\tilde{M}HM(X)) \to D_c(X;\mathbf{Q})$.

The duality functor $\mathbf{D}_X$ on $MHM(X)$ defines the duality functor

$$\mathbf{D}_X : \tilde{M}HM(X) \to \tilde{M}HM(X)^{\mathrm{op}} \tag{3.3.8}$$

by $\mathbf{D}_X \circ \varphi_X = \varphi_X \circ \mathbf{D}_X$. Then we have $i_X \circ \mathbf{D}_X = \mathbf{D}_X \circ i_X$, where $\mathbf{D}_X(K) = \mathbf{R}\mathcal{H}om(K, \mathbf{Q}_{X_{\mathrm{an}}})$ for $K \in \mathrm{Perv}(X;\mathbf{Q})$.

3.4. For a morphism $f: X \to Y$ of smooth **C**-schemes, we define functors

$$f^*, f^!: D^b(\tilde{M}HM(Y)) \to D^b(\tilde{M}HM(X)) \tag{3.4.1}$$

$$f_*, f_!: D^b(\tilde{M}HM(X)) \to D^b(\tilde{M}HM(Y)) \tag{3.4.2}$$

using those defined in [**S**] and the isomorphism (3.3.7). In particular, if $f: X \to Y$ is smooth and $M = (\mathfrak{M}, F, K, W)$ is an object of $\tilde{M}HM(Y)$, then we have

$$f^*(M) = (f^*\mathfrak{M}, F, f^*K, W) \in \mathrm{Ob}(\tilde{M}HM(X)), \tag{3.4.3}$$

where

$$F_p(f^*\mathfrak{M}) = f^*F_p(\mathfrak{M}), \tag{3.4.4}$$

$$W_k(f^*\mathfrak{M}) = f^*(W_k(\mathfrak{M})), \quad W_k(f^*K) = f^*(W_k(K)). \tag{3.4.5}$$

We extend this definition when X is essentially smooth, Y is smooth and f is weakly smooth. Hence in this case, f^* is a functor from $\tilde{M}HM(Y)$ into $\tilde{M}FW(X)$ defined by (3.4.3), (3.4.4) and (3.4.5).

3.5. For an essentially smooth **C**-scheme X, we define a full subcategory $\tilde{M}HM(X)$ of $\tilde{M}FW(X)$ as follows. An object M of $\tilde{M}FW(X)$ belongs to $\tilde{M}HM(X)$ if and only if X is covered by open subsets U such that there are a weakly smooth morphism $f: U \to Y$ to a smooth **C**-scheme Y and an object M' of $\tilde{M}HM(Y)$ satisfying $M|U \simeq f^*M'$. We can easily see that $\tilde{M}HM(X)$ is a stack. In this paper we call objects of $\tilde{M}HM(X)$ *mixed Hodge modules* on X. Note that $\tilde{M}HM(X)$ is an abelian category.

We can define the duality functor

$$\mathbf{D}_X: \tilde{M}HM(X) \to \tilde{M}HM(X)^{\mathrm{op}} \tag{3.5.1}$$

by $\mathbf{D}_X M|U \simeq f^*\mathbf{D}_Y M'$. It is an exact functor satisfying $\mathbf{D}_X \circ \mathbf{D}_X \simeq \mathrm{id}$. Hence this extends to

$$\mathbf{D}_X: D^b(\tilde{M}HM(X)) \to D^b(\tilde{M}HM(X))^{\mathrm{op}}. \tag{3.5.2}$$

3.6. Let X be an essentially smooth **C**-scheme satisfying (S) and let $\{S_n\}_{n\in\mathbf{N}}$ be a projective system as in §1.2. Then we have

$$\tilde{M}HM(X) \simeq \varinjlim_n \tilde{M}HM(S_n), \tag{3.6.1}$$

$$D^b(\tilde{M}HM(X)) \simeq \varinjlim_n D^b(\tilde{M}HM(S_n)) \tag{3.6.2}$$

(cf. Proposition 2.6.2).

3.7. For a morphism $f\colon X \to Y$ of essentially smooth **C**-schemes satisfying (S), we define

$$f^*\colon D^b(\tilde{M}HM(Y)) \to D^b(\tilde{M}HM(X)) \tag{3.7.1}$$

as follows. Let $X \simeq \varprojlim_n X_n$ and $Y \simeq \varprojlim_n Y_n$, where $\{X_n\}$ and $\{Y_n\}$ satisfy (1.2.1) and (1.2.2). We may assume that there are morphisms $f_n\colon X_n \to Y_n$ ($n \in \mathbf{N}$) such that $f = \varprojlim_n f_n$. Let $p_{X,n}\colon X \to X_n$ and $p_{Y,n}\colon Y \to Y_n$ be the projections. For a bounded complex $M^{\cdot}$ of mixed Hodge modules on Y, there exist some n and a bounded complex $M_n^{\cdot}$ of mixed Hodge modules on Y_n such that $M^{\cdot} \simeq (p_{Y,n})^* M_n^{\cdot}$. Then we define (3.7.1) by

$$f^* M^{\cdot} = (p_{X,n})^*((f_n)^* M_n^{\cdot}). \tag{3.7.2}$$

It is easy to check that this is well-defined.

3.8. Let $f\colon X \to Y$ be a morphism of essentially smooth **C**-schemes. Then, for each $i \in \mathbf{Z}$, we can define a functor

$$H^i f^*\colon \tilde{M}HM(Y) \to \tilde{M}HM(X). \tag{3.8.1}$$

In fact, locally on X, $(H^i f^*)(M)$ is defined as $H^i(f^*(M))$, and they can be patched together. It satisfies the following properties:

(3.8.2) If f is weakly smooth, then we have $(H^i f^*)(M) = 0$ for $i \neq 0$, and $(H^0 f^*)(M)$ is given by (3.4.4) and (3.4.5).

(3.8.3) If $(H^i f^*)(M) = 0$ for $i \neq p$ and if $g\colon W \to X$ is another morphism, then we have $(H^i g^*)(H^p f^*)(M) \simeq (H^{i+p}(f \circ g)^*)(M)$.

3.9. Let $f\colon X \to Y$ be a morphism of finite presentation. Assume that Y satisfies (S), so that X also satisfies (S). Then we define

$$f_*\colon D^b(\tilde{M}HM(X)) \to D^b(\tilde{M}HM(Y)), \tag{3.9.1}$$

$$f_!\colon D^b(\tilde{M}HM(X)) \to D^b(\tilde{M}HM(Y)), \tag{3.9.2}$$

$$f^!\colon D^b(\tilde{M}HM(Y)) \to D^b(\tilde{M}HM(X)) \tag{3.9.3}$$

as follows. Let $X \simeq \varprojlim_n X_n$ and $Y \simeq \varprojlim_n Y_n$, where $\{X_n\}$ and $\{Y_n\}$ satisfy (1.2.1) and (1.2.2), and let $M^{\cdot}$ be a bounded complex of mixed

Hodge modules on X (resp. Y). We may assume that there exists a morphism $f_0 : X'_0 \to Y_0$ such that $X \simeq X'_0 \times_{Y_0} Y$ and $f = f_0 \times_{Y_0} Y$. Set $f_n = f_0 \times_{Y_0} Y_n$. We may further assume that there exists $\{g_n\} : \{X'_0 \times_{Y_0} Y_n\} \to \{X_n\}$ (resp. $\{h_n\} : \{X_n\} \to \{X'_0 \times_{Y_0} Y_n\}$) such that $\varprojlim g_n = \mathrm{id}_X$ (resp. $\varprojlim h_n = \mathrm{id}_X$). Let $p_{X,n} : X \to X_n$ and $p_{Y,n} : Y \to Y_n$ be the projections. There exist some n and a bounded complex $M_n^{\cdot}$ of mixed Hodge modules on X_n (resp. Y_n) such that $M^{\cdot} \simeq (p_{X,n})^* M_n^{\cdot}$ (resp. $M^{\cdot} \simeq (p_{Y,n})^* M_n^{\cdot}$) . Then we define (3.9.1), (3.9.2) (resp. (3.9.3)) by

$$f_* M^{\cdot} = (p_{Y,n})^* \varphi_{Y_n} (f_n)_* (g_n)^* \varphi_{X_n}^{-1} (M_n^{\cdot}), \tag{3.9.4}$$

$$f_! M^{\cdot} = (p_{Y,n})^* \varphi_{Y_n} (f_n)_! (g_n)^* \varphi_{X_n}^{-1} (M_n^{\cdot}) \tag{3.9.5}$$

$$(\text{ resp. } f^! M^{\cdot} = (p_{X,n})^* \varphi_{X_n} (h_n)^* (f_n)^! \varphi_{Y_n}^{-1} (M_n^{\cdot})). \tag{3.9.6}$$

It is easy to check that they are well-defined. We have the following properties concerning them.

(3.9.7) The functor f_* (resp. $f^!$) is a right adjoint functor of f^* (resp. $f_!$).

(3.9.8) $f_* \circ \mathbf{D}_X \simeq \mathbf{D}_Y \circ f_!$.

(3.9.9) If f is proper, then we have $f_* = f_!$.

Note that (3.9.9) follows from the fact that f_n is proper for $n \gg 0$ if f is proper.

3.10. For an essentially smooth **C**-scheme X, we define

$$\mathbf{Q}_X^H = (\mathcal{O}_X, F, \mathbf{Q}_X, W, \iota) \in \mathrm{Ob}(\tilde{M}HM(X)) \tag{3.10.1}$$

by

$$F_p(\mathcal{O}_X) = \begin{cases} \mathcal{O}_X & (p \geqq 0) \\ 0 & (p < 0), \end{cases} \tag{3.10.2}$$

$$W_k(\mathcal{O}_X) = \begin{cases} \mathcal{O}_X & (k \geqq 0) \\ 0 & (k < 0), \end{cases} \tag{3.10.3}$$

$$W_k(\mathbf{Q}_X) = \begin{cases} \mathbf{Q}_X & (k \geqq 0) \\ 0 & (k < 0), \end{cases} \tag{3.10.4}$$

(3.10.5) $\iota\colon \mathbf{C}_X \otimes_{\mathbf{Q}_X} \mathbf{Q}_X \xrightarrow{\sim} \mathrm{Sol}(\mathcal{O}_X)$ is induced by

$$1 \mapsto \mathrm{id} \in \mathcal{H}om_{\mathcal{D}_X}(\mathcal{O}_X, \mathcal{O}_X) \subset \mathcal{H}om_{\mathcal{D}_X}(\mathcal{O}_X, \mathfrak{B}^{(0,0)}_{X_{an}}).$$

Set (pt) = Spec(**C**) and let $a_X : X \to (\mathrm{pt})$ be the projection. We shall identify $\tilde{M}HM((\mathrm{pt}))$ with the category MHS of mixed Hodge structures. Then we have $\mathbf{Q}_X^H = (a_X)^*\mathbf{Q}^H$, where $\mathbf{Q}^H$ is the trivial mixed Hodge structure on **Q**.

3.11. Let X and Y be essentially smooth **C**-schemes and let $j\colon Y \hookrightarrow X$ be an emmbeding of finite presentation. Then for any $M \in \tilde{M}HM(Y)$, there exists an object ${}^{\pi}M$ of $D^b(\tilde{M}HM(X))$ satisfying the following properties:

(3.11.1) ${}^{\pi}M[-\operatorname{codim} Y] \in \mathrm{Ob}(\tilde{M}HM(X))$,

(3.11.2) $\operatorname{Supp} {}^{\pi}M[-\operatorname{codim} Y] \subset \overline{Y}$,

(3.11.3) $j^*({}^{\pi}M) \simeq M$,

(3.11.4) ${}^{\pi}M[-\operatorname{codim} Y]$ has neither non-zero quotient nor non-zero subobject whose support is contained in $\overline{Y} - Y$.

Such ${}^{\pi}M$ is unique up to isomorphism, and we call it the *minimal extension* of M.

3.12. The following descent theorem being proven in a canonical way, we leave its proof to the readers.

PROPOSITION 3.12.1. *Let $f\colon X \to Y$ be a weakly smooth morphism of essentially smooth* **C***-schemes. Assume that f admits a section locally on Y. Let $p_i\colon X \times_Y X \to X$ $(i = 1, 2)$ and $p_{ij} : X \times_Y X \times_Y X \to X \times_Y X$ $(i, j = 1, 2, 3)$ be the obvious projections.*

(i) *For any $M, M' \in \mathrm{Ob}(\tilde{M}HM(Y))$ we have an exact sequence:*

$$0 \to Hom(M, M') \to Hom(f^*M, f^*M') \xrightarrow{(p_1)^* - (p_2)^*} Hom((p_1)^*f^*M, (p_1)^*f^*M').$$

(ii) *Let $M \in \mathrm{Ob}(\tilde{M}HM(X))$ and let $\alpha\colon (p_1)^*M \xrightarrow{\sim} (p_2)^*M$ be an isomorphism satisfying $(p_{23})^*\alpha \circ (p_{12})^*\alpha = (p_{13})^*\alpha$ (Note that we have $(p_{13})^*(p_1)^*M = (p_{12})^*(p_1)^*M$, $(p_{12})^*(p_2)^*M = (p_{23})^*(p_1)^*M$ and $(p_{13})^*(p_2)^*M = (p_{23})^*(p_2)^*M$). Then there exist some $N \in \mathrm{Ob}(\tilde{M}HM(Y))$ and an isomorphism $\beta\colon M \xrightarrow{\sim} f^*N$ satisfying $(p_1)^*\beta = (p_2)^*\beta \circ \alpha$.*

3.13. Let G be an essentially smooth affine group scheme acting on an essentially smooth **C**-scheme X. Let $\mu\colon G \times X \to X$ be the composition morphism, $pr\colon G\times X \to X$ the projection and $i\colon X \to G\times X$ the embedding by the identity element $e \in G$. We define morphisms $p_i\colon G \times G \times X \to G \times X$ $(i = 1,2,3)$ by $p_1(g_1, g_2, x) = (g_1, g_2 x)$, $p_2(g_1, g_2, x) = (g_1 g_2, x)$ and $p_3(g_1, g_2, x) = (g_2, x)$. For a mixed Hodge module M on X we have mixed Hodge modules $\mu^* M$ and $pr^* M$ since μ and pr are weakly smooth.

We define an abelian category $\tilde{M}HM^G(X)$ as follows. An object is a mixed Hodge module M on X, together with an isomorphism $\alpha_M : \mu^* M \xrightarrow{\sim} pr^* M$, satisfying the following conditions:

(3.13.1) $i^*\alpha_M : i^*\mu^* M \to i^* pr^* M$ coincides with $\mathrm{id}\colon M \to M$ under the identificaions $i^*\mu^* M = M$ amd $i^* pr^* M = M$.

(3.13.2) We have $(p_2)^*\alpha_M = (p_3)^*\alpha_M \circ (p_1)^*\alpha_M$ under the identifications $(p_2)^*\mu^* M = (p_1)^*\mu^* M$, $(p_2)^* pr^* M = (p_3)^* pr^* M$ and $(p_1)^* pr^* M = (p_3)^*\mu^* M$.

A morphism $\varphi\colon M \to N$ in $\tilde{M}HM^G(X)$ is a morphism of mixed Hodge modules satisfying $pr^*\varphi \circ \alpha_M = \alpha_N \circ \mu^*\varphi$. An object of $\tilde{M}HM^G(X)$ is called a *G-equivariant mixed Hodge module* on X.

Note that (3.13.1) is a consequence of (3.13.2) since α_M is an isomorphism.

If M is a mixed Hodge structue, then $(a_X)^* M$ is naturally endowed with a structure of G-equivariant mixed Hodge module, and we call it a *constant* G-equivariant mixed Hodge module. Here a_X is the morphism $X \to (\mathrm{pt})$.

LEMMA 3.13.1. *Any G-equivariant mixed Hodge module on G (with respect to the left multiplication) is constant.*

PROOF: Let $M \in \mathrm{Ob}(\tilde{M}HM^G(G))$. Let $\iota\colon (\mathrm{pt}) \to G$ be the embedding by e and let $\iota'\colon G \to G \times G$ be the morphism given by $g \mapsto (g, e)$. Since $\mu \circ \iota' = \mathrm{id}$ and $pr \circ \iota' = \iota \circ a_G$, we have $M = \iota'^*\mu^* M \simeq \iota'^* pr^* M = (a_G)^*\iota^* M$. We can easily check that the action of G on M coincides with the one on the constant mixed Hodge module $(a_G)^*\iota^* M$. □

Set $MHS^G = \tilde{M}HM^G((\mathrm{pt}))$.

LEMMA 3.13.2. *The abelian category MHS^G is naturally equivalent to MHS^{G/G^0}, where G^0 is the connected component of G containing the identity element e (Note that G/G^0 is a finite group by Lemma 1.6.4).*

PROOF: This follows from the fact that, for any $M \in \mathrm{Ob}(MHS^G)$, the restriction of $\alpha_M\colon (a_G)^* M \to (a_G)^* M$ to each connected component of G comes from an automorphism of M in MHS. □

The following theorem can be easily proven by Lemma 3.13.2 and Proposition 3.12.1.

THEOREM 3.13.3. *Let G be an affine group scheme and H an essentially smooth closed subgroup scheme of G. Assume that H acts locally freely on G and that G/H is separated and essentially smooth. Let i: (pt) $\to G/H$ be the embedding by $e \in G$. Then $M \mapsto i^*M$ gives an equivalence: $\tilde{M}HM^G(G/H) \xrightarrow{\sim} MHS^{H/H^0}$.*

4. Kac-Moody Lie algebras and flag varieties

4.1. Let $A = (a_{ij})_{1 \leqq i,j \leqq \ell}$ be a matrix of integers satisfying $a_{ii} = 2$, $a_{ij} \leqq 0\,(i \neq j)$, $a_{ij} \neq 0 \Leftrightarrow a_{ji} \neq 0$. Assume that we are given a finite-dimensional $\mathbf{C}$-vector space $\mathfrak{h}$, and elements $h_1, \ldots, h_\ell \in \mathfrak{h}$, $\alpha_1, \ldots, \alpha_\ell \in \mathfrak{h}^*$ satisfying the following conditions:

(4.1.1) $$\langle h_i, \alpha_j \rangle = a_{ij} \quad (i, j = 1, \ldots, \ell),$$

(4.1.2) $$\{\alpha_1, \ldots, \alpha_\ell\} \text{ is linearly independent,}$$

(4.1.3) $$\{h_1, \ldots, h_\ell\} \text{ is linearly independent.}$$

A Kac-Moody Lie algebra associated to these data is a Lie algebra $\mathfrak{g}$ over $\mathbf{C}$ which contains $\mathfrak{h}$ as an abelian subalgebra and which is generated by $\mathfrak{h}$ and elements $e_1, \ldots, e_\ell$, $f_1, \ldots, f_\ell$ satisfying the following relations:

(4.1.4) $$[h, e_i] = \alpha_i(h) e_i, \quad [h, f_i] = -\alpha_i(h) f_i \quad (h \in \mathfrak{h}, i = 1, \ldots, \ell),$$

(4.1.5) $$[e_i, f_j] = \delta_{ij} h_i \quad (i, j = 1, \ldots, \ell),$$

(4.1.6) $$(\operatorname{ad} e_i)^{1 - a_{ij}} e_j = 0 \quad (i \neq j),$$

(4.1.7) $$(\operatorname{ad} f_i)^{1 - a_{ij}} f_j = 0 \quad (i \neq j).$$

4.2. For $i = 1, \ldots, \ell$, let s_i be the linear automorphism of $\mathfrak{h}^*$ given by

(4.2.1) $$s_i(\lambda) = \lambda - \langle h_i, \lambda \rangle \alpha_i.$$

The Weyl group W of $(\mathfrak{g}, \mathfrak{h})$ is the subgroup of $\operatorname{Aut}(\mathfrak{h}^*)$ generated by $S = \{ s_1, \ldots, s_\ell \}$. It is well known that (W, S) is a Coxeter group with

$a_{ij}a_{ji}$	0	1	2	3	$\geqq 4$
$\operatorname{ord}(s_i s_j)$	2	3	4	6	∞

for $i \neq j$. We denote the length function and the Bruhat order on W by ℓ and $\geqq$, respectively.

Set

(4.2.2) $$\mathfrak{g}_\alpha = \{\, x \in \mathfrak{g}\,;\, [h, x] = \alpha(h)x\,(h \in \mathfrak{h})\,\} \quad (\alpha \in \mathfrak{h}^*),$$

(4.2.3) $$\Delta = \{\, \alpha \in \mathfrak{h}^* \setminus \{0\}\,;\, \mathfrak{g}_\alpha \neq 0\,\},$$

(4.2.4) $$\Delta^+ = \Delta \cap \sum_{i=1}^{\ell} \mathbf{Z}_{\geqq 0}\alpha_i, \quad \Delta^- = \Delta \cap \sum_{i=1}^{\ell} \mathbf{Z}_{\leqq 0}\alpha_i.$$

Let $\mathfrak{n}$ (resp. $\mathfrak{n}^-$) be the subalgebra of $\mathfrak{g}$ generated by $e_1, \dots, e_\ell$ (resp. $f_1, \dots, f_\ell$). Then we have

(4.2.5) $$\mathfrak{n} = \oplus_{\alpha \in \Delta^+} \mathfrak{g}_\alpha, \quad \mathfrak{n}^- = \oplus_{\alpha \in \Delta^-} \mathfrak{g}_\alpha,$$

(4.2.6) $$\mathfrak{g} = \mathfrak{n}^- \oplus \mathfrak{h} \oplus \mathfrak{n}.$$

Set

(4.2.7) $$\mathfrak{b} = \mathfrak{h} \oplus \mathfrak{n}, \quad \mathfrak{b}^- = \mathfrak{h} \oplus \mathfrak{n}^-,$$

(4.2.8) $$\mathfrak{g}_i = \mathfrak{h} \oplus \mathbf{C}e_i \oplus \mathbf{C}f_i \quad (i = 1, \dots, \ell),$$

(4.2.9) $$\mathfrak{n}_i = \oplus_{\alpha \in \Delta^+ \setminus \{\alpha_i\}} \mathfrak{g}_\alpha, \quad \mathfrak{n}_i^- = \oplus_{\alpha \in \Delta^- \setminus \{-\alpha_i\}} \mathfrak{g}_\alpha \quad (i = 1, \dots, \ell),$$

(4.2.10) $$\mathfrak{p}_i = \mathfrak{g}_i \oplus \mathfrak{n}_i \quad \mathfrak{p}_i^- = \mathfrak{g}_i \oplus \mathfrak{n}_i^- \quad (i = 1, \dots, \ell).$$

They are subalgebras of $\mathfrak{g}$.

4.3. We shall define groups corresponding to certain subalgebras of $\mathfrak{g}$ (see [**M**], [**K2**]). Fix a $\mathbf{Z}$-lattice P in $\mathfrak{h}^*$ satisfying

(4.3.1) $$\alpha_i \in P, \quad \langle h_i, P\rangle \subset \mathbf{Z} \quad (i = 1, \dots, \ell).$$

Let

(4.3.2) $T = \mathrm{Spec}(\mathbf{C}[P])$,

(4.3.3) $U = \varprojlim_k \exp(\mathfrak{n}/(\mathrm{ad}\,\mathfrak{n})^k\mathfrak{n})$, $U^- = \varprojlim_k \exp(\mathfrak{n}^-/(\mathrm{ad}\,\mathfrak{n}^-)^k\mathfrak{n}^-)$,

(4.3.4) B (resp. B^-) is the semi-direct product of T and U (resp. U^-),

(4.3.5) G_i is the algebraic group with $G_i \supset T, \mathrm{Lie}(G_i) = \mathfrak{g}_i, \mathrm{Lie}(T) = \mathfrak{h}$,

(4.3.6) $U_i = \varprojlim_k \exp(\mathfrak{n}_i/(\mathrm{ad}\,\mathfrak{n})^k\mathfrak{n}_i)$, $U_i^- = \varprojlim_k \exp(\mathfrak{n}_i^-/(\mathrm{ad}\,\mathfrak{n}^-)^k\mathfrak{n}_i^-)$,

(4.3.7) P_i (resp. P_i^-) is the semi-direct product of G_i and U_i (resp. U_i^-).

Here we denote by $\exp(\mathfrak{a})$ the unipotent algebraic group corresponding to a finite-dimensional nilpotent Lie algebra $\mathfrak{a}$. The groups defined above are naturally endowed with group scheme structures (see [**M**], [**K2**]).

4.4. In [**K2**] the first-named author has given a scheme theoretic construction of the flag variety of $(\mathfrak{g}, \mathfrak{h}, P)$. It is the quotient $X = G/B^-$, where G is the scheme defined in [**K2**] which has a locally free action of B^-. Let $x_0 = (1 \mod B^-) \in X$ and set $X_w = Bwx_0 \subset X$ for $w \in W$. As in the finite-dimensional case we have the following.

PROPOSITION 4.4.1 ([**K2**]).
(i) X_w *is an affine scheme with codimension* $\ell(w)$ *in* X.
(ii) $X = \sqcup_{w \in W} X_w$.
(iii) $\overline{X}_w = \sqcup_{z \geq w} X_z$.

For $i = 1, \ldots, \ell$ set $X^i = G/P_i^-, x_i = (1 \mod P_i^-) \in X^i$ (P_i^- acts on G locally freely). Let $q_i : X \to X^i$ be the natural morphism.

PROPOSITION 4.4.2 ([**K3**]).
(i) q_i *is a* $\mathbf{P}^1$*-bundle.*
(ii) $X^i = \sqcup_{\ell(ws_i) > \ell(w)} Bwx_i$.
(iii) $q_i^{-1}(Bwx_i) = X_w \sqcup X_{ws_i}$.
(iv) q_i *induces an isomorphism* $Bwx_0 \simeq Bwx_i$ *for* $\ell(ws_i) < \ell(w)$.
(v) q_i *induces an* $\mathbf{A}^1$*-bundle* $Bwx_0 \to Bwx_i$ *for* $\ell(ws_i) > \ell(w)$.

LEMMA 4.4.3. *Any* B*-invariant quasi-compact open subset of* X *or* X^i *satisfies* (S).

PROOF: The proof being similar, we shall prove the theorem only for X. Let Ω be a B-invariant quasi-compact open subset of X. Then there exists a finite subset J of W such that $\Omega = \cup_{w \in J} Bwx_0 = \cup_{w \in J} wBx_0$. Let Θ be a subset of Δ^+ such that $\Delta^+ \setminus \Theta$ is a finite set, $(\Theta + \Theta) \cap \Delta^+ \subset \Theta$ and $w^{-1}\Theta \subset \Delta^+$ for any $w \in J$. We denote by U_Θ the closed subgroup of U corresponding to $\mathfrak{n}_\Theta = \sum_{\alpha \in \Theta} \mathfrak{g}_\alpha$; i.e. $U_\Theta = \varprojlim_k \exp(\mathfrak{n}_\Theta / (\operatorname{ad} \mathfrak{n})^k \mathfrak{n}_\Theta)$. For $w \in J$ the action of U_Θ on wBx_0 is equivalent to the action of $w^{-1}U_\Theta w$ ($\subset U$) on Bx_0, and hence U_Θ acts on wBx_0 freely. Thus Ω/U_Θ exists and it is a quasi-compact smooth $\mathbf{C}$-scheme. Since $\Omega \simeq \varprojlim_\Theta \Omega/U_\Theta$, the assertion follows from Lemma 1.2.1. □

4.5. Let us recall the results of [**K3**]. Assume that $\mathfrak{g}$ is *symmetrizable* until the end of §4. For $\lambda \in P$, let $\mathcal{O}_X(\lambda)$ be the corresponding invertible $\mathcal{O}_X$-module. Set $\mathcal{D}_\lambda = \mathcal{O}_X(\lambda) \otimes_{\mathcal{O}_X} \mathcal{D}_X \otimes_{\mathcal{O}_X} \mathcal{O}_X(-\lambda)$ and $\mathcal{F}(\lambda) = \mathcal{O}_X(\lambda) \otimes_{\mathcal{O}_X} \mathcal{F}$ for an $\mathcal{O}_X$-module $\mathcal{F}$. Note that, if $\mathfrak{M}$ is a $\mathcal{D}_X$-module, then $\mathfrak{M}(\lambda)$ is a $\mathcal{D}_\lambda$-module. For $w \in W$ set $\mathfrak{B}_w = \mathcal{H}^{\ell(w)}_{X_w}(\mathcal{O}_X)$, where $\ell(w)$ is the length of w. Let $\mathfrak{M}_w$ be the dual of the $\mathcal{D}_X$-module $\mathfrak{B}_w$ and let $\mathfrak{L}_w$ be the image of the unique non-zero homomorphism $\mathfrak{M}_w \to \mathfrak{B}_w$. Then $\mathfrak{L}_w$ is the minimal extension of $\mathfrak{B}_w|wBx_0$.

4.6. For $\lambda \in \mathfrak{h}^*$ let $M(\lambda)$ be the Verma module with highest weight λ, $M^*(\lambda)$ the $\mathfrak{h}$-finite part of the dual of the Verma module with lowest weight $-\lambda$ and $L(\lambda)$ the image of the unique non-zero homomorphism $M(\lambda) \to M^*(\lambda)$. Then $L(\lambda)$ is the irreducible module with highest weight λ.

Set $w \circ \lambda = w(\lambda + \rho) - \rho$ for $w \in W$ and $\lambda \in \mathfrak{h}^*$, where ρ is an element of $\mathfrak{h}^*$ such that $\langle h_i, \rho \rangle = 1$ for any i.

4.7. Let $\lambda \in P_+ = \{\lambda \in P; \langle h_i, \lambda \rangle \geqq 0 \text{ for any } i\}$. For a B-equivariant $\mathcal{D}_\lambda$-module $\mathfrak{M}$ we set

$$\tilde{H}^n(X; \mathfrak{M}) = \oplus_{\mu \in P} \varprojlim_{\Omega} (H^n(\Omega; \mathfrak{M}))_\mu, \tag{4.7.1}$$

$$\tilde{\Gamma}(X; \mathfrak{M}) = \tilde{H}^0(X; \mathfrak{M}), \tag{4.7.2}$$

where Ω ranges over B-invariant quasi-compact open subsets of X, and for a semisimple $\mathfrak{h}$-module M the weight space with weight μ is denoted by M_μ. By [**K3**, Theorem 5.2.1] we have

$$\tilde{H}^n(X; \mathfrak{M}) = 0 \text{ for any } n \neq 0, \tag{4.7.3}$$

$$\tilde{\Gamma}(X; \mathfrak{M}_w(\lambda)) = M(w \circ \lambda), \tag{4.7.4}$$

$$\tilde{\Gamma}(X; \mathfrak{B}_w(\lambda)) = M^*(w \circ \lambda), \tag{4.7.5}$$

$$\tilde{\Gamma}(X; \mathfrak{L}_w(\lambda)) = L(w \circ \lambda). \tag{4.7.6}$$

4.8. Our main theorem is the following.

THEOREM 4.8.1. *Let $\mathfrak{g}$ be a symmetrizable Kac-Moody Lie algebra. Then, for $\lambda \in P_+$ and $w \in W$, we have:*

$$\operatorname{ch} L(w \circ \lambda) = \sum_{z \geqq w} (-1)^{\ell(z) - \ell(w)} Q_{w,z}(1) \operatorname{ch} M(z \circ \lambda),$$

where $Q_{w,z}$ is the inverse Kazhdan-Lusztig polynomial (see [**KL2**] *and* §5.3 *below).*

In order to prove this theorem, it is sufficient to show that, for any B-invariant quasi-compact open subset Ω, we have:

$$[\mathfrak{L}_w|\Omega] = \sum_{z \geqq w} (-1)^{\ell(z) - \ell(w)} Q_{w,z}(1) [\mathfrak{B}_z|\Omega] \tag{4.8.1}$$

in the Grothendieck group of the abelian category of B-equivariant holonomic $\mathcal{D}_\Omega$-modules. Note that $M(w \circ \lambda)$ and $M^*(w \circ \lambda)$ have the same characters. Since we have

$$\text{Sol}(\mathfrak{L}_w) = {}^\pi \mathbf{C}_{X_w}[-\ell(w)], \tag{4.8.2}$$

$$\text{Sol}(\mathfrak{B}_w) = \mathbf{C}_{X_w}[-\ell(w)], \tag{4.8.3}$$

this is again reduced to:

$$[{}^\pi \mathbf{C}_{X_w}[-\ell(w)]|\Omega] = \sum_{z \geqq w} (-1)^{\ell(z)-\ell(w)} Q_{w,z}(1)[\mathbf{C}_{X_z}[-\ell(z)]|\Omega] \tag{4.8.4}$$

in the Grothendieck group of the abelian category of B-equivariant perverse sheaves on Ω.

The last statement will be proven for any (not necessarily symmetrizable) Kac-Moody Lie algebras in §6 by the aid of mixed Hodge modules.

5. Hecke-Iwahori Algebras

5.0. In this section W denotes a Coxeter group with canonical generator system S. The length function and the Bruhat order on W are denoted by ℓ and $\geqq$, respectively.

5.1. The Hecke-Iwahori algebra $H(W)$ is the associative algebra over the Laurent polynomial ring $\mathbf{Z}[q, q^{-1}]$ which has a free $\mathbf{Z}[q, q^{-1}]$-basis $\{T_w\}_{w \in W}$ satisfying the following relations:

$$(T_s + 1)(T_s - q) = 0 \quad \text{for} \quad s \in S, \tag{5.1.1}$$

$$T_{w_1} T_{w_2} = T_{w_1 w_2} \quad \text{if} \quad \ell(w_1) + \ell(w_2) = \ell(w_1 w_2). \tag{5.1.2}$$

Let $h \mapsto \overline{h}$ be the automorphism of the ring $H(W)$ given by

$$\overline{q} = q^{-1}, \qquad \overline{T_w} = T_{w^{-1}}^{-1}, \tag{5.1.3}$$

and define $R_{y,w} \in \mathbf{Z}[q, q^{-1}]$ for $y, w \in W$ by

$$\overline{T_w} = \sum_{y \in W} \overline{R_{y,w}} q^{-\ell(y)} T_y. \tag{5.1.4}$$

The following is easily checked by direct calculations (see [**KL1**]).

(5.1.5) $R_{y,w} \neq 0$ if and only if $y \leqq w$.

(5.1.6) $R_{y,w}$ is a ploynomial in q with degree $\ell(w) - \ell(y)$ for $y \leqq w$.

(5.1.7) $R_{w,w} = 1$.

Following [**KL1**] we introduce a free $\mathbf{Z}[q, q^{-1}]$-basis $\{C_w\}_{w \in W}$ of $H(W)$.

PROPOSITION 5.1.1 ([**KL1**]). *For $w \in W$ there exists a unique element*

$$C_w = \sum_{y \leqq w} (-q)^{\ell(w)-\ell(y)} \overline{P}_{y,w} T_y \in H(W)$$

satisfying the following conditions:

(a) $P_{w,w} = 1$.
(b) *If $y < w$, $P_{y,w}$ is a polynomial in q with degree $\leqq (\ell(w) - \ell(y) - 1)/2$.*
(c) $\overline{C}_w = q^{-\ell(w)} C_w$.

We set $P_{y,w} = 0$ if $y \nleqq w$.

5.2. Set $H^*(W) = \mathrm{Hom}_{\mathbf{Z}[q,q^{-1}]}(H(W), \mathbf{Z}[q,q^{-1}])$. For $w \in W$ let S_w be the element of $H^*(W)$ determined by

$$\langle S_w, \overline{T}_y \rangle = \delta_{w,y^{-1}} q^{-\ell(w)}, \tag{5.2.1}$$

where $\langle\ ,\ \rangle$ denotes the natural paring of $H^*(W)$ and $H(W)$. Any element of $H^*(W)$ is uniquely written as a formal infinite sum $\sum_{w \in W} a_w S_w$ ($a_w \in \mathbf{Z}[q, q^{-1}]$).

Define an endomorphism $u \mapsto \overline{u}$ of the abelian group $H^*(W)$ by

$$\langle \overline{u}, h \rangle = \overline{\langle u, \overline{h} \rangle} \qquad (u \in H^*(W), h \in H(W)). \tag{5.2.2}$$

We also define a right $H(W)$-module structure on $H^*(W)$ by

$$\langle u \cdot h_1, h_2 \rangle = \langle u, h_1 h_2 \rangle \qquad (u \in H^*(W), h_1, h_2 \in H(W)). \tag{5.2.3}$$

We can check the following lemma by direct calculations.

LEMMA 5.2.1. (i) $\overline{\sum_{w \in W} a_w S_w} = \sum_{w \in W} q^{\ell(w)} (\overline{\sum_{y \leqq w} a_y R_{y^{-1}, w^{-1}}}) S_w$.

(ii) *For $s \in S$ we have*

$$(\sum_{w \in W} a_w S_w) \cdot T_s = \sum_{ws > w} ((q-1) a_w + a_{ws}) S_w + \sum_{ws < w} q a_{ws} S_w.$$

(iii) $\langle u, h \rangle = \epsilon(u \cdot h)$ $(u \in H^*(W), h \in H(W))$, *where $\epsilon \colon H^*(W) \to R$ is given by $\epsilon(\sum_{w \in W} a_w S_w) = a_e$.*

5.3. For $w \in W$ we define an element D_w of $H^*(W)$ by

$$\langle D_w, \overline{C}_y \rangle = \delta_{w,y^{-1}} q^{-\ell(w)}. \tag{5.3.1}$$

Set $D_w = \sum_{z \in W} Q_{w,z} S_z$ ($Q_{w,z} \in \mathbf{Z}[q, q^{-1}]$). It is easily seen that $Q_{w,z} = 0$ unless $z \geqq w$, and $Q_{y,w}$ for $y \leqq w$ are uniquely determined by

$$(5.3.2) \qquad \sum_{y \leqq w \leqq z} (-1)^{\ell(w)-\ell(y)} Q_{y,w} P_{w,z} = \delta_{y,z} \quad (y \leqq z).$$

By definition we have the following properties:

(5.3.3) $Q_{w,w} = 1$.

(5.3.4) If $z > w$, $Q_{w,z}$ is a polynomial in q with degree $\leqq (\ell(z) - \ell(w) - 1)/2$.

(5.3.5) $\overline{D}_w = q^{\ell(w)} D_w$.

Moreover these properties characterize the element D_w. We shall formulate this uniquenes in a more general setting.

Let R be a commutative ring with 1 containing $\mathbf{Z}[q, q^{-1}]$. Assume that we are given a grading $R = \oplus_{i \in \mathbf{Z}} R_i$ and an involutive automorphism $r \mapsto \overline{r}$ of the ring R satisfying

$$(5.3.6) \qquad R_i R_j \subset R_{i+j}, \quad q \in R_2, \quad \overline{R_i} = R_{-i}, \quad \overline{q} = q^{-1}.$$

Set $H_R(W) = R \otimes_{\mathbf{Z}[q,q^{-1}]} H(W)$ and $H_R^*(W) = \mathrm{Hom}_R(H_R(W), R)$. Similarly to (5.2.2) and (5.2.3), we have an involution $u \mapsto \overline{u}$ of $H_R^*(W)$ and a right $H_R(W)$-module structure on $H_R^*(W)$.

Proposition 5.3.1. *Let $w \in W$. If $D'_w = \sum_{z \geqq w} Q'_{w,z} S_z$ ($Q'_{w,z} \in R$) is an element of $H_R^*(W)$ satisfying the following conditions* (a), (b), (c), *then we have $D'_w = D_w$.*

(a) $Q'_{w,w} = 1$.

(b) $Q'_{w,z} \in \oplus_{i \leqq \ell(z) - \ell(w) - 1} R_i$ *for* $z > w$.

(c) $\overline{D'_w} = q^{\ell(w)} D'_w$.

Proof: We shall show $Q'_{w,z} = Q_{w,z}$ for $z \geqq w$ by induction on $\ell(z) - \ell(w)$. If $\ell(z) - \ell(w) = 0$, we have $w = z$, and the assertion is trivial. Assume that $z > w$. By Lemma 5.2.1 (i) we have

$$\overline{D'_w} = \sum_{v \geqq w} q^{\ell(v)} \Big(\sum_{v \geqq y \geqq w} \overline{Q'_{w,y} R_{y^{-1}, v^{-1}}} \Big) S_v,$$

and hence (c) implies :

$$Q'_{w,z} = q^{\ell(z)-\ell(w)} \Big(\overline{Q'_{w,z}} + \sum_{z > y \geqq w} \overline{Q'_{w,y} R_{y^{-1}, z^{-1}}} \Big).$$

By the inductive hypothesis we have

$$(5.3.7) \qquad Q'_{w,z} - q^{\ell(z)-\ell(w)}\overline{Q'_{w,z}} = q^{\ell(z)-\ell(w)} \sum_{z>y\geqq w} \overline{Q_{w,y}R_{y^{-1},z^{-1}}}.$$

On the other hand (b) implies

$$(5.3.8) \qquad Q'_{w,z} \in \oplus_{i\leqq \ell(z)-\ell(w)-1} R_i,$$

$$(5.3.9) \qquad q^{\ell(z)-\ell(w)}\overline{Q'_{w,z}} \in \oplus_{i\geqq \ell(z)-\ell(w)+1} R_i,$$

and hence the equation (5.3.7) uniquely determines $Q'_{w,z}$. Since $Q_{w,z}$ also satisfies the same equation, we have $Q'_{w,z} = Q_{w,z}$. □

6. Hodge modules on flag varieties

6.0. In this section we shall give a proof of (4.8.4) for any (not necessarily symmetrizable) Kac-Moody Lie algebra $\mathfrak{g}$. For an abelian category $\mathcal{A}$ we denote its Grothendieck group by $K(\mathcal{A})$.

6.1. Set $R = K(MHS)$. The abelian group R is naturally endowed with a structure of commutative ring with 1 via the tensor product. Since MHS is an Artinian category, R has a free $\mathbf{Z}$-basis consisting of simple objects. For $i \in \mathbf{Z}$ we denote by R_i the $\mathbf{Z}$-submodule of R generated by the elements corresponding to pure Hodge structures of weight i. Since any simple object of MHS is a pure Hodge structure, we have

$$(6.1.1) \qquad R = \oplus_{i\in\mathbf{Z}} R_i, \quad R_iR_j \subset R_{i+j}.$$

In the following we regard $\mathbf{Z}[q, q^{-1}]$ as a subring of R via $q^i = [\mathbf{Q}^H(-i)] \in R_{2i}$, where $\mathbf{Q}^H(-i)$ is the pure Hodge structure of weight $2i$ obtained by twisting the trivial Hodge structure $\mathbf{Q}^H$. Let $r \mapsto \overline{r}$ be the involutive automorphism of the ring R induced by the duality operation in MHS. Then we have

$$(6.1.2) \qquad \overline{R_i} = R_{-i}, \quad \overline{q} = q^{-1},$$

and hence the ring R satisfies the condition (5.3.6).

6.2. We have a natural R-module structure on $K(\tilde{M}HM^B(X_w))$ for $w \in W$.

LEMMA 6.2.1. (i) *Any object M of $\tilde{M}HM^B(X_w)$ is isomorphic to a constant B-equivariant mixed Hodge module $\mathbf{Q}^H_{X_w} \otimes L (= (a_{X_w})^*(L))$ for some $L \in \mathrm{Ob}(MHS)$.*

(ii) *$K(\tilde{M}HM^B(X_w))$ is a rank one free R-module with basis $[\mathbf{Q}^H_{X_w}]$.*

PROOF: This follows from Theorem 3.13.3 because the isotropy group with respect to the action of B on X_w is connected. □

6.3. We say that a subset J of W is *admissible* if J is a finite set satisfying the condition:

$$w \in J, \quad y \leqq w \Rightarrow y \in J. \tag{6.3.1}$$

We denote by $\mathcal{C}$ the set of admissible subsets of W. For a subset J of W set $\Omega_J = \sqcup_{w \in J} X_w$. By **[K2]** we see that Ω_J is a quasi-compact open subset of X if and only if J is admissible.

For admissible subsets J_1, J_2 satisfying $J_1 \subset J_2$, we have a natural functor and a natural homomorphism

$$\tilde{M}HM^B(\Omega_{J_2}) \to \tilde{M}HM^B(\Omega_{J_1}) \tag{6.3.2}$$

$$K(\tilde{M}HM^B(\Omega_{J_2})) \to K(\tilde{M}HM^B(\Omega_{J_1})) \tag{6.3.3}$$

by the restriction, and they give projective systems $\{\tilde{M}HM^B(\Omega_J)\}_{J \in \mathcal{C}}$ and $\{K(\tilde{M}HM^B(\Omega_J))\}_{J \in \mathcal{C}}$. Set

$$\tilde{M}HM^B(X) = \varprojlim_{J \in \mathcal{C}} \tilde{M}HM^B(\Omega_J), \tag{6.3.4}$$

$$K^B(X) = \varprojlim_{J \in \mathcal{C}} K(\tilde{M}HM^B(\Omega_J)), \tag{6.3.5}$$

and let $p_J : K^B(X) \to K(\tilde{M}HM^B(\Omega_J))$ be the projection. The R-module $K^B(X)$ may be regarded as a completion of the Grothendieck group of $\tilde{M}HM^B(X)$.

Let $i_w : X_w \to X$ be the inclusion. Let $w \in W$ and $J \in \mathcal{C}$ such that $w \in J$, and let $i_{w,J} : X_w \to \Omega_J$ be the inclusion. We define objects $(i_w)_! \mathbf{Q}^H_{X_w}$ and ${}^\pi\mathbf{Q}^H_{X_w}$ of $D^b(\tilde{M}HM^B(X))$ by

$$(i_w)_! \mathbf{Q}^H_{X_w} | \Omega_J = (i_{w,J})_! (\mathbf{Q}^H_{X_w}), \tag{6.3.6}$$

(6.3.7)

$${}^\pi\mathbf{Q}^H_{X_w} | \Omega_J = (\text{ the minimal extension of } \mathbf{Q}^H_{X_w} \text{ with respect to } i_{w,J}).$$

Set $[M]=\sum_{k\in\mathbf{Z}}(-1)^k[H^k(M)]$ for $M\in \mathrm{Ob}(D^b(\tilde{M}HM(\Omega_J)))$. We have elements $[(i_w)_!\mathbf{Q}^H_{X_w}]$ and $[{}^\pi\mathbf{Q}^H_{X_w}]$ of $K^B(X)$ satisfying

$$p_J([(i_w)_!\mathbf{Q}^H_{X_w}])=[(i_w)_!\mathbf{Q}^H_{X_w}|\Omega_J], \tag{6.3.8}$$

$$p_J([{}^\pi\mathbf{Q}^H_{X_w}])=[{}^\pi\mathbf{Q}^H_{X_w}|\Omega_J]. \tag{6.3.9}$$

We nextly define an R-homomorhism

$$(i_w)^*: K^B(X)\to K(\tilde{M}HM^B(X_w)) \tag{6.3.10}$$

as follows. For an admissible subset J such that $w\in J$, we have an R-homomorphism

$$(i_{w,J})^*: K(\tilde{M}HM^B(\Omega_J))\to K(\tilde{M}HM^B(X_w))$$

given by

$$(i_{w,J})^*([M])=\sum_{k\in\mathbf{Z}}(-1)^k[(H^k(i_{w,J})^*)(M)] \quad (M\in \mathrm{Ob}(K(\tilde{M}HM^B(\Omega_J))))\,,$$

and (6.3.10) is defined by

$$(i_w)^*(m)=(i_{w,J})^*(p_J(m)). \tag{6.3.11}$$

For $m\in K^B(X)$ and $w\in W$ we define $\varphi_w(m)\in R$ by

$$(i_w)^*(m)=\varphi_w(m)[\mathbf{Q}^H_{X_w}] \tag{6.3.12}$$

(see Lemma 6.2.1). We also define an R-homomorphism $\varphi\colon K^B(X)\to H^*_R(W)$ by

$$\varphi(m)=\sum_{w\in W}\varphi_w(m)S_w. \tag{6.3.13}$$

LEMMA 6.3.1. (i) $\varphi([(i_w)_!\mathbf{Q}^H_{X_w}])=S_w$
(ii) *φ is an isomorphism of R-modules.*

PROOF: (i) is clear. Let us show (ii). Let J be an admissible subset of W. Since $\tilde{M}HM^B(\Omega_J)$ is an Artinian category, its Grothendieck group has a free $\mathbf{Z}$-basis consisting of the simple objects. Since any simple object of $\tilde{M}HM^B(\Omega_J)$ is isomorphic to $({}^\pi\mathbf{Q}^H_{X_w}|\Omega_J)[-\ell(w)]\otimes L$ for some $w\in J$

and some simple object L of MHS, we see that $K(\tilde{M}HM^B(\Omega_J))$ is a free R-module with basis $\{[{}^\pi\mathbf{Q}^H_{X_w}|\Omega_J]\,;\, w \in J\}$. Since we have

$$[{}^\pi\mathbf{Q}^H_{X_w}|\Omega_J] \in [(i_w)_!\mathbf{Q}^H_{X_w}|\Omega_J] + \sum_{\substack{y\in J\\ y>w}} R[(i_y)_!\mathbf{Q}^H_{X_y}|\Omega_J]$$

for $w \in J$, $\{[(i_w)_!\mathbf{Q}^H_{X_w}|\Omega_J]\,;\, w \in J\}$ is also a free basis of the R-module $K(\tilde{M}HM^B(\Omega_J))$. Therefore the assertion follows from (i). □

6.4. We shall define an R-homomorphism

$$\tau_i\colon K^B(X) \to K^B(X) \tag{6.4.1}$$

for each $i = 1, \dots, \ell$ as follows. Let $\mathcal{C}_i$ be the set of admissible subsets J of W such that $ws_i \in J$ if $w \in J$. For $J \in \mathcal{C}_i$ let $q_{i,J}\colon \Omega_J \to q_i(\Omega_J)$ be the restriction of $q_i\colon X \to X^i$ and define an endomorphism $\tau_{i,J}$ of the R-module $K(\tilde{M}HM^B(\Omega_J))$ by

$$\tau_{i,J}([M]) = [(q_{i,J})^*(q_{i,J})_!M] \quad \text{for} \quad M \in \mathrm{Ob}(\tilde{M}HM^B(\Omega_J)).$$

Since $q_{i,J}$ is a B-equivariant $\mathbf{P}^1$-bundle, $\tau_{i,J}$ is well-defined. Then we define an endomorphism τ_i of $K^B(X) = \varprojlim_{J\in\mathcal{C}_i} K(\tilde{M}HM^B(\Omega_J))$ by $\tau_i = \varprojlim_J \tau_{i,J}$.

LEMMA 6.4.1. *$\varphi(\tau_i(m)) = \varphi(m)\cdot(T_{s_i}+1)$ for $m \in K^B(X)$.*

PROOF: By Lemma 5.2.1 (ii) and Lemma 6.3.1 it is sufficient to show

$$\tau_i([(i_w)_!\mathbf{Q}^H_{X_w}]) = \begin{cases} [(i_{ws_i})_!\mathbf{Q}^H_{X_{ws_i}}] + [(i_w)_!\mathbf{Q}^H_{X_w}] & (ws_i < w) \\ q([(i_{ws_i})_!\mathbf{Q}^H_{X_{ws_i}}] + [(i_w)_!\mathbf{Q}^H_{X_w}]) & (ws_i > w). \end{cases}$$

Let $J \in \mathcal{C}_i$ such that $w \in J$. Set $\tilde{X}_w = q_i^{-1}q_i(X_w) = X_w \sqcup X_{ws_i}$ and let $j_{w,J}\colon \tilde{X}_w \to \Omega_J$ be the inclusion. Since $\tilde{X}_w \to q_i(\tilde{X}_w)$ is a $\mathbf{P}^1$-bundle and since $X_w \to q_i(\tilde{X}_w)$ is an isomorphism (resp. $\mathbf{A}^1$-bundle) for $ws_i < w$ (resp. $ws_i > w$), we have

$$(q_{i,J})^*(q_{i,J})_!((i_{w,J})_!\mathbf{Q}^H_{X_w}) = \begin{cases} (j_{w,J})_!\mathbf{Q}^H_{\tilde{X}_w} & (ws_i < w) \\ (j_{w,J})_!\mathbf{Q}^H_{\tilde{X}_w}[-2](-1) & (ws_i > w). \end{cases}$$

On the other hand, if $ws_i > w$, we have an exact sequence:

$$0 \to (i_{ws_i,J})_! \mathbf{Q}^H_{X_{ws_i}}[-\ell(w)-1] \to (i_{w,J})_! \mathbf{Q}^H_{X_w}[-\ell(w)] \to (j_{w,J})_! \mathbf{Q}^H_{\tilde{X}_w}[-\ell(w)] \to 0$$

in $\tilde{M}HM^B(\Omega_J)$. Hence the assertion is clear. □

By Lemma 6.3.1 and Lemma 6.4.1 we can define a right $H(W)$-module structure on $K^B(X)$ by

$$m \cdot (T_{s_i} + 1) = \tau_i(m) \quad (m \in K^B(X)). \tag{6.4.2}$$

6.5. We denote by $m \mapsto m^*$ the endomorphisms of the abelian groups $K^B(X)$ and $K(\tilde{M}HM^B(X_e))$ induced by the duality operation of mixed Hodge modules.

LEMMA 6.5.1. *$\varphi(m^*) = \overline{\varphi(m)}$ for $m \in K^B(X)$.*

PROOF: We have to show $\langle \varphi(m^*), h \rangle = \overline{\langle \varphi(m), \overline{h} \rangle}$ for $m \in K^B(X), h \in H(W)$. By Lemma 5.2.1 (iii) and §§6.3, 6.4 this is equivalent to

$$(i_e)^*(m^* \cdot T_z) = ((i_e)^*(m \cdot \overline{T}_z))^* \quad (m \in K^B(X), z \in W), \tag{6.5.1}$$

where the right action of $H(W)$ on $K^B(X)$ is given by (6.4.2). Let us prove (6.5.1) by induction on $\ell(z)$. The case $z = e$ being trivial, we take $w \in W$ satisfying $s_i w > w$ and prove (6.5.1) for $z = s_i w$ assuming (6.5.1) for $z = w$.

Let $J \in \mathcal{C}_i$. Since $q_{i,J}$ is a $\mathbf{P}^1$-bundle, we have

$$(q_{i,J})^*(q_{i,J})_! \mathbf{D}_{\Omega_J}(M) = (\mathbf{D}_{\Omega_J}(q_{i,J})^*(q_{i,J})_!(M))[-2](-1) \tag{6.5.2}$$

$$(M \in \operatorname{Ob} \tilde{M}HM^B(\Omega_J)),$$

and hence we have

$$\tau_i(m^*) = (q^{-1}\tau_i(m))^* \quad (m \in K^B(X)). \tag{6.5.3}$$

Therefore we have

$$\begin{aligned}(i_e)^*(m^* \cdot T_{s_i w}) &= (i_e)^*((\tau_i(m^*) - m^*) \cdot T_w) \\ &= (i_e)^*((q^{-1}\tau_i(m) - m)^* \cdot T_w) \\ &= ((i_e)^*((q^{-1}\tau_i(m) - m) \cdot \overline{T}_w))^* \\ &= ((i_e)^*(m \cdot \overline{T}_{s_i w}))^*. \quad \square\end{aligned}$$

6.6. We shall determine $H^i((i_{z,J})^*({}^\pi\mathbf{Q}^H_{X_w}|\Omega_J))$ for any admissible subset J and $z, w \in J$. Since this does not depend on J, we simply denote it by $H^i((i_z)^*({}^\pi\mathbf{Q}^H_{X_w}))$.

We first give a weaker result.

PROPOSITION 6.6.1. *$\varphi([{}^{\pi}\mathbf{Q}^{H}_{X_w}]) = D_w$ for $w \in W$.*

PROOF: Setting $Q'_{w,z} = \varphi_z([{}^{\pi}\mathbf{Q}^{H}_{X_w}]) \in R$ and $D'_w = \sum_{z \geqq w} Q'_{w,z} S_z \in H^*_R(W)$, we have $\varphi([{}^{\pi}\mathbf{Q}^{H}_{X_w}]) = D'_w$. Hence by Proposition 5.3.1, it is sufficient to show the following conditions:

(6.6.1) $Q'_{w,w} = 1$.

(6.6.2) $Q'_{w,z} \in \oplus_{i \leqq \ell(z)-\ell(w)-1} R_i$ for $z > w$.

(6.6.3) $\overline{D'_w} = q^{\ell(w)} D'_w$.

(6.6.1) is trivial, and (6.6.3) follows from Lemma 6.5.1 and

$$\mathbf{D}_{\Omega_J}({}^{\pi}\mathbf{Q}^{H}_{X_w}|\Omega_J) = ({}^{\pi}\mathbf{Q}^{H}_{X_w}|\Omega_J)[-2\ell(w)](-\ell(w)). \tag{6.6.4}$$

Let us show (6.6.2). Let $z > w$. Since ${}^{\pi}\mathbf{Q}^{H}_{X_w}|\Omega_J$ is pure of weight 0, $(i_{z,J})^*({}^{\pi}\mathbf{Q}^{H}_{X_w}|\Omega_J)$ is of weight $\leqq 0$, and hence $H^i((i_{z,J})^*({}^{\pi}\mathbf{Q}^{H}_{X_w}|\Omega_J))$ is of weight $\leqq i$. On the other hand we have $H^i((i_{z,J})^*({}^{\pi}\mathbf{Q}^{H}_{X_w}|\Omega_J)) = 0$ for $i \geqq \ell(z) - \ell(w)$ by the definition. Therefore $Q'_{w,z} \in \oplus_{i \leqq \ell(z)-\ell(w)-1} R_i$. □

LEMMA 6.6.2. *Let Y be an irreducible closed subvariety of $\mathbf{C}^n$ such that there exist integers $a_1, \dots, a_n > 0$ satisfying*

$$z \in \mathbf{C}^*, \quad (z_1, \dots, z_n) \in Y \Rightarrow (z^{a_1} z_1, \dots, z^{a_n} z_n) \in Y, \tag{6.6.5}$$

and let $i\colon \{0\} \to Y$ be the inclusion. Then $H^j(i^*({}^{\pi}\mathbf{Q}^{H}_{Y}))$ is a pure Hodge structure of weight j.

The proof is similar to [**KL2**, Lemma 4.5].

LEMMA 6.6.3. *$H^j((i_z)^*({}^{\pi}\mathbf{Q}^{H}_{X_w}))$ is pure of weight j.*

PROOF: We may assume that $z > w$. Let $x \in X_z$. By [**K2**, Remark 4.5.14] we can take an open neighborhood V of x in X such that there exists a commutative diagram

$$\begin{array}{ccccc} X_z \cap V & \longrightarrow & \overline{X}_w \cap V & \longrightarrow & V \\ \downarrow & & \downarrow & & \downarrow \\ \{0\} \times \mathbf{A}^\infty & \longrightarrow & Y \times \mathbf{A}^\infty & \longrightarrow & \mathbf{C}^n \times \mathbf{A}^\infty \end{array} \tag{6.6.6}$$

where Y is an irreducible closed subvariety of $\mathbf{C}^n$ satisfying the assumption of Lemma 6.6.2, the horizontal arrows are the natural inclusions and the vertical arrows are isomorphisms. Hence the assertion follows from Lemma 6.6.2. □

Set $Q_{w,z} = \sum_j c_{w,z,j} q^j$ $(c_{w,z,j} \in \mathbf{Z})$ for $z, w \in W$ with $z \geqq w$.

THEOREM 6.6.4. *Let* $z \geqq w$.

(i) $H^{2j+1}((i_z)^*({}^\pi\mathbf{Q}^H_{X_w})) = 0$ *for any* $j \in \mathbf{Z}$.

(ii) *For any* $j \in \mathbf{Z}$ *we have* $c_{w,z,j} \geqq 0$, *and* $H^{2j}((i_z)^*({}^\pi\mathbf{Q}^H_{X_w}))$ *is isomorphic to* $(\mathbf{Q}^H_{X_z}(-j))^{\oplus c_{w,z,j}}$.

PROOF: By Theorem 3.13.3 there exist some $N_k \in \mathrm{Ob}(MHS)$ $(k \in \mathbf{Z})$ such that $H^k((i_z)^*({}^\pi\mathbf{Q}^H_{X_w})) = (a_{X_z})^*(N_k)$. Then we see from Proposition 6.6.1 that

$$\sum_j c_{w,z,j}q^j = \sum_{k\in\mathbf{Z}}(-1)^k[N_k]. \tag{6.6.7}$$

Since $[N_k] \in R_k$ by Lemma 6.6.3, we have $[N_{2j+1}] = 0$ and $[N_{2j}] = c_{w,z,j}q^j$, and this implies that $N_{2j+1} = 0$ and $N_{2j} = (\mathbf{Q}^H)^{\oplus c_{w,z,j}}$. □

It is easily seen that (4.8.4) is a consequence of Theorem 6.6.4 (or even Lemma 6.6.1).

REFERENCES

[EGA] A. Grothendieck and J. Dieudonné, *Eléments de Géométrie Algébrique, I-IV*, Publ. Math. IHES.

[BB] A. Beilinson, J. Bernstein, *Localisation de* $\mathfrak{g}$*-modules*, Comptes Rendus **292** (1981), 15–18.

[BBD] A. Beilinson, J. Bernstein and P. Deligne, *Faisceaux pervers*, Astérisque **100** (1983).

[BK] J.-L. Brylinski and M. Kashiwara, *Kazhdan-Lusztig conjecture and holonomic systems* , Invent. Math. **64** (1981), 387-410.

[DGK] V. Deodhar, O. Gabber and V. Kac, *Structure of some categories of representations of infinite-dimensional Lie algebras*, Adv. Math. **45** (1982), 92–116.

[K'] V. Kac, "Infinite dimensional Lie algebras," Progress in Math. **44**, Birkhäuser, Boston, 1983.

[K1] M. Kashiwara, *The Riemann-Hilbert problem for holonomic systems*, Publ. RIMS, Kyoto Univ. **20** (1984), 319–365.

[K2] M. Kashiwara, *The flag manifold of Kac-Moody Lie algebras*, Amer. J. Math. **111** (1989).

[K3] M. Kashiwara, *Kazhdan-Lusztig conjecture for symmetrizable Kac-Moody Lie algebra*, to be published in the volume in honor of A. Grothendieck's sixtieth birthday.

[KS] M. Kashiwara and P. Schapira, " *Sheaves on Manifolds*," Springer-Verlag.

[KL1] D. Kazhdan and G. Lusztig, *Representations of Coxeter groups and Hecke algebras*, Invent. Math. **53** (1979), 165–184.

[KL2] D. Kazhdan and G. Lusztig, *Schubert varieties and Poincaré duality*, Proc. Symp. in Pure Math. **36** (1980), 185–203.

[LV] G. Lusztig and D. Vogan, *Singularities of closures of K-orbits on flag manifolds*, Invent. Math. **71** (1983), 365–379.

[M] O. Mathieu, *Formules de caractères pour les algèbres de Kac-Moody générales*, Astérisque **159-160** (1988).

[S] M. Saito, *Mixed Hodge Modules*, Publ. RIMS, Kyoto Univ. **26** (1989).

[Sp] T. A. Springer, *Quelques applications de la cohomologie d'intersection*, Séminaire Bourbaki, exposé **589**, Astérisque **92-93** (1982), 249–273.

[T] T. Tanisaki, *Hodge modules, equivariant K-theory and Hecke algebras*, Publ. RIMS, Kyoto Univ. **23** (1987), 841–879.

Received October 10, 1989

Masaki Kashiwara
R.I.M.S.,
Kyoto University
Kyoto 606, Japan

Toshiyuki Tanisaki
College of General Education
Osaka University
Toyonaka 560, Japan

Standard Monomial Theory for $\hat{SL}_n$

V. LAKSHMIBAI*

Dedicated to Professor Jacques Dixmier on his 65th birthday

Abstract

Let $\mathfrak{g}$ be a Kac-Moody Lie algebra, G the associated Kac-Moody group. Let B be a Borel subgroup in G and X a Schubert variety in G/B. For $\mathfrak{g} = \hat{sl}(n, \mathbb{C})$, we construct a characteristic-free basis for $H^0(X, L)$, L being line bundles on X associated to dominant integral weights of $\mathfrak{g}$.

Introduction

Let G be a semi-simple algebraic group and B a Borel subgroup. Let X be a Schubert variety in G/B. Let L be an ample line bundle on G/B, as well as its restriction to X. A standard monomial theory for Schubert varieties in G/B is developed in [9], [11], [8], [10] as a generalization of the classical Hodge-Young theory (cf [3], [4]). This theory consists in the construction of a characteristic-free basis for $H^0(X, L)$. This theory is extended to Schubert varieties in the infinite dimensional flag variety $\hat{S}L_2/B$ in [12]. In this paper, we extend the theory to Schubert varieties in the infinite dimensional flag variety $\hat{S}L_n/B$.

Let A be a symmetrizable, generalized Cartan matrix (cf [5]). Let $\mathfrak{g}$ (resp. G) be the associated Kac-Moody Lie algebra (resp. Kac-Moody group). Let W be the Weyl group. Let U be the universal enveloping algebra of $\mathfrak{g}$ and $U_{\mathbb{Z}}^+$, the $\mathbb{Z}$-subalgebra of U generated by $X_\alpha^n/n!$, α a simple root. Let λ be a dominant, integral weight and V_λ the integrable, highest weight $\mathfrak{g}$ module (over $\mathbb{C}$) with highest weight λ. Let us fix a generator e for the highest weight space (note that e is unique up to scalars). For $\tau \in W$, let $e_\tau = \tau e$, $V_{\mathbb{Z},\tau} = U_{\mathbb{Z}}^+ e_\tau$. In [12], we gave a conjectural basis for $V_{\mathbb{Z},\tau}$. Using this, we construct an explicit basis for

*Partially supported by NSF Grant DMS-8701043

$V_{\mathbf{Z},\tau}$ for the case $\mathfrak{g} = \hat{sl}(n, \mathbf{C})$. We shall briefly describe below the results for $\hat{sl}(n, \mathbf{C})$.

Let us fix a fundamental weight ω_i, $0 \leq i \leq l$ $(= n - 1)$ and denote ω_i by just ω. Let P be the maximal parabolic subgroup of G associated to ω. Let V_ω be the irreducible $\mathfrak{g}$-module (over $\mathbf{C}$) with highest weight ω. Defining $e, e_\tau, V_{\mathbf{Z},\tau}$ etc. as above, for any field k, let $V_\tau = V_{\mathbf{Z},\tau} \otimes k$. Consider the canonical embedding $X(\tau) \hookrightarrow \mathbf{P}(V_\tau)$. Let L be the tautological line bundle on $\mathbf{P}(V_\tau)$. Let us denote $L|_{X(\tau)}$ by just L. We first construct a $\mathbf{Z}$–basis $\{Q_\Lambda\}$ for $V_{\mathbf{Z},\tau}$ (cf §3) indexed by "admissible Young diagrams on $X(\tau)$." Let $\{P_\Lambda\}$ be the basis of the $\mathbf{Z}$-dual of $V_{\mathbf{Z},\tau}$ dual to $\{Q_\Lambda\}$ and for any field k, let $p_\Lambda = P_\Lambda \otimes 1$. To an admissible Young diagram Λ on $X(\tau)$, we associate a pair of Weyl group elements $\rho(\Lambda)$, $\delta(\Lambda)$, where $\tau \geq \rho(\Lambda) \geq \delta(\Lambda)$. We then define a monomial $p_{\Lambda_1} \cdot p_{\Lambda_2} \cdots p_{\Lambda_m}$ to be *standard on* $X(\tau)$ if $\tau \geq \rho(\Lambda_1) \geq \delta(\Lambda_1) \geq \rho(\Lambda_2) \geq \delta(\Lambda_2) \geq \cdots \geq \delta(\Lambda_m)$. Then we prove

Theorem 1. *Let $\tau \in W/W_P$ and $X(\tau)$ the associated Schubert variety in G/P. The standard monomials on $X(\tau)$ of degree m form a basis of $H^0(X(\tau), L^m)$.*

Let now $X(\tau)$ be a Schubert variety in G/B and $L = \otimes L_i^{a_i}, a_i \in \mathbf{Z}^+$($L_i$ being the line bundle defined as above with respect to ω_i (or P_i)). *Let $F \in H^0(X(\tau), L)$. Further, let $F = f_0 f_1 f_2 \cdots f_l$, where $f_i = p_{\Lambda_{i1}}, p_{\Lambda_{i2}} \cdots p_{\Lambda_{ia_i}}$. We say, F is standard on $X(\tau)$ of multidegree $a = (a_0, \cdots, a_l)$ if there exists a sequence in W*

$$\tau \geq \theta_{01} \geq \varphi_{01} \geq \theta_{02} \geq \cdots \geq \varphi_{0a_1} \geq \theta_{11} \geq \cdots \geq \varphi_{la_l}$$

such that $\pi_i(\theta_{ij}) = \rho(\Lambda_{ij})$, $\pi_i(\varphi_{ij}) = \delta(\Lambda_{ij})$, $1 \leq j \leq a_i$, $0 \leq i \leq l$ (here π_i denotes the projection $\pi_i : G/B \to G/P_i$ (or same as $\pi_i : W \to W/W_{P_i}$)).

We prove

Theorem 2. *Standard monomials on $X(\tau)$ of degree a form a basis of $H^0(X(\tau), L)$.*

The philosophy of the proof of Theorems 1 and 2 is the same as in [11], namely, given $X = X(\tau)$, we fix a nice Schubert divisor Y in X. We then construct a proper birational morphism $\psi : Z \to X$ such that Z is a fiber space over $\mathbf{P}^1$ with fiber Y. By induction, we suppose the results to be true on Y, prove the results for Z and then make the results "go down to X."

As important consequences of the main theorem we obtain

(1) $X(\tau)$ is normal

(2) $H^i(X(\tau), L) = 0,\ i \geq 1$
(3) A character formula for $H^0(X(\tau), L)$.

§1. Preliminaries

Let $H = SL(n,k)$, k being the base field which we assume to be algebraically closed. Let $A = k[t,t^{-1}]$, $A^+ = k[t]$, $A^- = k[t^{-1}]$. Let D be the maximal torus in H consisting of all diagonal matrices and B the Borel subgroup consisting of all upper triangular matrices. Let N be the normalizer of D in H. The projection $\pi^+ : A^+ \to k$ sending t to 0, induces a map $\pi^+ : H(A^+) \to H$. Let $\mathcal{B} = (\pi^+)^{-1}(B)$, $W = N(A)/D$, $G = H(A)$. We have a cellular decomposition

$$(*) : G = \dot{\bigcup_{w \in W}} \mathcal{B} w \mathcal{B}$$

Let $\mathfrak{g}$ be the Kac-Moody lie algebra corresponding to the matrix

$$\begin{pmatrix} 2 & -1 & 0 & 0 & \cdots & 0 & -1 \\ -1 & 2 & -1 & 0 & \cdots & 0 & 0 \\ 0 & -1 & 2 & -1 & \cdots & 0 & 0 \\ \cdot & & & & & & \\ \cdot & & & & & & \\ \cdot & & & & & & \\ -1 & 0 & 0 & \cdots & \cdots & -1 & 2 \end{pmatrix}$$

With notation as in [5], let $S = \{\alpha_0, \alpha_1, \ldots, \alpha_{n-1}\}$ be the set of simple roots of $\mathfrak{g}$. For $w \in W$, let $X(w) = \bigcup_{\tau \leq w} \mathcal{B}\tau\mathcal{B}$ (mod $\mathcal{B}$) be the Schubert variety in $G/\mathcal{B}$ (see [6] for generalities on the infinite dimensional flag variety $G/\mathcal{B}$). Let us fix a fundamental weight ω_i, $0 \leq i \leq n-1$ and let P be the maximal parabolic subgroup of G obtained by "omitting α_i." Let W^P be the set of minimal representatives of W/W_P.

Definition 1.1 (cf [1]). Let $\tau \in W^P$ and let $X(\tau)$ be the associated Schubert variety in G/P. Let φ in W^P be such that $X(\varphi)$ is a divisor in $X(\tau)$, say $\varphi = \tau s_\beta$, for some positive root β. We define the *multiplicity of $X(\varphi)$ in $X(\tau)$ as*

$$m(\varphi, \tau) = (\omega, \beta^*) \left(= \frac{2(\omega, \beta)}{(\beta, \beta)} \right).$$

Definition 1.2. With notation as above, if $m(\varphi, \tau) > 1$, then we shall refer to $X(\varphi)$ as a *multiple divisor* in $X(\tau)$.

Definition 1.3. With notation as above, let $\varphi = s_\gamma \tau$, for some positive root γ. We say $X(\varphi)$ is a *moving* (resp. *nonmoving*) divisor in $X(\tau)$ if γ is simple (resp. nonsimple).

Lemma 1.4. (cf [11], Lemma 1.5). *Let $X(\varphi)$ be a moving divisor in $X(\tau)$, say $\varphi = s_\alpha \tau$, for some $\alpha \in S$. Then for any $\theta \leq \tau$, we have either $\theta \leq \varphi$ or $\theta = s_\alpha \theta'$, for some $\theta' \leq \varphi$.*

Lemma 1.5. ([11]). *With notation as in Lemma 1.5, let $X(\theta)$ be a divisor in $X(\tau)$ and let $\theta' = s_\alpha \theta$. Then $m(\theta, \tau) = m(\theta', \varphi)$.*

Lemma 1.6. *Let β_i, $1 \leq i \leq r+1 < n$, be consecutive simple roots. Let $(\tau(\omega), \beta_i^*) = x_i$, $1 \leq i \leq r+1$, where*

$$(i)\ x_1 + x_2 + \cdots + x_i > 0,\ 1 \leq i \leq r$$
$$(ii)\ x_1 + x_2 + \cdots + x_{r+1} = -m, \text{ for some } m \geq 1.$$

Let $\gamma = \sum_{i=1}^{r+1} \beta_i$ and $\eta = s_\gamma \tau$. Then $X(\eta)$ is a divisor in $X(\tau)$ with $m(\eta, \tau) = m$.

Proof (By induction on r).
(a) $r = 1$.
Let $\varphi = s_{\beta_2} \tau$, $\theta = s_{\beta_1} \theta$, $\eta = s_{\beta_2} \tau$. Then we have $m(\varphi, \tau) = m + x_1$, $m(\theta, \varphi) = m(\eta, \tau) = m$. Further, $\eta = s_\gamma \tau$, where $\gamma = \beta_1 + \beta_2$.
(b) $r > 1$.
Let $\theta_{r+2} = \tau$ and $\theta_i = s_i \cdots s_{r+1} \tau$, $1 \leq i \leq r+1$, where s_j denotes the reflection with respect to β_j. Let $\gamma_i = \beta_1 + \cdots + \beta_{i-1}$.

Claim $X(s_{\gamma_i} \theta_i)$ is a divisor in $X(\theta_i)$ occurring with multiplicity m, $2 \leq i \leq r+2$.

Proof of claim (by induction on i).
When $i = 2$, $\gamma_i = \beta_1$ and the claim is clear. When $i = 3$, the claim follows from (a). Let $i > 3$ and assume the result for $i-1$. For simplicity of notation, let us denote $\theta = \theta_i$, $\theta' = \theta_{i-1}$, $\alpha = \beta_{i-1}$, $\gamma = \gamma_{i-1}$, $\lambda = s_\gamma \theta'$. Now by induction, $X(\lambda)$ is a divisor in $X(\theta')$ with $m(\lambda, \theta') = m$. Hence $X(s_\alpha \lambda)$ is a divisor in $X(\theta)$, with $m(s_\alpha \lambda, \theta) = m$ (cf Lemma 1.5). But $s_\alpha \lambda = s_\alpha s_\gamma \theta' = s_{i-1} s_{i-2} \cdots s_1 \cdots s_{r+1} \tau = s_{\gamma_i} \theta_i$. This completes the proof of the claim and also of Lemma 1.6.

Conversely, we have

Lemma 1.7. *Any nonmoving divisor $X(\eta)$ in $X(\tau)$ is of the form $X(s_\gamma \tau)$ (with notation as in Lemma 1.6).*

Proof (by induction on dim $X(\tau)$).

Let $\eta = s_\varepsilon \tau$, for some nonsimple positive root ε. Let us fix a moving divisor $X(\varphi)$ in $X(\tau)$, say $\varphi = s_\alpha \tau$, for some $\alpha \in S$. Let $\theta = s_\alpha \eta$. Then $\theta \leq \varphi$ (cf Lemma 1.4). Let $\theta = s_\beta \varphi$, for some $\beta \in R^+$. We have (cf Lemma 1.5), $m(\eta, \tau) = m(\theta, \varphi) = m$ say. Let $m(\varphi, \tau) = t$, $m(\theta, \eta) = n'$. Then we have $m\varepsilon + n'\alpha = m\beta + t\alpha$.

Case 1. β is simple.

In this case, we have $\varepsilon = \alpha + \beta$, $n' = t - m$. Taking $\alpha = \beta_2$, $\beta = \beta_1$, we have $x_1 = t - m$, $x_2 = -t$ and $\varepsilon = \beta_1 + \beta_2$ (notation being as in Lemma 1.6).

Case 2. β is nonsimple.

By induction, there exist β_i, $1 \leq i \leq r$, such that if $(\varphi(\omega), \beta_i^*) = y_i$, then $y_1 + y_2 + \cdots + y_i > 0$, $1 \leq i \leq r-1$, $y_1 + y_2 + \cdots + y_r < 0$ and $\beta = \sum_{i=1}^{r} \beta_i$.

Subcase 2(a). $(\beta, \alpha^*) = 0$.

We have $\varepsilon = \beta$ and the result is clear.

Subcase 2(b). $(\beta, \alpha^*) = -1$.

We have, $\varepsilon = \beta + \alpha$. Hence taking $\alpha = \beta_{r+1}$ we have

$$(\tau(\omega), \beta_i^*) = y_i, \ i \neq r, \ r+1$$
$$(\tau(\omega), \beta_r^*) = y_r + t, \ (\tau(\omega), \beta_{r+1}^*) = -t$$

Also, in this case, $y_1 + \cdots + y_r = -m$ and $t = m + n'$. The result now follows from this.

Subcase 2(c). $(\beta, \alpha^*) = 1$.

This implies either β_1 or $\beta_r = \alpha$, say $\beta_r = \alpha$. We have, $\gamma = \beta - \alpha$, $t = n' - m$. Further

$$(\tau(\omega), \beta_i^*) = y_i, \ i \neq r-1, \ r, \ (\tau(\omega), \beta_{r-1}^*) = y_{r-1} + y_r, \ (\tau(\omega), \beta_r^*) = -y_r$$

(Note that $y_r = t$, in this case.) thus we have $\varepsilon = \beta_1 + \cdots + \beta_{r-1}$ and the result is now immediate (with $\gamma = \beta_1 + \cdots + \beta_{r-1}$).

Starting point of induction. Let $X(\tau)$ be of least dimension such that $X(\tau)$ has a nonmoving divisor $X(\eta)$. Then with notation as above, we obtain that β is necessarily simple. The rest of the proof is as in Case 1.

This completes the proof of Lemma 1.7.

§2. The Bases $\{Q(\lambda,\mu)_N\}$ and $\{P(\lambda,\mu)_N\}$

Let $U = U(\mathfrak{g})$ be the universal enveloping algebra of $\mathfrak{g}$. Let $U_{\mathbb{Z}}$ (resp. $U^+_{\mathbb{Z}}, U^-_{\mathbb{Z}}$) denote the $\mathbb{Z}$-subalgebra of U generated by $X^n_{\pm\alpha}/n!$ (resp. $X^n_{\alpha}/n!, X^n_{-\alpha}/n!$), $\alpha \in S$ and $n \in \mathbb{N}$. In the sequel, we shall denote $X^n_\alpha/n!$ by $X^{(n)}_\alpha$. Let U_α (resp. $U_{\alpha,\mathbb{Z}}$) denote the $\mathbb{Q}$-vector subspace of U generated by X^n_β (resp. $X^{(n)}_\beta$), $\beta = \pm\alpha$, $\alpha \in S$. Let $V = V_\omega$ be the integrable highest weight $\mathfrak{g}$-module (over $\mathbb{C}$) with highest weight ω. Let us fix a highest weight vector e (which is unique up to scalars). Let $V_{\mathbb{Z}} = U_{\mathbb{Z}}e$. For $\tau \in W^P$, let $e_\tau = \tau e$, $V_{\mathbb{Z},\tau} = U^+_{\mathbb{Z}} e_\tau$. For any field k, let $V_{k,\tau}$ (or just V_τ) be $V_{\mathbb{Z},\tau} \otimes k$.

A $\mathbb{Z}$-basis for $V_{\mathbb{Z},\tau}$:
Let

$$Z = \left\{(a_1, b_1, a_2, b_2, \dots, a_s, b_s) \;\middle|\; \begin{array}{l}(1)\ a_i, b_j \in \mathbb{Z}^+ \\ (2)\ a_i, b_j > 0,\ i \neq 1,\ j \neq s\end{array}\right\}$$

Definition 2.1.

(a) A is called a *head* if for every b_t, $\sum\limits_{i\le t} b_i \le \sum\limits_{i \le t} a_i$ (in particular, $a_1 \neq 0$)

(b) A is called a *tail* if for every a_t, $\sum\limits_{i\ge t} a_i \le \sum\limits_{i \ge t} b_i$ (in particular, $b_s \neq 0$).

Lemma 2.2. *Let A be not a tail. Then there exists a canonical tail associated to A.*

Proof. Let $1 \le i_1 < i_2 < \cdots < i_r \le s$ be the set of all indices such that for $1 \le t \le r$,

$$\sum_{i_t < j < i_{t+1}} a_j - b_j \le 0, \qquad \sum_{i_t \le j < i_{t+1}} a_j - b_j > 0$$

(here, $i_{r+1} = s+1$). Let us denote $\sum\limits_{i_t \le j < i_{t+1}} a_j - b_j$ by C_t. Note that $a_{i_t} > C_t$. Let $m = C_1 + \cdots + C_r$. We define A_j, $0 \le j \le m$ inductively as follows: Define $A_0 = A$. Let $j \ge 1$, say $j = C_1 + \cdots + C_{t-1} + k$, where $1 \le k \le C_t$ (here, if $t = 1$, then $j = k$). Then A_j is obtained from A_{j-k} by replacing a_{i_t} by $a_{i_t} - k$ and $b_{i_t - 1}$ by $b_{i_t-1} + k$ (here, if $i_1 = 1$, then b_{i_1-1} is understood as 0). Then A_m is obviously a tail.

Lemma 2.3. *Let A be not a head. Then there exists a canonical head associated to A.*

Proof. Let $1 \leq i_r < i_{r-1} < \cdots < i_1 \leq s$ be the set of all indices such that for $1 \leq t \leq r$,

$$\sum_{i_{t+1}<j<i_t} b_j - a_j \leq 0, \qquad \sum_{i_{t+1}<j\leq i_t} b_j - a_j > 0$$

(here, $i_{r+1} = 0$). Let us denote $\sum_{i_{t+1}<j\leq i_t} b_j - a_j$ by C_t. Note that $b_{i_t} > C_t$. Let $m = C_1 + \cdots + C_r$. We define A_j, $0 \leq j \leq m$ inductively as follows: Define $A_0 = A$. Let $j \geq 1$, say $j = C_1 + \cdots + C_{t-1} + k$, where $1 \leq k \leq C_t$ (here, if $t = 1$, then $j = k$). Then A_j is obtained from A_{j-k} by replacing b_{i_t} by $b_{i_t} - k$ and a_{i_t+1} by $a_{i_t+1} + k$ (here, if $i_1 = s$, then a_{i_1+1} is understood as 0). Then A_m is obviously a head.

Remark 2.4. Let $A = (a_1 b_1 a_2 b_2 \cdots)$ be a head. Let l be the least integer such that $a_l > b_l$. This implies $a_t = b_t$, $t < l$ and $\sum_{j\geq l} a_j \geq \sum_{j\geq l} b_j$ (since A i s a head). If $\sum_{i\geq 1} a_i > \sum_{j\geq 1} b_j$, then with notation as in Lemma 2.2, we have $m = \sum_{j\geq l}(a_j - b_j) = \sum_{j\geq 1}(a_j - b_j)$.

Definition 2.5. We call a subset Z_1 of Z *complete*, if either

(1) $Z_1 = \{A\}$, where A is both a head and a tail (in which case we call it trivial) or
(2) $Z_1 = \{A_0, A_1, \ldots, A_t\}$ where A_0 is a head, say

$$A_0 = (a_1, b_1, a_2, b_2, \ldots),\ t = \sum_{j\geq 1}(a_j - b_j)$$

and $A_1, \ldots, A_t$ are as in the proof of Lemma 2.2, with $A = A_0$.

Remark 2.6. Given $A = (a_1, b_1, a_2, b_2, \ldots)$, following the procedure in Lemmas 2.2 and 2.3, there exists a unique complete subset Z_1 to which A belongs.

Definition 2.6 (cf [2]). Let $\Lambda = (\lambda_1, \ldots, \lambda_r)$ be a Young diagram with λ_j boxes in the jth row. We say Λ is *admissible* if

(1) $\lambda_1 \geq \lambda_2 \geq \cdots \geq \lambda_r$
(2) Number of rows of same length is $\leq n - 1$.

Let $\omega = \omega_d$, for some d, $0 \le d \le n-1$. To an admissible Young diagram Λ, we want to associate a vector Q_Λ in V. To this end, we proceed as follows: We fill in Λ with integers. We fill in the first row with $\alpha_d, \alpha_{d+1}, \alpha_{d+2}, \cdots$ (or just $d, d+1, d+2, \cdots$ (modulo n)). Then we go down each column decreasing the value by 1 at a time (again modulo n).

Definition 2.7. A *corner box* in Λ is one such that

(1) it is a box at the end of a certain row, say the kth row r_k.

(2) $l(r_k) > l(r_{k+1})$.

Consider the right most corner box. It is filled with α_j say. Let us denote α_j by just α. Consider all the blank boxes of the form

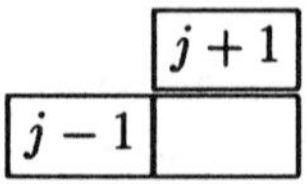

(and also the blank box 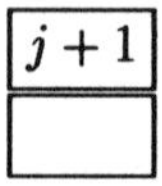

in the first column, in case the first column ends with $j+1$) adding all of which to Λ yields an admissible Young diagram. We shall refer to these as *blank corner* α*-boxes*. Consider

$$A(\Lambda) = (b_1, a_2, b_2, a_3, b_3, \dots)$$

where (going from top to bottom)

$$b_1 = \# \text{ first set of corner } \alpha\text{– boxes}$$
$$a_2 = \# \text{ first set of blank corner } \alpha\text{– boxes}$$
$$b_2 = \# \text{ second set of corner } \alpha\text{– boxes}$$

and so on.

Let us denote $A(\Lambda)$ by just A, so that $A = (a_1, b_1, a_2, b_2, \cdots)$, where $a_1 = 0$. Let Z_1 be the (unique) complete subset of Z such that $A \in Z_1$ (cf Remark 2.6). Note that Z_1 is not trivial (since A is not a head, as $a_1 = 0$). Let $Z_1 = (A_0, A_1, \dots, A_t)$ (notation as in Definition 2.5). Let $A = A_i$. We now associate admissible Young diagrams Λ_j, $0 \le j \le t$.

Case 1. $j > i$.

Let $m = t - i$. Then $m = \sum_{u=1}^{r} C_u$ (notation as in the proof of Lemma 2.2). Let $j - i = C_1 + \cdots + C_{u-1} + k$, where $1 \le k \le C_u$ (here, if $u = 1$, then $j - i = k$). Then Λ_j is obtained from Λ by filling in with α, the first D_ℓ blank corner α-boxes in the $(i_\ell - 1)^{th}$ set of blank corner α-boxes, $1 \le \ell \le u$, where

$$D_\ell = \begin{cases} C_\ell, \ell \ne u \\ k, \ell = u \end{cases}$$

(notation as in the proof of Lemma 2.2).

Case 2. $j < i$.

Then the m in the proof of Lemma 2.3 is simply i. Writing $m = \sum_{u=1}^{r} C_u$ (notation as in Lemma 2.3), let $j = C_1 + \cdots + C_{u-1} + k$, where $1 \leq k \leq C_u$ (here, if $u = 1$, then $j = k$). Then Λ_j is obtained from Λ by replacing the bottom D_ℓ corner α-boxes by blank corner α -boxes, in the i_ℓ^{th} set (from the bottom) of corner α-boxes, $1 \leq \ell \leq u$, where

$$D_\ell = \begin{cases} C_\ell, l \neq u \\ k, l = u \end{cases}$$

(notation as in the proof of Lemma 2.3).

Definition 2.8. With notation as above, set

$$Q_\Lambda = X_{-\alpha}^{(i)} Q_{\Lambda_0}$$

Remark 2.9. Q_Λ is a weight vector of weight $\omega - \sum_{t=0}^{n-1} m_t \alpha_t$, where $m_t = \#$ of boxes in Λ filled with t (or α_t).

Lemma 2.10. *Let $\chi(\Lambda)$ be the weight of Q_Λ and let $\alpha \in S$. Then $(\chi(\Lambda), \alpha^*) = x(\Lambda) - y(\Lambda)$, where $y(\Lambda)$ (resp. $x(\Lambda)$) is the number of corner (resp. blank corner) α-boxes.*

Proof (by induction on $n(\Lambda)$, the number of boxes in Λ).

When $n(\Lambda) = 1$, $\chi(\Lambda) = \omega_i - \alpha_i$. For $\alpha = \alpha_j$, $j \neq i-1, i+1 \pmod n$, $(\chi(\Lambda), \alpha^*) = 0$, $x(\Lambda) = 0 = y(\Lambda)$. For $\alpha = \alpha_j$, $j = i \pm 1$, $(\chi(\Lambda), \alpha^*) = 1$, $x(\Lambda) = 1$, $y(\Lambda) = 0$. For $\alpha = \alpha_i$, $(\chi(\Lambda), \alpha^*) = -1$, $x(\Lambda) = 0$, $y(\Lambda) = 1$. Let now $n(\Lambda) > 1$. Let β be a corner box in Λ and let Λ' be the admissible Young diagram obtained from Λ by deleting this corner β-box. We have $\chi(\Lambda) = \chi(\Lambda') - \beta$.

Case 1. $\alpha = \beta$.

We have, $\chi(\Lambda) = \chi(\Lambda') - \alpha$, $x(\Lambda) = x(\Lambda') - 1$, $y(\Lambda) = y(\Lambda') + 1$ and the result follows (by induction).

Case 2. $(\beta, \alpha^*) = 0$.

We have, $(\chi(\Lambda), \alpha^*) = (\chi(\Lambda'), \alpha^*)$, $x(\Lambda) = x(\Lambda')$, $y(\Lambda) = y(\Lambda')$.

Case 3. $(\beta, \alpha^*) = -1$, say $\alpha = \alpha_{j+1}$, $\beta = \alpha_j$.

We have $(\chi(\Lambda), \alpha^*) = (\chi(\Lambda'), \alpha^*) + 1$. Let the deleted corner β-box appear in the kth row r_k.

Subcase 3(a). $l(r_{k-1}) = l(r_k)$.

In this case, we have $x(\Lambda) = x(\Lambda')$, $y(\Lambda) = y(\Lambda') - 1$.

Subcase 3(b). $l(r_{k-1}) > l(r_k)$.

We have, $x(\Lambda) = x(\Lambda') + 1$, $y(\Lambda) = y(\Lambda')$.

Remark 2.11(a). As before, let $\alpha = \alpha_j$ be the right most corner box in Λ. We have $Q_\Lambda = X_{-\alpha}^{(i)} Q_{\Lambda_0}$ (notation as before). It is easily seen that Q_Λ is an extremal weight vector if and only if Q_{Λ_0} is an extremal weight vector and in Λ_0, $\#\{\text{corner } \alpha\text{– boxes}\} = 0$ and $\#\{\text{blank corner } \alpha\text{–boxes}\} = i$. Conversely, to an extremal weight vector Q_θ, we can associate an admissible Young diagram θ^* (in the obvious way).
(b) Given an extremal weight vector Q_θ and $\alpha \in S$, let $y = \#\{\text{corner } \alpha\text{-boxes in } \theta^*\}$, $x = \#\{\text{blank corner } \alpha\text{-boxes in } \theta^*\}$. Then we have, either

(1) $x = 0 = y$, in which case $(\theta(\omega), \alpha^*) = 0$
(2) $x \neq 0$, $y = 0$, in which case $(\theta(\omega), \alpha^*) = x$
(3) $x = 0$, $y \neq 0$, in which case $(\theta(\omega), \alpha^*) = -y$.

(c) Given θ, suppose β is a nonsimple positive root such that $X(\theta)$ is a divisor in $X(s_\beta\theta)$ with $m(\theta, s_\beta\theta) = r$, then we talk about corner β–boxes and it is clear that in θ^*, there are r corner blank β–boxes and no corner β–boxes (note that $\beta = \alpha_j + \alpha_{j+1} + \cdots + \alpha_t$, for some $0 \leq j < t \leq n-1$ (cf Lemmas 1.6 and 1.7)).

The elements $\rho(\Lambda)$ and $\delta(\Lambda)$

With notation as above let $Q_\Lambda = X_{-\alpha}^{(i)} Q_{\Lambda_0}$. Then we define $\rho(\Lambda)$ inductively as $\rho(\Lambda) = s_\alpha \rho(\Lambda_0)$. To define $\delta(\Lambda)$, let us write

$$Q_\Lambda = X_{-\beta_t}^{(m_t)} \cdots X_{-\beta_1}^{(m_1)} e$$

where β_i is the right most corner box at each step. Let l, $1 \leq l \leq t$ be the largest integer such that $X_{-\beta_l}^{(m_l)} \cdots X_{-\beta_1}^{(m_1)} e$ is an extremal weight vector, say Q_θ. Then from the way Q_Λ is defined, it is clear that Λ is obtained from θ^* in $t - l$ steps as follows:

We have $(\theta(\omega), \beta_{l+1}^*) \geq m_{l+1}$. Let Λ_1 be the admissible Young diagram obtained by filling in the first m_{l+1} blank corner β_{l+1} boxes in θ. Next we have $\chi(\Lambda_1), \beta_{l+2}^*) \geq m_{l+2}$ and let Λ_2 be the admissible Young diagram obtained by filling in the first m_{l+2} blank corner β_{l+2} boxes in Λ_1 and so on. Now, if in the above process we allow for nonsimple roots also, then it is clear that there exists a unique maximal element $\delta \in W^P$, positive roots γ_k, positive integers $n_k > 0$, $1 \leq k \leq s$ such that $X(s_k \cdots s_1 \delta)$ is a divisor in $X(s_{k+1} \cdots s_1 \delta)$ occurring with multiplicity $\geq n_k$ (here, s_i denotes the reflection with respect to γ_i). Further, let Λ_1 be the admissible Young diagram obtained by filling in the first n_1 blank corner γ_1–boxes in δ^* (cf Remark 2.11 (c)). We have $(\chi(\Lambda_1), \gamma_2^*) \geq n_2$. Let Λ_2 be the admissible Young diagram obtained by filling in the first n_2 blank corner γ_2–boxes in Λ_1 and so on. Then $\Lambda = \Lambda_s$. We set

$$\delta(\Lambda) = \delta.$$

Notation 2.12: Let us denote $\rho(\Lambda), \delta(\Lambda)$, by just ρ, δ and $s_k \cdots s_1 \delta$ by δ_k. We shall in the sequel denote Q_Λ by $Q(\rho, \delta)_N$, also, where

$$N = \{(\delta_0 = \delta < \delta_1 < \cdots < \delta_s = \rho);\ (n_1, \cdots, n_s)\}.$$

We shall have occasion to look at the following special vectors:
(1) Let $s_\alpha \rho = \varphi$, $\alpha \in S$ and let $r = m(\varphi, \rho)$.
If $\delta = \varphi$ and $n_1 = t$ for some $t \leq r$, we shall denote Q_Λ by just $Q(\rho, \varphi)_t$ (note that $Q_\Lambda = X_{-\alpha}^{(t)} Q_\varphi$).
(2) Let $\eta = s_\beta \rho$ be such that $X(\eta)$ is a divisor in $X(\rho)$, where $\beta \neq \alpha$. Let $\theta = s_\alpha \eta$. Let $m(\eta, \rho) = m$ (note that $m(\theta, \varphi)$ is also m). Corresponding to $\delta = \eta$, $n_1 = t$ for some $t \leq m$, we shall denote Q_Λ by just $Q(\rho, \eta)_t$ (note that $(\rho, \eta)_t = X_{-\alpha}^{(b)} (\varphi, \delta)_t$ where

$$b = \begin{cases} t & , \text{ if } (\beta, \alpha^*) = 0 \\ t + r' & , \text{ if } (\beta, \alpha^*) = 1 \\ r + m - t & , \text{ if } (\beta, \alpha^*) = -1 \end{cases}$$

where $r' = m(\theta, \eta)$ and $r = m(\varphi, \rho)$.

§3. First Basis Theorem

Generalized Demazure character formula (cf [7]).

Let $N = \{\text{integral weights of } \mathfrak{g}\}$. Let $\mathbf{Z}[N]$ denote the group ring of the multiplicative group $\exp N$. For a simple root α, let M_α be the linear operator $M_\alpha : \mathbf{Z}[N] \to \mathbf{Z}[N]$ defined by

$$M_{s_\alpha}(\exp \lambda) = \begin{cases} \exp \lambda + \exp(\lambda - \alpha) + \cdots + \exp(\lambda - m\alpha), \text{ if } m = (\lambda, \alpha^*) \geq 0 \\ -[\exp(\lambda + \alpha) + \cdots + \exp(\lambda + (m-1)\alpha), \text{ if } -m = (\lambda, \alpha^*) < 0 \end{cases}$$

Let $\tau = s_1 \cdots s_r$ (reduced expression) and let $M_\tau = M_{s_1} \circ M_{s_2} \circ \cdots \circ M_{s_r}$.
Let $V_\tau = V_{\mathbf{Z},\tau} \otimes \mathbf{C}$.

Theorem 3.1. (cf [7]).

$$char\, V_\tau = M_\tau(\exp(\omega)).$$

Notation 3.2: For τ in W, let

$$I_\tau = \{\text{admissible Young diagrams } \Lambda \mid \tau \geq \rho(\Lambda)\}$$

Theorem 3.3.

$$(a)\ char\, V_\tau = \sum_{\Lambda\in I_\tau} \exp(\chi(\Lambda))$$

$$(b)\ \dim V_\tau = \# I_\tau$$

(note that (b) is an immediate consequence of (a)*).*

Proof (by induction on $\dim X(\tau)$). If $\dim X(\tau) = 0$, then the result is obvious. Let then $\dim X(\tau) > 0$. Let the right most corner box in τ^* be an α-box (cf §2). Let $\varphi = s_\alpha \tau$. We have

$$\begin{aligned} \text{char}\, V_\tau &= M_{s_\alpha}(\text{char}\, V_\varphi) \\ &= M_{s_\alpha} \sum_{\Lambda \in I_\varphi} \exp(\chi(\Lambda)) \end{aligned}$$

Suppose Λ_0 in I_φ is such that $A(\Lambda_0)$ is a head. Let $A(\Lambda_0), A(\Lambda_1), \dots, A(\Lambda_t)$ be the associated complete subset of Z (cf §2). Then from our discussion in §2, it is clear that either $\Lambda_i \in I_\varphi$, $1 \le i \le t$ or $\Lambda_i \notin I_\varphi$, $1 \le i \le t$ according as $s_\alpha \rho(\Lambda_0) \le \varphi$ or $\not\le \varphi$. Hence I_φ breaks up as

$$I_\varphi = \dot{\bigcup_i} B_i \dot{\bigcup_j} C_j$$

where

$$B_i = \left\{ \Lambda_0, \dots, \Lambda_t \;\middle|\; \begin{array}{l} A(\Lambda_0), \dots, A(\Lambda_t) \text{ is a} \\ \text{complete subset of } Z \end{array} \right\}$$

and

$$C_j \text{ is a singleton } \{\Lambda\},$$

where $A(\Lambda)$ is a head and $s_\alpha \rho(\Lambda) \not\le \varphi$. We have for a B_i,

$$M_\alpha(\exp(\Lambda_0) + \cdots + \exp(\Lambda_t)) = \exp(\Lambda_0) + \cdots + \exp(\Lambda_t)$$

(note that $s_\alpha(\chi(\Lambda_i)) = \chi(\Lambda_{t-i})$). If a $C_j (= \Lambda)$ is such that $(\chi(\Lambda), \alpha^*) = 0$, again $M_\alpha(\exp(\chi(\Lambda))) = \exp(\chi(\Lambda))$. Thus the extra terms in char V_τ arise from C_j's such that $(\chi(C_j), \alpha^*) > 0$, and are given by $\Lambda_1, \dots, \Lambda_t$ (where $A(C_i), A(\Lambda_1), \dots, A(\Lambda_t)$ form a complete subset of Z). On the other hand any Λ in I_τ arises from an unique Λ_0 in I_τ such that $A(\Lambda_0)$ is a head. The result now follows:

Theorem 3.4 (First Basis Theorem). *Let* $\mathcal{B}_\tau = \{Q_\Lambda, \Lambda \in I_\tau\}$. *Then* $\mathcal{B}$ *is a* $\mathbb{Z}$*–basis for* $V_{\mathbb{Z},\tau}$.

Proof (by induction on $\dim X(\tau)$). In view of Theorem 3.3, it suffices to check that $\mathcal{B}_\tau$ generates $V_{\mathbb{Z},\tau}$. If $\dim X(\tau) = 0$, then the result is obvious. Let then $\dim X(\tau) > 0$. Let α and φ be as in the proof of Theorem 3.3. We have (cf [11]), $V_{\mathbb{Z},\tau} = U_{-\alpha,\mathbb{Z}} V_{\mathbb{Z},\varphi}$. Hence suffices to check $X_{-\alpha}^{(r)} Q_\Lambda$, $\Lambda \in I_\varphi$ belongs to the $\mathbb{Z}$-span of Q_Δ, $\Delta \in I_\tau$. Again, suffices to consider Q_Λ's such that $s_\alpha \rho(\Lambda) \not\leq \varphi$ (since $X_{-\alpha}^{(r)} Q_\Lambda \in V_{\mathbb{Z}, s_\alpha \rho(\Lambda)})$. As in the proof of Theorem 3.3, we have, for such a Λ, $A(\Lambda)$ is a head. Let $A(\Lambda_t)$ be the corresponding tail (note that $\Lambda_t = X_{-\alpha}^{(t)} Q_\Lambda$). Let $A(\Lambda_t) = (a_1 b_1 a_2 b_2 \cdots a_s b_s)$. Consider all elements B_j of Z obtained from $A(\Lambda_t)$ by replacing a_i by $x, 1, a_i - (x+1)$, for some i, $1 \leq i \leq s$. (Here $0 \leq x \leq a_i - 1$; if x is 0 (resp. $a_i - 1$) then the corresponding B_j has $b_{i-1}+1$, $a_i - 1$ (resp. $a_i - 1, b_i + 1$) in the place of b_{i-1}, a_i (resp. a_i, b_i).) Let $A(\Delta_j)$ be the corresponding head. Then it is clear that

$$X_{-\alpha} Q_{\Lambda_t} \in \mathbb{Z} \text{ span of } \{X_{-\alpha}^{(r)} Q_{\Delta_j}\}.$$

Thus we may assume Λ is such that $A(\Lambda_t) = (b_1)$. This implies $A(\Lambda) = (a_1)$. In this case, we have obviously

$$X_{-\alpha}^{(i)} Q_\Lambda = Q_{\Lambda_i},\ 1 \leq i \leq t$$

(note that $t = a_1$) and $X_{-\alpha}^{(i)} Q_\Lambda = 0$, $i > t$. This completes the proof of Theorem 3.4.

Definition 3.5. Let $\{P_\Lambda,\ \Lambda \in I_\tau\}$ be the $\mathbb{Z}$-basis of the $\mathbb{Z}$-dual of $V_{\mathbb{Z},\tau}$, dual to Q_Λ, $\Lambda \in I_\tau\}$.

Remark 3.6. Let $\theta \leq \tau$. Then $P_\Lambda|_{X(\theta)} \equiv 0$ if and only if $\theta \not\geq \rho(\Lambda)$.

§4. Standard monomials and their linear independence

Let k be the base field, $V_\tau = V_{\mathbb{Z},\tau} \otimes k$, $p_\Lambda = P_\Lambda \otimes 1$. If $\Lambda = \theta^*$, for some $\theta \in W$, then we shall denote p_Λ by just p_θ.

Definition 4.1. A monomial $p_{\Lambda_1} p_{\Lambda_2} \cdots p_{\Lambda_m}$ (in $S^m(V_\tau^*)$) is *called standard on* $X(\tau)$ if

$$\tau \geq \rho(\Lambda_1) \geq \delta(\Lambda_1) \geq \rho(\Lambda_2) \geq \cdots \geq \rho(\Lambda_m) \geq \delta(\Lambda_m).$$

Notation 4.2: Let Λ be given and let $\rho(\Lambda) = \tau$. Let $Q_\Lambda = X_{-\alpha}^{(t)} Q_{\Lambda'}$. We shall denote t by $t(\Lambda)$, δ_{s-1} by $\eta(\Lambda)$, n_s by $n(\Lambda)$ (cf Notation 2.12). In the sequel, we shall also denote Λ by $(\tau, \eta, \delta)_N$ (where $\eta = \eta(\Lambda), \delta = \delta(\Lambda)$) and p_Λ by $p(\tau, \eta, \delta)_N$. Also, if $\delta = \eta$ (so that $s = 1$), then we shall denote p_Λ by just $p(\tau, \eta)_{n_1}$.

Lemma 4.3. *Let $\tau \in W^P$. Let φ, α be as in §3.*

(1) *Let $\eta \neq \varphi$. Let $m(\eta, \tau) = m$. Then for $\eta' \neq \eta$, we have, on $X(\tau)$,*

(a) $p(\tau,\eta',\delta')_{N'}p(\tau,\eta)_{m-s} = p(\tau)p(\tau,\lambda,\lambda')_{M'}$ *where either* $\eta' \neq \varphi$ *or* $\eta' = \varphi$ *in which case* $t(\Lambda') > s$ *(here* $\Lambda' = (\tau,\eta',\delta')_{N'}$*). Further* $\lambda = \eta$ *or* η'.

(b) Let $\Lambda = (\tau,\eta,\delta)_N$. *If* $n(\Lambda) > s$, *then*

$$p(\tau,\eta,\delta)_N p(\tau,\eta)_{m-s} = p(\tau)p(\tau,\eta,\delta)_{N'}$$

for some N'.

(c) If $n(\Lambda) = s$, *then*

$$p(\tau,\eta,\delta)_N p(\tau,\eta)_{m-s} = p(\tau)F$$

where $F|_{X(\eta)} = F(\eta,\lambda,\delta)_{N'}$, *for some* λ *and* N'.

(2) *Let* $m(\varphi,\tau) = r$. *Let* $\Lambda = (\tau,\eta,\delta)_N$, $\eta \neq \varphi$.

(a) If $t(\Lambda) > a$, *then*

$$p(\tau,\eta,\delta)_N p(\tau,\varphi)_{r-a} = p(\tau)p(\tau,\lambda,\lambda')_{N'},$$

where $\lambda = \varphi$ *or* η.

(b) If $t(\Lambda) = a$ *and* $\delta \leq \varphi$, *then*

$$p(\tau,\eta,\delta)_N p(\tau,\varphi)_{r-a} = p(\tau)F$$

where $F|_{X(\varphi)} = p(\varphi,\varphi',\delta)_{N'}$ *for some* φ' *and* N'.

(c) Let $\Lambda = (\tau,\varphi,\delta)_N$. *If* $t(\Lambda) > s$, *then*

$$p(\tau,\eta,\delta)_N p(\tau,\varphi)_{r-a} = p(\tau)p(\tau,\varphi,\delta)_{N'},$$

for some N'.

(d) If $t(\Lambda) = s$, *then*

$$p(\tau,\varphi,\delta)_N p(\tau,\varphi)_{r-a} = p(\tau)F$$

where $F|_{X(\varphi)} = F(\varphi,\lambda,\delta)_{N'}$ *for some* λ *and* N'.

The above assertions are proved in the same spirit as in [10] and we skip the details.

Theorem 4.4. *Standard monomials on* $X(\tau)$ *of degree* m *are linearly independent.*

Proof (by induction on dim $X(\tau)$). If dim $X(\tau) = 0$, then p_τ^m is the only degree m standard monomial on $X(\tau)$ and the result is obvious. Let then dim $X(\tau) > 1$. Let φ, α be as in §3. Let

$$\sum a_i F_i = 0,\ a_i \in k \tag{I}$$

be a linear relation among standard monomials of degree m on $X(\tau)$. If there is a F_i such that F_i starts with a p_Λ, with $\rho(\Lambda) < \tau$, then by restricting (I) to $X(\rho(\Lambda))$, we conclude $a_i = 0$ (using induction hypothesis). Hence we may assume that each F_i starts with a p_Λ where $\rho(\Lambda) = \tau$.

Let us then rewrite (I) as

$$a_\tau p_\tau F_\tau + \sum a_\Lambda p_\Lambda F_\Lambda = 0 \tag{II}$$

where F_Λ is a standard monomial of the form $p_{\Lambda_1} p_{\Lambda_2} \cdots$ with $\rho(\Lambda_1) \leq \delta(\Lambda)$. Let $Q_\Lambda = X_{-\alpha}^{(t)} Q_\Lambda$, and $t(\Lambda) = t$ (as before). Let $t_0 = \min_\Lambda \{t(\Lambda)\}$.

Case 1. All Λ with $t(\Lambda) = t_0$ are such that $\eta(\Lambda) \neq \varphi$. Let us fix a Λ with $t(\Lambda) = t_0$. Let $\eta = \eta(\Lambda)$. Consider all Δ (appearing on the L.H.S. of II) with $\eta(\Delta) = \eta$. Let $b = \min_\Delta \{n(\Delta)\}$. We now multiply II throughout by $p(\tau, \eta)_{m-b}$ (where $m = m(\eta, \tau)$), use Lemma 4.3, (1), cancel $p(\tau)$, restrict (II) to $X(\eta)$ and conclude $a_\Delta = 0$ for all Δ such that $\eta(\Delta) = \eta$ and $n(\Delta) = b$.

Case 2. There are Λ's, with $\eta(\Lambda) = \varphi$ and $t(\Lambda) = t_0$. We divide this case into the following two subcases:

Subcase 2(a). All Δ's with $t(\Lambda) = t_0$ are such that $\delta(\Delta) \leq \varphi$. In this case, we multiply (II) throughout by $p(\tau, \varphi)_{r-t_0}$, use Proposition 4.3, (2), cancel $p(\tau)$, restrict (II) to $X(\varphi)$ and conclude $a_\Delta = 0$ for all Δ such that $t(\Lambda) = t_0$.

Subcase 2(b). There are Δ's with $t(\Lambda) = t_0$ and $\delta(\Delta) \not\leq \varphi$. We fix such a Δ. Let $\eta(\Delta) = \eta$. Now we consider *all* Θ's with $\eta(\Theta) = \eta$ and $n(\Theta) \leq n(\Delta)$ and proceed as in Case (1). Then we are reduced to Subcase 2(a). Thus proceeding, II reduces to

$$a_\tau p_\tau F_\tau = 0$$

from which we conclude $a_\tau = 0$. Thus II reduces to the trivial relation. This completes the proof of Theorem 4.4.

§5. Filtration for the ideal of $H(\tau)$.

Consider $X(\tau) \hookrightarrow \mathbf{P}(V_\tau)$. Let L denote the tautological line bundle on $\mathbf{P}(V_\tau)$. Let $R(\tau)$ denote the homogeneous coordinate ring of $X(\tau)$ under the above embedding. We shall show (cf Theorem 5.3) that the standard monomials on $X(\tau)$ of degree m form a basis of $H^0(X(\tau), L^m)$. Let $H(\tau)$ = zero set of p_τ in $X(\tau)$ and let $I(H(\tau))$ = the ideal $p_\tau R(\tau)$ in $R(\tau)$. We now take a total ordering on $K_\tau = \{$ admissible Young Diagrams Λ on $X(\tau)$ such that $\rho(\Lambda) = \tau$ $\}$.

Fix an integer a, $1 \leq a < r$, (here $r = m(\varphi, \tau)$, $\varphi = s_\alpha \tau$, $\alpha \in S$). Let

$$I_a = \{\Lambda | t(\Lambda) = a\}.$$

Let $I = K_\tau - \bigcup_{1 \leq a < r} I_a - \{\tau^*\}$. We index K_τ as follows. We keep τ^* as the first element, followed by I, then $I_{r-1}, I_{r-2}, \cdots I_1$. We now take a total ordering on I, I_a, $1 \leq a < r$ as follows:

1. Indexing of I: We fix a Λ in I and denote $\eta = \eta(\Lambda)$. We then arrange all Δ's in I with $\eta(\Delta) = \eta$ with decreasing order of $n(\Delta)$ in such a way that among the Δ's with the same $n(\Delta)$, if $\delta(\Delta_1) < \delta(\Delta_2)$, then Δ_2 will precede Δ_1.

2. Indexing of $\mathbf{I}_a$: We first consider $\{\Lambda \in I_a | \eta(\Lambda) = \varphi\}$ and arrange them as in (1). We then consider $\{\Lambda \in I_a | \eta(\Lambda) \neq \varphi \text{ and } \delta(\Lambda) \leq \varphi\}$ and arrange them as in (1). Finally we consider $\{\Lambda \in I_a | \delta(\Lambda) \not\leq \varphi\}$ and arrange them as in (1).

We write $K_\tau = \{\Lambda_0, \Lambda_1, \ldots, \Lambda_m\}$ where $\Lambda_0 = \tau^*$ and $m + 1 = \#K_\tau$. We now define the ideals I_t, $0 \leq t \leq m$ in $R(\tau)$ as

$$I_t = \sum_{0 \leq i \leq t} p_i R(\tau)$$

where $p_i = p_{\Lambda_i}$, $0 \leq i \leq m$ (note that $I_0 = p_\tau R(\tau)$) $(= I(H(\tau))$. Let $\underline{I}_t$ be the ideal sheaf in $\mathcal{O}_{X(\tau)}$ associated to I_t. Let us fix a simple root β such that if $w = s_\beta \tau$, then $X(\tau)$ is a divisor in $X(w)$. Proceeding as in [12], we have

Proposition 5.1. *Suppose that the standard monomials on $X(\tau)$ of degree s form a basis of $H^0(X(\tau), L^s)$, $s \in \mathbf{N}$.*

(1) *We have canonical isomorphisms (as $\mathcal{O}_{X(\tau)}$-modules)*

$$f_t : \underline{I}_t / \underline{I}_{t-1} \approx \mathcal{O}_{X(\delta)}(-1),\ 0 \leq t \leq m$$

where δ is given by $\delta = \delta(\Lambda)$, $p_\Lambda = p_t$ (here $I_{-1} = (0)$).

(2) *Let $B_\beta = B \cap SL(2, \beta)$, where $SL(2, \beta)$ is the "$SL(2)$" associated to the simple root β. We have B_β- isomorphisms*

$$\underline{I}_t / \underline{I}_{t-1} \approx \chi_t \otimes \mathcal{O}_{X(\delta)}(-1),\ 0 \leq t \leq m$$

where χ_t represents the 1-dimensional B_β-module associated to $-\chi(\Lambda)$ (where recall that $\chi(\Lambda) =$weight of Q_Λ).

(3) *The ideal sheaf $\underline{I}(H(\tau)_{\text{red}})$ in $\mathcal{O}_{X(\tau)}$ is precisely $\underline{I}_m$.*

With w, τ, β etc. as above, we define (as in [12]) Z_τ (or just Z) as

(i) $Z =$ the fiber product $SL(2, \beta) \times^{B_\beta} X(\tau)$.

(ii) We denote by ψ, the canonical morphism

$$\psi : Z \longrightarrow X(w),$$

defined by

$$(g, y) \longmapsto gy,\ g \in SL(2, \beta),\ y \in X(\tau)$$

(iii) For a coherent $\mathcal{O}_{X(\tau)}$-module F with a B_β action compatible with the action of B_β on $X(\tau)$, we set

$$\widetilde{F} = SL(2,\beta) \times^{B_\beta} F.$$

(iv) For a bundle N on Z, we set

$$N^{(\ell)} = N \otimes p^*(\mathcal{O}_{\mathbf{P}^1}(\ell)),\ \ell \in \mathbf{Z}$$

where p is the canonical map

$$p : Z \longrightarrow \mathbf{P}^1\ (= SL(2,\beta)/B_\beta).$$

(v) Let $H(Z)$ be the closed subscheme of Z with ideal sheaf $\underline{I}(H(Z)) = \underline{I}(H(\tau)) \otimes M$, where $M = \underline{I}(X(\tau))^{\otimes r}$, $r = (\tau(\omega), \beta^*)$ (here, we identify $X(\tau)$ as the closed subscheme $p^{-1}(\bar{e})$ of Z, where $\bar{e} =$ the coset eB_β).

Lemma 5.2. *With notation as above, suppose that standard monomials on $X(\tau)$ of degree s form a basis of $H^0(X(\tau), L^s)$, $s \in \mathbf{N}$. Let $(\tau(\omega), \beta^*) = t$. Let us define ideal sheaves K_i, $0 \le i \le m$, M_j, $1 \le j \le t$ on Z as follows:*

$$M_j = \tilde{\underline{I}}_0^{(-j)},\ 1 \le j \le t$$
$$K_j = \tilde{\underline{I}}_j^{(-1)},\ 0 \le j \le m$$

(where, recall $m+1 = \#K_\tau$ (cf §4)) so that we have

$$M_t \subset M_{t-1} \subset \cdots \subset M_1 = K_0 \subset K_1 \subset \cdots \subset K_m \subset \mathcal{O}_Z.$$

Then we have

$$(i)\ \ M_t = \psi^*(L^{-1})$$
$$(ii)\ \ M_j/M_{j+1} \approx L^{-1}|_{X(\tau)},\ 1 \le j \le t-1$$
$$(iii)\ K_i/K_{i-1} \approx \psi^*(L^{-1})^{(m_j-1)}|_{Z_\delta},\ 0 \le i \le m$$

where δ is given by $\delta = \delta(\Lambda)$, $p_\Lambda = p_i$ and $m_i = (\chi(\Lambda), \beta^*)$.

$$(iv)\ \ \mathcal{O}_Z/K_m = \mathcal{O}_{H(Z)_{\mathrm{red}}}$$

Proof. Similar to that of [12], Lemma 5.11.

Theorem 5.3. *Let $w \in W^P$. Then*

(a) $H^i(X(w), L^s) = 0$, $i \geq 1$, $s \in \mathbf{Z}^+$

(b) $X(w)$ *is normal*

(c) *Standard monomials on $X(w)$ of degree s form a basis for $H^0(X(w), L^s)$*

Proof. We prove the theorem by induction on $\dim X(w)$, the result being obvious if $\dim X(w) = 0$. Let then $\dim X(w) > 0$. Let us fix a moving divisor $X(\tau)$ in $X(w)$, say $\tau = s_\beta w$, for some simple root β. Let Z be as before. We now consider the following exact sequences on Z.

$$0 \to M_t \to \mathcal{O}_Z \to \mathcal{O}_Z/M_t \to 0$$
$$0 \to M_j/M_{j+1} \to \mathcal{O}_Z/M_{j+1} \to \mathcal{O}_Z/M_j \to 0, \ 1 \leq j \leq t-1$$
$$0 \to K_i/K_{i-1} \to \mathcal{O}_Z/K_{i-1} \to \mathcal{O}_Z/K_i \to 0, \ 0 \leq i \leq m$$

(where $K_{-1} = 0$). In view of Lemma 5.2, we have for $s \in \mathbf{N}$,

$$\begin{aligned} &(1) \quad \psi^*(L^s) \otimes M_t = \psi^*(L^{s-1}) \text{ (on } Z) \\ &(2) \quad \psi^*(L^s) \otimes M_j/M_{j+1} = L^{s-1}|_{X(\tau)} \\ &(3) \quad \psi^*(L^s) \otimes K_i/K_{i-1} = \psi^*(L^{s-1})^{(m_i-1)}|_{Z_\delta} \\ &(4) \quad \mathcal{O}_Z/K_m = \mathcal{O}_{H(Z)_{\text{red}}} \end{aligned}$$

Now tensoring the exact sequences in $(*)$ with $\psi^*(L^s)$, we obtain (denoting $\psi^*(L)$ by just L)

$$\left\{ \begin{aligned} \chi(Z, L^s) &= \chi(Z, L^{s-1}) + \chi(H(Z)_{\text{red}}, L^s) + (t-1)\chi(X(\tau), L^{s-1}) \\ &\quad + \textstyle\sum_{0 \leq i \leq m} \chi(Z_\delta, \mathcal{O}_{Z_\delta}^{m_i - 1} \otimes L^{s-1}) \end{aligned} \right. \qquad (**)$$

where χ is the Euler–Poincaré characteristic. From $(**)$ we obtain (proceeding as in [12])

$$h^0(Z, L^s) = h^0(Z, L^{s-1}) + h^0(H(w)_{\text{red}}, L^s) + \sum_{\theta < w} c_\theta h^0(X(\theta), L^{s-1}) \quad (***)$$

where $c_\theta = \#\{\Lambda \in K_w | \delta(\Lambda) = \theta\}$.

For any $\delta \in W^P$, let us denote $S(\delta, s) = \{$standard monomials on $X(\delta)$ of degree s$\}$ and $s(\delta, s) = \#S(\delta, s)$. We can write

$$S(w, s) = S_1 \dot{\cup} S_2 \dot{\cup} S_3,$$

where

$$S_1 = \{F \in S(w,s) | F \text{ starts with } p_\Lambda, \text{ where } \rho(\Lambda) < w\}$$
$$S_2 = \{F \in S(w,s) | F \text{ starts with } p_\Lambda, \text{ where } \rho(\Lambda) = w\}$$
$$S_3 = \{F \in S(w,s) | F \text{ starts with } p_w\}$$

We have

$$\#S_1 = \#\{\text{standard monomials on } H(W)_{\text{red}} \text{ of degree } s\}$$

(note that $H(w)_{\text{red}} = U$ all divisors $X(w')$ in $X(w)$),

$$\#S_2 = \sum_{\theta < w} c_\theta s(\theta, s-1), \text{ where } c_\theta = \#\{\Lambda \in K_w | \delta(\Lambda) = \theta\}$$

and

$$\#S_3 = s(w, s-1).$$

Thus we obtain

$$s(w,s) = s_3 + s_1 + s_2 \qquad (****)$$

where $s_i = \#S_i$. Now by induction on s, we may assume $H^0(Z, L^{s-1}) = H^0(X(w), L^{s-1})$ and hence we obtain

$$\text{R.H.S. of } (***) = \text{R.H.S. of } (****).$$

Hence we obtain $h^0(Z, L^s) = s(w,s)$. This implies that the canonical inclusions

$$H^0(X(w), L^s) \hookrightarrow H^0(\tilde{X}(w), L^s) \hookrightarrow H^0(Z, L^s)$$

(where $\tilde{X}(w)$ is the normalization of $X(w)$) are in fact isomorphisms. From this we obtain (b) and (c). The assertion (a) is proved as in [12], Theorem 5.14.

§6 Standard monomials for Schubert varieties in $\hat{Sl}_n/\mathcal{B}$.

Because of the symmetry involved, the results of §1 thru §5 give a standard mnomial theory for Schubert varieties in $\hat{S}L_n/P$, for every maximal parabolic subgroup P.

Let $\tau \in W$ and let $X(\tau^{(i)})$ be the projection of $X(\tau)$ under $\pi_i; G/\mathcal{B} \to G/P_i$, where $G = \hat{S}L_n$ and P_i, $0 \le i \le n-1$ are the maximal parabolic subgroups of G. Let $V_{\mathbb{Z},\tau^{(i)}}$ be the $\mathbb{Z}$–submodule of $V_{\mathbb{Z},\omega_i}$, associated to $\tau^{(i)}$ (cf §2).

For any field k, let $V_{\tau^{(i)}} = V_{Z,\tau^{(i)}} \otimes k$. Consider the canonical embedding $X(\tau^{(i)}) \hookrightarrow \mathbf{P}(V_{\tau^{(i)}})$. Let L_i be the tautological line bundle on $\mathbf{P}(V_{\tau^{(i)}})$.

Definition 6.1. Let $F \in H^0(X(\tau, L)$, where $L = \otimes L_i^{a_i}$, $a_i \in \mathbb{Z}^+$. Let $F = f_0 f_1 f_2 \cdots f_\ell$, $(\ell = n-1)$, where $f_i = p_{\Lambda_{i1}} p_{\Lambda_{i2}} \cdots p_{\Lambda_{ia_i}}$. F is said to be <u>standard on $X(\tau)$ of multidegree</u> $\underline{a} = (a_0, \dots, a_\ell)$, if there exists a sequence in W

$$\tau \geq \theta_{01} \geq \phi_{01} \geq \theta_{02} \geq \phi_{02} \geq \cdots \geq \theta_{0a_1} \geq \phi_{0a_1} \geq \theta_{11} \geq \cdots \geq \theta_{\ell a_\ell} \geq \phi_{\ell a_\ell}$$

such that $\pi_i(\theta_{ij}) = \rho(\Lambda_{ij})$, $\pi_i(\phi_{ij}) = \delta(\Lambda_{ij})$, $1 \leq j \leq a_i$, $0 \leq i \leq \ell$. Proceeding as in [11], we have

Theorem 6.2. *Standard monomials on* $X(\tau)$ *of degree* $\underline{a}$ *form a basis of* $H^0(X(\tau), L)$.

Remark 6.3. In Definition 6.1, the order $\{P_0, \dots, P_\ell\}$ is immaterial.

REFERENCES

[1] C.C. Chevalley, *Sur les Décompositions cellulaires des espaces G/B*, (manuscript non publie) (1958).

[2] E. Date, M. Jimbo, A. Kuniba, T. Miwa and M. Okado, *Paths, Maya diagrams and representations of $\hat{sl}(r, \mathbf{C})$*, Adv. Studies in Pure Math **19** (1989), 149-191.

[3] W.V.D. Hodge, *Some enumerative results in the theory of forms*, Proc. Camb. Phil. Soc. **39** (1943), 22-30.

[4] W.V.D. Hodge and C. Pedoe, *Methods of Algebraic Geometry, vol. II*, Cambridge University Press, 1952.

[5] V.G. Kac, *Infinite dimensional Lie Algebras*, Progress in Math **44** (1983), Birkäuser.

[6] M. Kashiwara, *The flag manifold of Kac-Moody Lie Algebra*, preprint.

[7] S. Kumar, *Demazure character formula in arbitrary Kac-Moody setting*, Inventiones Math **89** (1987), 395-423.

[8] V. Lakshmibai, *Standard Monomial Theory for G_2*, J. Algebra **98** (1986), 281-318.

[9] V. Lakshmibai, C. Musili and C.S. Seshadri, *Geometry of G/P–IV*, Proc. Ind. Acad. Sci. **99A** (1979), 279-362.

[10] V. Lakshmibai and K.N. Rajeswari, *Towards a Standard Monomial Theory for Exceptional Groups*, Contemporary Mathematics **88**, 449-578.

[11] V. Lakshmibai and C.S. Seshadri, *Geometry of G/P- V*, J. Algebra **100** (1986), 462-557.

[12] V. Lakshmibai and C. S. Seshadri, *Standard Monomial Theory for $\hat{S}L_2$, Infinite Dimensional Lie Algebras and Groups*, Advanced Series in Mathematical Physics **7**, 178-234.

Received March 12, 1990

Department of Mathematics
Northeastern University
Boston, MA 02215

Equivariant Cohomology of Wonderful Compactifications

PETER LITTELMANN *

CLAUDIO PROCESI**

Dedicated to J. Dixmier on his 65th birthday

Introduction

The theory of complete symmetric varieties has evolved in several papers with the aim to give a satisfactory framework for Schubert calculus on a particularly relevant class of non compact homogeneous spaces, the symmetric spaces over adjoint groups. It is an aspect of Hilbert's 15th problem and it includes some classical examples like the variety of quadrics.

For an automorphism σ of order 2 of a semisimple algebraic group G of adjoint type we denote by $H = G^\sigma$ the fixed point subgroup of σ. Let G/H be the corresponding symmetric space. We study the geometry of the canonical G–equivariant compactification X of G/H, and more generally the geometry of the wonderful compactifications of G/H (of which we shall recall some of the main features in sections 2,3 and 11). In the following we refer to the canonical compactification X as the minimal wonderful compactification.

In previous papers (cf. [6],[8],[9],[10],[11]) it has been carried out a detailed study of the wonderful compactifications, and it has been developed the interpretation of Schubert calculus as limit cohomology of equivariant compactifications. The computation of cohomology has been carried out in several different ways in ([6],[8],[11]).

The aim of this paper is to describe the cohomology of X extending

* Supported by C.N.R.

** Member of G.N.S.A.G.A. of C.N.R.

the results obtained in [8] for quadrics and at the same time to simplify the treatment by the use of equivariant cohomology (cf. [6]).

Let $T \subset G$ be a σ–stable maximal torus containing a maximal σ–split torus and let $B \supset T$ be a suitably chosen Borel subgroup (see 3.5). Denote by N_T the normalizer of T in G. We describe the T–fixed points X^T in terms of the Satake diagram of the symmetric space (3.7, 6.7). Moreover, we construct a canonical smooth N_T–stable subvariety $\mathcal{N}_T$ (containing X^T) with finitely many T–orbits. This variety has the following remarkable properties:

a) Every B–orbit of X meets $\mathcal{N}_T$ in a T–orbit, and the B–orbit is a product of the T–orbit and an affine space.

b) The G–equivariant cohomology of X restricts to the N_T–equivariant cohomology of $\mathcal{N}_T$ and its image is described by a natural combinatorial filtration: We group the connected components of $\mathcal{N}_T$ (which are torus embeddings) in N_T–orbits O_i (ordered in combinatorial way). The factors of the filtration of the G–equivariant cohomology of X are then the N_T–equivariant cohomology groups of O_i shifted by a suitable Euler class. We view this as a natural variation of the localization principle.

The strategy of the proof consists first in a careful geometric description of the variety $\mathcal{N}_T$ (which is of independent interest) and of a sequence of reductions through certain remarkable subvarieties of X associated to σ–stable Levi subgroups of G.

Acknowledgements: The first author would like to thank the University of Rome for the hospitality offered to him during his visit in the academic year 88/89.

1. Equivariant Cohomology

1.0. In this paragraph we wish to recall some results on equivariant cohomology which will be needed later. We refer to [2], [7] and [14] for the details.

The description of the equivariant cohomology of a smooth torus embedding given at the end of the paragraph can be found in [6].

1.1. For a topological group G let $E_G \to B_G$ be a universal principal G–fibration. The space E_G is contractible and G acts freely on E_G.

If X is a G–space with continuous G–action, then denote by X_G the fibre space $E_G \times_G X$ over B_G. The G–equivariant cohomology $H^*_G(X, F)$

of X, with coefficients in a ring F, is by definition

$$H_G^*(X,F) := H^*(X_G,F),$$

the cohomology of the space X_G. Note that $H_G^*(X,F)$ is in a canonical way a $H_G^*(pt,F)$–module.

We will restrict ourself to rational equivariant cohomology and simply write $H_G^*(X)$ for $H_G^*(X,\mathbf{Q})$.

1.2. If G is a Lie group with finitely many connected components and K is a maximal compact subgroup of G, then G/K is contractible and hence $H_G^*(X) = H_K^*(X)$.

If $H \subset G$ is a closed Lie subgroup, then $E_G \to E_G/H$ is a universal principal H–fibration. So if Z is a H–space, then the equivariant cohomology rings $H_H^*(Z)$ and $H_G^*(G \times_H Z)$ are canonically isomorphic.

1.3. Let T be a compact torus. Denote by $X(T_{\mathbf{C}})$ the character group of the complexification $T_{\mathbf{C}}$ of T. For a character λ let $c_\lambda \in H^2_{T_{\mathbf{C}}}(pt)$ be the Chern class of the corresponding line bundle on $B_{T_{\mathbf{C}}}$. The map $\lambda \to c_\lambda$ induces an isomorphism

$$\mathrm{Sym}(X(T_{\mathbf{C}}) \otimes_{\mathbf{Z}} \mathbf{Q}) \longrightarrow H_T^*(pt).$$

1.4. For a compact group K let K^0 be its connected component containing the identity and denote by Γ the quotient group K/K^0. Let $T \subset K^0$ be a maximal torus and let W be the Weyl group of K^0. If X is a K–space, then the morphisms $H_K^*(X) \to H_{K^0}^*(X)$ and $H_{K^0}^*(X) \to H_T^*(X)$ induce isomorphisms

$$H_K^*(X) \simeq H_{K^0}^*(X)^\Gamma \simeq (H_T^*(X)^W)^\Gamma.$$

In particular, $H_K^*(pt)$ is a graded algebra with elements only in even degree and without zero divisors.

1.5. If X is a K–space whose cohomology vanishes in odd degrees, then the spectral sequence corresponding to the map $X_K \to B_K$ collapses. So $H_K^*(X)$ is a free module over $H_K^*(pt)$ and

$$H^*(X) \cong H_K^*(X)/\mathcal{I}H_K^*(X),$$

where $\mathcal{I}$ is the ideal in $H_K^*(pt)$ generated by the elements of strictly positive degree.

A key result in the theory of equivariant cohomology is the following:

Localization Theorem 1.6. *If T is a compact torus acting smoothly on a compact manifold X, then the kernel as well as the cokernel of the canonical map*

$$i^* : H_T^*(X) \to H_T^*(X^T)$$

induced by the inclusion $i : X^T \hookrightarrow X$ are torsion modules over $H_T^*(pt)$. In particular, if $H_T^*(X)$ is a free module over $H_T^*(pt)$, then i^* is injective.

1.7. In our case we will study the action of a reductive algebraic group G on a smooth projective variety X such that X^T is finite for a maximal torus T of G. By the theory of Bialynicki–Birula (see [3]–[5]), the odd cohomology of X vanishes, and hence by 1.5 the G– as well as the T–equivariant cohomology is a free module over $H_G^*(pt)$ respectively $H_T^*(pt)$. Let W be the Weyl group of G. The Localization Theorem implies:

Proposition 1.8. *The odd G–equivariant cohomology of X vanishes and $H^*(X) = H_G^*(X)/\mathcal{I}H_G^*(X)$. Furthermore, the inclusion $i : X^T \hookrightarrow X$ induces an injective map*

$$i^* : H_G^*(X) \hookrightarrow H_T^*(X^T)^W .$$

1.9. We want now to recall the description of the ring $H_T^*(X)$, where T is an algebraic torus and X is a smooth T–embedding, i.e. X is a smooth T–variety with an open orbit isomorphic to T. For details we refer to [6], §4.

Denote by $X(T)$ the character group of T. Recall, the smooth torus embedding X of T is completely determined by its associated rational partial polyhedral decomposition (r.p.p.d.) of $\mathrm{Hom}_{\mathbf{Z}}(X(T), \mathbf{R})$.

Let $\{v_\alpha\}$ be the primitive integral vectors generating the one–dimensional cones in the decomposition. Consider the polynomial ring $\mathbf{Q}[x_\alpha]$ in the variables x_α corresponding to the vectors v_α. We define a morphism $\mathbf{Q}[x_\alpha] \to H_T^*(X)$ by sending each variable x_α to the equivariant Chern class of the T–stable divisor D_α corresponding to v_α.

Denote by I the ideal in $\mathbf{Q}[x_\alpha]$ generated by the monomials

$$\{\Pi_{i=1}^l x_{\alpha_i} \mid v_{\alpha_1}, \ldots, v_{\alpha_l} \text{ do not generate a cone in the given r.p.p.d.}\}.$$

Then I is contained in the kernel of the map defined above, so we have a well defined map from the "Reisner–Stanley" algebra $R_X := \mathbf{Q}[x_\alpha]/I$ to $H_T^*(X)$.

Theorem 1.10. ([6], Theorem 8) *The morphism $R_X \to H_T^*(X)$ is an isomorphism.*

Remark. The class c_χ in $H_T^2(X)$ associated to a character $\chi \in H_T^2(pt)$ under the map $H_T^*(pt) \to H_T^*(X)$ is $c_\chi := \sum_\alpha \chi(v_\alpha)x_\alpha$.

2. The minimal wonderful compactification of the group

2.0. As an example of our approach to compute the equivariant cohomology we consider in this paragraph the minimal wonderful compactification $\overline{G}$ of a semisimple algebraic group G of adjoint type.

2.1. Consider the group $G \times G$ and the involution $\theta(g_1, g_2) := (g_2, g_1)$ of $G \times G$. Then its fixed group $(G \times G)^\theta$ is the image of the diagonal embedding of G, and the symmetric space $G \times G/G$ is as $G \times G$–variety isomorphic to G.

Denote by $\mathcal{G}$ the Lie algebra of G. For $n = \dim G$, consider the Grassmann variety

$$Grass_n(\mathcal{G} \oplus \mathcal{G}) \subset \mathbf{P}(\bigwedge^n(\mathcal{G} \oplus \mathcal{G})).$$

Let $\bigwedge^n \mathcal{G}$ be the point in $Grass_n(\mathcal{G} \oplus \mathcal{G})$ corresponding to the subspace Lie $(G \times G)^\theta$ of $\mathcal{G}\oplus\mathcal{G}$. The orbit $(G\times G).\bigwedge^n \mathcal{G}$ is isomorphic to $G\times G/G$ and can be identified with G (as $G \times G$–variety) by the map $g \mapsto (g, id).\bigwedge^n \mathcal{G}$.

We define the *minimal wonderful compactification* $\overline{G}$ of G as the closure of the orbit $(G \times G).\bigwedge^n \mathcal{G}$ in $Grass_n(\mathcal{G} \oplus \mathcal{G})$.

The variety $\overline{G}$ has the following "wonderful" properties (see [9], §3): $\overline{G}$ is a smooth variety with finitely many $G \times G$–orbits and only one closed orbit. The complement of the open orbit $G \subset \overline{G}$ is the union of rk (G) smooth divisors intersecting transversely, and the closure of an orbit is obtained as the intersection of the codimension one orbit closures containing it.

2.2. Consider the closure $\overline{T}$ of a maximal torus $T \subset G$ in $\overline{G}$. This is a $T \times T$–stable variety. The Weyl group W of G acts on $\overline{T}$ via the diagonal embedding of N_T in $G \times G$, where N_T is the normalizer of T in G. The open orbit $(T \times T).id$ in $\overline{T}$ is isomorphic to the torus $S = T/H$, where H is the subgroup of T of elements of order 2.

The structure of this torus embedding is well understood ([10], §5): $\overline{T}$ is a smooth S–embedding. The r.p.p.d of $\mathrm{Hom}_{\mathbf{Z}}(X(T), \mathbf{R})$ corresponding to $\overline{T}$ is the decomposition into the Weyl chambers of the root system of G.

Furthermore, every $G \times G$–orbit in $\overline{G}$ intersects $\overline{T}$ in a union of $T \times T$–orbits, which are permuted transitively by W.

A precise description of the stabilizer of a point in $\overline{G}$ is given in [9], §5. We will need here only the fact that the unique closed orbit is isomorphic to $G/B \times G/B$ for a Borel subgroup B of G, and all $T \times T$–fixed points in $\overline{G}$ are contained in the closed orbit.

Theorem 2.3. *The inclusion* $\overline{T} \hookrightarrow \overline{G}$ *induces an isomorphism*

$$H^*_{G\times G}(\overline{G}) \xrightarrow{\sim} H^*_{T\times T}(\overline{T})^W \simeq (H^*_T(pt) \otimes H^*_S(\overline{T}))^W.$$

Remark. Let K be a maximal compact subgroup of G such that $T_c = K \cap T$ is a maximal torus in K. A consequence of the theorem is by §1 (see [11]):

$$H^*(\overline{G}) \simeq H^*((K \times K) \times_{T_c \times T_c} \overline{T})^W.$$

Proof. Consider the following inclusions:

$$\begin{array}{ccc} (\overline{T})^{T\times T} & \hookrightarrow & \overline{T} \\ \downarrow & & \downarrow \\ (\overline{G})^{T\times T} & \hookrightarrow & \overline{G} \end{array}$$

The $T \times T$–fixed points in $\overline{G}$ are in one $W \times W$–orbit, and the intersection of this orbit with $\overline{T}$ is one W–orbit. So by the Localization Theorem the inclusions induce the following commutative diagram:

$$\begin{array}{ccc} H^*_{T\times T}((\overline{T})^{T\times T})^W & \hookleftarrow & H^*_{T\times T}(\overline{T})^W \\ \| & & \uparrow \\ H^*_{T\times T}((\overline{G})^{T\times T})^{W\times W} & \hookleftarrow & H^*_{G\times G}(\overline{G}) \end{array}$$

which proves the morphism $H^*_{G\times G}(\overline{G}) \to H^*_{T\times T}(\overline{T})^W$ is injective.

Recall, $H^*_{G\times G}(\overline{G})$ is a free $H^*_{T\times T}(pt)^{W\times W}$–module. And $H^*_{T\times T}(\overline{T})$ is a free $H^*_{T\times T}(pt)$–module. It is well known, one can choose a graded $W \times W$–submodule R of $H^*_{T\times T}(pt)$, isomorphic to the regular representation of $W \times W$, such that

$$H^*_{T\times T}(pt) \simeq R \otimes H^*_{T\times T}(pt)^{W\times W}$$

as graded $H^*_{T\times T}(pt)^{W\times W}$–module. So $H^*_{T\times T}(\overline{T})^{W\times W}$ is in a natural way a free $H^*_{T\times T}(pt)^{W\times W}$–module.

Let $\mathcal{I}$ denote the ideal in $H^*_{T\times T}(pt)^{W\times W}$ of elements of strictly positive degree. Since the morphism we are considering is an injective, graded morphism of free $H^*_{T\times T}(pt)^{W\times W}$-modules, to prove its surjectivity it suffices to show the induced injection

$$H^*_{G\times G}(\overline{G})/\mathcal{I}H^*_{G\times G}(\overline{G}) \longrightarrow H^*_{T\times T}(\overline{T})^W/\mathcal{I}H^*_{T\times T}(\overline{T})^W$$

is an isomorphism. The dimension of the left hand side is by 1.8 the Euler characteristic of X, which is equal to the number of $T\times T$–fixed points in X and hence equal to $|W\times W|$.

To compute the dimension of the right hand side, note first we can find a graded W–stable submodule U of $H^*_{T\times T}(\overline{T})$ such that the morphism

$$U\otimes H^*_{T\times T}(pt) \longrightarrow H^*_{T\times T}(\overline{T})$$

is a W–equivariant isomorphism of graded $H^*_{T\times T}(pt)$–modules. And the dimension of U is by 1.8 equal to the Euler characteristic of $\overline{T}$, and hence equal to $|W|$, the number of $T\times T$–fixed points in $\overline{T}$. So

$$H^*_{T\times T}(\overline{T})^W/\mathcal{I}H^*_{T\times T}(\overline{T})^W$$

is isomorphic to $(U\otimes R)^W$ as W–representation. Since R decomposes into the direct sum of $|W|$–copies of the regular representation of W, the following simple lemma shows $\dim\,(U\otimes R)^W = |W\times W|$. And hence

$$\dim\, H^*_{T\times T}(\overline{T})^W/\mathcal{I}H^*_{T\times T}(\overline{T})^W = |W\times W| = \dim\, H^*_{G\times G}(\overline{G})/\mathcal{I}H^*_{G\times G}(\overline{G}),$$

which proves $H^*_{G\times G}(\overline{G}) \to H^*_{T\times T}(\overline{T})^W$ is an isomorphism.

Lemma. *If N is a finite group, and U and V are two finite dimensional representations of N such that V is the sum of copies of the regular representation of N, then*

$$\dim\,(V\otimes U)^N = \frac{\dim V\cdot \dim U}{|N|}$$

Since the image of the diagonal embedding $T\hookrightarrow T\times T$ acts trivially on $\overline{T}$, the isomorphism

$$H^*_{T\times T}(\overline{T}) \simeq H^*_T(pt)\otimes H^*_S(\overline{T})$$

follows by the exact sequence $1\to T\to T\times T\to S\to 1$

Q.E.D.

3. Minimal wonderful compactifications

3.0. This paragraph is entirely devoted to introduce some notation and recall the construction of the minimal wonderful compactification of a symmetric space and its orbit structure.

3.1. Let G be a semisimple group of adjoint type and denote by $\mathcal{G}$ its Lie algebra. Consider an involution σ of G and denote its fixed group G^σ by H, and the Lie algebra of H by $\mathcal{H}$.

If $M \subset G$ is a σ–stable torus, then M admitts a decomposition $M = M^\sigma.M^1$, where M^1 is the connected component containing the identity of the subgroup $\{m \in M|\sigma(m) = m^{-1}\}$, the so called σ–split part of M.

We fix a maximal σ–stable torus T of G such that T^1 is a maximal σ–split torus in G. The dimension l of T^1 is called the rank of the symmetric space G/H.

We fix further a σ–stable maximal compact subgroup K of G such that $T_c := T \cap K$ is a maximal torus in K.

3.2. To construct the minimal wonderful compactification X of G/H, consider first the compactification $\overline{G}$ of G defined in 2.1. Via the "twisted embedding"

$$G \hookrightarrow G \times G, \quad g \mapsto (g, \sigma(g))$$

we obtain an action of G on $Grass_n(\mathcal{G} \oplus \mathcal{G})$ such that the stabilizer in G of the point $\bigwedge^n \mathcal{G}$ is H. So the orbit map $g \mapsto g.\bigwedge^n \mathcal{G}$ induces an embedding $G/H \to G$.

The action of G on the open orbit $G \subset \overline{G}$ is given by $g.g' := gg'\sigma(g)^{-1}$ and the embedding of G/H into $G \subset \overline{G}$ is given by the map $gH \mapsto g\sigma(g)^{-1}$.

We define the *minimal wonderful compactification* X of G/H as the closure of the orbit $G.\bigwedge^n \mathcal{G}$ in $\overline{G}$.

Remark. For the definition of the minimal wonderful compactification X we refer to [9], §2 and §4. In [9], §6, it is proved that X is isomorphic to the closure of the orbit $G.\bigwedge^m \mathcal{H}$ in $Grass_m(\mathcal{G})$, $m = \dim H$. The proof that X is isomorphic to the closure of the orbit $G.\bigwedge^n \mathcal{G}$ in $\overline{G}$ is similar, so we omit it.

3.3. We can consider X as a connected component of the fixed points of an automorphism of $\overline{G}$: Let $d\sigma$ be the Lie algebra automorphism induced by σ. The linear automorphism

$$\phi : \mathcal{G} \oplus \mathcal{G} \to \mathcal{G} \oplus \mathcal{G}, \quad (g_1, g_2) \mapsto (d\sigma(g_2), d\sigma(g_1))$$

induces an automorphism of order two of $\mathbf{P}(\bigwedge^n(\mathcal{G} \oplus \mathcal{G}))$, which we denote by $\tilde{\sigma}$. Since $\tilde{\sigma}$ fixes $\bigwedge^n \mathcal{G}$, it is easy to see $\overline{G}$ is $\tilde{\sigma}$–stable and the restriction of $\tilde{\sigma}$ to the open orbit $G \subset \overline{G}$ is the morphism $g \mapsto \sigma(g^{-1})$.

Obviously we have $G/H \hookrightarrow (\overline{G})^{\tilde{\sigma}}$ and the $\tilde{\sigma}$–fixed point set $(\overline{G})^{\tilde{\sigma}}$ is G–stable (with respect to the G–action on $\overline{G}$ in 3.2). Since $d\tilde{\sigma}$ acts on the tangent space $T_{id}\overline{G} \simeq \mathcal{G}$ at the $\tilde{\sigma}$–fixed point $id \in \overline{G}$ by $g \mapsto -d\sigma(g)$, we see that

$$\dim T_{id}(\overline{G})^{\tilde{\sigma}} = \dim \mathcal{G}^{d\tilde{\sigma}} = \dim G/H.$$

So X is a connected component of $(\overline{G})^{\tilde{\sigma}}$. Since the fixed points of a finite order automorphism on a smooth variety form a smooth subvariety, we see in particular that X is smooth.

Remark. Note, $g \in G^{\tilde{\sigma}}$ if and only if $\mathrm{int}(g) \circ \sigma$ is an involution. If we identify G with the connected component $G(\sigma)$ of Aut (G) containing σ via the map $g \mapsto \mathrm{int}(g)\circ\sigma$, then $G^{\tilde{\sigma}}$ can be identified with the set of involutions in $G(\sigma)$. It is now easy to see each connected component of $G^{\tilde{\sigma}}$ is an orbit $G.\tau$ isomorphic to the symmetric space G/G^{τ}, and $X_{\tau} := \overline{G.\tau} \subset \overline{G}$ is the minimal wonderful compactification of the symmetric space G/G^{τ}.

3.4. The variety X has the following "wonderful" properties ([9], §3): X is smooth and contains exactly 2^l G–orbits. Further, the complement in X of the open orbit G/H is the union of l smooth divisors $S_1, \ldots, S_l$, which intersect transversely. The closure of any orbit in X is obtained as the intersection of the codimension one orbit closures containing the orbit. Finally, there exists only one closed orbit in X, namely $S_1 \cap \ldots \cap S_l$.

3.5. To describe the orbits and the stabilizers we have to introduce first some notation.

Since T is σ–stable, σ operates on the root system $\Phi = \Phi(G,T)$ of G. Let Φ_0 be the set of roots fixed by σ and let Φ_1 be the set of roots moved by σ. Choose a set of positive roots Φ^+ such that for $\Phi_1^+ := \Phi_1 \cap \Phi^+$ we have $\sigma(\Phi_1^+) = -\Phi_1^+$ and $\Phi_1 = \Phi_1^+ \cup -\Phi_1^+$. Denote by Δ the set of simple roots of Φ^+, and decompose Δ into $\Delta_0 \cup \Delta_1$ such that $\Delta_0 \subset \Phi_0$ and $\Delta_1 \subset \Phi_1$. Let $r : X(T) \to X(T^1)$ be the restriction map from the character group of T to the character group of T^1. Then $r(\Phi_1)$ gives rise to a root system $\overline{\Phi}_1$, called the restricted root system, with basis $\Sigma := r(\Delta_1)$. We denote by W the Weyl group of Φ and by W^1 the Weyl group of $\overline{\Phi}_1$. Finally, let B be the Borel subgroup of G corresponding to the choice of Φ^+.

3.6. The following results can be found in [10], §5:
Let S be the torus $S = T^1/T^1 \cap H$. The characters 2γ, $\gamma \in \Sigma$, are a basis of the character group $X(S)$. The closure $\overline{S}$ of S in X is a smooth S–embedding with a canonical action of W^1. The rational partial polyhedral decomposition (r.p.p.d) of $\mathrm{Hom}_{\mathbf{Z}}(X(T^1), \mathbf{R})$ associated to $\overline{S}$ consists of the Weyl chambers of $\overline{\Phi}_1$.

The negative Weyl chamber of this decomposition corresponds to a S–stable affine chart of $\overline{S}$, which can be identified with $\mathbf{A}^l$ in a canonical way. The embedding of S into $\mathbf{A}^l$ is given by

$$t \mapsto (t^{-2\gamma_1}, \dots, t^{-2\gamma_l}), \quad \text{where } \Sigma = \{\gamma_1, \dots, \gamma_l\}.$$

This affine chart is a fundamental domain for the action of W^1 on $\overline{S}$, and two elements x and y in this affine chart are conjugate under G if and only if they are conjugate under S.

For every subset Γ of Σ, denote by x_Γ the following element of $\mathbf{A}^l \subset X$:

$$x_\Gamma = (x_1, \dots, x_l), \quad x_i = 1 \text{ if } \gamma_i \in \Gamma \text{ and } x_i = 0 \text{ if } \gamma_i \notin \Gamma.$$

Then X is the disjoint union of the 2^l orbits $G.x_\Gamma$, where Γ runs over all possible subsets of Σ.

3.7. The description of the stabilizer in G of the point $x_\Gamma \in X$ we will give now can be found in [9], §5.

For $\Gamma \subset \Sigma$ denote by $P_\Gamma \supset B$ the parabolic subgroup of G associated to the set of simple roots $\Delta_\Gamma := \{\alpha \in \Delta \mid r(\alpha) = 0 \text{ or } r(\alpha) \in \Gamma\}$, and let L_Γ denote the Levi factor of P_Γ containing T. Note that L_Γ is σ–stable.

To avoid too many indices, set $P = P_\Gamma$ and $L = L_\Gamma$. Let C be the center of L and denote by C^1 its connected component containing the identity. Note that $C^1 \subset T^1$, so C^1 is a σ–split torus. And let P^u be the unipotent radical of P. The stabilizer in G of the point $x = x_\Gamma$ is

$$G_x = L^\sigma.C^1.P^u.$$

Note that $P = \mathrm{Nor}_G G_x^u$, the normalizer of the unipotent radical of G_x. So we can associate to any point $y \in X$ in a canonical way the parabolic subgroup $P(y) := \mathrm{Nor}_G G_y^u$.

Remark. 1) As an immediate consequence we see the orbit $G.x$ has a T–fixed point if and only if (L, L) contains a maximal torus contained in $(L, L)^\sigma$, i.e. the restriction of σ to (L, L) is an inner automorphism. This can be easily read off the Satake diagram corresponding to σ. (See for example the tables in [13]).

2) For $z \in X$ let l_z be the Lie algebra of the kernel of the representation of the stabilizer G_z on N_z, the normal space of the orbit $G.z$ in X at the point z. Then for the embedding of X in $Grass_m\mathcal{G}$ (see Remark 3.2), the image of z is the subspace l_z of $\mathcal{G}$ [6],§13. For the point $x = x_\Gamma$ is $l_x = \text{Lie } L^\sigma \oplus \text{Lie } P^u$.

3.8. For $L := L_\Gamma$ the subgroups $L_c := L \cap K$ and $C_c^1 := C^1 \cap K$ are maximal compact subgroups of L and C^1.

The stabilizer of $x = x_\Gamma$ in K is $K_x = L_c^\sigma.C_c^1$ ([6], §13).

3.9. The local structure of X around a T–fixed point $z \in \overline{S}$ can be described as follows (see [9], 2.3):

Let A be the affine chart of $\overline{S}$ corresponding to z, i.e. $A = \{y \in \overline{S} | z \in \overline{T.y}\}$. The stabilizer G_z is a parabolic subgroup of G. Let P^- be a parabolic subgroup of G opposite to G_z such that $T \subset G_z \cap P^-$. If L is the Levi part of P^- containing T, then the semisimple part of L acts trivially on A.

Let U^- be the unipotent radical of P^-. The canonical map

$$U^- \times A(z) \to X$$

is a P^-–equivariant isomorphism onto an open affine neighborhood of z.

4. Orbits with T–fixed points and T–stable subvarieties

4.0. We want now to use the construction of X as a subvariety of $\overline{G}$ to construct a canonical set of smooth T–stable projective torus embeddings, which contains all T–fixed points in X. We use throughout the notation $g.x = gx\sigma(g)^{-1}$ for the twisted action.

4.1. Let N_T be the normalizer of T in G, which is a σ–stable subgroup. The closure $\overline{N_T}$ of N_T in $\overline{G}$ is a $T \times T$–stable variety. In fact, let us write

$$N_T = \bigcup_{w \in W} Tn_w,$$

where n_w is a representative of $w \in W$. Consider $\overline{N_T} = \bigcup_{w \in W} \overline{Tn_w}$. Of course, $\overline{Tn_w}$ is isomorphic to $\overline{T}$, but the $T \times T$ action is twisted by w. I.e. there is an isomorphism $\psi : \overline{T} \to \overline{Tn_w}$ with $\psi((t_1, t_2)p) = t_1 p t_2^{-1} n_w = t_1 p n_w w^{-1}(t_2)^{-1}$. In any case, $\overline{Tn_w}$ contains the fixed points $(\overline{T}^{T \times T}) n_w$; the $T \times T$–fixed points lie all in the closed orbit $G/B \times G/B$ (cf. §2) and can be identified with $W \times W$. The fixed points $(\overline{T}^{T \times T})$ can be identified with $d(W)$, the image of the diagonal embedding of W into $W \times W$, and hence $W \times W = \bigcup_{w \in W} d(W)(1, w)$.

It follows that the varieties $\overline{Tn_w}$ are disjoint, otherwise the intersection of two of them should contain $T \times T$–fixed points. We have thus proved:

Proposition. *$\overline{N_T}$ is smooth and the disjoint union of $|W|$ irreducible varieties isomorphic to $\overline{T}$ (with the appropriate twist for the $T \times T$ action).*

Since N_T is σ–stable, its closure $\overline{N_T}$ is $\tilde{\sigma}$–stable and $(\overline{N_T})^{\tilde{\sigma}}$ is stable under the action of N_T (acting on $\overline{G}$ via the twisted action as a subgroup of G). Since

$$\mathcal{N}_T := X \cap (\overline{N_T})^{\tilde{\sigma}}$$

is an open subset of $(\overline{N_T})^{\tilde{\sigma}}$ by 3.3, we get

Proposition 4.2. *$\mathcal{N}_T$ is a N_T–stable disjoint union of smooth T–stable subvarieties of X.*

4.3. We will now investigate the irreducible components of $\mathcal{N}_T$. Let x denote the point x_Γ, $\Gamma \subset \Sigma$, in $\overline{S}$ (3.6). For the rest of this paragraph we fix Γ such that the orbit $G.x$ contains T–fixed points. Recall that this is equivalent to assume that for $L = L_\Gamma$ the derived subgroup (L, L) contains a maximal torus fixed by σ (Remark 3.7.1). As before we denote by L_c and C_c^1 the intersections of L and C^1 with the compact group K.

To prove all T–fixed points in X are contained in $\mathcal{N}_T$, we will choose a T–fixed point y in the orbit $G.x$, and describe the irreducible component of $\mathcal{N}_T$ containing y.

Let M be a maximal torus in L such that the intersection $M \cap (L_c, L_c)$ is a maximal torus in (L_c, L_c) fixed by σ, so $M \subset G_x$. If $k \in L_c$ is such that $kTk^{-1} = M$, then the point $y = k^{-1}.x$ is a T–fixed point.

And, since $M = kTk^{-1}$ is σ–stable by the choice of M, the element $n := k^{-1}\sigma(k)$ normalizes T. In fact, n is in the image of the embedding $G/H \hookrightarrow G$ (see 3.2), so we can view n as a point in $\mathcal{N}_T \subset X$. Denote by $\mathcal{T}$ the irreducible component of $\mathcal{N}_T$ containing n.

Note that $\phi := \mathrm{int}(n) \circ \sigma$ is an involution which stabilizes T. Moreover, since $K_x = L_c^\sigma . C_c^1$ we see $K_y = L_c^\phi . C_c^1$.

Concerning the choices made above, note that $y, \mathcal{T}, n$ and ϕ do not depend on the choice of M. And since $k'Tk'^{-1} = M$ implies $k^{-1}k'$ normalizes T, the possible choices for y and $\mathcal{T}$ are conjugate under the Weyl group of L, and the corresponding choices for n and ϕ are conjugate under the normalizer of T in L_c.

Fix a 1-parameter subgroup $\lambda : k^* \rightarrow T$ contained in the dominant Weyl chamber. It defines a Bialynicki–Birula cell decomposition in $\mathcal{N}_T$ and X. For a T-fixed point z in $\mathcal{T}$ denote by $\mathcal{C}_z$ the corresponding cell in X (i.e. $\mathcal{C}_z$ is the set of points p in X such that $\lim_{t\rightarrow 0} \lambda(t).p = z$) and by $\mathcal{C}'_z$ the corresponding cell in $\mathcal{T}$.

If α is a root, then we denote by U_α the T-stable unipotent subgroup of G such that Lie $U_\alpha = \mathcal{G}_\alpha$ is the T-eigenspace in $\mathcal{G}$ corresponding to α.

Recall, we denote by $P(z)$ the parabolic subgroup associated to the point $z \in X$ (see 3.7).

Proposition 4.4.

1) $\mathcal{T}$ is the closure of the orbit $C^1.n = k^{-1}.(C^1.id)$ and $y = k^{-1}.x$ is a T-fixed point in $\mathcal{T}$.

2) $\mathcal{T}$ is a smooth torus embedding for the torus $R := C^1/C^1 \cap H$ and is R-equivariantly isomorphic to $\overline{R}$, the closure of R in $\overline{S}$.

3) Let z be a T-fixed point in $\mathcal{T}$.

a) The intersection of the orbit $G.z$ with $\mathcal{T}$ is a union of T-fixed points permuted transitively by the normalizer of T in K. Moreover, the intersection is transversal at the point z.

b) Let $A(z)$ be the open T-stable affine chart of $\mathcal{T}$ containing z (3.9). The intersection of a G-orbit with $A(z)$ is either empty or one T-orbit.

c) L is a Levi part of $P(z)$, and the stabilizer of z admitts a decomposition $G_z = L^\phi.C^1.P(z)^u$. In particular, the compact stabilizer $K_z = L^\phi_c.C^1_c$ does not depend on the choice of z.

d) Let $\alpha_1, \ldots, \alpha_s$ be the set of positive roots such that $\mathcal{G}_{\alpha_i}$ is not contained in $\mathcal{G}_z$ and denote by $\mathcal{U}$ the product $\prod_{i=1}^{s} U_{\alpha_i}$. The multiplication map $\mu : \mathcal{U} \times \mathcal{C}'_z \rightarrow X$ gives an isomorphism of $\mathcal{U} \times \mathcal{C}'_z$ with the cell $\mathcal{C}_z$.

4) Every B-orbit in X meets $\mathcal{N}_T$ in one T-orbit.

5) The set of T-fixed points in X is included in $\mathcal{N}_T$.

Proof. Since $\sigma(n) = \sigma(k^{-1})k = n^{-1}$, the irreducible component $\overline{Tn}$ of $\overline{\mathcal{N}_T}$ containing n is stable under $\tilde{\sigma}$. Using the $G \times G$-action on $\overline{G}$ we see further that $\overline{Tn} = \overline{T}n = n\overline{T}$. But since $\sigma(n) = n^{-1}$ we know $\tilde{\sigma}(zn) = n\tilde{\sigma}(z)$ for $z \in \overline{G}$. So for $z \in \overline{T}$ one has $zn \in (\overline{T}n)^{\tilde{\sigma}}$ if and only if $n\tilde{\sigma}(z)n^{-1} = z$, and for $t \in T$ one has $tn \in (Tn)^{\tilde{\sigma}}$ if and only if $n\sigma(t)n^{-1} = t^{-1}$.

For $\phi = \mathrm{int}(n) \circ \sigma$ denote by $T^{1,\phi}$ the connected component containing the identity of the ϕ-split part of T. Note that $\phi(t) = t^{-1}$ if and only if $\sigma(ktk^{-1}) = kt^{-1}k^{-1}$. So $t \in T^{1,\phi}$ if and only if ktk^{-1} is a σ-split element in M. But by the choice of M this implies that $ktk^{-1} \in C^1$, so $ktk^{-1} = t$ and $T^{1,\phi} = C^1$.

Since $\mathcal{T}$ is the irreducible component of $(\overline{T}n)^{\tilde{\sigma}}$ containing n, the discussion above shows that dim $\mathcal{T}$ = dim $T^{1,\phi}$ = dim C^1. Moreover, by the definition of the twisted action is $t.n = n$ if and only if $t = n\sigma(t)n^{-1} = \phi(t)$, so $T_n = T^\phi$. Hence $\dim C^1.n = \dim \mathcal{T}$, which implies $\mathcal{T} = \overline{C^1.n}$. But k commutes with C^1, so $C^1.n = k^{-1}.(C^1.id) = k^{-1}.R$, where R denotes the torus $C^1/C^1 \cap H \subset S = T^1/T^1 \cap H$. We have thus proved

$$\mathcal{T} = \overline{T.n} = \overline{C^1.n} = k^{-1}.\overline{C^1.id} = k^{-1}.\overline{R}.$$

To see that $y = k^{-1}.x$ is contained in $\mathcal{T}$ we consider the action of C^1 on the affine chart A_{id} of $\overline{S}$ corresponding to the negative Weyl chamber $\mathcal{C}$ (3.6). Recall that A_{id} can be identified with $\mathbf{A}^l$ in a canonical way such that the point $id \in S \subset G/H$ corresponds to the point $p = (p_1, \dots, p_l)$, $p_i = 1$ for $i = 1, \dots, l$. Since C^1 is the connected component of $\bigcap_{\gamma_i \in \Gamma} \operatorname{Ker} 2\gamma_i$, the point $x = (x_1, \dots, x_l)$ with $x_i = 1$ for $\gamma_i \in \Gamma$ and $x_i = 0$ for $\gamma_i \in \Sigma - \Gamma$ is in the closure of $C^1.id$ in A_{id}. So $x \in \overline{R}$ and hence $y \in \mathcal{T}$.

As an immediate consequence we see that all T–fixed points in X are contained in $\mathcal{N}_T$, so it remains to prove 3) and 4).

Recall that T^ϕ operates trivially on $\mathcal{T}$ and k commutes with C^1, so there is a canonical bijection between the C^1–orbits in $\overline{R}$ and the T–orbits in $\mathcal{T}$. Hence, to prove 3) it suffices to prove the corresponding statements for the C^1–orbits in $\overline{R}$.

Denote by A_w, $w \in W^1$, the affine chart of $\overline{S}$ corresponding to the Weyl chamber $w\mathcal{C}$. Then, by the description of the action of T^1 on A_w in 3.6, $\overline{R}$ contains a C^1–fixed point z in A_w if and only if there exists a subset $\Lambda \subset \Sigma$ such that C^1 is the connected component of $\bigcap_{\gamma_i \in \Lambda} \operatorname{Ker} 2w(\gamma_i)$. But this means $C^1 = wC^1_\Lambda w^{-1}$, where C^1_Λ is the connected component of the σ–split part of the center of L_Λ, and hence $wL_\Lambda w^{-1} = L$. Moreover, if we identify A_w again with $\mathbf{A}^l$, then the unique C^1–fixed point in the closure of $C^1.id = C^1.(1, \dots, 1)$ in $\mathbf{A}^l$ is the point $z = (z_1, \dots, z_l)$, where $z_i = 0$ if $\gamma_i \in \Sigma - \Lambda$ and $z_i = 1$ if $\gamma_i \in \Lambda$. Hence we have $z = w.x_\Lambda$ and we obtain:

$$K_z = wL^\sigma_{\Lambda,c}C^1_{\Lambda,c}w^{-1} = L^\sigma_c C^1_c, \quad G_z = w(L^\sigma_\Lambda C^1_\Lambda P^u_\Lambda)w^{-1} = L^\sigma C^1 P(z)^u,$$

which proves c). Note further that $z = wx_\Lambda$ and $C^1 = wC^1_\Lambda w^{-1}$ implies $\dim T.x = \dim T.z$ and hence by 3.9 also $\dim G.x = \dim G.z$.

The affine chart $A(z)$ of $\overline{R}$ corresponding to the fixed point z can be identified in $\mathbf{A}^l$ with the set of points $p = (p_1, \dots, p_l)$ with $p_i = 1$ if $\gamma_i \in \Lambda$. By this description it is clear that the intersection of a T–orbit in A_w with $A(z)$ is either empty or one C^1–orbit. But since the intersection of a G–orbit with A_w is one T–orbit, this proves b).

Moreover, we see that the tangent space of A_w at z admitts a decomposition $T_z A_w = T_z T.z \oplus T_z A(z)$. But by 3.9 we have a decomposition

$T_zX = V \oplus T_zA_w$ such that $T_zG.z = V \oplus T_zT.z$. Hence the intersection of $G.z$ with $\overline{R}$ is transversal at the point z.

Let z' be an element in the intersection of $G.z$ with $\overline{R}$. The orbit $C^1.z'$ contains a C^1–fixed point z'' in its closure. We have already seen $\dim G.x = \dim G.z = \dim G.z''$, so $z'' \in G.z$. On the other hand is $G.z' \cap A(z'') = C^1.z'$ by b), so $z' = z''$.

Now let z be a T–fixed point in $\mathcal{T}$ and let z' be an element of $\mathcal{T} \cap G.z$. Since z' is a T–fixed point there exists an element $k \in K$ such that $k.z = z'$ and k normalizes T (and hence $k.\mathcal{N}_T \subset \mathcal{N}_T$). But $k.\mathcal{T} \cap \mathcal{T} \neq$ implies $k.\mathcal{T} \subset \mathcal{T}$, so k normalizes $\mathcal{T}$, which finishes the proof of a).

To prove d) note first that $\mathcal{C}'_z \subset \mathcal{C}_z$ and $\mathcal{C}_z$ is B–stable by the choice of λ, so the image of μ is contained in $\mathcal{C}_z$. Now the tangent space T_zX of X at z has a T–stable decomposition $T_zX = T_z\mathcal{T} \oplus T_zG.z$ by a). Moreover, if we denote by V^+ the set of elements in a representation space of T such that $\lim_{t\to 0} \lambda(t).v = 0$, then we have the corresponding T–stable decomposition $T_z^+X = T_z^+\mathcal{T} \oplus T_z^+G.z$, and $T_z^+X = T_z\mathcal{C}_z$, $T_z^+\mathcal{T} = T_z\mathcal{C}'_z$. By the construction of $\mathcal{U}$ the differential of the orbit map $\rho : \mathcal{U} \to G.z$, $\rho(u) := u.z$, is an isomorphism at id of $T_{id}\mathcal{U}$ with $T_z^+G.z$. So the differential of μ is an isomorphism at the point (id, z) of the tangent space of $\mathcal{U} \times \mathcal{C}'_z$ with the tangent space of $\mathcal{C}_z$ at z.

Moreover, the action of the 1–parameter subgroup λ on $\mathcal{U}$ as well as on $\mathcal{C}_z$ and $\mathcal{C}'_z$ are linearizable (see [3]). I.e., for the one–dimensional torus $\mathbf{G}_m \simeq k^*$ there exist representations $\rho_i : \mathbf{G}_m \to Gl(U_i)$, $i = 1, 2$, and $\mathbf{G}_m$–equivariant isomorphisms $U_1 \simeq \mathcal{U} \times \mathcal{C}'_z$ and $U_2 \simeq \mathcal{C}_z$. Note that the $\mathbf{G}_m$–equivariance implies that $\lim_{t\to 0} \rho_i(t)u_i = 0$ for any $u_i \in U_i$, $i = 1, 2$. Further, the morphism μ induces a $\mathbf{G}_m$–equivariant morphism $\tilde{\mu} : U_1 \to U_2$ such that $\tilde{\mu}(0) = 0$ and the differential $d\tilde{\mu}$ induces an isomorphism of the tangent spaces at 0. Now Lemma 4.4 below implies that $\tilde{\mu}$, and hence μ, is an isomorphism.

To conclude the proof of Proposition 4.4 it remains to prove 4). Every B–orbit in X is contained in a Bialynicki–Birula cell, so by 3d) every B–orbit meets $\mathcal{N}_T$. Let now $\mathcal{T}$ be an irreducible component of $\mathcal{N}_T$ and fix an element $z \in \mathcal{T}$. By 3d) there exists a T–fixed point $z' \in \mathcal{T}$ such that $B.z$ is contained in the cell $\mathcal{C}_{z'}$. In particular, $z' \in \overline{T.b.z}$ for any $b \in B$. So $B.z \cup \mathcal{T}$ is contained in the affine chart $A(z')$ of $\mathcal{T}$ and is hence one T–orbit by 3b).

Q.E.D.

Lemma 4.4. ([15], p.121) *Let $\rho_i : \mathbf{G}_m \to Gl(U_i)$, $i = 1, 2$, be two representations of the one–dimensional torus $\mathbf{G}_m \simeq k^*$ such that*

$$\lim_{t\to 0} \rho_i(t)u_i = 0$$

for any $u_i \in U_i$, $i = 1, 2$. If $\tilde{\mu} : U_1 \to U_2$ is a $\mathbf{G}_m$-equivariant morphism such that $\tilde{\mu}(0) = 0$ and the differential $\mathrm{d}\tilde{\mu}$ induces an isomorphism of the tangent spaces at 0, then $\tilde{\mu}$ is an isomorphism.

Q.E.D.

5. The subvariety Z of X

5.0. Let Γ be a subset of Σ and denote by $L = L_\Gamma$ the corresponding (σ–stable) Levi subgroup of P_Γ containing T. For a Levi subgroup of this type the variety

$$Z := \overline{L.id} \subset X$$

has been investigated in [1]:

Z is a smooth compactification of the variety L/L^σ. Every L–orbit in Z meets $\overline{S}$ in a union of S–orbits, which are permuted transitively by $W^1(\Gamma)$, the subgroup of W^1 generated by the simple reflections s_γ, $\gamma \in \Gamma$. The intersection of a G–orbit with Z is a finite union of L–orbits. For $z \in Z$ the L–orbit $L.z$ is closed if and only if the orbit $G.z$ is closed in X. The closure of a L–orbit in Z is smooth, and it is the transversal intersection of the codimension one orbit closures containing it.

Furthermore, let $z \in \overline{S}$ be a T–fixed point and let $A = \{y \in \overline{S} | z \in \overline{T.y}\}$ be the corresponding affine chart of $\overline{S}$. The stabilizer L_z is a parabolic subgroup of L. Let P^- be a parabolic subgroup of L opposite to L_z such that $T \subset P^- \cap L_z$, and let U^- be the unipotent radical of P^-. The canonical morphism

$$U^- \times A(z) \longrightarrow Z$$

is a P^-–equivariant isomorphism onto an open affine neighborhood of z.

5.1. Let C be the center of L and denote by C^1 its connected component containing the identity. Consider the semisimple group of adjoint type $L' := L/C$. Since L and C are σ–stable, we obtain an induced involution of L' which we denote also by σ. Let X_L be the minimal wonderful compactification of L'/L'^σ.

Proposition 5.2. *The canonical map $\pi : L/L^\sigma \to L'/L'^\sigma$ extends to a L-equivariant morphism $\pi : Z \to X_L$.*

Proof. To define π we use the realization of X and X_L as subvarieties of $Grass_m(\mathcal{G})$ and $Grass_{m'}(\mathcal{L}')$, where $\mathcal{L}' :=$ Lie L' and $m' =$ dim L'^σ (see Remark 3.2 and Remark 3.7.2).

We will see, Z is contained in the subset U of all subspaces $V \in Grass_m\mathcal{G}$, such that dim $(V \cap \mathcal{L}') = m'$. By L–equivariance and semicontinuity, it suffices to check this for a T–fixed point z in $\overline{S}$. Let Q be the stabilizer in G of z, which is a parabolic subgroup of G. Then $Q_L := Q \cap L$, the stabilizer in L of z, is a parabolic subgroup of L (since $T \subset L \cap Q$).

Denote by l_z the subspace of $\mathcal{G}$ corresponding to the point z. By Remark 3.7.2 is Lie Q = Lie $T^1 \oplus l_z$. Since Lie $C^1 \subset$ Lie T^1, we get Lie Q_L = Lie $T^1 \oplus (\mathcal{L}' \cap l_z)$. Furthermore, by the description of the local structure of Z in 5.0, we know T^1 has a dense orbit in the normal space N_z to the orbit $L.z \subset Z$ at the point z, and dim N_z = dim T^1. This implies dim (Lie Q_L/Lie T^1) = dim $\mathcal{L}^\sigma$. But $\mathcal{L}^\sigma = \mathcal{L}'^\sigma$, so dim $(l_z \cap \mathcal{L}')$ = dim $\mathcal{L}'^\sigma$.

But U can be considered in a canonical way as a fibre space with base space $Grass_{m'}\mathcal{L}'$ and fibre map $f : V \mapsto V \cap \mathcal{L}'$. Since the restriction of f to the open orbit in Z is the morphism $L/L^\sigma \to L'/L'^\sigma$, the restriction of f to Z is the desired extension.

Q.E.D.

5.3. If $z \in \pi^{-1}(\pi(\overline{S}))$, then we can find an element $z' \in \overline{S}$ such that $\pi(z') = \pi(z)$ and $l.z' = z$ for some $l \in L$. But this implies $l \in L_{\pi(z)}$. So $l = l_1.t$ with $l_1 \in L_z$ and $t \in C$, which implies $z \in \overline{S}$. Hence:

Lemma. $\pi^{-1}(\pi(\overline{S})) = \overline{S}$.

5.4. Note that the image T' of T in L' is a σ–stable maximal torus such that T'^1 is the image of T^1, and T'^1 is a maximal σ–split torus in L'. Furthermore, $\pi(\overline{S})$ is the closure of $T'^1/T'^1 \cap L'^\sigma$ in X_L.

In [6], Theorem 27, is proved $K.\overline{S} = X$. Applying this to the wonderful compactification X_L and $\pi(\overline{S})$, we obtain as an immediate consequence of the lemma above:

Corollary. $L_c.\overline{S} = Z$.

5.5. Throughout the rest of this paragraph we will again assume that $\Gamma \subset \Sigma$ is such that (L, L) contains a maximal torus contained in $(L, L)^\sigma$. We use the same notation as in §4, i.e. x denotes the point x_Γ, R is the image of C^1 in S and $\overline{R}$ denotes the closure of R in $\overline{S}$.

Consider the map $\pi : Z \to X_L$. The fibre over the point $id \in X_L$ is $\overline{R}$. By assumption the open orbit in X_L contains T–fixed points. We know by 4.3 that the T–fixed points in L'/L'^σ are contained in $L_c.id \subset X_L$. Let $k \in L_c$ be such that $k.id$ is a T–fixed point in X_L. Since x is a C^1–fixed point in the fibre $\overline{R}$ over the point $id \in X_L$, the point $y = k.x$ is a T–fixed point in the fibre $\mathcal{T} = k.\overline{R}$ over the point $k.id$. In fact, $\mathcal{T}$ is the irreducible component of $\mathcal{N}_T$ considered in 4.3: $k.id \in X_L$ is a T–fixed point if an only if $M = k^{-1}Tk$ is a maximal torus such that the intersection $M \cap (L_c, L_c)$ is a maximal torus in $(L_c, L_c)^\sigma$. So $n = k.id \in Z$ is contained in the T–stable fibre $\mathcal{T} = k.\overline{R}$ of π over the point $k.id \in X_L$, which implies $\mathcal{T}$ is the irreducible component of $\mathcal{N}_T$ containing n.

Denote by $Z^0 \subset Z$ the preimage of the dense orbit in X_L. Since the stabilizer in L of the point $id \in X_L$ is L_x and in L_c is K_x, we obtain:

Proposition 5.6.

1) The irreducible component of $\mathcal{N}_T$ containing a T–fixed point $y \in Z^0$ is the fibre $\pi^{-1}(\pi(y))$ of $\pi : Z \to X_L$ over the point $\pi(y)$.

2) The canonical maps $L \times_{L_x} \overline{R} \longrightarrow Z^0$ and $L \times_{L_y} \mathcal{T} \longrightarrow Z^0$ are isomorphisms.

3) The canonical maps $L_c \times_{K_x} \overline{R} \longrightarrow L_c.\overline{R}$ and $L_c \times_{K_y} \mathcal{T} \longrightarrow L_c.\mathcal{T}$ are isomorphisms.

5.7. As a consequence we see the map $S \times_R \overline{R} \longrightarrow S.\overline{R}$ is an isomorphism onto an open subset of $\overline{S}$. This implies the following description of the r.p.p.d. of $\overline{R}$ (and $\mathcal{T}$) as a R–embedding:

Corollary. *The r.p.p.d. of* $\mathrm{Hom}_{\mathbf{Z}}(X(R), \mathbf{R})$ *corresponding to $\overline{R}$ is obtained by intersecting the r.p.p.d. corresponding to $\overline{S}$ with the image of*

$$\mathrm{Hom}_{\mathbf{Z}}(X(R), \mathbf{R}) \subset \mathrm{Hom}_{\mathbf{Z}}(X(S), \mathbf{R}).$$

In particular, the intersection of $\overline{R}$ with the open orbit G/H in X is one C^1–orbit, so the intersection of $\mathcal{T}$ with G/H is the open T–orbit in $\mathcal{T}$.

5.8. Consider the group $\mathrm{Nor}_H L$, the normalizer of L in H. Since $\mathrm{Nor}_H L$ normalizes C^1, $\mathrm{Nor}_H L$ operates on $\overline{R}$. If z is a C^1–fixed point in $\overline{R}$ and z' is contained in $G.z \cap \overline{R}$, then $z' = w.z$ for some $w \in W^1$ (see proof of Proposition 4.4). Since w can be represented by an element $h \in H$, and h normalizes $L_z = L^\sigma.C^1$, we see h normalizes L^σ and C^1, and hence L. So $h \in \mathrm{Nor}_H L$, and

$$G.z \cap \overline{R} = N_L.z, \text{ where } N_L := \mathrm{Nor}_H L/L^\sigma.$$

Furthermore, N_L is isomorphic to $\mathrm{Nor}_{W^1} R/\mathrm{Cen}_{W^1} R$, the quotient of the normalizer in W^1 of R by the centralizer in W^1 of R.

Proposition. *L^σ is a normal subgroup of* $\mathrm{Nor}_H L$ *of index* $\sharp|G.x \cap \overline{R}|$. *The quotient group* $N_L = \mathrm{Nor}_H L/L^\sigma$ *is isomorphic to* $\mathrm{Nor}_{W^1} R/\mathrm{Cen}_{W^1} R$. *Furthermore,* $G.z \cap \overline{R} = N_L.z$. *for all R–fixed points in $\overline{R}$.*

5.9 Let $\mathcal{T}$ be the irreducible component of $\mathcal{N}_T$ considered in 5.5. Two T–fixed points in $\mathcal{T}$ are in the same G–orbit if and only if they are conjugate under N_T. Since for $n \in N_T$ either

$$n.\mathcal{T} = \mathcal{T} \quad \text{or} \quad n.\mathcal{T} \cap \mathcal{T} = \emptyset,$$

we see the normalizer $\mathrm{Nor}_{N_T}\mathcal{T}$ operates transitively on the intersection $G.z \cap \mathcal{T}$ for a T–fixed point z in $\mathcal{T}$. Denote by $W_\mathcal{T}$ the subgroup $W_\mathcal{T} := \mathrm{Nor}_{N_T}\mathcal{T}/T$ of W, and let W_y be the stabilizer in W of y. By Proposition 4.4 and conjugation it is now easy to see:

Lemma. *The following quotient groups are isomorphic to N_L:*

$$L.\mathrm{Nor}_H L/L, \quad \mathrm{Nor}_K \overline{R}/K_x, \quad \mathrm{Nor}_K \mathcal{T}/K_y, \quad W_\mathcal{T}/W_y.$$

5.10. By abuse of notation we will denote all these quotient groups by N_L, it will be clear from the context the quotient of which groups we are considering.

For example, the group $\tilde{L} := L.\mathrm{Nor}_H L$ operates naturally on Z^0. By Proposition 5.7 (and 1.2, 1.4) we obtain the following natural isomorphisms in equivariant cohomology:

$$H^*_{\tilde{L}}(Z^0) \simeq H^*_L(Z^0)^{N_L} \simeq H^*_{L_x}(\overline{R})^{N_L} \simeq (H^*_{L^\sigma}(pt) \otimes H^*_R(\overline{R}))^{N_L}$$

$$H^*_{\tilde{L}}(Z^0) \simeq H^*_{L_y}(\mathcal{T})^{N_L} \simeq H^*_T(\mathcal{T})^{W_\mathcal{T}}.$$

5.11. We want now to study the relation between the T–fixed points in Z and the varieties $Z_\Lambda := \overline{L_\Lambda.id}$, where Λ is a subset of Γ. Let C_Λ be the center of L_Λ, denote by C^1_Λ its connected component containing the identity, and let L'_Λ be the quotient L_Λ/C_Λ. Denote by Z^0_Λ the preimage of the open orbit in X_Λ under the map $\pi_\Lambda : Z_\Lambda \to X_\Lambda$, where X_Λ is the minimal wonderful compactification of $L'_\Lambda/(L'_\Lambda)^\sigma$.

Lemma. *Let $z \in Z$ be a T-fixed point not contained in Z^0. Then there exists a subset Λ of Γ such that (L_Λ, L_Λ) contains a maximal torus contained in $(L_\Lambda, L_\Lambda)^\sigma$, and z is conjugate under the Weyl group of L to a T-fixed point in Z^0_Λ.*

Proof. Consider the map $\pi : Z \to X_L$. Then $\pi(z)$ is a T'-fixed point (where T' is the image of T in $L' = L/C$) not contained in the open L'-orbit in X_L. Note that we can consider Γ in a canonical way as a basis of the restricted root system of the symmetric space L'/L'^σ. So by Proposition 4.4 and 5.7 we know there exists a proper subset Λ of Γ such that the derived subgroup (L'_Λ, L'_Λ) of the Levi subgroup L'_Λ of L' contains a maximal torus contained in $(L'_\Lambda, L'_\Lambda)^\sigma$. And $\pi(z)$ is conjugate under the Weyl group of L' to a T'-fixed point in the open subset ${Z'_\Lambda}^0$ of $Z'_\Lambda := \overline{L'_\Lambda.id}$, where ${Z'_\Lambda}^0$ is the preimage of the open orbit in X_Λ under the map $Z'_\Lambda \to X_\Lambda$.

Now if we consider the Levi subgroup L_Λ of G, then (L_Λ, L_Λ) obviously contains a maximal torus contained in $(L_\Lambda, L_\Lambda)^\sigma$. Since the map $Z_\Lambda \to X_\Lambda$ factors into the map $\pi : Z_\Lambda \to Z'_\Lambda$ and $Z'_\Lambda \to X_\Lambda$, we see $\pi^{-1}({Z'_\Lambda}^0) = Z^0_\Lambda$.

Since $\pi(z)$ is conjugate under the Weyl group of L' to a point in ${Z'_\Lambda}^0$, it follows z is conjugate under the Weyl group of L to a point in Z^0_Λ.

Q.E.D.

6. The Weyl group orbits in the set of irreducible components of $\mathcal{N}_T$

6.0. Let us now study the action of the Weyl group on the set of irreducible components of $\mathcal{N}_T$.

In 5.7 we have seen the intersection of an irreducible component $\mathcal{T}$ of $\mathcal{N}_T$ with the open orbit G/H in X is the open T-orbit in $\mathcal{T}$. So we can identify the set of W-orbits in the set of irreducible components of $\mathcal{N}_T$ with the points in $\mathcal{N}^0_T/N_T$, where $\mathcal{N}^0_T := \mathcal{N}_T \cap G/H$.

6.1. We wish to recall various different interpretations of the set $\mathcal{N}^0_T$. Denote by $\mathcal{T}_G$ the set

$$\mathcal{T}_G := \{M \subset G \mid M \text{ is a maximal torus }, \sigma(M) = M\}.$$

Then H acts on $\mathcal{T}_G$ by conjugation.

For a maximal torus $M = gTg^{-1}$ is $M \in \mathcal{T}_G$ if and only if $g^{-1}\sigma(g)$ is an element in N_T. Let us introduce the set

$$P := \{g \in G \mid g^{-1}\sigma(g) \in N_T\}.$$

We have a canonical action of H on P by left multiplication and of N_T by right multiplication. The map $\rho : P \to \mathcal{N}_T^0$ given by $\rho(g) = g^{-1}\sigma(g)$ is the quotient under H. Its image is $\mathcal{N}_T^0$, and ρ is equivariant for the appropriate actions.

The map $\nu : P \to \mathcal{T}_G$ given by $\nu(g) = gTg^{-1}$ is a quotient under N_T and H-equivariant. So we can identify $\mathcal{T}_G$ with P/N_T, and $\mathcal{T}_G/H$ with $H\backslash P/N_T$ and $\mathcal{N}_T^0/N_T$.

6.2. As in Remark 3.4 let us identify G with $G(\sigma)$, the connected component of Aut G containing σ. We can identify the conjugacy class $\mathcal{A}_\sigma$ of σ with G/H, where

$$\mathcal{A}_\sigma := \{\phi \in \text{Aut } G \mid \phi = \text{int}(g) \circ \sigma \circ \text{int}(g)^{-1} = \text{int}(g\sigma(g^{-1})) \circ \sigma\}.$$

Let $\mathcal{A}_\sigma(T)$ be the set of ϕ in $\mathcal{A}_\sigma$ such that $\phi(T) = T$. Since $\phi = \text{int}(g\sigma(g)^{-1}) \circ \sigma$ is in $\mathcal{A}_\sigma(T)$ if and only if $g\sigma(g)^{-1} \in N_T$, we can identify $\mathcal{A}_\sigma(T)$ with $\mathcal{N}_T^0$.

Summarizing, we have the following identifications:

Proposition 6.3.

1) $\mathcal{A}_\sigma(T) = H\backslash P = \mathcal{N}_T^0$.
2) $\mathcal{T}_G/H = \mathcal{A}_\sigma(T)/N_T = H\backslash P/N_T = \mathcal{N}_T^0/N_T$.

6.4. Let $\mathcal{T}$ be an irreducible component of $\mathcal{N}_T$ and let $z \in \mathcal{T}$ be a T-fixed point. If $P(z)$ is the parabolic subgroup associated to z (see 3.7), then the Levi part L of $P(z)$ containing T is independent of the choice of $z \in \mathcal{T}^T$ by Proposition 4.4. So if we denote by (L) the Weyl group conjugacy class of L and by $(n) \in \mathcal{N}_T^0/N_T$ the class corresponding to the Weyl group orbit of $\mathcal{T}$, then we can associate in a canonical way to (n) the class (L). Denote by LT the set

$$LT := \{(L_\Gamma) | \Gamma \subset \Sigma,\ (L_\Gamma, L_\Gamma) \text{ contains a maximal torus contained in } (L_\Gamma, L_\Gamma)^\sigma\}.$$

By abuse of notation we will sometimes say $\Gamma \in LT$ or $L \in LT$ if $(L_\Gamma) \in LT$ or $(L) \in LT$.

The map $(n) \mapsto (L)$ defined above induces by Remark 3.7 and Proposition 4.4 a surjective map $\phi : \mathcal{N}_T^0/N_T \longrightarrow LT$.

Lemma. The map $\phi : \mathcal{N}_T^0/N_T \longrightarrow LT$ is bijective.

Proof. To see ϕ is injective, let $L = L_\Gamma$ be such that $\Gamma \in LT$, and let M be a maximal torus in L such that the intersection $M \cap (L_c, L_c)$ is a maximal torus in $(L_c, L_c)^\sigma$. And choose $k \in L_c$ such that $M = kTk^{-1}$. We have seen in 4.3 that $n_\Gamma := k^{-1}\sigma(k)$ is an element of $\mathcal{N}_T^0$, and the possible choices of n_Γ are all conjugate under the normalizer $\mathrm{Nor}_{L_c}T$ of T in L_c. So we can canonically associate to L the class (n_Γ) of n_Γ in $\mathcal{N}_T^0/N_T$.

The following Lemma shows that if $\Lambda \in LT$ is such that $(L_\Lambda) = (L)$, then L and L_Λ are conjugate under $N_T \cap H$. But this implies the map which associates to L the class (n_Γ) defined above, induces a well defined map $\psi : LT \to \mathcal{N}_T^0/N_T$.

Note that the image of (n_Γ) in LT under ϕ is (L) by Proposition 4.4, so ψ is the inverse map to ϕ, and hence $\phi : \mathcal{N}_T^0/N_T \to LT$ is a bijective map.

Q.E.D.

Lemma 6.5. *Let $\Gamma, \Gamma' \in LT$. Then $(L_\Gamma) = (L_{\Gamma'})$ if and only if there exists an element $w \in W^1$ such that $w(\Gamma') = \Gamma$.*

Proof. Set $V = X(T) \otimes_{\mathbf{Z}} \mathbf{R}; V^1 = X(T^1) \otimes_{\mathbf{Z}} \mathbf{R}$ and let $\mathcal{C}$ and $\mathcal{C}^1$ denote the dominant Weyl chambers in V and V^1.

Assume there exists a $w \in W$ such that $wL_\Gamma w^{-1} = L_{\Gamma'}$. Denote by Δ_Γ the set of roots in Δ such that $r(\alpha) = 0$ or $r(\alpha) \in \Gamma$. Let F be the face of $\mathcal{C}$ defined by

$$F = \{v \in V \mid \alpha(v) > 0 \text{ for } \alpha \in \Delta_\Gamma, \alpha(v) = 0 \text{ for } \alpha \in \Delta - \Delta_\Gamma\}.$$

The faces $F^1 = F \cap V^1$ and $F_w^1 = wF \cap V^1$ are not empty. By Lemma 2 in [1] there exists a $w_1 \in W^1$ such that $w_1 F_w^1 = F^1$. And hence there exists a $w' \in W^1$ such that $w'w_1^{-1}(\Gamma') = \Gamma$.

Now assume there exists a $w_1 \in W^1$ such that $w_1(\Gamma) = \Gamma'$. Recall,

$$W^1 = \{w \in W \mid w(T^1) = T^1\}/\{w \in W \mid w(t) = t \ \forall t \in T^1\}.$$

If $w \in W$ is a representative of w_1, then obviously $wL_\Gamma w^{-1} = L_{\Gamma'}$.

Q.E.D.

6.6. Denote by O_T the set of all G–orbits in X containing T–fixed points. We associate to any T–fixed point z the Levi part $L(z)$ of $P(z)$ containing T and introduce on O_T the following relation:

$$G.z \sim G.z', \quad z, z' \in X^T, \quad \Longleftrightarrow \quad (L(z)) = (L(z')).$$

The bijection between LT and $\mathcal{N}_T^0/N_T$ implies now $G.z \sim G.z'$ if and only if $W.z$ and $W.z'$ meet the same irreducible components of $\mathcal{N}_T$.

Summarizing, we obtain:

Proposition 6.7. *The following sets are in a bijective correspondence to each other:*

1) *The set of N_T-orbits in $\mathcal{N}_T^0$.*
2) *The set of Weyl group orbits in the set of irreducible components of $\mathcal{N}_T$.*
3) *The set of H-orbits in the set $\mathcal{T}_G$ of maximal σ-stable tori in G.*
4) *The set of N_T-orbits in*

$$\mathcal{A}_\sigma(T) := \{\phi \in \mathrm{Aut}\ G \mid \phi = \mathrm{int}(g) \circ \sigma \circ \mathrm{int}(g)^{-1}, \phi(T) = T\}.$$

5) *The set of $H \times N_T$-orbits in $P := \{g \in G \mid g^{-1}\sigma(g) \in N_T\}$.*
6) *The set LT of Weyl group conjugacy classes (L_Γ), where $\Gamma \subset \Sigma$ is such that (L_Γ, L_Γ) contains a maximal torus contained in $(L_\Gamma, L_\Gamma)^\sigma$.*
7) *The set $O_T/$'$\sim$', where O_T is the set of G-orbits in X with T-fixed points and '$\sim$' is the relation*

$$G.z \sim G.z', \quad z, z' \in X^T, \quad \Longleftrightarrow \quad (L(z)) = (L(z')).$$

Remark. Consider again $X^T = \mathcal{N}_T^T \subset \mathcal{N}_T \subset X$. Fix a 1-parameter subgroup in the interior of the dominant Weyl chamber. It defines a Bialynicki-Birula cell decomposition in the two varieties $\mathcal{N}_T$ and X. So for every point $p \in X^T$ we have two cells C_p in $\mathcal{N}_T$ and C_p' in X. Then there exists a unipotent subgroup U_p of the Borel subgroup such that $C_p' = U_p \times C_p$, furthermore the B-orbits of X meet $\mathcal{N}_T$ in T-orbits. This follows from [12], and it gives (with a precise analysis of the torus embeddings forming $\mathcal{N}_T$) a complete description of the B-orbits in X.

7. The ring A_X

7.0. Let us first recall the notation introduced in the preceding paragraphs. For $\Gamma \subset \Sigma$ denote by L_Γ the Levi subgroup of P_Γ containing T. And let C_Γ^1 be the connected component containing the identity of the center C_Γ of L_Γ.

We assume that Γ is such that the Weyl group conjugacy class (L_Γ) of L_Γ is an element of

$$LT = \{(L_\Lambda) | \Lambda \subset \Sigma, (L_\Lambda, L_\Lambda) \text{ contains a maximal torus contained in } (L_\Lambda, L_\Lambda)^\sigma\}.$$

Denote by X_Γ the minimal wonderful compactification of $L'_\Gamma/(L'_\Gamma)^\sigma$, where $L'_\Gamma = L_\Gamma/C_\Gamma$, and let Z_Γ be the variety $\overline{L_\Gamma.id}$ (5.0). Consider the morphism $\pi : Z_\Gamma \to X_\Gamma$ (5.2). The fibre over the point $id \in X_\Gamma$ is $\overline{R}_\Gamma$, the closure of the image R_Γ of C^1_Γ in $\overline{S}$. We denote by $Z^0_\Gamma \subset Z$ the preimage of the open orbit in X_Γ. Finally, the finite group $\mathrm{Nor}_H L_\Gamma/L^\sigma_\Gamma$ is denoted by N_Γ.

Fix a system of representatives $\{\Gamma_1, \ldots, \Gamma_r\}$ of the set LT. To simplify the notation, we will write $L_i, Z_i, \ldots$ instead of $L_{\Gamma_i}, Z_{\Gamma_i}, \ldots$

Consider the ring $A = \bigoplus_{i=1}^r A_i$ defined by

$$A := \bigoplus_{i=1}^{r} H^*_{L_i}(Z^0_i)^{N_i}.$$

By Proposition 4.4 and Proposition 6.7 we know every Weyl group orbit of a T-fixed point meets some of the Z^0_i. So by the Localization Theorem the inclusions $Z^0_i \hookrightarrow X$ induce an injection $H^*_G(X) \hookrightarrow A$.

Our aim will be to describe $H^*_G(X)$ as a subring of A.

7.1. Let $\Lambda \subset \Gamma_i$ be such that $\Lambda \in LT$. Since $L_\Lambda \subset L_i$ and $\overline{R}_i \supset \overline{R}_\Lambda$, the set $L_\Lambda.\overline{R}_i$ is included in Z^0_i and in Z^0_Λ.

If Γ_j is W^1-conjugate to Λ, then one can find a $h \in H$ such that $h.Z^0_j = Z^0_\Lambda$. Note that the corresponding isomorphism

$$H^*_{L_i}(Z^0_i)^{N_i} \simeq H^*_{L_\Lambda}(Z^0_\Lambda)^{N_\Lambda}$$

does not depend on the choice of $h \in H$, since we are considering only N_i-invariants.

Definition 7.2. A triple (i, j, Λ) denotes a subset Λ of Γ_i which is conjugate under W^1 to Γ_j. We associate to each triple (i, j, Λ) the morphisms

$$\Phi(i,j,\Lambda) : A_i \longrightarrow H^*_{L_\Lambda}(L_\Lambda.\overline{R}_i) \quad \text{and} \quad \Psi(i,j,\Lambda) : A_j \longrightarrow H^*_{L_\Lambda}(L_\Lambda.\overline{R}_i)$$

induced by the isomorphism $H^*_{L_j}(Z^0_j)^{N_j} \simeq H^*_{L_\Lambda}(Z^0_\Lambda)^{N_\Lambda}$ and the inclusions

$$Z^0_i \supset L_\Lambda.\overline{R}_i \subset Z^0_\Lambda.$$

7.3. For an element $\xi \in H^*_G(X)$ let $(a^\xi) = (a^\xi_1, \ldots, a^\xi_r)$ be its image in A. Since the morphisms Ψ and Φ are given by restrictions, we have

$$\Psi(i,j,\Lambda)(a^\xi_j) = \Phi(i,j,\Lambda)(a^\xi_i) \quad \text{for any triple} \quad (i,j,\Lambda).$$

This motivates the following

Definition 7.4. Let A_X be the subring of all elements $(a) = (a_1, \ldots, a_r)$ of A such that

$$\Psi(i,j,\Lambda)(a_j) = \Phi(i,j,\Lambda)(a_i) \quad \text{for any triple} \quad (i,j,\Lambda).$$

Remark. 1) Denote by x_i the point x_{Γ_i}. By 5.8 is $\mathrm{Nor}_{W^1} R_i/\mathrm{Cen}_{W^1} R_i$ isomorphic to N_i. And by the isomorphism $H^*_{L_i}(Z_i^0)^{N_i} \simeq H^*_{L_{x_i}}(\overline{R}_i)^{N_i}$ (see 5.10), we see if (i,j,Λ) and (i,j,Λ') are triples such that Λ and Λ' are conjugate under $\mathrm{Nor}_{W^1} R_i$, then

$$\Psi(i,j,\Lambda)(a_j) = \Phi(i,j,\Lambda)(a_i) \Longleftrightarrow \Psi(i,j,\Lambda')(a_j) = \Phi(i,j,\Lambda')(a_i).$$

So instead of considering in the definition of A_X all subsets of Γ_i which are conjugate to Γ_j, it suffices to consider a system of representatives of the $\mathrm{Nor}_{W^1} R_i$–orbits in the set of W^1–conjugates of Γ_j contained in Γ_i.

2) Since the morphisms Φ are defined by restrictions, one has furthermore the following compatibility: Let $\Lambda \subset \Gamma_i$ be W^1–conjugate to Γ_j and let $\Upsilon \subset \Lambda$ be W^1–conjugate to Γ_k. If $(a) \in A$ is an element such that

$$\Psi(i,j,\Lambda)(a_j) = \Phi(i,j,\Lambda)(a_i), \text{ and } \Psi(j,k,\Upsilon')(a_k) = \Phi(j,k,\Upsilon')(a_j)$$

for any triple (j,k,Υ'), then

$$\Psi(i,k,\Upsilon)(a_k) = \Phi(i,k,\Upsilon)(a_i).$$

7.5. We want to describe now the morphisms $\Phi(i,j,\Lambda)$ and $\Psi(i,j,\Lambda)$ more explicitly. Since for $x_i = x_{\Gamma_i}$ the subvariety Z_i^0 is isomorphic to $L_i \times_{L_{i,x_i}} \overline{R}_i$ by 5.6, we get

$$A_i \simeq (H^*_{L_i^\sigma}(pt) \otimes H^*_{R_i}(\overline{R}_i))^{N_i}.$$

Furthermore, $L_\Lambda.\overline{R}_i \simeq L_\Lambda \times_{L^\sigma_\Lambda . C^1_\Lambda} (C^1_\Lambda \times_{C^1_i} \overline{R}_i)$, so that

$$H^*_{L_\Lambda}(L_\Lambda.\overline{R}_i) \simeq H^*_{L^\sigma_\Lambda}(pt) \otimes H^*_{R_i}(\overline{R}_i).$$

The morphism $\Phi(i,j,\Lambda)$ can be now described as the restriction to the N_i–invariants of the tensor product of the map

$$H^*_{L_i^\sigma}(pt) \to H^*_{L^\sigma_\Lambda}(pt),$$

induced by the inclusion $L^\sigma_\Lambda \hookrightarrow L^\sigma_i$, with the identity map on $H^*_{R_i}(\overline{R}_i)$. And for

$$A_j \simeq (H^*_{L^\sigma_j}(pt) \otimes H^*_{R_j}(\overline{R}_j))^{N_j}$$

the map $\Psi(i,j,\Lambda)$ can be described as the restriction to the N_j-invariants of the tensor product of the isomorphism $H^*_{L^\sigma_j}(pt) \simeq H^*_{L^\sigma_\Lambda}(pt)$ with the morphism

$$H^*_{R_j}(\overline{R}_j) \longrightarrow H^*_{R_i}(\overline{R}_i)$$

induced by the isomorphism $\overline{R}_j \simeq \overline{R}_\Lambda$ and the inclusion $\overline{R}_i \subset \overline{R}_\Lambda$.

8. The equivariant cohomology of X

8.0. By the definition of the ring A_X we see the image of $H^*_G(X)$ in A is contained in A_X.

Theorem A. *The inclusion $H^*_G(X) \hookrightarrow A$ induces an isomorphism of rings*

$$H^*_G(X) \simeq A_X.$$

Example 1. If the closed orbit contains all T-fixed points in X, then $\Gamma_1 = \emptyset$ is a system of representatives for the set LT. Denote by L the corresponding Levi subgroup of G. The group N_1 is in this case W^1, so we obtain by Theorem A an isomorphism

$$H^*_G(X) \simeq (H^*_{L^\sigma}(pt) \otimes H^*_S(\overline{S}))^{W^1}.$$

To recover the result in §2 for the compactification $\overline{G}$, note that in this case the group L is $T \times T$, $L^\sigma \simeq T$, $\overline{S}$ is the closure of T in $\overline{G}$ and W^1 is isomorphic to the Weyl group W of G.

Example 2. Consider the minimal wonderful compactification X of the nondegenerate quadrics in $\mathbf{P}^3$. The group G is PSl_4 and the restricted root system $\overline{\Phi}_1$ is of type $\mathtt{A}_3$. Let $\gamma_1, \gamma_2, \gamma_3$ be a basis of $\overline{\Phi}_1$. The set $\{\Gamma_1, \Gamma_2, \Gamma_3\}$ with $\Gamma_1 = \emptyset$, $\Gamma_2 = \{\gamma_1\}$ and $\Gamma_3 = \{\gamma_1, \gamma_3\}$ is a system of representatives for the set LT.

For Γ_1 we have $L_1 = T$, $\overline{R}_1 = \overline{S}$, $N_1 = W^1$ and

$$A_1 = H^*_S(\overline{S})^{W^1} \simeq \mathbf{Q}[x_1, x_2, x_3],$$

where x_1, x_2, x_3 are generators of degree 2.

For Γ_2 we have L_2^g is isomorphic to O_2, the group N_2 is isomorphic to $\mathbf{Z}/2\mathbf{Z}$ and acts trivially on L_2^g, and

$$H^*_{R_2}(\overline{R}_2)^{N_2} \simeq \mathbf{Q}[u_1, u_2, u_3, u_4]/(u_1u_4, u_1u_3, u_2u_4),$$

so

$$A_2 \simeq \mathbf{Q}[v, u_1, u_2, u_3, u_4]/I,$$

where v is a generator of degree 4, the u_i are of degree 2 and I is the ideal generated by the monomials u_iu_j, $|i-j| > 1$.

For Γ_3 we have $L_3^g \simeq O_2 \times O_2$, $\overline{R}_3 \simeq \mathbf{P^1}$, the group N_3 is isomorphic to $\mathbf{Z}/2\mathbf{Z}$, and N_3 permutes the simple factors of L_3^g and the R_3–fixed points in $\overline{R}_3$. In this case A_3 can be described as the subring of $\mathbf{Q}[r, s, w_1, w_2]/(w_1w_2)$ generated by the elements $r+s$, rs, w_1+w_2, and rw_1+sw_2. The generators r and s are of degree 4, and the generators w_1 and w_2 are of degree 2.

To determine A_X it is by Remark 7.4 sufficient to consider the morphisms

$$\Psi(1,2,\emptyset) : u_i \mapsto u_i, i = 1, \ldots 4;\ v \mapsto 0$$

$$\Phi(1,2,\emptyset) : x_1 \mapsto u_3; x_2 \mapsto u_1 + u_4; x_3 \mapsto u_2$$

$$\Psi(2,3,\gamma_1) : r+s \mapsto v; rs \mapsto 0; w_1 + w_2 \mapsto w_1 + w_2; rw_1 + sw_2 \mapsto vw_1$$

$$\Phi(2,3,\gamma_1) : v \mapsto v; u_1 \mapsto w_1; u_4 \mapsto w_2; u_2, u_3 \mapsto 0.$$

It is now easy to see

$$H^*_{PSl_4}(X) \simeq \mathbf{Q}[a, b, c, d, e, f]/(ad, af, bd, f(ce-b), bce-b^2),$$

where d, e, f are generators of degree 2, and a, b, c are generators of degree $8, 6, 4$.

To obtain $H^*(X)$, one has further to divide by the relations

$$I_1 = -8c - 3d^2 - 4de - 2df - 4e^2 - 4ef - 3f^2$$

$$I_2 = -8ce + 16b + 4cf - 4cd + d^3 + 2d^2e + d^2f - df^2 - 2ef^2 - f^3$$

$$I_3 = 256a - 64ce^2 - 64ce(d+f) + 160cdf + 96c(f^2+d^2) - 3d^4 - 8d^3e + 8d^2e^2$$
$$+32de^3 + 16e^4 - 4d^3f + 8d^2ef + 48de^2f + 32e^3f + 14d^2f^2 + 8def^2 + 8e^2f^2$$
$$-4df^3 - 8ef^3 - 3f^4.$$

8.3. Assume that the enumeration of the Γ_i is such that if Γ_j is W^1–conjugate to a subset of Γ_i, then $i > j$. For $\xi \in H^*_G(X)$ denote by (a^ξ)

its image in A_X. By Theorem A we have a canonical graded filtration of $H_G^*(X)$

$$0 = F^{r+1}H_G^*(X) \subset F^r H_G^*(X) \subset \ldots \subset F^1 H_G^*(X) = H_G^*(X)$$

with $F^i H_G^*(X) = \{\xi \in H_G^*(X) | a_j^\xi = 0 \text{ for all } j < i\}$. Denote by $Gr\ H_G^*(X)$ the associated graded group.

We have seen in Proposition 5.7 the set $L_{c,i}.\overline{R}_i$is isomorphic to $L_{c,i} \times_{K_{x_i}} \overline{R}_i$ and hence is a smooth, compact submanifold of Z_i. We will see later that the Euler class E_i of its normal bundle in Z_i is a nonzero divisor in $H^*_{L_{c,i}}(L_{c,i}.\overline{R}_i)$. Denote the degree of E_i by d_i.

Theorem B.

$$Gr\ H_G^k(X) \simeq \bigoplus_{i=1}^{r} \Big(\bigoplus_{a+b=k-d_i} (H^a_{L_i^\sigma}(pt) \otimes H^b_{R_i}(\overline{R}_i))^{N_i} \Big)$$

Remark. Let y_i be a T-fixed point in Z_i^0 and denote by $\mathcal{T}_i$ the irreducible component of $\mathcal{N}_T$ containing y_i. Denote by W_i the subgroup of W of elements normalizing $\mathcal{T}_i$. Another way to state Theorem B is by 5.10:

$$Gr\ H_G^*(X) = \bigoplus_{i=1}^{r} H_T^{*-d_i}(\mathcal{T}_i)^{W_i}.$$

9. The proof of Theorem A and B

9.0. The strategy of the proof of the theorems will be to define a filtration of A_X compatible with the inclusion $H_G^*(X) \subset A_X$, and then to prove the associated graded groups are isomorphic.

We assume throughout the following that the enumeration of the sets $\Gamma_1, \ldots, \Gamma_r$ is such that if Γ_j is W^1-conjugate to a subset of Γ_i, then $i > j$.

9.1. We will first define a filtration of A_X. For $i = 1, \ldots, r$ denote by $F^i A_X$ the subspace

$$F^i A_X := \{(a) \in A_X | a_j = 0 \quad \text{for all} \quad j < i\}$$

We obtain a filtration

$$0 = F^{r+1}A_X \subset F^r A_X \subset \ldots \subset F^1 A_X = A_X$$

By 7.5, $(a) \in A_X$ is an element in $F^i A_X$ only if a_i is an element of the kernel $\mathcal{K}_i$ of the map

$$\mathcal{K}_i := \operatorname{Ker}\{(H^*_{L_i^\sigma}(pt) \otimes H^*_{\overline{R}_i}(\overline{R}_i))^{N_i} \to \bigoplus_{\substack{\Lambda \in LT \\ \Lambda \subset \Gamma_i, \Lambda \neq \Gamma_i}} H^*_{L_\Lambda^\sigma}(pt) \otimes H^*_{\overline{R}_i}(\overline{R}_i)\}.$$

So the morphisms $F^i A_X \to \mathcal{K}_i$, $(a) \mapsto a_i$, induce an inclusion of the associated graded group $Gr\ A_X \hookrightarrow \bigoplus_{i=1}^r \mathcal{K}_i$.

9.2. We will now define a filtration of $H^*_G(X)$. If $x \in X$ is a T–fixed point, then let $L(x)$ be the Levi subgroup of the associated parabolic subgroup $P(x)$ (3.7), such that $T \subset L(x)$. And denote by $(L(x))$ the Weyl group conjugacy class of $L(x)$. For $i = 2, \ldots, r+1$ denote by $F^i X^T$ the subset

$$F^i X^T := \{x \in X^T | (L(x)) = (L_j) \text{ for some } j < i\}.$$

And let $F^i H^*_G(X)$ be the subspace

$$F^i H^*_G(X) := \operatorname{Ker}\{H^*_G(X) \to H^*_T(F^i X^T)^W\}.$$

We obtain a filtration

$$0 = F^{r+1} H^*_G(X) \subset F^r H^*_G(X) \subset \ldots \subset F^1 H^*_G(X) = H^*_G(X)$$

For $\xi \in H^*_G(X)$ let (a^ξ) be the image of ξ in A. If $x \in X^T$ is a T–fixed point with $(L(x)) = (L_j)$, then the Weyl group orbit of x meets Z_j^0. By the injection $H^*_{L_j}(Z_j^0) \subset H^*_T((Z_j^0)^T)$ we see

$$\xi \in F^i H^*_G(X) \iff a_j^\xi = 0 \text{ for all } j < i \iff (a^\xi) \in F^i A_X$$

and hence $F^i H^*_G(X) = H^*_G(X) \cap F^i A_X$. This implies for the associated graded groups $Gr\ A_X$ and $Gr\ H^*_G(X)$:

Lemma. *The filtrations of $H^*_G(X)$ and R_X induce inclusions*

$$Gr\ H^*_G(X) \hookrightarrow Gr\ A_X \hookrightarrow \bigoplus_{i=1}^r \mathcal{K}_i$$

Let y_i be a T–fixed point in Z_i^0 and denote by $\mathcal{T}_i$ the irreducible component of $\mathcal{N}_T$ containing y_i. Let W_i be the normalizer in W of this component, and denote by d_i the degree of the Euler class of the normal bundle of $L_{c,i}.\mathcal{T}_i$ in Z_i (see Proposition 5.6).

Proposition 9.3. *The kernel $\mathcal{K}_i$ is isomorphic to $H_T^{*-d_i}(\mathcal{T}_i)^{W_i}$.*

This is in fact the main point in the proof. Since the proof of the proposition is rather technical, we postpone it to the next chapter and proceed with the proof of the theorems.

9.4. The next step will be to show the inclusion

$$Gr\ H_G^*(X) \hookrightarrow \bigoplus_{i=1}^{r} H_T^{*-d_i}(\mathcal{T}_i)^{W_i}$$

is an isomorphism. Note, this implies $Gr\ H_G^*(X) \simeq Gr\ A_X$ and hence $H_G^*(X) \simeq A_X$, which proves Theorem A. And, by 5.10,

$$H_T^*(\mathcal{T}_i)^{W_i} \simeq (H_{L_i^\sigma}^*(pt) \otimes H_{R_i}^*(\overline{R}_i))^{N_i},$$

so this implies also Theorem B.

To prove the map $Gr\ H_G^*(X) \hookrightarrow \bigoplus_{i=1}^{r} H_T^{*-d_i}(\mathcal{T}_i)^{W_i}$ is an isomorphism we proceed as in §2.

9.5. Recall that $H_G^*(X) \simeq H_T^*(X)^W$ is a free $H_T^*(pt)^W$–module. Let $\mathcal{I}$ be the ideal of all elements of strictly positive degree in $H_T^*(pt)^W$. Since for $i < j$ $F^j H_G^*(X) \cap \mathcal{I} F^i H_G^*(X) \subset \mathcal{I} F^j H_G^*(X)$, it is easy to see one can find a graded $H_T^*(pt)^W$–submodule U of $H_G^*(X)$, such that the intersections $F^i U := U \cap F^i H_G^*(X)$ induce a filtration

$$0 = F^{r+1} U \subset F^r U \subset \ldots \subset F^1 U = U$$

and $Gr\ H_G^*(X) \simeq H_T^*(pt)^W \otimes Gr\ U$. In particular, $Gr\ H_G^*(X)$ is a free $H_T^*(pt)^W$–module. And dim $Gr\ H_G^*(X)/\mathcal{I}\ Gr\ H_G^*(X)$ is the Euler characteristic of X, which is equal to $\sharp X^T$.

On the other hand, $H_T^*(\mathcal{T}_i)$ is a free $H_T^*(pt)$–module. So we can find a graded W_i–stable $H_T^*(pt)$–submodule U_i of $H_T^*(\mathcal{T}_i)$ such that $U_i \otimes H_T^*(pt) \to H_T^*(\mathcal{T}_i)$ is a W_i–equivariant isomorphism of graded $H_T^*(pt)$–modules.

Let now R be a graded W–stable $H_T^*(pt)^W$–submodule of $H_T^*(pt)$ such that $H_T^*(pt) \simeq R \otimes H_T^*(pt)^W$. Then R is isomorphic to the regular representation of W. So $H_T^*(\mathcal{T}_i)^{W_i} \simeq (R \otimes U_i)^{W_i} \otimes H_T^*(pt)^W$ as $H_T^*(pt)^W$–module.

Since dim U_i is the Euler characteristic of $\mathcal{T}_i$ and hence equal to $\sharp\mathcal{T}_i^T$, we see by Lemma 2.3 dim $(U_i \otimes R)^{W_i}$ is equal to $\sharp\mathcal{T}_i^T|W|/|W_i|$. But W_i operates transitively on the intersection $W.z \cap \mathcal{T}_i$ for a T–fixed point $z \in \mathcal{T}_i$. So dim $H_T^*(\mathcal{T}_i)^{W_i}/\mathcal{I}H_T^*(\mathcal{T}_i)^{W_i}$ is equal to the number of T–fixed points in X whose Weyl group orbit meet $\mathcal{T}_i$.

Since the inclusion $Gr\ H_G^*(X) \hookrightarrow \bigoplus_{i=1}^r H_T^{*-d_i}(\mathcal{T}_i)^{W_i}$ is a graded morphism of free $H_T^*(pt)^W$–modules, to prove its surjectivity it suffices to prove that the induced inclusion

$$Gr\ H_G^*(X)/\mathcal{I}Gr\ H_G^*(X) \hookrightarrow \bigoplus_{i=1}^{r} H_T^{*-d_i}(\mathcal{T}_i)^{W_i}/\mathcal{I}\bigoplus_{i=1}^{r} H_T^{*-d_i}(\mathcal{T}_i)^{W_i}$$

is an isomorphism. But the dimension of both sides is by Proposition 6.7 and the previous considerations equal to $\sharp X^T$.

Q.E.D.

10. Proof of Proposition 9.3

10.0. We use the same notation as before, only we denote by N_L the group N_i and by W_T the group W_i, and we drop the index i else. We want to study the kernel

$$\mathcal{K} = \mathrm{Ker}\{(H_{L^\sigma}^*(pt) \otimes H_R^*(\overline{R}))^{N_L} \to \bigoplus_{\substack{\Lambda \in LT \\ \Lambda \subset \Gamma, \Lambda \neq \Gamma}} H_{L_\Lambda^\sigma}^*(pt) \otimes H_R^*(\overline{R})\},$$

which is isomorphic (see 7.5) to the kernel of the map

$$H_L^*(Z^0)^{N_L} \to \bigoplus_{\substack{\Lambda \in LT \\ \Lambda \subset \Gamma, \Lambda \neq \Gamma}} H_{L_\Lambda}^*(L_\Lambda \times_{L_\Lambda^\sigma . C_\Lambda^1} (C_\Lambda^1 \times_{C^1} \overline{R})).$$

10.1. To prove Proposition 9.3, we will first investigate

$$Q := \mathrm{Ker}\{H_L^*(Z) \to \bigoplus_{\substack{\Lambda \in LT \\ \Lambda \subset \Gamma, \Lambda \neq \Gamma}} H_{L_\Lambda}^*(Z_\Lambda^0)\}.$$

By Lemma 5.10, every T–fixed point in Z, which is not contained in Z^0, is conjugate under the Weyl group of L to a T–fixed point in Z_Λ^0 for some $\Lambda \subset \Gamma$, $\Lambda \in LT$. Furthermore, by Proposition 4.4, all T–fixed points in Z^0 are contained in $L_c.\overline{R}$. So we get by the Localization Theorem:

Lemma. $Q = \mathrm{Ker}\{H^*_L(Z) \to H^*_T((Z - L_c.\overline{R})^T)\}$

10.2. The set $L_c.\overline{R}$ is a smooth compact manifold by Proposition 5.6.

Lemma. *The Euler class E of the normal bundle of $L_c.\overline{R}$ in Z is a nonzero divisor in $H^*_{L_c}(L_c.\overline{R})$.*

Proof. Consider the morphism $Z \to X_L$ defined in 5.2, and let N_1 be the normal bundle of $L_c.id \simeq L_c/L_c^\sigma.C_c^1$ in X_L. Let $M_c \subset L_c^\sigma.C_c^1$ be a maximal torus. Since its fixed points in X_L are isolated, all characters of M_c occuring in the fibre $N_1(id)$ of N_1 at the point id are nontrivial. So the Euler class of $N_1(id)$ in $H^*_{M_c}(pt)$ is nonzero, and hence by the injection $H^*_{L_c^\sigma}(pt) \otimes H^*_{C_c^1}(pt) \hookrightarrow H^*_{M_c}(pt)$ the Euler class of N_1 in $H^*_{L_c}(L_c.id) \simeq H^*_{L_c^\sigma}(pt) \otimes H^*_{C_c^1}(pt)$ is nonzero. And, since C_c^1 acts trivially on X_L, we see furthermore the Euler class of N_1 is of the form $E_1 \otimes 1$.

The normal bundle N of $L_c.\overline{R}$ in Z is the pull back of N_1, so its Euler class E is equal to $E_1 \otimes 1$ in $H^*_{L_c}(L_c.\overline{R}) \simeq H^*_{L_c^\sigma}(pt) \otimes H^*_{\overline{R}}(\overline{R})$. Since $E_1 \neq 0$ and $H^*_{L_c^\sigma}(pt)$ is a domain, this implies E is a nonzero divisor.

Q.E.D

10.3. Let d be the degree of E. Since E is a nonzero divisor and the odd equivariant cohomology of Z and $\overline{R}$ (and hence of $L_c.\overline{R}$) vanish, we obtain:

Corollary. *The Thom–Gysin sequence splits into short exact sequences*

$$0 \longrightarrow H^{j-d}_{L_c}(L_c.\overline{R}) \longrightarrow H^j_{L_c}(Z) \longrightarrow H^j_{L_c}(Z - L_c.\overline{R}) \longrightarrow 0$$

and all odd terms vanish.

10.4. Consider the following diagram

$$\begin{array}{ccccccc}
0 \to & H^{j-d}_{T_c}(L_c.\overline{R}) & \to & H^j_{T_c}(Z) & \to & H^j_{T_c}(Z - L_c.\overline{R}) & \to 0 \\
 & \downarrow & & \downarrow & & \downarrow & \\
0 \to & H^{j-d}_{T_c}((L_c.\overline{R})^{T_c}) & \to & H^j_{T_c}((Z)^{T_c}) & \to & H^j_{T_c}((Z - L_c.\overline{R})^{T_c}) & \to 0
\end{array}$$

The sequence in the first row is exact, one replaces in Lemma 10.2 and Corollary 10.3 just L_c by T_c. The second row is obviously exact. The first two vertical arrows are injective by the Localization Theorem and Lemma 10.2.

Since $H^*_{T_c}(Z - L_c.\overline{R})$ has no odd parts, it is a free $H^*_{T_c}(pt)$–module by 1.4. Furthermore, the cokernel of the first vertical arrow is a torsion module over $H^*_{T_c}(pt)$ by the Localization Theorem, which implies the kernel of the last vertical arrow is a torsion module as well, and hence all vertical arrows are injective.

Since the map $H^*_{L_c}(Z - L_c.\overline{R}) \to H^*_{T_c}(Z - L_c.\overline{R})$ is injective, we see the kernel Q is by Lemma 10.1 equal to the kernel of the map

$$H^*_{L_c}(Z) \longrightarrow H^*_{L_c}(Z - L_c.\overline{R}).$$

So by Corollary 10.3: $Q \simeq H^{*-d}_{L_c}(L_c.\overline{R})$.

Consider now the map $\pi : Z \to X_L$. Choose a T–fixed point y' in the open orbit in X_L, let $\mathcal{T} = \pi^{-1}(y')$ be the fibre and let $y \in \mathcal{T}$ be a T–fixed point. We obtain by Proposition 5.6:

Proposition 10.5. $Q \simeq H^{*-d}_{K_y}(\mathcal{T})$

10.6. Note that Q is by the description in Lemma 10.1 in a canonical way a N_L–module, and the isomorphism above is in fact a N_L–isomorphism. We get by 5.10:

Corollary. $Q^{N_L} \simeq H^{*-d}_T(\mathcal{T})^{W_T}$.

10.7. The next step will be to construct an isomorphism between $\mathcal{K}$ and Q^{N_L}. The inclusions

$$Z^0 \subset Z \quad \text{and} \quad L_\Lambda \times_{L^\sigma_\Lambda . C^1_\Lambda} (C^1_\Lambda \times_{C^1} \overline{R}) \subset Z^0_\Lambda$$

induce a morphism $Q \to Q^0$, where Q^0 is the kernel of the map

$$H^*_L(Z^0) \to \bigoplus_{\substack{\Lambda \in LT \\ \Lambda \subset \Gamma, \Lambda \neq \Gamma}} H^*_{L_\Lambda}(L_\Lambda \times_{L^\sigma_\Lambda . C^1_\Lambda} (C^1_\Lambda \times_{C^1} \overline{R})).$$

Note that the map $Q \to Q^0$ is injective, for an element in the kernel would vanish at all T–fixed point in Z, which is not possible by the Localization Theorem.

Proposition 10.8. *The map $Q \to Q^0$ is an isomorphism.*

10.9. This implies that the image of Q^{N_L} is $Q^0 \cap H^*_L(Z^0)^{N_L} = \mathcal{K}$. Hence we obtain by Corollary 10.6:

Proposition 9.3. $\mathcal{K} \simeq H_T^{*-d}(T)^{W_T}$.

10.10. The proof of Proposition 10.8 needs some preparation, for we have first to recall the results in [6] on the equivariant cohomology of regular embeddings.

Let G be an affine algebraic group acting on a connected algebraic variety X. We say X is a regular embedding, if each orbit closure is smooth and it is the transversal intersection of the codimension one orbit closures containing it. We require furthermore that for any $x \in X$ the stabilizer G_x has a dense orbit in the normal space to $G.x \subset X$ at x (see [6],§3).

Let $O_1, \ldots, O_t$ be an enumeration of the G–orbits in X such that for all $i = 1, \ldots, r$ the union $X_i := O_1 \cup \ldots \cup O_i$ is open in X. Then the maps

$$H_G^*(X_i) \longrightarrow \mathcal{R}_i = \bigoplus_{k=1}^{i} H_G^*(O_k)$$

are injective, and the restriction maps $H_G^*(X_i) \longrightarrow H_G^*(X_{i-1})$ are surjective. A description of $H_G^*(X)$ as a subalgebra of $\mathcal{R}_t$ can be found in [6], §7. We will use this result to describe $H_L^*(Z)$.

10.11. Note first that the L–variety Z is a regular embedding. By 5.0, the first condition is satisfied, and we see by the description of the local structure of Z in 5.0 that L_z has an open orbit in the normal space N_z to the orbit $L.z$ in Z at z.

We associate to Z a simplicial complex $C = (V, \mathcal{S})$, whose vertices are the orbits of codimension one, and $F \subset V$ is a simplex whenever $\bigcap_{v \in F} \overline{O_v}$ is not empty. We include the empty set among the vertices, so the set of simplexes is in one–to–one correspondence with the L–orbits in Z (see [6],§3). For $F \in \mathcal{S}$ denote by O_F the corresponding orbit, which is the unique open L–orbit in $\bigcap_{v \in F} \overline{O_v}$.

Let $x_v \in H_L^*(Z)$, $v \in V$, be the equivariant Chern class corresponding to the (linearized) line bundle $\mathcal{L}_v$ associated to the divisor $\overline{O_v}$.

For $F \in \mathcal{S}$ choose $z \in O_F$ such that $z = w.x_\Lambda$ for some subset Λ of Σ and $w \in W^1$. Denote by $L_F := L_z$ its stabilizer in L. By the choice of z the intersection $K_F := L_F \cap L_c$ is a maximal compact subgroup of L_F (3.8). By the description of the stabilizer K_{x_Λ} in 3.8 one knows K_F is σ–stable and has a decomposition $K_F = K_F^\sigma . C_F$, where C_F is a σ–split compact torus, and

$$K_F^\sigma = (\mathrm{Cen}_K C_F)^\sigma \cap L_c.$$

The subgroup K_F^σ operates trivially on the normal space $N(z)$ of O_F at z. Denote by $\mathcal{L}_v(z)$ the fibre of the line bundle $\mathcal{L}_v$ at z. Since O_F is the transversal intersection of the O_v, $v \in F$, is

$$N(z) \simeq \bigoplus_{v \in F} \mathcal{L}_v(z)$$

as C_F representation and $\dim_{\mathbf{R}} C_F = \dim N(z)$. So there is a canonical isomorphism

$$H_L^*(O_F) \simeq H_{K_F}^*(pt) \simeq H_{K_F^\sigma}^*(pt) \otimes \mathbf{Q}[x_v]_{v \in F}.$$

If $F' \in \mathcal{S}$ is such that $F' \subset F$ (which is equivalent to $O_F \subset \overline{O_{F'}}$), then we can find elements $z \in O_F$ and $z' \in O_{F'}$, such that $z = w.x_\Lambda$, $z' = w.x_\Upsilon$, where $\Upsilon \supset \Lambda$ and $w \in W^1$. The inclusions (see 3.8) $K_F^\sigma \subset K_{F'}^\sigma$ and $C_F \supset C_{F'}$ induce canonical morphisms

$$\Phi(F, F') : H_L^*(O_F) \longrightarrow H_{K_F^\sigma}^*(pt) \otimes \mathbf{Q}[x_v]_{v \in F}/(x_v, v \notin F')$$

$$\Psi(F, F') : H_L^*(O_{F'}) \longrightarrow H_{K_F^\sigma}^*(pt) \otimes \mathbf{Q}[x_v]_{v \in F}/(x_v, v \notin F').$$

Remark. A geometrical interpretation of these morphisms in terms of a tubular neighborhood of the orbit O_F is given in [6], §7.

Definition 10.12. Let $\mathcal{R}_Z$ be the subalgebra of $\mathcal{R} := \bigoplus_{F \in \mathcal{S}} H_L^*(O_F)$ of all elements $(r_F)_{F \in \mathcal{S}}$ such that

$$\Phi(F, F')(r_F) = \Psi(F, F')(r_{F'}) \quad \text{for all} \quad F, F' \in \mathcal{S},\ F' \subset F.$$

Proposition 10.13. ([6],§7, §14) *The image of $H_L^*(Z) \hookrightarrow \mathcal{R}$ coincides with $\mathcal{R}_Z$.*

10.14. Since every T–fixed point in Z is conjugate under the Weyl group of L to a T–fixed point in Z_Λ^0 for some $\Lambda \subset \Gamma$, $\Lambda \in LT$, by the Localization Theorem we have an injection

$$H_L^*(Z) \hookrightarrow \bigoplus_{\Lambda \subset \Gamma, \Lambda \in LT} H_{L_\Lambda}^*(Z_\Lambda^0).$$

A connection to the inclusion $H_L^*(Z) \hookrightarrow \mathcal{R}$ is given by the following

Lemma. *For every L-orbit O_F in Z there exists a subset Λ of Γ and a point $z \in O_F$ such that $z \in Z^0_\Lambda$ and the restriction map*

$$H^*_L(O_F) \longrightarrow H^*_{L_\Lambda}(L_\Lambda.z)$$

is injective.

Proof. Choose $z \in O_F$ such that $z = w.x_\Upsilon$ for some subset Υ of Σ. Let

$$K_F = K^\sigma_F.C_F$$

be a decomposition of the stabilizer K_F of z in L_c as in 10.11. Consider the Levi subgroup $\mathrm{Cen}_L C_F$ of L. By 10.11 we have

$$(\mathrm{Cen}_L C_F)^\sigma \cap L_c = K^\sigma_F \tag{2}$$

Let M be a maximal σ–stable torus in $\mathrm{Cen}_L C_F$ such that $M \cap K^\sigma_F$ is a maximal torus in K^σ_F. We can assume that the connected component containing the identity M^1 of the σ–split part of M is contained in T^1. Note that $C_F \subset M^1$ by construction.

Recall that $W^1(\Gamma)$ is the subgroup of W^1 generated by the reflections s_γ, $\gamma \in \Gamma$. Replacing z by a $W^1(\Gamma)$–conjugate if necessary, we can furthermore assume that $\mathrm{Cen}_L M^1 = L_\Lambda$ for some subset Λ of Γ. By construction, we have $M^\sigma \cap (L_\Lambda, L_\Lambda)$ is a maximal torus of (L_Λ, L_Λ) , so $\Lambda \in LT$.

And (2) implies $L^\sigma_{\Lambda,c}$ is a subgroup of maximal rank of K^σ_F. Since by the choice of z the stabilizer $(L_{\Lambda,c})_z$ is a maximal compact subgroup of $(L_\Lambda)_z$, the inclusion

$$(L_{\Lambda,c})_z = L^\sigma_{\Lambda,c}.(C^1_{\Lambda,c})_z \subset K_F = K^\sigma_F.C_F$$

induces an injection in equivariant cohomology

$$H^*_L(O_F) \simeq H^*_{K_F}(pt) \hookrightarrow H^*_{L^\sigma_{\Lambda,c}.(C^1_{\Lambda,c})_z}(pt) \simeq H^*_{L_\Lambda}(L_\Lambda.z).$$

Consider the morphism $\pi_\Lambda : Z_\Lambda \to X_\Lambda$, where X_Λ is the minimal wonderful compactification of the symmetric space $L'_\Lambda/{L'_\Lambda}^\sigma$, and $L'_\Lambda := L_\Lambda/C_\Lambda$. Since $\pi_\Lambda(z)$ contains ${L'_\Lambda}^\sigma$ in its stabilizer, we see that z is contained in the preimage of the dense L'_Λ–orbit in X_Λ, which implies $z \in Z^0_\Lambda$.

Q.E.D.

10.15. We want now to prepare the proof of Proposition 10.8. Consider a L-orbit O_F in Z such that $O_F \subset Z^0$. Since $\pi(O_F)$ is the dense L'-orbit in Y, every codimension one orbit in Z which contains O_F in its closure is contained in Z^0. Furthermore, by 5.6 the intersection of a L-orbit with $\overline{R}$ is one R-orbit. So if we still denote by x_v the class in $H^*_R(\overline{R})$ of the line bundle $\mathcal{L}_v|_{\overline{R}}$, then $H^*_R(\overline{R})$ is generated by the x_v, $O_v \subset Z^0$. And the inclusion $O_F \subset Z^0$ induces an isomorphism

$$H^*_L(O_F) \simeq H^*_{L^\sigma}(pt) \otimes (H^*_R(\overline{R})/(x_v | v \notin F, O_v \subset Z^0)) \tag{3}$$

Let $O_{F'}$ be a L-orbit such that $F \subset F'$ and $O_{F'}$ is not contained in Z^0. Let $\Lambda \subset \Gamma$, $\Lambda \in LT$, and $z' = w.x_\Upsilon \in O_{F'} \cap Z^0_\Lambda$ be such that restriction map

$$H^*_L(O_{F'}) \longrightarrow H^*_{L_\Lambda}(L_\Lambda.z')$$

is injective (Lemma 10.14). The injection $L_\Lambda.\overline{R} \subset Z^0$ induces by 7.5 a morphism $H^*_{L^\sigma}(pt) \otimes H^*_R(\overline{R}) \to H^*_{L^\sigma_\Lambda}(pt) \otimes H^*_R(\overline{R})$, and hence a morphism

$$\begin{aligned} &H^*_{L^\sigma}(pt) \otimes (H^*_R(\overline{R})/(x_v | v \notin F, O_v \subset Z^0)) \\ &\qquad \longrightarrow H^*_{L^\sigma_\Lambda}(pt) \otimes H^*_R(\overline{R})/(x_v | v \notin F, O_v \subset Z^0) \end{aligned} \tag{4}$$

Since we can choose $z' = w.x_\Upsilon \in O_{F'} \cap \overline{R}_\Lambda$ and $z = w.x_{\Upsilon'} \in O_F$, by the isomorphism (3) and the description of $\Psi(F', F)$ in 10.11, we see the morphism in (4) is the composition of the morphism $\Psi(F', F)$ and the morphism induced by the inclusion $L^\sigma_\Lambda \subset L^\sigma_{z'}$

$$H^*_{L^\sigma_{z'}}(pt) \longrightarrow H^*_{L^\sigma_\Lambda}(pt).$$

We have seen in 10.14 that the last map is injective. So $\Psi(F', F)(\xi) = 0$ for any element ξ^0 in the kernel of the map $H^*_L(Z^0) \to H^*_{L_\Lambda}(L_\Lambda.\overline{R})$.

For $\xi \in H^*_L(Z)$ denote by ξ^0 its image in $H^*_L(Z^0)$ and by ξ_F, $F \in \mathcal{S}$, its image in $H^*_L(O_F)$. We obtain as consequence of the discussion above:

Lemma 10.16. *If $\xi \in H^*_L(Z)$ is such that $\xi^0 \in Q^0$, then $\Psi(F', F)(\xi_F) = 0$ for all $F, F' \in \mathcal{S}$, with $F \subset F'$ and $O_F \subset Z^0$, $O_{F'} \not\subset Z^0$.*

10.17. *Proof of Proposition 10.8.* It remains to prove the surjectivity of the morphism $Q \to Q^0$.

Since Z^0 is an open subset we can choose an enumeration $O_1, \ldots, O_r$ of the L–orbits, such that $Z_i = O_1 \cup \ldots \cup O_i$ is open in Z for all $i = 1, \ldots, r$ and $Z^0 = Z_j$ for some j. So by 10.10 the restriction map

$$H_L^*(Z) \longrightarrow H_L^*(Z^0)$$

is surjective. For $\xi^0 \in Q^0$, let $\xi \in H_L^*(Z)$ be such that its restriction to Z^0 is equal to ξ^0. Denote by ξ_F its image in $H_L^*(O_F)$ for $F \in \mathcal{S}$. By Lemma 10.16, we know $\Psi(F', F)(\xi_F) = 0$, whenever $O_F \subset Z^0$ and $O_{F'} \not\subset Z^0$. So the element (ξ'_F) of $\mathcal{R}$ with $\xi'_F = \xi_F$ if $O_F \subset Z^0$ and $\xi'_F = 0$ else, is in $\mathcal{R}_Z$.

Hence by Proposition 10.13, one can assume that the restriction of ξ to any orbit not contained in Z^0 vanishes. But then the restriction of ξ to any T–fixed point not contained in Z^0 vanishes, which by Lemma 10.1 implies that $\xi \in Q$. Hence the morphism $Q \to Q^0$ is surjective.

Q.E.D.

11. Equivariant cohomology of wonderful compactifications

11.0. We want now to reformulate Theorem A and Theorem B in §8 for the general case of a wonderful compactification.

11.1. A *wonderful compactification* Y of G/H is a complete, smooth G–variety Y with an embedding $G/H \hookrightarrow Y$ and a G–equivariant morphism $\phi : Y \to X$ to the minimal wonderful compactification X of G/H (see 3.2), which extends the identity on G/H (see [10]).

11.2. We use the notation as in §3. Let $\overline{S}_Y$ be the closure of $S = T^1/T^1 \cap H$ in Y. Then $\overline{S}_Y$ is a smooth, complete W^1–equivariant torus embedding of S (see [10]).

Let $\overline{S}_X$ be the closure of S in X. The restriction of ϕ to $\overline{S}_Y$ is a W^1–equivariant morphism of S–embeddings and $\phi^{-1}(\overline{S}_X) = \overline{S}_Y$ (see [10]).

Furthermore, Y is completely determined by $\overline{S}_Y$. This means that for every W^1–equivariant, smooth, complete S–embedding $\mathcal{S}$ with a W^1–equivariant morphism $\psi : \mathcal{S} \to \overline{S}_X$ there exists (up to isomorphism) a unique wonderful compactification Y, such that $\overline{S}_Y$ is isomorphic to $\mathcal{S}$ and the restriction of $\phi : Y \to X$ to $\overline{S}_Y$ is ψ (see [10]).

11.3. For a subset $\Gamma \subset \Sigma$ let L_Γ be the Levi subgroup of P_Γ containing T, let C_Γ be its center and let C_Γ^1 be the connected component containing the identity of C_Γ (§3). We assume that its Weyl group conjugacy class

(L_Γ) is an element of

$LT = \{(L_\Lambda) | \Lambda \subset \Sigma, (L_\Lambda, L_\Lambda)$ contains a maximal torus contained in

$$(L_\Lambda, L_\Lambda)^\sigma\}$$

By X_Γ we denote the minimal wonderful compactification of $L'_\Gamma/L'^{\sigma}_\Gamma$, where $L'_\Gamma = L_\Gamma/C_\Gamma$, and by Z_Γ the variety $\overline{L_\Gamma.id}$. Recall that Z^0_Γ is the preimage in Z_Γ of the open orbit in X_Γ under the map $\pi : Z_\Gamma \to X_\Gamma$ (5.2). And the fibre $\pi^{-1}(id)$ is the closure $\overline{R}_\Gamma$ of $R_\Gamma = C^1_\Gamma/C^1_\Gamma \cap H$ in $\overline{S}_X$. Let y be a T-fixed point in Z^0_Γ and let $\mathcal{T}_\Gamma$ be the irreducible component of $\mathcal{N}_T$ containing y, which is by Proposition 5.6 the fibre $\pi^{-1}(\pi(y))$.

We index the corresponding preimages in Y by Y, i.e. $Z_{Y,\Gamma} = \phi^{-1}(Z_\Gamma)$, $Z^0_{Y,\Gamma} = \phi^{-1}(Z^0_\Gamma)$, $\overline{R}_{Y,\Gamma} = \phi^{-1}(\overline{R}_\Gamma)$ and $\mathcal{T}_{Y,\Gamma} = \phi^{-1}(\mathcal{T}_\Gamma)$. We will see later (see 11.12, 11.13) that these are smooth varieties, and the canonical morphism:

$$L_\Gamma \times_{L^\sigma_\Gamma C^1_\Gamma} \overline{R}_{Y,\Gamma} \longrightarrow Z^0_{Y,\Gamma}$$

is an isomorphism. Finally, N_Γ is the quotient $\mathrm{Nor}_H L_\Gamma / L^\sigma_\Gamma$, and W_Γ is the subgroup of elements in W normalizing $\mathcal{T}_\Gamma$.

11.4. Let $\{\Gamma_1, \ldots, \Gamma_r\}$ be a set of representatives of the set LT. We index the corresponding groups and varieties just by i.

Consider the ring $A = \bigoplus_{i=1}^r A_i$ defined by

$$A := \bigoplus_{i=1}^{r} H^*_{L_i}(Z^0_{Y,i})^{N_i}.$$

Since $\phi : Y \to X$ is G-equivariant, the Weyl group orbit of every T-fixed point meets some $Z^0_{Y,i}$ by Proposition 6.7. So the inclusions $Z^0_{Y,i} \subset Y$ induce an injection

$$H^*_G(Y) \hookrightarrow A.$$

11.5. If $\Lambda \subset \Gamma_i$, $\Lambda \in LT$, then $L_\Lambda \subset L_i$ and $\overline{R}_{Y,i} \subset \overline{R}_{Y,\Lambda}$. So $L_\Lambda.\overline{R}_{Y,i}$ is contained in $Z^0_{Y,\Lambda}$ and in $Z^0_{Y,i}$. And if Λ is conjugate to Γ_j, this induces a canonical isomorphism

$$H^*_{L_j}(Z^0_{Y,j})^{N_j} \simeq H^*_{L_\Lambda}(Z^0_{Y,\Lambda})^{N_\Lambda}.$$

11.6. By a triple (i, j, Λ) we mean a subset Λ of Γ_i which is conjugate under W^1 to Γ_j. To each triple (i, j, Λ) we associate the morphisms

$$\Phi(i,j,\Lambda) : A_i \longrightarrow H^*_{L_\Lambda}(L_\Lambda.\overline{R}_{Y,i}) \text{ and } \Psi(i,j,\Lambda) : A_j \longrightarrow H^*_{L_\Lambda}(L_\Lambda.\overline{R}_{Y,i})$$

induced by the isomorphism $H^*_{L_j}(Z^0_{Y,j})^{N_j} \to H^*_{L_\Lambda}(Z^0_{Y,\Lambda})^{N_\Lambda}$ and the inclusions

$$Z^0_{Y,i} \supset L_\Lambda \times_{L^\sigma_\Lambda C^1_\Lambda} (C^1_\Lambda \times_{C^1_i} \overline{R}_{Y,i}) \subset Z^0_{Y,\Lambda}.$$

Definition 11.7. Let A_Y be the subring of all elements $(a) = (a_1, \ldots, a_r)$ of A, such that

$$\Phi(i,j,\Lambda)(a_i) = \Psi(i,j,\Lambda)(a_j) \text{ for any triple } (i,j,\Lambda).$$

Remark. The comments on the definition on the ring A_X in Remark 7.4 hold as well for the definition of A_Y.

Theorem A'. *The inclusion $H^*_G(Y) \hookrightarrow A$ induces an isomorphism of rings*

$$H^*_G(Y) \simeq A_Y.$$

11.8. Assume that the elements Γ_i are enumerated such that if Γ_j is W^1-conjugate to a subset of Γ_i, then $i > j$. For $\xi \in H^*_G(Y)$ denote by (a^ξ) its image in A_Y. As in 8.3 consider the graded filtration

$$0 = F^{r+1}H^*_G(Y) \subset F^r H^*_G(Y) \subset \ldots \subset F^1 H^*_G(Y) = H^*_G(Y)$$

with $F^i H^*_G(Y) := \{\xi \in H^*_G(Y) | a^\xi_j = 0 \text{ for all } j < i\}$. And denote by $Gr\ H^*_G(Y)$ the associated graded group.

We will see later (11.15) the set $L_{c,i}.\overline{R}_{Y,i}$ is a smooth submanifold of $Z_{Y,i}$. Let d_i be the degree of the Euler class in $H^*_{L_{c,i}}(L_{c,i}.\overline{R}_{Y,i})$ of the corresponding normal bundle.

Theorem B'.

$$Gr\ H^k_G(Y) \simeq \bigoplus_{i=1}^r \Big(\bigoplus_{a+b=k-d_i} (H^a_{L^\sigma_i}(pt) \otimes H^b_{R_i}(\overline{R}_{Y,i}))^{N_i}\Big) \simeq \bigoplus_{i=1}^r H^{k-d_i}_T(\mathcal{T}_{Y,i})^{W_i}$$

11.9. The proof of the Theorems is essentially the same as in the case of the minimal compactification, so we omit it. It remains to prove the properties of the varieties $Z_{Y,i}$, $Z^0_{Y,i}$ and $\overline{R}_{Y,i}$ stated above. We wish first to recall some results about the structure of the wonderful compactifications.

11.10. Let A be the affine chart of $\overline{S}_X$ corresponding to the negative Weyl chamber (see 3.6), and denote by V its preimage $\phi^{-1}(A)$ in Y. Then V is a smooth torus embedding, and every G–orbit in Y meets V in a unique S–orbit.

Furthermore, let P be the stabilizer of the unique T–fixed point in A. Let P^- be a parabolic subgroup opposite to P such that $T \subset P \cap P^-$, and denote by U^- its unipotent radical. The canonical map

$$U^- \times V \longrightarrow Y$$

is a P^-–equivariant isomorphism onto an open subset of Y.

As a consequence we see that the closure of every G–orbit in Y is smooth, and it is the transversal intersection of the codimension one orbit closures containing it (see [10]).

11.11. To describe the stabilizer in G of a point $y \in Y$ we may assume that $\phi(y) = x := x_\Gamma$ for some subset Γ of Σ (see 3.6). Set $P = P_\Gamma$, $L = L_\Gamma$ and $C^1 = C^1_\Gamma$. The stabilizer of y is (see [6],§9)

$$G_y = L^\sigma . P^u . (C^1)_y .$$

11.12. For $\Gamma \subset \Sigma$ and $L = L_\Gamma$ consider now $Z = \overline{L.id}$ and $Z_Y = \phi^{-1}(Z)$. Since $\phi^{-1}(\overline{S}_X) = \overline{S}_Y$, it follows by Corollary 5.3:

$$Z_Y = L_c . \overline{S}_Y .$$

Hence every L–orbit meets $\overline{S}_Y$ in a union of T–orbits, which are permuted transitively by $W^1(\Gamma)$, the subgroup of W^1 generated by the reflections s_γ, $\gamma \in \Gamma$.

Let A be the affine chart of $\overline{S}_X$ considered in 11.10, and let z be the unique T–fixed point in A. The stabilizer L_z is a parabolic subgroup of L. Let P_L^- be a parabolic subgroup of L opposite to L_z, such that $T \subset L_z \cap P_L^-$, and let U_L^- be its unipotent radical. We obtain by 11.10 for $V := \phi^{-1}(A)$: The canonical map

$$U_L^- \times V \longrightarrow Z_Y$$

is a P_L^-–equivariant isomorphism onto an open subset of Z_Y. As a consequence we see Z_Y is a regular embedding in the sense of 10.10.

11.13. Consider a C^1–fixed point y in $\phi^{-1}(x_\Gamma)$. The center C of L is contained in L_y, so the variety $Y_L := \overline{L.y}$ is a smooth compactification of

L'/L'^{σ}, where $L' = L/C$. Denote by X_L the minimal wonderful compactification of L'/L'^{σ}. Since the image of y is $id \in X_L$ under the morphism $\pi \circ \phi : Y_L \to X_L$, Y_L is in fact a wonderful compactification of L'/L'^{σ}.

The image $T^{1'}$ of T^1 in L' is a maximal split torus, and

$$S' = T^{1'}/T^{1'} \cap L'^{\sigma}$$

is isomorphic to S/S_y. Since $\overline{S'.y} = \overline{S.y}$, the r.p.p.d. of the closure $\overline{S}'_Y$ of S' in Y_L is the image of the r.p.p.d. of $\overline{S}_Y$ under the morphism

$$\mathrm{Hom}_Z(X(S), \mathbf{R}) \longrightarrow \mathrm{Hom}_Z(X(S'), \mathbf{R}).$$

This implies further we have well defined morphism $\overline{S}_Y \to \overline{S}'_Y$. Denote by $\overline{S}'_X$ the closure of S' in X_L.

Proposition 11.14. *The morphism* $L/L^{\sigma} \to L'/L'^{\sigma}$ *extends to a morphism* $Z_Y \to Y_L$.

Proof. The map $L/L^{\sigma} \to L'/L'^{\sigma}$ induces a rational L–equivariant map $Z_Y \to Y_L$. The set of points in Z_Y where this map is defined is an open, L–stable subset of Z_Y.

Let now y be a T–fixed point in $\overline{S}_Y$, x its image in $\overline{S}_X$, x' its image in $\overline{S}'_X$, and let y' be its image in $\overline{S}'_Y$ under the morphism $\overline{S}_Y \to \overline{S}'_Y$. Note that x' and y' are $T' = T/C$–fixed points, and $\phi(\pi(y')) = x'$. Let $A(x)$ and $A(x')$ be the affine charts of the torus embeddings $\overline{S}_X$ and $\overline{S}'_X$ corresponding to the T– respectively T'–fixed point, and denote by $V(y)$ and $V(y')$ the preimages in Z_Y and Y_L. The morphism $\overline{S}_Y \to \overline{S}'_Y$ induces a well defined morphism $V(x) \to V(y')$. Furthermore, let $P \subset L$ and $P' \subset L'$ be parabolic subgroups opposite to L_y respectively $L'_{y'}$, such that $T \subset L_y \cap P$ and $T' \subset P' \cap L'_{y'}$. Then the unipotent radicals U and U' of P and P' are canonically isomorphic, and the restriction of the morphism

$$U^- \times V(y) \to U'^- \times V(y')$$

to the intersection with L/L^{σ} coincides with the morphism $L/L^{\sigma} \to L'/L'^{\sigma}$.

Since every L–orbit meets such an open subset for some T–fixed point in $\overline{S}_Y$, the morphism $L/L^{\sigma} \to L'/L'^{\sigma}$ extends all of Z_Y. Q.E.D.

11.15. We suppose for the rest of the paragraph that $\Gamma \in LT$. As an immediate consequence of Proposition 11.14 we see:

1) Set $x = x_\Gamma$, let y be a T–fixed point in Z^0, and denote by $\mathcal{T} = \pi^{-1}(\pi(y))$ the irreducible component of $\mathcal{N}_T$ containing y. Then $Z^0_Y = \phi^{-1}(Z^0)$ is isomorphic to the fibre spaces

$$L \times_{L_x} \overline{R}_Y \text{ and } L \times_{L_y} \mathcal{T}_Y, \quad \text{where } \mathcal{T}_Y = \phi^{-1}(\mathcal{T}).$$

The subset $L_c.\overline{R}_Y$ is a smooth compact submanifold of Z.

2) $\overline{R}_Y$ is a smooth R–embedding. The corresponding r.p.p.d. is the intersection of the image of the inclusion

$$\mathrm{Hom}_{\mathbf{Z}}(X(R), \mathbf{R}) \subset \mathrm{Hom}_{\mathbf{Z}}(X(S), \mathbf{R})$$

with the r.p.p.d. of $\overline{S}_Y$.

3) For a T–fixed point $y \in \mathcal{T}_Y$ the intersection $G.y \cap \mathcal{T}_Y$ is a union of T–fixed points permuted transitively by $\mathrm{Nor}_K \mathcal{T} = \mathrm{Nor}_K \mathcal{T}_Y$.

4) $\mathcal{N}_{T,Y} := \phi^{-1}(\mathcal{N}_T)$ is N_T–stable, and it is the disjoint union of smooth T–stable varieties. All T–fixed points in Y are contained in $\mathcal{N}_{T,Y}$. The Weyl group orbits in the set of irreducible components of $\mathcal{N}_{T,Y}$ is in one–to–one correspondence with the set LT.

5) The normalizer $W_{\mathcal{T}}$ in W of the irreducible component $\mathcal{T}$ in W is equal to the normalizer in W of $\mathcal{T}_Y$. Denote by N_L the quotient group $L.\mathrm{Nor}_H L^\sigma / L$. There are canonical isomorphisms in equivariant cohomology

$$H_L^*(Z_Y^0)^{N_L} \simeq (H_{L^\sigma}^*(pt) \otimes H_R^*(\overline{R}_Y))^{N_L} \simeq H_T^*(\mathcal{T}_Y)^{W_{\mathcal{T}}}.$$

Remark. Using 5) one can show that the morphisms $\Phi(i, j, \Lambda)$ and $\Psi(i, j, \Lambda)$ can be described as in §7 as maps induced by the inclusion $L_\Lambda \subset L_i$ and $\overline{R}_{Y,i} \subset \overline{R}_{Y,\Lambda}$.

References

[1] S. Abeasis, On a remarkable class of subvarieties of a symmetric variety, *Adv. in Math.* **71** (1988) , 113-129.

[2] M. F. Atiyah, R. Bott, The moment map and equivariant cohomology, *Topology* **23** (1984) , 1–28.

[3] A. Bialynicki–Birula, Some theorems on actions of algebraic groups, *Ann. of Math.* **98** (1973) , 480–497.

[4] A. Bialynicki–Birula, On fixed points of torus actions on projective varieties, *Bull. Acad. Polon. Sci. Sér. Sci. Math. Astronom et Phys.* **22** (1974) , 1097–1101.

[5] A. Bialynicki–Birula, Some properties of the decomposition of algebraic varieties determined by actions of a torus, *Bull. Acad. Polon. Sci. Sér. Sci. Math. Astronom et Phys.* **24** (1976) , 667–674.

[6] E. Bifet, C. De Concini, C. Procesi, Cohomology of regular embeddings, *Adv. in Math., to appear* (1989) .

[7] A. Borel *et al.*, "Seminar on Transformation Groups," Ann. of Math. Studies **46**, Princeton, (1961) .

[8] C. De Concini, M. Goresky, R. MacPherson, C. Procesi, On the geometry of quadrics and their degenerations, *Comment. Math. Helvetici* **63** (1988) , 337–413.

[9] C. De Concini, C. Procesi, Complete symmetric varieties *in*," Invariant Theory," Springer Lecture Notes in Mathematics, **996** (1983) , 1–44.

[10] C. De Concini, C. Procesi, Complete symmetric varieties II *in*," Algebraic Groups and Related Topics, Advanced studies in Pure Math.," **6** (1985) , 481–513.

[11] C. De Concini, C. Procesi, Cohomology of compactifications of algebraic groups, *Duke Math. J.* **53** (1986) , 585–594.

[12] C. De Concini, A. Springer, Betti numbers of complete symmetric varieties *in*," Geometry of Today," Birkhäuser–Boston, (1984) , 87–107.

[13] S. Helgason, "Differential Geometry, Lie groups and symmetric spaces," Academic Press, (1978) .

[14] W. Hsiang , "Cohomology Theory of Topological Transformation Groups," Springer Verlag, (1975) .

[15] P. Slodowy, "Simple singularities and simple algebraic groups," Springer Lecture Notes in Mathematics, **815** (1980) .

Received November 20, 1989

Peter Littelmann
Istituto Guido Castelnuovo
Università di Roma "La Sapienza"
Piazzale Aldo Moro, 2
00185 Roma
Italia

Claudio Procesi
Istituto Guido Castelnuovo
Università di Roma "La Sapienza"
Piazzale Aldo Moro, 2
00185 Roma
Italia

Nilpotent Orbital Integrals in a Real Semisimple Lie Algebra and Representations of Weyl Groups

W. ROSSMANN[1]

Dedicated to Jacques Dixmier on his 65th birthday

0. Introduction

Let $g_{\mathbf{R}}$ be a semisimple real Lie algebra, $G_{\mathbf{R}} = Ad(g_{\mathbf{R}})$, $h_{\mathbf{R}}$ a real Cartan subalgebra. For $\lambda \in h^*_{\mathbf{R}}$, let μ_λ denote the canonical invariant measure on the $G_{\mathbf{R}}$-orbit of λ in $g^*_{\mathbf{R}}$. A well-known theorem of Harish-Chandra [8] says that

$$\lim_{\lambda \to 0} \varpi_\lambda \mu_\lambda = \kappa \mu_0, \text{ with } \varpi = \prod_{\alpha \in \Delta^+} \partial_\alpha;$$

the limit is taken through regular λ, and κ is a constant, which is non-zero if and only if $h_{\mathbf{R}}$ is fundamental.

The differential operator ϖ transforms according to *sgn* under the Weyl group W of (g, h). This is the irreducible character associated to $\{0\}$ under Springer's correspondence between nilpotent orbits in g^* and irreducible characters of W.

This paper deals with the problem of finding an analogous formula for arbitrary nilpotent $G_{\mathbf{R}}$-orbits. The problem is solved in theorem 5.3 only under an additional hypothesis. The correspondence between real nilpotent orbits and certain representations of W is given in theorem 3.3, based on the theory for complex groups developed in [16], which is recalled in §2. The general framework for the study of Fourier transforms of orbital contour-integrals is given in §1.

Barbasch and Vogan [3] solved the problem for complex orbits in the classical groups and formulated the result as a conjecture for complex orbits in general [4]. Their conjecture was proved in [10] and in [16]. For $G_{\mathbf{R}} = U(p, q)$ the solution was given by Barbasch and Vogan in [5].

I thank Michèle Vergne for correcting a mistake in the proof theorem 4.1.

[1]Supported by a grant from NSERC Canada

1. Coherent families of contours and invariant eigendistributions

Let g be a complex semisimple Lie algebra, $g_{\mathbf{R}}$ a real form of g. Let $G := Ad(g)$, $G_{\mathbf{R}} := Ad(g_{\mathbf{R}})$. Fix a Cartan subalgebra h of g and let W be the Weyl group of (g, h).

We recall some definitions and results from [15], [16]. We shall be interested in integrals of the form

$$\frac{1}{(-2\pi i)^n n!} \int_{\Gamma(\lambda)} \varphi \sigma_\lambda{}^n \tag{1}$$

where $\lambda \in h^*_{reg}$ is a regular element in the dual of the complex Cartan subalgebra h,

$$\Gamma(\lambda) \subset \Omega_\lambda := G \cdot \lambda \subset g^*$$

is an $2n$-cycle with $2n = dim_{\mathbf{C}} \Omega_\lambda$, $\sigma_\lambda{}^n$ is the n-th exterior power of the canonical holomorphic 2-form on Ω_λ,

$$\sigma_\lambda(x \cdot \xi, y \cdot \xi) := \xi([x, y]), \quad \xi \in G \cdot \lambda, \quad x, y \in g,$$

and

$$\varphi(\xi) := \int_{g_{\mathbf{R}}} f(x)\, e^{\xi(x)} dx, \quad f \in C_c^\infty(g_{\mathbf{R}})$$

is the Fourier transform of a compactly supported C^∞-function on $g_{\mathbf{R}}$. We note that φ is an entire holomorphic function on g^* satisfying for all N an estimate of the form

$$|\, \varphi(\xi) \,| \leq \frac{A e^{B\|Re\, \xi\|}}{1 + \|\xi\|^N} \tag{2}$$

The cycles $\Gamma(\lambda)$ must be restricted so that (2) guarantees the convergence of (1): one considers locally finite sums of singular m-simplices on Ω_λ,

$$\gamma = \sum_k c_k \sigma_k, \quad c_k \in \mathbb{Z},$$

which satisfy

(a) $\| Re\, \sigma_k \| \leq C$ for all k,

(b) $\displaystyle\sum_k |\, c_k \,|\, max \frac{1}{1 + \|\sigma_k\|^N} < \infty$ for some N;

$\partial\gamma$ is required to have the same properties. Using such m-chains γ one defines a homology group ${}_{\prime}H_m(\Omega_\lambda)$ as usual. An oriented, closed, m-dimensional, real submanifold of Ω_λ which admits a triangulation satisfying (a) and (b) defines an element of ${}_{\prime}H_m(\Omega_\lambda)$ independent of the triangulation.

The integral (1) depends only on the class of $\Gamma(\lambda)$ in ${}_{\prime}H_{2n}(\Omega_\lambda)$ so that we shall take $\Gamma(\lambda) \in {}_{\prime}H_{2n}(\Omega_\lambda)$.

Let

$$\mathfrak{B} := \{\text{Borel subalgebras } b \text{ of } g\},$$

the flag manifold of g,

$$\mathfrak{B}^* := \{(b, \nu) \mid b \in \mathfrak{B}, \nu \in b^{\perp} \subset g^*\},$$

the cotangent bundle of $\mathfrak{B}$, and

$$\mathfrak{S} := \{(b, \nu) \mid b \in \mathfrak{B}, \nu \in ib_{\mathbf{R}}^{\perp} := b^{\perp} \cap ig_{\mathbf{R}}^*\},$$

the conormal variety (union of conormal bundles) of the $G_{\mathbf{R}}$-orbits on $\mathfrak{B}$. For $\lambda \in h^*_{reg}$, define a map

$$p_\lambda \colon \mathfrak{B}^* \to \Omega_\lambda$$

by the formula

$$p_\lambda(u \cdot (b_1, \nu)) := u \cdot (\lambda + \nu),\ u \in U,\ \nu \in b_1^{\perp}; \tag{3}$$

$b_1 \in \mathfrak{B}$ is a fixed base-point so that $b_1 \supset h$; $U \subset G$ is a compact real form for which $h \cap u$ is a maximal torus in the Lie algebra u of U. The map p_λ is a U-equivariant, real-analytic bijection. By [15], this map induces an isomorphism

$$p_\lambda \colon H_{2n}(\mathfrak{S}) \overset{\approx}{\to} {}_{\prime}H_{2n}(\Omega_\lambda). \tag{4}$$

$H_{2n}(\mathfrak{S})$ is the (Borel-Moore) homology with arbitrary supports.

1. 1 *Remark.* The image of a contour $\Gamma(\lambda)$ on Ω_λ under the inverse map $p_\lambda^{-1}\colon \Omega_\lambda \to \mathfrak{B}^*$ does generally not lie in $\mathfrak{S}$. Rather it lies in a homology group ${}_{\prime}H_{2n}(\mathfrak{B}^*)$ defined by the same conditions (a), (b) above, with $\| Re(b, \nu) \| := \| Re\, \nu \|$ and $\| (b, \nu) \| := \| \nu \|$ for $(b, \nu) \in \mathfrak{B}^*$. $H_{2n}(\mathfrak{S})$ is free of rank equal to the number of $G_{\mathbf{R}}$-orbits in $\mathfrak{B}$ [15]. The isomorphism (4) asserts that a class in ${}_{\prime}H_{2n}(\mathfrak{B}^*)$ has a representative in $\mathfrak{S}$, which defines a class in $H_{2n}(\mathfrak{S})$. $\mathfrak{S}$ is the union of the conormal bundles on the $G_{\mathbf{R}}$-orbits on $\mathfrak{B}$, which are smooth (but not closed) real analytic submanifolds of $\mathfrak{B}^*$ of dimension

$$dim_{\mathbf{R}}\, \mathfrak{S} = dim_{\mathbf{C}}\, \mathfrak{B}^* = 2\, dim_{\mathbf{C}}\, \mathfrak{B} = 2n.$$

This means that a class in $H_{2n}(\mathcal{S})$ is represented by a finite $\mathbb{Z}$-linear combination of the chains defined by a triangulation of these submanifolds. It should be noted however that arbitrary chains constructed in this way are generally not closed in the sense of homology. In particular, the chains constructed from the conormal bundles themselves in this way are generally not cycles.

An element $\Gamma(\lambda) \in {}_{\prime}H_{2n}(\Omega_\lambda)$ will be referred to as a *contour* on Ω_λ; a *coherent family of contours* is a family of the form

$$\Gamma(\lambda) = p_\lambda \Gamma, \quad \Gamma \in H_{2n}(\mathcal{S}) \text{ fixed.}$$

$\Gamma(\lambda)$ is considered a function of $\lambda \in h^*_{reg}$.

For fixed $\Gamma(\lambda) \in H_{2n}(\Omega_\lambda)$, the integral (1), considered as a function of $f \in C_c^\infty(g_{\mathbf{R}}^*)$, defines a distribution θ on $g_{\mathbf{R}}$, which is easily seen to be an invariant eigendistribution [16]. When $\Gamma(\lambda) = p_\lambda \Gamma$, then $\theta = \theta(\Gamma, \lambda)$ is called a *coherent family of invariant eigendistributions* on $g_{\mathbf{R}}^*$; these form a $\mathbb{Z}$-module denoted $CH(g_{\mathbf{R}})$, isomorphic with $H_{2n}(\mathcal{S})$ by the map

$$H_{2n}(\mathcal{S}) \to CH(g_{\mathbf{R}}), \quad \Gamma \to \theta(\Gamma)$$

defined by the integral (1). Write this as

$$\theta(\Gamma, \lambda) = \frac{1}{(-2\pi i)^n n!} \int_{\xi \in p_\lambda \Gamma} e^\xi \sigma_\lambda{}^n, \tag{5}$$

the integral being understood in the distribution sense as explained above. We set $CH(g_{\mathbf{R}}, \mathbb{C}) := CH(g_{\mathbf{R}}) \otimes \mathbb{C}$.

We introduce the following notation.

$\Delta :=$ the root system of (g, h),

$\Delta^+ :=$ a system of positive roots for Δ,

$W :=$ the Weyl group of (g, h),

$$\pi := \prod_{\alpha \in \Delta^+} \alpha,$$

The subscipt "c" in the following symbols indicates conjugation by an element $c \in G$: $h_c := c \cdot h$, $b_c := c \cdot b_1$, Δ_c, Δ_c^+, W_c, π_c, λ_c. If $h_{c,\mathbf{R}} := h_c \cap g_{\mathbf{R}}$ is a real Cartan subalgebra of $g_{\mathbf{R}}$ we let

$$W_{c,\mathbf{R}} := Norm_{G_{\mathbf{R}}}(h_{c,\mathbf{R}}) / Cent_{G_{\mathbf{R}}}(h_{c,\mathbf{R}}),$$

$\Delta_{c,\mathbf{R}}, \Delta_{c,\mathbf{I}} :=$ real, imaginary roots in Δ_c,

$W(\Delta_{c,\mathbf{R}}) :=$ the Weyl group of $\Delta_{c,\mathbf{R}}$,

$\epsilon_{c,\mathbf{R}} := sgn \prod_{\alpha\epsilon\Delta^+_{c,\mathbf{R}}} \alpha.$

Note that

$$w\cdot(\epsilon_{c,\mathbf{R}}\pi_c) := sgn_{c,\mathbf{I}}(w)(\epsilon_{c,\mathbf{R}}\pi_c) \text{ for } w \in W_{c,\mathbf{R}}$$

where $sgn_{c,\mathbf{I}}(w) = \pm 1$ may be defined by this equation.

A real Cartan subalgebra decomposes as

$$h_c = t_c + a_c$$

so that the roots of h_c are imaginary on $t_{c,\mathbf{R}}$ and real on $a_{c,\mathbf{R}}$. Introduce a partial order on the (conjugacy classes of) real Cartan subalgebras by stipulating that

$$h_c < h_{c\prime} \text{ if } t_c \subsetneq t_{c\prime}$$

after suitable conjugation by $G_{\mathbf{R}}$.

When $\mu \in h^*_{reg}$ and $\mu_c \in ig^*_{\mathbf{R}}$, then the real orbit $G_{\mathbf{R}}\cdot\mu_c$ defines a contour $[G_{\mathbf{R}}\cdot\mu_c] \in {}_\prime H_{2n}(\Omega_\mu)$: the orientation on $G_{\mathbf{R}}\cdot\mu_c$ is specified by the form $(-i\sigma_\mu)^n$, which is real-valued and non-vanishing on $G_{\mathbf{R}}\cdot\mu_c$.

1. 2 Lemma. Let $\mu \in h^*_{reg}, c \in G, \mu_c \in ig^*_{\mathbf{R}}$. For $w \in W(\Delta_{c,\mathbf{R}})$

$$p_{w\mu}{}^{-1} \circ p_\mu[G_{\mathbf{R}}\cdot\mu_c] = [G_{\mathbf{R}}\cdot\mu_c].$$

Proof. Let $M = Cent_G(a)$, $p = m + b_1$. Then

$$G_{\mathbf{R}}\cdot\mu_c = (U \cap G_{\mathbf{R}})\cdot(M_{c,\mathrm{R}}\cdot\mu_c + ip_{c,\mathrm{R}}{}^\perp).$$

On $G_{\mathbf{R}}\cdot\mu_c$ the map $p_{w\mu}{}^{-1} \circ p_\mu$ is given by

$$k\cdot(u_M\cdot(\mu_c + \nu_M) + \nu_N). \to k\cdot(u_M\cdot(w_c\mu_c + \nu_M) + \nu_N)$$

Here $k \in U \cap G_{\mathbf{R}}$, $u_M \in U \cap M$, $\nu_M \in b_1{}^\perp \cap m$, $\nu_N \in ip_{c,\mathrm{R}}{}^\perp$, and $u_M\cdot(\mu_c + \nu_M) = m_c\cdot\mu_c$ with $m_c \in M_{c,\mathbf{R}}$. Since M_c fixes $w_c\mu_c - \mu_c$, $u_M\cdot(\mu_c + \nu_M) = m_c\cdot\mu_c$ gives $u_M\cdot(w_c\mu_c + \nu_N) = m_c w_c\cdot\mu_M$. Thus on $G_{\mathbf{R}}\cdot\mu_c$ the map $p_{w\mu}{}^{-1} \circ p_\mu$ is given by

$$k\cdot(m_c\cdot\mu_c + \nu_N) \to k\cdot(m_c w_c\cdot\mu_c + \nu_N)$$

This transformation maps $G_{\mathbf{R}}\cdot\mu_c = G_{\mathbf{R}}w_c\mu_c$ into itself. The map

$M_{c,\mathbf{R}} \cdot \mu_c \to M_{c,\mathbf{R}} w_c \mu_c$ preserves the orientations induced by the restrictions of $i\sigma_\mu$ (because $sgn_{\mathrm{I}}(w) = 1$), hence the transformation $p_{w\mu}{}^{-1} \circ p_\mu$ of $G_{\mathbf{R}}\mu_c$ preserves orientation as well. Q.E.D.

1. 3 *Remark.* It may be shown that for *any* $w \in W_{c,\mathbf{R}}$

$$p_{w\mu}{}^{-1} \circ p_\mu[G_{\mathbf{R}} \cdot \mu_c] = sgn_{\mathrm{I}}(w)[G_{\mathbf{R}} \cdot \mu_c] + \cdots$$

where the dots indicate a $\mathbb{Z}$-linear combination of contours $[G_{\mathbf{R}} \cdot \mu_{c'}]$ with $h_{c'} < h_c$.

The contours $[G_{\mathbf{R}} \cdot \mu_c]$ fit into coherent families as follows.

1. 4 Theorem. (a) *Let S be a $G_{\mathbf{R}}$-orbit on $\mathfrak{B}$. There is a unique coherent family $\Gamma(S, \cdot\,)$ so that*

$$\Gamma(S, \mu) = [G_{\mathbf{R}} \cdot \mu_c]$$

whenever $c \in G$ and $\mu \in h^*_{reg}$ satisfy the following conditions:

(1) $\mu_c \in i\mathfrak{g}^*_{\mathbf{R}}$
(2) $b_c \in S$
(3) $\mu_c(H_{\alpha_c}) > 0$ if α_c *is an imaginary root of* h_c *on* b_c.

(b) The coherent families of eigendistributions $\theta(S, \cdot\,)$ corresponding to the $\Gamma(S, \cdot\,)$, S running over the $G_{\mathbf{R}}$-orbits on $\mathfrak{B}$, form a $\mathbb{Z}$-basis for the $\mathbb{Z}$-module $CH(\mathfrak{g}_{\mathbf{R}})$.

(c) Let $(h_c)_{\mathbf{R}}$, $c \in G$, be a real Cartan subalgebra of $\mathfrak{g}_{\mathbf{R}}$. Let S be a $G_{\mathbf{R}}$-orbit on $\mathfrak{B}$ so that $b_c \in S$. Then

$$\theta(S, \lambda) \equiv \frac{1}{\epsilon_{c,\mathbf{R}}\pi_c} \sum_{w \in W_{c,\mathbf{R}}} sgn_{c,\mathrm{I}}(w)\, e^{w^{-1}\lambda_c} \tag{6}$$

on $(h_c)_{\mathbf{R}}$ and vanishes on a real Cartan subalgebra $(h_{c'})_{\mathbf{R}}$ unless $h_{c'} \leq h_c$.

Proof. (a) Fix S and a pair (c, μ) with the indicated properties. Let $\Gamma = \Gamma(c, \mu, S) \in {}'H_{2n}(\mathfrak{B}^*)$ be defined by

$$\Gamma = p_\lambda \circ p_\mu{}^{-1}[G_{\mathbf{R}} \cdot c\mu]. \tag{7}$$

We show that Γ is independent of (c, μ). Let (c', μ') be another such pair. In view of (2) we may assume that

$$b_c = b_{c'} =: b_S.$$

In view of (1), h_c and $h_{c\prime}$ are both conjugation-stable Cartan subalgebras in b_S. It is well-known that such Cartan subalgebras are $G_{\mathbf{R}}$-conjugate. We may therefore assume that

$$h_c = h_{c\prime} =: h_S.$$

Thus $c = c\prime h$ for some $h \in Cent_G(h)$. We may as well assume $c = c\prime$. Observe: There is $w \in W(\Delta_{c,\mathbf{R}})$ so that

$$c \cdot \mu \text{ and } wc \cdot \mu\prime \text{ lie in the same connected component of } (h_{S,\mathbf{R}})^*_{reg}. \quad (8)$$

Reason: Such a connected component is a chamber cut out by the real and imaginary roots of h_S. In view of (3), only reflections in real roots are needed to transform the chamber containing $c \cdot \mu$ into the chamber containing $c\mu\prime$.

Fix $w \in W(\Delta_{c,\mathbf{R}})$ so that $c \cdot \mu$ and $wc \cdot \mu\prime$ lie in the same connected component of $(h_{S,\mathbf{R}})^*_{reg}$. There is a continuous curve $\mu(t)$, $0 \leq t \leq 1$, in h^* so that

$$\mu(0) = \mu,\ \mu(1) = c^{-1} wc\mu\prime,\ c \cdot \mu(t) \in (h_{S,\mathbf{R}})^*_{reg},\ 0 \leq t \leq 1.$$

Consider the following one-parameter family of cycles on $\mathfrak{B}^*$:

$$p_{\mu(t)}{}^{-1}[G_{\mathbf{R}} \cdot c\mu(t)],\quad 0 \leq t \leq 1.$$

This is a homotopy, which satisfies the conditions (a) and (b) defining $\prime H_{2n}(\mathfrak{B}^*)$ uniformly in t, hence represents the same class in $\prime H_{2n}(\mathfrak{B}^*)$. Thus

$$p_{\mu}{}^{-1}[G_{\mathbf{R}} \cdot c\mu] = p_{c^{-1} wc\mu\prime}[G_{\mathbf{R}} \cdot c\mu\prime].$$

By lemma 1.2,

$$p_{c^{-1} wc\mu\prime}{}^{-1}[G_R \cdot c\mu\prime] = p_{\mu\prime}[G_{\mathbf{R}} \cdot c\mu\prime].$$

This proves that Γ is independent of (c, μ).

As remarked in 1.1, the class $\Gamma \in \prime H_{2n}(\mathfrak{B}^*)$ has a representative which lies on $\mathcal{S}$ and defines an element $\Gamma_S \in H_{2n}(\mathcal{S})$. Let

$$\Gamma(S, \lambda) = p_\lambda \Gamma_S.$$

This is the required coherent family.

(b) Let S be a $G_{\mathbf{R}}$-orbit on $\mathfrak{B}$, $\mathcal{S}_S$ the part of $\mathcal{S}$ over S, i.e.

$$\mathcal{S}_S := \{(b, \nu) \in \mathcal{S} \mid b \in S\},$$

(the conormal bundle of the $G_{\mathbf{R}}$-orbit S). The element $\Gamma_S \in H_{2n}(\mathcal{S})$ corresponding to $\Gamma(S, \cdot\,)$ is of the form

$$\Gamma_S = [\mathcal{S}_S] + \cdots$$

where $[\mathcal{S}_S]$ is the $2n$-chain on $\mathcal{S}$ defined by an orientation on $\mathcal{S}_S$ and the dots indicate a $\mathbb{Z}$-linear combination of $[\mathcal{S}_{S'}]$'s with $S' \subset \partial S := \overline{S} - S$ ([15]). This implies that the Γ_S form a $\mathbb{Z}$-basis for $H_{2n}(\mathcal{S})$, hence (b).

(c) For $(h_c)_{\mathbf{R}}$ of compact type, this follows from [13]. The general case is proved by a familiar reduction thereto using parabolic subalgebras. ([22], *Part* I, §3.7. The positive constant in Theorem 32 of [22] becomes $c = 1$ with the normalization of forms used here. This may be seen as in §2 of [15].)

Q.E.D.

1. 5 *Remarks.* We mention some supplementary facts about coherent families.

(a) By a well-known result of Harish-Chandra, *any* invariant eigendistribution θ with regular infinitesimal character $\lambda \in h^*_{reg}$ is given by a locally L^1-function, whose restriction to a real Cartan subalgebra $h_{c,\mathbf{R}}$ is given by a formula

$$\theta \equiv \frac{1}{\epsilon_{c,\mathbf{R}}\pi_c} \sum_{y \in W_c} m_{c,y}\, e^{y^{-1}\lambda_c} \quad on\ (h_c)_{\mathbf{R}},$$

where the $m_{c,y}$ are $\mathbb{C}$-valued, locally constant functions on $h_{\mathbf{R},reg}$. A family $\theta(\,\cdot\,)$ of the above type is *coherent* (i.e. $\in CH(g_{\mathbf{R}}, \mathbb{C})$) if and only if the $m_{c,y}$ are independent of λ. (Compare the definition of "coherent familiy of invariant eigendistributions" given in [17].) In particular, a coherent family $\theta(\,\cdot\,)$ extends to a holomorphic function on all of h^* with values in the space of distributions on $g_{\mathbf{R},\,reg}$.

(b) A family $\Gamma(\lambda) \in {}'H_{2n}(\Omega_\lambda)$ of contours parametrized by $\lambda \in h^*_{reg}$ is *coherent* in the sense defined above if and only if the integral (1) is a holomorphic function of $\lambda \in h^*_{reg}$ (for arbitrary φ).

(c) For any $\lambda_o \in h^*_{reg}$ the map $\theta(\,\cdot\,) \to \theta(\lambda_o)$ induces a linear isomorphism from $CH(g_{\mathbf{R}}, \mathbb{C})$ to the space of invariant eigendistributions with infinitesimal character λ_o.

2. Representations of Weyl groups

We require some constructions from [16]. Recall the bijection p_λ: $\mathfrak{B}^* \to \Omega_\lambda$, $\lambda \in h^*_{reg}$, defined in (3). For any $w \in W$, define a transformation $a_\lambda(w)$ of $\mathfrak{B}^*$ by

$$a_\lambda(w) := p_{w\lambda}{}^{-1} \circ p_\lambda : \mathfrak{B}^* \to \mathfrak{B}^*. \tag{9}$$

Let $\mathcal{N}$ be the nilpotent cone in $\boldsymbol{g}^*$ and

$$p: \mathfrak{B}^* \to \mathcal{N},\ (b, \nu) \to \nu$$

the *Springer map*. For any subset V of $\mathcal{N}$, let

$$\mathfrak{B}^*(V) := p^{-1}(V) \subset \mathfrak{B}^*.$$

Choose $\epsilon > 0$ and let

$$U = \{\nu \in \mathcal{N} \mid \|\nu - \nu\prime\| < \epsilon \text{ for some } \nu\prime \in V\}$$

Assume that for sufficiently small $\epsilon > 0$ the inclusion $i: \mathfrak{B}^*(V) \to \mathfrak{B}^*(U)$ admits a proper homotopy inverse in the sense that there is a continuous map $j: \mathfrak{B}^*(U) \to \mathfrak{B}^*(V)$ so that

$$j \circ i \sim 1 \text{ on } \mathfrak{B}^*(V), \quad i \circ j \sim 1 \text{ on } \mathfrak{B}^*(U)$$

with "$\sim$" meaning "properly homotopic". Then for $\lambda \in h^*_{reg}$ in some ball around 0, the map

$$a(w) := j \circ a_\lambda(w) \circ i$$

induces a transformation of $\mathfrak{B}^*(V)$ whose proper homotopy class is independent of λ. This gives a representation of W on $H_*(\mathfrak{B}^*(V))$.

Applied to $V = \{\nu\}$, $\nu \in \mathcal{N}$, the above construction produces a representation of W on the homology of $\mathfrak{B}^*(\nu)$; $\mathfrak{B}^*(\nu)$ may be identified with

$$\mathfrak{B}^\nu := \{b \in \mathfrak{B} \mid \nu \in b^\perp\}.$$

The component group

$$A_\nu := G_\nu/(G_\nu)_o$$

of the stabilizer G_ν of ν in G acts naturally on $H_*(\mathfrak{B}^\nu)$ by W-automorphisms, so that one has a representation of $W \times A_\nu$ thereon. In highest degree $2e := 2\, dim_{\mathbf{C}}\, \mathfrak{B}^\nu$, this representation decomposes as

$$H_{2e}(\mathfrak{B}^\nu) \approx \sum_{\varphi \epsilon \Phi_\nu} \chi_{\nu,\varphi} \otimes \varphi$$

where Φ_ν is a set of irreducible characters of A_ν and $\chi_{\nu,\varphi}$ an irreducible character of W. Every irreducible character of W is of this form $\chi_{\nu,\varphi}$, and two pairs (ν, φ), $(\nu\prime, \varphi\prime)$ correspond to the same character if and only if they are conjugate by G. These results are due to Springer [19],

[20] with a different construction of the representations. The construction outlined here is given in [16].

3. Decomposition of the coherent continuation representation

We apply the construction of representations of W outlined in §2 with $V = \mathcal{N}_{\mathbf{R}} := \mathcal{N} \cap i\mathfrak{g}_{\mathbf{R}}^{*}$. (As explained in [16], it follows from general facts in algebraic topology that $V = \mathcal{N}_{\mathbf{R}}$ satisfies the conditions required for the construction.) Note that for $V = \mathcal{N}_{\mathbf{R}}$, $\mathfrak{B}^*(V) = \mathcal{S}$, so that we get a representation of W on $H_*(\mathcal{S})$. In top degree $2n = dim_{\mathbf{R}}\mathcal{S}$, we get a representation of W on on $H_{2n}(\mathcal{S})$. Essentially by definition:

3. 1 Lemma. *The isomorphism* $H_{2n}(\mathcal{S}) \to CH(\mathfrak{g}_{\mathbf{R}})$, $\Gamma \to \theta(\Gamma)$, *satisfies*

$$\theta(w\Gamma, \lambda) := \theta(\Gamma, w^{-1}\lambda).$$

Proof. Let $\Gamma \in H_{2n}(\mathcal{S})$. By definition (9) of the action of W on $H_{2n}(\mathcal{S})$,

$$w \cdot \Gamma \sim {p_{w\lambda}}^{-1} \circ p_{\lambda}(\Gamma)$$

for any $\lambda \in \mathfrak{h}_{reg}^{*}$ in some ball around 0. Here " $\sim$ " means "properly homotopic". Thus

$$p_{w\lambda} w\Gamma \sim p_{\lambda}\Gamma$$

which gives

$$\theta(w\Gamma, w\lambda) = \theta(\Gamma, \lambda)$$

as required. Q.E.D.

Thus under the isomorphism $H_{2n}(\mathcal{S}) \to CH(\mathfrak{g}_{\mathbf{R}})$, $\Gamma \to \theta(\Gamma, \cdot\ )$, the representation of W on $H_{2n}(\mathcal{S})$ becomes the *coherent continuation representation* of W on the $\mathbb{Z}$-module $CH(\mathfrak{g}_{\mathbf{R}})$ of coherent families $\theta(\ \cdot\ )$ of invariant eigendistribution given by

$$(w \cdot \theta)(\lambda) = \theta(w^{-1} \cdot \lambda).$$

3. 2 Corollary. *As representation of* W,

$$H_{2n}(\mathcal{S}) \approx \sum_{c} ind_{W_{c,\mathbf{R}}}^{W} sgn_{c,\mathbf{I}} \tag{10}$$

where $c \in G$ *is chosen so that* $\mathfrak{h}_{c,\mathbf{R}}$ *runs over a complete set of representatives for the* $G_{\mathbf{R}}$*-conjugacy classes of Cartan subalgebras of* $\mathfrak{g}_{\mathbf{R}}$, *and* $W_{c,\mathbf{R}}$ *is considered as a subgroup of* W *via* c.

Proof. This follows from the lemma and theorem 1.4 (c). Q.E.D.

3. 3 Theorem. *The representation of* W on $H_{2n}(\mathcal{S})$ *decomposes as*

$$H_{2n}(\mathcal{S}) \approx \sum_{\nu,\varphi} m(\nu,\varphi)\,\chi_{\nu,\varphi}$$

where ν *runs over a set of representatives for the nilpotent* $G_{\mathbf{R}}$*-orbits,* φ *over the set of irreducible representations of* A_ν *in* Φ_ν *which contain a vector fixed by* $A_{\nu,\mathbf{R}}$*, and* $m(\nu,\varphi)$ *is the dimension of the space of these vectors.*

The proof of theorem 3.3 requires some preliminary lemmas. Consider the Springer map $p\colon \mathcal{S} \to \mathcal{N}_{\mathbf{R}}$. Let

$$\mathcal{N}_{\mathbf{R}} = \bigcup_{O} O$$

be the decomposition into $G_{\mathbf{R}}$-orbits, and correspondingly

$$\mathcal{S} = \bigcup_{O} \mathcal{S}(O)$$

with $\mathcal{S}(O) := p^{-1}(O)$.

3. 4 Lemma. *For all* O,

$$dim_{\mathbf{R}}\mathcal{S}(O) = 2n.$$

Furthermore, the restriction of $\mathcal{S} \to \mathcal{N}_{\mathbf{R}}$ to $\mathcal{S}(O)$ *is a* $G_{\mathbf{R}}$*-equivariant fibration with fibre* $\mathcal{B}^\nu$ *over* $\nu \in O$:

$$\mathcal{B}^\nu \overset{\subset}{\rightarrow} \mathcal{S}(O) \rightarrow O. \tag{11}$$

Proof. The second assertion is evident. The first is a consequence of results of Spaltenstein and Steinberg, as follows. According to [18], the irreducible components of the complex variety $\mathcal{B}^\nu$ have the same dimension, say

$$dim_{\mathbf{C}}\,\mathcal{B}^\nu = e.$$

According to [21], the components of $\mathcal{B}^\nu \times \mathcal{B}^\nu$are the intersections of the components of the variety

$$\mathcal{Z} := \{(b, b\prime, \nu) \mid \nu \in b^{\perp} \cap b\prime^{\perp}\}$$

with the fibre $\mathcal{B}^\nu \times \mathcal{B}^\nu$ of the map $\mathcal{Z} \to \mathcal{N}$. Furthermore, the components of $\mathcal{Z}$ all have the same C-dimension, namely $2n$, because they are the closures of the conormal bundles of the G-orbits on $\mathcal{B} \times \mathcal{B}$ (under the diagonal action). From the fibration

$$\mathfrak{B}^\nu \times \mathfrak{B}^\nu \hookrightarrow \mathfrak{Z}(\mathcal{O}) \to \mathcal{O}$$

induced by $\mathfrak{Z} \to \mathcal{N}$ over a fixed G-orbit $\mathcal{O} = G \cdot \nu$, one sees that

$$dim_{\mathbf{C}}\, \mathcal{O} + 2\ dim_{\mathbf{C}} \mathfrak{B}^\nu = 2n.$$

For a $G_{\mathbf{R}}$-orbit $O = G_{\mathbf{R}} \cdot \nu$, $\nu \in \mathcal{N}_{\mathbf{R}}$, the fibration

$$\mathfrak{B}^\nu \hookrightarrow \mathcal{S}(O) \to O$$

mentioned above now gives

$$\begin{aligned} dim_{\mathbf{R}}\, \mathcal{S}(O) &= dim_{\mathbf{R}} O + dim_{\mathbf{R}} \mathfrak{B}^\nu \\ &= dim_{\mathbf{C}} O_{\mathbf{C}} + 2\ dim_{\mathbf{C}} \mathfrak{B}^\nu \\ &= 2n, \end{aligned}$$

as required. Q.E.D.

For $\nu \in \mathcal{N}_{\mathbf{R}}$ set

$$A_{\nu,\mathbf{R}} := G_{\nu,\mathbf{R}} / G_{\nu,\mathbf{R}} \cap (G_\nu)_o,$$

a subgroup of

$$A_\nu := G_\nu / (G_\nu)_o.$$

3. 5 Lemma. *For any $G_{\mathbf{R}}$-orbit $O = G_{\mathbf{R}} \cdot \nu$ on $\mathcal{N}_{\mathbf{R}}$,*

$$H_{2n}(\mathcal{S}(O)) \approx H_{2e}(\mathfrak{B}^\nu)^{A_{\nu,\mathbf{R}}},$$

where $e = dim_{\mathbf{C}} \mathfrak{B}^\nu$ and the right side denotes the $A_{\nu,\mathbf{R}}$-invariants under the natural action of A_ν on $H_{2e}(\mathfrak{B}^\nu)$.

Proof. The fibration

$$\mathfrak{B}^\nu \hookrightarrow \mathcal{S}(O) \to O \tag{11}$$

leads to an isomorphism

$$H_{2n}(\mathcal{S}(O)) \approx H_{2e}(\mathfrak{B}^\nu)^{\pi_1(O,\nu)}, \tag{12}$$

which may be described as follows. A continuous path from ν to ν' in O gives an isomorphism $H_*(\mathfrak{B}^\nu) \to H_*(\mathfrak{B}^{\nu'})$ by trivialization of the fibration (11) along the path. On the invariants of the monodromy action of $\pi_1(O, \nu)$ in $H_*(\mathfrak{B}^\nu)$, this isomorphism is independent of the path. An element $\Gamma \in H_{2n}(\mathcal{S}(O))$ decomposes into its fibres

$$\Gamma_\nu \in H_{2e}(\mathfrak{B}^\nu)^{\pi_1(O,\, \nu)}$$

under $\mathcal{S}(O) \to O$:

$$\Gamma_\nu \hookrightarrow \Gamma \to O.$$

The Γ_ν and $\Gamma_{\nu'}$ belonging to ν and ν' are related by the isomorphism mentioned. $\Gamma \leftrightarrow \Gamma_\nu$ gives the isomorphism (12).

An element $[\gamma]$ of $\pi_1(O, \nu)$ may be realized in the form $\gamma(t) = g(t)\cdot\nu$, where $g(t)$ $(0 \leq t \leq 1)$ is a continuous path in $G_{\mathbf{R}}$ with $g(0) = 1$, $g(1) \in G_{\nu\mathbf{R}}$. The monodromy action of $[\gamma]$ on $H_*(\mathcal{B}^\nu)$ coincides with the natural action of $g(1)$. If $g(1) \in (G_\nu)_o$, then $g(1)$ acts trivially, hence the action of $\pi_1(O, \nu)$ on $H_*(\mathcal{B}^\nu)$ factors though the homomorphism $\pi_1(O, \nu) \to A_{\nu,\mathbf{R}}$, $[\gamma] \to [g(1)]$. Q.E.D.

Fix a $G_{\mathbf{R}}$-orbit $O = G_{\mathbf{R}}\cdot\nu$ on $\mathcal{N}_{\mathbf{R}}$. Let $\overline{O}$ be its closure, $\partial O := \overline{O} - O$ its topological boundary.

3. 6 Lemma. *The natural maps*

$$0 \to H_{2n}(\mathcal{S}(\partial O)) \to H_{2n}(\mathcal{S}(\overline{O})) \to H_{2n}(\mathcal{S}(O)) \to 0$$

form an exact sequence of W-modules.

Proof. The couple $\partial O \subset \overline{O}$ of closed $\mathbb{R}$-subvarietes of $\mathcal{N}_{\mathbf{R}}$ gives a long-exact sequence of W-modules

$$\cdots \to H_{i+1}(\mathcal{S}(O)) \to H_i(\mathcal{S}(\partial O)) \to H_i(\mathcal{S}(\overline{O})) \to$$

$$\to H_i(\mathcal{S}(O)) \to H_{i-1}(\mathcal{S}(\partial O)) \to \cdots$$

In top degree $i = 2n$ this gives

$$0 \to H_{2n}(\mathcal{S}(\partial O)) \to H_{2n}(\mathcal{S}(\overline{O})) \to H_{2n}(\mathcal{S}(O)) .$$

The surjectivity of the map on the right follows from the fact that $\overline{O}$ and ∂O admit compatible CW-decompositions [7] and standard facts about the homology of CW-complexes ([12], *Theorem 4.1*). Q.E.D.

Proof of theorem 3. 1. The map $H_{2n}(\mathcal{S}(\overline{O})) \to H_{2n}(\mathcal{S})$ is injective, as one sees from the exact sequence associated to the inclusion $\mathcal{S}(\overline{O}) \subset \mathcal{S}$. $H_{2n}(\mathcal{S}(\overline{O}))$ may therefore be considered a W-submodule of $H_{2n}(\mathcal{S})$, and $H_{2n}(\mathcal{S}(O))$ a W-subquotient of $H_{2n}(\mathcal{S})$:

$$H_{2n}(\mathcal{S}(O)) \approx H_{2n}(\mathcal{S}(\overline{O}))/\sum_{O' < O} H_{2n}(\mathcal{S}(O)) ,$$

where $O' < O$ means $O' \subset \partial O$.

The filtration of $\mathcal{N}_{\mathbf{R}}$ according to the dimension of the $G_{\mathbf{R}}$-orbits,

$$\mathcal{N}_{i,\mathbf{R}} := \bigcup_{\dim_{\mathbf{R}} O \leq i} O$$

induces a filtration of $\mathcal{S}$,

$$\mathcal{S}_i := \bigcup_{\dim_{\mathbf{R}} O \leq i} \mathcal{S}(O)$$

and hence of $H_{2n}(\mathcal{S})$:

$$H_{2n}(\mathcal{S}_i) = \sum_{\dim_{\mathbf{R}} O \leq i} H_{2n}(\mathcal{S}(\bar{O})).$$

The corresponding graded group is then

$$gr\, H_{2n}(\mathcal{S}) = \sum_{O} H_{2n}(\mathcal{S}(O)). \tag{13}$$

From lemma 3.5

$$H_{2n}(\mathcal{S}(O)) \approx H_{2e}(\mathcal{B}^{\nu})^{A_{\nu}(\mathbf{R})} \tag{14}$$

and from Springer's theory (§2)

$$H_{2e}(\mathcal{B}^{\nu}) \approx \sum_{\varphi \epsilon \Phi_{\nu}} \chi_{\nu,\varphi} \otimes \varphi$$

as representation of $W \times A_{\nu}$. Putting these pieces together one gets the decomposition of theorem 3.3, since $gr\, H_{2n}(\mathcal{S}) \approx H_{2n}(\mathcal{S})$ as representations of W. Q.E.D.

We point out the following consequence of theorem 3.3 and corollary 3.2.

3. 7 Corollary. *Let* $\nu \in \mathcal{N}_{\mathbf{R}}$ *and* $G \cdot \nu \cap i\mathfrak{g}_{\mathbf{R}} = G_{\mathbf{R}} \cdot \nu_1 \cup \cdots \cup G_{\mathbf{R}} \cdot \nu_m$ *the decomposition of* $G \cdot \nu \cap i\mathfrak{g}_{\mathbf{R}}$ *into distinct* $G_{\mathbf{R}}$*-orbits. Then*

$$m = \sum_{c} [\, ind_{W_{c,\mathbf{R}}}^{W} sgn_{c,\mathbf{I}} : \chi_{\nu,1}].$$

Proof. By theorem 3.3

$$m = [H_{2n}(\mathcal{S}) : \chi_{\nu,\, 1}];$$

by corollary 3.2

$$[H_{2n}(\mathcal{S}) : \chi_{\nu,\, 1}] = \sum_{c} [\, ind_{W_{c,\mathbf{R}}}^{W} sgn_{c,\mathbf{I}} : \chi_{\nu,1}].$$

Q.E.D.

3. 8 *Remark.* For $G_{\mathbf{R}} = U(p, q)$, Barbasch and Vogan [5] prove the decomposition of the coherent continuation representation (defined in terms of characters) according to nilpotent $G_{\mathbf{R}}$-orbits by direct verification based on (10). (In that case A_{ν} is trivial for all $\nu \in \mathcal{N}$.)

4. Nilpotent orbital integrals

We return to the integral

$$\theta(\Gamma, \lambda) = \frac{1}{(-2\pi i)^n n!} \int_{\xi \in p_\lambda \Gamma} e^{\xi} {\sigma_\lambda}^n, \tag{5}$$

with $\Gamma \in H_{2n}(\mathcal{S})$. Recall the natural isomorphisms

$$gr\, H_{2n}(\mathcal{S}) \approx \sum_{O} H_{2n}(\mathcal{S}(O)), \tag{13}$$

$$H_{2n}(\mathcal{S}(O)) \approx H_{2e}(\mathcal{B}^{\nu})^{A_\nu(\mathbf{R})}. \tag{14}$$

The inclusion $\mathcal{B}^{\nu} \overset{\subset}{\rightarrow} \mathcal{B}$ induces a W-injection

$$H_{2e}(\mathcal{B}^{\nu})^{A_\nu} \overset{\subset}{\rightarrow} H_{2e}(\mathcal{B}) \tag{15}$$

([16], *Lemma* 3.1.) The character of W on $H_{2e}(\mathcal{B}^{\nu})^{A_\nu}$ is $\chi_\nu := \chi_{\nu,1}$.

According to Borel [6], the cohomology ring of $\mathcal{B}$ can be described as follows. For $\lambda \in h^*$, let τ_λ denote the U-invariant 2-form on $\mathcal{B}$ which at the base point b_1 is given by

$$\tau_\lambda([x \cdot b_1, y \cdot b_1]) := \lambda([x, y]) \text{ for } x, y \in u.$$

Let I^+ denote the ideal in the ring $\mathbf{C}[h]$ of polynomial function on h generated by the W-invariants without constant term. There is a unique isomorphism of rings

$$\mathbf{C}[h]/I^+ \rightarrow H^*(\mathcal{B}) \tag{16}$$

so that

$$[\lambda] \rightarrow [\tfrac{1}{-2\pi i} \tau_\lambda].$$

The dual space of $\mathbf{C}[h]/I^+$ is the space $\mathcal{H}(h^*)$ of W-harmonic polynomials on h^*, and the transpose of the isomorphism (16) is a W-isomorphism

$$H_*(\mathcal{B}) \rightarrow \mathcal{H}(h^*),\ \gamma \rightarrow c_\gamma \tag{17}$$

given by

$$c_\gamma(\lambda) = \frac{1}{(-2\pi i)^e} \int_\gamma {\tau_\lambda}^e$$

for $\gamma \in H_{2e}(\mathcal{B})$.

For $\nu \in \mathcal{N}$, let $\mathcal{H}_\nu(h^*)$ denote the space of W-harmonic polynomials on h^* which are homogeneous of degree $e = dim_{\mathbf{C}}\, \mathcal{B}^{\nu}$ and transform according to the irreducible character χ_ν of W. From (13), (14), (15), and

(17), together with

$$H_{2n}(\mathcal{S}(O)) \approx H_{2n}(\mathcal{S}(\overline{O}))/\sum_{O_\prime < O} H_{2n}(\mathcal{S}(O)),$$

there results a map

$$H_{2n}(\mathcal{S}(\overline{O})) \to \mathcal{H}_\nu(h^*),\ \Gamma \to c_\Gamma.$$

We introduce the following notation.

$O := G_{\mathbf{R}} \cdot \nu$, a $G_{\mathbf{R}}$-orbit on $\mathcal{N}_{\mathbf{R}} \subset i\mathfrak{g}_{\mathbf{R}}^*$;
$2d := dim_{\mathbf{R}}\, O = dim_{\mathbf{C}} G\cdot \nu$;
$e := dim_{\mathbf{C}}\, \mathcal{B}^\nu$;
$\sigma_\nu :=$ the canonical holomorphic 2-form on $G \cdot \nu$;
$\mu_\nu :=$ the canonical measure on O, i.e.

$$\mu_\nu(\varphi) = \frac{1}{(-2\pi i)^d} \int_{\xi\, \in\, O} \varphi {\sigma_\nu}^d\,;$$

$\theta_\nu :=$ the Fourier transform of μ_ν, i.e.

$$\theta_\nu(f) := \mu_\nu(\varphi) \text{ where } \varphi(\xi) := \int_{\mathfrak{g}_{\mathbf{R}}} f(x)\, e^{\xi(x)}\, dx.$$

With this notation:

4. 1 Theorem. Let $\Gamma \in H_{2n}(\mathcal{S}(\overline{O}))$, $\theta(\Gamma, \cdot\,)$ *the corresponding coherent family of invariant eigendistributions. Then*

$$\theta(\Gamma, \lambda) = c_\Gamma(\lambda)\theta_\nu + o(\|\lambda\|^e). \tag{18}$$

Proof. Let $\mathcal{O} = G\cdot \nu$ be the complex orbit containing $O = G_{\mathbf{R}}\cdot \nu$, $\mathcal{B}^*(\mathcal{O})$ the inverse image of $\mathcal{O}$ under $p\colon \mathcal{B}^* \to \mathcal{N}$. Since $\mathcal{B}^*(\mathcal{O}) \subset \mathcal{B} \times \mathcal{O}$, we may consider $\tau_\lambda + \sigma_\nu$ as 2-form on $\mathcal{B}^*(\mathcal{O})$. By [16], this form agrees on $\mathcal{B}^*(\mathcal{O})$ with the pull-back ${p_\lambda}^*\sigma_\lambda$ of the form σ_λ on Ω_λ by the map p_λ: $\Omega_\lambda \to \mathcal{B}^*$. Note that

$$d + e = n$$

because of the fibration

$$\mathcal{B}^\nu \overset{\subset}{\to} \mathcal{S}(O) \to O \tag{11}$$

of lemma 3.4.

Let Γ by a $2n$-cycle on $\mathcal{S}(O)$. As noted in the proof of lemma 3.5, the fibration (11) induces a fibre-decomposition

$$\Gamma_\nu \overset{\subset}{\to} \Gamma \to O, \tag{19}$$

with Γ_ν a $2e$-cycle on $\mathcal{B}^\nu$.

Now compute the integral (1) of §1 as a repeated integral according to (19):

$$\frac{1}{(-2\pi i)^n n!}\int_{p_\lambda\Gamma}\varphi\sigma_\lambda{}^n$$

$$= \frac{1}{(-2\pi i)^n n!}\int_\Gamma (\varphi\circ p_\lambda)(\tau_\lambda+\sigma_\nu)^n$$

$$= \frac{1}{(-2\pi i)^n n!}\int_\Gamma \sum_{r+s=n}(\varphi\circ p_\lambda)\ \frac{n!}{r!\,s!}\tau_\lambda{}^r\sigma_\nu{}^s$$

$$= \frac{1}{(-2\pi i)^n e!d!}\int_{\mu\,\in\,O}\left\{\int_{b\,\in\,\Gamma_\mu}\varphi(p_\lambda(b,\mu))\,\tau_\lambda{}^e\right\}\sigma_\nu{}^d$$

$$+ \mathrm{o}(\|\lambda\|^e)$$

Write $(b, \mu) = u\cdot(b_1, \mu')$ with $u \in U$, and $\mu' \in b_1{}^\perp$. Then

$$\varphi(p_\lambda(b,\mu)) = \varphi(u\cdot(\lambda+\mu'))$$

$$= \varphi(u\cdot\mu') + o(\|u\cdot\lambda\|)$$

$$= \varphi(\mu) + o(\|\lambda\|).$$

Thus the integral

$$= \frac{1}{(-2\pi i)^d d!}\int_{\mu\,\in\,O}\varphi(\mu)\left\{\frac{1}{(-2\pi i)^e e!}\int_{\Gamma_\mu}\tau_\lambda{}^e\right\}\sigma_\nu{}^d + o(\|\lambda\|^e).$$

The expression in parentheses is independent of μ and equals $c_\Gamma(\lambda)$. Thus the integral

$$= c_\Gamma(\lambda)\,\frac{1}{(-2\pi i)^d d!}\int_{\mu\,\in\,O}\varphi(\mu)\,\sigma_\nu{}^d + o(\|\lambda\|^e),$$

which is the desired formula when Γ is a $2n$-cycle on $\mathcal{I}(O)$. Since $\partial\bar{O} = \bigcup O'$ with $e' > e$, the same formula holds for $\Gamma \in H_{2n}(\mathcal{I}(\bar{O}))$, with c_Γ depending only on the image of Γ under the map $H_{2n}(\mathcal{I}(\bar{O})) \to H_{2n}(\mathcal{I}(O))$. Q.E.D.

5. A limit formula

We fix a real Cartan subalgebra of $g_{\mathbf{R}}$, which we may as well assume to be $h_{\mathbf{R}}$ for the previously fixed h. It will be convenient to use another parametization of the $\Gamma(S,\ \cdot\ )$ with S containing a fixed-point of h, as follows.

For $y \in W$, let $b_y := y \cdot b_1$, $S_y := G_{\mathbf{R}} \cdot b_y$. Then $\Gamma(S_y, \cdot\,)$ is the unique coherent family of contours satisfying

$$\Gamma(S_y, \mu) = [G_{\mathbf{R}} \cdot y\mu] \text{ if } y\mu(H_\alpha) > 0 \text{ for all } \alpha \in (y\Delta^+) \cap \Delta_{\mathbf{I}}.$$

Let

$$C_y := \{\mu \in h^*_{\mathbf{R},reg} \mid \mu(H_\alpha) > 0 \text{ for } \alpha \in (y\Delta^+) \cap \Delta_{\mathbf{I}}\}$$

so that $\Gamma(S_y, \cdot\,)$ is the unique coherent family satisfying

$$\Gamma(S_y, y^{-1}\mu) = [G_{\mathbf{R}} \cdot \mu] \text{ if } \mu \in C_y.$$

Define

$$\Gamma(C, \cdot\,) := y^{-1}\Gamma(S_y, \cdot\,) \text{ if } C = C_y.$$

$\Gamma(C, \cdot\,)$ is a coherent family depending only on the chamber C for $\Delta_{\mathbf{I}}$. As (C, y) runs over a set of representatives of the $W_{\mathbf{R}}$-orbits of pairs satisfying $C_y = C$, the corresponding families $y \cdot \Gamma(C, \cdot\,)$ run through $\{\Gamma(S, \cdot\,) \mid h \subset b \text{ for some } b \in S\}$. Write $\theta(C, \cdot\,)$ for the coherent family of invariant eigendistributions corresponding to $\Gamma(C, \cdot\,)$.

The chamber C for $\Delta_{\mathbf{I}}$ will remain fixed from now on. Let

$$\mathcal{N}_C := \bigcap_{\mu \in C} \left(\mathcal{N} \cap \overline{G_{\mathbf{R}} \cdot \mathbb{R}^{\times}_{+}\mu}\right).$$

Let $O = G_{\mathbf{R}} \cdot \nu$ be a nilpotent orbit in $ig^*_{\mathbf{R}}$, $\mathcal{O} = G \cdot \nu$ the complex orbit containing O. We shall repeatedly refer to the following *hypothesis* (O):

$$O = \mathcal{O} \cap \mathcal{N}_C. \qquad (O)$$

5. 1 *Remarks.* (a) It seems likely that $\mathcal{N} \cap \overline{G_{\mathbf{R}} \cdot \mathbb{R}^{\times}_{+}\mu}$ is in fact the same for all $\mu \in C$.

(b) It follows from the theory of sl_2-triples ([1] *Proposition 3.1*) that for *any* nilpotent orbit $O = G_{\mathbf{R}} \cdot \nu$ in $ig^*_{\mathbf{R}}$ there is an elliptic (not necessarily regular) element $\mu \in ig^*_{\mathbf{R}}$ so that $O \subset \overline{G_{\mathbf{R}} \cdot \mathbb{R}^{\times}_{+}\mu}$.

(c) It is known that the hypothesis (O) is satisfied for nilpotent orbits of dimension $2n$ with C a chamber in the fundamental Cartan subalgebra [14]. It is trivally satisfied at the opposite extreme, $O = \{0\}$, and for all O when $g_{\mathbf{R}}$ is complex.

For *any* $\xi \in ig^*_{\mathbf{R}}$, let μ_ξ be the canonical measure on $G_{\mathbf{R}} \cdot \xi$ and θ_ξ its Fourier transform (defined as in §4 for $\xi = \nu$). In particular

$$\theta(C, \mu) = \theta_\mu \text{ if } \mu \in C.$$

A polynomial $p \in \mathbb{C}[h]$ on h may be considered a differential operator on h^*, denoted $p(\partial)$ or $p(\partial_\lambda)$ with λ indicating the variable of differentiation.

5. 2 **Lemma.** *Let* $p \in \mathbb{C}[h]$, $\theta \in CH(g_{\mathbf{R}}, h)$. *Then*

$$\lim_{\lambda \to 0} p(\partial_\lambda)\theta(\lambda)$$

exits as a distribution on $g_{\mathbf{R},reg}$ *and defines a W-invariant pairing on* $\mathbb{C}[h] \otimes CH(g_{\mathbf{R}}, h]$.

Proof. This is clear, since $\theta(\ \cdot\ )$ extends to a holomorphic function on all of h^* with values in the space of distributions on $g_{\mathbf{R},reg}$ (1.4(a)). Q.E.D.

5. 3 **Theorem.** *Assume O satisfies hypothesis (O). Let p be any polynomial on h, homogeneous of degree e and transforming according to* χ_ν *by W. Then*

$$\lim_{\lambda \to 0(C)} p(\partial_\lambda)\mu_\lambda = \kappa\mu_\nu$$

for some constant $\kappa = \kappa(C, p, \nu)$. *The constant* $\kappa \neq 0$ *for some p if and only if* χ_ν *occurs in the W-module generated by* $\theta(C, \ \cdot\ )$.

Proof. Let $\theta_\nu(C, \ \cdot\ )$ be the component of $\theta(C, \ \cdot\ )$ transforming according to χ_ν,

$$\theta_\nu(C, \lambda) = \frac{\deg \chi_\nu}{|W|} \sum_{y \in W} \chi_\nu(y)\theta(C, y^{-1}\lambda),$$

$\Gamma_\nu(C) \in H_{2n}(\mathcal{I})$ the element correspondig to $\theta_\nu(C, \ \cdot\ ) \in CH(g_{\mathbf{R}})$.

According to theorem 3.3, the subspace of $H_{2n}(\mathcal{I})_\nu$ of $H_{2n}(\mathcal{I})$ of type χ_ν can be described as follows. Let $\mathcal{O} := G\cdot\nu$ the complex orbit containing $O = G_{\mathbf{R}}\cdot\nu$, and write

$$\mathcal{O} \cap \mathcal{N}_{\mathbf{R}} = \bigcup_{O'} O'$$

with $O' = G_{\mathbf{R}}\cdot\nu'$. (We choose $\nu' = \nu$ for $O' = O$.) Then

$$H_{2n}(\mathcal{I})_\nu \approx \sum_{\nu'} H_{2n}(\mathcal{I}(O'))^{A_{\nu'}} \approx \sum_{\nu'} H_{2e}(\mathcal{B}^{\nu'})^{A_{\nu'}}. \tag{20}$$

Applied to $\Gamma_\nu(C) \in \sum_{\nu'} H_{2n}(\mathcal{I}(O'))^{A_{\nu'}}$, theorem 4.1 gives

$$\theta_\nu(C, \lambda) = \sum_{\nu'} c_{\nu'}(\lambda)\theta_{\nu'} + o(\| \lambda \|^e). \tag{21}$$

We record that

$$c_{\nu'}(\,\cdot\,) \in \mathcal{H}(h^*) \textit{ represents the component of } \Gamma_\nu(C) \textit{ in } H_{2e}(\mathfrak{B}_{\nu'})^{A_{\nu'}} \textit{ according to the decomposition } (20). \tag{22}$$

Suppose $p \in \mathbb{C}[h]$ is homogeneous of degee e and transforms by χ_ν. Then (21) and lemma 5.2 give

$$\lim_{\lambda \to 0} p(\partial_\lambda)\theta(C, \lambda) = \sum_{\nu'} (p, c_{\nu'})\theta_{\nu'} \tag{23}$$

where

$$(p, c) := p(\partial)c(0)$$

is the natural pairing. Take the limit in (23) from within C and apply the inverse Fourier transform to find that

$$\lim_{\lambda \to 0(C)} p(\partial_\lambda)\mu_\lambda = \sum_{\nu'} (p, c_{\nu'})\mu_{\nu'}.$$

The left side of this equation has support in $\mathcal{N} \cap \overline{G_{\mathbf{R}} \cdot \mathbb{R}_+^\times \lambda}$ for any fixed $\lambda \in C$, hence in $\mathcal{N}_C$. The hypothesis (O) implies that

$$(p, c_{\nu'}) = 0 \text{ for } \nu' \neq \nu.$$

Since this holds for all p

$$c_{\nu'} = 0 \text{ for } \nu' \neq \nu, \tag{24}$$

i.e.

$$\theta_\nu(C, \lambda) = c_\nu(\lambda)\theta_\nu + o(\| \lambda \|^e). \tag{25}$$

This gives the desired formula

$$\lim_{\lambda \to 0(C)} p(\partial_\lambda)\mu_\lambda = (p, c_\nu)\mu_\nu,$$

and it follows from (22) and (24) that $(p, c_\nu) \neq 0$ for some p if and only if $\theta_\nu(C, \,\cdot\,) \neq 0$. *Q.E.D.*

It would be desirable to have for each O an explicit formula for a corresponding polynomial p. Under an additional hypothesis, this can be done as follows.

5. 4 **Corollary.** *Assume O satisfies hypothesis (O) and assume the restriction of* χ_ν *to* $W_{\mathbf{R}}$ *contains the character* $sgn_{\mathbf{I}}$ *of* $W_{\mathbf{R}}$ *with multiplicity one. Let* p_ν *be the (up to scalars unique) W-harmonic polynomial on* h *which is homogeneous of degee* $e = dim_{\mathbf{C}}\mathfrak{B}^\nu$, *transforms by* χ_ν *under W, and transforms by* $sgn_{\mathbf{I}}$ *under* $W_{\mathbf{R}}$. *Then*

$$\lim_{\lambda \to 0(C)} p_\nu(\partial_\lambda)\mu_\lambda = \kappa\mu_\nu$$

with $\kappa \neq 0$.

Proof. As in theorem 1.4 (c), write

$$\theta(C, \lambda) \equiv \frac{1}{\epsilon_{\mathbf{R}}\pi} \sum_{w \in W_{\mathbf{R}}} sgn_{\mathbf{I}}(w)\, e^{w^{-1}\lambda} \quad on\ h_{\mathbf{R}}.$$

$(sgn_{\mathbf{I}}(w))_{w \in W_{\mathbf{R}}}$ can be considered as an element in $ind_{W_{\mathbf{R}}}^{W} sgn_{\mathbf{I}}$; its component in the the irreducible subspace of type χ_ν is non-zero if and only if the restriction of χ_ν to $W_{\mathbf{R}}$ contains the character $sgn_{\mathbf{I}}$ of $W_{\mathbf{R}}$. This is the case, by assumption. Let $m = (m(w))_{w \in W}$ be the (up to scalars unique) non-zero element of $ind_{W_{\mathbf{R}}}^{W}$ transforming according to χ_ν under W and according to $sgn_{\mathbf{I}}$ under $W_{\mathbf{R}}$. We assume m normalized so that

$$\sum_{w \in W} m(w)\, w^{-1}$$

operates as the projection on the space of such elements. Then

$$\theta_\nu(C, \lambda) \equiv \frac{1}{\epsilon_{\mathbf{R}}\pi} \sum_{w \in W} m(w)\, e^{w^{-1}\lambda} \quad on\ h_{\mathbf{R}}. \tag{26}$$

By what has just been said, the expression on the right is non-zero.

Expand the exponentials in (26):

$$\theta_\nu(C, \lambda) \equiv \sum_{k=0}^{\infty} \frac{1}{k!} \frac{1}{\epsilon_{\mathbf{R}}\pi} \sum_{w \in W} m(w)\, w^{-1}\lambda^k. \tag{27}$$

Compare (27) with (25):

$$\theta_\nu(C, \lambda) = c_\nu(\lambda)\theta_\nu + o(\|\lambda\|^e), \quad c_\nu(\,\cdot\,) \neq 0.$$

One finds that

$$\sum_{w \in W} m(w)\, w^{-1}\lambda^k = 0 \ \text{ for } k < e$$

$$\frac{1}{\epsilon_{\mathbf{R}}\pi} \sum_{w \in W} m(w)\, w^{-1}\lambda^e \equiv c_\nu(\lambda)\, \theta_\nu.$$

Since $m(yw) = sgn_{\mathrm{I}}(y)\, m(w)$ for $y \in W$, also $c_\nu(y\lambda) = sgn_{\mathrm{I}}(y)\, c_\nu(y^{-1}\lambda)$. Hence $c_\nu(\,\cdot\,)$ is the (up to scalars unique) W-harmonic polynomial on h^* which is homogeneous of degee e, transforms by χ_ν under W, and transforms by sgn_{I} under $W_{\mathbf{R}}$. Hence $(p_\nu, c_\nu) \neq 0$. Q.E.D.

The "multiplicity-one" hypothesis, when it holds, may be verified with the help of the following criterion.

5. 5 Lemma. *Let G be a locally compact group, σ an automorphism of G, H a σ-stable, compact subgroup of G. Assume that for g in some set of $H{:}H$ double-coset representatives (and hence for all $g \in G$),*

$$g^\sigma = h_g g^{-1} k_g$$

with $h_g, k_g \in H$. Let $\rho\colon H \to \mathbb{C}^\times$ be a one-dimensional representation of H so that $\rho(h^\sigma) = \rho(h)$ for all $h \in H$ and $\rho(h_g) = \rho(k_g)$ for all $g \in G$. Then $ind_H^K \rho$ decomposes with multiplicity one.

Proof. The algebra of continuous G-endomorphisms of $ind_H^G \rho$ contains as a dense subalgebra the convolution algebra of compactly supported, continuous, $\mathbb{C}$-valued functions φ on G satisfying

$$\varphi(hgk) = \rho(h)\varphi(g)\rho(k)^{-1} \tag{28}$$

for all $g \in G$, $h, k \in H$. If suffices to show that this convolution algebra is abelian.

Let $g^\tau = (g^\sigma)^{-1}$. If $\varphi(g)$ satisfies (28), so does $\varphi^\tau(g) := \varphi(g^\tau)$. Since τ is an anti-automorphism of G,

$$(\varphi * \psi)^\tau = \psi^\tau * \varphi^\tau. \tag{29}$$

On the other hand, a function φ as in (28) satisfies $\varphi^\tau = \varphi$:

$$\varphi^\tau(g) = \rho(k_g{}^{-1})\varphi(g)\rho(h_g) = \varphi(g).$$

Applied to the functions in (29), this gives $\varphi * \psi = \psi * \varphi$ as required.

Q.E.D.

5. 6 *Remark.* This kind of argument goes back at least to Gelfand [11]. (See also [9] *Ch. X, Theorem 4.1.*)

The following result is due to E. Neher (unpublished).

5. 7 Lemma. *Assume $h_{\mathbf{R}}$ is a Cartan subalgebra of compact type in $g_{\mathbf{R}}$. Then any left (or right) coset of $W_{\mathbf{R}}$ in W has a representative $s \in W$ satisfying $s^2 = 1$.*

Proof. Any element of W is a product of reflections, say

$$\cdots s_\alpha s_\beta \cdots. \tag{30}$$

Generally

$$s_\alpha s_\beta = s_\gamma s_\alpha, \text{ where } \gamma = s_\alpha \beta. \tag{31}$$

If α is compact and β is non-compact, then $s_\alpha \beta$ is non-compact. So (31) may be used to bring all the compact roots in (30) to the left (or to the right). The coset representative may therefore be chosen of the form (30) with non-compact roots only. It suffices to verify the following statement.

(32)

If α, β are non-orthogonal, non-compact roots, then $s_\alpha \beta$ is compact.

Assuming this, any representative of the form (30) with a minimal number of reflections must consist of non-compact orthogonal reflections: otherwise (32) and (31) could be used to reduce the numbers of reflections without changing the coset.

The statement (32) concerns only the subalgebra of g generated by root-vectors for α, β and its real form obtained by intersection with $g_{\mathbf{R}}$. This algebra has rank two and (32) may be verified by inspection of the rank-two root systems. Q.E.D.

5. 8 Corollary. *Assume $h_{\mathbf{R}}$ is a Cartan subalgebra of compact type in $g_{\mathbf{R}}$. Then $ind_{W_{\mathbf{R}}}^{W} sgn_{\mathbf{I}}$ decomposes with multiplicity one.*

Proof. One can take $\sigma = identity$ in lemma 5.5. Q.E.D.

5. 9 *Example: Harish-Candra's Limit Formula* [8]. Take $\nu = 0$. Then $\chi_\nu = sgn$. Let

$$\varpi = \prod_{\alpha \in \Delta^+} \partial_\alpha,$$

with $\partial_\alpha \varphi = d\varphi(\alpha)$ as operator on h^*. ϖ transforms according to $\chi_0 = sgn$. Theorem 5.3 becomes

$$\lim_{\lambda \to 0} \varpi_\lambda \mu_\lambda = \kappa \mu_0 .$$

The constant κ is non-zero if and only if $h_{\mathbf{R}}$ is fundamental.

The last assertion is seen as follows.

$$[sgn \mid_{W_{\mathbf{R}}} : sgn_{\mathbf{I}}] \begin{array}{l} = 1, \text{ if } h \text{ is fundamental} \\ = 0, \text{ otherwise.} \end{array}$$

It follows that $\chi = sgn$ occurs exactly once in

$$H_{2n}(\mathcal{Y}) \approx \sum_{c} ind_{W_{c,\mathbf{R}}}^{W} sgn_{c,\mathbf{I}}$$

namely in the summand for which h_c is fundamental. Hence $\chi = sgn$ occurs in the W-module generated by $\theta(C, \cdot\,)$ if only if C lies in the fundamental Cartan subalgebra. ("If" because of the formula for $\theta(C, \cdot\,)$ in theorem 1.4 (c).)

REFERENCES

[1] D. Barbasch, *Fourier inversion for unipotent invariant integrals.* Trans. Amer. Math. Soc. 249, 51-83 (1979).

[2] D. Barbasch and D. Vogan, Jr., *The local structure of characters.* J. Functional Analysis 37 (1980), 27-55.

[3] D. Barbasch and D. Vogan, Jr., *Primitive ideals and orbital integrals in complex classical groups.* Math. Ann. 259 (1982), 153-199.

[4] D. Barbasch and D. Vogan, Jr., *Primitive ideals and orbital integrals in complex exceptional groups.* J. of Algebra 80 (1983), 350-382.

[5] D. Barbasch and D. Vogan, Jr., *Weyl group representations and nilpotent orbits.* In *Representations of Reductive Groups*, P. Trombi, ed. Birkhäuser, 1983.

[6] A. Borel, *Sur la cohomologie des espaces fibrés principaux et des espaces homogènes des groupes de Lie compactes.* Ann. Math. 57 (1953), 115-207 .

[7] R. Hardt, *Topological properties of subanalytic sets.* Trans. Amer. Math. Soc. 211 (1975), 57-70.

[8] Harish-Chandra, *Some results on the invariant integral on a semisimple Lie algebra.* Ann. Math. 80 (1964), 551-593.

[9] S. Helgason, *Differential Geometry and Symmetric Spaces.* Academic Press, New York and London, 1962.

[10] R. Hotta and M. Kashiwara, *The invariant holonomic system on a semisimple Lie algebra.* Inventiones math. 75 (1984), 327-358.

[11] I.M. Gelfand, *Analogue of the Plancherel formula for real semi-simple Lie groups.* Doklady Akad. Nauk. S.S.S.R. 92 (1953), 461-464.

[12] W.S. Massey, *Singular Homology Theory.* Springer-Verlag, 1980.

[13] W. Rossmann, *Kirillov's character formula for reductive Lie groups.*

Inventiones math. 48 (1978), 207-220.

[14] W. Rossmann, *Limit orbits in reductive Lie algebras.* Duke Math. J. 49 (1982), 215-229.

[15] W. Rossmann, *Characters as contour integrals.* Lecture Notes in Math. 1077, 375-388 , Springer-Verlag, 1984.

[16] W. Rossmann, *Invariant eigendistributions on a complex Lie algebra and homology classes on the conormal variety I: an integral formula; II: representations of Weyl groups.* Preprint (1988).

[17] W. Schmid, *Two character identities for semisimple Lie groups.* Lecture Notes in Math. 587, 129-154, Springer-Verlag, 1977.

[18] N. Spaltenstein, *On the fixed point set of a unipotent element on the variety of Borel subgroups.* Topology 16 (1977), 203-204.

[19] T. A. Springer, *Trigonometric sums, Green functions of finite groups and representations of Weyl groups.* Inventiones Math. 36 (1976), 173-207.

[20] T. A. Springer, *A construction of representations of Weyl groups.* Inventiones Math. 44 (1978), 279-293.

[21] R. Steinberg, *On the desingularization of the unipotent variety.* Inventiones Math. 36 (1976), 209-312 .

[22] V. S. Varadarajan, *Harmonic Analysis on Real Reductive Groups.* Lecture Notes in Math. 576, Springer-Verlag, 1977.

W. Rossmann
Department of Mathematics
University of Ottawa
Ottawa, Canada

Received October 27, 1989

Twisted Ideals of the Nullcone

RANEE KATHRYN BRYLINSKI

Dedicated to Professor J. Dixmier on his 65th birthday

1. Introduction

This paper is a continuation of [B1]. Again, the goal is to generalize Kostant's fundamental work ([K1],[K2]) on the coordinate rings of maximal dimension adjoint orbits and the actions of the principal TDS on adjoint group representations, for a complex semisimple Lie algebra $\mathfrak{g}$. The generalization is to an analogous theory for *twisted functions* on the maximal dimension semisimple and nilpotent adjoint orbits, which then relates to the action of the TDS on all finite-dimensional representations of $\mathfrak{g}$.

A main point is that the generalization ultimately comes from the invariant theory of the adjoint representation, in much the same spirit as Kostant's original ideas. To do this, we study the orbits as varieties fibered over the flag variety X ([B1]). The fibers then lie naturally in the adjoint representation- for a regular semisimple orbit Q, the fibers are translates of the nilradicals of Borel subalgebras, and for the principal nilpotent orbit N^o, the fibers are open subvarieties of those nilradicals. So the geometry ultimately reduces the study of twisted functions on the orbits to the study of certain functions on these fibers.

The key construction in this paper is a (non-linear) mapping Ω which "degenerates" twisted functions on Q to twisted functions on N^o, by essentially picking out the "leading term". This works because the points of the fibers of $N^o \to X$ identify with direction vectors in the fibers of $Q \to X$.

Twisted functions on $Q = G/T$ are determined by the characters of T- these are the functions exp^{μ} on T, wheres μ is an integral weight. Twisted functions on $N^o = G/G^e$ are determined just by characters γ of the center $Z(G)$ (see §2 and §5 for more precise information).Here G is the connected and simply connected semisimple Lie group with Lie algebra $\mathfrak{g}$, T is a maximal torus of G, and G^e is the centralizer group of a principal nilpotent e. We call the restriction γ_μ of exp^μ to $Z(G)$ the "central character of μ". Our map Ω is fully compatible with twisting in the following way: Ω degenertes exp^μ-twisted functions on Q to γ_μ-twisted functions on N^o.

As our mapping Ω is defined geometrically, it rather automatically has a lot of nice properties, like G-equivariance. To describe its effect, we first need to identify the twisted functions.

Once we fix the weight $-\mu$, with central character $\gamma = \gamma_{-\mu}$, we can build homogeneous line bundles $\underline{L}^{-\mu}$ on Q and $\underline{E}_\gamma$ on N^o. Then the spaces $\Gamma(Q, \underline{L}^{-\mu})$ and

$$R_\gamma(N^o) := \Gamma(N^o, \underline{E}_\gamma)$$

of global sections are the spaces of twisted functions. If the weight μ lies in the root lattice, then γ is trivial, and $R_1(N^o)$ is simply the coordinate ring $R(N^o) = R(N)$, where N is the cone of all nilpotents in $\mathfrak{g}$ (see [K2] for the fact that all regular functions on N^o extend to the closure N). (Note that N itself supports no non-trivial homogeneous line bundle.)

Our map Ω degenerates the whole space $\Gamma(Q, \underline{L}^{-\mu})$ to a certain subspace $I_{-\mu}$ of $R_\gamma(N^o)$. We now summarize our main results.

Let $-\mu$ be an integral weight, with central character γ. We construct a graded $R(N)$-submodule $I_{-\mu}$ of $R_\gamma(N^o)$; if μ lies in the root lattice, then $I_{-\mu}$ is just an ideal of the ring $R(N)$. $I_{-\mu}$ is the image of a (non-linear) map $\Omega : \Gamma(Q, \underline{L}^{-\mu}) \to R_\gamma(N^o)$.

The multiplicity in $I_{-\mu}$ of each irreducible finite-dimensional representation V_λ^* of $\mathfrak{g}$ is the weight multiplicity $dim(V_\lambda^\mu)$. (I.e., as a G-representation, $I_{-\mu}$ coincides with the "induced representation" $\Gamma(Q, \underline{L}^{-\mu})$.) Moreover, under a certain vanishing hypothesis (satisfied, for instance, when μ is regular or $\mathfrak{g}$ is classical), the graded multiplicity $\sum_{p \geq 0} \langle V_\lambda^*, I_{-\mu}^p \rangle q^p$ is equal to Lusztig's polynomial $m_\lambda^\mu(q)$, the q-analog of weight multiplicity. With respect to the fiber degree grading introduced in [B1], we get a natural G-linear graded isomorphism $gr\ \Gamma(Q, \underline{L}^{-\mu}) \xrightarrow{\sim} I_{-\mu}$.

Now $I_{-\mu}$ contains a unique copy of V_μ^* (since $dim(V_\mu^\mu) = 1$); call this subspace $K_{-\mu}$. Then $K_{-\mu}$ is the highest degree copy of V_μ^* occurring in $R_\gamma(N^o)$; this copy was first studied by Kostant ([K2],[K3]). Moreover, $K_{-\mu}$ is the pullback to N^o of the space of global $exp^{-\mu}$-twisted functions on the flag variety of G.

At $\mu = 0$, we have $I_0 = R(N)$, and we recover Kostant's picture.

It turns out that the $I_{-\mu}$ have interesting properties. For instance, it is well known that the basic lattice relation among dominant integral weights is reflected by a dimension inequality of weight spaces. I.e., take μ,ν dominant integral, so that $\mu \preceq \nu \iff \mu$ occurs as a weight of V_ν. Then $\mu \preceq \nu$ implies that $dim(V^\nu) \leq dim(V^\mu)$ for every representation V. We prove something much stronger, in the presence of the vanishing hypothesis: *The basic lattice relation among dominant integral weights is reflected by an inclusion of our modules: if μ,ν are dominant integral with $\mu \preceq \nu$, then $I_{-\nu} \subseteq I_{-\mu}$ (Proposition 7.2). Moreover, we also obtain (Corollary 7.5) a term-by-term coefficient inequality for the q-analogs of weight multiplicities.*

In the case μ lies in the root lattice, the most natural invariant of the ideal $I_{-\mu}$ is its zero-locus, the locus of points of N where all $f \in I_{-\mu}$ vanish. (We will consider only the set-theoretic zero-locus, but cf. Conjecture 13.3.) It turns out that the zero-locus depends only very coarsely on μ- it depends only on the list of simple components of $\mathfrak{g}$ relative to which μ has zero projection. See Theorem 9.1 for the precise statement. In particular, we prove: *For $\mathfrak{g}$ simple and $\mu \neq 0$, the zero-locus of $I_{-\mu}$ is precisely the boundary of N^o in N, i.e., the closure of the subregular nilpotent orbit.* Theorem 9.1 was motivated by an unpublished result of Kostant, and I am grateful to him for explaining it to me and letting me reproduce it here as Theorem 8.1..

We conjecture (see §13) that, for μ dominant, $I_{-\mu}$ is generated as an $R_\gamma(N^o)$-module by its unique copy $K_{-\mu}$ of V_μ^*. At $\mu = 0$ this is true: $R(N)$ is generated over itself by the constants. Some evidence is given in §13.

The connection of the $I_{-\mu}$ with the principal TDS action comes (by Frobenius reciprocity, following [K2]) in §6, see in particular Corollary 6.5. In later sections, statements about the $I_{-\mu}$ are most often translated into statements about representations.

Let us explain the relation of this paper to the previous one. In [B1], we degenerated twisted functions on Q to twisted function on the cotangent bundle $\underline{T}_X^*$ of the flag variety $X = G/B$. In this paper we complete the process to reach twisted functions on N^o by just restricting functions from $\underline{T}_X^*$ to N^o. It turns out that this process is somewhat subtle, as most of the twisting (in fact, all save the central twisting) is lost in the restriction. We describe this in §4, though the main work, analyzing the *degree shift* that occurs, is done in §3.

There should be some strong connection between our twisted ideals and the work of Gelfand and Kirillov ([Ge-Ki]) on the modules R_μ of regular differential operators of weight μ on the basic affine space. The multiplicities are the same, and one can see similarities in some of the constructions and calculations. The connection should become clearer in light of [J-S,

Proposition 5.5]. Kirillov and Joseph have, respectively, pointed out these references to me, but I have so far been unable to find the connection. However, it also seems clear that our Proposition 9.2 should be compared with Shapovalov's earlier divisibility result [S, Theorem 3(3)].

Ginsburg has recently informed me that he proved in [Gi] (by completely different means) [B1, Theorem 6.4], but *without the assumption on vanishing of cohomology*. Thus, the vanishing hypothesis can be removed from at least some of my results here.

Acknowledgements. I thank Jens Jantzen for suggesting that I compute the Hilbert series of the ideals $I_{-\mu}$ (§10) to check my ideas on their zero-loci. The proofs in §9 were in fact found during the week of Colloque Dixmier, and I thank the organizers for arranging the conference and inviting me. I thank Bert Kostant for teaching me all about the invariant theory of the adjoint representation, and for countless inspiring conversations.

2. Twisted Functions on N^o

We set notations for the paper. $\mathfrak{g}$ is a complex semisimple Lie algebra of rank ℓ. G is a connected and simply connected complex semisimple algebraic group with Lie algebra $\mathfrak{g}$ and adjoint group G^{ad}. We write ad_g for the adjoint action of $g \in G$ on $\mathfrak{g}$.

Fix a pair $T \subseteq B$ of a maximal torus inside a Borel subgroup, and let $\mathfrak{t}$ and $\mathfrak{b}$ be the Lie algebras. The pair $(\mathfrak{t}, \mathfrak{b})$ determines the sets Φ^+ and Δ of positive and simple roots, and the cone $\mathcal{P}^{++}$ of dominant integral weights in the lattice $\mathcal{P}$ of integral weights. Let $\mathcal{Q}$ be the root lattice, and set $\mathcal{Q}^{++} = \mathcal{Q} \cap \mathcal{P}^{++}$. Set $\mathcal{P}^{++}_{\gamma} := \{\lambda \in \mathcal{P}^{++} \mid \gamma_\lambda = \gamma\}$, for $\gamma \in \widehat{Z(G)}$; here γ_λ is the restriction to the center $Z(G)$ of exp^λ.

Let ρ be the half-sum of the positive roots; its "dual" is the vector $h_\rho \in \mathfrak{t}$ on which all simple roots take unit value.

View the flag variety $X = G/B$ as the variety of Borel subalgebras of $\mathfrak{g}$; we write $x_{\mathfrak{b}_1}$ when considering the subalgebra $\mathfrak{b}_1$ as a point of X. Take $x_0 := x_{\mathfrak{b}}$ as the base point of X.

A nilpotent $e \in \mathfrak{g}$ is *principal* if $dim\ \mathfrak{g}^e = \ell$. Let N^o be the variety of principal nilpotents in $\mathfrak{g}$. N^o forms a single dense adjoint orbit in the cone N of all nilpotents.

We fix as the *base point* of N^o a principal nilpotent e_0 in $\mathfrak{b}$ satisfying $[h_\rho, e_0] = e_0$; so e_0 is any linear combination $\sum_{\alpha \in \Delta} c_\alpha X_\alpha$, with all c_α non-zero, of the simple root vectors X_α (see [B1, 2.3]). Let $\eta : N^o \to X$ be the G-projection (i.e., G-equivariant projection) sending e_0 to $x_{\mathfrak{b}}$; then η sends each principal nilpotent e to the unique Borel subalgebra $\mathfrak{b}^{(e)}$ containing it.

The choice of e_0 identifies, once and for all, N^o with G/A, where $A := G^{e_0}$. A_{conn}, the connected component of A through the identity, is a subgroup of the unipotent radical U of B ([K2]).

Form the covering variety $N^{\sim} := G/A_{conn}$ with natural G-projection $\sigma : N^{\sim} \to N^o$.

A semisimple $h \in \mathfrak{g}$ is *regular* if $dim\ \mathfrak{g}^h = \ell$. Let (Q, π) be a fixed *regular semisimple adjoint orbit over* X; i.e., Q is a regular semisimple adjoint orbit and $\pi : Q \to X$ is a G-projection. The corresponding base point of (Q, π) is $h_0 := \pi^{-1}(x_{\mathfrak{b}}) \cap \mathfrak{t}$. Let Q_ρ be the adjoint orbit through h_ρ, with $\pi_\rho : Q_\rho \to X$ the G-projection sending h_ρ to $x_{\mathfrak{b}}$. Then h_ρ is the base point of (Q_ρ, π_ρ).

The purpose of this section is to set the stage for §3. We will look at "twisted functions" on N^o, i.e., sections of G-homogeneous line bundles on N^o. Twisted functions on N^o may also be regarded as functions on $N^{\sim}$. We will define an "internal" $\mathbf{C}^*$-action on $N^{\sim}$ compatible (see Lemma 2.4) with the usual *dilation* $\mathbf{C}^*$-action on N^o given by scalar multiplication on the vector space $\mathfrak{g}$. Homogeneity with respect to this internal $\mathbf{C}^*$-action will then give us a grading of twisted functions on N^o generalizing the usual grading of untwisted functions on N^o.

Lemma 2.1. *For any $e \in N^o$, we have $G^e = G^e_{conn} \times Z(G)$.*

Proof: The projection $G \longrightarrow G/Z(G) \xrightarrow{\sim} G^{ad}$ carries G^e_{conn} onto $(G^{ad})^e$, as both are connected ([K2, Proposition 14, pg. 362]) groups with Lie algebra $\mathfrak{g}^e$. So $G^e = Z(G)G^e_{conn}$. Now $Z(G) \cap G^e_{conn} = \{1\}$ since G^e_{conn} lies in the unipotent radical of B ([K2]), and $Z(G)$ is central, so the product is direct. Q.E.D.

G-homogeneous line bundles on N^o correspond exactly to the characters of A. Since unipotent groups have no non-trivial characters, a character of A is just a character γ of $Z(G)$, extended trivially over A_{conn}. For such γ, we build over N^o the G-homogeneous locally trivial algebraic vector bundle

$$\underline{E}^\gamma := G \times^A \mathbf{C}^\gamma,$$

with base fiber $\mathbf{C}^\gamma$. Let $\mathcal{E}^\gamma$ be the sheaf of algebraic sections of $\underline{E}^\gamma$ over N^o. The space

$$R_\gamma(N^o) := \Gamma(N^o, \underline{E}^\gamma)$$

of global sections is naturally a module over the ring $R(N^o) = R(N)$ of global functions.

Lemma 2.1 implies that each line bundle $\underline{E}^\gamma$ trivializes upon pullback to $N^{\sim}$, and we have

Lemma 2.2. *$\sigma_* \mathcal{O}_{N^{\sim}} = \oplus_{\gamma \in \widehat{Z(G)}} \mathcal{E}^\gamma$; hence, $R(N^{\sim}) = \oplus_{\gamma \in \widehat{Z(G)}} R_\gamma(N^o)$, as $R(N^o)$-modules and G-representations.*

$R_\gamma(N^o)$ is the sum inside $R(N^{\sim})$ of all G-irreducibles on which the center acts by the character γ; $R_1(N^o) = R(N^o)$ is the subspace of $Z(G)$-invariants in $R(N^{\sim})$. Clearly, every G-irreducible occurs in $R(N^{\sim})$, so

$$V_\lambda \text{ occurs in } R_\gamma(N^o) \Leftrightarrow \lambda \in \mathcal{P}_\gamma^{++}.$$

We can G-equivariantly lift the square of the dilation $\mathbf{C}^*$-action on N^o to a $\mathbf{C}^*$-action on $N^{\sim}$ as follows. First define a one-parameter subgroup $c : \mathbf{C}^* \to T$ by

$$c_t := exp(-2(log\ t)h_\rho);$$

so "$c_t = t^{-2h_\rho}$". Although *log* is only determined modulo $2\pi i$, c_t is well-defined, on account of

Lemma 2.3. *For any $h \in Q_\rho$, $exp(2\pi i h)$ is a central element of G of order 1 or 2, and its order is independent of h.*

Proof: Decomposing $\mathfrak{g}$ into h-eigenspaces, we find that $a := exp(2\pi i h)$ acts trivially on $\mathfrak{g}$, hence lies in $Z(G)$. But also a lies in the exponential of a principal TDS in $\mathfrak{g}$, hence in a subgroup of G isomorphic to SL_2 or PGL_2. Then a has order 2 in the former case (by direct calculation), and order 1 in the latter. As h varies, the subgroups in question are all G-conjugate, so isomorphic. Q.E.D.

As all multiples of h_ρ preserve the line in $\mathfrak{g}$ through e_0, all c_t normalize A. So we have the action of c on $N^{\sim}$ through the right multiplication:

$$c_t \star gA_{conn} := gc_t^{-1}A_{conn}.$$

Call this the *internal* $\mathbf{C}^*$*-action on* $N^{\sim}$. A simple calculation gives

Lemma 2.4. *The internal $\mathbf{C}^*$-action on $N^{\sim}$ lifts the square of the dilation $\mathbf{C}^*$-action on N^o.*

Remark 2.5. c_t is completely determined, modulo A_{conn}, by the choice of e_0; c_t does not depend on the choices of $\mathfrak{b}$, $\mathfrak{t}$, and h_ρ. To see this, consider the elements $c_{t,h} := exp(-2(log\ t)h)$, for $h \in \mathfrak{g}$ satisfying $[h, e] = e$, and $t \in \mathbf{C}^*$. For each t, the fact $ad_{c_{t,h}} e = t^{-2}e$ implies right away that, at least modulo G^e, $c_{t,h}$ is independent of h. But h varies in a linear coset of $\mathfrak{g}^e$, so the continuity of *exp* insures that, as h varies (and t is held fixed), the $c_{t,h}$ all lie in a connected component of a coset of G^e.

We will say a regular function f on a $\mathbf{C}^*$-stable open set U of $N^\sim$ is homogeneous of *internal degree* p if

$$(2.6) \qquad f(c_t \star z) = t^{2p} f(z),$$

for all $z \in U$, $t \in \mathbf{C}^*$. Thus we get a notion of *internal degree* for any section of $\mathcal{E}^\gamma$, $\gamma \in \widehat{Z(G)}$, over a dilation stable open set of N^o.

Equivalently, one can define the internal degree of a section s of $\mathcal{E}^\gamma$ directly by using the "internal $\mathbf{C}^*$-action on $\underline{E}^\gamma$ (again, by the right multiplication action) and the following variant of (2.6):

$$(2.7) \qquad c_t^{-1} \star s(t^2 z) = t^{2p} s(z).$$

3. N^o fibered over the flag variety

In this section, we compare, in Proposition 3.3, two notions of degree for a twisted function f on N^o: (1) the *internal degree*, measuring how f transforms under the $\mathbf{C}^*$-action constructed in the last section, and (2) the *fiber degree*, measuring the homogeneous degree of f upon restriction to the fibers of η.

As the fibers of $\sigma\eta$ are dilation stable in N^o, the construction of "internal degree" in §2 produces on the pushdown sheaf $\eta_*\sigma_*\mathcal{O}_{N^\sim}$ a *internal grading* over the half-integers by subsheaves of $\mathcal{O}_X$-modules; $\eta_*\sigma_*\mathcal{O}_{N^\sim} = \oplus_{2p \in \mathbf{Z}} (\eta_*\sigma_*\mathcal{O}_{N^\sim})^p$. Thus we get the *internal gradings*:

$$\eta_*\mathcal{E}^\gamma = \oplus_{2p\in\mathbf{Z}} (\eta_*\mathcal{E}^\gamma)^p, \qquad R_\gamma(N^o) = \oplus_{2p\in\mathbf{Z}} R_\gamma^p(N^o),$$

$\gamma \in \widehat{Z(G)}$. When γ is trivial, we recover the familiar grading $\eta_*\mathcal{O}_{N^o} = \oplus_{p\in\mathbf{Z}} (\eta_*\mathcal{O}_{N^o})^p$ coming from the dilation action. As the global regular functions on N^o all extend to N ([K2]), the global functions on N^o occur only in non-negative degrees.

Lemma 3.1. *Let γ be any character of $Z(G)$. Then*

(1) *$\eta_*\mathcal{E}^\gamma$ is $\mathbf{Z}$-graded or $(\mathbf{Z}+\frac{1}{2})$-graded, according to the sign of $\gamma(c_{-1})$.*
(2) *The space of global sections of $\underline{E}^\gamma$ is graded in non-negative degrees.*

Proof: (1) According to Lemma 2.2, the group element c_{-1} is central. So c_{-1} acts on $\underline{E}^{\gamma}$ by dilating each fiber of η by a factor of $\gamma(c_{-1})$. So if s is a non-zero local section of $\mathcal{E}^{\gamma}$ and is homogeneous of degree p, then (2.6) forces that $\gamma(c_{-1}) = (-1)^{2p}$. Thus the integer $2p$ is even if $\gamma(c_{-1}) = 1$, and odd if $\gamma(c_{-1}) = -1$. (2) As $Z(G)$ is finite, some finite power of a global section of $\underline{E}^{\gamma}$ is a global function on N^o, and hence has non-negative degree. Q.E.D.

The G-line bundles over N^o all arise through pullback from X, but not uniquely. Given an integral weight μ of $\mathfrak{t}$, form the homogeneous line bundle

$$\underline{F}^{\mu} := G \times^{B} \mathbf{C}^{exp(\mu)}$$

over X, with base fiber $\mathbf{C}^{exp(\mu)}$ and sheaf of sections $\mathcal{F}^{\mu}$. (As usual, we extend $exp(\mu)$ trivially over the unipotent radical of B to get a character of B.)

Then $\eta^{*}\underline{F}^{\mu}$ and $\underline{E}^{\gamma_{\mu}}$ are equivalent as algebraic G-line bundles on N^o, since γ_{μ} is the character of $Z(G)$ obtained by restricting $exp(\mu)$. For

$$\eta^{*}\underline{F}^{\mu} = N^o \times_X \underline{F}^{\mu} = G/A \times_{G/B} (G \times^{B} \mathbf{C}^{exp(\mu)}) \xrightarrow{\sim} G \times^{A} \mathbf{C}^{\gamma_{\mu}} = \underline{E}^{\gamma_{\mu}};$$

the map is given by $(gA, f, v) \mapsto (g, (g^{-1}f)\cdot v)$, where $g, f \in G$ satisfy $gB = fB$ and v is a vector in the B-representation $\mathbf{C}^{exp(\mu)}$. Hence, we obtain a G-equivariant isomorphism of $\mathcal{O}_{N^o}$-modules

$$\Theta^{\mu} : \eta^{*}\mathcal{F}^{\mu} \xrightarrow{\sim} \mathcal{E}^{\gamma_{\mu}}.$$

What is happening here is that most of the twisted structure of $\eta^{*}\underline{F}^{\mu}$ collapses G-equivariantly in this isomorphism. This collapsing is effected by the action of G. When μ lies in the root lattice, the G-action G-equivariantly trivializes $\eta^{*}\underline{F}^{\mu}$. (To see this happen for general μ, we need to pullback $\underline{F}^{\mu}$ all the way to $N^{\sim}$.)

Our goal right now is to understand the collapsing map

$$\Theta^{\mu} : \Gamma(N^o, \eta^{*}\mathcal{F}^{\mu}) \to R_{\gamma_{\mu}}(N^o)$$

on global sections. The target $R_{\gamma_{\mu}}(N^o)$ consists of those functions f on $N^{\sim}$ satisfying $f(grA_{conn}) = \gamma_{\mu}(r)^{-1}f(gA_{conn})$, for all $g \in G$, $r \in Z(G)$. Recall that our very construction of $\underline{F}^{\mu}$ equips the base fiber $\underline{F}^{\mu}_{x_0}$, and hence the base fiber $(\eta^{*}\underline{F}^{\mu})_{e_0}$, with an identification to $\mathbf{C}$.

Lemma 3.2. *Let $\mu \in \mathcal{P}$. Then, for $s \in \Gamma(N^o, \eta^* \mathcal{F}^\mu)$, the function $f = \Theta^\mu(s)$ on $N^\sim$ is given by*

$$f(gA_{conn}) = g^{-1} \cdot s(g \cdot e_0) \in (\eta^* \underline{F}^\mu)_{e_0} = \mathbf{C}.$$

We can now describe the degree shift between the induced grading $\eta_* \sigma_* \sigma^* \eta^* \mathcal{F}^\mu = \oplus_{2p \in \mathbf{Z}} \mathcal{F}^\mu \otimes (\eta_* \sigma_* \mathcal{O}_{N^\sim})^p$ and the internal grading on $\eta_* \mathcal{E}^{\gamma_\mu}$. The former is the *fiber degree* grading (in the sense of [B1,§4§5]) coming from the fibering of N^o over X.

Proposition 3.3. *Let $\mu \in \mathcal{P}$. Then Θ^μ gives, for each p, an isomorphism*

$$\mathcal{F}^\mu \otimes (\eta_* \sigma_* \mathcal{O}_{N^\sim})^p \xrightarrow{\sim} (\eta_* \mathcal{E}^{\gamma_\mu})^{p - \langle \mu, \rho \rangle}.$$

In particular, if $s \in \Gamma(N^o, \eta^* \mathcal{F}^\mu)$, and $f = \Theta^\mu(s) \in R_{\gamma_\mu}(N^o)$, then s is of homogeneous fiber degree p iff f is of homogeneous internal degree $p - \langle \mu, \rho \rangle$.

Proof: Set $\underline{F} = \underline{F}^\mu$ and $\gamma = \gamma_\mu$. Suppose $U \subseteq X$ and s is a section over $\eta^{-1}(U)$ of $\eta^* \underline{F}$, and $f = \Theta^\mu(s)$. For $x \in X$ and $e \in \eta^{-1}(x)$, we have the fiber projection $\eta_* : (\eta^* \underline{F})_e \to \underline{F}_x$.

Now s is of homogeneous fiber degree d iff for all $x \in U$, the equality

$$\eta_*(s(te)) = t^d \eta_*(s(e))$$

holds in $\underline{F}_x$ for all $e \in \eta^{-1}(x)$ and $t \in \mathbf{C}^*$.

On the other hand, f is of homogeneous internal degree n iff for all $e \in \eta^{-1}(U)$, the equality

$$c_t^{-1} \star s(t^2 e) = t^{2n} s(e)$$

holds in $(\eta^* \underline{F})_e$, for all $t \in \mathbf{C}^*$. Hence, f is of homogeneous internal degree n iff for all $x \in U$, the equality

$$\eta_*(c_t^{-1} \star s(t^2 e)) = t^{2n} \eta_*(s(e))$$

holds in $\underline{F}_x$ for all $e \in \eta^{-1}(x)$ and $t \in \mathbf{C}^*$. Now

$$\eta_*(c_t^{-1} \star s(t^2 e)) = c_t^{-1} \star \eta_*(s(t^2 e)) = t^{-2\langle \mu, \rho \rangle} \eta_*(s(t^2 e)),$$

since the action of $c_t^{-1} \star$ on the entire bundle $\underline{F}$ is just multiplication by the scalar $exp^\mu(c_t^{-1}) = t^{-2\mu(h_\rho)} = t^{-2\langle \mu, \rho \rangle}$. So f is of homogeneous internal degree n iff

$$\eta_*(s(t^2 e)) = t^{2\langle \mu, \rho \rangle + 2n} \eta_*(s(e)).$$

Comparing with the condition for fiber degree, we find the notions of homogeneity coincide, save for the degree shift $d = n + \langle \mu, \rho \rangle$. Q.E.D.

Kostant proved ([K2]) that, for $\mu \in \mathcal{Q}^{++}$, V_μ^* occurs exactly once in degree $\langle \mu, \rho \rangle$ in $R(N)$, and this is the highest degree copy. The same argument works for $R(N^{\sim})$ and all $\mu \in \mathcal{P}^{++}$. Evaluation at e_0 gives a linear isomorphism $ev_{e_0} : Hom_G(V_\mu^*, R_\gamma(N^o)) \to V_\mu^{\mathfrak{g}^{e_0}}$, $\gamma = \gamma_{-\mu}$. Under ev_{e_0}, the grading on $R_\gamma(N^o)$ translates into the grading on $V_\mu^{\mathfrak{g}^{e_0}}$ by the eigenspaces of h_ρ. The highest weight space V_μ^μ is the highest graded component, with degree $\langle \mu, \rho \rangle$. So we make

Definition 3.4. Let $K_{-\mu}$ be the unique highest degree copy of V_μ^* in $R_{\gamma_{-\mu}}(N^o)$, for $\mu \in \mathcal{P}^{++}$.

Corollary 3.5. *Suppose $\mu \in \mathcal{P}^{++}$ and $\gamma = \gamma_{-\mu}$. Then the pullback of $\Gamma(X, \underline{F}^{-\mu})$ through η to $R_\gamma(N^o)$ is $K_{-\mu}$.*

Proof: 3.3 says in particular that if s has fiber degree 0, i.e., if s is a section of $\mathcal{F}^\mu$, then $f = \Theta^\mu(s)$ has internal degree equal to $-\langle \mu, \rho \rangle$. But also one knows (Borel-Weil) that $\Gamma(X, \underline{F}^{-\mu})$ carries the G-representation to V_μ^*. So its pullback is a copy of V_μ^* in $R_\gamma^{\langle \mu, \rho \rangle}(N^o)$, hence it is $K_{-\mu}$. Q.E.D.

Remark 3.6. We can calculate in a more direct way the degree of $f = \Theta^{-\mu}(s) \in R_\gamma(N^o)$ for $s \in \Gamma(X, \underline{F}^{-\mu})$. Suppose $e = \sum_{\alpha \in \Delta} X_\alpha$. Then $t \cdot e = \sum_{\alpha \in \Delta} exp^\alpha(t) X_\alpha$. So $X_{-\alpha}(t \cdot e) = exp^\alpha(t)$. Thus

$$f(t \cdot e) = (t^{-1} \cdot f)(e) = exp^{-\mu}(t^{-1}) f(e) = exp^\mu(t) f(e).$$

Now $exp^\mu(t)$ is a polynomial of total degree $\sum_{\alpha \in \Delta} p_\alpha = \langle \mu, \rho \rangle$ in the $exp^\alpha(t)$, where $\mu = \sum_{\alpha \in \Delta} p_\alpha \alpha$.

4. N^o as a subvariety of the cotangent bundle

Let $\iota : N^o \to \underline{T}_X^*$ be the natural G-equivariant inclusion over X of $N^o = G \times^B \mathfrak{m}^o$ as an open dense subset of the cotangent bundle $\underline{T}_X^* = G \times^B \mathfrak{m}$ of X. As $\underline{T}_X^*$ identifies canonically with the variety of pairs $(e_1, \mathfrak{b}_1)$ of a nilpotent vector e_1 inside a Borel subalgebra $\mathfrak{b}_1$, we write $\xi_{e_1, \mathfrak{b}_1}$ for the corresponding cotangent vector, or simply ξ_{e_1} when $e_1 \in N^o$. Let $\tau : \underline{T}_X^* \to X$ be the natural projection, and let $\mathcal{T}_X$ be the tangent sheaf of X.

So we have the picture

$$\begin{array}{ccccc} N^{\sim} & & & & \\ & \overset{\sigma}{\searrow} & & & \\ & & N^o & \xrightarrow{\iota} & \underline{T}^*_X \\ & & & \eta \searrow \quad \swarrow \tau & \\ & & & X & \end{array}$$

The diagram is equivariant for not only the action of G, but also for the internal $\mathbf{C}^*$-action, which we continue to denote by "$\star$".

Restriction of functions gives a natural inclusion $\iota^* : S(\mathcal{T}_X) = \tau_* \mathcal{O}_{\underline{T}^*_X} \to \eta_* \mathcal{O}_{N^o}$, of sheaves of $\mathcal{O}_X$-modules. Twisting, with respect to some $\mu \in \mathcal{P}$, and composing with the collapsing map of §3, we get an $\mathcal{O}_X$-module inclusion

$$j^\mu : \mathcal{F}^\mu \otimes S(\mathcal{T}_X) = \tau_* \tau^* \mathcal{F}^\mu \xrightarrow{\iota^*_\mu} \eta_* \eta^* \mathcal{F}^\mu \xrightarrow{\Theta^\mu} \eta_* \mathcal{E}^{\gamma_\mu}.$$

On global sections, this gives the inclusion

$$j^\mu_X : \Gamma(X, \mathcal{F}^\mu \otimes S(\mathcal{T}_X)) = \Gamma(\underline{T}^*_X, \tau^* \mathcal{F}^\mu) \to R_{\gamma_\mu}(N^o).$$

It is well known that the restriction map $j^0_X : R(\underline{T}^*_X) \to R(N^o)$ is an isomorphism.

The grading of $\mathcal{F}^\mu \otimes S(\mathcal{T}_X)$ induced by the grading of the symmetric algebra is again a *fiber degree grading* in the sense of [B1, §4,§5]. So ι^*_μ is graded of degree 0.

Now 3.3 gives

Proposition 4.1. *Let $\mu \in \mathcal{P}$. Then j^μ and j^μ_X are G-linear, graded injections of degree $-\langle \mu, \rho \rangle$. j^μ is a module homomorphism over the sheaf of algebras $S(\mathcal{T}_X) \xrightarrow{\sim} \iota^*(S(\mathcal{T}_X))$; j^μ_X is a module homomorphism over the ring $R(\underline{T}^*_X) = R(N^o)$.*

Corollary 4.2. *Let d be any positive integer and suppose $\alpha_1, \ldots, \alpha_d$ are d (not necessarily distinct) simple roots. Then*

$$\Gamma(X, \mathcal{F}^{\alpha_1 + \cdots + \alpha_d} \otimes S^d(\mathcal{T}_X)) = \mathbf{C}.$$

Furthermore, if $\mu \in \mathcal{P}^{++}$ and $\nu \in \mathcal{P}$ with $\mu - \nu = \alpha_1 + \cdots + \alpha_d$, then $\Gamma(X, \mathcal{F}^{-\nu} \otimes S^d(\mathcal{T}_X))$ contains a unique copy of V^*_μ.

Proof: Let $\theta = \alpha_1 + \cdots + \alpha_d$. We construct a G-equivariant inclusion $\phi : \mathcal{F}^{-\mu} \hookrightarrow \mathcal{F}^{-\nu} \otimes S^d(\mathcal{T}_X)$ of $\mathcal{O}_X$-modules as follows.

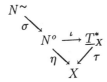

For any simple root α, form the parabolic subalgebra $\mathfrak{p}_\alpha = \mathfrak{b} + \mathbf{C}X_{-\alpha}$ of $\mathfrak{g}$. The inclusion $\mathfrak{p}_\alpha/\mathfrak{b} \hookrightarrow \mathfrak{g}/\mathfrak{b}$ is $\mathfrak{b}$-linear, and hence induces an inclusion $\mathcal{F}^{-\alpha} \to \mathcal{T}_X$. Tensoring these maps, we build the inclusion $\mathcal{F}^{-\theta} \hookrightarrow S^d(\mathcal{T}_X)$ of $\mathcal{O}_X$-modules; tensoring next with $\mathcal{F}^\theta$ we get ϕ. Thus $\Gamma(X, \mathcal{F}^{-\nu} \otimes S^d(\mathcal{T}_X))$ contains a copy of $\Gamma(X, \mathcal{F}^{-\mu}) = V_\mu^*$.

On the other hand, $j_X^{-\nu}$ injects $\Gamma(X, \mathcal{F}^{-\nu} \otimes S^d(\mathcal{T}_X))$ into $R^{d+(\nu,\rho)}(N^o) = R^{(\mu,\rho)}(N^o)$, which contains just one copy of V_μ^*. Moreover at $\mu = 0$, simply $R^0(N^o) = \mathbf{C}$. Q.E.D.

We will make a more precise statement about $j = j_X^{-\mu}$ in Proposition 9.2.

5. Degeneration from Q to N^o of twisted functions

Recall (Q, π) is a regular semisimple adjoint orbit over X. For each integral weight μ, form the pullback bundle $\underline{L}^\mu = \pi^* \underline{F}^\mu$ on Q, with sheaf of sections $\mathcal{L}^\mu$.

In [B1, §4 and §5], we constructed the G-equivariant *symbol map*

$$\Delta^\mu : \pi_* \mathcal{L}^\mu \longrightarrow \tau_* \tau^* \mathcal{F}^\mu = \mathcal{F}^\mu \otimes S(\mathcal{T}_X)$$

of sheaves in the following way. Let U be an open set in X, and let s be a section of $\underline{L}^\mu$ over $\pi^{-1}(U)$. Take $x = x_{\mathfrak{b}_1} \in U$, and let $\underline{F}^\mu_x$ be the fiber of $\underline{F}^\mu$ at x. Consider the map $f(s,x) : \pi^{-1}(x) \longrightarrow \underline{F}^\mu_x$ defined by evaluating s on $\pi^{-1}(x)$ and then projecting these values down to $\underline{F}^\mu$. The fiber $\pi^{-1}(x)$ is an affine space, in fact a translate $h_1 + \mathfrak{m}_1$ of the nilpotent radical $\mathfrak{m}_1$ of $\mathfrak{b}_1$. So $f(s,x)$ has a well-defined *affine degree*. The *fiber degree* of s is the maximum of the affine degrees of the maps $f(s,x)$, as x varies in U. Now, the value of the symbol $s^+ = \Delta^\mu_U(s) \in \Gamma(\tau^{-1}(U), \tau^* \mathcal{F}^\mu)$ at any point $\xi = \xi_{e_1, \mathfrak{b}_1}$ of $\tau^{-1}(U)$ is given by, for p the fiber degree of s,

$$s^+(\xi) = \lim_{t \to \infty} \tau^* \pi_* s(h_1 + te_1)/t^p \tag{5.1}$$

Here, π_* gives the fiber projection $\underline{L}^\mu_{h_1+te_1} \to \underline{F}^\mu_x$, while τ^* gives the fiber lifting $\underline{F}^\mu_x \to (\tau^* \underline{F}^\mu)_\xi$. Note the limit may be computed in either $\underline{F}^\mu_x$ or $(\tau^* \underline{F}^\mu)_\xi$.

In this setting, we proved

Theorem 5.2. ([B1, Theorem 5.5]) *Let (Q, π) be a regular semisimple adjoint orbit over X. Let $\mathcal{F}^\mu$ be the sheaf of $exp(\mu)$-twisted functions on X, for $\mu \in \mathcal{P}$. Then the sheaf $\pi_* \mathcal{L}^\mu$ is filtered by the G-stable $\mathcal{O}_X$-submodules*

$(\pi_*\mathcal{L}^\mu)^{\leq p}$, *the sheaves of sections of fiber degree at most p, for* $p \geq 0$. *The symbol map induces a graded G-linear isomorphism*

$$gr\ \Delta^\mu : gr\ \pi_*\mathcal{L}^\mu \xrightarrow{\sim} \mathcal{F}^\mu \otimes S(\mathcal{T}_X)$$

of modules over the $\mathcal{O}_X$*-algebra* $gr\ \pi_*\mathcal{O}_Q \xrightarrow{\sim} S(\mathcal{T}_X)$.

Set, for $p \geq 0$, $\Gamma^{\leq p}(Q,\mathcal{L}^\mu) := \Gamma(X,(\pi_*\mathcal{L}^\mu)^{\leq p})$,

$$\Gamma^p(Q,\mathcal{L}^\mu) := \Gamma^{\leq p}(X,\pi_*\mathcal{L}^\mu)/\Gamma^{\leq p-1}(X,\pi_*\mathcal{L}^\mu),$$

and $gr\ \Gamma(Q,\mathcal{L}^\mu) = \oplus_{p\geq 0}\Gamma^p(Q,\mathcal{L}^\mu)$.

We call the composition of the symbol map Δ^μ with j^μ the *special symbol map*

$$\Omega^\mu : \pi_*\mathcal{L}^\mu \to F^\mu \otimes S(\mathcal{T}_X) \to \eta_*\mathcal{E}^{\gamma_\mu}.$$

Theorem 5.2 together with Proposition 4.1 gives at once:

Theorem 5.3. *Keep the notations of 5.2. Then the special symbol map induces a graded, of degree* $\langle\mu,\rho\rangle$, *G-linear injection*

$$gr\ \Omega^\mu : gr\ \pi_*\mathcal{L}^\mu \to \eta_*\mathcal{E}^{\gamma_\mu}$$

of modules over the $\mathcal{O}_X$*-algebra* $gr\ \pi_*\mathcal{O}_Q \xrightarrow{\sim} \iota_0^*(S(\mathcal{T}_X))$. *In particular, the induced map on global sections is a graded, of degree* $\langle\mu,\rho\rangle$, *G-linear injection*

$$gr\ \Omega^\mu_X : gr\ \Gamma(Q,\mathcal{L}^\mu) \to R_{\gamma_\mu}(N^o)$$

of modules over gr $R(Q) \xrightarrow{\sim} R(N^o)$.

Remark 5.4. Let H be any graded G-stable complement in $S(\mathfrak{g}^*)$ to the ideal of the nullcone N; so H is also a complement to the ideal of Q ([K2]). Then the untwisted special symbol map Ω^0_X on global functions can be characterized as the unique map such that, for all $p \geq 0$, the triangle below, with the vertical maps given by restriction, is commutative.

$$\begin{array}{ccc} & H^p & \\ \swarrow & & \searrow \\ R^{\leq p}(Q) & \xrightarrow{\Omega^0_X} & R^p(N^o) \end{array}$$

Commutativity is easy to check. For, given $e \in N^o$, take any $h \in \pi^{-1}(x_{\mathfrak{b}(e)})$. Then, for any non-zero $F \in H^p$, $f = F|_Q$ has fiber degree p (see proof of [B1, 5.4]), so that

$$\Omega^0_X(f)(e) = \lim_{t\to\infty} f(h+te)/t^p = \lim_{t\to\infty} F(h/t+e) = F(e).$$

Following [B1, §2], define the e_0-*order* of a vector v in a G-representation V to be the least non-negative integer p such that $e_0^{p+1} \cdot v = 0$. Construct a (non-linear) map $\delta_{e_0} : V \to V^{e_0}$ by

$$\delta_{e_0}(v) = \frac{(-1)^p}{p!} e_0^p \cdot v,$$

where p is the e_0-order of v.

Proposition 5.5. *Suppose μ is a weight of V_λ, $\lambda \in \mathcal{P}^{++}$, and $\gamma = \gamma_{-\mu}$. Then the following diagram, where the vertical maps are the linear isomorphisms given by evaluation at h_ρ and e_0, is commutative.*

$$\begin{array}{ccc} Hom_G(V_\lambda^*, \Gamma(Q_\rho, \mathcal{L}^{-\mu})) & \xrightarrow{\Omega_X^{-\mu}} & Hom_G(V_\lambda^*, R_\gamma(N^o)) \\ {\scriptstyle ev_{h_\rho}} \downarrow & & \downarrow {\scriptstyle ev_{e_0}} \\ V_\lambda^\mu & \xrightarrow{\delta_{e_0}} & V_\lambda^{\mathfrak{g}^{e_0}} \end{array} \tag{5.6}$$

Proof: The commutativity of the diagram follows at once from the proof of [B1, Theorem 5.8]. Q.E.D.

Remark 5.7. This gives another proof of the fact [B1, Proposition 2.6] that $\delta_{e_0}(V^\mu)$ is annihilated by $\mathfrak{g}^{e_0}$.

6. The twisted ideals $J_{-\mu}$ and $I_{-\mu}$

Define, for any weight $\mu \in \mathcal{P}$,

$$J_{-\mu} := j_X^{-\mu}(\Gamma(\underline{T}_X^*, \tau^*\mathcal{F}^\mu)) \subseteq R_{\gamma_\mu}(N^o),$$

$$I_{-\mu} := \Omega_X^{-\mu}(gr\ \Gamma(Q, \mathcal{L}^{-\mu})) \subseteq R_{\gamma_\mu}(N^o).$$

So $J_0 = I_0 = R(N^o)$. Note that $I_{-\mu}$ (a priori) depends on our fixed pair (Q, π); as usual, we suppress this dependence in the notation.

The homogeneous components of $I_{-\mu}$ are the images, for $p \geq 0$, of the sequences of G-linear inclusions, via $\Delta = \Delta_X^{-\mu}$ and $j = j_X^{-\mu}$:

$$\Gamma^p(Q, \mathcal{L}^{-\mu}) \overset{\Delta}{\hookrightarrow} \Gamma(X, \mathcal{F}^{-\mu} \otimes S^p(\mathcal{T}_X)) \overset{j}{\hookrightarrow} R_{\gamma_{-\mu}}^{p+(\mu,\rho)}(N^o). \tag{6.1}$$

Set $m_\lambda^\mu := |V_\lambda^\mu|$, $\lambda \in \mathcal{P}^{++}$, $\mu \in \mathcal{P}$. We call a polynomial $f(q)$ with non-negative integral coefficients a q-analog of its particular value $f(1)$.

The results of the last two sections imply

Proposition 6.2. *Let $\mu \in \mathcal{P}$. Then $J_{-\mu} = \oplus_{p\geq 0} J^p_{-\mu}$ and $I_{-\mu} = \oplus_{p\geq 0} I^p_{-\mu}$ are graded $G-$ stable $R(N^o)$-submodules of $R_{\gamma_{-\mu}}(N^o)$, with $I_{-\mu} \subseteq J_{-\mu}$. We have $\langle V^*_\lambda, I_{-\mu}\rangle = m^\mu_\lambda$, for all $\lambda \in \mathcal{P}^{++}$. The polynomial*

$$r^\mu_\lambda(q) := q^{\langle\mu,\rho\rangle} \sum_{p\geq 0} \langle V^*_\lambda, \Gamma^p(Q, \mathcal{L}^{-\mu})\rangle q^p = \sum_{p\geq 0} \langle V^*_\lambda, I^p_{-\mu}\rangle q^p$$

is a q-analog of m^μ_λ.

Lemma 6.3. *Let $\mu \in \mathcal{P}^{++}$. Then $I^p_{-\mu} = J^p_{-\mu} = 0$ for $0 \leq p < \langle\mu,\rho\rangle$, and $I^{\langle\mu,\rho\rangle}_{-\mu} = J^{\langle\mu,\rho\rangle}_{-\mu} = K_{-\mu}$ (see 3.4).*

Proof: Immediate from Corollary 3.5 and (6.1), since, for all $\nu \in \mathcal{P}$, $\Gamma^{\leq 0}(Q, \mathcal{L}^{-\nu}) = \Gamma(X, \mathcal{F}^{-\nu})$. Q.E.D.

Example 6.4. Suppose $G = SL_2$, and μ is the highest or lowest weight of $S^p(\mathbb{C}^2)$. Then $R(N^\sim)$ contains each irreducible G-representation exactly once. As a representation, $I_{-\mu} = \oplus_{i\geq p} S^i(\mathbb{C}^2)$. If p is even, then $I_{-\mu}$ is the ideal of $R(N^o)$ generated by the copy of $S^p(\mathbb{C}^2)$. If p is odd, then $I_{-\mu}$ is the $R(N^o)$-submodule of $R_{-1}(N^o)$ generated by the copy of $S^p(\mathbb{C}^2)$. $J_{-\mu} = I_{-\mu}$ if μ is dominant, while $J_{-\mu} = R_{\gamma_{-\mu}}(N^o)$ if μ is anti-dominant.

Through Proposition 5.5, we can explicitly relate computing the $I_{-\mu}$ to doing certain calculations on representations. Recall from [B1, §2] that the *e_0-limit $lim_{e_0}(V^\mu)$ of V^μ* is the subspace generated by $\delta_{e_0}(V^\mu)$.

Corollary 6.5. *In diagram (5.6), ev_{e_0} carries $Hom_G(V^*_\lambda, I_{-\mu})$ isomorphically onto $lim_{e_0}(V^\mu_\lambda)$.*

In particular, Lemma 6.3 is equivalent to the obvious fact: $lim_{e_0}(V^\mu_\mu) = V^\mu_\mu$.

Vanishing of the higher sheaf cohomology of $\mathcal{F}^{-\mu} \otimes S(\mathcal{T}_X)$ simplifies the situation in various ways. The two relevant conditions are

Hypotheses 6.6. *We have the following two conditions on the pair (μ, G), for $\mu \in \mathcal{P}$.*

1. *FVH (First Vanishing Hypothesis): $H^1(X, \mathcal{F}^{-\mu} \otimes S(\mathcal{T}_X)) = 0$.*
2. *HVH (Higher Vanishing Hypothesis): $H^i(X, \mathcal{F}^{-\mu} \otimes S(\mathcal{T}_X)) = 0$, for all $i \geq 1$.*

Lemma 6.7. *Let $p \geq 0$. If $H^1(X, \mathcal{F}^{-\mu} \otimes S^i(\mathcal{T}_X)) = 0$ for $i = 0, \ldots, p-1$, then $I^{p+\langle\mu,\rho\rangle}_{-\mu} = J^{p+\langle\mu,\rho\rangle}_{-\mu}$. In particular, if (μ, G) satisfies the FVH, then $I_{-\mu} = J_{-\mu}$, and hence $I_{-\mu}$ is independent of the choice of (Q, π).*

Proof: The hypothesized vanishing of cohomology implies that the first map in (6.1) is surjective. (Cf. [B1, Proof of Corollary 5.6].) Q.E.D.

Lusztig ([Lu]) introduced, for $\mu \in \mathcal{P}, \lambda \in \mathcal{P}^{++}$, the polynomials

$$m_\lambda^\mu(q) := \sum_{w \in W} sgn(w) \wp_q(w(\lambda+\rho)-\mu-\rho), \tag{6.8}$$

where $W = N(T)/T$ is the Weyl group, and $\wp_q$ is the usual q-analog to Kostant's partition function. Then $m_\lambda^\mu(1) = m_\lambda^\mu$, so that $m_\lambda^\mu(q)$ is a $q-$ *analog* of weight multiplicity, whenever $m_\lambda^\mu(q)$ has non-negative coefficients. This happens for all dominant μ.

The comparison of the two q-analog polynomials is

Theorem 6.9. ([B1]) *If (μ, G) satisfies the HVH, then, for all $\lambda \in \mathcal{P}^{++}$, $r_\lambda^\mu(q) = q^{\langle \mu, \rho \rangle} m_\lambda^\mu(q)$.*

In [B1, Theorem 6.3], we observed that results of Andersen and Jantzen and of Griffiths say (respectively) that(μ, G) satisfies the HVH if

(6.10)
(1) $\mathfrak{g}$ *is of classical type, and $\mu+\rho$ is dominant, or*
(2) μ *is regular and dominant.*

7. Lattice properties of the twisted ideals

Proposition 7.1. *Suppose $\mu, \nu \in \mathcal{P}$. Then $J_\mu J_\nu \subseteq J_{\mu+\nu}$ and $I_\mu I_\nu \subseteq I_{\mu+\nu}$, where the multiplication is inside $R(N^\sim)$.*

Proof: This follows formally. Q.E.D.

We have the usual partial ordering on weights by $\mu \preceq \nu \Leftrightarrow \nu - \mu$ is a sum of simple roots.

Proposition 7.2. *Let $\mu, \nu \in \mathcal{P}$, $\mu \preceq \nu$. Then*

(1) $J_{-\nu} \subseteq J_{-\mu}$.
(2) *Assume also that (μ, G) satisfies the FVH. Then $I_{-\nu} \subseteq I_{-\mu}$.*

Proof: Say $\nu - \mu = \alpha_1 + \cdots + \alpha_d$, a sum of d simple roots.

Choose $g \in \Gamma(X, \mathcal{F}^{\nu-\mu} \otimes S^d(\mathcal{T}_X))$ such that $j_X^{-\mu}(g) = 1$ (see Corollary 4.4). Then g is a degree d twisted function on $\underline{T}_X^*$ with values in $\tau^* \underline{F}^{\mu-\nu}$, so multiplication by g gives a degree d G-linear injection $\Gamma(\underline{T}_X^*, \tau^* \mathcal{F}^{-\nu}) \to \Gamma(\underline{T}_X^*, \tau^* \mathcal{F}^{-\mu})$. Set $\gamma = \gamma_{-\mu} = \gamma_{-\nu}$.

Thus the following diagram is commutative, for each $p \geq 0$:

$$
\begin{array}{ccc}
\Gamma^p(Q,\mathcal{L}^{-\nu}) & & \Gamma^{d+p}(Q,\mathcal{L}^{-\mu}) \\
\Delta_X^{-\nu}\big\downarrow & & \big\downarrow\Delta_X^{-\mu} \\
\Gamma(X,\mathcal{F}^{-\nu}\otimes S^p(\mathcal{T}_X)) & \xrightarrow{\times\, g} & \Gamma(X,\mathcal{F}^{-\mu}\otimes S^{d+p}(\mathcal{T}_X)) \\
j_X^{-\nu}\big\downarrow & & \big\downarrow j_X^{-\mu} \\
R_\gamma^{p+\langle\nu,\rho\rangle}(N^o) & \xrightarrow{\times\, 1} & R_\gamma^{d+p+\langle\mu,\rho\rangle}(N^o)
\end{array}
\tag{7.3}
$$

Note all maps in the diagram are injective.

So $J_{-\nu} \subseteq J_{-\mu}$. To establish the inclusion $I_{-\nu}^{p+\langle\nu,\rho\rangle} \subseteq I_{-\mu}^{p+\langle\nu,\rho\rangle}$, it is enough to verify the condition

$$\times\, g \text{ carries } \Delta_X^{-\nu}(\Gamma^p(Q,\mathcal{L}^{-\nu})) \text{ into } \Delta_X^{-\mu}(\Gamma^{d+p}(Q,\mathcal{L}^{-\mu})). \tag{7.4}$$

In general, (7.4) fails; note that, for $d \neq 0$, $g \notin \Delta_X^{\nu-\mu}(gr\ \Gamma(Q,\mathcal{L}^{\nu-\mu}))$. However, (7.4) obviously goes through when $\Delta_X^{-\mu}$ is surjective in degree $d+p$, so as long as $H^1(X,\mathcal{F}^{-\mu}\otimes S^i(\mathcal{T}_X)) = 0$ for $0 \leq i \leq d+p-1$, by Lemma 6.7. Q.E.D.

Let us write "$f(q) \leq g(q)$" for two polynomials $f(q) = a_0 + a_1 q + \dots$ and $g(q) = b_0 + b_1 q + \dots$ over the integers if $a_i \leq b_i$ for all $i \geq 0$.

Corollary 7.5. *Keep the conditions of 7.2(2). Take any $\lambda \in \mathcal{P}^{++}$. Then $r_\lambda^\nu(q) \leq r_\lambda^\mu(q)$. In particular, if (μ, G) and (ν, G) satisfy the HVH (e.g., they satisfy (6.10)), then*

$$q^{\langle\nu-\mu,\rho\rangle} m_\lambda^\nu(q) \leq m_\lambda^\mu(q). \tag{7.6}$$

In the case $\mathfrak{g} = \mathfrak{sl}_n$, Lascoux and Schutzenberger ([La-S]) have already proven the inequality (7.6), and, in fact, a stronger result on alternating sums.

Corollary 7.7. *Keep the conditions of 7.2(2), and assume further that $\nu \in \mathcal{P}^{++}$. Then $I_{-\mu}$ contains $K_{-\nu}$.*

Translating into statements about representations via Proposition 5.5 and Corollary 6.5, we find also

Corollary 7.8. *Keep the conditions of 7.2(2). Then, for each $\lambda \in \mathcal{P}^{++}$, $lim_{e_0}(V_\lambda^\nu) \subseteq lim_{e_0}(V_\lambda^\mu)$.*

8. Kostant's Theorem on $K_{-\mu}$.

We now present Kostant's result discussed in the introduction.

Let $\mathfrak{m}$ be the nilpotent radical of $\mathfrak{b}$.

Given a simple Lie algebra, the boundary of the principal nilpotent adjoint orbit in the nullcone contains an open dense orbit, which is called the *subregular nilpotent adjoint orbit* ([K1]).

Consider the $\mathfrak{b}$-linear ring homomorphism

$$\psi : R(N) \to S(\mathfrak{m}^*) = S(\mathfrak{m}^-)$$

by restriction of functions from N to $\mathfrak{m}$. The Killing form on $\mathfrak{g}$ induces a $\mathfrak{b}$-linear identification of the dual of $\mathfrak{m}$ with $\mathfrak{m}^-$, the nilradical of the Borel subalgebra $\mathfrak{b}^-$ opposite to $\mathfrak{b}$ with respect to $\mathfrak{t}$.

For $\mu \in Q$, define coefficients $p_\alpha(\mu)$ for α simple by $\mu = \sum_{\alpha\in\Delta} p_\alpha(\mu)\alpha$. Set

$$k_{-\mu} := \prod_{\alpha\in\Delta} X_{-\alpha}^{p_\alpha(\mu)}.$$

This is a rational function on $\mathfrak{m}$, and regular if $\mu \in Q^{++}$.

Theorem 8.1 ([K3]). *Assume $\mathfrak{g}$ is simple. Let $0 \neq \mu \in Q^{++}$. Then*
(1) *The unique highest degree copy $K_{-\mu}$ in $R(N)$ of V_μ^* vanishes on the subregular adjoint orbit in N. Moreover,*
(2) *Suppose $f \in K_{-\mu}$ is a weight vector. Then f is vanishes on $\mathfrak{m}$, unless f is a lowest weight vector, in which case $f|_{\mathfrak{m}}$ is a non-zero scalar multiple of $k_{-\mu}$.*

Proof: (2) Suppose $g \in L$ is a weight vector of weight $-\nu$. Then $\psi(g)$ is a linear combination of monomials M in the negative root vectors, with each M of degree $\langle \mu, \rho\rangle$ and weight $-\nu$. The degree and weight conditions force the relation $\langle \mu, \rho\rangle \leq \langle \nu, \rho\rangle$, with equality if and only if each M is a product of negative simple root vectors, hence if and only if $M = k_{-\mu}$. But we know $\langle \mu, \rho\rangle \geq \langle \nu, \rho\rangle$, since $-\nu$ is a weight of V_μ^*. So $\psi(g) = 0$ if $\nu \neq \mu$, while $\psi(f)$ is a scalar multiple of $k_{-\mu}$ if f is a lowest weight vector. Moreover, ψ is injective on lowest weight vectors, so $\psi(f) \neq 0$.

(1) Let $\mathfrak{m}^{subreg}$ be the subregular locus in $\mathfrak{m}$. The closure in $\mathfrak{m}$ of $\mathfrak{m}^{subreg}$ is exactly the union of hyperplanes described by $\prod_{\alpha\in\Delta} X_{-\alpha} = 0$ ([K1]). So $\mathfrak{m}^{subreg}$ meets some (in fact, each) hyperplane $X_{-\alpha} = 0$. Thus $k_{-\mu}$ vanishes at some point p on $\mathfrak{m}^{subreg}$. By (2) then, all of $K_{-\mu}$ vanishes at p. Since $K_{-\mu}$ is G-stable, $K_{-\mu}$ must vanish on the adjoint orbit through p. Q.E.D.

Remark 8.2. Alternatively, one can observe in the proof of 8.1(1) that $k_{-\mu}$ vanishes on all of $\mathfrak{m}^{subreg}$, since all the numbers $p_\alpha(\mu)$, $\alpha \in \Delta$, are non-zero. Then (1) follows without considering the other weight vectors of $K_{-\mu}$, since the set $B^- \cdot \mathfrak{m}^{subreg}$ is dense in the subregular orbit.

9. The zero-loci of the ideals $J_{-\mu}$ and $I_{-\mu}$

In this section, we assume that μ lies in the root lattice Q, so that $J_{-\mu}$ and $I_{-\mu}$ are ideals of $R(N)$.

Let us fix a decomposition $\mathfrak{g} = \mathfrak{g}_1 \oplus \cdots \oplus \mathfrak{g}_k$ of $\mathfrak{g}$ into simple Lie algebras. Relative to this, we have $\mu = \mu_1 + \cdots + \mu_k$, a sum of weights for the $\mathfrak{g}_i$.

Let S_i be the product of the subregular nilpotent adjoint orbit in $\mathfrak{g}_i$ with the principal nilpotent adjoint orbits in each of the $\mathfrak{g}_j$ with $j \neq i$, for $1 \leq i, j \leq k$. We may regard S_i as the "ith subregular nilpotent adjoint orbit in N". Call the union $S_1 \cup \cdots \cup S_k$ the *subregular locus in* N. More generally, define the *subregular locus in* N *relative to* μ to be the union of the S_i for those i such that $\mu_i \neq 0$. Call μ *strongly non-zero* if and only if $\mu_i \neq 0$ for all $i = 1, \ldots, k$.

Theorem 9.1. *Let $0 \neq \mu \in Q^{++}$. Then $J_{-\mu}$ and $I_{-\mu}$ have the same zero-locus $Z_{-\mu}$ inside N, namely the closure of subregular locus relative to μ. In particular, if μ is strongly non-zero, then $Z_{-\mu}$ is the boundary of N^o in N.*

We generalize Theorem 8.1(2) in the following way; this is purely a statement about $j = j_X^{-\mu}$.

Proposition 9.2. *Let $0 \neq \mu \in Q^{++}$. Then, for all $f \in J_{-\mu}$, and hence for all $f \in I_{-\mu}$, $f|_{\mathfrak{m}}$ is a multiple, in $S(\mathfrak{m}^-)$, of $k_{-\mu} = \prod_{\alpha\in\Delta} X_{-\alpha}^{p_\alpha}$, where $\mu = \sum_{\alpha\in\Delta} p_\alpha \alpha$.*

Proof: We have $e_0 = \sum_{\alpha\in\Delta} c_\alpha X_\alpha$, with all c_α non-zero; put $e = e_0$. We can easily calculate some coordinate functions on $\mathfrak{m}^{reg} := \mathfrak{m} \cap N^o = B \cdot e$. Since $B = T \cdot U$ (semi-direct product), where U is the unipotent radical of B, each $b \in B$ has a unique factoring $b = ut$ such that $u \in U$ and $t \in T$. Then $b \cdot e = (\sum_{\alpha\in\Delta} exp^\alpha(t) X_\alpha) + \ldots$, where the remaining terms only involve X_ϕ for ϕ positive but not simple. So

$$X_{-\alpha}(b \cdot e) = exp^\alpha(t).$$

Next we compute the values on $\mathfrak{m}^{reg}$ of $f = j(s) \in R(N)$ for a section $s \in \Gamma(\underline{T}_X^*, \tau^* \underline{F}^{-\mu})$. Set $x = x_{\mathfrak{b}}$ and $\underline{F} = \underline{F}^{-\mu}$. From §3, we see that (up to a non-zero scalar factor which is immaterial here), f is given on $\mathfrak{m}^{reg}$ by

$$f(b \cdot e) = b^{-1} \cdot s(\xi_{b \cdot e}) \in b^{-1} \cdot (\tau^* \underline{F})_{\xi_{b \cdot e}} = (\tau^* \underline{F})_{\xi_e} = \underline{F}_x,$$

once we fix an identification of the fiber $\underline{F}_x$ with $\mathbf{C}$. Now for any point ξ in the fiber $\tau^{-1}(x)$, we have the B-equivariant fiber projection $\tau_* : (\tau^* \underline{F})_\xi \to \underline{F}_x$. So

$$(9.3) \quad f(b \cdot e) = \tau_*(b^{-1} \cdot s(\xi_{b \cdot e})) = b^{-1} \cdot \tau_*(s(\xi_{b \cdot e})) = exp^{\mu}(t) \tau_*(s(\xi_{b \cdot e}))$$

since b^{-1} acts on the fiber $\underline{F}_x$ by dilation by the factor $exp^{-\mu}(t^{-1}) = exp^{\mu}(t)$ (note U acts trivially).

Our final answer in (9.3) is a product of two regular functions on $\mathfrak{m}^{reg}$. The first factor is

$$exp^{\mu}(t) = \prod_{\alpha \in \Delta} exp^{\alpha}(t)^{p_\alpha} = \prod_{\alpha \in \Delta} X_{-\alpha}(b \cdot e)^{p_\alpha} = k_{-\mu}(b \cdot e).$$

The second factor is $g(b \cdot e)$, where g is the regular function on $\mathfrak{m}$ given by

$$g(z) = \tau_*(s(\xi_{z,\mathfrak{b}})) \in \underline{F}_x,$$

$z \in \mathfrak{m}$. Note g is defined because *every* $z \in \mathfrak{m}$ is the direction of some tangent vector to X at x.

Since the equation

$$(9.4) \qquad f = k_{-\mu} g$$

holds on $\mathfrak{m}^{reg}$, it holds true on its closure $\mathfrak{m}$. Q.E.D.

Proof of Theorem 9.1: Proposition 9.2 and the proof of 8.1(1) give 9.1 in the case $\mathfrak{g}$ is simple. The generalization to the semi-simple case is immediate. Q.E.D.

Remark 9.5. Since $\mathfrak{m}^{reg}$ is the non-vanishing locus in $\mathfrak{m}$ of $\prod_{\alpha \in \Delta} X_{-\alpha}$, each function $k_{-\mu}$, $\mu \in Q$, is actually regular on $\mathfrak{m}^{reg}$. In particular, equation (9.4) on $\mathfrak{m}^{reg}$ is valid for all $\mu \in Q$, so that (9.4) gives a tangible form of the degree shift discussed in §3 and §4. Cf. Remark 3.6.

10. $I_{-\mu}$ as a representation

Let Λ be the character ring of $\mathfrak{g}$; put $\chi_\lambda := ch(V_\lambda)$, $\lambda \in \mathcal{P}^{++}$. The *graded character* of a representation A graded over the non-negative integers is the formal series $ch_q(A) = \sum_{p\geq 0} ch(A^p)q^p \in \Lambda[[q]]$. So

$$ch_q(I_{-\mu}) = \sum_{\lambda\in\mathcal{P}^{++}} r_\lambda^\mu(q)\chi_\lambda.$$

Suppose $\mu \in \mathcal{P}^{++}$. Let $t_\mu(q) := \sum_{w\in W^\mu}(-q)^{\ell(w)}$, where W^μ is the Weyl group stabilizer of μ, and ℓ is the usual length function on the Weyl group relative to the Bruhat order corresponding to the pair $(\mathfrak{b}, \mathfrak{t})$. The *Hall-Littlewood character* $P_\mu \in \Lambda[q]$ is defined by

$$P_\mu := t_\mu(q)^{-1} \sum_{w\in W} w(e^\mu \prod_{\phi>0} \frac{1-qe^{-\phi}}{1-e^{-\phi}});$$

see [Ka] and [G2].

Proposition 10.1. *Assume $\mu \in \mathcal{P}^{++}$ and (μ, G) satisfies the HVH (cf., (6.9)). Then*

$$ch_q(I_{-\mu}) = q^{\langle\mu,\rho\rangle} t_\mu(q)t_0(q)^{-1}P_\mu ch_q(R(N)). \tag{10.2}$$

Proof: Let $\mu \in \mathcal{P}$. Set $E_\mu := \sum_{\lambda\in\mathcal{P}^{++}} m_\lambda^\mu(q)\chi_\lambda$. If (μ, G) satisfies the HVH, then $ch_q(I_{-\mu}) = q^{\langle\mu,\rho\rangle}E_\mu$, by Theorem 6.8. On the other hand, for μ dominant, we know $E_\mu = t_\mu(q)t_0(q)^{-1}P_\mu E_0$, by [G2, (4.6) in Theorem 4.4]. But also $E_0 = ch_q(R(N))$ ([Hs]). Q.E.D.

Example 10.3. Suppose $\mathfrak{g} = \mathfrak{sl}_2$ and $\mu \neq 0$ is dominant. So $\mu = r\rho$, for some $r > 0$; let $\chi_r = \chi_{r\rho}$, etc. Then $P_r = \chi_r - \chi_{r-2}q$, and (10.2) says (cf Example 6.4):

$$\chi_r q^r + \chi_{r+2}q^{r+2} + \cdots = q^r(\chi_r - \chi_{r-2}q)(\chi_0 + \chi_2 q + \ldots)/(1+q).$$

Let $\overline{P}_\mu$ be the image of P_μ under the natural $\mathbf{Z}[q]$-linear projection $dim : \Lambda[q] \to \mathbf{Z}[q]$ obtained by replacing each irreducible character by its dimension.

Theorem 9.1 and Proposition 10.1 give

Corollary 10.4. *Keep the hypotheses of 10.1, and assume also $\mu \neq 0$. Then the polynomial*

$$f_\mu(q) := t_0(q) - q^{\langle \mu,\rho\rangle} t_\mu(q)\overline{P}_\mu$$

vanishes to order exactly 2 at $q = 1$.

Proof: 10.1 gives us a Hilbert series formula:

$$\begin{aligned} HS(R(N)/I_{-\mu}) &= HS(R(N)) - HS(I_{-\mu}) \\ &= (1 - q^{\langle \mu,\rho\rangle} t_\mu(q) t_0(q)^{-1}\overline{P}_\mu) dim(ch_q(R(N))) \end{aligned}$$

Clearly $dim(ch_q(R(N))) = HS(R(N))$. So we may apply the fact: for any graded ideal I of the coordinate ring R of an irreducible affine variety A with $\mathbb{C}^*$-action, the order of the zero at $q = 1$ in the rational function $HS(R/I)/HS(R)$ is equal to the codimension of the zero-locus in A of I. Now from Theorem 9.1, we know the zero-locus of $I_{-\mu}$, $\mu \neq 0$, has codimension exactly 2 in N. So we conclude that the rational function $(1 - q^{\langle \mu,\rho\rangle} t_\mu(q) t_0(q)^{-1}\overline{P}_\mu)$ must vanish at $q = 1$ to order exactly 2. As $t_0(1) = |W| \neq 0$, nothing is changed by multiplying by $t_0(q)$. Q.E.D.

Examples 10.5. Suppose $\mathfrak{g} = \mathfrak{sl}_n$. Let $\omega_1, \dots, \omega_{n-1}$ be the fundamental dominant weights, in the usual order, and let ϑ be the highest root. Then

$$f_\vartheta(q) = \tfrac{(1-q^{n+1})(1-q^{n-1})}{(1-q)^2} - (n^2 - 1)q^{n-1}.$$

Other cases are:

$$\begin{aligned} f_{2\omega_2}(q) &= 1 + q + 2q^2 + q^3 - 19q^4 + 15q^5 - q^7, n = 4 \\ f_{2\omega_1+\omega_3}(q) &= 1 + 2q + 4q^2 + 5q^3 + 6q^4 + 5q^5 + 4q^6 - 127q^7 + 100q^8 \\ &\quad - q^9 - q^{10} - q^{11}, n = 5 \\ f_{8\omega_1}(q) &= 1 + q + \cdots + q^7 - 330q^{14} + 630q^{15} - 378q^{16} + 70q^{17}, n = 8 \end{aligned}$$

Remark 10.6. It might be interesting to understand how to compute these polynomials $f_\mu(q)$, for even say $\mathfrak{sl}_n$, and to make a combinatorial proof of 10.4. More generally, it might be interesting to look at the polynomials $f_{\mu,\lambda}(q) = q^{\langle \lambda,\rho\rangle} t_\lambda(q)\overline{P}_\lambda - q^{\langle \mu,\rho\rangle} t_\mu \overline{P}_\mu$. Since $\overline{P}_\lambda|_{q=1} = |W/W^\lambda|$, it is obvious that $f_{\mu,\lambda}$ vanishes at $q = 1$. However, no other vanishing (even when $\lambda = 0$) seems apparent from the algebra.

11. Some more consequences of §9

Proposition 11.1. *Let $\mu \in \mathcal{Q}^{++}$. Then $I_{-\mu}$, $J_{-\mu}$, and the ideal generated by $K_{-\mu}$, all have the same restriction to $\mathfrak{m}$, namely the principal ideal $A_{-\mu}$ of $S(\mathfrak{m}^-)$ generated by $k_{-\mu}$.*

Proof: Theorem 8.1(2) implies that $\psi(K_{-\mu}R(N)) = A_{-\mu}$, while Lemma 6.3 and Proposition 9.2 imply the $\psi(I_{-\mu}) = \psi(J_{-\mu}) = A_{-\mu}$. Q.E.D.

Remarks 11.2. (1) Theorem 8.1(2) implies that $K_{-\mu}R(N)$ is the smallest ideal of $R(N)$ with restriction to $\mathfrak{m}$ equal to $A_{-\mu}$.
(2) In general, restriction to $\mathfrak{m}$ is not injective on ideals. For example, suppose $\mathfrak{g} = \mathfrak{sl}_3$, and let α and β be the simple roots. Let J be the ideal of $R(N)$ generated by the the lowest degree copy L of $V_{2(\alpha+\beta)}$ (the copy with lowest weight vector $X^2_{-\alpha-\beta}$). Let I be the ideal generated by L together with the top degree copy (so the degree 2 copy) of the adjoint representation $V_{\alpha+\beta}$. Then $J \subseteq I$, $J \neq I$, but $\psi(J) = \psi(I)$. The latter holds since $X_{-\alpha}X_{-\beta} \in \psi(L)$; in fact, $\psi(J)$ is the ideal of $S(\mathfrak{m}^-)$ generated by $X_{-\alpha}X_{-\beta}$, $X_{-\alpha}X_{-\alpha-\beta}$, $X_{-\beta}X_{-\alpha-\beta}$, and $X^2_{-\alpha-\beta}$.

Let ϑ be the highest root of $\mathfrak{g}$.

Proposition 11.3. *Let $\mathfrak{g} = \mathfrak{sl}_n$. Then $I_{-\vartheta} = J_{-\vartheta}$ is the defining prime ideal of $N - N^o$, the closure of the nilpotent subregular orbit, and is generated by $K_{-\vartheta}$.*

Proof: It is known that $K_{-\vartheta}$ generates the defining ideal of the closure of the nilpotent subregular orbit S. But $I_{-\vartheta}$ and $J_{-\vartheta}$ vanish on S (Theorem 9.1) and both these ideals contain $K_{-\vartheta}$ (Lemma 6.3). So $I_{-\vartheta}$ and $J_{-\vartheta}$ both equal the defining ideal. Q.E.D.

Corollary 11.4. *Let $\mathfrak{g} = \mathfrak{sl}_n$. Form the Lie subalgebra $\mathfrak{l} := \mathfrak{t} \oplus \mathbb{C}X_\vartheta \oplus \mathbb{C}X_{-\vartheta}$ (so $\mathfrak{l}$ is isomorphic to the Levi factor of a supraminimal parabolic subalgebra). Then for all $\lambda \in \mathcal{P}^{++}$, $|V_\lambda^0| = |V_\lambda^\vartheta| + |V_\lambda^{\mathfrak{l}}|$.*

Proof: Multiplicities are preserved in the coordinate rings of adjoint orbits through elements in the "subregular sheet" ([Bo-Kr]). So $|V_\lambda^0| - |V_\lambda^\vartheta|$, the multiplicity of V_λ^* in the nilpotent subregular orbit, is equal to $|V_\lambda^{\mathfrak{l}}|$, the multiplicity of V_λ^* in the semi-simple subregular orbit. Q.E.D.

Remarks 11.5. (1) The equality in 11.4 is easy to check in the case that V_λ is a "first layer representation" (i.e., a component in $(\mathbb{C}^n)^{\otimes n}$, see [B2, 8.1]). For then, one finds that $|V_\lambda^0|$ is the number of standard Young tableaux of shape λ, while $|V_\lambda^\vartheta|$ counts those tableaux in which "2" occurs in its first

row, and $|V_\lambda^{\mathfrak{l}}|$ counts those tableaux in which "2" occurs in its first column. It would be interesting to have a combinatorial proof of the general fact.
(2) It should be possible to prove a stronger assertion than 11.4: $V_\lambda^{\mathfrak{g}^{e_0}} = lim_{e_0} V_\lambda^{\vartheta} \oplus lim_{e_0} V_\lambda^{\mathfrak{l}}$.

12. Multiplication of twisted functions.

Probably the simplest result on multiplication in our setting is

Lemma 12.1. *Let $\mu \in \mathcal{P}$. Then $\Gamma(Q, \mathcal{L}^{-\mu})$ is generated over $R(Q)$ by any non-zero G-stable subspace.*

Proof: $\mathcal{L}^{-\mu}$ is a locally free sheaf of rank 1 on Q. So any set $\{s_i\}_{i \in I}$ of global sections generates $\mathcal{L}^{-\mu}$, as long as for each point $h \in Q$, some s_i is non-vanishing at h. Thus, if s is a non-zero global section, then the set $\{g \cdot s \mid g \in G\}$ generates. Q.E.D.

We can describe the generation of $\Gamma(Q, \mathcal{L}^{-\mu})$ more precisely by computing the multiplication map $M \otimes R(Q) \xrightarrow{\times} \Gamma(Q, \mathcal{L}^{-\mu})$ on each G-isotypic component.

For any two vector spaces U and V and $u \in U$, let $u^\vee : Hom(U, V) \to V$ be evaluation at u. If χ is a character of an algebraic group H, i.e., $\chi \in \hat{H}$, let $V^{H,\chi}$ be the space of χ-semi-invariants in an H-representation V.

Lemma 12.2. *Let $\pi, \lambda \in \mathcal{P}^{++}$. Let O be an adjoint orbit, with $p \in O$, and $H = G^p$. Let $\mathcal{L}^i$ be the sheaf of sections of the homogeneous line bundle $G \times^H \mathbb{C}_{\chi_i^{-1}}$ ($\chi_i \in \hat{H}$) on O, $i = 1, 2$. Suppose $0 \neq r \in Hom_G(V_\pi^*, \Gamma(O, \mathcal{L}^1))$, with $\overline{r} = ev_p \circ r \in V_\pi^{H,\chi_1}$ and $M = r(V_\pi^*)$. Then, for α the natural map induced by r, the following diagram is commutative, and all vertical maps linear isomorphisms.*

$$
\begin{array}{cccc}
 & Hom_G(V_\lambda^*, M \otimes \Gamma(O, \mathcal{L}^2)) & \xrightarrow{\times} & Hom_G(V_\lambda^*, \Gamma(O, \mathcal{L}^1 \otimes \mathcal{L}^2)) \\
 & \downarrow \alpha & & \\
(12.3) & Hom_G(V_\lambda^* \otimes V_\pi, \Gamma(O, \mathcal{L}^2)) & & \downarrow ev_p \\
 & \downarrow ev_p & & \\
 & (V_\pi^* \otimes V_\lambda)^{H,\chi_2} & \xrightarrow{\overline{r}^\vee} & V_\lambda^{H,\chi_1\chi_2}
\end{array}
$$

Proof: Here α is the composition of natural maps

$$Hom_G(V_\lambda^*, M \otimes U) \to Hom_G(V_\lambda^*, V_\pi^* \otimes U) \to Hom_G(V_\pi \otimes V_\lambda^*, U),$$

where $U = \Gamma(O, \mathcal{L}^2)$, and the first map is induced by r^{-1}.

Consider the "larger" diagram obtained by dropping, for each of the 5 spaces, the invariance and semi-invariance conditions. (I.e., we replace $Hom_G(V_\lambda^*, M \otimes \Gamma(O, \mathcal{L}^2))$ by $Hom(V_\lambda^*, M \otimes \Gamma(O, \mathcal{L}^2))$, etc.) All the maps still make sense. We will show that this "larger" diagram is commutative.

We compute the two images in V_λ^μ of $F \in Hom(V_\lambda^*, M \otimes \Gamma(O, \mathcal{L}^2))$. Suppose $F = v \otimes s \otimes f$, for $v \in V_\lambda$, $s \in M$, $f \in \Gamma(O, \mathcal{L}^2)$. Then $(ev_p \circ \times)(F) = s(p)f(p)v$. Following the diagram in the other direction, we find $F \mapsto v \otimes r^{-1}(s) \otimes f \mapsto f(p)v \otimes r^{-1}(s) \mapsto f(p)\langle r^{-1}(s), \overline{r}\rangle v$. Clearly, $\langle r^{-1}(s), \overline{r}\rangle = s(p)$. Q.E.D.

Remark 12.4. We can apply Lemma 12.2 in the case where $O = Q$ and $\mathcal{L} = \mathcal{L}^{-\mu}$, $\mu \in \mathcal{P}$, to get another proof of Lemma 12.1. For then $0 \neq r \in V_\pi^\mu$, and the bottom map in (12.3) is $(V_\pi^* \otimes V_\lambda)^0 \xrightarrow{\overline{r}^\vee} V_\lambda^\mu$. The latter is obviously surjective, which means that $r(V_\pi^*)$ generates the V_λ^*-isotypic component in $\Gamma(Q, \mathcal{L}^{-\mu})$.

Set $e := e_0$ The following is an "e-analog" to the fact: $f \in Hom_{\mathfrak{t}}(V_\pi, V_\lambda) \Rightarrow f(V_\pi^\mu) \subseteq V_\lambda^\mu$.

Lemma 12.5. *Let $\pi, \lambda \in \mathcal{P}^{**}$, $\mu \in \mathcal{P}$. Suppose $f \in Hom_{\mathfrak{g}^e}(V_\pi, V_\lambda)$. Then $f(lim_e V_\pi^\mu) \subseteq lim_e V_\lambda^\mu$.*

Proof: Set $\gamma = \gamma_\pi$. Suppose $r \in Hom_G(V_\pi^*, I_{-\mu})$. Then $\overline{r} = ev_e \circ r \in lim_e V_\pi^\mu$ (Corollary 6.5); $lim_e V_\pi^\mu$ is spanned by such vectors. Applying Lemma 12.2 when $O = N^o$, we find the following diagram, with maps as in (12.3), is commutative:

$$
\begin{array}{ccc}
Hom_G(V_\pi \otimes V_\lambda^*, R(N)) & \longrightarrow & Hom_G(V_\lambda^*, I_{-\mu}) \\
\Big\downarrow{\scriptstyle ev_e} & & \Big\downarrow{\scriptstyle ev_e} \\
(V_\pi^* \otimes V_\lambda)^{\mathfrak{g}^e} & \xrightarrow{r^\vee} & lim_e V_\lambda^\mu
\end{array}
\tag{12.6}
$$

So $\overline{r^\vee}(f) = f(\overline{r})$ lies in $lim_e V_\lambda^\mu$. Q.E.D.

Remark 12.7. One can also prove 12.5 more directly by using the fact

$$lim_e V_\lambda^\mu = \lim_{t \to \infty} exp(te) \cdot V_\lambda^\mu.$$

Proposition 12.8. *Suppose that for some $\pi_i \in \mathcal{P}^{++}$, $i = 1, \ldots, m$, we have $r_i \in Hom_G(V_{\pi_i}^*, I_{-\mu})$. Set $\overline{r_i} = ev_e \circ r_i \in lim_e V_{\pi_i}^\mu$. Then: the spaces $r_i(V_{\pi_i}^*)$, $i = 1, \ldots, m$, together generate $I_{-\mu}$ over $R(N)$ if and only if for each $\lambda \in \mathcal{P}_{\gamma_\mu}^{++}$, the map*

$$\overline{r_1}^\vee \oplus \cdots \oplus \overline{r_m}^\vee : ((V_{\pi_1}^* \oplus \cdots \oplus V_{\pi_m}^*) \otimes V_\lambda)^{\mathfrak{g}^e} \to lim_e V_\lambda^\mu$$

is surjective.

Proof: Immediate from diagram (12.6). Q.E.D.

Since $I_{-\mu}$ is finitely generated, we derive immediately a fact about representations:

Corollary 12.9. *Let $\mu \in \mathcal{P}$. Then there exists finitely many pairs (V_{π_i}, v_i), $i = 1, \ldots, m$, with $v_i \in \lim_e V^{\mu}_{\pi_i}$, such that, for all $\lambda \in \mathcal{P}^{++}_{\gamma_\mu}$, the map*

$$v_1^{\vee} \oplus \cdots \oplus v_m^{\vee} : ((V^*_{\pi_1} \oplus \cdots \oplus V^*_{\pi_m}) \otimes V_\lambda)^{\mathfrak{g}^e} \to lim_e V^{\mu}_{\lambda}$$

is surjective.

13. On the generation of the $I_{-\mu}$

Let us now filter $\Gamma(Q, \mathcal{L}^{-\mu})$, for $\mu \in \mathcal{P}^{++}$, by the spaces

$$C^{\leq p}(Q, \mathcal{L}^{-\mu}) := R^{\leq p}(Q) \cdot \Gamma^{\leq 0}(Q, \mathcal{L}^{-\mu}),$$

$p \geq 0$; call this the *induced filtration* of $\Gamma(Q, \mathcal{L}^{-\mu})$.

Proposition 13.1. *Let $\mu \in \mathcal{P}^{++}$, with v a highest weight vector of V_μ. Then the following are equivalent:*

(1) *$I_{-\mu}$ is generated over $R(N)$ by $K_{-\mu}$.*
(2) *For all $p \geq 0$, $R^{\leq p}(Q) \cdot \Gamma^{\leq 0}(Q, \mathcal{L}^{-\mu}) = \Gamma^{\leq p}(Q, \mathcal{L}^{-\mu})$; i.e., the fiber degree filtration coincides with the induced filtration.*
(3) *For all $\lambda \in \mathcal{P}^{++}$, the map $(V^*_\mu \otimes V_\lambda)^{\mathfrak{g}^e} \to lim_e V^{\mu}_{\lambda}$ given by evaluation at v is surjective.*

Proof: While $I_{-\mu}$ is the associated graded object of $\Gamma(Q, \mathcal{L}^{-\mu})$ with respect to its fiber degree filtration, $K_{-\mu} \cdot R(N)$ is the graded object

$$\oplus_{p \geq 0} C^{\leq p}(Q, \mathcal{L}^{-\mu}) / (C^{\leq p}(Q, \mathcal{L}^{-\mu}) \cap \Gamma^{\leq p-1}(Q, \mathcal{L}^{-\mu})).$$

So (2) $\Longrightarrow$ (1). Since also $C^{\leq p}(Q, \mathcal{L}^{-\mu}) \subseteq \Gamma^{\leq p}(Q, \mathcal{L}^{-\mu})$, all $p \geq 0$, also (1) $\Longrightarrow$ (2). On the other hand, Proposition 12.8 implies that, when $Q = Q_\rho$, (3) $\Longleftrightarrow$ (2). But clearly, (1) holds for a particular pair (Q, π) $\Leftrightarrow$ (1) holds for every pair (Q, π). Q.E.D.

Lemma 13.2. *Let $\mu \in \mathcal{P}^{++}$. Then the following are equivalent:*

(1) *$J_{-\mu}$ is generated over $R(N)$ by $K_{-\mu}$.*
(2) *$\Gamma(\underline{T}^*_X, \tau^* \mathcal{F}^{-\mu})$ is generated over $R(\underline{T}^*_X)$ by $\Gamma(X, \mathcal{F}^{-\mu})$.*

Proof: Immediate from Proposition 4.1 and Corollary 3.5. Q.E.D.

Conjecture 13.3. *Let $\mu \in \mathcal{P}^{++}$. Then the 5 statements in Proposition 13.1 and Lemma 13.2 are all true. In particular, $I_{-\mu} = J_{-\mu}$.*

Note that 13.2(1) implies $I_{-\mu} = J_{-\mu}$ and 13.1(1). Our conjecture is true if $G = SL_2$, and of course Theorem 9.1 (in view of Theorem 8.1) strongly supports the conjecture. Moreover, the next result is a piece of quantitative evidence.

We have seen above that $K_{-\mu} \otimes R(N)$ contains $I_{-\mu}$ as a representation; i.e., each irreducible V_λ occurs at least many times in $K_{-\mu} \otimes R(N)$ as it does in $I_{-\mu}$. This means that, as a representation, $K_{-\mu} \otimes R(N)$ is "large enough" to G-linearly project onto $I_{-\mu}$. The *graded* version of this fact is

Proposition 13.4. *Suppose $\mu \in \mathcal{P}^{++}$ and (μ, G) satisfies the HVH (cf. (6.10)). Form the tensor grading:*

$$(K_{-\mu} \otimes R(N))^p = K_{-\mu} \otimes R(N)^{p-\langle\mu,\rho\rangle},$$

$p \geq 0$. Then, as graded representations, $K_{-\mu} \otimes R(N)$ contains $I_{-\mu}$.

Proof: We need to show for all $\lambda \in \mathcal{P}^{++}_{\gamma-\mu}$, the polynomial

$$f_{\lambda,\mu}(q) := \sum_{p\geq 0} \langle V_\lambda^*, (K_{-\mu} \otimes R(N))^p \rangle q^p$$

satisfies the inequality $f_{\lambda,\mu}(q) \geq r_\lambda^\mu(q)$.

Now

$$f_{\lambda,\mu}(q) = q^{\langle\mu,\rho\rangle} \sum_{p\geq 0} \langle V_\lambda^* \otimes V_\mu, R(N)^p \rangle q^p = q^{\langle\mu,\rho\rangle} \sum_{\pi\in\mathcal{P}^{++}} m_\lambda^\pi(q) m_\mu^\pi(q) \frac{t_0(q)}{t_\pi(q)}.$$

The second equality is exactly [G2, Corollary 2.4]. The "first term" in the last summation occurs at $\pi = \mu$; the term is just $q^{\langle\mu,\rho\rangle} m_\lambda^\mu(q) t_0(q)/t_\mu(q)$. Since $t_0(q)/t_\mu(q)$ has constant term 1, we have shown that $f_{\lambda,\mu}(q) \geq q^{\langle\mu,\rho\rangle} m_\lambda^\mu(q)$. The vanishing assumption insures that $q^{\langle\mu,\rho\rangle} m_\lambda^\mu(q) = r_\lambda^\mu(q)$ (Theorem 6.9). Q.E.D.

REFERENCES

[Bo-Kr] W. Borho and H. Kraft, *Uber Bahnen und deren Deformationen bei linearen Aktionen reduktiver Gruppen*, Comment. Math. Helv. **54** (1979), 61-104.

[B1] R.K. Brylinski, *Limits of weight spaces, Lusztig's q-analogs, and fiberings of adjoint orbits*, J. Amer. Math. Soc. **2** (1989).

[B2] __________, *The stable calculus of the mixed tensor character I*, in "Séminaire d'Algèbre Paul Dubreil et Marie-Paule Malliavin", M.-P. Malliavin ed., p 35–94, Lecture Notes in Mathematics 1404, Springer-Verlag..

[Ge-Ki] I. M. Gelfand and A.A. Kirillov, *The structure of the Lie field connected with a split semisimple Lie algebra*, Funct. Anal **3** (1969), 7–26.

[Gi] V. Ginsburg, **Пучки на группе петель и Двоиственность Ленглендса** *preprint.*

[G1] R.K. Gupta (now Brylinski), *Generalized exponents via Hall-Littlewood symmetric functions*, Bull. A.M.S. **16** (1987), 287–291.

[G2] ________________, *Characters and the q-analog of weight multiplicity*, Jour. London Math. Soc. (2) **36** (1987), 68–76.

[Hs] W. Hesselink, *Characters of the nullcone*, Math Ann. **252** (1980), 179–182.

[J-S] A. Joseph and J.T. Stafford, *Modules of $\mathfrak{k}$-finite vectors over semi-simple Lie algebras*, Proc. Lond. Math. Soc. **49** (1984), 361–384.

[Ka] S. Kato, *Spherical functions and a q-analog of Kostant's weight multiplicity formula*, Inv. Math. **66** (1982), 461-468.

[K1] B. Kostant, *The principal three-dimensional subgroup and the Betti numbers of a complex semi-simple Lie group*, Amer. Jour. Math. **81** (1959), 973–1032.

[K2] __________, *Lie group representations on polynomial rings*, Amer. J. Math. **85** (1963), 327–404.

[K3] __________, *personal communication.*

[La-S] A. Lascoux and M.P. Schutzenberger, *Croissance des polynomes de Foulkes-Green*, C.R. Acad. Sci. Paris **288**, 95-98.

[Lu] G. Lusztig, *Singularities, character formulas, and a q-analog of weight multiplicities*, in "Analyse et Toplogie sur les Espaces Singuliers (II-III)", Asterisque **101-102** (1983), 208-227.

[S] N.N. Shapovalov, *Structure of the algebra of regular differential operators on a basic affine space*, Funct. Anal **8** (1974), 43-54.

Received September 29, 1989

Department of Mathematics
McAllister Building 312
The Pennsylvania State University
University Park, PA 16802
E-mail: RKB1@ PSUVM.BITNET

Representations with Maximal Primitive Ideal

DAN BARBASCH[1]

Dedicated to Prof. Dixmier on his 65th birthday

1. Introduction

Let G be a connected Lie group. An irreducible representation is a strongly continuous homomorphism

$$\pi : G \longrightarrow Aut(\mathcal{H}),$$

where $\mathcal{H}$ is a Hilbert space, such that there is no non–trivial closed invariant subspace. For a reductive group this is studied via $(\mathfrak{g}, K)$–modules. These are representations of the enveloping algebra,

$$\pi_\infty : \mathcal{U}(\mathfrak{g}) \longrightarrow End(\mathcal{H}^\infty)$$

with a certain natural compatibility with respect to the action of the maximal compact subgroup K. There are many aspects of the study of these modules. In particular, I mention the following.

(1) **Classify all irreducible representations,**
(2) **Find a character theory,**
(3) **Classify all irreducible unitary modules.**

In his book, "Enveloping algebras", Dixmier suggests in the introduction that for some of these problems it should be easier to study the corresponding problems for the primitive ideals. In this note I want to illustrate this for the problem of computing the characters of the irreducible spherical representations in the case of complex groups. This is a subproblem of (2).

[1] Supported by NSF grant DMS–8803500

These repesentations are intimately connected with the maximal primitive ideals. Most cases were done in [BV1] and [B]. In particular all the classical groups were done in [B], so only the exceptional groups will be treated here. The main difficulty is that the representations of the Weyl group on the left cells do not decompose with multiplicity 1. I will rely heavily on the results in [BV1] and [B].

In the case of left cells which decompose with multiplicity one, Theorem 3.5 in this note is equivalent to Theorem 3.8 in [J1], where a different approach is used. In the case of left cells which do not decompose with multiplicity 1, Theorem 3.14 gives character formulas for a certain set of irreducible representations including the spherical representation. (The corresponding result in [J1] is Theorem 4.6 which gives some character identities for the same set of representations).

The character formulas allow one to compute K–type multiplicities which are necessary in dealing with questions of unitarity.

2. Notation

Let G be a connected simply connected complex Lie group. We denote its Lie algebra viewed as a real algebra by $\mathfrak{g}_{\Re}$. Denote by K a fixed maximal compact subgroup and by B^+ a Borel subgroup. Let $\mathfrak{k}_{\Re}$ and $\mathfrak{b}_{\Re}^+$ be their respective Lie algebras. Then $T = K \cap B^+$ is a maximal torus in K. If $\mathfrak{t}_{\Re}$ is its Lie algebra, then $\mathfrak{a}_{\Re} = \sqrt{-1}\mathfrak{t}_{\Re}$ is a maximally split component. Let $A = exp\ \mathfrak{a}_{\Re}$ and $H = T \cdot A$ be a Cartan subgroup. Its Lie algebra is denoted by $\mathfrak{h}_{\Re}$. Fix an automorphism σ that preserves $\mathfrak{k}_{\Re}$ and such that $\sigma|_{\mathfrak{h}_{\Re}} = -Id$. Define

$$\check{X} = -X, \qquad {}^tX = -\sigma X, \qquad \text{for } X \in \mathfrak{g}_{\Re}. \tag{2.1}$$

Then $\check{}$ and ^{t} extend uniquely to antiautomorphisms of $\mathcal{U}(\mathfrak{g}_{\Re})$ by the formulas

$$\begin{aligned} (X_1 \dots X_n)\check{} &= (-1)^n X_n \dots X_1, \\ {}^t(X_1 \dots X_n) &= {}^tX_n \dots {}^tX_1. \end{aligned} \tag{2.2}$$

Let $\mathfrak{g} = (\mathfrak{g}_{\Re})_c$. Write j for the action of $\sqrt{-1}$ coming from the complexification and i for the action of $\sqrt{-1}$ coming from the original complex structure on $\mathfrak{g}_{\Re}$. Then $\mathfrak{g}$ can be written as the direct sum of two ideals each isomorphic to $\mathfrak{g}_{\Re}$,

$$\begin{aligned} \mathfrak{g}^L &= \{1/2(X + jiX) | X \in \mathfrak{g}_{\Re}\}, \\ \mathfrak{g}^R &= \{1/2(X - jiX) | X \in \mathfrak{g}_{\Re}\}. \end{aligned} \tag{2.3}$$

They identify with $\mathfrak{g}_{\Re}$ via the maps

$$\begin{aligned}\phi^L(X) &= 1/2(X + jiX), \\ \phi^R(X) &= 1/2(-{}^t\overline{X} + ji{}^t\overline{X}),\end{aligned} \tag{2.4}$$

for $X \in \mathfrak{g}_{\Re}$(the bar denotes complex conjugation with respect to the real form $\mathfrak{k}_{\Re}$ of $\mathfrak{g}_{\Re}$ viewed as a complex algebra this time). Then

$$\mathcal{U}(\mathfrak{g}) = \mathcal{U}(\mathfrak{g}^L) \otimes \mathcal{U}(\mathfrak{g}^R). \tag{2.5}$$

The complexification of $\mathfrak{k}_{\Re}$ can be identified as the image of $\mathfrak{g}_{\Re}$ under the map

$$\begin{aligned}\phi^D &: \mathfrak{g}_{\Re} \longrightarrow \mathfrak{g}, \\ \phi^D(X) &= \phi^L(-{}^tX) + \phi^R(X).\end{aligned} \tag{2.6}$$

Thus we can identify $\mathfrak{k} = (\mathfrak{k}_{\Re})_c$, $\mathfrak{t} = (\mathfrak{t}_{\Re})_c$, $\mathfrak{a} = (\mathfrak{a}_{\Re})_c$ with the following objects in $\mathfrak{g}_{\Re} \oplus \mathfrak{g}_{\Re}$.

$$\begin{aligned}\mathfrak{k} &= \{(-{}^tX, X) | X \in \mathfrak{g}_{\Re}\}, \\ \mathfrak{t} &= \{(H, -H) | H \in \mathfrak{h}_{\Re}\}, \\ \mathfrak{a} &= \{(H, H) | \mathfrak{h} \in \mathfrak{h}_{\Re}\}.\end{aligned} \tag{2.7}$$

Furthermore $j(X,Y) = (-\ {}^t\overline{Y}, X)$ so that $\mathfrak{g}_{\Re}$ identifies with

$$\mathfrak{g}|_{\Re} = \{(-\ {}^t\overline{X}, X) | X \in \mathfrak{g}_{\Re}\}. \tag{2.8}$$

We now describe the classification of irreducible $(\mathfrak{g}, K)$ modules. Denote by Δ the roots and by Δ^+ the roots of the fixed Borel subgroup $B^+ = H \cdot N^+$.

Let $\lambda, \lambda' \in \mathfrak{h}^*_{\Re}$ be such that $\lambda - \lambda' = \mu$ is the weight of a finite dimensional holomorphic representation of G. Let $\nu = \lambda + \lambda'$. Define a character $\mathbb{C}_{(\lambda,\lambda')}$ of H via

$$\begin{aligned}\mathbb{C}_{(\lambda,\lambda')}|_T &= \mathbb{C}_{\mu}, \\ \mathbb{C}_{(\lambda,\lambda')}|_A &= \mathbb{C}_{\nu}.\end{aligned} \tag{2.9}$$

$\mathbb{C}_{(\lambda,\lambda')}$ can be extended to a character of B^+ by making it trivial on N^+. Let

$$\begin{aligned}X(\lambda, \lambda') &= [Ind_{B^+}^G\ \mathbb{C}_{(\lambda,\lambda')}]_{\text{K-finite}}, \\ \overline{X}(\lambda, \lambda') &= \text{the unique irreducible subquotient of } X(\lambda, \lambda') \\ &\qquad \text{containing the K-type with extremal weight } \mu.\end{aligned} \tag{2.10}$$

THEOREM. *(Parthasarathy–Rao–Varadarajan, Zhelobenko) Fix* (λ, λ') *and* (λ_1, λ'_1) *as before. The following are equivalent.*

(1) $X(\lambda, \lambda')$ *and* $X(\lambda_1, \lambda'_1)$ *have the same composition factors with multiplicities.*
(2) $\overline{X}(\lambda, \lambda') \simeq \overline{X}(\lambda_1, \lambda'_1)$.
(3) *There is* $w \in W$ *such that* $w\lambda = \lambda_1$ *and* $w\lambda' = \lambda'_1$.

Any irreducible $(\mathfrak{g}, K)$ *module is isomorphic to an* $\overline{X}(\lambda, \lambda')$.

Proof: See [D1]. Q.E.D.

In particular given a primitive ideal $\mathcal{I}$, we can form the $(\mathfrak{g}, K)$–module $\mathcal{U}(\mathfrak{g})/\mathcal{I}$ with the action

$$(a \times b) \cdot v = {}^t\check{a} \cdot v \cdot \check{b}. \tag{2.11}$$

This is spherical and has an infinitesimal character. We want to study the particular case when $\mathcal{I}$ is maximal so that in fact this module is irreducible.

3. Spherical Modules

3.1. For each infinitesimal character χ there is a unique maximal primitive ideal $\mathcal{I}(\chi) \subset \mathcal{U}(\mathfrak{g})$. Then the unique irreducible spherical module satisfies

$$\overline{X}(\lambda, \lambda) \approx \mathcal{U}(\mathfrak{g})/\mathcal{I}(\chi). \tag{3.1.1}$$

There are two general results that are useful in reducing the problem to a smaller number of cases.

Translation Principle. *(Jantzen–Zuckerman)*

For character formulas it is enough to find such formulas for a representative of $\lambda \in (\mathfrak{h}^*/\Lambda)^+$, *where* Λ *is the weight lattice.*

Reduction to Integral Infinitesimal Character. *(Kazhdan–Lusztig conjecture for non–integral infinitesimal character)*

To each λ *one can attach the integral root system,*

$$\Delta(\lambda) = \{\alpha | (\check{\alpha}, \lambda) \in \mathbb{Z}\}.$$

Then it is enough to prove such formulas in the case $\Delta(\lambda) = \Delta$.

We can describe the representations in terms of data involving the nilpotent orbits as follows.

3.3. Given a representation (π, V) one can attach to it an $Ad\ G$ invariant set in the nilpotent cone of the Lie algebra called $WF(\pi)$. In particular, this is the closure of one nilpotent orbit by the work of [BV2], [BV3], [J2], [BB1] and [BB2]. Let

$$\mathcal{O}(\lambda) = WF(\overline{X}(\lambda, \lambda)), \quad \mathcal{X}(\lambda) = \{\overline{X}(\lambda, w\lambda) | WF(\overline{X}(\lambda, w\lambda)) \subset \overline{\mathcal{O}(\lambda)}\}, \tag{3.3.1}$$

and $\mathcal{G}(\lambda)$ be the Grothendieck group of virtual characters generated by the elements in $\mathcal{X}(\lambda)$.

THEOREM.

(1) *There is a left cell $V(\lambda)$ of Weyl group representations in the sense of Joseph–Kazhdan–Lusztig such that*

$$card\ \mathcal{X}(\lambda) = dim\ \mathcal{G}(\lambda) = dim\ Hom_W[V(\lambda) : V(\lambda)].$$

(2) *$\mathcal{G}(\lambda)$ is generated by*

$$\{R_{\sigma,w}(\lambda) = \sum_{x \in W} tr\ \sigma(x) X(\lambda, xw\lambda)\}$$

for $\sigma \in V(\lambda)$, $w \in W$.

This is essentially proved in sections 5 and 6 in [BV1].

3.4. Assume (as we may) that the infinitesimal character is integral. Then the spherical representations are parametrized by standard parabolic subalgebras $\mathfrak{p} = \mathfrak{m} + \mathfrak{n} \supset \mathfrak{b}$. Each such parabolic subalgebra determines a unique λ which is dominant for the fixed positive system Δ^+ and has inner product 0 with each simple root in $\mathfrak{m}$, 1 with each simple root not in $\mathfrak{m}$. The nilpotent orbit $\mathcal{O}(\lambda)$ satisfies

$$^L\mathcal{O}(\lambda) = ind_{^L\mathfrak{p}}^{^L\mathfrak{g}}[Triv], \tag{3.4.1}$$

where $^L\mathfrak{g}$ is the dual algebra and $^L\mathcal{O}$ the dual nilpotent as in [BV1]. Finally, the cell $V(\lambda)$ satisfies

$$V(\lambda) \approx J_{W(\mathfrak{m})}^{W}[Sgn] \otimes Sgn. \tag{3.4.2}$$

where J is truncated induction as defined in [L].

Thus, to describe the character theory of the spherical representations, we can try to find a basis formed of $R_{\sigma,w}(\lambda)$ and then express the irreducible representations in terms of that. This works very well when $V(\lambda)$ decomposes with multiplicity 1.

3.5. THEOREM. *Assume that $V(\lambda)$ has multiplicity 1. Then there is a finite group $M(\mathcal{O})$ and correspondences*

$$[x] \in [M(\mathcal{O})] \mapsto \sigma_x \in V(\lambda)$$
$$\chi \in \hat{M}(\mathcal{O}) \mapsto X_\chi \in \mathcal{X}(\lambda)$$

such that at the level of characters,

$$X_\chi = \frac{1}{|M(\mathcal{O})|} \sum_{[x]\in[M(\mathcal{O})]} tr\ \chi(x) R_{\sigma_x}(\lambda)$$
$$R_{\sigma_x}(\lambda) = \sum_{\chi\in M(\mathcal{O})} tr\ \chi(x) X_\chi.$$

Here

$$R_{\sigma_x}(\lambda) = \frac{1}{|M(\mathcal{O})|} \sum_{w\in W} tr \sigma_x(w) X(\lambda, w\lambda).$$

For the classical groups this is proved in [B]. The group $M(\mathcal{O}) \approx \overline{A}(\mathcal{O})$ is defined in [L] chapter 4. Except for the case of special unipotent, the correspondence

$$[x] \in [M(\mathcal{O})] \mapsto \sigma_x \in V(\lambda) \tag{3.5.1}$$

is different (see [B] section 5).

As mentioned in the introduction, this is equivalent to Theorem 3.8 in [J1].

Note first that in this case, $\{R_\sigma(\lambda)\}$ form a basis of $\mathcal{G}(\lambda)$. Then all the reductions in sections 7 and 8 in [BV1] apply. Thus we can reduce to the case when neither $(\mathcal{O}, \lambda)$ nor $({}^L\mathcal{O}, \lambda)$ are smoothly induced (hypotheses 7A, 7B, 7C, and hypotheses 8A, 8B in [B]). Furthermore, it is enough to consider only one representative of ${}^L\mathfrak{m}$ of linked Levi components as in definition 7.3 in [BV1].

Then we are reduced to the following cases.

	$\mathfrak{g}$	$\mathcal{O}$	${}^L\mathfrak{m}$	$V(\lambda)$
(1a)	G_2	A_2	A_1^s	V, V', ϵ_2
(1b)			A_1^l	V, V', ϵ_1
(2a)	F_4	$F_4(a_3)$	$A_1^l A_2^s$	$12_1, 9_2, 6_1, 4_3, 16_1$
(2b)			$A_1^s A_2^l$	$12_1, 9_3, 6_1, 4_4, 16_1$
(3a)	E_6	$D_4(a_1)$	$2A_2A_1$	$80_s, 60_s, 10_s$
(3b)			A_3A_1	$80_s, 60_s, 90_s$

(4a)	E_8	$2A_4$	A_4A_3	$4480_y, 3150_y, 4200_y, 420_y$ $7168_w, 1344_w, 2016_w$

Cases (1a), (2a), (3a) and (4a) are done in [BV1] and are known as special unipotent. Cases (1b), (2b) and (3b) are entirely similar. The correspondence alluded to earlier is as in the following table. In cases (a), the correspondence is as in chapter 4 of [L]. When more than one representation corresponds to one conjugacy class, the second one refers to case (b).

G2. $1 \leftrightarrow V;\ g_3 \leftrightarrow \epsilon_2, \epsilon_1;\ g_2 \leftrightarrow V'$.

F4. $1 \leftrightarrow 12_1;\ g_2' \leftrightarrow 9_2, 9_3;\ g_4 \leftrightarrow 4_3, 4_4;\ g_3 \leftrightarrow 6_1;\ g_2 \leftrightarrow 16_1$.

E6. $1 \leftrightarrow 80_s\ g_2 \leftrightarrow 60_s\ g_3 \leftrightarrow 90_s$.

In cases (1b) and (2b) the argument is the same as in the case of special unipotents with the long and short roots reversed. The proof will be given in a series of Lemmas. We follow [BV1] section 9 closely. The details will be useful in the case when $V(\lambda)$ does not decompose with multiplicity 1.

3.6. LEMMA. *There is a parabolic subalgebra* $\mathfrak{p}(\lambda) = \mathfrak{m}(\lambda) + \mathfrak{n}(\lambda) \supset \mathfrak{b}$ *satisfying the following.*

Let

$$P(\lambda) = \{w\lambda | w \in W',\ w\lambda|_{\mathfrak{m}} \text{ dominant regular}\}.$$

Then

(1) $P(\lambda)| = [M(\mathcal{O})]$.
(2) *There is a unique* λ^1 *such that for any* $\mu \in P(\lambda)$ $\mu - \lambda^1$ *is a sum of negative roots and* $(\check{\alpha}, \lambda^1) = 1$ *for all* $\alpha \in \Delta^+(\mathfrak{m})$ *simple.*
(3) *Suppose* μ *and* μ' *are in* $P(\lambda)$. *If* $\mu' = w\mu$ *for some* $w \in W(\lambda^1)$ *then* $\mu' = \mu$.
(4) *The elements in* $P(\lambda)$ *can be ordered so that*

$$|\lambda^i - \lambda^1| < |\lambda^{i+1} - \lambda^1|.$$

Proof: This is done by direct calculation.

G2. $\Delta(\mathfrak{m}) = \{\alpha_{short}\}$.

λ^i	$\lambda^i - \lambda^1$	**Dim**	**Length**
$(1, 0, -1)$	$(0, 0, 0)$	1	0
$(0, 1, -1)$	$(1, -1, 0)$	2	2
$(-1, 1, 0)$	$(-2, 1, 1)$	2	6

F4. $\Delta(\mathfrak{m}) = \Delta - \{\alpha_{short}\}$.

λ^i	$\lambda^i - \lambda^1$	**Dim**	**Length**
$(2,1,0,-1)$	$(0,0,0,0)$	1	0
$(1,1,0,-2)$	$(1,0,0,1)$	3	2
$(1,0,-1,-2)$	$(1,1,1,1)$	2	4
$(0,1,-1,-2)$	$(2,0,1,1)$	3	6
$(-1,0,-1,-2)$	$(3,1,1,1)$	1	12

E6. Let the labelling of the roots be as in [Bou]. Then we take $\Delta(\mathfrak{m}) = \{\alpha_1, \alpha_4, \alpha_5, \alpha_6\}$.

λ^i	**Dim**	**Length**
$(-\frac{1}{2}, -\frac{3}{2}, -\frac{1}{2}, \frac{1}{2}, \frac{3}{2}, -\frac{5}{6}, -\frac{5}{6}, \frac{5}{6})$	1	0
$(-1, -2, -1, 0, 1, -\frac{1}{3}, -\frac{1}{3}, \frac{1}{3})$	1	2
$(1, -2, -1, 0, 1, -\frac{1}{3}, -\frac{1}{3}, \frac{1}{3})$	2	4

Q.E.D.

3.7. For $\mu \in P(\lambda)$ let $F_\mu^{\mathfrak{m}}$ be the finite dimensional representation of $\mathfrak{m}$ of highest weight $\mu - \rho(\mathfrak{m})$. Set

$$\begin{aligned} E_\mu &= F_\mu^{\mathfrak{m}} \otimes F_{\lambda^1}^{\mathfrak{m}}, \\ I_\mu &= Ind_P^G[E_\mu], \\ X_\mu &= \text{ unique irreducible subquotient of } I_\mu \\ &\qquad \text{containing the K–type } \mu - \lambda \end{aligned} \tag{3.7.1}$$

LEMMA.

(1) *All I_μ and X_μ have infinitesimal character (λ, λ).*
(2) *The X_μ are distinct and exhaust $\mathcal{X}(\lambda)$.*
(3) *There is a constant $c(\lambda)$ depending only on λ such that*

$$c(I_\mu) = c(\lambda) dim\ F_\mu^{\mathfrak{m}}.$$

(4) *$c(X_\mu)$ is an integral multiple of $c(\lambda)$.*

Proof: This is Lemma 9.10 in [BV1]. Q.E.D.

3.8 PROPOSITION. *I_μ is irreducible.*

The proof is the same as for Proposition 9.11 in [BV1]. It follows from the following Lemmas.

3.9. Let

$$M_\mu = \mathcal{U}(\mathfrak{g}) \underset{\mathfrak{p}}{\otimes} [F_\mu^{\mathfrak{m}} \otimes \mathbf{C}_{\rho(\mathfrak{n})}],$$
$$L_\mu = \text{ unique irreducible quotient of } M_\mu \tag{3.9.1}$$

Write M_i for M_{λ^i}.

LEMMA. *Assume $\mathfrak{g}$ is of exceptional type.*

(1) *L_μ exhaust the irreducible representations in category $\mathcal{O}$ of Bernstein–Gelfand–Gelfand with infinitesimal character λ and GK–dimension $dim_{\mathbf{C}}\, \mathfrak{n}$.*
(2) *The multiplicity of M_μ is dim $F_\mu^{\mathfrak{m}}$.*
(3) *$M_i = L_i + \sum_{j>i} a_{ij} L_j$ on the level of characters.*

Proof: This is 9.15 in [BV1]. Q.E.D.

LEMMA. *Suppose $i < j$ and L_i has a nonsplit extension with L_j. Then L_j is a composition factor of M_i.*

Proof: This is 9.16 in [BV1]. Q.E.D.

3.10 LEMMA. *M_i is irreducible.*

Proof: This is Lemma 9.17 in [BV1]. Q.E.D.

3.11 Proof of Proposition 3.8. Let $\mathcal{F}_i = F_i \otimes \mathbf{C}$, where F_i has extremal vector $\lambda^i - \lambda^1$, and

$$T_i(X) = P_{(\lambda,\lambda)}(X \otimes \mathcal{F}_i) \tag{3.11.1}$$

This is a functor in the category of Harish–Chandra modules.

Because of multiplicity 1, I_{λ^1} is irreducible. Then

$$T_i(I_{\lambda^1}) = I_{\lambda^i} \oplus S \tag{3.11.2}$$

where S comes from the corresponding formula for the M_i from 3.9 and 3.10. If I_{λ^i} is not irreducible, then one of

$$Hom_G(I_{\lambda^i}, X_{\lambda^j}), \quad Hom_G(X_{\lambda^j}, I_{\lambda^i}) \tag{3.11.3}$$

is not 0 for some $j > i$. Say the first one is nonzero. Then

$$Hom_G(I_{\lambda^1} \otimes \mathcal{F}_i, X_{\lambda^j}) \neq 0. \tag{3.11.4}$$

Then

$$Hom_G(I_{\lambda^1}, \mathcal{F}^* \otimes X_{\lambda^j}) \neq 0, \tag{3.11.5}$$

so since I_{λ^1} is spherical irreducible,

$$Hom_K(Triv, \mathcal{F}_i^* \otimes X_{\lambda^j}) \approx Hom_K(F_i, X_{\lambda^j}) \neq 0. \tag{3.11.6}$$

This is a contradiction. The proof follows because by 3.2,

$$|\lambda^i - \lambda^1| < |\lambda^j - \lambda^1| \tag{3.11.7}$$

for $i < j$. Q.E.D.

REMARK. In case (3b), the proof of 3.8 is much simpler. I_3 is irreducible because its lowest K–type has the biggest length. I_1, I_2 are irreducible because of their multiplicity.

3.12 Proof of Theorem 3.5. We give details in case (3a) only. Observe first that R_{σ_x} is a joint eigenvector for the T_i ([BV1], Lemma 6.3). We can compute these eigenvectors in terms of the X_i. We get

$$\begin{aligned} T_2(X_1) &= 2X_1 + X_2, & T_3(X_1) &= 3X_1 + X_2 + X_3, \\ T_2(X_2) &= X_1 + 2X_2, & T_3(X_2) &= X_1 + 3X_2 + X_3, \\ T_2(X_3) &= 3X_3 & T_3(X_3) &= X_1 + X_2 + 5X_3. \end{aligned} \tag{3.12.1}$$

The common eigenvectors are

$$X_1 - X_2, \quad X_1 + X_2 - X_3, \quad X_1 + X_2 + 2X_3. \tag{3.12.2}$$

Then $X_1 + X_2 + 2X_3$ identifies with 80_s and $X_1 + X_2 - X_3$ identifies with 60_s as in section 10 of [BV1], by observing that

$$\begin{aligned} J_{W(A_2A_2A_1)}^{W(E_6)}[Sgn] &= 80_s + 60_s + 10_s, \\ J_{W(A_3A_2A_1)}^{W(E_6)}[Sgn] &= 60_s, \end{aligned} \tag{3.12.4}$$

and computing the eigenvectors of the corresponding induced representations. Thus 90_s must correspond to $X_1 - X_2$.

The proof follows. Q.E.D.

3.13. We consider the case when $V(\lambda)$ does not decompose with multiplicity 1. There are two cases.

$\mathfrak{g}$	$\mathcal{O}$	${}^L\mathfrak{m}$	$V(\lambda)$
F_4	$F_4(a_3)$	B_2	$12_1, 9_2, 9_3, 4_1, 6_2,$ $2 \times 16_1$

E_8	$2A_4$	A_5A_1	$4480_y, 3150_y, 4536_y, 5670_y,$ $5600_w, 4200_y, 2688_y, 2016_w,$ $2 \times 7168_w$

3.14. THEOREM. *For each of the cases above there is a parabolic subalgebra $\mathfrak{p} = \mathfrak{m} + \mathfrak{n}$ and finite dimensional representations $\mathcal{F}_i$ such that*

$$I_i = Ind_{\mathfrak{m}}^{\mathfrak{g}}[\mathcal{F}_i]$$

are irreducible and form a basis of $\mathcal{G}(\lambda)$.

Proof: With notation as in 3.6, we find that $P(\lambda)$ has the following properties.

(1) $|P(\lambda)| = |\mathcal{X}(\lambda)|$.
(2) There is a unique λ^1 such that for any $\mu \in P(\lambda)$, $\mu - \lambda^1$ is a sum of negative roots.
(3) The elements of $P(\lambda)$ can be ordered such that $\lambda^{i+1} - \lambda^i$ is either a sum of negative roots or neither a sum of negative, nor a sum of positive roots.
(4) $|\lambda^i - \lambda^1| \leq |\lambda^{i+1} - \lambda^1|$ and $X_i \approx X_j$ if and only if $i = j$.

This is done case by case and follows from the following lists.

F4. We use the labelling and realization of the simple roots in [Bou]. Let $\Delta(\mathfrak{m}) = \{\alpha_2, \alpha_3\}$ so that $\mathfrak{m}$ is of type B_2. The roots can be written as $(0,0,1,-1)$, $(0,0,0,1)$. We get

Label	Weight	Dim	Length
F_1	$(\frac{3}{2}, \frac{1}{2}, \frac{3}{2}, \frac{1}{2})$	1	0
F_2	$(\frac{3}{2}, -\frac{1}{2}, \frac{3}{2}, \frac{1}{2})$	1	1
F_3	$(\frac{1}{2}, \frac{3}{2}, \frac{3}{2}, \frac{1}{2})$	1	2
F_4	$(0,0,2,1)$	4	4
F_5	$(\frac{1}{2}, -\frac{3}{2}, \frac{3}{2}, \frac{1}{2})$	1	5
F_6	$(-\frac{1}{2}, \frac{3}{2}, \frac{3}{2}, \frac{1}{2})$	1	5
F_7	$(-\frac{1}{2}, -\frac{3}{2}, \frac{3}{2}, \frac{1}{2})$	1	8
F_8	$(-\frac{3}{2}, \frac{1}{2}, \frac{3}{2}, \frac{1}{2})$	1	9
F_9	$(-\frac{3}{2}, -\frac{1}{2}, \frac{3}{2}, \frac{1}{2})$	1	10

E8. Let $\Delta(\mathfrak{m}) = \{\alpha_1, \alpha_3, \alpha_4, \alpha_5, \alpha_6, \alpha_8\}$. Then the $F_i^{\mathfrak{m}}$ are

	Weight	**Dim**	**Length**
F_1	$(-\frac{5}{2}, -\frac{3}{2}, -\frac{1}{2}, \frac{1}{2}, \frac{3}{2}, -\frac{3}{2}, \frac{1}{2}, \frac{7}{2})$	2	0
F_2	$(-3, -2, -1, 0, 1, -1, 1, 3)$	2	2
F_3	$(-3, -2, -1, 0, 2, -2, 0, 2)$	12	4
F_4	$(-\frac{5}{2}, -\frac{3}{2}, -\frac{1}{2}, \frac{1}{2}, \frac{3}{2}, -\frac{5}{2}, -\frac{3}{2}, \frac{5}{2})$	6	6
F_5	$(-\frac{5}{2}, -\frac{3}{2}, -\frac{1}{2}, \frac{1}{2}, \frac{3}{2}, -\frac{7}{2}, \frac{1}{2}, \frac{3}{2})$	4	8
F_6	$(-\frac{7}{2}, -\frac{5}{2}, -\frac{3}{2}, -\frac{1}{2}, \frac{1}{2}, -\frac{3}{2}, -\frac{1}{2}, \frac{3}{2})6$	10	
F_7	$(-3, -2, -1, 0, 1, -3, 1, 1)$	4	10
F_8	$(-\frac{5}{2}, -\frac{3}{2}, -\frac{1}{2}, \frac{1}{2}, \frac{5}{2}, -\frac{5}{2}, -\frac{3}{2}, \frac{3}{2})$	6	10
F_9	$(-3, -2, -1, 0, 1, -3, -1, 1)$	12	12
F_{10}	$(-\frac{7}{2}, -\frac{5}{2}, -\frac{3}{2}, -\frac{1}{2}, \frac{3}{2}, -\frac{3}{2}, -\frac{1}{2}, \frac{1}{2})$	6	14
F_{11}	$(-\frac{5}{2}, -\frac{3}{2}, -\frac{1}{2}, \frac{1}{2}, \frac{3}{2}, -\frac{7}{2}, -\frac{3}{2}, -\frac{1}{2})$	2	24
F_{12}	$(-3, -2, -1, 0, 1, -3, -1, -1)$	2	26

We now follow the same argument as in Proposition 3.8.

In case F_4 we note that all I_i for $i \neq 4$ are irreducible because of multiplicity considerations. For I_4 we can apply the arguments in 3.6–3.11.

In case E_8 the arguments in 3.6–3.11 do not apply for I_6, I_7 and I_8. We first show that M_6, M_7 and M_8 are irreducible. We have

$$T_6(M_1) = M_6 + M_7 + M_8 \oplus \text{ lower terms.} \tag{3.14.1}$$

None of the $\lambda^i - \lambda^j$ with $i \neq j$ for $i, j = 6, 7, 8$ is a sum of negative roots. The claim follows as before, by observing that because M_1 is self–dual then so is $T_6(M_1)$. Thus L_6, L_7, L_8 are direct summands, a contradiction unless $M_i = L_i$ for $i = 6, 7, 8$.

To check that I_i are irreducible we note that (3.14.3) implies

$$T_6(I_1) = I_6 \oplus I_7 \oplus I_8 \oplus \text{ lower order terms.}$$

Then the argument follows as before. Q.E.D.

4. An Example of Unitary Representations

4.1. In this section we give an example how to use the character formulas to decide the unitarity of some nontrivial examples. In particular we collect evidence for Conjecture 11.3 in [B1].

DEFINITION. An infinitesimal character χ_λ will be called half-integral if

$$\check{\Delta}(\lambda) = \{\check{\alpha} | (\check{\alpha}, \lambda) \in \mathbf{Z}\}$$

is a maximal proper system of roots in the co-roots.

These infinitesimal characters are of particular interest because if $\mathcal{O}$ is cuspidal (rigid in the terminology of [Sp]) then any representation $\overline{X}$ satisfying $WF(\overline{X}) = \overline{\mathcal{O}}$ must have integral or half-integral infinitesimal character.

We recall Conjecture 11.3 in [B1].

CONJECTURE. *Let $\overline{X}(\lambda, \lambda)$ be such that $WF(\overline{X}(\lambda, \lambda)) = \overline{\mathcal{O}}$ cuspidal. Then $\overline{X}(\lambda, \lambda)$ is unitary if and only if $||\lambda||$ is minimal subject to the condition that $WF(\overline{X}(\lambda, \lambda)) = \mathcal{O}$.*

In particular, for $\mathcal{O}$ special in the sense of Lusztig [L], $\overline{X}(\lambda, \lambda)$ must be special unipotent.

We examine this conjecture in the case when $\mathfrak{g}$ is type F_4. In the notation of [Sp], the cuspidal nilpotents are

$\mathcal{O}$	s s l l	Sp/nSp	Degree	$\check{\Delta}(\lambda)$
$\emptyset$	0 0 0 0	1_4 *sp*	24	F_4
A_1	0 0 0 1	2_4 *nsp*	16	$B_4(1^9)$
$\tilde{A}_1$	1 0 0 0	4_5 *sp*	13	F_4
$A_1 + \tilde{A}_1$	0 1 0 0	9_4 *sp*	10	F_4,
$A_2 + \tilde{A}_1$	0 0 1 0	4_3 *nsp*	7	$A_3(1^4)A_1(1^2)$
$\tilde{A}_2 + A_1$	0 1 0 1	6_1 *nsp*	6	$A_2(1^3)A_2(1^3)$

Consider the case of $\mathcal{O} = \tilde{A}_2 + A_1$, corresponding to the system A_2A_2. Then the minimal length parameter is obtained by taking the 3-dimensional representation on A_2^s, and the one dimensional representation on A_2^l. In coordinates, the parameter is $(3/2, 7/6, 1/6, -5/6) \approx (-1, 5/3, 2/3, -1/3)$. We now consider the induced representation from $A_1^s \times A_2^l$ with parameter

$$\lambda_t = t(3/2, 1/2, 1/2, 1/2) + (1/4, -1/4, -1/4, -1/4) + (0, 1, 0, -1). \quad (4.1)$$

This is irreducible for $0 \leq t < 1/2$. At $t = 1/2$ the infinitesimal character is integral and the other factor is

$$(-1/2, 3/2, 1/2, 1/2) \times (1/2, 3/2, 1/2, 1/2). \quad (4.2)$$

This has lowest K–type $(1,0,0,0)$. This is induced irreducible from a representation on $A_1^l \times A_2^s$ and clearly unitary. The next reducibility point is at $t = 5/6$. The parameter is integral for $A_2^s \times A_2^l$ and the induced representation has a second factor

$$(-1,5/3,2/3,-1/3) \times (1,5/3,2/3,-1/3), \tag{4.3}$$

with lowest K–type $(2,0,0,0)$. Thus $(1,0,0,0)$ occurs in the spherical factor only; and there it occurs with multiplicity 1. At the next reducibility point, we have $t = 1$. The parameter is $(7/4,5/4,1/4,-3/4) \approx (1/2,2,1,0)$, with integral system $A_1 \times B_3$ The induced representation has only one other factor with lowest K–type $(3,1,1,1)$. The spherical factor is induced irreducible from the spherical representation with parameter $(1/2,2,1)$ in B_3. We note that the K–type $(1,0,0)$ does not occur. This is because the induced representation from B_2 with parameter $(t,2,1)$ is irreducible up to $t = 1/2$, where it has one other factor, $(1/2,2,1) \times (-1/2,2,1)$ with K–type exactly $(1,0,0)$; and this K–type occurs only once in the induced representation. On the other hand,

$$(1,0,0,0)|_{B_3} = (\pm 1;0,0,0) + (0;1,0,0) + (\pm 1/2;1/2,1/2,1/2) + (0,0,0,0), \tag{4.4}$$

so the signature on $(1,0,0,0)$ comes only from the trivial representation.

Thus the claim follows.

References

[B] D. Barbasch, *The unitary dual for complex classical Lie groups*, Invent. Math. **96** (1989), 103–176.

[BV1] D. Barbasch , D. Vogan, *Unipotent representations of complex semisimple Lie groups*, Ann. of Math. **121** (1985), 41–110.

[BV2] D. Barbasch , D. Vogan, *Primitive ideals and orbital integrals in complex classical groups*, Math. Ann. **259** (1982), 153–199.

[BV3] D. Barbasch , D. Vogan, *Primitive ideals and orbital integrals in complex exceptional groups*, J. of Algebra **80** (1983), 350–382.

[BB1] J.L. Brylinski, W. Borho, *Differential Operators on Homogeneous Spaces I*, Inv. Math. **69** (1982), 437–476.

[BB2] J.L. Brylinski, W. Borho, *Differential Operators on Homogeneous Spaces III*, Inv. Math. **80** (1985), 1–68.

[BOU] N. Bourbaki, "Groupes et algebres de Lie," Hermann, 1968.

[D1] M. Duflo, *Representations irreductibles des groupes semisimples complexes*, SLN **497** (1975), 26–88.

[D2] M. Duflo, *Representations unitaires irreductibles des groupes semisimples complexes de rang deux*, Bull. Soc. Math. France **107** (1979), 55–96.

[**J1**] A. Joseph, *Characters of Unipotent Representations*, preprint (1988),.

[**J2**] A. Joseph, *On the Associated Variety of a Primitive Ideal*, J. of Algebra **93** (1985), 509–523.

[**L**] G. Lusztig, "Characters of reductive groups over a finite field, Annals of Math. Studies," Princeton University Press.

[**Z**] G. Zuckerman, *Tensor products of finite and infinite dimensional representations of semisimple Lie groups*, Ann. of Math. **106** (1977).

Received January 29, 1990

Dan Barbasch
Department of Mathematics
White Hall
Cornell University
Ithaca NY 14853

Dixmier Algebras, Sheets, and Representation Theory

DAVID A. VOGAN, JR.*

Dedicated to Professor Jacques Dixmier on his sixty-fifth birthday

1. Introduction.

One of the grand unifying principles of representation theory is the method of coadjoint orbits. After impressive successes in the context of nilpotent and solvable Lie groups, however, the method encountered serious obstacles in the semisimple case. Known examples (like $SL(2, \mathbf{R})$) suggested a strong connection between the structure of the unitary dual and the geometry of the orbits, but it proved very difficult to formulate any precise general conjectures that were entirely consistent with these examples.

In the late 1960's, Dixmier suggested a way to avoid some of these problems. Motivated in part by the theory of C^*-algebras, he suggested that one should temporarily set aside a direct study of unitary representations and concentrate instead on their annihilators in the universal enveloping algebra. Classification of the annihilators would be a kind of approximation to the classification of the unitary representations themselves. The hope was that this approximation would be crude enough to be tractable, and yet precise enough to provide useful insight into the unitary representations themselves. This hope has been abundantly fulfilled: the two classification problems are now inextricably intertwined, and they continue constantly to shed new light on each other.

To be more precise, suppose $G_{\mathbf{R}}$ is a connected real Lie group with Lie algebra $\mathfrak{g}_{\mathbf{R}}$. The set of irreducible unitary representations of $G_{\mathbf{R}}$ is written Unit $G_{\mathbf{R}}$. Write $\mathfrak{g}$ for the complexification of $\mathfrak{g}_{\mathbf{R}}$, and $U(\mathfrak{g})$ for its universal enveloping algebra. If $(\pi, \mathcal{H})$ is a unitary representation of $G_{\mathbf{R}}$, then $U(\mathfrak{g})$ acts on the dense subspace $\mathcal{H}^\infty$ of smooth vectors in $\mathcal{H}$. We define $\mathrm{Ann}(\pi)$ to be the annihilator in $U(\mathfrak{g})$ of $\mathcal{H}^\infty$. Then $\mathrm{Ann}(\pi)$ is a two-sided ideal in

* Supported in part by NSF grant DMS-8805665.

$U(\mathfrak{g})$. (Our ideals will always be two-sided unless the contrary is explicitly stated.)

An ideal I in any ring R with unit is called *(left) primitive* if it is the annihilator of a simple (left) R-module. (This says exactly that I is the largest two-sided ideal contained in some maximal left ideal.) A maximal ideal is always primitive, but a primitive ideal need not be maximal. The ideal I is called *prime* if whenever J and J' are ideals with $JJ' \subset I$, then either $J \subset I$ or $J' \subset I$. A primitive ideal is necessarily prime, but a prime ideal need not be primitive. We say that I is *completely prime* if the quotient ring R/I has no zero divisors. A completely prime ideal is prime, but a prime ideal need not be completely prime. We write

$$\begin{aligned} &\text{Spec}\, R = \text{set of prime ideals in } R \\ &\text{Prim}\, R = \text{set of primitive ideals in } R \\ &\text{Spec}_1 R = \text{set of completely prime ideals in } R \\ &\text{Prim}_1 R = \text{set of completely prime primitive ideals in } R\,. \end{aligned} \tag{1.1}$$

Suppose now that $(\pi, \mathcal{H})$ is an irreducible representation of G. Because the irreducibility is topological rather than algebraic, the space of smooth vectors $\mathcal{H}^\infty$ will not be a simple module for $U(\mathfrak{g})$ (unless the representation is finite-dimensional). Nevertheless, Dixmier proved

Theorem 1.2 ([7]). *Suppose π is an irreducible unitary representation of a connected Lie group $G_\mathbb{R}$. Then* $\text{Ann}(\pi)$ *is a primitive ideal in* $U(\mathfrak{g})$.

To explain the connection with the orbit method, suppose that $\mathcal{O}_\mathbb{R}$ is an orbit of $G_\mathbb{R}$ on $\mathfrak{g}^*_\mathbb{R}$. Let G be a complex connected Lie group with Lie algebra $\mathfrak{g}$, and let $\mathcal{O}_\mathbb{C} = G \cdot \mathcal{O}_\mathbb{R}$. Roughly speaking, the method of coadjoint orbits seeks to attach to $\mathcal{O}_\mathbb{R}$ an irreducible unitary representation $\pi(\mathcal{O}_\mathbb{R})$ of $G_\mathbb{R}$. (Actually one needs to restrict attention to certain orbits, called "admissible," and one needs some additional data beyond the orbit itself.) We will call a correspondence from orbits to representations an *orbit correspondence*, and denote it KK (for Kirillov-Kostant). Under favorable circumstances (for example if $G_\mathbb{R}$ is an algebraic group) it appears that almost all interesting unitary representations should appear in the image of an orbit correspondence. The problem of constructing an orbit correspondence is therefore one of the most important unsolved problems in representation theory. We consider next what Dixmier's ideas can tell us about this problem.

One of Dixmier's insights was that one should attach to the complexified orbit a primitive ideal in $U(\mathfrak{g})$. This may be formulated as

Conjecture 1.3 (Dixmier). Suppose G is a complex connected Lie group with Lie algebra $\mathfrak{g}$, and $\mathcal{O}_{\mathbf{C}}$ is an orbit of G on $\mathfrak{g}^*$. Then there is attached to $\mathcal{O}_{\mathbf{C}}$ a completely prime primitive ideal $I(\mathcal{O}_{\mathbf{C}})$ in $U(\mathfrak{g})$.

In fact the ideal should depend only on the Zariski closure of $\mathcal{O}_{\mathbf{C}}$ in $\mathfrak{g}^*$. A correspondence of the form in the conjecture is called a *Dixmier map*, and often written Dix. In its strongest original form (never formulated by Dixmier) the "Dixmier conjecture" asks that Dix should be a bijection from Zariski closures of orbits to completely prime primitive ideals. (Work of Borho and others has shown that this is not possible if the Dixmier map is to have other reasonable properties.)

If one had both an orbit correspondence and a Dixmier map, there would be (roughly speaking) a diagram

$$\begin{array}{ccc} \mathcal{O}_{\mathbf{R}} & \longrightarrow & \pi \\ \downarrow & & \downarrow \\ \mathcal{O}_{\mathbf{C}} & \longrightarrow & I \end{array} \tag{1.4)(a}$$

of maps among the sets

$$\begin{array}{ccc} \mathfrak{g}^*_{\mathbf{R}}/G_{\mathbf{R}} & \xrightarrow{\text{KK}} & \text{Unit } G_{\mathbf{R}} \\ \downarrow & & \downarrow \text{Ann} \\ \mathfrak{g}^*/G & \xrightarrow{\text{Dix}} & \text{Prim } U(\mathfrak{g}) \end{array} \tag{1.4)(b}$$

A natural requirement to impose on KK and Dix is that this diagram ought to commute. Unfortunately this is not a reasonable condition. The Dixmier map is supposed to take values in $\text{Prim}_1 U(\mathfrak{g})$ (the completely prime primitive ideals); but the annihilator of a unitary representation need not be completely prime. (The simplest example is the defining representation of $SU(2)$. The corresponding quotient $U(\mathfrak{g})/\text{Ann}(\pi)$ is isomorphic to the algebra of 2 x 2 matrices, and therefore has zero divisors.)

Since the diagram (1.4) cannot commute, we look for slightly weaker requirements. The simplest one consistent with examples like $SU(2)$ is

$$\text{Ann}(\pi(\mathcal{O}_{\mathbf{R}})) \subset I(\mathcal{O}_{\mathbf{C}}) \tag{1.5}$$

Equivalently,

$$\pi(\mathcal{O}_{\mathbf{R}}) \text{ is a } U(\mathfrak{g})/I(\mathcal{O}_{\mathbf{C}})\text{-module.} \tag{1.5)'$$

Properly understood, (1.5)′ casts representation theory in an entirely new light. We want to interpret it as a program for constructing an orbit correspondence. The first step is to construct a Dixmier map: more precisely, to construct and understand the algebra $U(\mathfrak{g})/I(\mathcal{O}_{\mathbf{C}})$. A unitary representation $\pi(\mathcal{O}_{\mathbf{R}})$ attached to $\mathcal{O}_{\mathbf{R}}$ —that is, the image of the orbit correspondence — should then be constructed and understood as a module for this algebra.

From now on we will confine our attention almost exclusively to reductive groups. To a large extent the point of view just described is the one adopted by Beilinson and Bernstein in their fundamental paper [1]. (It is not difficult to find similar ideas in much earlier work — for instance in the theory of C^*-algebras, or in the work of Gelfand and Kirillov on quotient rings of enveloping algebras. What is unique to Beilinson and Bernstein is the successful application of a general structure theorem for quotients of $U(\mathfrak{g})$ to representation theory.) They showed that if I is any minimal primitive ideal in $U(\mathfrak{g})$, then $U(\mathfrak{g})/I$ is isomorphic to an algebra of differential operators on a flag variety. Consequently any irreducible $\mathfrak{g}$-module may be regarded as a module for a differential operator algebra. This perspective has proven to be tremendously illuminating in a wide range of contexts: for the classification of representations, for the construction of intertwining operators, for analysis on symmetric spaces, and for primitive ideal theory, for example.

Nevertheless, the Beilinson-Bernstein approach has some limitations. In the context of (1.5), it amounts to looking for $\pi(\mathcal{O}_{\mathbf{R}})$ as a module not for the natural algebra $U(\mathfrak{g})/I(\mathcal{O}_{\mathbf{R}})$, but rather for some much larger algebra (of which $U(\mathfrak{g})/I(\mathcal{O}_{\mathbf{R}})$ is a quotient). The modules we want are certainly present in the Beilinson-Bernstein picture, but so are many extraneous ones. Putting more precise constraints on the primitive ideal should put more precise constraints on the modules, and so (one hopes) help to suggest the definition of the orbit correspondence. For this reason, it is still worthwhile to pursue the program described after (1.5).

There is already a tremendous amount of information available about the construction of a Dixmier map. Most of it is based on the notion of "parabolic induction" in one form or another. Roughly speaking, the idea is that most coadjoint orbits for G can be constructed in a simple way from coadjoint orbits for a Levi subgroup L. Parabolic induction also provides a way to construct primitive ideals for G from those for L. If we already know something about a Dixmier map on L, then we can hope to specify part of a Dixmier map for G by requiring that induction of orbits should correspond to induction of primitive ideals under the Dixmier maps for L and G. (One can make exactly parallel remarks about representations and orbit correspondences.) In the case of $SL(n)$, Borho in [2] used exactly this idea to define a Dixmier map completely.

For reductive groups not of type A, Borho discovered a fundamental obstruction to this approach. It can happen that the same coadjoint orbit $\mathcal{O}_{\mathbf{C}}$ for G arises in two different ways by induction, and that the corresponding induced primitive ideals I_1 and I_2 are different; apparently both ought to be attached to $\mathcal{O}_{\mathbf{C}}$. One of the goals of the theory of "Dixmier algebras" initiated in [20] and [15] is to circumvent this problem. Roughly speaking, the idea is this. In Conjecture 1.3, the orbit $\mathcal{O}_{\mathbf{C}}$ is replaced by an "orbit datum," consisting of some additional algebro-geometric structure on an orbit (Definition 2.2). The primitive quotient $U(\mathfrak{g})/I$ is replaced by a "Dixmier algebra" (Definition 2.1), which is an extension ring of a quotient of $U(\mathfrak{g})$. Conjecture 1.3 is replaced by a conjectural map from orbit data to Dixmier algebras (Conjecture 2.3). (Such a map would automatically descend to a multi-valued Dixmier map in the sense of Conjecture 1.3.)

The primary purpose of this paper is to extend to orbit data and Dixmier algebras the notions of parabolic induction discussed above. This is accomplished in Proposition 3.15 and Corollary 4.17 respectively. These results suggest a way to define a Dixmier map on induced orbit data (in terms of Dixmier maps on Levi subgroups). In order to justify this definition, one would have to show that if an orbit datum is induced in two different ways, then the corresponding induced Dixmier algebras must coincide. This we have not been able to prove; it would follow from Conjectures 3.24 and 4.18. A more complete discussion of the status of Conjecture 2.3 may be found in section 5.

These ideas of course focus attention on the orbit data that are *not* induced. (In fact the primary motivation for this paper was not so much to say something about induced orbits (or primitive ideals, or Dixmier algebras) as to understand by a process of elimination those that are not.) We call these non-induced orbit data *rigid*, for reasons that will be clearer in section 3 (cf. Definition 3.22 and Proposition 3.23). An interesting point is that rigidity is a property of the full orbit datum, and not just of the underlying orbit; it may be possible to deform the orbit but not the orbit datum. In section 5 we recall from [21] a conjectural construction of Dixmier algebras attached to rigid orbit data.

The program described after (1.5) suggests that one should turn next to the description of modules for induced Dixmier algebras, seeking among these candidates for representations attached by the orbit correspondence to real forms of $\mathcal{O}_{\mathbf{C}}$. In this direction we do only a little. Induced Dixmier algebras are generalizations of Beilinson and Bernstein's twisted differential operator algebras. It should therefore be possible to analyze their modules by the kind of geometric "localization" familiar in the differential operator case. We prove here only a few of the basic facts about such a localization theory (notably Corollary 6.16 and Theorem 7.9).

Here is a more detailed outline of the contents of this paper. Section 2 recalls from [20] and [15] the definition of Dixmier algebras and orbit data, and a corresponding refinement of Conjecture 1.3. (One of McGovern's results in [15] is that the main conjecture in [20] is false; and McGovern has since found counterexamples for a revision circulated in an earlier version of this paper. Conjecture 2.3 appears to be consistent with all of his work to date.) Section 3 outlines the extension to orbit data of some of the basic structure theory for coadjoint orbits: Jordan decomposition, parabolic induction, and sheets. In the theory of sheets we find some strong (conjectural!) geometric evidence for the correctness of the general approach to the Dixmier conjecture in [20]: Conjecture 3.24 says that distinct sheets of "orbit data" should be disjoint. The failure of the corresponding fact for sheets of orbits is at the heart of the non-uniqueness problems discovered by Borho and discussed above. Section 4 presents the construction of Dixmier algebras by parabolic induction. Section 5 outlines how these ingredients should fit together to define a Dixmier map for G.

The rest of the paper is devoted to related technical results. In section 6 (following [5]), we relate induction of Dixmier algebras to ordinary induction of Harish-Chandra bimodules. (Recall that Harish-Chandra bimodules are closely related to infinite-dimensional representations of G regarded as a real Lie group. By "ordinary induction" we mean the bimodule construction corresponding to parabolic induction of group representations (in the sense of Mackey and Gelfand-Naimark).) Perhaps the most important consequence is a cohomology vanishing theorem (Corollary 6.16). This generalizes the fact that the higher cohomology of G/Q with coefficients in the sheaf of differential operators is zero. Section 7 considers the translation principle for induced Dixmier algebras and their modules.

A key tool in all the induction constructions (both for orbit data and for Dixmier algebras) is the notion of equivariant bundles on homogeneous spaces (in the algebraic category). These help to formalize the idea that G-equivariant constructions on G/H are equivalent to H-equivariant constructions at a point. A few of the basic definitions and results are summarized in an appendix for the convenience of the reader.

2. Dixmier algebras.

Suppose for the balance of this paper that G is a connected complex reductive algebraic group with Lie algebra $\mathfrak{g}$.

Definition 2.1 (cf. [15]). A *Dixmier algebra* for G is a pair (A, ϕ) satisfying the following conditions.

i) A is an algebra over $\mathbb{C}$, equipped with a locally finite algebraic action (called Ad) of G on A by algebra automorphisms.

ii) The map ϕ is an algebra homomorphism of $U(\mathfrak{g})$ into A, respecting the two adjoint actions of G. The differential of the action Ad of G on

A is the difference of the left and right actions of $\mathfrak{g}$ defined by ϕ.

iii) A is a finitely generated $U(\mathfrak{g})$-module.

iv) Each irreducible G-module occurs at most finitely often in the adjoint action of G on A.

The simplest example of a Dixmier algebra is any quotient $U(\mathfrak{g})/I$ of $U(\mathfrak{g})$ by a primitive ideal.

Definition 2.2. An *orbit datum* for G is a pair (R, ψ) satisfying the following conditions.

i) R is an algebra over $\mathbb{C}$, equipped with a locally finite algebraic action (called Ad) of G on R by algebra automorphisms.

ii) The map ψ is an algebra homomorphism of $S(\mathfrak{g})$ into the center of R, respecting the two adjoint actions of G.

iii) R is a finitely generated $S(\mathfrak{g})$-module.

iv) Each irreducible G-module occurs at most finitely often in the adjoint action of G on R.

The *support* Σ of the orbit datum is the algebraic variety $\mathcal{V}(\ker\psi) \subset \mathfrak{g}^*$. (Thus Σ is the image of the *moment map* $\psi^* : \operatorname{Spec} R \to \mathfrak{g}^*$.) The orbit datum is called *completely prime* if R is completely prime, and *commutative* if R is commutative. It is called *geometric* if R is commutative, completely prime, and normal. It is called *pre-unipotent* if G is semisimple, and Σ is contained in the nilpotent cone. Finally, it is called *unipotent* if it is pre-unipotent and geometric.

The simplest example of an orbit datum is the quotient $S(\mathfrak{g})/J$ of $S(\mathfrak{g})$ by the ideal of functions vanishing on an orbit. In this case the support Σ is the closure of the orbit. (An example not of this kind appears as Example 3.21 below.) For general orbit data, condition (iv) implies that Σ is a finite union of orbit closures. (To see this, consider the algebra $Z = S(\mathfrak{g})^G$ of invariants in the symmetric algebra. The maximal ideals of Z parametrize the semisimple orbits of G on $\mathfrak{g}^*$: if $\mathfrak{m}$ is such a maximal ideal, then the associated variety $\mathcal{V}(\mathfrak{m})$ consists of all elements of $\mathfrak{g}^*$ for which the semisimple part of the Jordan decomposition belongs to the corresponding orbit. Consequently each $\mathcal{V}(\mathfrak{m})$ is a finite union of coadjoint orbits; in fact Kostant's theorem on the principal nilpotent element implies that it is the closure of a single coadjoint orbit. On the other hand, the image $\psi(Z)$ is contained in the G-invariants of R, which form a finite-dimensional algebra by (2.2)(iv). It follows that the kernel of ψ contains an ideal $\mathcal{I}$ of finite codimension in Z. Consequently Σ is contained in the (finite) union of the various $\mathcal{V}(\mathfrak{m})$, with $\mathfrak{m}$ a maximal ideal in Z containing $\mathcal{I}$.) If the orbit datum is completely prime, then Σ is necessarily the closure of a single coadjoint orbit. A geometric orbit datum (R, ψ) is the same thing as a normal irreducible affine algebraic variety X (namely Spec R) equipped with a G action and an equivariant finite morphism ψ^* from X to an orbit closure in $\mathfrak{g}^*$. A unipotent orbit datum is therefore exactly a unipotent Poisson variety in the sense of [21].

Before formulating the Dixmier conjecture, we should say a little bit about filtrations. Suppose A is an algebra filtered by $\frac{1}{2}\mathbf{N}$. This means that we are given an increasing family of subspaces

$$A_0 \subset A_{\frac{1}{2}} \subset A_1 \subset \cdots$$

so that

$$\cup_j A_j = A, \qquad A_p A_q \subset A_{p+q}.$$

Then the associated graded space $\operatorname{gr} A$ is a graded algebra. (Our basic example is the standard filtration $U_n(\mathfrak{g})$; in this case $\operatorname{gr} U(\mathfrak{g}) = S(\mathfrak{g})$.) An increasing filtration on an A-module M is called *compatible* if

$$\cup_j M_j = M, \qquad A_p M_q \subset M_{p+q}.$$

In this case the associated graded space $\operatorname{gr} M$ is in a natural way a graded module for $\operatorname{gr} A$. A compatible filtration of M is called *good* if $\operatorname{gr} M$ is finitely generated as a module for $\operatorname{gr} A$.

Here is a version of the Dixmier conjecture for reductive groups. It is taken from [20], but modified in accordance with the requirements of [15].

Conjecture 2.3. Suppose G is a complex connected reductive algebraic group. Then there is a natural injection Dix from the set of geometric orbit data for G (Definition 2.2) into the set of completely prime Dixmier algebras for G. This correspondence should have the following properties. Fix an orbit datum (R, ψ), and write (A, ϕ) for the corresponding Dixmier algebra.

i) The Gelfand-Kirillov dimensions of A and R are equal.
ii) A and R are isomorphic as G-modules.
iii) A and R admit filtrations indexed by $\frac{1}{2}\mathbf{N}$, with the following properties.
   a) The filtration of A is good (and therefore by definition compatible) for A regarded as a $U(\mathfrak{g})$-module. In particular, $\phi(U_n(\mathfrak{g})) \subset A_n$.
   b) The filtration of R is good for R regarded as an $S(\mathfrak{g})$-module. In particular, $\psi(S^n(\mathfrak{g})) \subset R_n$.
   c) The associated graded algebras gr A and gr R are completely prime.
   d) There is a G-equivariant isomorphism ξ : gr $A \rightarrow$ gr R carrying gr ϕ to gr ψ.

In (a), $U_n(\mathfrak{g})$ is the nth level of the standard filtration of $U(\mathfrak{g})$; and in (b), $S^n(\mathfrak{g})$ is the nth level of the standard gradation. (Of course we could equally well use the standard filtration of $S(\mathfrak{g})$ in (b).)

Conditions (i) and (ii) are included only for expository purposes; they are consequences of (iii). That the filtration in (iii) ought to be indexed by $\frac{1}{2}\mathbb{N}$ (rather than some $\frac{1}{k}\mathbb{N}$) is suggested by [16]. The map Dix should extend to a bijection from some larger set of completely prime orbit data onto all completely prime Dixmier algebras. McGovern has pointed out that it cannot be defined on *all* completely prime orbit data, however.

Orbit data are close enough to coadjoint orbits to admit Jordan decompositions, which we now describe. Suppose (R, ψ) is a completely prime orbit datum. Fix λ in $\mathfrak{g}^*$ so that $\ker \psi$ is the ideal of functions vanishing on $\mathcal{O} = G \cdot \lambda$. Write

$$\lambda = \lambda_s + \lambda_u \tag{2.4)(a}$$

for the Jordan decomposition of λ, and

$$L = \{\, g \in G \mid \mathrm{Ad}^*(g)\lambda_s = \lambda_s \,\} \tag{2.4)(b}$$

(a Levi subgroup of G). L is again a connected reductive algebraic group; we want to relate (R, ψ) to a completely prime orbit datum (R_L, ψ_L) for L. Define

$$\lambda_L = \lambda \mid_L; \tag{2.4)(c}$$

we want the support of (R_L, ψ_L) to be the closure Σ_L of $\mathcal{O}_L = L \cdot \lambda_L$. Let $\mathfrak{s}$ be the unique $\mathrm{Ad}(L)$-invariant complement for $\mathfrak{l}$ in $\mathfrak{g}$. We identify $\mathfrak{l}^*$ with the linear functionals on $\mathfrak{g}$ vanishing on $\mathfrak{s}$. By Proposition A.2(b), the natural inclusion of Σ_L in Σ induces a G-equivariant morphism

$$G \times_L \Sigma_L \to \Sigma. \tag{2.4)(d}$$

Every point $\sigma_L \in \Sigma_L$ has semisimple part λ_s. By the Jordan decomposition, the stabilizer of σ_L in G is contained in L. By (A.3), the map (2.4)(d) is one-to-one. A slightly more careful analysis, which we omit, shows that (2.4)(d) is actually an isomorphism of varieties. By Proposition A.5, the category of finitely generated modules with G-action on Σ is equivalent (by passage to geometric fibers) to the category of finitely generated modules with L-action on Σ_L. A little more explicitly, let $J_L \subset S(\mathfrak{g})$ be the defining ideal for Σ_L. Define

$$R_L = R/R\psi(J_L). \tag{2.5)(a}$$

Obviously R_L is an algebra equipped with an action of L and a map

$$\psi_L : S(\mathfrak{l}) \to R_L. \tag{2.5)(b}$$

(The map comes from ψ by restriction on the domain and passage to the quotient on the range.) The preceding discussion implies that (R_L, ψ_L) is a completely prime orbit datum for L, and that

$$R \simeq G \times_L R_L \tag{2.5)(c}$$

(cf. (A.4)). By (A.4)(c), this last formula says that R may be identified with the space of algebraic maps ρ from G to R_L, subject to the condition

$$\rho(gl) = \mathrm{Ad}(l^{-1})\rho(g) \tag{2.5)(d}$$

for g in G and l in L. The algebra structure on R is just pointwise multiplication.

Finally, one can check easily that the adjoint action of $Z(L)_0$ on R_L must be trivial. Consequently (R_L, ψ_L) gives rise to a pre-unipotent orbit datum (R_u, ψ_u) for $L/Z(L)_0$. The algebra R_u is just R_L, and the map ψ_u is the restriction of ψ_L to $[\mathfrak{l}, \mathfrak{l}]$, composed with the natural isomorphism

$$Lie(L/Z(L)_0) \simeq [\mathfrak{l}, \mathfrak{l}]. \tag{2.5)(e}$$

Conversely, one can recover ψ_L from ψ_u by the requirement

$$\psi_L(A) = \lambda_s(A) \qquad (A \in \mathfrak{z}(\mathfrak{l})). \tag{2.5)(f}$$

The following theorem summarizes this discussion.

Theorem 2.6 (Jordan decomposition for orbit data). *There is a natural bijection between the set of completely prime orbit data (R, ψ) for G (Definition 2.2) and the set of G-conjugacy classes of triples $(L, \lambda_s, (R_u, \psi_u))$. Here*

i) L is a Levi subgroup of G.
ii) $\lambda_s : \mathfrak{l} \to \mathbb{C}$ is a G-regular Lie algebra homomorphism.
iii) (R_u, ψ_u) is a completely prime pre-unipotent orbit datum for $L/Z(L)_0$.

The correspondence is specified by (2.4)-(2.5). In this bijection, R is commutative (respectively normal or geometric) if and only if R_u is.

In (ii), the "G-regular" hypothesis is just the condition (2.4)(b) above.

In light of the classification of unipotent orbit data in [21], Theorem 5.3, we get

Corollary 2.7. *The following three sets are in natural one-to-one correspondence:*

i) geometric orbit data (R, ψ) *for* G*;*

ii) G*-conjugacy classes of pairs* (λ, S)*, with* λ *in* $\mathfrak{g}^*$ *and* S *a subgroup of the "*G*-equivariant fundamental group"* G^λ/G_0^λ *of* $G \cdot \lambda$*;*

iii) G*-conjugacy classes of triples* $(L, \lambda_s, (R_u, \psi_u))$*, with* L *a Levi subgroup of* G*,* λ_s *a* G*-regular character of* $\mathfrak{l}$*, and* (R_u, ψ_u) *unipotent orbit data for* $L/Z(L)_0$*.*

If G *is simply connected, we can add*

iv) finite coverings of coadjoint orbits.

Unfortunately, no analogous theorems are known for completely prime primitive Dixmier algebras. There is, however, a construction (by parabolic induction) of a Dixmier algebra associated to an orbit datum, assuming that such an algebra associated to (R_u, ψ_u) is already available. It will be described in section 4, after some geometric preliminaries in section 3.

We conclude this section with some useful formal ideas; for more background, see [17] and [8].

Definition 2.8. Suppose (A, ϕ) is a Dixmier algebra for G. The *opposite Dixmier algebra* is the pair (A^{op}, ϕ^{op}) defined as follows. The algebra A^{op} is the opposite algebra to A, with the same underlying vector space and multiplication

$$(a) \cdot_{op} (b) = ba.$$

The action of G on A^{op} is the same as on A. The map ϕ^{op} is characterized by

$$\phi^{op}(X) = \phi(-X) \qquad (X \in \mathfrak{g}).$$

A *transpose antiautomorphism* of (A, ϕ) is an isomorphism (usually written $a \mapsto {}^t a$) of A with its opposite Dixmier algebra.

Definition 2.9. Suppose (R, ψ) is an orbit datum for G with support Σ. The *opposite orbit datum* is the pair (R^{op}, ψ^{op}) defined in obvious analogy with Definition 2.8; it is an orbit datum with support $-\Sigma$. We can also define a *transpose antiautomorphism*.

The Dixmier correspondence of Conjecture 2.3 should respect passage to opposite algebras in the sense of these definitions. Now a geometric unipotent orbit datum is easily seen to be isomorphic to its opposite. The corresponding Dixmier algebras should therefore admit transpose antiautomorphisms. Such automorphisms will play a role in the theory of unipotent representations (cf. [23], Theorem 8.7(ii)).

3. Induction, sheets, and rigid orbit data.

In this section we consider parabolic induction for orbit data. Suppose L is a Levi subgroup of G. It is well known that there is a codimension-preserving map from coadjoint orbits for L to coadjoint orbits for G. This map sends a G-regular semisimple orbit $L \cdot \lambda$ to $G \cdot \lambda$, and sends $\{0\}$ to a Richardson nilpotent orbit. The purpose of this section is to extend this map to orbit data. Because the construction is not a very easy one to grasp, we will begin by recalling the theory on the level of orbits.

Definition 3.1 (cf. [14]). Suppose L is a Levi factor in G, and $\mathcal{O}_L$ is a nilpotent coadjoint orbit in $\mathfrak{l}^*$. We will construct from $\mathcal{O}_L$ a nilpotent coadjoint orbit for G. Although this orbit turns out to depend only on L, its construction requires the choice of a parabolic subgroup Q with Levi factor L. Write U for the unipotent radical of Q; then $L \simeq Q/U$. This quotient map gives rise to an injection

$$i_Q : \mathfrak{l}^* \hookrightarrow \mathfrak{q}^* \tag{3.1)(a}$$

identifying linear functionals on $\mathfrak{l}$ with linear functionals on $\mathfrak{q}$ that vanish on $\mathfrak{u}$. Define

$$\mathcal{O}_Q = i_Q(\mathcal{O}_L) \subset \mathfrak{q}^*. \tag{3.1)(b}$$

(Thus $\mathcal{O}_Q$ is isomorphic to $\mathcal{O}_L$.) Next, consider the restriction map

$$\pi_{G/Q} : \mathfrak{g}^* \twoheadrightarrow \mathfrak{q}^*. \tag{3.1)(c}$$

This exhibits $\mathfrak{g}^*$ as an affine bundle over $\mathfrak{q}^*$, with vector space $(\mathfrak{g}/\mathfrak{q})^*$. (That is, each fiber of $\pi_{G/Q}$ is a principal homogeneous space for the vector space $(\mathfrak{g}/\mathfrak{q})^*$ — a copy of the vector space with the origin forgotten.) Define

$$\tilde{\mathcal{O}}_{G/Q} = \pi_{G/Q}^{-1}(\mathcal{O}_Q) \subset \mathfrak{g}^*, \tag{3.1)(d}$$

an affine bundle over $\mathcal{O}_Q$. In particular, we have

$$\dim \tilde{\mathcal{O}}_{G/Q} = \dim \mathcal{O}_L + \dim G/Q. \tag{3.1)(e}$$

The action of Q on $\mathfrak{g}^*$ evidently preserves $\tilde{\mathcal{O}}_{G/Q}$. By Proposition A.2(b), we get a G-equivariant morphism

$$G \times_Q \tilde{\mathcal{O}}_{G/Q} \to G \cdot \tilde{\mathcal{O}}_{G/Q} \subset \mathfrak{g}^*. \tag{3.1)(f}$$

The induced bundle on the left has dimension equal to

$$\dim G/Q + \dim \tilde{\mathcal{O}}_{G/Q} = \dim \mathcal{O}_L + 2 \cdot \dim G/Q = \dim \mathcal{O}_L + \dim G/L. \tag{3.1)(g}$$

It follows that any G-orbit in the induced bundle — and *a fortiori* any G-orbit in $G \cdot \tilde{\mathcal{O}}_{G/Q}$ — has dimension at most equal to $\dim \mathcal{O}_L + \dim G/L$.

The *induced orbit*

$$\mathcal{O}_G = \mathrm{Ind}_{orb}(L \uparrow G)(\mathcal{O}_L)$$

is the unique nilpotent coadjoint orbit in $\mathfrak{g}^*$ satisfying any of the following equivalent conditions.

i) The orbit $\mathcal{O}_G$ is contained in $G \cdot \tilde{\mathcal{O}}_{G/Q}$; and the dimension of $\mathcal{O}_G$ is $\dim \mathcal{O}_L + \dim G/L$.

ii) There is a λ in $\mathcal{O}_G$ such that the restriction of λ to $\mathfrak{q}$ is trivial on $\mathfrak{u}$ and belongs to $\mathcal{O}_L$ on $\mathfrak{l}$; and the codimension of $\mathcal{O}_G$ in $\mathfrak{g}^*$ is equal to the codimension of $\mathcal{O}_L$ in $\mathfrak{l}^*$.

iii) Fix a Cartan subalgebra $\mathfrak{h}$ of $\mathfrak{l}$, and write W_L and W for the Weyl groups of L and G. Then the Springer representation of W (on W-harmonic polynomials on $\mathfrak{h}$) for $\mathcal{O}_G$ is generated by the Springer representation for $\mathcal{O}_L$.

Several remarks are in order here. First, the equivalence of (i) and (ii) is immediate from the definitions. In (iii), the Springer correspondence must be normalized to take the principal orbit to the trivial representation. We have included (iii) only because it shows that the induced orbit is independent of the choice of Q. The proof of its equivalence with (i) and (ii) ([14], Theorem 3.5) would require a substantial digression, so we omit it.

Second, there is (given Q) at most one orbit $\mathcal{O}_G$ satisfying (i) and (ii). To see this, notice that $\mathcal{O}_L$ (as a homogeneous space for a connected group) is an irreducible algebraic variety. As an affine bundle over $\mathcal{O}_L$, the variety $\tilde{\mathcal{O}}_{G/Q}$ is irreducible as well. Since G is connected, it follows that $G \cdot \tilde{\mathcal{O}}_{G/Q}$ is irreducible. We have already observed that the dimension of this set is at most $\dim \mathcal{O}_L + \dim G/L$. It can therefore contain at most one G-orbit of the desired dimension. For the proof that one exists, we refer to [14].

Third, we should check that $\mathcal{O}_G$ is nilpotent. For this it suffices to show that $\tilde{\mathcal{O}}_{G/Q}$ consists of nilpotent elements. Now it is an elementary exercise to show that a linear functional λ on a reductive Lie algebra is nilpotent if and only if there is a Borel subalgebra $\mathfrak{b}$ such that $\lambda\,|_{\mathfrak{b}} = 0$. So suppose $\lambda \in \tilde{\mathcal{O}}_{G/Q}$. Then $\lambda\,|_{\mathfrak{u}} = 0$, and $\lambda\,|_{\mathfrak{l}} = \lambda_L \in \mathcal{O}_L$. By hypothesis λ_L is nilpotent, so there is a Borel subalgebra $\mathfrak{b}_L$ of $\mathfrak{l}$ on which λ_L is zero. Since Q is parabolic, $\mathfrak{b} = \mathfrak{b}_L + \mathfrak{u}$ is a Borel subalgebra of $\mathfrak{g}$; and clearly $\lambda\,|_{\mathfrak{b}} = 0$.

A nilpotent orbit is called *rigid* if it is not induced from a proper Levi subgroup, and *induced* otherwise. It is called *Richardson* if it is induced from the zero orbit on a Levi subgroup. (Thus the zero orbit is called Richardson but not induced.)

To get a little feeling for the notion of induced orbit, we consider some examples in the classical groups. If G is $GL(n)$ or $SL(n)$, then every nilpotent orbit except 0 is induced. (To describe this explicitly, one can replace 2 by 1 everywhere in the discussion of other classical groups below.) If G_n is any other classical group of n by n matrices, then any nilpotent coadjoint orbit $\mathcal{O}$ gives rise (via Jordan blocks) to a partition

$$\pi = (p_1, ..., p_r) \tag{3.2}$$

of n. The sequence of non-negative integers p_i is (weakly) decreasing, and the sum of the p_i is n. (It is often convenient to allow some of the p_i to be zero; we identify partitions when their non-zero parts agree.) Suppose that there is a jump of 2 in this sequence; say

$$p_m - 2 \geq p_{m+1}. \tag{3.3)(a}$$

Then $\mathcal{O}$ is induced, in the following way. G has a Levi factor

$$L = G_{n-2m} \times GL(m), \tag{3.3)(b}$$

with the first factor a classical group of $n-2m$ by $n-2m$ matrices. There is a nilpotent orbit $\mathcal{O}'$ for G_{n-2m} corresponding to the partition

$$\pi' = (p_1 - 2, ..., p_m - 2, p_{m+1}, ..., p_r). \tag{3.3)(c}$$

(In the case of $SO(2k)$ the orbit $\mathcal{O}'$ may not be uniquely determined by π'; one must make an appropriate choice of $\mathcal{O}'$.) Define $\mathcal{O}_L$ to be the orbit $(\mathcal{O}', 0)$ in L. Then

$$\mathcal{O} = \mathrm{Ind}_{orb}(L \uparrow G)(\mathcal{O}_L). \tag{3.3)(d}$$

There are other ways for a nilpotent orbit to be induced in the classical groups, but the preceding special case captures most of the general flavor. The first example of a non-zero rigid orbit is the minimal (4-dimensional) nilpotent in $Sp(4)$; it cannot be induced because its dimension is not of the form $\dim \mathcal{O}_L + \dim G/L$ for any orbit in proper Levi subgroup L. The corresponding partition is $(2,1,1)$, which of course has no jumps of 2. (Viewed within the locally isomorphic group $SO(5)$, this orbit gives rise to the partition $(2,2,1)$, which still has no jumps.)

We now extend Definition 3.1 to include induction from non-nilpotent orbits.

Definition 3.4. Suppose L is a Levi subgroup of G, and $\mathcal{O}_L \subset \mathfrak{l}^*$ is a coadjoint orbit for L. The *induced orbit*

$$\mathcal{O}_G = \mathrm{Ind}_{orb}(L \uparrow G)(\mathcal{O}_L)$$

is defined precisely as in Definition 3.1(i) or (ii), as the unique dense orbit in $G \cdot (\pi_{G/Q}^{-1}(i_Q(\mathcal{O}_L)))$. It follows easily from the construction that the codimension of $\mathcal{O}_G$ in $\mathfrak{g}^*$ is equal to the codimension of $\mathcal{O}_L$ in $\mathfrak{l}^*$. (Again the construction depends on a choice of parabolic, but the orbit constructed does not. We will not stop to prove this independence; it follows from the special case of Definition 3.1 by Proposition 3.6.)

We have already seen one instance of this more general induction in (2.4). Here is the connection.

Lemma 3.5. *Suppose $\lambda = \lambda_s + \lambda_u$ is the Jordan decomposition of a linear functional in $\mathfrak{g}^*$. Write L for the centralizer in G of λ_s, a Levi subgroup of G, and $\lambda_L = (\lambda_L)_s + (\lambda_L)_u$ for the restriction of λ to L. Set*

$$\mathcal{O}_L = L \cdot \lambda_L = (\lambda_L)_s + L \cdot (\lambda_L)_u.$$

Then the orbit for G induced by $\mathcal{O}_L$ is

$$\mathrm{Ind}_{orb}(L \uparrow G)(\mathcal{O}_L) = G \cdot \lambda.$$

Proof. Choose a parabolic $Q = LU$ as in Definition 3.4. Define $\mathcal{O}_Q \subset \mathfrak{q}^*$ and $\tilde{\mathcal{O}}_{G/Q} \subset \mathfrak{g}^*$ as in Definition 3.1. It is immediate from the definitions that $\lambda \in \tilde{\mathcal{O}}_{G/Q}$. It remains only to check that $G \cdot \lambda$ has the correct dimension. But this follows from the fact (part of the Jordan decomposition) that the centralizer in G of λ is just the centralizer in L of λ_L. Q.E.D.

In general, the induction operation of Definition 3.4 can be expressed in two steps: the first an induction of nilpotent orbits in the sense of Definition 3.1, and the second the "Jordan decomposition" case considered in Lemma 3.5. Here is a precise statement.

Proposition 3.6. *Suppose L is a Levi subgroup of G, and $\mathcal{O}_L \subset \mathfrak{l}^*$ is a coadjoint orbit for L; write*

$$\mathcal{O}_G = \mathrm{Ind}_{orb}(L \uparrow G)(\mathcal{O}_L).$$

Then the Jordan decomposition of an element of $\mathcal{O}_G$ may be computed as follows. Fix an element λ_L of $\mathcal{O}_L$, and write its Jordan decomposition as

$$\lambda_L = \lambda_s + (\lambda_L)_u.$$

Let L' be the centralizer in G of λ_s, and put $M = L' \cap L$. Then λ_L is zero on the natural complement of $\mathfrak{m}$ in $\mathfrak{l}$, and so may be identified with an element

$$\lambda_L = \lambda_M = \lambda_s + (\lambda_M)_u.$$

of $\mathfrak{m}^*$. Similarly, λ_s may be regarded as a semisimple element of $(\mathfrak{l}')^*$ or of $\mathfrak{g}^*$.

Write

$$(\mathcal{O}_M)_u = M \cdot (\lambda_M)_u$$

$$(\mathcal{O}_{L'})_u = \mathrm{Ind}_{orb}(M \uparrow L')((\mathcal{O}_M)_u).$$

Fix a representative $(\lambda_{L'})_u \in (\mathcal{O}_{L'})_u$. Then

$$\lambda = \lambda_s + (\lambda_{L'})_u$$

is the Jordan decomposition of an element of $\mathcal{O}_G$. We have

$$\mathcal{O}_G = \mathrm{Ind}_{orb}(L' \uparrow G)(\lambda_s + (\mathcal{O}_{L'})_u).$$

Proof. Because L' fixes λ_s, the sets

$$\mathcal{O}_M = \lambda_s + (\mathcal{O}_M)_u, \qquad \mathcal{O}_{L'} = \lambda_s + (\mathcal{O}_{L'})_u$$

are coadjoint orbits for M and L' respectively. By Lemma 3.5 applied to L,

$$\mathcal{O}_L = \mathrm{Ind}_{orb}(M \uparrow L)(\mathcal{O}_M).$$

By an easy "induction by stages" fact, we deduce

$$\mathcal{O}_G = \mathrm{Ind}_{orb}(M \uparrow G)(\mathcal{O}_M).$$

Now we induce in stages first from M to L', and then to G. This gives

$$\mathcal{O}_G = \mathrm{Ind}_{orb}(L' \uparrow G)(\mathcal{O}_{L'}).$$

This is the last formula in the proposition. The description of the Jordan decomposition follows from Lemma 3.5 applied to G. Q.E.D.

We can now describe sheets of coadjoint orbits. Recall that a nilpotent coadjoint orbit is called rigid if it is not induced.

Definition 3.7 (cf. [3], [4]). Suppose L is a Levi subgroup of G, and $\mathcal{O}_u$ is a rigid nilpotent coadjoint orbit for L. The *sheet* of coadjoint orbits attached to $(L, \mathcal{O}_u)$ is the collection of orbits

$$\{\operatorname{Ind}_{orb}(L \uparrow G)(\lambda_s + \mathcal{O}_u) \mid \lambda_s \in (\mathfrak{l}/[\mathfrak{l}, \mathfrak{l}])^* \}.$$

Two sheets are identified if they contain exactly the same orbits; by Proposition 3.6, this amounts to conjugacy of the pair $(L, \mathcal{O}_u)$ under G. A *Dixmier sheet* is one attached to the zero orbit in a Levi subgroup.

Here are some of the most important facts about sheets.

Proposition 3.8 (cf. [3]). *In the notation of Definition 3.7,*

i) Each coadjoint orbit belongs to at least one sheet.
ii) Each sheet contains exactly one nilpotent orbit (namely $\operatorname{Ind}_{orb}(L \uparrow G)(\mathcal{O}_u)$*).*
iii) All of the orbits in a sheet have the same dimension (namely $\dim \mathcal{O}_u + \dim G/L$*).*

Proof. By induction by stages, every nilpotent orbit is induced from a rigid nilpotent orbit. Now (i) follows from Proposition 3.6. Part(ii) is also clear from Proposition 3.6, which says that induction preserves "semisimple part." Part (iii) follows from Definition 3.4. (In fact the original definition of a sheet is as a component of the variety of coadjoint orbits of a fixed dimension; from this point of view the description in Definition 3.7 is something to be proved.) Q.E.D.

In the case of $SL(n)$, distinct sheets are actually disjoint. This partition of the orbits is the key to the definition of a Dixmier map in that case (cf. [2]). Even for the other classical groups, however, distinct sheets can overlap: for $Sp(4)$, the (Dixmier) sheets corresponding to $(GL(2), \{0\})$ and $(GL(1) \times Sp(2), \{0\})$ share the same nilpotent orbit. As was explained in the introduction, this failure of disjointness is the main reason that one cannot have a nice bijection between orbits and completely prime primitive ideals for the other groups.

We turn now to the problem of inducing orbit data. Each step in the geometric construction of Definition 3.1 (or Definition 3.4) has an algebraic analogue, provided by the standard dictionary between algebraic geometry and commutative algebra.

Definition 3.9. Fix a pair $(L,(R_L,\psi_L))$, where

L is a Levi subgroup of G, and
(R_L,ψ_L) is an orbit datum for L.

(Definition 2.2). Notice that these hypotheses are much weaker than those of Theorem 2.6: the main difference is that R_L (or rather its restriction to the commutator subgroup) is not required to be pre-unipotent. We will construct from these data (and a parabolic subgroup) an orbit datum for G. Define

$$\Sigma_L = \text{Supp } R_L, \tag{3.10)(a}$$

the image of the morphism ψ_L^* from Spec R_L to $\mathfrak{l}^*$. If R_L is completely prime, then

$$\Sigma_L = \overline{\mathcal{O}}_L, \tag{3.10)(b}$$

the closure of a single coadjoint orbit for L. If R_L is geometric, we write

$$X_L = \text{Spec } R_L \tag{3.10)(c}$$

for the corresponding variety (a ramified finite cover of Σ_L). Our orbit datum for G is going to be supported on the closure of

$$\mathcal{O}_G = \text{Ind}_{orb}(L \uparrow G)(\mathcal{O}_L). \tag{3.10)(d}$$

Now fix a parabolic subgroup $Q = LU$ of G with Levi factor L. (Recall that L is most naturally regarded as the quotient of Q by its unipotent radical, rather than as a subgroup. The reader may observe that it is this structure that we will actually use.) We will use the notation of Definitions 3.1 and 3.4. Define

$$R_Q = R_L, \tag{3.11)(a}$$

and let Q act on R_Q by making U act trivially. Define a map ψ_Q from $S(\mathfrak{q})$ to R_Q by

$$\psi_Q(A) = \begin{cases} \psi_L(A) & (A \in \mathfrak{l}) \\ 0 & (A \in \mathfrak{u}). \end{cases} \tag{3.11)(b}$$

The image of ψ_Q^* is

$$\Sigma_Q = i_Q(\Sigma_L) \simeq \Sigma_L; \tag{3.11)(c}$$

here $\mathfrak{l}^*$ is identified with $(\mathfrak{q}/\mathfrak{u})^* \subset \mathfrak{q}^*$. If R_L is completely prime, then Σ_Q is the closure of $\mathcal{O}_Q \subset \mathfrak{q}^*$. If R_L is geometric, then $X_Q = \text{Spec } R_Q$ is a ramified finite cover of Σ_Q as before.

The algebraic version of (3.1)(d) is the definition

$$R_{G/Q} = S(\mathfrak{g}) \otimes_{S(\mathfrak{q})} R_Q. \tag{3.12)(a}$$

This algebra carries a natural action Ad of Q by algebraic automorphisms. It is the tensor product of the adjoint action on $S(\mathfrak{g})$ with the action on R_Q. There is an $\mathrm{Ad}(Q)$-equivariant homomorphism

$$\psi_{G/Q} : S(\mathfrak{g}) \to R_{G/Q}, \qquad p \mapsto p \otimes 1. \tag{3.12)(b}$$

The image of $\psi^*_{G/Q}$ is

$$\begin{aligned} \Sigma_{G/Q} &= \pi^{-1}_{G/Q}(\Sigma_Q) \\ &= \{\, \lambda \in \mathfrak{g}^* \mid \lambda\,|_{\mathfrak{q}} \in \Sigma_Q \,\}. \end{aligned} \tag{3.12)(c}$$

This exhibits $\Sigma_{G/Q}$ as an affine bundle over Σ_Q with fiber $(\mathfrak{g}/\mathfrak{q})^*$. If R_L is geometric, we write $X_{G/Q} = \mathrm{Spec}\ R_{G/Q}$; again this is an affine bundle over X_Q. Suppose that R_L is completely prime, so that Σ_Q and therefore $\Sigma_{G/Q}$ are irreducible. The existence of the induced orbit $\mathcal{O}_G$ is equivalent (by (A.3) and the definitions) to the fact that Q has a (unique) open orbit $\mathcal{O}_{G/Q}$ on $\Sigma_{G/Q}$.

We define a sheaf of algebras on G/Q by

$$\mathcal{R}_G = G \times_Q R_{G/Q}. \tag{3.13)(a}$$

If V is an open set in G, then the space of sections of $\mathcal{R}_G$ over VQ is the space of algebraic maps ρ from VQ to $R_{G/Q}$, satisfying

$$\rho(gq) = \mathrm{Ad}(q^{-1})\rho(g) \tag{3.13)(b}$$

(cf. (2.5)(d)). (It might be slightly more natural first to identify $R_{G/Q}$ with a sheaf of algebras $\mathcal{R}_{G/Q}$ over $\Sigma_{G/Q}$, and then to define

$$\mathcal{R}'_G = G \times_Q \mathcal{R}_{G/Q} \tag{3.13)(a)'}$$

as a sheaf of algebras over $G \times_Q \Sigma_{G/Q}$ (cf. (A.4)). Because $\Sigma_{G/Q}$ is affine, the two approaches are interchangeable; the one we have chosen seems slightly simpler.) The group G acts on the sheaf $\mathcal{R}_G$; an element g defines an algebra homomorphism from the sections over VQ to the sections over $(g^{-1})VQ$ by right translation. The *induced orbit datum*

$$(R_G, \psi_G) = \mathrm{Ind}_{orb}(Q \uparrow G)(R_L, \psi_L) \tag{3.14)(a}$$

is defined by

$$R_G = \text{global sections of } \mathcal{R}_G; \tag{3.14)(b}$$

these are just the algebraic maps from G to $R_{G/Q}$ satisfying (3.13)(b) (cf. (A.4)(c)). The algebra structure is pointwise multiplication, and the action of G is $(g \cdot \rho)(x) = \rho(g^{-1}x)$. Finally,

$$\psi_G : S(\mathfrak{g}) \to R_G, \qquad (\psi_G(p))(x) = (\mathrm{Ad}(x^{-1})p) \otimes 1 \in R_{G/Q}. \quad (3.14)(c)$$

Proposition 3.15. *Suppose $Q = LU$ is a parabolic subgroup of G, and (R_L, ψ_L) is an orbit datum for L. Then the pair*

$$(R_G, \psi_G) = \mathrm{Ind}_{orb}(Q \uparrow G)(R_L, \psi_L)$$

(Defintion 3.9) is an orbit datum for G. It is completely prime (respectively geometric) if (R_L, ψ_L) is. The support of R_G is

$$\Sigma_G = Ad^*(G) \cdot (\Sigma_{G/Q}) = Ad^*(G) \cdot (\pi_{G/Q}^{-1}(i_Q(\Sigma_L)))$$

Proof. Conditions (i) and (ii) in Definition 2.2 follow easily from the definitions. We consider (iii). By hypothesis, R_L is a finitely generated $S(\mathfrak{l})$-module supported on Σ_L. It follows easily from the definitions that $R_{G/Q}$ is a finitely generated $S(\mathfrak{g})$-module supported on Σ_Q. We may therefore identify it with a coherent sheaf $\mathcal{R}_{G/Q}$ on $\Sigma_{G/Q}$. By Proposition A.5,

$$\mathcal{R}'_G = G \times_Q \mathcal{R}_{G/Q}$$

is a coherent sheaf on $G \times_Q \Sigma_{G/Q}$. By Proposition A.2(c), the natural map

$$G \times_Q \Sigma_{G/Q} \to G \cdot \Sigma_{G/Q} \qquad (3.16)$$

is proper. (In fact the proof shows that the morphism is projective.) By [9], Corollary II.5.20, the space of global sections of $\mathcal{R}'_G$ is a finitely generated $S(\mathfrak{g})$-module supported on $G \cdot \Sigma_{G/Q}$. By the remark at (3.13)(a)′, this space of global sections is precisely R_G.

We know that Σ_L is a finite union of orbits of L (see the remarks after Definition 2.2). The theory of induced orbits recalled at the beginning of this section therefore implies that $G \cdot \Sigma_{G/Q}$ is a finite union of orbits of G. Condition (iv) in Definition 2.2 follows.

That the support of R_G is all of $G \cdot \Sigma_{G/Q}$ (rather than some proper subvariety) follows from the fact that the map (3.16) is generically finite; this in turn is a consequence of the the theory of induced orbits (cf. (3.1) and (A.3)).

That R_G is completely prime (or commutative) whenever R_L is follows from the definitions. Suppose that R_L is geometric; write

$$X_L = \text{Spec } R_L, \qquad X_{G/Q} = \text{Spec } R_{G/Q} \tag{3.17)(a}$$

as in Definition 3.9. Then the sheaf of algebras $\mathcal{R}_G$ is just the structure sheaf of the bundle

$$\mathcal{X}_G = G \times_Q X_{G/Q}, \tag{3.17)(b}$$

over G/Q. Since $X_{G/Q}$ is normal, so is $\mathcal{X}_G$. It follows that R_G, as the algebra of global functions on $\mathcal{X}_G$, is normal as well. Consequently $X_G =$ Spec R_G is normal, so the orbit datum (R_G, ψ_G) is geometric. Q.E.D.

Just as in the case of coadjoint orbits (Proposition 3.6), induction of orbit data is closely related to the Jordan decomposition.

Proposition 3.18. *Suppose $Q = LU$ is a parabolic subgroup of G, and (R_L, ψ_L) is a completely prime orbit datum for L. Suppose that this orbit datum corresponds in the Jordan decomposition for L (Theorem 2.6) to $(M, \lambda_s, (R_u, \psi_u))$. Let L' be the centralizer in G of λ_s. Then $Q' = Q \cap L' = MU'$ is a parabolic subgroup of L', and*

$$(R'_u, \psi'_u) = \text{Ind}_{orb}(Q' \uparrow L')(R_u, \psi_u)$$

is a completely prime pre-unipotent orbit datum for $L'/Z(L')_0$. The induced orbit datum

$$(R_G, \psi_G) = \text{Ind}_{orb}(Q \uparrow G)(R_L, \psi_L)$$

has Jordan decomposition $(L', \lambda_s, (R'_u, \psi'_u))$.

The proof is parallel to that of Proposition 3.6, and we omit it. (In the statement, we have been a little sloppy about identifying orbit data for (for example) L' and $L'/Z(L')_0$.)

We know that the support of an induced orbit datum does not depend on the choice of parabolic Q; and we know (Corollary 2.7, for example) that an orbit datum is nearly determined by its support. This suggests

Conjecture 3.19. In the setting of Definition 3.9, the orbit datum $\text{Ind}_{orb}(Q \uparrow G)(R_L, \psi_L)$ is independent of the choice of parabolic subgroup with Levi factor L.

An interesting special case (both of the construction and of the conjecture) is when R_L is $\mathbb{C}$ and ψ_L is zero on $\mathfrak{l}$. Then $R_{G/Q}$ is the algebra $S(\mathfrak{g}/\mathfrak{q})$ of functions on the cotangent space at eQ to G/Q, and $\mathcal{X}_G$ is just $T^*(G/Q)$.

The algebra R_G is the algebra of regular functions on $T^*(G/Q)$. If Q' is another parabolic subgroup with Levi factor L, then $T^*(G/Q)$ need not be isomorphic to $T^*(G/Q')$. The conjecture asks for an isomorphism between the algebras of global regular functions on these (non-affine) varieties (respecting the G actions and the maps ψ_G). Such an isomorphism does exist, but I do not know any very satisfactory proof of the fact. (Because of normality, it suffices to show that the unique dense G-orbits in the two cotangent bundles are isomorphic as homogeneous spaces. Fix elements

$$\lambda \in (\mathfrak{g}/\mathfrak{q})^* \simeq T^*_{eQ}(G/Q),$$
$$\lambda' \in (\mathfrak{g}/\mathfrak{q}')^* \simeq T^*_{eQ'}(G/Q')$$

representing these orbits. Then λ and λ' are conjugate as elements of $\mathfrak{g}^*$, so the isotropy groups $G(\lambda)$ and $G(\lambda')$ are conjugate. The identity components of these groups are contained in Q and Q' respectively. The desired isomorphism is equivalent to the fact that $G(\lambda) \cap Q$ is conjugate to $G(\lambda') \cap Q'$. This can be verified by more or less explicit computation on a case-by-case basis.) The simplest non-trivial example is for $G = GL(3)$ and $L = GL(2) \times GL(1)$. We can take G/Q to be the variety of lines in $\mathbb{C}^3$, and G/Q' to be the variety of planes in $\mathbb{C}^3$. Of course these varieties are isomorphic, but the isomorphism cannot be made to respect the G actions. Nevertheless the algebras of regular functions on the respective cotangent bundles *are* isomorphic in a G-invariant way: they are both isomorphic to the algebra of regular functions on the corresponding (subregular) nilpotent orbit in $\mathfrak{g}^*$.

Because of Proposition 3.18, a completely prime orbit datum for a semisimple group that is *not* pre-unipotent must be induced from a proper parabolic subgroup. It therefore makes sense to concentrate on pre-unipotent orbit data.

Definition 3.20. Suppose (R, ψ) is a completely prime pre-unipotent orbit datum for the semisimple group G. We say that (R, ψ) is *rigid* if it is not induced (in the sense of Definition 3.9) from a completely prime orbit datum on a proper Levi subgroup. Otherwise we say that (R, ψ) is *induced.*

This notion of rigidity is much broader than that for orbits (defined after Definition 3.1). That is, an orbit datum will be rigid if its support is rigid; but it may be rigid even if its support is not. If G is $GL(n)$, every non-zero completely prime orbit datum is induced; but already for $SL(2)$ this is not true.

Example 3.21. Suppose G is $SL(2)$. We define a unipotent orbit datum (R, ψ) for G as follows. The algebra R is $\mathbb{C}[p, q]$, with the action of

G induced by linear change of variables. We define ψ from $S(\mathfrak{g})$ to R on the standard basis (H, E, F) of $\mathfrak{g}$ by

$$\psi(H) = 2pq, \qquad \psi(E) = p^2, \qquad \psi(F) = -q^2.$$

Obviously R is finitely generated as an $S(\mathfrak{g})$ module (by 1, p, and q). It is easy to check that ψ respects the action of G. Every irreducible representation of G appears in R exactly once. The image of ψ^* is the cone $H^2 + 4EF = 0$, which is the nilpotent cone in $\mathfrak{g}^*$. It follows that (R, ψ) is a unipotent orbit datum. One can show that it is rigid for example by computing all the induced orbit data; this is not difficult since there is only one proper Levi subgroup. In the parametrization of Corollary 2.7, R corresponds to the double cover of the principal nilpotent orbit.

Iwan Pranata has observed that the preceding example can be modified to produce many non-geometric completely prime commutative orbit data. To do that, fix a non-negative integer k, and define R_k to be the subring of R generated by the image of ψ and the polynomials of degree $2k + 1$. (Pranata also found the Dixmier algebras associated to these orbit data.) Together with $(\mathbb{C}, 0)$, the various (R_k, ψ) exhaust the rigid completely prime orbit data for $SL(2)$.

Here is the analogue of Definition 3.7.

Definition 3.22. Suppose $Q = LU$ is a parabolic subgroup of G, and (R_u, ψ_u) is a rigid completely prime pre-unipotent coadjoint orbit datum for $L/Z(L)_0$. To every character λ_s of $\mathfrak{l}$ we can associate a completely prime orbit datum $(R_L, \psi_L(\lambda_s))$ for L as in (2.5)(e) and (f):

$$R_L = R_u, \qquad \psi_L(A) = \psi_u(A + \mathfrak{z}(\mathfrak{l})) + \lambda_s(A) \quad (A \in \mathfrak{l}).$$

The *sheet* of completely prime orbit data attached to $(L, (R_u, \psi_u))$ is

$$\{ \operatorname{Ind}_{orb}(Q \uparrow G)(R_L, \psi_L(\lambda_s)) \mid \lambda_s \in (\mathfrak{l}/[\mathfrak{l}, \mathfrak{l}])^* \}$$

(Definition 3.9). It is called a *Dixmier sheet* if $R_u = \mathbb{C}$. Two sheets are identified if they contain exactly the same orbit data; according to Conjecture 3.19, this amounts to conjugacy of the pair $(L, (R_u, \psi_u))$ under G. (By considering regular λ_s in Proposition 3.18, one sees that this conjugacy is a *necessary* condition for the sheets to coincide.)

We have at once an analogue of Proposition 3.8.

Proposition 3.23. *In the notation of Definitions 3.9 and 3.22,*

i) *Each completely prime coadjoint orbit datum belongs to at least one sheet.*

ii) *If G is semisimple, each sheet contains exactly one preunipotent orbit datum (corresponding to $\lambda_s = 0$).*

iii) *The supports of the orbit data in a single sheet are contained in a single sheet of coadjoint orbits.*

This follows from Proposition 3.18. In (iii), the supports may not constitute an entire sheet of coadjoint orbits. This happens exactly when the support of the rigid orbit datum R_u is not a rigid coadjoint orbit.

Suppose that $G = Sp(4)$, $L = GL(2)$, and $L' = GL(1) \times Sp(2)$. Fix parabolic subgroups Q and Q' with Levi factors L and L'. As was pointed out after Proposition 3.8, the Dixmier sheets of coadjoint orbits attached to L and L' share the same nilpotent orbit. Let us now consider the corresponding sheets of orbit data. The unipotent orbit data for these sheets are the rings of regular functions on $T^*(G/Q)$ and $T^*(G/Q')$ respectively. They are *not* isomorphic as orbit data: the second contains the five-dimensional representation of G, and the first does not. It follows that these two sheets of orbit data are disjoint. (The non-unipotent orbit data are distinguished by their supports.) Many similar examples suggest

Conjecture 3.24. Two sheets of orbit data are disjoint or they coincide. The first possibility occurs exactly when the pairs $(L, (R_u, \psi_u))$ defining the sheets are conjugate by G.

Conjecture 3.19 is more or less subsumed in this formulation. In the case of geometric orbit data, Conjecture 3.24 amounts to an assertion about the behavior of fundamental groups under induction (compare Corollary 2.7 and the example after Conjecture 3.19). It could in principle be verified by a finite calculation for each group. For the classical groups this should not be too difficult to carry out, but I have not done so. For non-geometric sheets (in particular when the normality hypothesis is dropped) the number of essentially different cases becomes infinite, and some more conceptual approach is probably necessary.

4. Parabolic induction of Dixmier algebras.

The main purpose of this section is to give a construction of Dixmier algebras parallel to the construction of orbit data described in section 3.

Fix a pair $(Q,(A_L,\phi_L))$, where

$$Q \text{ is a parabolic subgroup of } G, \text{ and } (A_L,\phi_L) \text{ is a Dixmier algebra for } L. \tag{4.1}$$

We want to construct an "induced Dixmier algebra" (A_G,ϕ_G) for G. Of course we will imitate and extend the construction of induced primitive ideals (cf. [6]).

Before we embark on the rather convoluted construction of A_G, it is worth pausing to explain why the problem is not trivial. Suppose $V_{\mathfrak{l}}$ is a faithful module for A_L. Let $V_{\mathfrak{g}} = U(\mathfrak{g}) \otimes_{\mathfrak{q}} V_{\mathfrak{l}}$ be a parabolically induced module for $\mathfrak{g}$. (We are neglecting "ρ-shifts" — that is, the twist by the character δ in (4.2)(b) — for this discussion.) The ideal I in $U(\mathfrak{g})$ induced by the annihilator of $V_{\mathfrak{l}}$ is by definition the annihilator of $V_{\mathfrak{g}}$; so the quotient ring $U(\mathfrak{g})/I$ is naturally embedded in End $V_{\mathfrak{g}}$. The Dixmier algebra A_G is supposed to be some sort of nice ring extension of $U(\mathfrak{g})/I$, so it is natural to look for this extension in End $V_{\mathfrak{g}}$. Now it is an important theme of Joseph's work that the algebra of G-finite endomorphisms of $V_{\mathfrak{g}}$ is a natural and well-behaved ring; so it is an obvious candidate for A_G. To see that it is not the right one, consider the case $L = G$. In that case we had better have $A_L = A_G$ (if our notion of induction is to be related to induction of orbit data). Our "obvious candidate" for A_G will have this property only if the Dixmier algebra A_L is already the *full* algebra of L-finite endomorphisms of $V_{\mathfrak{l}}$. So this approach from the outset runs into a rather difficult technical problem: is every Dixmier algebra for L the full algebra of L-finite endomorphisms of some $\mathfrak{l}$-module? (The answer in this generality is "no;" if A_L is required to be reasonable (say prime, for example) the answer may be "yes," but I have no idea how to prove it.)

What we actually do is this. We show that a large class of endomorphisms of $V_{\mathfrak{g}}$ (including all the G-finite ones) can be described essentially as matrices with entries in the endomorphism algebra of $V_{\mathfrak{l}}$ (cf. Definition 4.6 and (4.9)). We can then define A_G to consist of those G-finite endomorphisms of $V_{\mathfrak{g}}$ whose "matrix entries" belong to A_L (Definition 4.7). This definition is not so difficult. What is slightly less easy is proving that A_G satisfies the finiteness requirements in the definition of Dixmier algebras, and relating the definition to induction of orbit data. To do that, we have to give a rather different description of A_G (Corollary 4.16). (We cannot easily use Corollary 4.16 as the definition of A_G, because it is difficult to see the algebra structure from this point of view.)

Especially for the material on the symbol calculus, the reader should keep in mind the case $A_L = \mathbb{C}$; in that case A_G is the algebra of (holomorphic linear) differential operators on G/Q. (Recall that we are still neglecting ρ shifts!) That case has been thoroughly analyzed in [Borho-

Brylinski], and most of the serious ideas we use may be found there.

Suppose then that

$$V_{\mathfrak{l}} \text{ is a faithful module for } A_L. \tag{4.2)(a}$$

(By Proposition 6.0 of [15], any prime Dixmier algebra is primitive; if A_L is prime, we could therefore even choose $V_{\mathfrak{l}}$ to be simple as an A_L-module. This makes no difference in the construction, however.) Let 2δ denote the character of L on the top exterior power of $\mathfrak{g}/\mathfrak{q}$. Then δ is a well-defined character of $\mathfrak{l}$. Twisting the map ϕ_L by δ we get a new Dixmier algebra (A'_L, ϕ'_L): the algebra A'_L is equal to A_L, and

$$\phi'_L(X) = \phi_L(X) + \delta(X) \qquad (X \in \mathfrak{l}). \tag{4.2)(b}$$

This algebra has a faithful module

$$V'_{\mathfrak{l}} = V_{\mathfrak{l}} \otimes \mathbf{C}_\delta. \tag{4.2)(c}$$

Make Q act on A'_L by making U act trivially; to emphasize this structure, we call the algebra A_Q, and write

$$\mathrm{Ad} : Q \to \mathrm{Aut}(A_Q). \tag{4.2)(d}$$

Extend ϕ'_L to

$$\phi_Q : U(\mathfrak{q}) \to A_Q \tag{4.2)(e}$$

by sending $\mathfrak{u}$ to 0. When we regard $V'_{\mathfrak{l}}$ as a module for A_Q, we call it $V_{\mathfrak{q}}$. Write $\mathfrak{u}^-$ for the nil radical of the parabolic subalgebra of $\mathfrak{g}$ opposite to Q. Now define

$$V_{\mathfrak{g}} = U(\mathfrak{g}) \otimes_{\mathfrak{q}} V_{\mathfrak{q}} \simeq U(\mathfrak{u}^-) \otimes_{\mathbf{C}} V_{\mathfrak{q}}. \tag{4.2)(f}$$

We are going to construct A_G as an algebra of endomorphisms of $V_{\mathfrak{g}}$. We begin by constructing a larger algebra. To motivate the definition of this algebra, we will study the adjoint action of $\mathfrak{z}(\mathfrak{l})$ on the endomorphisms of $V_{\mathfrak{g}}$. Our immediate goal is Lemma 4.5 below.

Because A_L is a Dixmier algebra, the kernel of ϕ_L must contain an ideal of finite codimension in $U(\mathfrak{z}(\mathfrak{l}))$. To simplify the notation, we assume that this ideal is maximal (as it must be if A_L is prime); the reader can easily modify the constructions that follow to cover the general case. Then $\mathfrak{z}(\mathfrak{l})$ acts by a character λ on $V_{\mathfrak{q}}$. If α is any character of $\mathfrak{z}(\mathfrak{l})$, write $U(\mathfrak{u}^-)_\alpha$ for the (finite-dimensional) subspace of $U(\mathfrak{u}^-)$ on which $\mathfrak{z}(\mathfrak{l})$ acts by α. Then $V_{\mathfrak{g}}$ is the direct sum of its weight spaces

$$(V_{\mathfrak{g}})_\beta = U(\mathfrak{u}^-)_{\beta-\lambda} \otimes V_{\mathfrak{q}}. \tag{4.3}$$

Next, recall that the algebra End $V_{\mathfrak{g}}$ is a bimodule for $U(\mathfrak{g})$. The left action of an element u on an endomorphism T is on the range of T, and the right action is on the domain. Write ad for the corresponding diagonal action of $\mathfrak{g}$: if T is an endomorphism, X is in $\mathfrak{g}$, and v is in $V_{\mathfrak{g}}$, then

$$(\mathrm{ad}(X)T)(v) = X \cdot (Tv) - T(X \cdot v). \tag{4.4}$$

An endomorphism T has weight γ for $\mathrm{ad}(\mathfrak{z}(\mathfrak{l}))$ if and only if it carries $(V_{\mathfrak{g}})_\beta$ to $(V_{\mathfrak{g}})_{\beta+\gamma}$ for every β. Using (4.3), one deduces immediately

Lemma 4.5. *Suppose T is an ad($\mathfrak{z}(\mathfrak{l})$)-finite endomorphism of $V_{\mathfrak{g}}$. Then for every u in $U(\mathfrak{u}^-)$ there are finitely many elements $u_1, \ldots, u_n$ in $U(\mathfrak{u}^-)$, and endomorphisms $E_1, \ldots, E_n$ of $V_{\mathfrak{q}}$, so that for each v in $V_{\mathfrak{q}}$,*

$$T(u \otimes v) = \sum u_i \otimes E_i v.$$

Here is a first approximation to the induced Dixmier algebra.

Definition 4.6. Suppose we are in the setting (4.2). An endomorphism T of $V_{\mathfrak{g}}$ is said to be *of type A_L* if and only if for every u in $U(\mathfrak{u}^-)$ there are finitely many elements $u_1, \ldots, u_n$ in $U(\mathfrak{u}^-)$, and endomorphisms $E_1, \ldots, E_n$ in A_Q, so that for each v in $V_{\mathfrak{q}}$,

$$T(u \otimes v) = \sum u_i \otimes E_i v. \tag{4.6)(a}$$

Using the defining relations for $V_{\mathfrak{g}}$, one checks that it is equivalent to require this condition for every u in $U(\mathfrak{g})$, or to require only that u_i belong to $U(\mathfrak{g})$. Write

$$A_{\mathfrak{g}} = \{\, T \in \mathrm{End}(V_{\mathfrak{g}}) \mid T \text{ is of type } A_L \,\}$$

It is easy to check that $A_{\mathfrak{g}}$ is an algebra.

If w is in $U(\mathfrak{g})$, define $\phi_{\mathfrak{g}}(w)$ to be the endomorphism of $V_{\mathfrak{g}}$ defined by the action of w. This is always of type A_L, so we get

$$\phi_{\mathfrak{g}} : U(\mathfrak{g}) \to A_{\mathfrak{g}}, \tag{4.6)(b}$$

a homomorphism of algebras.

In (4.9) below we will give another description of $A_{\mathfrak{g}}$ from which one can see that it depends only on the data (4.1) (and not on the choice of $V_{\mathfrak{l}}$).

Because of (4.4) and the existence of the map $\phi_{\mathfrak{g}}$, we see that the algebra $A_{\mathfrak{g}}$ is ad($\mathfrak{g}$)-stable. The reason $A_{\mathfrak{g}}$ is not a Dixmier algebra is that this Lie algebra adjoint action does not exponentiate to an algebraic action of G. We remedy this in the simplest possible way.

Definition 4.7. In the setting of Definition 4.6, define

$$\begin{aligned} A_G &= \mathrm{Ad}(G)\text{-finite subalgebra of } A_{\mathfrak{g}} \\ &= \mathrm{Ad}(G)\text{-finite endomorphisms of } V_{\mathfrak{g}} \text{ of type } A_L \end{aligned} \tag{4.7)(a}$$

To understand this definition, recall that an element T of $A_{\mathfrak{g}}$ is called Ad(G)*-finite* if it belongs to a finite-dimensional ad($\mathfrak{g}$)-invariant subspace F of $A_{\mathfrak{g}}$, on which there exists an algebraic action Ad_F of G with differential ad. The elements in the image of $\phi_{\mathfrak{g}}$ certainly have this property: one can take for F the image of some level $U_n(\mathfrak{g})$ of the standard filtration of $U(\mathfrak{g})$. Restricting the range of $\phi_{\mathfrak{g}}$ therefore defines

$$\phi_G : U(\mathfrak{g}) \to A_G, \tag{4.7)(b}$$

a homomorphism of algebras. It is easy to check that A_G is a subalgebra of $A_{\mathfrak{g}}$ containing the identity element. Since G is connected, the actions Ad_F are uniquely determined, and can be assembled into an algebraic action

$$\mathrm{Ad} : G \to \mathrm{Aut}(A_G). \tag{4.7)(c}$$

The action is by algebra automorphisms since ad($\mathfrak{g}$) acts by derivations. We define the *induced Dixmier algebra* to be

$$(A_G, \phi_G) = \mathrm{Ind}_{Dix}(Q \uparrow G)(A_L, \phi_L). \tag{4.7)(c}$$

To show that A_G is a Dixmier algebra, we only need to check the finiteness conditions (iii) and (iv) of Definition 2.1. To do that, and to get a clearer picture of the structure of A_G, we need first to study $A_{\mathfrak{g}}$ much more carefully.

Lemma 4.8. *The set $A_{\mathfrak{g}}$ is an algebra of endomorphisms of $V_{\mathfrak{g}}$. If A_L consists of all L-finite endomorphisms of $V_{\mathfrak{l}}$, then $A_{\mathfrak{g}}$ contains all the L-finite endomorphisms of $V_{\mathfrak{g}}$.*

This is straightforward.

Here is another description of $A_{\mathfrak{g}}$. Fix characters β and γ of $\mathfrak{z}(\mathfrak{l})$. Define $(\mathrm{End}\ V_{\mathfrak{g}})_{\beta\gamma}$ to consist of those endomorphisms transforming by the character β by the left action of $\mathfrak{z}(\mathfrak{l})$, and by γ under the right action. These are precisely the endomorphisms vanishing on all the weight spaces of $V_{\mathfrak{g}}$ except the one for γ, and whose image is contained in the β weight space:

$$\begin{aligned}(\text{End } V_{\mathfrak{g}})_{\beta\gamma} &\simeq \text{Hom}((V_{\mathfrak{g}})_{\gamma}, (V_{\mathfrak{g}})_{\beta}) \\ &\simeq \text{Hom}(U(\mathfrak{u}^-)_{\gamma-\lambda} \otimes V_{\mathfrak{q}}, U(\mathfrak{u}^-)_{\beta-\lambda} \otimes V_{\mathfrak{q}}) \qquad (4.9)(a) \\ &\simeq \text{Hom}(U(\mathfrak{u}^-)_{\gamma-\lambda}, U(\mathfrak{u}^-)_{\beta-\lambda}) \otimes \text{End } V_{\mathfrak{q}}.\end{aligned}$$

The corresponding double weight space for $A_{\mathfrak{g}}$ is

$$(A_{\mathfrak{g}})_{\beta\gamma} \simeq \text{Hom}(U(\mathfrak{u}^-)_{\gamma-\lambda}, U(\mathfrak{u}^-)_{\beta-\lambda}) \otimes A_Q. \qquad (4.9)(b)$$

The algebra $A_{\mathfrak{g}}$ is built from these pieces in a slightly subtle way:

$$A_{\mathfrak{g}} \simeq \prod_{\gamma} \left(\sum_{\beta} (A_{\mathfrak{g}})_{\beta\gamma} \right). \qquad (4.9)(c)$$

The reason we insist on a direct product outside is to ensure that the identity operator belongs to $A_{\mathfrak{g}}$. The direct sum inside ensures that the resulting algebra acts on $V_{\mathfrak{g}}$.

In the description (4.9) of $A_{\mathfrak{g}}$, the algebra structure is the obvious one (induced by the algebra structure on A_Q and the "composition maps"

$$\text{Hom}(U(\mathfrak{u}^-)_{\gamma}, U(\mathfrak{u}^-)_{\beta}) \times \text{Hom}(U(\mathfrak{u}^-)_{\delta}, U(\mathfrak{u}^-)_{\gamma}) \to \text{Hom}(U(\mathfrak{u}^-)_{\delta}, U(\mathfrak{u}^-)_{\beta}).$$

The structure that is subtle in this picture is the homomorphism $\phi_{\mathfrak{g}}$ of (4.7)(b). At any rate, (4.9) shows that $A_{\mathfrak{g}}$ depends only on A_L (and Q), not on the choice of module $V_{\mathfrak{l}}$.

We need two more descriptions of $A_{\mathfrak{g}}$. For each of them, it is convenient to introduce an auxiliary space.

Definition 4.10. In the setting of (4.2), define

$$A_{G/Q} = U(\mathfrak{g}) \otimes_{\mathfrak{q}} A_Q. \qquad (4.10)(a)$$

This space is clearly analogous to the algebra $R_{G/Q}$ of (3.12), but I do not know any very simple motivation for its introduction. Clearly $A_{G/Q}$ carries a left action of $U(\mathfrak{g})$ and a commuting right action of $U(\mathfrak{q})$. There is also an algebraic action

$$\text{Ad} : Q \to End(A_{G/Q}), \qquad (4.10)(b)$$

which is the tensor product of the adjoint actions on $U(\mathfrak{g})$ and A_Q. The differential of Ad is the difference of the left and right actions of $\mathfrak{q}$.

The first of our final descriptions of $A_{\mathfrak{g}}$ is

Lemma 4.11. *In the setting of Definition 4.10, define*

$$A^1_{\mathfrak{g}} = \mathrm{Hom}_{\mathfrak{q}(right,right)}(U(\mathfrak{g}), A_{G/Q}). \qquad (i)$$

(Here $\mathfrak{q}$ *acts on the first* $U(\mathfrak{g})$ *and on* $A_{G/Q}$ *on the right to define the* $\mathrm{Hom}_{\mathfrak{q}}$*.) Then there is an isomorphism*

$$\alpha^1 : A_{\mathfrak{g}} \to A^1_{\mathfrak{g}}. \qquad (ii)$$

In this isomorphism, the left action of $U(\mathfrak{g})$ *on* $A_{\mathfrak{g}}$ *corresponds to the left action on* $A_{G/Q}$*; and the right action of* $U(\mathfrak{g})$ *on* $A_{\mathfrak{g}}$ *corresponds to the left action on the domain* $U(\mathfrak{g})$ *in the* Hom. *Explicitly,*

$$\alpha^1(xTy)(u) = x\left(\alpha^1(T)(yu)\right) \qquad (x, y, u \in U(\mathfrak{g}), T \in A_{\mathfrak{g}}. \qquad (iii)$$

For $x \in U(\mathfrak{g})$*, the element* $\phi_{\mathfrak{g}}(x)$ *of (4.6)(b) satisfies*

$$\alpha^1(\phi_{\mathfrak{g}}(x))(u) = xu \otimes 1. \qquad (iv)$$

We use the cumbersome subscript in (i) to distinguish this realization from that of Lemma 4.14 below.

Proof. Fix $T \in A_{\mathfrak{g}}$; we must define $\alpha^1(T)$ as a map from $U(\mathfrak{g})$ to $A_{G/Q} = U(\mathfrak{g}) \otimes A_Q$. So fix u in $U(\mathfrak{g})$, and write

$$T(u \otimes v) = \sum u_i \otimes E_i v \qquad (4.12)(a)$$

(with $u_i \in U(\mathfrak{g})$ and $E_i \in A_Q$) as in Definition 4.6. Then put

$$\alpha^1(T)(u) = \sum u_i \otimes E_i. \qquad (4.12)(b)$$

We leave to the reader the straightforward verification that $\alpha^1(T)$ is a well-defined element of $A^1_{\mathfrak{g}}$, and that α^1 is an isomorphism. The assertion (iii) follows from (4.12). For (iv), one can use (iii) and the fact that $\phi_{\mathfrak{g}}(x) = x \cdot 1_A$ (the left action of x on the identity element of $A_{\mathfrak{g}}$). Q.E.D.

Using (4.12), we can compute the algebra structure on $A^1_{\mathfrak{g}}$ induced by the isomorphism α^1. Suppose S' and S are in $A^1_{\mathfrak{g}}$. To evaluate $S'S$ at an element u of $U(\mathfrak{g})$, first write

$$S(u) = \sum u_i \otimes E_i. \qquad (4.13)(a)$$

Next, write

$$S'(u_i) = \sum_j u_{ij} \otimes E_{ij}. \tag{4.13)(b}$$

Then

$$(S'S)(u) = \sum_{i,j} u_{ij} \otimes E_{ij} E_i. \tag{4.13)(c}$$

We leave the straightforward verification of this to the reader.

The last description of $A_{\mathfrak{g}}$ is the least transparent of all, but it will be crucial to the symbol calculus.

Lemma 4.14. *In the setting of Lemma 4.11, define*

$$A_{\mathfrak{g}}^2 = \operatorname{Hom}_{\mathfrak{q}(left,ad)}(U(\mathfrak{g}), A_{G/Q}) \tag{i}$$

to be the space of maps Σ from $U(\mathfrak{g})$ to $A_{G/Q}$ with the property that

$$\Sigma(Xu) = ad(X)\Sigma(u) \qquad (X \in \mathfrak{q}, u \in U(\mathfrak{g})). \tag{ii}$$

Then there is an isomorphism

$$\alpha^2 : A_{\mathfrak{g}} \to A_{\mathfrak{g}}^2. \tag{iii}$$

Under the isomorphism α^2, the adjoint action of $U(\mathfrak{g})$ on $A_{\mathfrak{g}}$ corresponds to the right action on the domain $U(\mathfrak{g})$ term in $A_{\mathfrak{g}}^2$:

$$\alpha^2(\mathrm{ad}(x)T)(u) = \alpha^2(T)(ux). \tag{iv}$$

The map $\phi_{\mathfrak{g}}$ is computed in this picture by

$$\alpha^2(\phi_{\mathfrak{g}}(x))(u) = ad(u)x \otimes 1. \tag{v}$$

Before proving this lemma, we deduce the consequences we want.

Definition 4.15. The *complete symbol sheaf* (for A_L and Q) is the quasicoherent sheaf

$$\mathcal{A}_G = G \times_Q A_{G/Q}$$

on G/Q. Here we use the action Ad of Q on $A_{G/Q}$.

It should be fairly easy to make $\mathcal{A}_G$ into a sheaf of algebras on G/Q, but I have not done this.

Corollary 4.16. *The algebra $A_{\mathfrak{g}}$ may be identified with the space of formal power series sections of $\mathcal{A}_G$ at the identity coset eQ. The subalgebra A_G then corresponds to the global sections of $\mathcal{A}_G$; the left action of G on sections corresponds to the adjoint action on A_G. In particular,*

$$\begin{aligned} A_G &\simeq \mathrm{Ind}_{alg}(Q \uparrow G)(A_{G/Q}) \\ &= \mathrm{Ind}_{alg}(Q \uparrow G)(U(\mathfrak{g}) \otimes_{\mathfrak{q}} A_Q). \end{aligned}$$

Here we use induction of algebraic representations (cf. (A.8)); Q acts on the inducing representation by $\mathrm{Ad} \otimes \mathrm{Ad}$.

Sketch of proof. This is a consequence of Lemma 4.14 and general facts about homogeneous vector bundles. A formal power series section Σ of the "vector bundle" $G \times_Q A_{G/Q}$ is specified by a map from $U(\mathfrak{g})$ to $A_{G/Q}$, sending u to the value at eQ of the derivative $\partial_u(\Sigma)$. Since Σ must transform (infinitesimally) under Q in a certain way, this map must respect the action of $\mathfrak{q}$; that is, it must belong to the space $A^2_{\mathfrak{g}}$ of Lemma 4.14. This gives the first assertion. For the rest, one needs to know that a G-finite formal power series section must come from a globally defined section; this is easy. Q.E.D.

Corollary 4.17. *In the setting of Definition 4.7, the pair*

$$(A_G, \phi_G) = \mathrm{Ind}_{Dix}(Q \uparrow G)(A_L, \phi_L)$$

is a Dixmier algebra. The kernel of ϕ_G in $U(\mathfrak{g})$ is the ideal induced (via $\mathfrak{q}$) from the kernel of ϕ_L in $U(\mathfrak{l})$.

Proof. Since A_L is a Dixmier algebra for L, the kernel of ϕ_L contains an ideal of finite codimension in the center of $U(\mathfrak{l})$. Since A_G is an algebra of endomorphisms of $V_{\mathfrak{g}}$, it follows from the theory of the Harish-Chandra homomorphism that the kernel of ϕ_G contains an ideal of finite codimension in the center of $U(\mathfrak{g})$. By the theory of Harish-Chandra modules for G, condition (iii) in the definition of Dixmier algebra (Definition 2.1) is a consequence of condition (iv). But condition (iv) is an elementary consequence of the description of A_G as an induced module in Corollary 4.16; we leave the calculation to the reader.

For the statement about induced ideals, notice first that the annihilator of $V_{\mathfrak{l}}$ in $U(\mathfrak{l})$ is the kernel of ϕ_L (since $V_{\mathfrak{l}}$ is assumed to be faithful for A_L). By definition the induced ideal is therefore the annihilator in $U(\mathfrak{g})$ of $V_{\mathfrak{g}}$. Since A_G is an algebra of endomorphisms of $V_{\mathfrak{g}}$, the claim follows. Q.E.D.

Conjecture 4.18. In the setting of Definition 4.7, the induced Dixmier algebra is independent of the choice of Q.

Of course this conjecture is analogous to Conjecture 3.19. The corresponding assertion for induced ideals is true (cf. [8]). Beyond this, there is very little evidence. It may be that one has to impose some extra condition on A_L, such as complete primality.

Proof of Lemma 4.14. Rather than passing directly from $A_{\mathfrak{g}}$ to $A^2_{\mathfrak{g}}$, we will construct an isomorphism

$$\alpha^{12} : A^1_{\mathfrak{g}} \to A^2_{\mathfrak{g}}. \qquad (4.19)(a)$$

In light of Lemma 4.11, it will suffice to show that α^{12} has nice properties; then the isomorphism

$$\alpha^2 = \alpha^{12} \circ \alpha^1 \qquad (4.19)(b)$$

will satisfy the requirements of Lemma 4.14. So fix $S \in A^1_{\mathfrak{g}}$, and define

$$\alpha^{12}(S)(u) = (\mathrm{ad}(u)S)(1). \qquad (4.19)(c)$$

That the map $\alpha^{12}(S)$ satisfies condition (ii) in Lemma 4.14, and that

$$\alpha^{12}(S)(uv) = \alpha^{12}(\mathrm{ad}(v)S)(u) \qquad (4.19)(d)$$

(which implies (4.14)(iv)) are easy calculations. The formula (4.14)(v) follows from (4.11)(iv) and the fact that the map $\phi_{\mathfrak{g}}$ is ad-equivariant.

The main point is therefore to show that the map α^{12} is a linear isomorphism. To see that, we will write an explicit inverse. Write $u \to {}^t u$ for the transpose antiautomorphism of $U(\mathfrak{g})$; we have ${}^t X = -X$ for X in $\mathfrak{g}$. Write

$$h : U(\mathfrak{g}) \to U(\mathfrak{g}) \otimes U(\mathfrak{g})$$

for the Hopf map; this is the algebra homomorphism sending X in $\mathfrak{g}$ to $X \otimes 1 + 1 \otimes X$. If M is a bimodule for $U(\mathfrak{g})$, then the adjoint action of $U(\mathfrak{g})$ may be computed as follows. Fix u in $U(\mathfrak{g})$, and write

$$h(u) = \sum u_i \otimes v_i. \qquad (4.20)(a)$$

Then

$$\mathrm{ad}(u)m = \sum (u_i)m({}^t v_i). \qquad (4.20)(b)$$

We have described the left and right actions of $U(\mathfrak{g})$ in $A^1_{\mathfrak{g}}$ explicitly; we deduce that for $S \in A^1_{\mathfrak{g}}$,

$$(\mathrm{ad}(u)S)(x) = \sum u_i[S({}^t v_i x)]. \qquad (4.20)(c)$$

Consequently

$$\alpha^{12}(S)(u) = \sum u_i(S({}^t v_i)). \tag{4.20)(d}$$

Now suppose that Σ is an element of $A^2_{\mathfrak{g}}$. Define a map $\alpha^{21}(\Sigma)$ from $U(\mathfrak{g})$ to $A_{G/Q}$ by

$$\alpha^{21}(\Sigma)(u) = \sum u_i(\Sigma({}^t v_i)). \tag{4.21)(a}$$

One way to see that α^{21} is well-defined is to regard $A^2_{\mathfrak{g}}$ as a subspace of $\mathrm{Hom}_{\mathbf{C}}(U(\mathfrak{g}), A_{G/Q})$. This larger space carries commuting left and right $U(\mathfrak{g})$ actions, from the left actions on $A_{G/Q}$ and $U(\mathfrak{g})$ respectively. Consequently there is an adjoint action ad; and

$$\alpha^{21}(\Sigma)(u) = (ad(u)\Sigma)(1). \tag{4.21)(b}$$

To check that $\alpha^{21}(\Sigma)$ belongs to $A^1_{\mathfrak{g}}$, fix u in $U(\mathfrak{g})$ and X in $\mathfrak{q}$; we must show that

$$\alpha^{21}(\Sigma)(uX) = (\alpha^{21}(\Sigma)(u))X.$$

Now $h(uX) = \sum(u_i X \otimes v_i + u_i \otimes v_i X)$, so

$$\alpha^{21}(\Sigma)(uX) = \sum(u_i X(\Sigma({}^t v_i)) + u_i(\Sigma(-X{}^t v_i))).$$

By (4.14)(ii), the second terms can be rewritten to give

$$\begin{aligned}\alpha^{21}(\Sigma)(uX) &= \sum(u_i X(\Sigma({}^t v_i)) - u_i(ad(X)\Sigma({}^t v_i)))\\ &= \sum(u_i X(\Sigma({}^t v_i)) - u_i(Xs({}^t v_i) - \Sigma({}^t v_i)X)).\end{aligned}$$

The first two terms in each summand cancel, leaving

$$\begin{aligned}\alpha^{21}(\Sigma)(uX) &= \sum(u_i s({}^t v_i)X)\\ &= (\alpha^{21}(\Sigma)(u))X,\end{aligned}$$

as we wished to show.

Finally, we must show that the correspondences defined by (4.20) and (4.21) are mutual inverses. Fix S in $A^1_{\mathfrak{g}}$; we will show that

$$\alpha^{21}(\alpha^{12}(S)) = S. \tag{$*$}$$

(The proof that $\alpha^{12}(\alpha^{21}(\Sigma)) = \Sigma$ is identical.) To do this, fix $X_1, \ldots, X_n$ in $U(\mathfrak{g})$; we will compute the left side of $(*)$ at $u = X_1 \ldots X_n$. If A is any subset of $\{1, \ldots, n\}$, write X_A for the corresponding product (with the indices arranged in increasing order). The complement of A is written A^c. Then the Hopf map is given by

$$h(u) = \sum_{A \subset \{1,\ldots,n\}} X_A \otimes X_{A^c}.$$

Therefore

$$\begin{aligned}\alpha^{21}(\alpha^{12}(S))(u) &= \sum_A X_A(\alpha^{12}(S)({}^tX_{A^c})) \\ &= \sum_A X_A\left(\sum_{B\subset A^c} {}^tX_B S(X_{A^c-B})\right) \\ &= \sum_{\substack{A \\ B\subset A^c}} X_A{}^tX_B S(X_{A^c-B}).\end{aligned}$$

This last expression may be regarded as a sum over partitions of $\{1,\ldots,n\}$ into three disjoint subsets A, B, C; it is

$$\begin{aligned}&\sum_{A,B,C} X_A{}^tX_B S(X_C) \\ &= \sum_C (\mathrm{ad}(X_{C^c})\cdot 1) S(X_C).\end{aligned}$$

Of course $\mathrm{ad}(X_{C^c})$ acts by zero on 1 unless C^c is empty. The only non-zero term is therefore

$$(ad(1)\cdot 1)S(X_{\{1,\ldots,n\}}),$$

which is $S(u)$ as we wished to show. Q.E.D.

We conclude this section with a symbol calculus for the induced Dixmier algebra.

Definition 4.22. Suppose we are in the setting of Definition 4.10. The standard filtration of $U(\mathfrak{g})$ induces an $\mathrm{Ad}(Q)$-stable filtration

$$A_{G/Q,n} = \text{ image of } U_n(\mathfrak{g}) \otimes A_Q \tag{4.22)(a}$$

of $A_{G/Q}$. The left action of $U(\mathfrak{g})$ is compatible with this filtration in the usual sense (see the discussion preceding Conjecture 2.3). Define $S_{G/Q}$ to be the associated graded object:

$$S^n_{G/Q} = S^n(\mathfrak{g}/\mathfrak{q}) \otimes A_Q. \tag{4.22)(b}$$

This is a graded algebra with a graded algebraic action (still called Ad) of Q by automorphisms. (In fact it is the pushforward to a point of a constant sheaf of $\mathcal{O}$-algebras on $(\mathfrak{g}/\mathfrak{q})^*$.) The *principal symbol sheaf* is the graded sheaf of algebras on G/Q

$$\mathcal{S}_G = G \times_Q S_{G/Q} \tag{4.22)(c}$$

(cf. (3.13)). (This is the pushforward to G/Q of a sheaf of $\mathcal{O}$-algebras on $T^*(G/Q)$.) The space of global sections of $\mathcal{S}_G$ is a graded algebra S_G that we call the *principal symbol algebra*; explicitly,

$$\begin{aligned} S_G &= \mathrm{Ind}_{alg}(Q \uparrow G)(S(\mathfrak{g}/\mathfrak{q}) \otimes A_Q) \\ &= \text{functions on } G \text{ with values in } S(\mathfrak{g}/\mathfrak{q}) \otimes A_Q, \\ &\qquad \text{transforming by Ad under } Q. \end{aligned} \tag{4.22)(d}$$

An immediate consequence is that if A_Q (or, equivalently, A_L) is completely prime, then S_G is as well. We write Ad for the action of G on S_G by left translation of sections.

Now define the *weak filtration* on $A_{\mathfrak{g}}$ by

$$A^1_{\mathfrak{g},p,wk} = \{\, S \in A^1_{\mathfrak{g}} \mid S(U_n(\mathfrak{g})) \subset (A_{G/Q})_{p+n} \}. \tag{4.23)(a}$$

(Here and below we will use the isomorphisms α^1 and α^2 to transfer filtrations among $A_{\mathfrak{g}}$, $A^1_{\mathfrak{g}}$, and $A^2_{\mathfrak{g}}$.) This makes $A^1_{\mathfrak{g}}$ a filtered algebra (as one checks using (4.13(c)), but not every element of $A^1_{\mathfrak{g}}$ belongs to some $A^1_{\mathfrak{g},p}$. It follows from (4.11)(iv) that

$$\phi_{\mathfrak{g}}(U_n(\mathfrak{g})) \subset (A_{\mathfrak{g}})_{n,wk}. \tag{4.23)(b}$$

Using (4.20) and (4.21), one can describe the weak filtration directly on $A^2_{\mathfrak{g}}$ as well: it is

$$A^2_{\mathfrak{g},p,wk} = \{\, \Sigma \in A^2_{\mathfrak{g}} \mid \Sigma(U_n(\mathfrak{g})) \subset (A_{G/Q})_{p+n} \}. \tag{4.23)(c}$$

The difficulty with the weak filtration is that it is not ad($\mathfrak{g}$)-invariant. Given any filtered algebra A and a family $\mathcal{D}$ of derivations of A, we can define a smaller filtration of A by

$$A_{n,sm} = \{\, a \in A \mid D_1 \dots D_k(a) \in A_n, \text{ all } \{D_i\} \subset \mathcal{D} \,\}.$$

This gives A a new structure of filtered algebra, this one preserved by $\mathcal{D}$. In our case, we set

$$A_{\mathfrak{g},p} = \{\, T \in A_{\mathfrak{g}} \mid ad(u)T \in A_{\mathfrak{g},p,wk}, \text{ all } u \in U(\mathfrak{g}) \,\}. \tag{4.24)(a}$$

This is not very easy to understand directly; but Lemma 4.14(iv) and (4.23)(c) show immediately that the corresponding filtration of $A^2_{\mathfrak{g}}$ is

$$A^2_{\mathfrak{g},p} = \{\, \Sigma \in A^2_{\mathfrak{g}} \mid \Sigma(U_n(\mathfrak{g})U(\mathfrak{g})) \subset (A_{G/Q})_{p+n}, \text{ all } n \,\}. \qquad (4.24)(a)$$

Clearly this is

$$A^2_{\mathfrak{g},p} = \{\, \Sigma \mid \Sigma(U(\mathfrak{g})) \subset (A_{G/Q})_p \,\}. \qquad (4.24)(b)$$

Using Corollary 4.16, we deduce a description of the restriction of this filtration to A_G:

$$A_{G,p} = \mathrm{Ind}_{alg}(Q \uparrow G)(A_{G/Q,p}). \qquad (4.24)(c)$$

Regarded as an algebraic map from G to $A_{G/Q}$, every element of A_G has its image contained in some finite-dimensional subspace. Consequently

$$\bigcup_p A_{G,p} = A_G. \qquad (4.24)(d)$$

On the level of sheaves, Definition 4.22 gives rise to an exact sequence

$$0 \to \mathcal{A}_{G,p-1} \to \mathcal{A}_{G,p} \to \mathcal{S}^p_G \to 0. \qquad (4.25)(a)$$

(Here we have used and extended slightly the notation of Definition 4.15 and (4.22)(c).) Taking global sections gives an exact sequence

$$0 \to A_{G,p-1} \to A_{G,p} \to S^p_G. \qquad (4.25)(b)$$

The last map in this sequence is called a *principal symbol map*, and is denoted π_p. Because of (4.24)(c), these maps can be assembled to a graded injection

$$\pi : \mathrm{gr}\ A_G \hookrightarrow S_G. \qquad (4.25)(c)$$

Clearly π respects the action Ad of G. What is less obvious is

Proposition 4.26. *The principal symbol map π of (4.25) is an algebra homomorphism. In particular, if A_L is completely prime, then the induced Dixmier algebra A_G is as well.*

If $H^1(G/Q, \mathcal{S}_G) = 0$, then π is an isomorphism of graded algebras.

Proof. Suppose $S \in A^1_{\mathfrak{g},p}$. Write $\Sigma = \alpha^{12}(S)$ for the corresponding element of $A^2_{\mathfrak{g},p}$. If u belongs to $U_n(\mathfrak{g})$, then (4.21) shows that

$$S(u) \equiv u\Sigma(1) \pmod{A_{G/Q,n+p-1}}.$$

By the definition of α^{12}, this is the same as

$$S(u) \equiv uS(1) \pmod{A_{G/Q,n+p-1}}. \tag{4.27)(a}$$

Write

$$S(1) = \sum u_i \otimes E_i, \tag{4.27)(b}$$

with u_i in $U_p(\mathfrak{g})$. It is immediate from the definition that the value at the identity of the principal symbol is given by

$$\pi_p(S)(e) = \sum p_i \otimes E_i; \tag{4.27)(c}$$

here p_i is the image of u_i in $S^p(\mathfrak{g}/\mathfrak{q})$.

Suppose S' belongs to $A_{\mathfrak{g},q}$. Write

$$S'(1) = \sum v_j \otimes E'_j, \tag{4.27)(d}$$

with v_j in $U_q(\mathfrak{g})$. If q_j is the image of v_j in $S^q(\mathfrak{g}/\mathfrak{q})$, then

$$\pi_q(S')(e) = \sum q_j \otimes E'_j. \tag{4.27)(e}$$

By (4.23) applied to S',

$$S'(u_i) = \sum_j u_i v_j \otimes E'_j \qquad \pmod{A_{G/Q,p+q-1}}. \tag{4.27)(f}$$

By Lemma 4.11(iv),

$$(S'S)(1) = \sum_{i,j} u_i v_j \otimes E'_j E_i \qquad \pmod{A_{G/Q,p+q-1}}. \tag{4.27)(g}$$

Consequently

$$\begin{aligned} \pi_{p+q}(S'S)(e) &= \sum_{i,j} p_i q_j \otimes E'_j E_i \\ &= \pi_q(S')(e)\pi_p(S)(e). \end{aligned} \tag{4.27)(h}$$

The proposition follows from (4.27)(h) and the Ad(G)-equivariance of π. Q.E.D.

There is a variant of this symbol calculus that is of interest for the Dixmier conjecture. (Because we will never need both at the same time, we will not use a different notation for the filtrations in the variant.) Suppose that A_L is endowed with a good filtration indexed by $1/2\mathbf{N}$. Recall from the discussion before Conjecture 2.3 that this amounts to the following requirements:

$$\begin{gathered}A_{L,p}A_{L,q} \subset A_{L,p+q};\\ A_{L,p} \text{ is } \mathrm{Ad}(L)\text{-invariant};\\ \bigcup_p A_{L,p} = A_L;\\ \phi_L(U_n(\mathfrak{l})) \subset (A_L)_n;\\ (R_L, \psi_L) = (\mathrm{gr}\ A_L, \mathrm{gr}\ \phi_L) \text{ is a graded orbit datum.}\end{gathered} \tag{4.28}$$

We want to construct a good filtration of the induced Dixmier algebra (A_G, ϕ_G).

To begin, filter $A_{G/Q}$ by

$$A_{G/Q,p} = \sum_n U_n(\mathfrak{g}) \otimes A_{Q,p-n}. \tag{4.29)(a}$$

(Of course the underlying algebra of A_Q is just A_L, so A_Q inherits the filtration of A_L.) This filtration is Ad(Q)-stable, and

$$U_n(\mathfrak{g}) \cdot A_{G/Q,p} \subset A_{G/Q,n+p}. \tag{4.29)(b}$$

The associated graded object therefore carries an action Ad of Q and a graded $S(\mathfrak{g})$-module structure. A little thought shows that it is the graded tensor product

$$\mathrm{gr}\ A_{G/Q} = S(\mathfrak{g}) \otimes_{S(\mathfrak{q})} \mathrm{gr}\ A_Q. \tag{4.29)(c}$$

This is exactly the algebra $R_{G/Q}$ attached to R_L in the construction of induced orbit data (cf. (3.12)). We call

$$\mathcal{R}_G = G \times_Q R_{G/Q} \tag{4.29)(d}$$

a *good symbol sheaf.* The space of global sections is a graded algebra R_G that we call a *good symbol algebra.* By (3.13) and (3.14),

$$R_G = Ind_{orb}(Q \uparrow G)(R_L). \tag{4.29)(e}$$

Of course R_G is an orbit datum for G by Proposition 3.15.

Now we can filter $A^2_{\mathfrak{g}}$ by

$$A^2_{\mathfrak{g},p} = \{\, \Sigma \in A^2_{\mathfrak{g}} \mid \Sigma(U(\mathfrak{g})) \subset A_{G/Q,p}\,\}. \tag{4.30)(a}$$

Exactly as in (4.23) and (4.24), one shows that this defines a filtered algebra structure on $A^2_{\mathfrak{g}}$. The induced filtration on A_G is

$$A_{G,p} = \mathrm{Ind}_{alg}(Q \uparrow G)(A_{G/Q,p}), \tag{4.30)(b}$$

and

$$\bigcup_p A_{G,p} = A_G. \tag{4.30)(c}$$

Again we get an exact sequence of sheaves on G/Q

$$0 \to \mathcal{A}_{G,p-1} \to \mathcal{A}_{G,p} \to \mathcal{R}^p_G \to 0. \tag{4.31)(a}$$

This gives rise to an exact sequence

$$0 \to A_{G,p-1,gd} \to A_{G,p,gd} \to (R_G)^p. \tag{4.31)(b}$$

The last map in this sequence is called a *good symbol map*, and is denoted γ_p. We therefore have a graded G-equivariant injection

$$\gamma : \mathrm{gr}\, A_G \to R_G. \tag{4.31)(c}$$

Exactly as in Proposition 4.26, one shows that γ is a homomorphism of algebras. We have proved

Proposition 4.32. *Suppose the Dixmier algebra (A_L, ϕ_L) is endowed with a good filtration indexed by $1/2\mathbb{N}$ (cf. (4.28) with associated graded orbit datum (R_L, ψ_L). Write (R_G, ψ_G) for the induced orbit datum for G. Then the induced Dixmier algebra (A_G, ϕ_G) has a natural good filtration, and the associated graded orbit datum injects into (R_G, ψ_G). If the cohomology $H^1(G/Q, \mathcal{R}_G)$ vanishes, then this injection is an isomorphism.*

Evidently this result has some relationship to Conjecture 2.3. We will discuss this in detail in section 5.

5. Sheets of Dixmier algebras.

We can now say something about what a "sheet of Dixmier algebras" ought to look like. Of course such a sheet ought to consist of the Dixmier algebras corresponding via Conjecture 2.3 to a sheet of geometric orbit data. Our goal, however, is to make a little progress toward proving that conjecture. We fix therefore a sheet of geometric orbit data attached to a Levi subgroup L of G and a rigid unipotent orbit datum (R_u, ψ_u) for $L/Z(L)_0$ (Definition 3.20). Thus

$$X_u = \text{Spec } R_u \text{ is a unipotent Poisson variety for } L/Z(L)_0 \qquad (5.1)(a)$$

(cf. [21]). One consequence is that the action of $\mathbb{C}^\times$ on the support Σ_u must lift to X_u, at least after passing to a finite cover of $\mathbb{C}^\times$. This means that R_u is graded by $k^{-1}\mathbb{N}$ for some positive integer k. As was already mentioned in section 2, Moeglin has observed that k is at most 2. (This is a consequence of the Jacobson-Morozov theorem.) Consequently

$$R_u \text{ is graded by } 1/2\mathbb{N}; \qquad (5.1)(b)$$

this gradation is compatible with the standard one on $S(\mathfrak{g})$ and the map ψ_u.

The first serious step is to attach to (R_u, ψ_u) a Dixmier algebra for $L/Z(L)_0$. There is a conjecture for how to do this in section 5 of [21]; we reproduce a strengthened version here.

Definition/Conjecture 5.2. Suppose G is a semisimple group, and (R, ψ) is a rigid geometric (and therefore unipotent) orbit datum (Definition 2.2 and Definition 3.20). Recall the natural grading of R by $1/2\mathbb{N}$, and the natural Poisson structure $\{,\}$ on R. A *(rigid) unipotent Dixmier algebra* associated to (R, ψ) is a Dixmier algebra (A, ϕ) (Definition 2.1) having an $\mathrm{Ad}(G)$-invariant filtration by $1/2\mathbb{N}$, subject to the following conditions.

i) $\phi(U_n(\mathfrak{g})) \subset A_n$.
ii) The pair (gr A, gr ϕ) is isomorphic as an algebra with G-action to (R, ψ).

These two properties conjecturally determine exactly one Dixmier algebra. An immediate consequence is

iii) A is completely prime.

As a conjectural consequence of (i) and (ii), one should have four additional properties.

iv) The infinitesimal character of A corresponds to a weight in the rational span of the roots.
v) The kernel of ϕ is a weakly unipotent primitive ideal in $U(\mathfrak{g})$.
vi) Suppose a and b in A_p and A_q have images r and s in R^p and R^q. Then the Poisson bracket $\{r, s\}$ is the image of $ab - ba$.
vii) A admits a transpose antiautomorphism ($a \to {}^t a$) of order 1, 2, or 4 (Definition 2.8). This preserves the filtration, and the associated graded antiautomorphism acts by $\exp(ip\pi)$ on R^p.

(The notion of *weakly unipotent* primitive ideal may be found in [19], section 8. The condition is that the infinitesimal character cannot be made shorter by tensoring with a finite-dimensional representation. The prototypical example is the augmentation ideal.)

This conjecture is very easy if $R = \mathbb{C}$ (corresponding to the zero nilpotent orbit). If R corresponds to a minimal non-zero nilpotent orbit (which is rigid except in type A) then the conjecture is true because of Joseph's work on the Joseph ideal (as supplemented by Garfinkle). The complete conjecture is known only in a handful of other cases, although there is some evidence for it in many other infinite families of examples. Some of the most powerful results in the direction of (ii) are those in [16].

Now assume that we are in the setting (5.1), and that we have a unipotent Dixmier algebra

$$(A_u, \phi_u) \qquad (5.3)(a)$$

for $L/Z(L)_0$ attached to (R_u, ϕ_u) in the sense of Conjecture 5.2. To every character λ of $\mathfrak{z}(\mathfrak{l})$ we can attach an orbit datum

$$(R_L(\lambda), \psi_L(\lambda)) \qquad (5.3)(b)$$

for L; here $R_L(\lambda)$ is isomorphic to R_u, and $\psi_L(\lambda)$ is given as in (2.5)(e) and (f). Similarly, we can construct a Dixmier algebra

$$(A_L(\lambda), \phi_L(\lambda)) \qquad (5.3)(c)$$

for L. Here $A_L(\lambda)$ is just A_u as an algebra with L action, and

$$\begin{aligned} \phi_L(\lambda)(X) &= \lambda(X) \qquad (X \in \mathfrak{z}(\mathfrak{l})), \\ \phi_L(\lambda)(Y) &= \phi_u(Y) \qquad (Y \in [\mathfrak{l}, \mathfrak{l}]). \end{aligned} \qquad (5.3)(d)$$

Definition 5.4. Suppose we are in the setting (5.1), and that the Dixmier algebra (A_u, ϕ_u) for $L/Z(L)_0$ exists. Fix a parabolic subgroup $Q = LU$ with Levi factor L. Form a sheet of orbit data

$$(R_G(\lambda), \psi_G(\lambda)) = \mathrm{Ind}_{orb}(Q \uparrow G)(R_L(\lambda), \psi_L(\lambda)) \qquad (5.4)(a)$$

(Definition 3.22). (Recall that, according to Conjecture 3.19, this sheet is independent of the choice of Q.) The *sheet of Dixmier algebras for G attached to* $(Q, (A_u, \phi_u))$ is

$$(A_G(\lambda), \phi_G(\lambda)) = \mathrm{Ind}_{Dix}(Q \uparrow G)(A_L(\lambda), \phi_L(\lambda))$$

(Definition 4.7). According to Conjecture 4.18, these algebras do not depend on the choice of Q.

By Proposition 4.32 (and hypothesis (ii) in Definition/Conjecture 5.2) each $A_G(\lambda)$ carries a natural good filtration indexed by $1/2\mathbb{N}$. These filtrations have the property that there is natural injection of graded orbit data

$$(\text{gr } A_G(\lambda), \text{gr } \phi_G(\lambda) \hookrightarrow (R_G(0), \psi_G(0)). \tag{5.5)(a}$$

An elementary argument also provides natural good filtrations of the orbit data in the sheet, with the property that

$$(\text{gr } R_G(\lambda), \text{gr } \psi_G(\lambda)) \hookrightarrow (R_G(0), \psi_G(0)). \tag{5.5)(b}$$

Both of these injections ought to be isomorphisms. Because of Proposition 4.32 (and an easier analogue for orbit data) this would be a consequence of

Conjecture 5.6. In the setting of Definition 5.4, form the sheaf $\mathcal{R}_G(0)$ on G/Q as in (3.13). Then the higher cohomology of G/Q with coefficients in $\mathcal{R}_G(0)$ vanishes.

If $R_u = \mathbb{C}$, then the sheaf $\mathcal{R}_G(0)$ is the sheaf of functions on the cotangent bundle of G/Q. In that case Conjecture 5.6 is true, by a well-known result of Elkik.

The correspondence taking $(R_G(\lambda), \psi_G(\lambda))$ to $(A_G(\lambda), \phi_G(\lambda))$ is our candidate for a Dixmier correspondence on geometric orbit data (cf. Conjecture 2.3). To define it, we need to know the existence of unipotent Dixmier algebras attached to rigid orbit data (Conjecture 5.2). For it to be well-defined on a single sheet, we need both $R_G(\lambda)$ and $A_G(\lambda)$ to be independent of Q (Conjectures 3.19 and 4.18). For it to be well-defined on all geometric orbit data, we need the geometric sheets to be disjoint (Conjecture 3.24). For it to satisfy Conjecture 2.3, we need a cohomology vanishing result (Conjecture 5.6).

6. Relation with bimodule induction.

In this section we will resume our study of induced Dixmier algebras, focusing on special conditions on the inducing algebra that permit a further analysis of their structure. The main tool is the idea of induction of Harish-Chandra bimodules, which we now recall.

Definition 6.1. Suppose $Q = LU$ is a parabolic subgroup of G, and B_L is a Harish-Chandra bimodule for L. Write $Q^{op} = LU^{op}$ for the opposite parabolic subgroup. Define a new bimodule B'_L by subtracting from the left and right actions of $\mathfrak{l}$ the character δ defined after (4.2)(a). Make B'_L into a left $\mathfrak{q}$-module by making $\mathfrak{u}$ act by zero, and into a right $\mathfrak{q}^{op}$-module by making $\mathfrak{u}^{op}$ act by zero; we denote this object $B_{\mathfrak{q},\mathfrak{q}^{op}}$. Consider

$$B_{\mathfrak{g},bi} = B_{\mathfrak{g}} = \mathrm{Hom}_{\mathfrak{q},\mathfrak{q}^{op}}(U(\mathfrak{g}) \otimes U(\mathfrak{g}), B_{\mathfrak{q},\mathfrak{q}^{op}}).$$

Here $\mathfrak{q}$ acts on the left on the first $U(\mathfrak{g})$ factor and $\mathfrak{q}^{op}$ on the right on the second to define the Hom. $B_{\mathfrak{g}}$ has the structure of a $\mathfrak{g}$-bimodule: the left action of $\mathfrak{g}$ comes from the right action on the first $U(\mathfrak{g})$, and the right action from the left action of $\mathfrak{g}$ on the second $U(\mathfrak{g})$. Put

$$B_{G,bi} = B_G = \mathrm{Ad}(G)\text{-finite part of } B_{\mathfrak{g}};$$

this is the Harish-Chandra bimodule for G *induced* by B_L. We write

$$B_G = \mathrm{Ind}_{bi}(Q \uparrow G)(B_L).$$

Induction of Harish-Chandra bimodules is a very simple and well-understood process. It is an exact functor, and (as algebraic representations under Ad) we have

$$B_G \simeq \mathrm{Ind}_{alg}(L \uparrow G)(B_L) \tag{6.2}$$

(notation (A.8)). Because this is induction from a reductive subgroup, it is much better behaved than the induction appearing in (say) Corollary 4.16. The reader may wonder why we did not use it in section 4 to construct induced Dixmier algebras. The reason is that bimodule induction does not in general take algebras to algebras.

Nevertheless, there is much to be gained (computability, for example) whenever we can relate induction of Dixmier algebras (which are, among other things, Harish-Chandra bimodules) to Ind_{bi}. Examples for SL(2) show that the relationship cannot be quite trivial. To understand it, we will first interpret bimodule induction in terms of endomorphisms between modules (Corollary 6.5). Once that is done, it is convenient to extend the induction construction of Definition 4.7 from Dixmier algebras to arbitrary Harish-Chandra bimodules (Definition 6.7). Once the two kinds of induction are described in parallel, it becomes clear that they are related by something like the Shapovalov form on a Verma module (Definition 6.11). Establishing an isomorphism between them therefore comes down to proving some irreducibility results for (appropriately generalized) Verma modules (Theorem 6.12). The idea of this argument comes from [5]; much more sophisticated incarnations of it appear in Joseph's work relating primitive ideals to highest weight modules (cf. [11] and [12], section 1.3).

Suppose $V_{\mathfrak{l}}$ and $W_{\mathfrak{l}}$ are modules for $\mathfrak{l}$. Assume that

$$B_L \subset \mathrm{Hom}(V_{\mathfrak{l}}, W_{\mathfrak{l}}) \tag{6.3)(a}$$

is a Harish-Chandra bimodule of maps. Here the left action of $\mathfrak{l}$ comes from the action on $W_{\mathfrak{l}}$, etc. (It is easy to see that any Harish-Chandra bimodule arises in this way; we do not require $V_{\mathfrak{l}}$ and $W_{\mathfrak{l}}$ to be particularly nice.) Write $V_{\mathfrak{l}}'$ for $V_{\mathfrak{l}}$ with the action of $\mathfrak{l}$ twisted by $-\delta$, and then $V_{\mathfrak{q}^{op}}$ for the extension to $\mathfrak{q}^{op}$ on which $\mathfrak{u}^{op}$ acts by zero. Similarly define $W_{\mathfrak{q}}$. Then

$$B_{\mathfrak{q},\mathfrak{q}^{op}} \subset \mathrm{Hom}(V_{\mathfrak{q}^{op}}, W_{\mathfrak{q}}). \tag{6.3)(b}$$

Define

$$V_{\mathfrak{g},ind} = U(\mathfrak{g}) \otimes_{\mathfrak{q}^{op}} V_{\mathfrak{q}^{op}} \simeq U(\mathfrak{u}) \otimes_{\mathbf{C}} V_{\mathfrak{q}^{op}} \tag{6.3)(c}$$

$$W_{\mathfrak{g},pro} = \mathrm{Hom}_{\mathfrak{q}}(U(\mathfrak{g}), W_{\mathfrak{q}}) \simeq \mathrm{Hom}_{\mathbf{C}}(U(\mathfrak{u}^{op}, W_{\mathfrak{q}}). \tag{6.3)(d}$$

Lemma 6.4. *In the setting (6.3), there is a natural identification of $U(\mathfrak{g})$-bimodules*

$$\begin{aligned}\mathrm{Hom}_{\mathbf{C}}(V_{\mathfrak{g},ind}, W_{\mathfrak{g},pro}) &\simeq \mathrm{Hom}_{\mathfrak{q},\mathfrak{q}^{op}}(U(\mathfrak{g}) \otimes U(\mathfrak{g}), \mathrm{Hom}(V_{\mathfrak{q}^{op}}, W_{\mathfrak{q}}))\\ &\simeq \mathrm{Hom}_{\mathbf{C}}(U(\mathfrak{u}^{op}) \otimes U(\mathfrak{u}), \mathrm{Hom}(V_{\mathfrak{q}^{op}}, W_{\mathfrak{q}})).\end{aligned}$$

We will not give the (easy and well-known) formal proof, but it is helpful to recall the form of the isomorphism. If ϕ on the left corresponds to Φ on the right, then for u and u' in $U(\mathfrak{g})$ and v in $V_{\mathfrak{q}}^{op}$,

$$[\phi(u \otimes v)]\,(u') = (\Phi(u' \otimes u))v.$$

Here $u \otimes v$ is an element of $V_{\mathfrak{g},ind}$. The term in square brackets on the left is therefore an element of $W_{\mathfrak{g},pro}$; that is, it is a map from $U(\mathfrak{g})$ to $W_{\mathfrak{q}}$. We specify it by specifying its value at u'.

Definition 6.5. Suppose we are in the setting of (6.3). A map

$$\phi : V_{\mathfrak{g},ind} \longrightarrow W_{\mathfrak{g},pro}$$

is said to be of *type* B_L if the corresponding map

$$\Phi : U(\mathfrak{g}) \otimes U(\mathfrak{g}) \longrightarrow \mathrm{Hom}(V_{\mathfrak{q}^op}, W_{\mathfrak{q}}))$$

(Lemma 6.4) takes values in $B_{\mathfrak{q},\mathfrak{q}^{op}}$. By Definition 6.1,

$$B_{\mathfrak{g},bi} \simeq \{\, \phi : V_{\mathfrak{g},ind} \to W_{\mathfrak{g},pro} \mid \phi \text{ is of type } B_L \,\}.$$

Definitions 6.1 and 6.5 combine (like all good definitions) to give a result.

Proposition 6.6. *Suppose we are in the setting (6.3). Then the induced Harish-Chandra bimodule* $B_G = \mathrm{Ind}_{bi}(Q \uparrow G)(B_L)$ *consists precisely of the* $\mathrm{Ad}(G)$*-finite maps from* $V_{\mathfrak{g},ind}$ *to* $W_{\mathfrak{q},pro}$ *that are of type* B_L*.*

We now outline the extension to bimodules of the construction of section 4.

Definition 6.7. In the setting of (6.3), extend $W'_{\mathfrak{l}}$ to a module $W_{\mathfrak{q}^{op}}$ for $\mathfrak{q}^{op}$ by making $\mathfrak{u}^{op}$ act by zero. Define

$$W_{\mathfrak{g},ind} = U(\mathfrak{g}) \otimes_{\mathfrak{q}^{op}} W_{\mathfrak{q}^{op}}. \tag{6.7)(a}$$

In analogy with (4.2), identify B_L with a $U(\mathfrak{q}^{op})$-bimodule

$$B_{Q^{op}} \subset \mathrm{Hom}(V_{\mathfrak{q}^{o}p}, W_{\mathfrak{q}^{op}}). \tag{6.7)(b}$$

As in Definition 4.6, we say that a map T from $V_{\mathfrak{g},ind}$ to $W_{\mathfrak{g},ind}$ is of *type* B_L if for every u in $U(\mathfrak{g})$ there are elements u_i in $U(\mathfrak{g})$ and E_i in $B_{\mathfrak{q}^{op}}$ such that for every v in $V_{\mathfrak{q},op}$,

$$T(u \otimes v) = \sum u_i \otimes E_i v. \tag{6.7)(c}$$

The collection of all such maps is a $U(\mathfrak{g})$-bimodule

$$B_{\mathfrak{g},Dix} \subset \mathrm{Hom}(V_{\mathfrak{g},ind}, W_{\mathfrak{g},ind}). \tag{6.7)(d}$$

Define

$$B_{G,Dix} = \mathrm{Ad}(G)\text{-finite maps from } V_{\mathfrak{g},ind} \text{ to } W_{\mathfrak{g},ind} \text{ of type } B_L. \tag{6.7)(e}$$

The proof of Corollary 4.17 shows that $B_{G,Dix}$ is a Harish-Chandra bimodule for G depending only on Q and B_L (and not on the choices of $V_{\mathfrak{l}}$ and $W_{\mathfrak{l}}$). We call it the *Dixmier induced Harish-Chandra bimodule*, and write

$$B_{G,Dix} = \mathrm{Ind}_{Dix}(Q^{op} \uparrow G)(B_L). \tag{6.7)(f}$$

As in Corollary 4.16, one gets an isomorphism of G-modules for Ad

$$B_{G,Dix} \simeq \mathrm{Ind}_{alg}(Q^{op} \uparrow G)(U(\mathfrak{g}) \otimes_{\mathfrak{q}^{op}} B_{Q^{op}}). \tag{6.7)(g}$$

Here $\mathfrak{q}^{op}$ acts on $B_{Q^{op}}$ on the left to define the tensor product, and Q^{op} acts by $\mathrm{Ad} \otimes \mathrm{Ad}$ on the inducing representation.

Recall that what we want is to compare $B_{G,bi}$ and $B_{G,Dix}$. We will do this by comparing the larger bimodules $B_{\mathfrak{g},bi}$ and $B_{\mathfrak{g},Dix}$. To do that, we determine their bimodule structure under the center $\mathfrak{z}(\mathfrak{l})$ of $\mathfrak{l}$. As in section 4, it is harmless and convenient to assume that $\mathfrak{z}(\mathfrak{l})$ acts by characters λ and ρ on $W'_{\mathfrak{l}}$ and $V'_{\mathfrak{l}}$ respectively. We assume also that the Harish-Chandra bimodule B_L has finite length. Fix characters β and γ of $\mathfrak{z}(\mathfrak{l})$. Then the subspace of $B_{\mathfrak{g},Dix}$ transforming on the left by β and on the right by γ is (cf. (4.9))

$$(B_{\mathfrak{g},Dix})_{\beta\gamma} \simeq \mathrm{Hom}(U(\mathfrak{u})_{\gamma-\rho}, U(\mathfrak{u})_{\beta-\lambda}) \otimes B_{Q^{op}}. \tag{6.8)(a}$$

The whole bimodule is assembled from these pieces by the prescription

$$B_{\mathfrak{g},Dix} \simeq \prod_{\gamma} \left(\sum_{\beta} (B_{\mathfrak{g},Dix})_{\beta\gamma} \right). \tag{6.8)(b}$$

Similarly, Lemma 6.4 gives

$$(B_{\mathfrak{g},bi})_{\beta\gamma} \simeq \mathrm{Hom}(U(\mathfrak{u}^{op})_{-\beta+\lambda} \otimes U(\mathfrak{u})_{\gamma-\rho}, B_{\mathfrak{q},\mathfrak{q}^{op}}. \tag{6.8)(c}$$

Now the indicated weight spaces (for ad($\mathfrak{z}(\mathfrak{l})$ in $U(\mathfrak{u})$ and $U(\mathfrak{u}^{op})$ are finite-dimensional. As an $\mathfrak{l}$-bimodule, $B_{\mathfrak{q},\mathfrak{q}^{op}}$ is equal to $B_{Q^{op}}$. We may therefore rewrite this last equation as

$$(B_{\mathfrak{g},bi})_{\beta\gamma} \simeq \mathrm{Hom}(U(\mathfrak{u})_{\gamma-\rho}, (U(\mathfrak{u}^{op})_{-\beta+\lambda})^*) \otimes B_{Q^{op}}. \tag{6.8)(d}$$

By Definition 6.5,

$$B_{\mathfrak{g},bi} \simeq \prod_{\gamma} (\prod_{\beta} (B_{\mathfrak{g},bi})_{\beta\gamma}). \tag{6.8)(e}$$

Lemma 6.9. *Suppose B_L is a Harish-Chandra bimodule of finite length for L. Then as algebraic representations of* $\mathrm{Ad}(L)$, *the bimodule weight spaces $(B_{\mathfrak{g},bi})_{\beta\gamma}$ and $(B_{\mathfrak{g},Dix})_{\beta\gamma}$ for $\mathfrak{z}(\mathfrak{l})$ contain the same representations with the same (finite) multiplicities.*

Proof. This follows from (6.8)(a) and (d), using the isomorphism of L-modules $\mathfrak{u} \simeq (\mathfrak{u}^{op})^*$ provided by the Killing form. Q.E.D.

Our next task is the construction of a bimodule map from $B_{\mathfrak{g},Dix}$ to $B_{\mathfrak{g},bi}$. We need a little more notation in the setting of (6.3). We have

$$U(\mathfrak{g}) = U(\mathfrak{q}) \oplus U(\mathfrak{g})\mathfrak{u}^{op}; \tag{6.10)(a}$$

write

$$\pi : U(\mathfrak{g}) \to U(\mathfrak{q}) \tag{6.10)(b}$$

for the corresponding projection on the first factor. Write

$$I_{\mathfrak{q},\mathfrak{q}^{op}} : W_{\mathfrak{q}^{op}} \to W_{\mathfrak{q}} \tag{6.10)(c}$$

for the "identity map" of the underlying $\mathfrak{l}$-modules.

Definition 6.11. Suppose $\mathfrak{q} = \mathfrak{l} + \mathfrak{u}$ and $\mathfrak{q}^{op} = \mathfrak{l} + \mathfrak{u}^{op}$ are opposite parabolic subalgebras of $\mathfrak{g}$. Suppose $V_{\mathfrak{l}}$ and $W_{\mathfrak{l}}$ are $\mathfrak{l}$-modules; we use other notation as in (6.3) and (6.10). Define a homomorphism of $\mathfrak{q}^{op}$ modules

$$j_{\mathfrak{q}^{op}} : W_{\mathfrak{q}^{op}} \to W_{\mathfrak{g},pro}, \qquad (j_{\mathfrak{q}^{op}}(w))(x) = \pi(x) \cdot (I_{\mathfrak{q},\mathfrak{q}^{op}} w). \tag{6.11)(a}$$

Here x is in $U(\mathfrak{g})$ and w is in $W_{\mathfrak{q}^{op}}$. It is easy to check that $j_{\mathfrak{q}^{op}}(w)$ really belongs to $W_{\mathfrak{g},pro}$, and that the map is a $\mathfrak{q}^{op}$-module injection. By the universality property of induction, $j_{\mathfrak{q}^{op}}$ induces a $\mathfrak{g}$-module map

$$j : W_{\mathfrak{g},ind} \to W_{\mathfrak{g},pro}, \qquad (j(u \otimes w))(x) = \pi(xu) \cdot (I_{\mathfrak{q},\mathfrak{q}^{op}} w). \tag{6.11)(b}$$

We call j the *canonical intertwining operator*; it is closely connected to the Shapovalov form on a Verma module (see [6] or [10]).

Composition with j induces a $\mathfrak{g}$-bimodule map

$$J : \mathrm{Hom}_{\mathbb{C}}(V_{\mathfrak{g},ind}, W_{\mathfrak{g},ind}) \to \mathrm{Hom}_{\mathbb{C}}(V_{\mathfrak{g},ind}, W_{\mathfrak{g},pro}). \tag{6.11)(c}$$

We claim that this composition sends maps of type B_L to maps of type B_L (Definitions 6.5 and 6.7) and therefore restricts to

$$J : B_{\mathfrak{g},Dix} \to B_{\mathfrak{g},bi}. \tag{6.11)(d}$$

(This will be proved in a moment.) As a bimodule map, J respects the action ad, and so restricts to

$$J : \mathrm{Ind}_{Dix}(Q^{op} \uparrow G)(B_L) \to \mathrm{Ind}_{bi}(Q \uparrow G)(B_L); \tag{6.11)(e}$$

this map we call the *canonical bimodule intertwining operator.*

To prove that J preserves type, suppose that $T \in \mathrm{Hom}_{\mathbf{C}}(V_{\mathfrak{g},ind}, W_{\mathfrak{g},ind})$ is of type B_L; we want to prove that $J(T) = j \circ T$ is as well. Write τ for the map from $U(\mathfrak{g}) \otimes U(\mathfrak{g})$ corresponding to $J(T)$ (Lemma 6.4), and fix u and u' in $U(\mathfrak{g})$. Then $\tau(u' \otimes u)$ is a map from $V_{\mathfrak{q}^{op}}$ to $W_{\mathfrak{q}}$; we have to show that it belongs to $B_{\mathfrak{q},\mathfrak{q}^{op}}$. To compute it, fix v in $V_{\mathfrak{q}^{op}}$. Now (6.6)(c) provides elements u_i in $U(\mathfrak{g})$ and E_i in $B_{\mathfrak{q}^{op}}$ (depending only on u) so that

$$T(u \otimes v) = \sum u_i \otimes E_i v.$$

Tracing through the definitions, we find

$$\begin{aligned}(\tau(u' \otimes u))v &= (J(T)(u \otimes v))(u') \qquad \text{(by Lemma 6.4)}\\ &= (j(\sum u_i \otimes E_i v))(u')\\ &= \sum \pi(u'u_i) \cdot (I_{\mathfrak{q},\mathfrak{q}^{op}} E_i v) \qquad \text{(by (6.11)(b))}.\end{aligned}$$

Now composition with $I_{\mathfrak{q},\mathfrak{q}^{op}}$ clearly defines an isomorphism from $B_{\mathfrak{q}^{op}}$ to $B_{\mathfrak{q},\mathfrak{q}^{op}}$. Write F_i for the image of E_i under this map. The action of $\pi(u'u_i)$ on $F_i v$ just corresponds to the left action of $U(\mathfrak{q})$ on $B_{\mathfrak{q},\mathfrak{q}^{op}}$; so we get

$$T(u \otimes v) = \left[\sum \pi(u'u_i)F_i\right] v.$$

Now the term in square brackets belongs to $B_{\mathfrak{q},\mathfrak{q}^{op}}$, as we wished to show.

Theorem 6.12. *Suppose $Q = LU$ is a parabolic subgroup of G, and B_L is a Harish-Chandra bimodule of finite length for L; say*

$$B_L \subset \mathrm{Hom}(V_{\mathfrak{l}}, W_{\mathfrak{l}}).$$

Define $W_{\mathfrak{g},ind}$ and $W_{\mathfrak{g},pro}$ as in (6.3) and Definition 6.7, and the canonical intertwining operator j between them as in Definition 6.11. If j is one-to-one, then the canonical bimodule intertwining operator

$$J : \mathrm{Ind}_{Dix}(Q^{op} \uparrow G)(B_L) \to \mathrm{Ind}_{bi}(Q \uparrow G)(B_L)$$

is an isomorphism.

Proof. Because j is one-to-one, the map

$$J : B_{\mathfrak{g},Dix} \to B_{\mathfrak{g},bi}$$

is obviously one-to-one. By Lemma 6.9, the restriction of J to each $\mathfrak{z}(\mathfrak{l})$-bimodule weight space $(B_{\mathfrak{g},Dix})_{\beta\gamma}$ is an isomorphism. The map J respects the decompositions (6.8)(b) and (e) in the obvious way. (This is easy, but it is not a formal consequence of linearity, since infinite direct products are involved. One has to recall how the decompositions on the level of endomorphisms arise from decompositions of modules. These latter decompositions are respected by the module intertwining operator j.) It follows at once that J is injective. More precisely, J carries the ad($\mathfrak{z}(\mathfrak{l})$)-finite part of the domain isomorphically onto the corresponding part of the range. *A fortiori* the restriction of J to the Ad(G)-finite part is an isomorphism. Q.E.D.

It is worth recording the slightly stronger statement that we actually proved: under the same hypotheses as in Theorem 6.12,

$$(B_{\mathfrak{g},Dix})_{\text{ad}(\mathfrak{z}(\mathfrak{l}))\text{-finite}} \simeq (B_{\mathfrak{g},bi})_{\text{ad}(\mathfrak{z}(\mathfrak{l}))\text{-finite}}. \tag{6.13}$$

One can find in section 8 of [19] various sufficient conditions for the map j of Theorem 6.12 to be an isomorphism. Here is one of them.

Proposition 6.14 ([19], Proposition 8.17). *Suppose $\mathfrak{q} = \mathfrak{l} + \mathfrak{u}$ is a parabolic subalgebra of $\mathfrak{g}$, and $W_\mathfrak{l}$ is an $\mathfrak{l}$-module. Define $W_{\mathfrak{g},ind}$ and $W_{\mathfrak{g},pro}$ as in (6.3) and (6.6), and the canonical intertwining operator between them as in (6.11). Assume that*

i) The annihilator in $U([\mathfrak{l},\mathfrak{l}])$ of $W_\mathfrak{l}$ is a weakly unipotent primitive ideal.
ii) The center $\mathfrak{z}(\mathfrak{l})$ of $\mathfrak{l}$ acts by a character λ on $W_\mathfrak{l}$.
iii) If α is a weight of $\mathfrak{z}(\mathfrak{l})$ in $\mathfrak{u}$, then

$$\text{Re} < \alpha, \lambda > \geq 0.$$

Then j is injective.

We can combine this with Theorem 6.12 to get

Corollary 6.15. *Suppose $Q = LU$ is a parabolic subgroup of G, and B_L is a Harish-Chandra bimodule of finite length for L. Assume that*

i) The annihilator in $U([\mathfrak{l},\mathfrak{l}])$ of the left action on B_L is a weakly unipotent primitive ideal.
ii) The center $\mathfrak{z}(\mathfrak{l})$ of $\mathfrak{l}$ acts on the left on B_L by a character λ.
iii) If α is a weight of $\mathfrak{z}(\mathfrak{l})$ in $\mathfrak{u}$, then

$$\text{Re} < \alpha, \lambda > \geq 0.$$

Then the canonical bimodule intertwining operator

$$J : \mathrm{Ind}_{Dix}(Q^{op} \uparrow G)(B_L) \to \mathrm{Ind}_{bi}(Q \uparrow G)(B_L)$$

is an isomorphism.

Corollary 6.16. *Suppose $Q = LU$ is a parabolic subgroup of G, and (A_L, ϕ_L) is a Dixmier algebra for L. Assume that*

i) The kernel of ϕ_L in $U([\mathfrak{l}, \mathfrak{l}])$ is a weakly unipotent primitive ideal.
ii) The center $\mathfrak{z}(\mathfrak{l})$ of $\mathfrak{l}$ acts on A_L by a character λ.
iii) If α is a weight of $\mathfrak{z}(\mathfrak{l})$ in $\mathfrak{u}$, then

$$\mathrm{Re} < \alpha, \lambda > \geq 0.$$

Then the induced Dixmier algebra (A_G, ϕ_G) is isomorphic as a bimodule to $\mathrm{Ind}_{bi}(Q \uparrow G)(A_L)$. In particular, the induced algebra is isomorphic to $\mathrm{Ind}_{alg}(L \uparrow G)(A_L)$ as an algebraic representation under Ad. *The higher cohomology groups of G/Q^{op} with coefficients in the complete symbol sheaf $\mathcal{A}_G$ (Definition 4.15) are zero.*

The point of the cohomology vanishing result is primarily that if it were *not* true, then our definition of induced Dixmier algebras would be flawed: the higher cohomology groups would have to be taken into account somehow. One expects that in interesting cases $\mathcal{A}_G$ has a filtration with associated graded sheaf of the form $\mathcal{R}_G(0)$ (cf. (5.5)). The present vanishing theorem would in such cases be a consequence of Conjecture 5.6; so we may regard it a as a kind of evidence for that conjecture.

Proof. Only the last assertion requires comment. Write Γ for Zuckerman's functor from $(\mathfrak{g}, L)$-modules to G-modules (passage to the G-finite part). Write Γ^i for its derived functors ([18], Chapter 6). If Z is any algebraic representation of Q^{op}, write

$$\mathcal{Z} = G \times_{Q^{op}} Z$$

for the corresponding sheaf on G/Q^{op}. We have already observed (in the proof of Corollary 4.16) that

$$H^0(G/Q^{op}, \mathcal{Z}) \simeq \Gamma(\mathrm{Hom}_{\mathfrak{q}^{op}}(U(\mathfrak{g}), Z)_{L-finite}).$$

It is easy to extend this fact to higher cohomology:

$$H^i(G/Q^{op}, \mathcal{Z}) \simeq \Gamma^i(\mathrm{Hom}_{\mathfrak{q}^{op}}(U(\mathfrak{g}), Z)_{L-finite}).$$

(One reason for this is that the same $\mathfrak{u}^{op}$-cohomology spaces can be used to compute either side.) To compute the higher cohomology with coefficients in $\mathcal{A}_G$, we must therefore apply Γ^i to

$$\mathrm{Hom}_{\mathfrak{q}^{op}}(U(\mathfrak{g}), U(\mathfrak{g}) \otimes_{\mathfrak{q}^{op}} A_{Q^{op}})_{L-finite}.$$

(We are considering only the ad action of $\mathfrak{g}$.) This is just the L-finite part of $A_{\mathfrak{g},Dix}$. By (6.13), it is isomorphic to the L-finite part of $A_{\mathfrak{g},bi}$. By Definition 6.1, this is

$$\mathrm{Hom}_{\mathfrak{q},\mathfrak{q}^{op}}(U(\mathfrak{g}) \otimes U(\mathfrak{g}), A_{\mathfrak{q},\mathfrak{q}^{op}})_{L-finite}.$$

Again we are interested only in the adjoint action of $\mathfrak{g}$. As a module for this action, $A_{\mathfrak{g},bi}$ is isomorphic to $\mathrm{Hom}_{\mathfrak{l}}(U(\mathfrak{g}), A_L)$; here $\mathfrak{l}$ acts on $U(\mathfrak{g})$ on the left and on A_L by ad. (This elementary fact is the basis of (6.2).) Therefore

$$H^i(G/Q^{op}, \mathcal{A}_G) \simeq \Gamma^i(\mathrm{Hom}_{\mathfrak{l}}(U(\mathfrak{g}), A_L))_{L-finite}. \tag{6.17}$$

The module to which Γ^i is applied is a standard injective $(\mathfrak{g}, L)$-module, so the right side is zero for positive i. Q.E.D.

Geometrically, the proof shows that the sheaf $\mathcal{A}_G$ is the pushforward of a sheaf on G/L. Since G/L is affine, the cohomology vanishing follows.

7. The translation principle for Dixmier algebras.

As mentioned in the introduction, there ought to be a theory of modules for induced Dixmier algebras entirely analogous to the Beilinson-Bernstein localization theory. I have not developed such a theory, but this section describes one of its basic results (Corollary 7.14). It is included mostly as evidence that the definition of induced Dixmier algebras is reasonable and interesting. The experts will find no surprises here; but such readers may wish to examine carefully the hypotheses in Theorem 7.9, which are substantially weaker than in some formulations of the translation principle.

We begin by recalling from section 3 of [22] the notion of translation functors for Dixmier algebras. Write $\mathfrak{Z}(\mathfrak{g})$ for the center of $U(\mathfrak{g})$. Suppose α is a character of $\mathfrak{Z}(\mathfrak{g})$ and I_α is the associated maximal ideal. If M is a $\mathfrak{Z}(\mathfrak{g})$-finite (left) $\mathfrak{g}$-module, write

$$_\alpha M = \{\, m \in M \mid \text{ for some } n,\ (I_\alpha)^n \cdot m = 0 \,\}. \tag{7.1)(a}$$

This is an exact functor on the category of $\mathfrak{Z}(\mathfrak{g})$-finite $\mathfrak{g}$-modules, and

$$M = \sum_\alpha {_\alpha M}, \tag{7.1)(b}$$

the sum running over all characters of $\mathfrak{Z}(\mathfrak{g})$. We use analogous notation for right modules.

Suppose (A, ϕ) is a Dixmier algebra. Then A is a $\mathfrak{Z}(\mathfrak{g})$-finite bimodule, so there is a finite direct sum decomposition of A

$$A = \sum_{\alpha,\beta} {}_{\alpha}A_{\beta}. \tag{7.2)(a}$$

This decomposition is preserved by the $U(\mathfrak{g})$-bimodule structure and (as a consequence) by $\mathrm{Ad}(G)$. The multiplication satisfies

$$({}_{\alpha}A_{\beta})({}_{\gamma}A_{\delta}) \subset \begin{cases} {}_{\alpha}A_{\delta} & \text{if } \beta = \gamma. \\ 0 & \text{if } \beta \neq \gamma \end{cases} \tag{7.2)(b}$$

In particular, each ${}_{\alpha}A_{\alpha}$ is a subalgebra. More generally, if M is an A-module,

$$({}_{\alpha}A_{\beta})({}_{\gamma}M) \subset \begin{cases} {}_{\alpha}M & \text{if } \beta = \gamma \\ 0 & \text{if } \beta \neq \gamma. \end{cases} \tag{7.2)(c}$$

As a formal consequence of these facts, we get

Lemma 7.3. *Suppose A is a Dixmier algebra and M is a simple A-module. With the notation (7.1) and (7.2), each non-zero ${}_{\alpha}M$ is an irreducible ${}_{\alpha}A_{\alpha}$-module.*

Conversely, suppose N is an irreducible ${}_{\alpha}A_{\alpha}$-module. Define

$$M' = A \otimes_{{}_{\alpha}A_{\alpha}} N$$

Then ${}_{\alpha}M' \simeq N$. Consequently M' has a unique maximal proper submodule S. The quotient $M = M'/S$ is a simple A-module, and ${}_{\alpha}M \simeq N$.

These constructions establish a bijection between the set of simple A-modules M with ${}_{\alpha}M$ non-zero, and the set of simple ${}_{\alpha}A_{\alpha}$-modules.

Next, suppose (π, F) is a finite-dimensional representation of $\mathfrak{g}$. If M is a $\mathfrak{Z}(\mathfrak{g})$-finite $\mathfrak{g}$-module, then so is $F \otimes M$ (cf. [13]). To make the corresponding construction for Dixmier algebras, fix a Dixmier algebra (A, ϕ_A) and form the algebra

$$B = \mathrm{End}(F) \otimes A. \tag{7.4)(a}$$

We can define an algebra homomorphism ϕ_B of $U(\mathfrak{g})$ into B by

$$\phi_B(X) = \pi(X) \otimes 1 + 1 \otimes \phi_A(X). \tag{7.4)(b}$$

To make a Dixmier algebra, we need an action Ad of G. This will exist whenever the adjoint action of $\mathfrak{g}$ on $\mathrm{End}(F)$ exponentiates to G (as it always does if F is irreducible, for example). In that case we can define

$$\mathrm{Ad}(g)\cdot(T\otimes a) = (\mathrm{Ad}(g)\cdot T)\otimes(\mathrm{Ad}(g)\cdot a). \qquad (7.4)(c)$$

These definitions make B into a Dixmier algebra.

Lemma 7.5. *Suppose A is a Dixmier algebra and F is a finite-dimensional representation of $\mathfrak{g}$. Assume that*

$$\textit{the adjoint action of } \mathfrak{g} \textit{ on } \mathrm{End}(F) \textit{ exponentiates to } G. \qquad (i)$$

Form the Dixmier algebra $B = \mathrm{End}(F)\otimes A$ as in (7.4). Then the map $M \to F\otimes M$ is an equivalence of categories from A-modules to B-modules. The inverse functor is $N \to \mathrm{Hom}_{\mathrm{End}(F)}(F, N)$.

This well-known result is an elementary exercise. (The Dixmier algebra structure is purely decorative; B is really just the ring of $n \times n$ matrices with entries in the ring A.)

Fix now a character α of $\mathfrak{Z}(\mathfrak{g})$ and a finite-dimensional representation F of $\mathfrak{g}$. The *elementary translation functor for modules* attached to α and F is the functor

$$TM = {}_{\alpha}(F\otimes M) \qquad (7.6)(a)$$

on $\mathfrak{Z}(\mathfrak{g})$-finite $\mathfrak{g}$-modules. Assume in addition that F satisfies condition (i) of Lemma 7.5. The *elementary translation functor for Dixmier algebras* attached to α and F is

$$\mathcal{T}A = {}_{\alpha}(\mathrm{End}(F)\otimes A)_{\alpha}. \qquad (7.6)(b)$$

Proposition 7.7 (Translation Principle for Dixmier algebras). *In the setting of (7.6), suppose A is a Dixmier algebra. Then the translation functor T takes modules for A to modules for $\mathcal{T}A$. This functor has the following additional properties.*

a) T is exact.

b) If M is an irreducible A-module, then TM is irreducible or zero as a $\mathcal{T}A$-module.

c) If N is an irreducible $\mathcal{T}A$-module, then there is a unique irreducible A-module M such that $N = TM$.

This proposition is an immediate consequence of Lemmas 7.3 and 7.5.

The expert reader may be wondering what has been hidden, since results of this form about the translation principle usually require more hypotheses and more proof. The difficulty arises if we want to speak about modules for $U(\mathfrak{g})$ instead of for some Dixmier algebra. If $A = U(\mathfrak{g})/I$, then an A-module is just a $U(\mathfrak{g})$-module annihilated by I. There will always be a natural inclusion

$$\phi : U(\mathfrak{g})/J \to \mathcal{T}A. \tag{7.8}$$

If ϕ is surjective, then a $\mathcal{T}A$-module is just a $U(\mathfrak{g})$-module annihilated by J, and Proposition 7.7 becomes a result about $\mathfrak{g}$-modules. It is exactly such surjectivity results that require additional hypotheses and more difficult proofs.

Since we have cast off the shackles of $\mathfrak{g}$-modules, we take a slightly different view of what is required to make Proposition 7.7 interesting. We will begin with a known Dixmier algebra A, and try to understand the translated algebra $\mathcal{T}A$. Here is such a result.

Theorem 7.9. *Suppose $Q = LU$ is a parabolic subgroup of G, and (A_u, ϕ_u) is a Dixmier algebra for $L/Z(L)_0$. For each character ξ of $\mathfrak{z}(\mathfrak{l})$ define a Dixmier algebra $(A_L(\xi), \phi_L(\xi))$ by (5.3)(d). Define*

$$(A_G(\xi), \phi_G(\xi)) = \mathrm{Ind}_{Dix}(Q \uparrow G)(A_L(\xi), \phi_L(\xi)).$$

Fix a one-dimensional character μ of $\mathfrak{l}$, and assume that μ occurs in the restriction to $\mathfrak{l}$ of a finite-dimensional representation of $\mathfrak{g}$. Let F be the unique irreducible finite-dimensional representation of $\mathfrak{g}$ containing μ as an extremal weight. (This means that F has a $\mathfrak{q}'$-invariant line of weight μ, for some parabolic subalgebra $\mathfrak{q}'$ having Levi factor $\mathfrak{l}$.) We sometimes identify μ with its restriction to $\mathfrak{z}(\mathfrak{l})$.

Fix a character λ of $\mathfrak{z}(\mathfrak{l})$. Assume that

i) $\ker(\phi_u)$ *is a weakly unipotent primitive ideal in $U([\mathfrak{l}, \mathfrak{l}])$.*
ii) If β is any weight of $\mathfrak{z}(\mathfrak{l})$ in $\mathfrak{g}/\mathfrak{l}$ and $\langle \beta, \mu \rangle > 0$, then

$$\mathrm{Re}\,\langle \beta, \lambda \rangle \geq 0.$$

In particular, (i) implies that $A_G(\lambda)$ has an infinitesimal character, which we denote by α. Let $\mathcal{T}$ be the elementary translation functor associated to α and F^*. Then

$$\mathcal{T}(A_G(\lambda + \mu)) = A_G(\lambda).$$

Proof. We argue as in [19], section 8. Fix a faithful module V_u for A_u. For every character ξ of $\mathfrak{z}(\mathfrak{l})$, V_u becomes a faithful module $V_{\mathfrak{l}}(\xi)$ for the Dixmier algebra $A_L(\xi)$. (The point is that the underlying algebra of $A_L(\xi)$ is just A_u; only the map $\phi_L(\xi)$ is changing.) Just as in (4.3) we can construct $V_{\mathfrak{q}}(\xi)$ and

$$V_{\mathfrak{g}}(\xi) = U(\mathfrak{g}) \otimes_{\mathfrak{q}} V_{\mathfrak{q}}(\xi). \tag{7.10}$$

By Lemma 7.5, $F^* \otimes V_{\mathfrak{g}}(\lambda+\mu)$ is a faithful module for $\mathrm{End}(F^*) \otimes A_G(\lambda+\mu)$. Consequently

$$T(V_{\mathfrak{g}}(\lambda+\mu)) = {}_{\alpha}[F^* \otimes V_{\mathfrak{g}}(\lambda+\mu)] \tag{7.11)(a}$$

is a faithful module for $\mathcal{T}(A_G(\lambda+\mu))$. We will show that

$$T(V_{\mathfrak{g}}(\lambda+\mu)) = V_{\mathfrak{g}}(\lambda). \tag{7.11)(b}$$

This will show that $\mathcal{T}(A_G(\lambda+\mu))$ and $A_G(\lambda)$ are both represented as endomorphism algebras of $V_{\mathfrak{g}}(\lambda)$. It is then very easy to check from the definitions that they are exactly the same endomorphisms; we leave this to the reader.

It therefore remains to check (7.11)(b). By (7.11)(a) and the algebraic version of Mackey's tensor product theorem, the right side of (7.11)(b) is

$$_{\alpha}\left[U(\mathfrak{g}) \otimes_{\mathfrak{q}} ((F^* \mid_{\mathfrak{q}}) \otimes V_{\mathfrak{q}}(\lambda+\mu))\right]. \tag{7.11)(c}$$

Now F^* has a $\mathfrak{q}$-stable filtration whose subquotients are irreducible representations of $\mathfrak{l}$. We get a filtration of (7.11)(c) whose subquotients are

$$_{\alpha}\left[U(\mathfrak{g}) \otimes_{\mathfrak{q}} (E^* \otimes V_{\mathfrak{q}}(\lambda+\mu))\right], \tag{7.11)(d}$$

where E can be any irreducible constituent of $F\mid_{\mathfrak{l}}$.

If E is the μ weight space of F, then (7.11)(d) is $V_{\mathfrak{g}}(\lambda)$. We must therefore show that the other terms are all zero. Explicitly, this means the following. Suppose E is a representation of $\mathfrak{l}$ occurring in F, other than the weight μ. Then we must show that the infinitesimal character α does not occur in

$$U(\mathfrak{g}) \otimes_{\mathfrak{q}} (E^* \otimes V_{\mathfrak{q}}(\lambda+\mu)). \tag{7.12)(a}$$

To prove this, fix a Cartan subalgebra $\mathfrak{t}$ of $[\mathfrak{l},\mathfrak{l}]$. Then $\mathfrak{h} = \mathfrak{t} + \mathfrak{z}(\mathfrak{l})$ is a Cartan subalgebra of $\mathfrak{g}$. Fix a weight $\alpha_0 \in \mathfrak{t}^*$ corresponding to the infinitesimal character of A_u in the Harish-Chandra correspondence. Then α corresponds to

$$(\alpha_0, \lambda) \in \mathfrak{t}^* + \mathfrak{z}(\mathfrak{l})^* = \mathfrak{h}^*. \tag{7.12)(b}$$

Let μ_1 be the weight of $\mathfrak{z}(\mathfrak{l})$ on E, and α_1 an infinitesimal character for $[\mathfrak{l},\mathfrak{l}]$ in $E^* \otimes V_u$. Then a typical infinitesimal character in (7.12)(a) is

$$(\alpha_1, \lambda+\mu-\mu_1). \tag{7.12)(c}$$

What we are trying to show is that this cannot be equal to α; that is, that the weights (7.12)(b) and (7.12)(c) are not conjugate under the Weyl group. To do that, it is obviously sufficient to show that

$$\mathrm{Re}\,\langle(\alpha_0,\lambda),(\alpha_0,\lambda)\rangle < \mathrm{Re}\,\langle(\alpha_1,\lambda+\mu-\mu_1),(\alpha_1,\lambda+\mu-\mu_1)\rangle. \quad (7.13)(a)$$

To prove (7.13)(a), we first use the hypothesis (i) of the Theorem. This implies that the infinitesimal characters occurring in $E^* \otimes V_u$ are all longer than α_0:

$$\langle\alpha_0,\alpha_0\rangle \leq \langle\alpha_1,\alpha_1\rangle. \quad (7.13)(b)$$

Next, any weight of $\mathfrak{z}(\mathfrak{l})$ in F must be of the form

$$\mu_1 = \mu - \sum n_\beta \beta.$$

Here the sum is over weights β of $\mathfrak{z}(\mathfrak{l})$ on $\mathfrak{g}/\mathfrak{l}$ having positive inner product with μ, and n_β is a non-negative integer. If μ_1 is different from μ, then the sum is non-empty. By hypothesis (ii),

$$\mathrm{Re}\,\langle\sum n_\beta\beta,\lambda\rangle \geq 0.$$

Consequently

$$\begin{aligned}\mathrm{Re}\,\langle\lambda+\mu-\mu_1,\lambda+\mu-\mu_1\rangle &= \mathrm{Re}\,\langle\lambda+\sum n_\beta\beta,\lambda+\sum n_\beta\beta\rangle \\ &= \mathrm{Re}\,\langle\lambda,\lambda\rangle + 2\cdot\mathrm{Re}\,\langle\sum n_\beta\beta,\lambda\rangle \\ &\quad + \langle\sum n_\beta\beta,\sum n_\beta\beta\rangle \\ &> \mathrm{Re}\,\langle\lambda,\lambda\rangle.\end{aligned} \quad (7.13)(c)$$

Adding (7.13)(b) and (7.13)(c) gives (7.13)(a) and completes the proof. Q.E.D.

It is very easy to refine the argument so that the positivity hypothesis (ii) is needed only for those weights β that are restrictions of roots integral on the infinitesimal character.

Corollary 7.14. *Under the hypotheses of Theorem 7.9, every irreducible module N for $A_G(\lambda)$ is of the form TM, with M a unique irreducible module for $A_G(\lambda+\mu)$. Conversely, if M is an irreducible $A_G(\lambda+\mu)$-module, then TM is an irreducible $A_G(\lambda)$-module or zero.*

The point of this corollary is that it allows one to study problems of irreducibility at "very regular" parameters, where they are typically much easier.

Appendix. Induced bundles.

We assemble here some basic definitions used throughout the paper. Suppose throughout this appendix that G is an (affine) algebraic group, and H is a closed (and therefore affine algebraic) subgroup. Recall that the homogeneous space G/H is a quasiprojective algebraic variety. As a point set, G/H is just the coset space. Its topology is the quotient topology from G: a subset U of G/H is open if and only if its preimage V in G is open. A regular function on such an open set U is by definition a (right) H-invariant regular function on V. The main point in the construction of G/H is that for small enough V there are many such functions.

Suppose now that Z_H is any algebraic variety on which H acts. The *induced bundle* $G \times_H Z_H$ is a bundle over G/H whose fiber at the identity coset eH is Z_H:

$$G \times_H Z_H \to G/H \qquad (A.1)(a)$$

This bundle is constructed from the product $G \times Z_H$ in the same way that G/H is constructed from G. That is, we define an equivalence relation $\sim$ on (closed points of) $G \times Z_H$ by

$$(x, z) \sim (xh^{-1}, h \cdot z) \qquad (x \in G, z \in Z_H, h \in H). \qquad (A.1)(b)$$

As a point set, $G \times_H Z_H$ is the set of equivalence classes for this relation. That is, it is the set of orbits of an action of H on $G \times Z_H$ (called the *right action*) defined by

$$h \cdot_R (x, z) = (xh^{-1}, h \cdot z) \qquad (A.1)(c)$$

The subscript R is included to distinguish this action from the (left) action of G defined by

$$g \cdot (x, z) = (gx, z) \qquad (x \in G, z \in Z_H, g \in G). \qquad (A.1)(d)$$

This action of G commutes with the right action of H, so it is inherited by $G \times_H Z_H$.

A subset U of $G \times_H Z_H$ is defined to be open if and only if its preimage V in $G \times Z_H$ is open. The regular functions on U are defined to be the regular functions on V that are invariant under the right action of H:

$$\mathcal{O}(U) = \{ f \in \mathcal{O}(V) \mid f(h \cdot_R v) = f(v) \quad (v \in V, h \in H) \} \qquad (A.1)(e)$$

This makes sense because the preimage in $G \times Z_H$ of any subset of $G \times_H Z_H$ is by definition a union of orbits of the right action of H. That it makes $G \times_H Z_H$ into an algebraic variety with a G action is proved just as for the case of G/H itself.

Here are some properties of the induced bundle construction; all follow fairly easily from the definitions.

Proposition A.2. *Suppose G is an algebraic group, H is a closed subgroup, and Z_H is an algebraic variety on which H acts.*

a) The induced bundle $G \times_H Z_H$ is an algebraic fiber bundle over G/H. The fiber over eH is naturally identified with Z_H, and the isotropy action of H on the fiber is the original action of H on Z_H.

b) Suppose Y is an algebraic variety on which G acts, and $f : Z_H \to Y$ is a morphism respecting the actions of H. Then there is a natural morphism $F = G \times_H f$ from $G \times_H Z_H$ to Y, respecting the actions of G. On the level of equivalence classes,

$$F(x, z) = x \cdot f(z).$$

c) In the setting of (b), suppose in addition that f is proper, and that G/H is a projective variety. Then the map F is proper.

d) Suppose $Z_H \simeq H/K$ is a homogeneous space for H. Then

$$G \times_H Z_H \simeq G/K.$$

In the setting of (b) in the proposition, one can easily describe the fibers (over closed points) of the map F. Of course the image of F is $G \cdot f(Z_H)$, so the fibers outside that set are empty. By G-invariance, it suffices to understand the fiber over a point $y \in f(Z_H)$. Write $K \subset G$ for the isotropy group of the G-action at y. Then

$$F^{-1}(y) \simeq K \times_{H \cap K} f^{-1}(y). \qquad (A.3)$$

In particular, F is injective if and only if the following two conditions are satisfied: f is injective; and the stabilizer of every point in $f(Z_H)$ is contained in H.

Suppose now that $\mathcal{M}_H$ is a quasicoherent sheaf on Z_H with an action of H (compatible with the action on Z_H). We can define the *induced sheaf* $\mathcal{M}_G = G \times_H \mathcal{M}_H$ on $Z_G = G \times_H Z_H$ as follows. Consider first the quasicoherent sheaf

$$\mathcal{N} = \mathcal{O}_G \otimes \mathcal{M}_H \qquad (A.4)(a)$$

on $G \times Z_H$. (This can be thought of informally as "functions on G with values in $\mathcal{M}_H$;" that description is accurate on any open set that is a product of an open set in G with one in Z_H.) The sheaf $\mathcal{N}$ carries a right action of H compatible with the right action of H on $G \times Z_H$, and a commuting (left) action of G. With notation as in (A.1)(e), we define

$$\mathcal{M}_G(U) = \{ m \in \mathcal{N}(V) \mid h \cdot_R m = m \quad (h \in H). \qquad (A.4)(b)$$

The space M_G of global sections of $\mathcal{M}_G$ has a simple description. Write M_H for the space of global sections of $\mathcal{M}_H$. This is a vector space (generally infinite-dimensional) carrying an algebraic action of H. We may speak of the algebraic functions on G with values in M_H; such a function is required to take values in a finite-dimensional subspace, and to be algebraic (in the obvious sense) as a map to that subspace. That is, it should belong to $(\mathcal{O}_G(G)) \otimes M_H$. Now it is clear from the definitions that

$$M_G = \{\, f : G \to M_H \mid f \text{ is algebraic, and } f(xh) = h^{-1} f(x) \,\}. \quad (A.4)(c)$$

Proposition A.5. *Suppose G is an algebraic group, H is a closed subgroup, and Z_H is an algebraic variety with G-action. Write $Z_G = G \times_H Z_H$. Then the category of quasicoherent sheaves on Z_H with H-action is equivalent to the category of quasicoherent sheaves on Z_G with G-action, by the induction construction $\mathcal{M}_H \mapsto G \times_H \mathcal{M}_H$ of (A.4). This equivalence identifies the subcategories of coherent sheaves.*

We omit the proof. It is worthwhile to describe the inverse functor, however. Let $\mathcal{I}_H$ be the sheaf of ideals defining the subvariety Z_H of $Z_G = G \times_H Z_H$. Then $\mathcal{O}_{Z_H}$ may be identified as a sheaf of $\mathcal{O}_{Z_G}$-modules with $\mathcal{O}_{Z_G}/\mathcal{I}_H$. If $\mathcal{M}_G$ is any quasicoherent sheaf on Z_G, then the "geometric fiber"

$$\mathcal{M}_H = \mathcal{O}_{Z_H} \otimes_{\mathcal{O}_{Z_G}} \mathcal{M}_G \simeq \mathcal{M}_G/\mathcal{I}_H\mathcal{M}_G \quad (A.6)$$

is a quasicoherent sheaf on Z_H. If $\mathcal{M}_G$ carries an action of G, then this fiber inherits an action of H (the largest subgroup of G preserving the ideal $\mathcal{I}_H$). As the notation indicates, the geometric fiber provides a natural inverse for the induced sheaf construction of (A.4). Again we omit the proof.

We conclude by considering the relationship between the induced sheaf construction of (A.4) and the notion of induced representation for algebraic groups. Recall first of all that an *algebraic representation* of G is a pair (π, V) with V a vector space and

$$\pi : G \to GL(V) \quad (A.7)(a)$$

a homomorphism. We require in addition that π be algebraic, in the following sense. For every $v \in V$, there should be a finite-dimensional subspace $E \subset V$, containing v, with the property that

$$\pi(G)E \subset E, \quad (A.7)(b)$$

and the resulting homomorphism

$$\pi_E : G \to GL(E) \quad (A.7)(c)$$

is a morphism of algebraic groups.

Suppose now that (π_H, V_H) is an algebraic representation of H. The *algebraically induced representation of* G is the algebraic representation

$$\mathrm{Ind}_{alg}(H \uparrow G)(\pi_H, V_H) = (\pi_G, V_G) \tag{A.8)(a}$$

defined by

$$\begin{aligned} V_G = \{\, f : G \to V_H \mid f \text{is algebraic, and } f(xh) \\ = \pi_H(h^{-1}) f(x)\, (x \in G, h \in H) \,\} \end{aligned} \tag{A.8)(b}$$

$$(\pi_G(g) f)(x) = f(g^{-1} x). \tag{A.8)(c}$$

(We will drop the maps π from the notation when no confusion can result.) The analogy with (A.4) is clear, and in fact the connection is very close.

Proposition A.9. *Suppose G is an algebraic group, H is a closed subgroup, and (π_H, V_H) is an algebraic representation of H.*

a) The induced representation $V_G = \mathrm{Ind}_{alg}(H \uparrow G)(V_H)$ is an algebraic representation of G.

b) Identify V_H with a quasicoherent sheaf $\mathcal{M}_H$ with an H-action on a point. Then V_G may be identified with the space of global sections of $G \times_H \mathcal{M}_H$ (cf. (A.4)).

c) Suppose that V_H is the space of global sections of a quasicoherent sheaf $\mathcal{M}_H$ on some Z_H as in (A.4). Then V_G may be identified with the space of global sections of the induced sheaf $G \times_H \mathcal{M}_H$ on $G \times_H Z_H$.

d) (Frobenius reciprocity.) Suppose W is any algebraic representation of G. Then there is a natural isomorphism

$$\mathrm{Hom}_G(W, V_G) \simeq \mathrm{Hom}_H(W\,|_H, V_H).$$

e) Suppose $\dim V_H < \infty$, so that $G \times_H V_H$ (defined as in (A.1)) is a vector bundle over G/H. Then V_G may be identified with the space of sections of this vector bundle.

This is very well-known, and we omit the straightforward proof.

REFERENCES

[1] A. Beilinson and J. Bernstein, "Localisation de $\mathfrak{g}$-modules," C. R. Acad. Sci. Paris **292** (1981), 15–18.

[2] W. Borho, "Definition einer Dixmier-Abbildung für $\mathfrak{sl}(n, \mathbb{C})$," Inventiones math. **40**(1977), 143–169.

[3] W. Borho, "Über Schichten halbeinfacher Lie-Algebren," Invent. math. **65** (1981), 283–317.

[4] W. Borho and H. Kraft, "Über Bahnen und deren Deformationen bei linearen Aktionen reduktiver Gruppen," Comment. Math. Helvetici **54** (1979), 61–104.

[5] N. Conze-Berline and M. Duflo, "Sur les représentations induites des groupes semi-simples complexes," Compositio math. **34** (1977), 307–336.

[6] J. Dixmier, *Algèbres Enveloppantes.* Gauthier-Villars, Paris-Brussels-Montreal 1974.

[7] J. Dixmier, "Ideaux primitifs dans les algèbres enveloppantes," J. Algebra **48** (1978) 96–112.

[8] M. Duflo, "Sur les idéaux induits dans les algèbres enveloppantes," Inventiones math. **67** (1982), 385–393.

[9] R. Hartshorne, *Algebraic Geometry.* Springer-Verlag, New York, Heidelberg, Berlin, 1977.

[10] J. C. Jantzen, *Moduln mit einem Höchsten Gewicht*, Lecture Notes in Mathematics **750**. Springer-Verlag, Berlin-Heidelberg-New York, 1979.

[11] A. Joseph, "Dixmier's problem for Verma and principal series modules," J. London Math. Soc.(2) **20** (1979), 193–204.

[12] A. Joseph, "The primitive spectrum of an enveloping algebra," 13–53 in *Orbites Unipotentes et Représentations III. Orbites et Faisceaux pervers*, Astérisque **173–174** (1989).

[13] B. Kostant, "On the tensor product of a finite and an infinite dimensional representation," J. Func. Anal. **20** (1975), 257–285.

[14] G. Lusztig and N. Spaltenstein, "Induced unipotent classes," J. London Math. Soc.(2) **19** (1979), 41–52.

[15] W. McGovern, "Unipotent representations and Dixmier algebras," Compositio math. **69** (1989), 241–276.

[16] C. Moeglin, "Modèles de Whittaker et idéaux primitifs complètement premiers dans les algèbres de lie semi-simples complexes II," Math. Scand. **63** (1988), 5–35.

[17] R. Rentschler, "Comportement de l'application de Dixmier par rapport à l'anti-automorphism principal pour les algèbres de Lie résolubles," C. R. Acad. Sci. Paris **282** (1976), 555–557.

[18] D. Vogan, *Representations of Real Reductive Lie Groups.* Birkhauser, Boston-Basel-Stuttgart, 1981.

[19] D. Vogan, "Unitarizability of certain series of representations," Ann. of Math. **120** (1984), 141–187.

[20] D. Vogan, "The orbit method and primitive ideals for semisimple Lie algebras," in *Lie Algebras and Related Topics*, CMS Conference Pro-

ceedings, volume 5, D. Britten, F. Lemire, and R. Moody, eds. American Mathematical Society for CMS, Providence, Rhode Island, 1986.

[21] D. Vogan, "Noncommutative algebras and unitary representations," in *The Mathematical Heritage of Hermann Weyl*, R. O. Wells, Jr., ed. Proceedings of Symposia in Pure Mathematics, volume 48. American Mathematical Society, Providence, Rhode Island, 1988.

[22] D. Vogan, "Irreducibility of discrete series representations for semisimple symmetric spaces," 191–221 in *Representations of Lie groups, Kyoto, Hiroshima, 1986*, K. Okamoto and T. Oshima, editors. Advanced Studies in Pure Mathematics, volume 14. Kinokuniya Company, Ltd., Tokyo, 1988.

[23] D. Vogan, "Associated varieties and unipotent representations," preprint.

Received October 12, 1989

David A. Vogan, Jr.
Department of Mathematics
Massachusetts Institute of Technology
Cambridge, MA 02139

Dixmier Algebras and the Orbit Method

WILLIAM M. McGOVERN

Dedicated to Professor Jacques Dixmier on his 65th birthday

Abstract

Let G be a complex Lie group with Lie algebra $\mathfrak{g}$, enveloping algebra $U(\mathfrak{g})$, and symmetric algebra $S(\mathfrak{g})$. The Nullstellensatz provides a tight link between ideal theory in $S(\mathfrak{g})$ and geometry in $\mathfrak{g}^*$. One naturally wants to generalize it to the slightly non-commutative setting of $U(\mathfrak{g})$. The fundamental work of Dixmier and others does this for solvable $\mathfrak{g}$. Here we give an historical survey of the various attempts to generalize Dixmier's results to the semisimple case, showing how these attempts have led to the notion of what we call a Dixmier algebra. We give an exposition of some recent work on Dixmier algebras. We conclude by giving a number of interesting examples of Dixmier algebras, some of them geometrically nice while others are not.

1. Introduction

Let $\mathfrak{g}$ be a complex Lie algebra with enveloping algebra $U(\mathfrak{g})$ and symmetric algebra $S(\mathfrak{g})$. Let G be a simply connected group with Lie $G = \mathfrak{g}$. The Nullstellensatz provides a tight link between ideal theory in $S(\mathfrak{g})$ and geometry in $\mathfrak{g}^*$; more precisely, it yields a bijection between prime quotients of $S(\mathfrak{g})$ and irreducible affine varieties in $\mathfrak{g}^*$. A basic goal in any reasonable theory of non-commutative algebraic geometry is to extend this bijection to a (slightly) non-commutative setting; a natural one to consider is that of $U(\mathfrak{g})$. For solvable $\mathfrak{g}$, Dixmier extended the bijection to $U(\mathfrak{g})$ as follows: he showed that primitive ideals of $U(\mathfrak{g})$ correspond 1 - 1 to G-orbits in $\mathfrak{g}^*$. We call this the orbit correspondence. It reduces the classification of the primitive spectrum of $U(\mathfrak{g})$ to a purely geometric question which in many cases admits a complete solution. In view of the power and generality of the orbit correspondence in the solvable case, it was natural to believe that it should extend to the semisimple case without too much difficulty. Twenty years of effort have unfortunately shown that there are fundamental obstacles to its implementation for semisimple $\mathfrak{g}$.

In this paper, we will describe these obstacles in some detail and show how the attempt to overcome them has led to the notion of a Dixmier algebra, roughly a finite algebra extension of a primitive quotient of $U(\mathfrak{g})$. We will then survey the work done so far on Dixmier algebras, briefly reviewing the methods currently available for constructing them. We will conclude the paper by giving a number of interesting examples of Dixmier algebras. Most of these algebras are completely prime and have nice geometric properties, but a few of them are not. The latter algebras are, however, closely related to certain algebras which we call unipotent, so we call them mock unipotent. Mock unipotent Dixmier algebras help to illustrate a basic principle of any orbit correspondence: only completely prime Dixmier algebras can hope to have a nice geometric theory.

2. Dixmier algebras

Henceforth, assume that $\mathfrak{g}$ is semisimple. One of the most important differences between primitive ideals of $U(\mathfrak{g})$ and primitive ideals in the solvable case, is that the former need not be completely prime. As we stated in the introduction, a basic tenet of any orbit correspondence is that it should involve only completely prime ideals of $U(\mathfrak{g})$. For example, $U(\mathfrak{g})$ has irreducible finite-dimensional modules of arbitrarily large dimension. The image of $U(\mathfrak{g})$ in the endomorphism ring E of any such module V is always E itself, whence it has zero divisors if V is nontrivial. Since any such image of $U(\mathfrak{g})$ is finite-dimensional, it must correspond to a 0-dimensional subvariety of $\mathfrak{g}^*$ under any reasonable implementation of the orbit correspondence. But the only G-stable such variety is the single point 0. We can attach this point to the augmentation ideal of $U(\mathfrak{g})$ but then it is clear that we cannot attach any G-orbit to the other primitive ideals of $U(\mathfrak{g})$ of finite codimension. We can also see the necessity of restricting to completely prime primitive quotients of $U(\mathfrak{g})$ in a less expected way. Take $\mathfrak{g} = \mathfrak{sl}(n)$, the algebra of traceless complex matrices, and consider primitive ideals of $U(\mathfrak{g})$ of a fixed regular infinitesimal character. It is well known that these are parametrized by the set of all standard Young tableaux with n boxes [15]. But there is no reasonable injective map from standard Young tableaux to coadjoint orbits; the best we can do is to attach to a Young *diagram* the nilpotent orbit whose Jordan form is given by the corresponding partition. Thus we need a way to cut down from the set of Young tableaux to the set of Young diagrams. Moeglin's beautiful characterization [30] of the completely prime primitive ideals as induced from codimension-one ideals in a parabolic subalgebra in effect does this. More precisely, her main result implies that the Dixmier map [6] yields a bijection between the set $\mathfrak{g}^*/G$ of coadjoint orbits and $\mathrm{Prim}^1(U(\mathfrak{g}))$, the sets of completely prime primitive ideals of $U(\mathfrak{g})$.

Resuming to general semisimple $\mathfrak{g}$, one is thus readily led to make

Conjecture 2.1. *There is a natural bijection between* $Prim^1(U(\mathfrak{g}))$ *and* $\mathfrak{g}^*/G$. *This bijection sends an ideal of Gelfand-Kirillov dimension d to an orbit of the same dimension.*

(For generalities on Gelfand-Kirillov dimension, see [13], Chapter 8.)

The first thing to say about this conjecture is that it is false in general for any reasonable interpretation of the word "natural". It is true and easily verified for the smallest example $\mathfrak{g} = \mathfrak{sl}(2)$, by a direct calculation. For $\mathfrak{g} = \mathfrak{sl}(n)$ we have already mentioned the Dixmier map from $\mathfrak{g}^*/G$ to $\mathrm{Prim}^1 U(\mathfrak{g}))$ which is a bijection by [8] and [31]. The smallest $\mathfrak{g}$ not isomorphic to a product of $\mathfrak{sl}(n)$'s is $\mathfrak{so}(5)$, the algebra of 5×5 complex skew-symmetric matrices. In this case, Conjecture 2.1 was first studied by Borho [5]. He parametrized the sets $\mathrm{Prim}^1(U(\mathfrak{g}))$ and $\mathfrak{g}^*/G$ explicitly. It is clear from his parametrization that there is a bijection between these sets, but it is not clear how to implement this bijection canonically. Any straightforward attempt to generalize the Dixmier map leads to a correspondence that is not well defined. Thus Conjecture 2.1 already stood on shaky ground shortly after it was first formulated and studied. It was given a death blow by Joseph, who studied $\mathrm{Prim}^1(U(\mathfrak{g}))$ for $\mathfrak{g}$ of type G_2. He showed that there are two elements $U(\mathfrak{g})/J_1$, $U(\mathfrak{g})/J_2$ of this set of Gelfand-Kirillov dimension 8, but only one orbit in $\mathfrak{g}^*/G$ of this dimension [17]. Vogan was the first to produce a geometric interpretation of these first failures of the orbit correspondence. More specifically, he showed that two geometric phenomena in $\mathfrak{g}^*$ have algebraic consequences for $U(\mathfrak{g})$: nilpotent orbits can admit nontrivial covers, and they can have nonnormal closures. The first of these phenomena accounts for the trouble in type B_2, while the second causes the problems in type G_2. For more details, see [34]; the bad orbits in question are the 6-dimensional subregular nilpotent orbit in $\mathfrak{so}(5)$ and the 8-dimensional nilpotent orbit in G_2.

It follows that objects slightly larger than orbits can affect ideal theory in $U(\mathfrak{g})$. To introduce this in the orbit correspondence, one may attach to each orbit not just the ring of regular functions on its closure but also certain extensions of this ring. Thus we are led to attach to each primitive ideal I certain extensions of the primitive quotient $U(\mathfrak{g})/I$. More generally, we are led to the notion of a Dixmier algebra and to Vogan's reformulation of Conjecture 2.1.

Definition 2.2. (cf. [24]) A *Dixmier algebra* A over (G and) $U(\mathfrak{g})$ is an algebra of finite type over a quotient U of $U(\mathfrak{g})$ equipped with an algebraic G-action extending the adjoint action of G on U and realizing A as an admissible G-module.

Definition 2.3. A ramified orbit cover X for G is an irreducible affine variety equipped with an algebraic G-action and a G-equivariant finite morphism $X \rightarrow \mathfrak{g}^*$ whose image is the closure of a single G-orbit.

Conjecture 2.4. [34] *There is a natural bijection between completely prime Dixmier algebras and ramified orbit covers.*

This formulation disposes of the difficulties mentioned above in types B_2 and G_2. We will see later that there is a partial implementation of the orbit method correspondence using rings of twisted differential operators. Vogan has generalized this implementation using the theory of endomorphism rings of induced modules [37]. If we supplement Vogan's generalized implementation by attaching the minimal 4-dimensional nilpotent orbit in type B_2 to $U(\mathfrak{g})$ modulo the Joseph ideal, then we get a bijection between $\mathfrak{g}^*/G$ and completely prime quotients of $U(\mathfrak{g})$, which may be fairly easily extended to a bijection from the set of orbit covers to a certain set of completely prime Dixmier algebras. (This extension uses Vogan's theory of induced Dixmier algebras in [37]. It does not unfortunately prove Conjecture 2.4 for $\mathfrak{g} = \mathfrak{so}(5)$, because neither *ramified* orbit covers nor completely prime Dixmier algebras have been classified in this case.) In type G_2, one can attach the closure $\overline{\mathcal{O}}_8$ of the 8-dimensional nilpotent orbit to one of the $U(\mathfrak{g})/I_i$, say $U(\mathfrak{g})/I_1$, and the normalization $N(\overline{\mathcal{O}}_8)$ of this closure to $U(\mathfrak{g})/I_2$. (By an old result of Borho and Kraft [9], the normalization $N(\overline{\mathcal{O}})$ of any orbit closure $\overline{\mathcal{O}}$ may be identified with the orbit $\mathcal{O}$.)

Unfortunately, Conjecture 2.4 is still false, due to an algebraic phenomenon with geometric consequences: it is possible for distinct Dixmier algebras to be isomorphic as $U(\mathfrak{g})$ -bimodules. Ramified orbit covers are not subtle enough to distinguish two such algebras. The simplest example has $\mathfrak{g} = \mathfrak{sl}(2) \times \mathfrak{sl}(2)$. The second Weyl algebra A_2, generated by two commuting copies of the first Weyl algebra A_1, is a Dixmier algebra over $U(\mathfrak{g})$. This can be seen by embedding $\mathfrak{sl}(2)$ into A_1 in the standard way as operators of total degree 2. Now form the algebra A_2' (= "twisted A_2"), generated by two copies of A_1 with anticommuting generators. It is easy to see that A_2', like A_2, is a completely prime Dixmier algebra. We may attach the universal (fourfold) cover of the principal nilpotent orbit in $\mathfrak{g}^*$ to A_2, but then there is no ramified orbit cover which can reasonably be attached to A_2'. This example first appears in [24]. There we study Dixmier algebras A with the kernel of the map $U(\mathfrak{g}) \rightarrow A$ a special unipotent primitive ideal (in the sense of [3]). It is shown that under certain circumstances completely prime Dixmier algebras having a fixed bimodule structure are parametrized by $H^2(F, C^\times)$, the Schur multiplier of a suitable finite group F. For the bimodule structure of A_2 and A_2', we have $F = Z_2 \times Z_2$, so that $H^2(F, C^\times) \cong Z_2$.

It is fairly clear why A_2' causes problems in the orbit correspondence: it is a sort of super version of A_2 and so ought to correspond on the geometric side to a super-commutative, rather than a commutative, algebra. More precisely, A_2' should correspond to the polynomial ring $C[x_1, x_2, y_1, y_2]$ in four variables x_1, x_2, y_1, y_2 such that the x's commute with each other and anticommute with the y's, and the y's commute with each other. In attempting to repair Conjecture 2.4, Vogan was led by this example and many others to the following definition.

Definition 2.5. [37] An *orbit datum* (R, ψ) for G is a pair satisfying the following conditions.

(a) R is a $\mathbb{C}$-algebra on which G acts algebraically by automorphisms.
(b) ψ is a homomorphism from $S(\mathfrak{g})$ into the center of R which respects both G-actions.
(c) R is finitely generated as a module over $S(\mathfrak{g})$.
(d) R is G-admissible.

Note that any orbit cover gives rise to a unique orbit datum via the ring $R(X)$ of regular functions on X and the comorphism $S(\mathfrak{g}) \rightarrow R(X)$. We identify the former with the latter. Vogan then reformulated Conjecture 2.4 as

Conjecture 2.6. *There is a natural bijection between completely prime Dixmier algebras and completely prime orbit data (orbit data (R, ψ) with R completely prime).*

The known examples (including the ones in [24]) confirm this latest version. The most interesting orbit data are (identified with) covers of nilpotent coadjoint orbits and are called unipotent (in [36], they are called unipotent Poisson varieties). The Dixmier algebras attached to them by Conjecture 2.6 are also called unipotent. In [27] and [28] we call the process of attaching a Dixmier algebra A to a unipotent orbit datum X via Conjecture 2.6 *quantization* and we say that A quantizes X. Given X, we give criteria that the A quantizing it should satisfy (which should specify A uniquely). The most important of these state that A and $R(X)$, the regular functions on X, should both admit G-stable good filtrations making gr $A \cong$ gr $R(X)$ as Poisson algebras, $S(\mathfrak{g})$-modules, and G-modules. For X corresponding to a trivial or universal cover, we construct an A which conjecturally satisfies these criteria. Using the techniques of [37], we make substantial progress towards proving this conjecture.

There is only one algebra $\mathfrak{g}$ for which the complete list of completely prime Dixmier algebras over $U(\mathfrak{g})$ is known, namely $\mathfrak{sl}(2)$. This list is given in [33], and it satisfies Conjecture 2.6. The precise situation is as follows. Every coadjoint orbit is either semisimple or nilpotent. The nonzero semisimple orbits are normal, closed, and simply connected. They are attached to quotients $U(\mathfrak{g})/I$ with I a minimal primitive ideal (such quotients are automatically completely prime). The 0 orbit is of course attached to $U(\mathfrak{g})/\mathfrak{g}U(\mathfrak{g})$. The only other orbit $\mathcal{O}$ is principal nilpotent. It admits a double cover $\widetilde{\mathcal{O}}$, but has normal closure (by a result of Kostant [19]). $\widetilde{\mathcal{O}}$ is attached to the Weyl algebra A_1, but $\mathcal{O}$ is *not* attached to the subalgebra of even operators in A_1 (the image of $U(\mathfrak{g})$). This fact can be deduced from the differential operator construction in the next section, as we shall see; we will discuss its implications later. Instead, $\mathcal{O}$ is attached to $U(\mathfrak{g})$ modulo the minimal primitive of infinitesimal character half the positive root. (The infinitesimal character of A_1, on the other hand, is one-fourth of the positive root.) There is also a very interesting family of orbit data lying between $\mathcal{O}$ and $\widetilde{\mathcal{O}}$ in the following sense: the underlying algebras contain $R(\mathcal{O})$ and are commutative and nonnormal, with normalizations all isomorphic to $R(\widetilde{\mathcal{O}})$. This family is indexed by odd integers ≥ 3; the n-th orbit datum has as its underlying algebra the subalgebra of the polynomial ring $\mathbb{C}[x, y]$ generated by the homogeneous polynomials of degrees 2 and n. The corresponding Dixmier algebra may be realized as the ring of differential operators on the *singular* variety $y^2 = x^n$; note that for $n = 1$ this variety is smooth and has A_1 as its ring of differential operators. There are no other completely prime Dixmier algebras or orbit data for $U(\mathfrak{g})$.

Just as one tries to understand quotients $U(\mathfrak{g})/I$ of $U(\mathfrak{g})$ by realizing them in a different way, so one wants nice realizations of Dixmier algebras as well. Most of the existing realizations of Dixmier algebras are actually constructions of the latter, since it is rather difficult to give abstract existence proofs for Dixmier algebras. These constructions are of two closely related types. One either looks at the ring of $\mathcal{L}(M, M)$ of G-finite endomorphisms of some one-sided $U(\mathfrak{g})$ -module M of finite length (the case where M lies in category $\mathcal{O}$ being of particular interest), or one looks at rings of differential operators on a (possibly singular) G-variety. In the former case, Joseph and Stafford [18] have powerful results showing that $\mathcal{L}(M, M)$ is to a large extent determined by its ring of fractions Fract $\mathcal{L}(M, M)$, which is in turn severely restricted by Goldie's Theorem and the Gelfand-Kirillov conjecture. If M is simple, then under certain hypotheses we can even assert that $\mathcal{L}(M, M) \cong \mathcal{L}(L, L)$ for some simple highest weight module L having the same annihilator as M. Most of the techniques used and results obtained in studying the ring $\mathcal{L}(M, M)$ carry over to the module $\mathcal{L}(M, N)$. The algebras $\mathcal{L}(M, M)$ are particularly important because of an embedding theorem of Joseph and Stafford: any prime Dixmier algebra A with an infinitesimal character embeds in

$\mathcal{L}(M, M)$ for some M in category $\mathcal{O}$ (but M need not be simple). Moreover A is a direct summand of $\mathcal{L}(M, M)$ as a bimodule [32].

We have already observed that certain Dixmier algebras over $U(\mathfrak{sl}(2))$ can be realized as rings of differential operators on a singular variety. Such rings were studied in [22] and [23]; these papers point out a very striking connection between the singularity of certain G-varieties and the finite generation of their rings of differential operators as $U(\mathfrak{g})$ -modules. In [22], it is shown that if $\mathfrak{g}$ is of type B_n, C_n, D_n, E_6, or E_7, then $U(\mathfrak{g})$ modulo the Joseph ideal may be realized as differential operators on an appropriate irreducible component of $\mathcal{O} \cap \mathfrak{n}$, where $\mathcal{O}$ is the minimal nilpotent adjoint orbit and $\mathfrak{n}$ is the nilradical of some Borel subalgebra. In [23], differential operators are studied over certain classical rings of invariants studied by Weyl. These are shown to take the form $U(\mathfrak{g})/J$ with J a completely prime maximal ideal; some of the ideals J arising in this way are special unipotent in the sense of [3]. Once again the varieties in question (i.e., the spectra of the rings of invariants) are singular (and consist of matrices satisfying a rank condition).

In the embedding result $A \hookrightarrow \mathcal{L}(M, M)$ it may not be possible to choose $M \in Ob\mathcal{O}$ simple. If A is a primitive ring then by definition it admits a simple faithful module N. One can ask if N can be chosen simple over $U(\mathfrak{g})$. The work of Moeglin ([29], [30]) strongly suggests that one can even choose N to be a generalized Whittaker module. Let I be a primitive ideal with associated variety $\overline{\mathcal{O}} = \overline{G \cdot e} \subset \mathfrak{g}^*$, $\mathcal{O}$ a nilpotent orbit. Then we have the comorphism $\pi : S(\mathfrak{g}) \to R(\widetilde{\mathcal{O}})$ of the natural map $\widetilde{\mathcal{O}} \to \mathcal{O} \hookrightarrow \mathfrak{g}^*$, with $\widetilde{\mathcal{O}}$ the universal cover of $\mathcal{O}$. Suppose that $U(\mathfrak{g})/I$ admits a filtration Filt such that π factors through the associated graded algebra Gr $U(\mathfrak{g})/I$ and the map Gr $U(\mathfrak{g})/I \to R(\widetilde{\mathcal{O}})$ is injective. Then there is a faithful simple Whittaker module M for $U(\mathfrak{g})/I$ and a filtration on M such that the associated graded module gr M has an algebra structure and injects as an algebra into $R(N \cdot e)$ for some unipotent subgroup N of G. The converse of this statement is also true. Moreover, in this situation, any Dixmier algebra A over $U(\mathfrak{g})/I$ equipped with a filtration agreeing with Filt on the copy of $U(\mathfrak{g})/I$ inside A and an embedding Gr $A \hookrightarrow R(\widetilde{\mathcal{O}})$ embeds in $\mathcal{L}(M, M)$. The hypothesis imposed on A is a weaker version of the requirements given above for quantization (from [27] and [28]).

3. Rings of twisted differential operators and related Dixmier algebras

The remainder of this paper is devoted to discussing various interesting examples of Dixmier algebras. In this section, we give the general constructions which account for most of these examples (one of them arises in an ad hoc way from a very special fact about G_2). We begin by recalling the definition and basic properties of the Beilinson-Bernstein rings of twisted differential operators. We also discuss their role in the orbit method correspondence. For more details see [4], [7], and [34]. Let $\mathfrak{g}$ be reductive with G as above. Let $\mathfrak{h}$ be a Cartan subalgebra of $\mathfrak{g}$. Let $P \supset \exp \mathfrak{h}$ be a parabolic subgroup of G with commutator subgroup P_0; then $T = P/P_0$ is a torus. Choose $\Delta^+(\mathfrak{g}, h)$, a set of positive roots of $\mathfrak{h}$ in $\mathfrak{g}$, in such a way that all $\alpha \in \Delta^+(\mathfrak{g}, h)$ occur in Lie P. Write

$$\begin{aligned}
P_0 &= M_0 N, \quad \text{a Levi decomposition,} \\
\mathfrak{m} &= \text{ Lie } M_0, \\
\mathfrak{t} &= \text{ Lie } T, \\
\Delta^+(\mathfrak{m}, \mathfrak{t}) &= \text{ induced choice of positive roots of } \mathfrak{t} \text{ in } \mathfrak{m} \text{ from } \Delta^+(\mathfrak{g}, \mathfrak{h}), \\
\rho_0 &= \frac{1}{2} \sum_{\alpha \in \Delta^+(\mathfrak{m}, \mathfrak{t})} \alpha, \\
\rho &= \frac{1}{2} \sum_{\alpha \in \Delta^+(\mathfrak{g}, \mathfrak{h})} \alpha = \rho_0 + \rho_1, \\
Z &= G/P, \\
Y &= G/P_0, \\
\mathfrak{n} &= T_e^* Z, \\
\mathfrak{u} &= T_e^* Y.
\end{aligned}$$

Then, using standard identifications, we may identify $\mathfrak{t}^*$ with $\mathfrak{u}/\mathfrak{n}$. Given $\xi \in \mathfrak{t}^*$, attach a Dixmier algebra A_ξ to ξ as follows. Start with the ring Diff G/P_0 of algebraic differential operators on G/P_0. There is a left G-action and a commuting right T-action on G/P_0. Differentiating these actions, we get a map $U(\mathfrak{g}) \otimes U(\mathfrak{t}) \to$ Diff G/P_0. Set

$$A = \text{centralizer of } T \text{ in Diff } G/P_0;$$

then we get a map

$$\Phi : U(\mathfrak{g}) \to A.$$

Now bring ξ into the picture by letting I_ξ be the ideal of A generated by all $H + (\xi - \rho_1)(H)$ for $H \in \mathfrak{t}$. These elements are central in A, since ρ_1 is trivial on Lie P_0. Finally, we set

$$A_\xi = A/I_\xi = \text{Diff}_\xi G/P,$$

the ring of ξ-twisted differential operators on G/P [4]. The map Φ induces a map $\Phi_\xi : U(\mathfrak{g}) \to A_\xi$. The basic properties of the algebras A_ξ are given by

Theorem 3.1. [34]

(a) A_ξ is finitely generated as a $U(\mathfrak{g})$ -module

(b) A_ξ has infinitesimal character $\xi + \rho_0$, so that it is indeed a Dixmier algebra.

(c) A_ξ is primitive and completely prime.

(d) If $\xi + \rho_0$ is nonnegative (or, more generally, never a negative rational) on any positive root, then Φ_ξ is a surjection.

Parts (a) and (b) are rather easy. The complete primality of (c) follows from the existence of a filtration on A_ξ with gr A_ξ completely prime (we will see this in a moment). Then the primitivity in (c) follows from primality and finite length of A_ξ. Part (d) is the key to the localization theory of Beilinson and Bernstein; they study quotients of $U(\mathfrak{g})$ and their modules by realizing these quotients as rings of differential operators. We are, however, primarily interested in the opposite situation, where A_ξ is strictly larger than the image of $U(\mathfrak{g})$ inside it.

From the point of view of the orbit correspondence, the algebras A_ξ are interesting chiefly because to each of them there corresponds a ramified orbit cover X_ξ, which we now define. Set

$$\mathfrak{n} + \xi = \{y \in \mathfrak{u} : y \text{ maps to } \xi \text{ in } \mathfrak{t}^*\},$$

an affine space, and

$$Y_\xi = G \times_P (\mathfrak{n} + \xi)$$

a homogeneous affine bundle over G/P. (If ξ is 0, this is just the cotangent bundle.) Now set

$$X_\xi = \text{ affinization of } Y_\xi = \text{ Spec } R(Y_\xi),$$

where $R(\ )$ as usual denotes the regular functions. The basic properties of X_ξ and Y_ξ are given by

Proposition 3.2. [34]

(a) Y_ξ is equipped with a proper moment map $\pi_\xi : Y_\xi \longrightarrow \mathfrak{g}^*$ sending (g, y) to $g \cdot y$.

(b) The image of π_ξ is the closure $\overline{\mathcal{O}}_\xi$ of a single G-orbit $\mathcal{O}_\xi$ in $\mathfrak{g}^*$. Over $\mathcal{O}_\xi$, π_ξ is a finite covering, so that X_ξ is a ramified orbit cover.

(c) $\mathcal{O}_\xi$ is nilpotent if and only if $\xi = 0$. In this case it is a Richardson orbit.

(d) $\mathcal{O}_\xi$ is semisimple if and only if $(\alpha, \xi) \neq 0$ for any root of $\mathfrak{h}$ in $\mathfrak{n}$. In this case π_ξ is an isomorphism onto $\mathcal{O}_\xi = \overline{\mathcal{O}}_\xi$ and $\mathcal{O}_\xi$ is simply connected

Here part (a) follows from the projectivity and simple connectedness of G/P. The remaining parts are well known and easy to verify. The fundamental link between A_ξ and X_ξ is given by

Theorem 3.3. [34] *A_ξ and $R(Y_\xi)$ admit natural G-stable filtrations making gr $A_\xi \cong$ gr $R(Y_\xi)$ as algebras and G-modules.*

The filtrations come from the symbol calculus; the isomorphism follows from the pair of isomorphisms

$$\text{gr } A_\xi \cong R(Y_0) \cong \text{ gr } R(Y_\xi)$$

which in turn follow from a vanishing theorem in sheaf cohomology due to Elkik (see [34]). Thus we get a partial implementation of the orbit correspondence by attaching X_ξ to A_ξ. It turns out, however, that this implementation has some rather surprising properties. We will discuss these in the next section.

There is another construction of Dixmier algebras which can produce algebras that are not completely prime. Let $\mathfrak{p}$ be any parabolic subalgebra of $\mathfrak{g}$ and $\mathfrak{b} \subset \mathfrak{p}$ a Borel subalgebra. Choose $\mathfrak{h} \subset \mathfrak{b}$ a Cartan subalgebra. For every $\lambda \in \mathfrak{h}^*$ such that there is a finite dimensional irreducible representation $L(\lambda)$ of $\mathfrak{p}$ with highest weight λ, form the generalized Verma module $M_\mathfrak{p}(\lambda) = U(\mathfrak{g}) \otimes_{U(\mathfrak{p})} L(\lambda)$. Then we have the algebra $A_\mathfrak{p}(\lambda) = \mathcal{L}(M_\mathfrak{p}(\lambda), M_\mathfrak{p}(\lambda))$ of all G-finite endomorphisms of $M_\mathfrak{p}(\lambda)$. Its basic properties are given by

Theorem 3.4.

(a) *$A_\mathfrak{p}(\lambda)$ is prime with infinitesimal character $\lambda + \rho$, ρ the half-sum of the roots in $\mathfrak{b}$. (Thus $A_\mathfrak{p}(\lambda)$ is a primitive Dixmier algebra.)*
(b) *The multiplicity of any irreducible G-module in $A_\mathfrak{p}(\lambda)$ is bounded below by its multiplicity in the principal series representation $\mathcal{L}(M_\mathfrak{p}(\lambda), M_\mathfrak{p}(\lambda)^d)$, d the standard contravariant duality functor in category $\mathcal{O}$.*
(c) *The Gelfand-Kirillov dimension of $A_p(\lambda)$ is $2\dim_\mathbf{C}(\mathfrak{g}/\mathfrak{p})$.*
(d) *The Goldie rank of $A_\mathfrak{p}(\lambda)$ is $\dim_\mathbf{C} L(\lambda)$.*

Here parts (a) and (c) follow from [16], part (b) from [10], and part (d) from [13]. It should be noted that parts (a) and (d) can fail for the image of $U(\mathfrak{g})$ inside $A_\mathfrak{p}(\lambda)$.

The algebras $A_\mathfrak{p}(\lambda)$ seem at first glance rather different from the algebras A_ξ; for example, the former algebras do not have obvious ramified orbit covers attached to them. Nevertheless, it is not difficult to show that every A_ξ is isomorphic to some $A_\mathfrak{p}(\lambda)$ with $L(\lambda)$ one-dimensional. This observation was made by a number of people, including Vogan, who has also, as we mentioned above, generalized the twisted differential operator construction to produce a theory of induced Dixmier algebras. He has also shown how to induce orbit data from a parabolic subgroup P to the group G [37]. One expects the correspondence of Conjecture 2.6 to respect induction of Dixmier algebras and of orbit data, but limitations on current technology of cohomology vanishing theorems make verifying this difficult.

4. Examples

We begin by looking at the simplest example of an algebra A_ξ. Set $G = SL(2)$, $P = B$, the Borel subgroup of upper triangular matrices. Put $\xi = 0$. It is easy to see that A_ξ is isomorphic to $U(\mathfrak{g})$ modulo the minimal primitive I of infinitesimal character ρ, while Y_ξ identifies with the principal nilpotent orbit $\mathcal{O}$. Thus $\mathcal{O}$ is attached to $U(\mathfrak{g})/I$. We see that the Dixmier algebra quantizing $\mathcal{O}$ does not embed in the Weyl algebra A_1 quantizing its double cover $\widetilde{\mathcal{O}}$. Thus we cannot expect the bijection of Conjecture 2.6 to have good functorial properties. We now give some further examples, which will show how this lack of functoriality leads to a rather loose connection between the algebraic map Φ_ξ of Theorem 3.1 and the geometric map π_ξ of Proposition 3.2. Let G be of type G_2. Let P_{short} be the parabolic subgroup of G with the short simple root in its Levi factor, P_{long} the other proper non-minimal parabolic subgroup. First set $P = P_{short}$, $\xi = 0$. One may calculate that $\mathcal{O}_\xi$ is $\mathcal{O}_{10}$, the 10-dimensional subregular nilpotent orbit in $\mathfrak{g}^*$, and that π_ξ restricts to an isomorphism over this orbit. Coordinatize the root system and choose the positive root system of $\mathfrak{g}$ in the standard way, so that the short simple root is $(1,-1,0)$, the long simple root is $(-2,1,1)$ and $\rho = (-1,-2,3)$ [12]. Then the infinitesimal character of A_ξ is $\frac{1}{2}(1,-1,0)$ and Φ_ξ is surjective with kernel a maximal ideal. (The surjectivity obtains because the image of Φ_ξ admits no bimodule other than itself, as is easily seen.) Thus $\mathcal{O}_{10}$ gets attached to A_0. Next set $P = P_{long}$, $\xi = 0$. Once again, $\mathcal{O}_\xi$ is $\mathcal{O}_{10}$, but this time π_ξ restricts to a triple cover over $\mathcal{O}_\xi$. The infinitesimal character of A_ξ is $\frac{1}{2}(2-1,-1)$, and once again we deduce that Φ_ξ is a surjection. The algebra quantizing $\mathcal{O}_{10}$ is thus not a subalgebra of the one quantizing the triple cover of $\mathcal{O}_{10}$. We also see that it is possible for Φ_ξ to be a surjection while π_ξ is not an isomorphism.

The converse phenomenon can also occur. Let $P = P_{long}$, $\xi = \frac{1}{2}(0,1,-1)$. Now π_ξ is an isomorphism onto a semisimple orbit but Φ_ξ is not surjective. To see the last assertion, note first that the infinitesimal character of A_ξ is $(1,-1,0)$ and that Φ_ξ has maximal kernel. This kernel is special unipotent, and we may apply the character formulas of [3, Theorem III; see also Section 9] to it. These formulas yield explicit K-type decompositions of the irreducible bimodules over the image of Φ_ξ. Then we may combine Theorem 3.3, the well-known formula for the ring of regular functions on a semisimple orbit as a G-module [19], and these decompositions, to deduce that A_ξ splits into two pieces as a bimodule.

It turns out that there is an extension of this last algebra A_ξ which gets attached to the universal (sixfold) cover $\widetilde{\mathcal{O}}_{10}$ of $\mathcal{O}_{10}$. To construct it, let J be the Joseph ideal of $U(\mathfrak{so}(8))$. Put $A_{Jos} = U(\mathfrak{so}(8))/J$.

Theorem 4.1. *Take g simple of type G_2. A_{Jos} is a Dixmier algebra over $U(\mathfrak{g})$ which admits an action of the symmetric group S_3 on three letters by $U(\mathfrak{g})$ -bimodule and algebra automorphisms. As a $(U(\mathfrak{g}), S_3)$-bimodule A_{Jos} takes the form*

$$\bigoplus_{\pi \in S_3^{\wedge}} (V_\pi \otimes E_\pi)$$

where E_π is the irreducible S_3-module corresponding to π (on which $U(\mathfrak{g})$ acts trivially) and V_π is an irreducible $U(\mathfrak{g})$ -bimodule (on which S_3 acts trivially). The map $\pi \to V_\pi$ is a bijection from $S_3^{\wedge}$ to the set of irreducible Harish-Chandra $U(\mathfrak{g})$ -bimodules having the same annihilator as A_{Jos}.

Proof. There is an action of S_3 on $\mathfrak{so}(8)$ by automorphisms of the Dynkin diagram; the fixed subalgebra is well known to be isomorphic to $\mathfrak{g}$. The extended action of $U(\mathfrak{so}(8))$ preserves J as it is well known to be the unique completely prime primitive with Gelfand-Kirillov dimension 10 [14]. It is also well known that the K-types of A_{Jos} as an $\mathfrak{so}(8)$-module are just the Cartan powers of the adjoint representation, each occurring once. Decomposing these first over $\mathfrak{so}(7)$ by a classical branching law and then over G_2 by the branching law in [26], we see that A_{Jos} contains the trivial K-type only once as a $\mathfrak{g}$-module. It follows at once that the kernel K of the map $U(\mathfrak{g}) \to A_{Jos}$ has an infinitesimal character. Using the realization of J as the annihilator of a simple highest weight module in [11], we can calculate that the $\mathfrak{g}$-infinitesimal character of A is $(1, -1, 0)$. Using the complete primality of A, we see that K is prime and then by highest weight considerations can be shown to be maximal. Now the character formulas of [3] take over; we easily see that every π belonging to $S_3^{\wedge}$ occurs in the decomposition of A_{Jos} over S_3, and we know what all the irreducible $U(\mathfrak{g})/K$-bimodules look like. We can compute that the π-isotypic component of A_{Jos} takes the form $V_\pi \otimes E_\pi$ with V_π irreducible, and all assertions in the theorem drop out. (See also [26]). QED

Thus A_{Jos} has a reductive dual pair decomposition (as do all the Dixmier algebras arising in [27] and [28]). To see the role of A_{Jos} in the orbit method correspondence, note first that it is known by [11] that if A_{Jos} is given the filtration induced by the standard one on $U(\mathfrak{so}(8))$, then the associated graded algebra gr A_{Jos} is isomorphic to $R(\mathcal{O}'_{10})$, the regular functions on the minimal nilpotent orbit $\mathcal{O}'_{10}$ in $\mathfrak{so}(8)^*$. The moment map $\mathfrak{so}(8)^* \to \mathfrak{g}^*$ yields a universal covering map $\mathcal{O}'_{10} \to \mathcal{O}_{10}$ when restricted to $\mathcal{O}'_{10}$; the fundamental group of $\mathcal{O}_{10}$ is isomorphic to S_3 [1]. We therefore conjecture that $\mathcal{O}'_{10} = \widetilde{\mathcal{O}}_{10}$ should be attached to A_{Jos} in the orbit correspondence for $U(\mathfrak{g})$, even though A_{Jos}, when regarded as a Dixmier algebra over $U(\mathfrak{g})$, is not an induced Dixmier algebra (and cannot be realized as $\mathcal{L}(M,M)$ for any simple $M \in Ob\mathcal{O}$). We observe that the algebra A_ξ with $\xi = \frac{1}{2}(0,1,-1)$ mentioned earlier is a subalgebra of A_{Jos} (but corresponds to a semisimple orbit). Setting $P = P_{long}$, $\xi = \frac{1}{2}(0,-3,3)$, we obtain an algebra A_ξ isomorphic to the image of $U(\mathfrak{g})$ in A_{Jos} (which corresponds to yet another semisimple orbit). It turns out that there is another proper subalgebra of A_{Jos} which is not realizable as a ring of twisted differential operators; it has length two as a $U(\mathfrak{g})$-bimodule and presumably corresponds to the double cover of $\mathcal{O}_{10}$ [26].

We conclude this section by constructing the mock unipotent Dixmier algebras mentioned in the introduction, together with certain auxiliary unipotent Dixmier algebras which we will need in the next section. First let G be of type C_2. Define P_{long} andP_{short} as for G_2 and denote their Lie algebras by $\mathfrak{p}_{long}$ and $\mathfrak{p}_{short}$. Coordinatize the root system and choose the positive root system in the usual way, so that the short simple root is $(1,-1)$, the long simple root is $(0,2)$, and $\rho = (2,1)$. Set $A_{m,C} = A_{\mathfrak{p}_{short}}(-1,-2)$, $A_{u,C} = A_{\mathfrak{p}_{long}}(-2,0)$. Here m, u, C stand for "mock," "unipotent," and "type C_2," respectively. $A_{m,C}$ is a mock unipotent Dixmier algebra; we see from Theorem 3.4 that it has infinitesimal character $(1,1)$, Gelfand-Kirillov dimension 6, and Goldie rank 2. It is also easy to see that $\mathrm{Ker}\,(U(\mathfrak{g})) \to A_{m,C})$ is maximal. $A_{u,C}$ is a unipotent Dixmier algebra; it is a ring of twisted differential operators and corresponds to the double cover $\widetilde{\mathcal{O}}_6$ of the 6-dimensional nilpotent orbit of $\mathcal{O}_6$ in $\mathfrak{g}^*$.

Next let G be of type G_2 and define $\mathfrak{p}_{short}$, $\mathfrak{p}_{long}$ as in the last paragraph. Set $A_{m1,G} = A_{\mathfrak{p}_{short}}(3,1,-4)$, $A_{m2,G} = A_{\mathfrak{p}_{short}}(-1,3,-2)$, $A_{u,G} = A_{Jos}$. Both m's stand for "mock," u stands for "unipotent," and G stands for "type G_2." The kernel of both maps $U(\mathfrak{g}) \to A_{m1,G}$, $U(\mathfrak{g}) \to A_{m2,G}$ is maximal of infinitesimal character $(-2,1,1)$. The dimension of $A_{m1,G}$, $A_{m2,G}$, and $A_{u,G}$ is 10, while their respective Goldie ranks are 3, 2, and 1. We saw above that $A_{u,G}$ is a unipotent Dixmier algebra.

Finally let G be of type F_4. Once again, coordinatize the root system and choose a positive root system in the standard way [12], so that the simple roots are $(0,1,-1,0)$, $(0,0,1,-1)$, $(0,0,0,1)$, and $\frac{1}{2}(1,-1,-1,-1)$. Let $\mathfrak{p}_1$ be the parabolic subalgebra with Levi factor of type B_2 and $\mathfrak{p}_2$ the maximal parabolic with Levi factor containing the long simple roots and of type $A_2 \times A_1$. Put $A_{m,F} = A_{\mathfrak{p}_2}\frac{1}{2}(-11,-3,-5,-5)$, $A_{u,F} = A_{\mathfrak{p}_1}(-5,-2,0,0)$. Then the kernels of $U(\mathfrak{g}) \to A_{m,F}$, $U(\mathfrak{g}) \to A_{u,F}$ are maximal of infinitesimal character $(2,1,1,0)$, $\frac{1}{2}(3,1,1,1,)$, respectively. $A_{m,F}$ and $A_{u,F}$ have dimension 40 and Goldie ranks 6,1, respectively. $A_{u,F}$ is a ring of twisted differential operators corresponding to a sixfold cover of $\mathcal{O}_{40}$, the 40-dimensional nilpotent orbit in $\mathfrak{g}^*$. This orbit has fundamental group S_4, the symmetric group on four letters [1].

5. Mock unipotent Dixmier algebras

We conclude this paper by studying the algebras $A_{m,C}$, $A_{m1,G}$, $A_{m2,G}$, and $A_{m,F}$ in more detail and giving a precise account of their relationship to the unipotent algebras $A_{u,C}$, $A_{u,G}$, and $A_{u,F}$. For this we need to recall some of the constructions of [3]. The central notion of that paper is that of a special unipotent representation, defined to be an irreducible Harish-Chandra bimodule whose (left and right) annihilator is maximal with infinitesimal character belonging to a certain specified list. This list is given as follows. Given a semisimple $\mathfrak{g}$, let ${}^L\mathfrak{g}$ denote its dual Lie algebra. Let ${}^L\mathcal{O}$ be an even nilpotent orbit in ${}^L\mathfrak{g}$ with representative e. By the Jacobson-Morozov theorem, e embeds in a standard triple $\{h, e, f\} \subset^L g$ satisfying the usual bracket relations in $\mathfrak{sl}(2)([he] = 2e$, etc.). Then h belongs to some Cartan subalgebra of $\mathfrak{g}$, which may be naturally identified with a Cartan subalgebra dual of $\mathfrak{g}$. Set $\lambda_{\mathcal{O}} = \frac{1}{2}h$; we may regard $\lambda_{\mathcal{O}}$ as an infinitesimal character of $\mathfrak{g}$. As an infinitesimal character, $\lambda_{\mathcal{O}}$ is well defined thanks to the conjugacy of all standard triples containing e. The evenness of ${}^L\mathcal{O}$ implies, by definition, that $\lambda_{\mathcal{O}}$ is integral. Then the list of infinitesimal characters in the definition of special unipotent representation consists precisely of the $\lambda_{\mathcal{O}}$ for each even ${}^L\mathcal{O}$.

Given such a $\lambda_{\mathcal{O}}$, Barbasch and Vogan give character formulas for the special unipotent representations of infinitesimal character $\lambda_{\mathcal{O}}$. They prove these formulas by an elaborate induction argument on ${}^L\mathcal{O}$ (and its "dual orbit" $\mathcal{O} \subset \mathfrak{g}^*$, which they define). The orbits ${}^L\mathcal{O}$ serving as base cases in the induction are called triangular in [2]. The character formulas for the corresponding special unipotent representations are proved by realizing these representations as induced from finite-dimensional representations on a parabolic. This realization is in turn obtained from the following combinatorial lemma (see the proof of Proposition 9.11 in [3]).

Lemma 5.1. ([3], **Lemma 9.7**) *Let $\lambda_{\mathcal{O}}$ be the infinitesimal character attached to a triangular orbit ${}^L\mathcal{O}$. Choose a set of positive roots, and let γ be the unique dominant weight conjugate to $\lambda_{\mathcal{O}}$. Let S denote the set of simple roots on which γ is nonzero. Then we may list the Weyl group conjugates of γ that are strictly positive on S as $(\gamma^1, \dots, \gamma^j)$ in such a way that*

(a) $(\gamma^1, \alpha^\vee) = 2(\gamma^1, \alpha)/(\alpha, \alpha) = 1$ *for* $\alpha \in S$,
(b) *no two distinct* γ^i *are conjugate by a Weyl group element fixing* γ^1,
(c) $\gamma^1 - \gamma^k$ *is a sum of positive roots for* $k \leq j$ *and is strictly shorter than* $\gamma^1 - \gamma^{k+1}$ *for* $k \leq j - 1$.

This is proved by a case-by-case verification, using the list of triangular orbits given in [3, Section 9]. Now it turns out that the orbits $\mathcal{O}_6$, $\mathcal{O}_{10}$, and $\mathcal{O}_{40}$ defined in the last section are triangular. The corresponding infinitesimal characters $\lambda_6 = \lambda_{\mathcal{O}_6}$, $\lambda_{10} = \lambda_{\mathcal{O}_{10}}$, and $\lambda_{40} = \lambda_{\mathcal{O}_{40}}$ are exactly those of $A_{u,C}$, $A_{u,G}$, and $A_{u,F}$, respectively. Hence these infinitesimal characters satisfy the properties of Lemma 5.1.

The first link between the mock unipotent Dixmier algebras and the $A_{u,*}$ is that the infinitesimal characters of $A_{m,C}$, $A_{m1,G}$, $A_{m2,G}$ and $A_{m,F}$, also satisfy the properties o f Lemma 5.1, except for (a) in the case of $A_{m,F}$. [One may verify this directly; the relevant lists $(\gamma^1, \ldots, \gamma^j)$ are, respectively, $((1,1),(1,-1))$, $((-1,-1,2),(-2,1,1),(-1,2,-1))$ and $((2,1,0,-1)$, $(1,1,0,-2)$, $(1,0,-1,-2)$, $(0,1,-1,-2)$, $(-1,0,-1,-2))$.] Thus the proof of Proposition 9.11 in [3] carries over to the mock unipotent infinitesimal characters; any irreducible Harish-Chandra bimodule having the same annihilator as $A_{m,C}$, $A_{m1,G}$, or $A_{m2,G}$ is induced from a finite-dimensional representation of the parabolic subgroup corresponding to the set S of simple roots in Lemma 5.1. More precisely there are exactly j such irreducible bimodules, with highest weights of the inducing representations given by $(\gamma^1 - \rho_S, \gamma^1 - \rho_S)$, $(\gamma^2 - \rho_S, \gamma^1 - \rho_S), \ldots,$ $(\gamma^j - \rho_S, \gamma^1 - \rho_S)$, where ρ_S is defined by the requirement that each of these inducing representations have infinitesimal character γ^1. (The proof of [3] that these induced representations are irreducible breaks down in the case of $A_{m,F}$, since Lemma 5.1(a) fails. The most we can say for certain is that the irreducible bimodules having the same annihilator as $A_{m,F}$ are the Langlands subquotients of the representations induced from finite-dimensional representations of highest weights $(\gamma^1 - \rho_S, \gamma^1 - \rho_S), \ldots,$ $(\gamma^j - \rho_S, \gamma^1 - \rho_S)$ as above). The arguments of [3, Sections 9, 10] then carry over to show that the character theories of irreducible bimodules with annihilators Ann $A_{u,C}$, Ann $A_{m,C}$ agree. The same result holds for type G_2 (and also for type F_4, provided that the induced representations occurring there are irreducible.)

There is another link between the mock unipotent Dixmier algebras and the $A_{u,*}$. To describe it, we need a construction of Lusztig and some primitive ideal theory. Given a nilpotent orbit $\mathcal{O}$ which is special in Lusztig's sense, he attaches a left cell $\mathcal{C}_{\mathcal{O}}$ of the corresponding Weyl group W to $\mathcal{O}$. Recall that any left cell of W may be viewed either as a subset of W or as a submodule of the group algebra $\mathbb{C}W$ of W under the left regular representation. When regarded as a W-module, it decomposes as a direct sum of irreducible W-modules. Lusztig's cells $\mathcal{C}_{\mathcal{O}}$ have the property that this decomposition is always multiplicity-free. Moreover, we can say how many irreducible W-modules occur in $\mathcal{C}_{\mathcal{O}}$. Lusztig defines a certain quotient $\overline{A(\mathcal{O})}$ of the fundamental group $\pi_1(\mathcal{O})$ and then sets up a bijection from $\overline{A(\mathcal{O})}^\wedge$ to the set of W-modules occurring in $\mathcal{C}_{\mathcal{O}}$. He also shows that if ${}^L\mathcal{O}$ is the Barbasch-Vogan dual of $\mathcal{O}$, then there is a natural isomorphism $\overline{A(\mathcal{O})} \cong \overline{A({}^L\mathcal{O})}$. For all this see [21, Chapter 4].

Barbasch and Vogan have applied Lusztig's results on left cells to primitive ideal theory, as follows. Given any integral infinitesimal character λ, let W^λ denote the stabilizer of λ in W, and denote by $\sigma_1, \ldots, \sigma_s$ the special representations of W in Lusztig's sense. For $1 \leq i \leq s$, let τ_i denote the space of W^λ-fixed vectors on σ_i. Recall that primitive ideals of infinitesimal character λ are parametrized by the basis of $\sum_{n=1}^{s} \tau_i$ obtained via Joseph's Goldie rank polynomials [13, Chapter 14]. Given one such primitive ideal I_λ, we may obtain I_λ from a primitive ideal $I_{\lambda'}$ of some *regular* integral infinitesimal character λ' by the Jantzen-Zuckerman translation principle. Then there is a unique left cell $\mathcal{C}$ of W corresponding to $I_{\lambda'}$ (by definition). As usual, regard $\mathcal{C}$ as a W-module and list the irreducible submodules of it according to multiplicity as $\nu_1, \ldots, \nu_t$. Let ν_i' denote the subspace of W^λ-fixed vectors in ν_i. Then the set of irreducible Harish-Chandra bimodules with left and right annihilator I_λ is parametrized by a basis of $\sum_{i=1}^{t} \nu_i'$. (This may be deduced from [13, Kapital 14], or from Joseph's work, but to the best of my knowledge it is stated explicitly only in an unpublished preprint of Barbasch and Vogan.) Now let ${}^L\mathcal{O}$ be an even nilpotent orbit; then it is automatically special. Recall that the Weyl group of ${}^L\mathfrak{g}$ identifies with that of $\mathfrak{g}$, so that left cells for $\mathfrak{g}$ and ${}^L\mathfrak{g}$ coincide. Barbasch and Vogan show in [3, Proposition 5.28] that Lusztig's cell $\mathcal{C}_{L_\mathcal{O}}$ is the relevant cell to use in counting irreducible Harish-Chandra bimodules of infinitesimal character $\lambda_\mathcal{O}$ and maximal annihilator. Moreover, they show that each W-representation occurring in $\mathcal{C}_{L_\mathcal{O}}$ has a unique $W^{\lambda_\mathcal{O}}$-fixed vector. Hence, there is a bijection between $\overline{A({}^L\mathcal{O})}^\wedge$ or $\overline{A(\mathcal{O})}^\wedge$ and special unipotent representations of infinitesimal character $\lambda_\mathcal{O}$. Denote it by $r \leftrightarrow V_r$. The trivial representation of $\overline{A(\mathcal{O})}$ corresponds to the spherical special unipotent representation, a quotient of $U(\mathfrak{g})$ ([3], Corollary 5.30).

It is natural to ask for the relationship between the left cells $\mathcal{C}_\mathcal{O}, \mathcal{C}_{L_\mathcal{O}}$ attached to Barbasch-Vogan dual orbits $\mathcal{O}, {}^L\mathcal{O}$. As usual, regard these left cells as W-modules. From Lusztig's algorithms for computing $\mathcal{C}_\mathcal{O}$ in the classical cases and his tables giving $\mathcal{C}_\mathcal{O}$ in the exceptional cases [21, Chapter 4], one can compute that $\mathcal{C}_{L_\mathcal{O}}$ is just $\mathcal{C}_\mathcal{O}$ tensored with the sign representation, with three exceptions. These exceptions occur exactly for the orbits $\mathcal{O} = \mathcal{O}_6, \mathcal{O}_{10}, \mathcal{O}_{40}$ of the last section. Each of these orbits is self-dual, and yet the corresponding left cell is not stable under tensoring with sign. Thus we obtain three new left cells $\mathcal{C}_{\mathcal{O}_i} = \mathcal{C}_{\mathcal{O}_i} \otimes sgn$ for $\mathcal{O}_i = \mathcal{O}_6, \mathcal{O}_{10}$, and $\mathcal{O}_{40}$. Each of the infinitesimal characters $\lambda_6', \lambda_{10}'$, and λ_{40}' of the mock unipotent Dixmier algebras $A_{m,C}$, $A_{m1,G}$, and $A_{m,F}$ differs from that of the corresponding unipotent Dixmier algebra, as we have seen. One may verify directly for each λ'_i that the relevant left cell to use in counting irreducible Harish-Chandra bimodules with infinitesimal character λ'_i and maximal annihilators is $\mathcal{C}'_{\mathcal{O}_i}$. Moreover, each W- representation occurring in $\mathcal{C}'_{\mathcal{O}_i}$ has a unique $W^{\lambda'_i}$-fixed vector. Thus the bijection between $\overline{A(\mathcal{O}_i)^\wedge}$ and the set of special unipotent representations attached to $\mathcal{O}_i(= 6, 10, 40)$ induces one between $\overline{(A(\mathcal{O}_i)^\wedge}$ and what we may call the set of mock unipotent representations attached to $\mathcal{O}_i$. Denote it by $r \leftrightarrow W_r$. Each $\overline{A(\mathcal{O}_i)}$, by the way, is isomorphic to $\pi_1(\mathcal{O}_i)$, which is in turn isomorphic to S_2, S_3, S_4 for $i = 6, 10, 40$ [21].

Now it can be seen why we restricted attention to $\mathcal{O}_6$, $\mathcal{O}_{10}$, and $\mathcal{O}_{40}$ in defining mock unipotent Dixmier algebras. The infinitesimal characters λ'_6, λ'_{10}, λ'_{40} were obtained from the Barbasch-Vogan infinitesimal characters λ_6, λ_{10}, λ_{40} as follows. For each self-dual semisimple Lie algebra g, there is an involution ι on the set of simple roots of g induced by the isomorphism $\mathfrak{g} \cong {}^L\mathfrak{g}$. Thus ι interchanges short and long roots. It induces an involution π on $\mathfrak{h}^*$ via the requirement $(\pi(H), \alpha^\vee) = (H, (\iota\alpha)^\vee)$ for α and $H \in \mathfrak{h}^*$ (but the map π does *not* extend ι). For $i = 6, 10, 40$, we have $\lambda'_i = \pi(\lambda_i)$. (We make π act on infinitesimal characters by restricting it to the dominant Weyl chamber.)

We conclude this section by giving the bimodule decompositions of the unipotent and mock unipotent Dixmier algebras of the preceding section. Denote the representations of S_2 by $\{1, sgn\}$; those of S_3 by $\{1, sgn, refl\}$ with $refl$ of dimension two; and those of S_4 by $\{1, sgn, two, refl, refl \otimes sgn\}$, where two is two-dimensional and $refl$ is realized on a Cartan subalgebra dual of $\mathfrak{sl}(4)$ by the action of its Weyl group. We have seen that to each of these representations r there corresponds a special unipotent representation V_r and a mock unipotent representation W_r. One can verify in the case of S_3 that this notation V_r is consistent with that of Theorem 4.1. If we combine Theorems 3.3 and 4.1 with the character formulas of [3], we can compute that the respective bimodule decompositions of $A_{u,C}$, $A_{u,G}$, and $A_{u,F}$ are

$$V_1 \oplus V_{sgn},\ V_1 \oplus V_{sgn} \oplus 2.V_{refl},\ V_1 \oplus V_{two} \oplus V_{refl}.$$

The decomposition of the mock unipotent algebras are more difficult to work out, but it may be computed from Theorem 3.4(b) that $A_{m,C}$, $A_{m1,G}$, $A_{m2,G}$, $A_{m,F}$ respectively contain at least the pieces

$$W_1 \oplus W_{sgn},\ W_1 \oplus W_{refl},\ W_1 \oplus W_{sgn} \oplus W_{refl},\ W_1 \oplus W_{two} \oplus W_{refl}.$$

I conjecture that there is a third mock unipotent algebra $A_{m3,G} \supset A_{m1,G}$, $A_{m2,G}$ in type C_2 having the bimodule decomposition

$$W_1 \oplus W_{sgn} \oplus 2.W_{refl}$$

and Goldie rank 6. Finally, note that by looking at the Dixmier algebras $A_{\mathfrak{p}_{long}}(-1,0)$ in type C_2 and $A_{\mathfrak{p}_{long}}(0,1,-1)$ in type G_2 one can show that the images of $U(\mathfrak{g})$ in $A_{m,C}$, $A_{m1,G}$, $A_{m2,G}$ are completely prime, though the algebras themselves are not. (I do not know whether this is also true of $A_{m,F}$.) We thus obtain two counterexamples to Theorem 3.4(d) for the image of $U(\mathfrak{g})$ inside $A_{\mathfrak{p}}(\lambda)$.

REFERENCES

[1] A. V. Alexeevsky, *Component groups of centralizer for unipotent elements in semisimple algebraic groups* [in Russian], Trudy Tbilissi. Mat. Inst. Razmazde Akad. Nauk. Gruzin. SSR **62** (1979), 5-27.

[2] D. Barbasch, *The unitary dual for complex classical Lie groups*, Inv. Math. **96** (1989), 103-176.

[3] D. Barbasch and D. A. Vogan, *Unipotent representations of complex semisimple groups*, Ann. of Math. **121** (1985), 41-110.

[4] A. Beilinson and J. Bernstein, *Localisation de $\mathfrak{g}$-modules*, C. R. Acad. Sci. Paris **292** (1981), 15-18.

[5] W. Borho, *Primitive vollprime Ideale in der Einhüllenden von* $\mathfrak{so}(5,C)$, J. Alg. **43** (1976), 619-654.

[6] W. Borho, *Definition einer Dixmier-Abbildung für* $\mathfrak{sl}(n,\mathbb{C})$, Inv. Math. **40** (197 7), 143-169.

[7] W. Borho and J.-L. Brylinski, *Differential operators on homogeneous spaces I: irreducibility of the associated variety*, Inv. Math. **69** (1982), 437-476.

[8] W. Borho and J. G. Jantzen, *Über primitive Ideale in der Einhüllenden einer halbeinfachen Lie-Algebra*, Inv. Math. **39** (1977), 1-53.

[9] W. Borho and H. Kraft, *Über Bahnen und deren Deformationen bei linearen Aketionen reduktiver Gruppen*, Comm. Math. Helv. **54** (1979), 61-104.

[10] N. Conze-Berline and M. Duflo, *Sur les représentations induites des groupes semisimples complexes*, Comp. Math. **34** (1977), 307-336.

[11] D. Garfinkle, *A new construction of the Joseph ideal*, Ph.D. dissertation, M.I.T., 1982.

[12] S. Helgason, *Differential Geometry, Lie Groups, and Symmetric Spaces*, Academic Press, New York, 1978.

[13] J. C. Jantzen, *Einhüllende Algebren halbeinfacher Lie-Algebren*, Ergebnisse der Mathematik und ihrer Grenzgebiete, Band 3, Springer-Verlag, Berlin/Heidelberg/New York, 1983.

[14] A. Joseph, *The minimal orbit in a simple Lie algebra and the associated maximal ideal*, Ann. Sci. Ec. Norm. Sup. (4) **9** (1976), 1-30.

[15] A. Joseph, *Towards the Jantzen conjecture II*, Comp. Math. **40** (1980), 69-78.

[16] A. Joseph, *Gelfand-Kirillov dimension for annihilators of simple quotients of Verma modules*, J. London Math. Soc (2) **18** (1978), 50-60.

[17] A. Joseph, *Kostant's problem and Goldie rank*, Non-commutative Harmonic Analysis and Lie Groups, Lecture Notes in Mathematics, **880**, Springer-Verlag, Berlin/Heidelberg/New York, 1981, 249-266.

[18] A. Joseph and J. Stafford, *Modules of $\mathfrak{k}$-finite vectors over semisimple Lie algebras*, Proc. London Math. Soc. **49** (1984), 361-384.

[19] B. Kostant, *Lie group representations on polynomial rings*, Amer. J. Math. **85** (1963), 327-404.

[20] H. Kraft and C. Procesi, *On the geometry of conjugacy classes in classical groups*, Comm. Math. Helv. (4) **57** (1982), 539-602.

[21] G. Lusztig, *Characters of Reductive Groups over a Finite Field*, Annals of Mathematics Studies, no. 107, Princeton University Press, Princeton NJ, 1984.

[22] T. Levasseur S. P. Smith and J. Stafford, *The minimal nilpotent orbit, the Joseph ideal, and differential operators*, J. Alg. **116** (1988), 480-501.

[23] T. Levasseur and J. Stafford, *Rings of differential operators on classical rings of invariants*, Mem. A.M.S., no. 412, A.M.S., Providence RI, 1989.

[24] W. McGovern, *Unipotent representations and Dixmier algebras*, Comp. Math. (3) **69** (1989), 241-276.

[25] W. McGovern, *Rings of regular functions on nilpotent orbits and their covers*, Inv. Math. **97** (1989), 209-217.

[26] W. McGovern, *A branching law for* $\mathrm{Spin}(7, C) \to G_2$ *and its applications to unipotent representations*, to appear in J. Alg.

[27] W. McGovern, *Quantization of nilpotent orbit covers in complex classical groups*, preprint, Yale University, 1989.

[28] W. McGovern Quantization of nilpotent orbit covers in complex exceptional groups, preprint, Yale University, 1989.

[29] C. Moeglin, *Modèles de Whittaker et idéaux primitifs complètement premiers dans les algèbres enveloppantes*, C. R. Acad. Sci. **303** (1986), 845-848.

[30] C. Moeglin, *Modèles de Whittaker et idéaux primitifs complètement premiers dans les algèbres enveloppantes des algèbres de Lie semisimples complexes II*, Math. Scand. **63** (1988), 5-35.

[31] C. Moeglin, *Idéaux complètements premiers de l'algèbre enveloppante de* $\mathfrak{gl}(n, C)$, J. Alg. **106** (1987), 287-366.

[32] I. Pranata, *Structure of Dixmier algebras*, Ph.D. dissertation, M.I.T., 1989.

[33] S. P. Smith, *Overrings of primitive factor rings of* $U(\mathfrak{sl}(2))$, to appear in J. Pure and Appl. Alg.

[34] D. Vogan, *The orbit method and primitive ideals for semisimple Lie algebras*, in Lie Algebras and Related Topics, F. W. Lemire, D. Britten and R. V. Moody, eds. **5** Canadian Mathematical Society for A.M.S., Providence RI, 1986.

[35] D. Vogan., *Unitary Representations of Reductive Lie Groups*, Ann. of Math. Studies, **118**, Princeton University Press, Princeton, 1987.

[36] D. Vogan, *Noncommutative algebras and unitary representations*, in The Mathematical Heritage of Herman Weyl, R. O. Wells, Jr., ed., A.M.S., Providence RI, 1988.

[37] D. Vogan, *Dixmier algebras, sheets, and representation theory*, to appear in Proc. Colloque Jacques Dixmier, Paris, 1989.

Received October 12, 1989

Department of Mathematics
GN 50
University of Washington
Seattle, Washington 98195

Trace in Categories

JOSEPH BERNSTEIN[1]

To Jacques Dixmier on his 65th birthday

0. Introduction

In this note we will introduce a trace morphism $tr_V : \mathcal{Z}(\mathfrak{g}) \longrightarrow \mathcal{Z}(\mathfrak{g})$ and give an explicit formula for it (formula $(*)$ in proposition 2). This is a beautiful formula, which I think has many applications. We present one such application – namely, we reprove the description of the algebra of endomorphisms of the big projective module in the category $\mathcal{O}$, due to W. Soergel (See [S]).

1. Definition of trace

1.1. Trace. Let V be a finite-dimensional vector space over a field k. For every endomorphism $a \in \mathrm{End}(V)$ we define its trace by $tr(a) = \sum a_{ii}$, where (a_{ij}) is the matrix of a in some basis v_i of V.

In order to see that this definition does not depend on a choice of the basis, let us write it in a more invariant form. Consider canonical morphism $\mu : V \otimes V^* \to \mathrm{End}(V)$, $\mu(v \otimes v^*)\xi = (v^*, \xi)v$. Since V is finite dimensional, this is an isomorphism, and we denote by ν the inverse morphism $\nu : \mathrm{End}(V) \to V \otimes V^*$. We also consider natural morphisms $i : k \to \mathrm{End}(V)$ and $p : V \otimes V^* \to k$, where $i(1) = 1_V$ and $p(v \otimes v^*) = (v^*, v)$.

Now for any endomorphism $a \in \mathrm{End}(V)$ we define $tr(a)$ as the composition $p \circ a' \circ \nu \circ i : k \to \mathrm{End}(V) \to V \otimes V^* \to V \otimes V^* \to k$, where $a' = a \otimes 1_{V^*}$.

1.2. Relative trace. The definition above can be immediately generalized. Namely, let M be another vector space over k. Then we define the trace morphism $tr_V : \mathrm{End}(V \otimes M) \to \mathrm{End}(M)$ as follows: for every $a \in \mathrm{End}(V \otimes M)$ we denote by $tr_V(a)$ the endomorphism of M given by composition $p' \circ a' \circ \nu' \circ i' : M \to M \otimes \mathrm{End}(V) \to M \otimes V \otimes V^* \to M \otimes V \otimes V^* \to M$, where $i' = 1_M \otimes i$ with similar formulae for p', v', and where $a' = a \otimes 1_{V^*}$.

---

[1] This research is supported by an NSF grant.

Let $\{v_i\}$ be some basis of V. Then we can write an endomorphism $a \in \mathrm{End}(V \otimes M)$ as a matrix with entries $a_{ij} \in \mathrm{End}(M)$. It is easy to see, that $tr_V(a) = \sum a_{ii} \in \mathrm{End}(M)$. Using this formula it is easy to establish functorial properties of the morphism tr_V. We will need the following

LEMMA. *If M is finite-dimensional, then $tr(tr_V(a)) = tr(a)$.*

2. Formula for trace morphism

Let us assume that the field k is algebraically closed of characteristic 0. We fix a reductive Lie algebra $\mathfrak{g}$ over k and denote by $U(\mathfrak{g})$ its universal enveloping algebra. The center of this algebra we call **central algebra** and denote by $\mathcal{Z}(\mathfrak{g})$.

We denote by $M(\mathfrak{g})$ the category of $\mathfrak{g}$-modules. Let us fix a finite-dimensional $\mathfrak{g}$-module V and denote by $F_V : M(\mathfrak{g}) \to M(\mathfrak{g})$ the functor $F_V(M) = V \otimes M$. Using functorial properties of morphism tr_V one can easily check that $tr_V : \mathrm{End}(V \otimes M) \to \mathrm{End}(M)$ defines a morphism $tr_V : \mathrm{End}(F_V) \to \mathrm{End}(Id)$ from endomorphisms of functor F_V to endomorphisms of identity functor on $M(\mathfrak{g})$. We would like to have an explicit description of this morphism.

In order to do this we need an explicit description of source and target of the morphism tr_V. The target space $\mathrm{End}(Id)$ is isomorphic to $\mathcal{Z}(\mathfrak{g})$. It can be explicitly described using Harish-Chandra's theorem. Namely, fix a Cartan subalgebra $\mathfrak{h} \subset \mathfrak{g}$ and denote by L the dual space and by W the corresponding Weyl group. Then $\mathrm{End}(Id) = \mathcal{Z}(\mathfrak{g})$ can be identified with $\mathcal{F}(L)^W$, where $\mathcal{F}(L)$ is the algebra of polynomial functions on L and $\mathcal{F}(L)^W$ its W-invariant part.

Probably there exists a similar description for $\mathrm{End}(F_V)$, but I do not know it. However we have a natural morphism $\mathcal{Z}(\mathfrak{g}) \to \mathrm{End}(F_V)$, which sends each $z \in \mathcal{Z}(\mathfrak{g})$ to an endomorphism of the functor F_V, given by the action of z on $V \otimes M \in M(\mathfrak{g})$. Composing it with the trace map tr_V, we get a trace map

$$tr_V : \mathcal{Z}(\mathfrak{g}) \longrightarrow \mathcal{Z}(\mathfrak{g}).$$

We will give an explicit formula for this morphism, as a morphism of $F(L)^W$ into itself.

Let $P(V)$ be the set of weights of V with multiplicities counted. We define a convolution $f \mapsto P(V) * f$ on the space of functions on L by $(P(V) * f)(x) = \sum f(x + \mu)$, the sum being taken over all weights $\mu \in P(V) \subset L$ with multiplicities counted. We denote by Λ the elementary W-skew symmetric function on L, that is $\Lambda(x) = \prod_{\alpha>0} h_\alpha(x)$.

Proposition. *Suppose we identified $\mathcal{Z}(\mathfrak{g})$ with $F(L)^W$ using Harish-Chandra isomorphism. Then the trace morphism $tr_V : \mathcal{Z}(\mathfrak{g}) \to \mathcal{Z}(\mathfrak{g})$ satisfies the following identity*

$$\Lambda \cdot tr_V(f) = P(V) * (\Lambda \cdot f). \qquad (*)$$

This proposition gives a formula for tr_V, namely

$$tr_V(f) = \Lambda^{-1}(P(V) * (\Lambda \cdot f)).$$

Proof. For each integral dominant regular weight $\lambda \in L$ we denote by V_λ an irreducible finite dimensional $\mathfrak{g}$-module with highest weight $\lambda - \rho$, where ρ is half sum of positive roots. It is known that the action of an element $z \in \mathcal{Z}(\mathfrak{g})$ on V_λ is given by multiplication on $f(\lambda)$, where f is the function on L corresponding to z. Also dim $(V_\lambda) = C \cdot \Lambda(\lambda)$, where C is some constant, which implies, that $tr(z|V_\lambda) = C \cdot (\Lambda f)(\lambda)$.

To prove formula $(*)$ it is enough to check that the functions on both sides coincide for integral dominant λ which are sufficiently regular. Choose such a λ and let us compute the trace of the operator $tr_V(z)$ on V_λ. On the one hand it equals $C \cdot (\Lambda \cdot tr_V(f))(\lambda)$. On the other hand by lemma in 1.2, it equals $tr(z|V \otimes V_\lambda)$. It is known, that $V \otimes V_\lambda$ is isomorphic to $\bigoplus_\mu V_{\lambda+\mu}$, where sum is taken over weights $\mu \in P(V)$ with multiplicities. Hence $tr(z|V \otimes V_\lambda) = \sum_\mu tr(z|V_{\lambda+\mu}) = C \cdot \sum_\mu \Lambda f(\lambda+\mu) = C \cdot (P(V) * (\Lambda f))(\lambda)$. This proves formula $(*)$. □

3. Trace in categories

The definition of the trace morphism is a special case of a general construction in category theory. This construction is interesting in itself, so I would like to describe it, though we will not use it explicitly in application described in section 4.

Let A and B be two categories and $F : A \longrightarrow B$ a functor. Suppose that

(a) The functor F has a left adjoint functor $E : B \longrightarrow A$ and a right adjoint functor $G : B \longrightarrow A$.
(b) We have fixed a morphism of functors $\nu : G \longrightarrow E$.

Then for all objects $X, Y \in A$ we define morphism

$$tr : \mathrm{Hom}_B(F(X), F(Y)) \longrightarrow \mathrm{Hom}_A(X, Y) \text{ by}$$

$$tr(a) = i_Y \circ E(a) \circ \nu_{F(X)} \circ j_X : X \to GF(X) \to EF(X) \to EF(Y) \to Y,$$

where $j_X : X \to GF(X)$ and $i_Y : EF(Y) \to Y$ are adjunction morphisms, $\nu_{F(X)} : GF(X) \to EF(X)$ ν-morphism, corresponding to the object $F(X)$, $a : F(X) \to F(Y)$ any morphism in B and $E(a) : EF(X) \to EF(Y)$ the corresponding morphisms in A.

In particular, for any object $X \in A$ we get a morphism $tr : \mathrm{End}(F(X)) \to \mathrm{End}(X)$. It is easy to see that this morphism has natural functorial properties. In particular, it defines a morphism

$$tr : \mathrm{End}(F) \longrightarrow \mathrm{End}(Id_A).$$

The case discussed in section 2 corresponds to $A = B = M(\mathfrak{g}), F = F_V, E = G = F_{V^*}$.

4. An application. Description of the endomorphism algebra of the big projective module

Fix a maximal nilpotent subalgebra $\mathfrak{n}$ normalized by $\mathfrak{h}$ and denote by $\mathcal{O}$ the corresponding category of highest weight modules. For every weight $\lambda \in L = \mathfrak{h}^*$ we denote by M_λ the corresponding Verma module of highest weight $\lambda - \rho$, by L_λ its irreducible quotient, and by P_λ the projective cover of L_λ in the category $\mathcal{O}$ (see[BGG1]).

Fix a regular integral antidominant weight λ. Then M_λ is irreducible and P_λ is what we call the "big" projective module. We want to describe the algebra $\mathrm{End}(P_\lambda)$ of its endomorphisms in category $\mathcal{O}$.

Theorem. *The natural morphism $\eta : \mathcal{Z}(\mathfrak{g}) \to End(P_\lambda)$ is an epimorphism. Its kernel coincides with the ideal J_λ, described below. In particular, the algebra $End(P_\lambda)$ is isomorphic to $\mathcal{Z}(\mathfrak{g})/J_\lambda$ which in turn is isomorphic to the cohomology algebra of the flag variety X of algebra $\mathfrak{g}$.*

Let us describe the ideal J_λ. We identify $\mathcal{Z}(\mathfrak{g})$ with the algebra $F(L)^W$ and consider a linear functional $\nu = \nu_\lambda : F(L)^W \to k$, given by

$$\nu(f) = [\sum_w T(w\lambda)(\Lambda f))/\Lambda](0) = [\sum_w \epsilon(w) \cdot w(T(\lambda)(\Lambda f))/\Lambda](0).$$

Here $T(\mu)$ is a translation operator on $F(L)$, $T(\mu)(h)(x) = h(x+\mu)$, $\epsilon(w)$ is the sign of element $w \in W$. In order to see that these two expressions coincide, we use that $w(f) = f$, $w(\Lambda) = \epsilon(w)\Lambda$ and $T(w\lambda) = wT(\lambda)w^{-1}$.

Remark. In order to compute $\nu(f)$ we use the fact that f is a polynomial, i.e. we computed $\nu(f)$ using some kind of limit.

In terms of the functional ν the ideal J_λ is described as

$$J_\lambda = \{f \in F(L)^W \mid \nu(f \cdot F(L)^W) = 0\}.$$

In order to prove the theorem it is enough to check the following three lemmas.

Lemma 1. $\dim(End(P_\lambda)) \leq \#(W)$.

Lemma 2. *Set* $J = Ann_{\mathcal{Z}(\mathfrak{g})}(P_\lambda)$, *i.e.* J *is the kernel of morphism* $\eta : F(L)^W \longrightarrow End(P_\lambda)$. *Then* $\nu(f) = 0$ *for all* $f \in J$. *In particular,* $J \subset J_\lambda$.

Lemma 3. *The algebra* $F(L)^W/J_\lambda$ *has dimension equal to* $\#(W)$ *and is isomorphic to the cohomology algebra of the flag variety* X.

Remark. Lemmas 1 and 3 are more or less straightforward exercises on category $\mathcal{O}$ and cohomology algebra of flag variety respectively. From the point of view of this note the key statement is lemma 2.

Proofs. 1. Let V be an irreducible finite dimensional $\mathfrak{g}$-module with lowest weight λ, θ the corresponding character of $\mathcal{Z}(\mathfrak{g})$. Consider Verma module M_0, corresponding to weight 0. Then P_λ is the direct summand of $F_\lambda(M_0)$, corresponding to the character θ of $\mathcal{Z}(\mathfrak{g})$. This implies that P_λ has a composition series whose factors are isomorphic to Verma modules M_μ with μ of the form $w\lambda$ for $w \in W$; moreover $[P_\lambda : M_{w\lambda}] \leq [F_V(M_0) : M_{w\lambda}] = 1$. Thus $\dim \mathrm{Hom}(P_\lambda, P_\lambda) \leq \sum_w \dim \mathrm{Hom}(P_\lambda, M_{w\lambda}) = \sum_w [M_{w\lambda} : L_\lambda] \leq \#(W)$.

2. Let $P(V)_e$ be the set of extremal weights in $P(V)$. Clearly $P(V)_e = \{w\lambda \mid w \in W\}$. Choose a W-invariant polynomial p on L with the following properties:

a) The polynomial p vanishes up to order $\geq \#(W)$ at all non extremal points of $P(V)$.

b) The polynomial $1-p$ vanishes up to order $\geq \#(W)$ at all extremal points of $P(V)$.

Clearly, the corresponding element $z(p) \in \mathcal{Z}(\mathfrak{g})$, acting on the module $F_V(M_0)$, gives a projection onto the submodule P_λ. This shows, that a function $f \in F(L)^W$ lies in the ideal $J = \mathrm{Ann}(P_\lambda)$ iff $z(f) \cdot z(p) = 0$ on $F_V(M_0)$. In this case clearly $tr_V(z(f) \cdot z(p)) = 0$ on the module M_0.

We claim that the action of the operator $tr_V(z(f) \cdot z(p))$ on M_0 is given by multiplication by $\nu(f)$, which implies that $\nu(f) = 0$ for $f \in J$.

Using formula $(*)$ from section 2 we see that the operator $tr_V(z(f) \cdot z(p))$ acts on M_0 as a scalar $[\sum_\mu (\Lambda f p)(x+\mu)/\Lambda(x)](0)$.

Using properties of p we can rewrite this sum as $[\sum_\mu (\Lambda f)(x+\mu)/\Lambda(x)](0)$, where the sum is over extremal weights μ. Since Λf is skew-symmetric under the action of W, and extremal weights are of the form $w\lambda$, this sum equals to $\nu(f)$.

3. Given a commutative k-algebra B and a linear map $\nu : B \rightarrow k$ we denote by $J(B, \nu)$ the ideal $J(B, \nu) = \{b \in B \mid \nu(bB) = 0\}$ and by

$Q(B,\nu)$ the quotient algebra $B/J(B,\nu)$. By definition $J_\lambda = J(F(L)^W,\nu)$. Our aim is to compute the algebra $Q = Q(F(L)^W,\nu)$.

Set $A = F(L)^W$. Clearly ν vanishes on some power of ideal $J_\theta \subset A$ corresponding to the character θ. Hence the algebra Q will not be changed if we replace A by its completion $\hat{A}$ at θ.

Since λ is regular point of L, the algebra $\hat{A}$ is naturally isomorphic to the completion $\hat{F}_\lambda$ of $F(L)$ at point λ.

Translation operator $T(\lambda)$, $T(\lambda)f(x) = f(x+\lambda)$, identifies $\hat{F}_\lambda$ with the algebra $\hat{F}_0$–completion of $F(L)$ at 0. Let us identify $\hat{A}$ with $\hat{F}_0$ using $T(\lambda)$. Then the functional ν on $\hat{A}$ corresponds to the following functional ν' on $\hat{F}_0$

$$\nu'(f) = \nu(T_\lambda^{-1}(f)) = [(\sum_w \epsilon(w)w(T(\lambda)\Lambda \cdot f))/\Lambda](0).$$

In other words, if we define a linear map $\tau : \hat{F}_0 \to k$ by

$$\tau(h) = [\mathrm{Alt}(h)/\Lambda](0)) = [(\sum_w \varepsilon(w)\cdot w(h))/\Lambda](0)$$

then $\nu'(f) = \tau(T_\lambda(\Lambda)\cdot f)$. Since the function $T_\lambda(\Lambda)$ is invertible in $\hat{F}_0$, we have $Q(A,\nu) = Q(\hat{A},\nu) = Q(\hat{F}_0,\nu') = Q(\hat{F}_0,\tau) = Q(F(L),\tau)$.

In order to describe this last algebra let us consider an ideal J_+ in $F(L)$, generated by W-invariant polynomials of positive degree, and denote by H the quotient algebra $F(L)/J_+$. It is easy to see that $\tau(J_+) = 0$, i.e. τ can be considered as a functional on H, and $Q(F(L),\tau) = Q(H,\tau)$.

By well known result of A. Borel (see [BGG2] or [D]) H is isomorphic to cohomology algebra of flag variety X and functional τ on H is given by evaluation on fundamental class of X. This implies, that the bilinear form $< h, f >= \tau(hf)$ on H is non degenerate and hence $Q(H,\tau) = H$ (direct algebraic proof of the fact that this form is non degenerate see in [D], Prop. 4). This proves lemma 3.

Remark. Slightly modifying above arguments one can prove the following more general result

Theorem. *Let $\lambda \in L$ be any antidominant weight, L_λ an irreducible module with highest weight $\lambda - \rho$ and P_λ its projective cover in category $\mathcal{O}$. Then the natural morphism $\eta : \mathcal{Z}(\mathfrak{g}) \to End(P_\lambda)$ is an epimorphism. Its image is isomorphic to $F(L)^{W(\lambda)}/J(W(\lambda/R))$, where $W(\lambda) = \{w \in W \mid w\lambda = \lambda\}$, $W(\lambda/R) = \{w \in W \mid w\lambda - \lambda \in$ Root lattice $R\}$, $F(L)^{W(\lambda)}$ is the algebra of $W(\lambda)$–invariant polynomial functions on L and $J(W(\lambda/R))$ is an ideal, generated by $W(\lambda/R)$–invariant polynomials of positive degree.*

This finite–dimensional algebra can be realized as cohomology algebra of some partial flag variety.

REFERENCES

[BGG1] J.N.Bernstein, I.M. Gelfand, S.I. Gelfand, *Category of $\mathfrak{g}$-modules*, Funct. Anal. Appl **10** (1976), 87–92.

[BGG2] J.N.Bernstein, I.M. Gelfand, S.I. Gelfand, *Schubert cells and cohomology of the spaces G/P*, Russian Math. Surveys 28 **3** (1973), 1–26.

[D] M. Demazure, *Invariants symetriques des groupes de Weyl et torsion*, Invent. Math. **21** (1973), 287–301.

[S] W. Soergel, *Kategorie O, perverse Garben und Moduln über den Koinvarianten zur Weylgruppe*, preprint.

Received April 30, 1990

Department of Mathematics
Harvard University
Cambridge, MA 02138

Filtered Auslander–Gorenstein Rings

J.-E. BJÖRK AND E.K. EKSTRÖM

Dedicated to Jacques Dixmier on his 65th birthday

Introduction

In this paper we construct a class G_M of good filtrations on a pure module M over a filtered noetherian ring whose associated Rees ring is Auslander–Gorenstein such that the graded modules $\mathrm{gr}_{\Omega}(M)$ are pure for every Ω in the class G_M. We refer to G_M as the class of Gabber filtrations.

Our work has its origin in lectures by O. Gabber at the Université Paris VI in 1981 where he used homological methods to prove in particular that the associated variety of $U(\mathfrak{g})/\mathfrak{p}$ is *equi-dimensional* when $\mathfrak{p}$ is a primitive ideal of an enveloping algebra. The main idea in Gabber's work was to extend results about *pure modules* over regular commutative noetherian rings to *filtered non-commutative noetherian rings.* For non-commutative noetherian rings it turns out that finiteness of the global injective dimension is not sufficient for a nice duality theory. Therefore certain conditions were introduced by M. Auslander in the late sixties. These conditions will play an important role and are recalled in Section 1 where we also define *Auslander–Gorenstein rings.* In Section 2 we define and study pure modules over an Auslander–Gorenstein ring. Using a construction due to R. Fossum we find an affirmative answer to a question raised in [Bj, p. 144]. To be precise, the purity of certain Ext-modules in Proposition 2.11 was only known for Auslander-regular rings in [Bj]. So called *tame pure extensions* are studied in §3 and will later on be used in §4.

In order to make this paper reasonably self-contained we have given a fairly detailed presentation of filtered noetherian rings in §4. The systematic use of *Rees modules* should be noticed. The material about the $\mathcal{E}$-functor in §4 has been inspired from lectures by V. Oystaeyen at a conference in Antwerpen in June 1988. At the end of §4 we give a recent example due to G.M. Bergman which shows that it is essential to use Rees modules in order to define good filtrations over non-positively filtered noetherian rings.

The main results of this paper occur in §5 where Theorem 5.23 gives the conclusive result about *Gabber filtrations* on a pure module.

1. Auslander's Condition

By a noetherian ring we mean a (not necessarily commutative) ring with a multiplicative unit which is both left and right noetherian. Let A be a noetherian ring. The category of finitely generated left, respectively right A-modules is denoted by $\mathrm{Mod}_f(A)$, resp. $\mathrm{Mod}_f(A^0)$. If $M \in \mathrm{Mod}_f(A)$, then right multiplication in the ring A gives a right A-module structure on $\mathrm{Ext}^v_A(M,A)$ for every $v \geq 0$. Moreover, using projective resolutions we see that $\mathrm{Ext}^v_A(M,A)$ belongs to $\mathrm{Mod}_f(A^0)$. Similarly, if $N \in \mathrm{Mod}_f(A^0)$, then $\mathrm{Ext}^v_A(N,A) \in \mathrm{Mod}_f(A)$. Keeping the ring A fixed we simplify the notations for Ext-groups having A as a second factor and write $E^v(M) = \mathrm{Ext}^v_A(M,A)$. More generally, if v and k is a pair of integers then $E^{v,k}(M)$ denotes $\mathrm{Ext}^v_A(\mathrm{Ext}^k_A(M,A),A)$.

1.1 *Definition.* A finitely generated left or right A-module M satisfies Auslander's condition if the following hold: For every $v \geq 0$ and every submodule N of $E^v(M)$, it follows that $E^i(N) = 0$ for every $i < v$.

1.2 *Remark.* The condition in Definition 1.1 occurs in [F-G-R] and [A-B]. These works also contain various equivalent definitions based upon flatness and syzygies. We shall not discuss this, but only rely on Definition 1.1.

1.3 *Injective dimensions.* A noetherian ring A has a finite injective dimension if there exists an integer w such that $E^v(M) = 0$ for every left or right A-module M and $v > w$. We refer to [Z] for the proof that the injective left dimension of A is equal to the injective right dimension when both are finite.

Suppose now that A is a noetherian ring with a finite injective dimension. If M is a finitely generated left or right A-module we construct the derived dual $\mathbf{R}\mathrm{Hom}_A(M,A)$ whose cohomology groups are $E^v(M)$. We have the equality $M = \mathbf{R}\mathrm{Hom}_A(\mathbf{R}\mathrm{Hom}_A(M,A),A)$, called the biduality formula. We refer to [Le 2] for the proof, based upon the use of the *Ischebeck complex.* The biduality formula implies that if M is a non-zero finitely generated A-module, then there exists at least one integer $v \geq 0$ such that $E^v(M) \neq 0$. This gives

1.4 *Definition.* Let M be a non-zero and finitely generated A-module. The unique smallest integer k such that $E^k(M) \neq 0$ is denoted by $j_A(M)$ and called the grade number of M.

1.5. *Definition.* A noetherian ring A with a finite injective dimension is called an Auslander–Gorenstein ring if every finitely generated A-module satisfies Auslander's condition.

1.6 *Auslander regular rings.* A noetherian ring A with a finite *global homological dimension* is called an Auslander regular ring if Auslander's condition holds for every finitely generated A-module. Since finiteness of gl.dim(A) implies that A has a finite injective dimension, it follows that Auslander regular rings is a subfamily of Auslander–Gorenstein rings. So all the subsequent results which are announced for Auslander–Gorenstein rings also hold for Auslander regular rings.

1.7 *Remark.* In [Re: Example 2.4.6] occurs an example of a noetherian ring A with gl.dim$(A) = 2$ but Auslander's condition does not hold for every finitely generated A-module. So A is not Auslander regular. See also [Bj: p. 138] for Reiten's example.

1.8 *The case when A is commutative.* Let A be a commutative noetherian ring with a finite injective dimension. Then the work by H. Bass in [Ba] shows that Auslander's condition holds and hence A is an Auslander–Gorenstein ring.

2. Pure Modules

Throughout this section A is an Auslander–Gorenstein ring. We shall perform a construction due to R. Fossum in [Fo]. Let M be a non-zero finitely generated A-module and put $n = j_A(M)$. We stop a projective resolution of M after $n+1$ steps. Thus, we consider a complex

$$P_{n+1} \to \cdots \to P_0 \to M \to 0 \tag{2.1}$$

where $P_0, \ldots, P_{n+1}$ are finitely generated projective A-modules. The map $P_{n+1} \to P_n$ is in general not injective. Now we consider the complex $\mathbf{P}^*$, defined by

$$\mathrm{Hom}_A(P_0, A) \to \cdots \to \mathrm{Hom}_A(P_{n+1}, A). \tag{2.2}$$

2.3 *Definition.* We set $H^{n+1}(\mathbf{P}^z) = \mathfrak{F}(M)$ and refer to $\mathfrak{F}(M)$ as a Fossum module associated with M.

2.4 *Remark.* Let $P_\bullet$ and $Q_\bullet$ be two projective resolutions of M which both stop after $n+1$ steps. By a well known result in homological algebra it follows that the *projective equivalence* classes of $H^v(\mathbf{P}^*)$ and $H^v(Q^*)$ are the same for every integer v.

In particular this holds if $v = n+1$. We conclude that the projective equivalence class of a Fossum module associated with M is unique. So $E^v(\mathfrak{F}(M))$ depend only upon M if $v \geq 1$.

2.5 **Proposition.** *Let M be a non-zero finitely generated A-module. Then there exists an exact sequence*

$$0 \to E^{n+1}(\mathfrak{F}(M)) \to M \to E^{n,n}(M) \to E^{n+2}(\mathfrak{F}(M)) \to 0$$

and if $v > n$, then $E^{v,n}(M) = E^{v+2}(\mathfrak{F}(M))$.

Proof. Consider the derived Hom-complex $\mathbf{R}\mathrm{Hom}_A(\mathbf{P}^*, A)$, where $P_\bullet$ is a projective resolution of M from 2.1. We notice that $H^0(\mathbf{R}\mathrm{Hom}_A(\mathbf{P}^*, A)) = M$ while $H^{-(n+1)}(\mathbf{R}\mathrm{Hom}_A(\mathbf{P}^*, A)) = \mathrm{Ker}(P_{n+1} \to P_n)$. Next, we have a spectral sequence with

$$E_2^{p,q} = E^p(H^{-q}(\mathbf{P}^*))$$

which abuts to $\mathbf{R}\mathrm{Hom}_A(\mathbf{P}^*, A)$. Moreover, the definition of the grade number $j_A(M)$ shows that $H^i(\mathbf{P}^*) = 0$ for $i \neq n,\ n+1$; $H^n(\mathbf{P}^*) = E^n(M)$ and $H^{n+1}(\mathbf{P}^*) = \mathfrak{F}(M)$. It follows that the spectral sequence above gives the exact sequence in Proposition 2.5 when we identify M with $H^0(\mathbf{R}\mathrm{Hom}_A(\mathbf{P}^*, A))$.

2.6 *Remark.* The exact sequence in Proposition 2.5 is due to R. Fossum in [Fo: Proposition 6].

2.7 *Definition.* A finitely generated A-module M is pure if $j_A(M') = j_A(M)$ for every non-zero submodule M'.

In Theorem 2.12 we establish equivalent conditions in order that an A-module is pure. But first we need some preliminary results.

2.8 **Lemma.** *Let M be a non-zero finitely generated A-module. Then, if $n = j_A(M)$, it follows that $E^n(M)$ is non-zero and its grade number is equal to n.*

Proof. Auslander's condition gives $j_A(E^{n+1}(\mathfrak{F}(M)) \geq n+1$. Then Proposition 2.5 shows that $E^{n,n}(M) \neq 0$. It follows that $E^n(M) \neq 0$ and

$j_A(E^n(M)) \leq n$. Finally, Auslander's condition also gives $j_A(E^n(M)) \geq n$ and Lemma 2.8 is proved.

2.9 **Lemma.** *Let* $0 \to M \to M' \to M'' \to 0$ *be an exact sequence of finitely generated A-modules. Then*

$$\inf\{j_A(M), j_A(M'')\} = j_A(M').$$

Proof. We have the long exact sequence

$$E^v(M'') \to E^v(M') \to E^v(M) \to E^{v+1}(M'').$$

Now Lemma 2.8 and Auslander's condition easily give Lemma 2.9.

2.10 **Lemma.** *Let M be a finitely generated A-module such that* $j_A(E^v(M)) > v$ *for every* $v > j_A(M)$. *Then M is pure.*

Proof. Let $0 \neq M' \subset M$ and suppose that $j_A(M') = v > j_A(M)$. The long exact sequence from Lemma 2.9 applied to $0 \to M' \to M \to M/M' \to 0$ gives $j_A(E^v(M')) > v$. This contradicts Lemma 2.8 and hence M is pure.

2.11 **Proposition.** *Let M be a finitely generated A-module. Then, with* $n = j_A(M)$ *it follows that* $E^n(M)$ *is pure.*

Proof. Set $M' = E^n(M)$. Lemma 2.8 gives $j_A(M') = n$. If $v > n$ we have $E^v(M') = E^{v+2}(\mathfrak{F}(M))$ by Proposition 2.5. This gives $j_A(E^v(M')) \geq v + 2 > v$ and hence M' is pure by Lemma 2.10.

2.12 **Theorem.** *Let M be a finitely generated A-module. Then, with* $n = j_A(M)$ *the following are equivalent:*

(1): *M is pure,* (2): $E^{n+1}(\mathfrak{F}(M)) = 0$, (3): $E^{v,v}(M) = 0$ *for every* $v > n$.

Proof. Lemma 2.10 gives (3) $\Rightarrow$ (1). Next, if (1) holds then the exact sequence in Proposition 2.5 shows that $E^{n+1}(\mathfrak{F}(M))$ is a submodule of the pure A-module M. Auslander's condition gives $j_A(E^{n+1}(\mathfrak{F}(M)) \geq n + 1$ and hence the purity of M implies that $E^{n+1}(\mathfrak{F}(M)) = 0$. This proves that (1) $\Rightarrow$ (2). There remains to show (2) $\Rightarrow$ (3). Put $M' = E^n(M)$ and then (2) and Proposition 2.5 give the exact sequence $0 \to M \to E^n(M') \to E^{n+2}(\mathfrak{F}(M)) \to 0$. Moreover, $v > n$ gives $E^v(M') = E^{v+2}(\mathfrak{F}(M))$ and hence $j_A(E^v(M')) \geq v + 2$. Then, using the long exact sequence from the proof of Lemma 2.9 we get $j_A(E^v(M)) > v$ for every $v > n$. This gives $E^{v,v}(M) = 0$ when $v > n$ and hence (2) $\Rightarrow$ (3).

2.13 *The case when A is commutative.* Let A be a commutative Auslander–Gorenstein ring. If M is a finitely generated A-module then the radical of the annihilating ideal $(0 : M) = \{a \in A : aM = 0\}$ is denoted by $J(M)$ and called the *characteristic ideal* of M. The radical ideal $J(M)$ has a unique prime decomposition, i.e. we have $J(M) = \mathfrak{p}_1 \cap \cdots \cap \mathfrak{p}_s$, where $\mathfrak{p}_1, \ldots, \mathfrak{p}_s$ are the minimal prime divisors of $J(M)$. Next, for every prime ideal $\mathfrak{p}$ of the ring A we denote by $A_\mathfrak{p}$ the localization of A with respect to $\mathfrak{p}$. Recall that $A_\mathfrak{p}$ is a local ring whose maximal ideal is $\mathfrak{p}A_\mathfrak{p}$.

2.14 **Theorem.** *Let A be a commutative Auslander–Gorenstein ring. Then, for every finitely generated A-module M we have*

$$j_A(M) = \inf\{\text{inj. dim}(A_\mathfrak{p})$$

with inf *taken over the family of minimal prime divisors of* $J(M)$.

Proof. The work by H. Bass in [Ba] gives the following for every prime ideal $\mathfrak{p}$ of the ring A:

$$\text{depth}(A_\mathfrak{p}) = \text{inj.dim}(A_\mathfrak{p}) = ht(\mathfrak{p}).$$

Moreover, the study of grade numbers in [Ba] gives the equality

$$j_A(M) = \inf_{\mathfrak{p} \supseteq J(M)} \{\text{depth}(A_\mathfrak{p})\}.$$

Then we easily get Theorem 2.14.

2.15 *Remark.* Let $\mathfrak{p}$ be a prime ideal in the ring A and consider the cyclic A-module $M = A/\mathfrak{p}$. Then $J(M) = \mathfrak{p}$ and Theorem 2.14 gives $j_A(M) = \text{inj.dim}(A_\mathfrak{p})$. Using this formula and the well known fact that every non-zero submodule N of M contains a submodule which is isomorphic with $A/\mathfrak{p}$, it follows that $j_A(N) = j_A(M)$ and hence the A-module M is pure. More generally, if $\mathfrak{g}$ is a primary ideal and $M = A/\mathfrak{g}$, then $J(M)$ is the prime radical $\sqrt{\mathfrak{g}}$. By a similar reasoning as above we find that M is a pure A-module. Next, using Theorem 2.14 one easily obtains the following:

2.16 **Proposition.** *Let A be a commutative Auslander–Gorenstein ring. Then, if M is a pure A-module, it follows that* $\text{inj.dim}(A_\mathfrak{p}) = j_A(M)$ *for every minimal prime divisor of* $J(M)$.

2.17 *Remark.* Proposition 2.16 may be expressed by saying that the radical ideal $J(M)$ if a pure A-module M is equi-dimensional in the sense that inj.dim($A_{\mathfrak{p}}$) are equal for all minimal prime divisors $\mathfrak{p}$ of $J(M)$.

2.18 *A final remark about purity.* Using Theorem 2.14 we find a necessary and sufficient condition in order that a cyclic A-module A/I is pure. Namely, let I be an ideal of the commutative Auslander–Gorenstein ring. Then A/I is a pure A-module if and only if I has a primary decomposition without embedded primary components and $\sqrt{I}$ is equi-dimensional. In other words, A/I is pure if and only if $I = \mathfrak{g}_1 \cap \cdots \cap \mathfrak{g}_s$, where $\sqrt{\mathfrak{g}_1}, \ldots, \sqrt{\mathfrak{g}_s}$ are the minimal prime divisors of the radical ideal $\sqrt{I}$ and moreover, inj.dim$(A_{\sqrt{\mathfrak{g}_i}}) = j_A(A/I)$ holds for every $1 \leq i \leq s$.

3. Tame Pure Extensions

Throughout this section A denotes an Auslander–Gorenstein ring. When we refer to a pure A-module we always assume that the module is finitely generated and satisfies the condition in Definition 2.7.

3.1 *Definition.* Let M be a pure A-module. A tame pure extension of M consists of a pair (M', α), where $\alpha: M \to M'$ is an injective A-linear map, M' is a pure A-module and $j_A(M'/\alpha(M)) \geq j_A(M) + 2$.

Remark. Let (M', α) be a tame pure extension. Lemma 2.9 gives $j_A(M') = j_A(M)$.

Example. Let M be a pure A-module and construct a Fossum module $\mathfrak{F}M$. Set $n = j_A(M)$. Theorem 2.12 gives $E^{n+1}(\mathfrak{F}M) = 0$ and then Proposition 2.5 gives the exact sequence

$$0 \to M \to E^{n,n}(M) \to E^{n+2}(\mathfrak{F}M) \to 0.$$

Lemma 2.8 and Proposition 2.11 imply that $E^{n,n}(M)$ is pure. Next, Auslander's condition gives $j_A(E^{n+2}(\mathfrak{F}M)) \geq n + 2$. Hence the injective map $M \to E^{n,n}(M)$ gives a tame pure extension of M.

In Theorem 3.6 we are going to show that if M is a pure A-module with $n = j(M)$, then $E^{n,n}(M)$ is a *maximal tame pure extension* of M. But first we need some preliminary results.

3.3 *Definition.* A finitely generated A-module M which satisfies $j(E^v(M)) \geq v + 2$ for every $v > j(M)$ is called a strongly pure A-module.

Remark. A strongly pure module is pure by Lemma 2.10.

3.4 **Proposition.** *Let M be a strongly pure A-module. Then, for any tame pure extension (M', α) it follows that $\alpha(M) = M'$.*

Proof. We have the exact sequence

$$0 \longrightarrow M \xrightarrow{\alpha} M' \longrightarrow M'/M \longrightarrow 0.$$

Suppose that M'/M is non-zero and put $v = j(M'/M)$. The long exact sequence contains

$$(i) \qquad \cdots \longrightarrow E^{v-1}(M) \longrightarrow E^{v}(M'/M) \longrightarrow E^{v}(M') \longrightarrow \cdots$$

Since M is strongly pure and $v - 1 \geq j(M) + 2 - 1 > j(M)$, it follows that $j(E^{v-1}(M)) \geq v + 1$. Also, since M' is pure we have $j(E^{v}(M')) > v$. Now (i) and Lemma 2.8 imply that $j(E^{v}(M'/M)) > v$. But this contradicts Lemma 2.9 and we conclude that M'/M is zero and hence α is surjective.

3.5 **Proposition.** *Let M be a finitely generated A-module. Then $E^{j(M)}(M)$ is strongly pure.*

Proof. Put $n = j(M)$. If $v > n$ we have $E^{v}(E^{n}(M)) = E^{v+2}(\mathfrak{J}M)$ by Proposition 2.6. This gives $j(E^{v}(E^{n}(M)) \geq v + 2$ and hence $E^{n}(M)$ is strongly pure.

3.6 **Theorem.** *Let M be a pure A-module with $n = j(M)$. Then, if (M', α) is a tame pure extension there exists an injective A-linear map $\alpha' \colon M' \longrightarrow E^{n,n}$ such that the diagram below commutes*

$$\begin{array}{ccc} M & \xrightarrow{\alpha} & M' \\ & \searrow \quad \swarrow {\scriptstyle \alpha'} & \\ & E^{n,n}(M) & \end{array}$$

Before we prove Theorem 3.6 we insert some remarks about Fossum modules. Consider an injective A-linear map $\alpha \colon M \longrightarrow M'$ where $j(M) = j(M') = n$.

We can construct a commutative diagram

$$\begin{array}{ccccccccc} P_{n+1} & \longrightarrow & \cdots & \longrightarrow & P_0 & \longrightarrow & M & \longrightarrow & 0 \\ \downarrow & & & & \downarrow & & \downarrow & & \\ Q_{n+1} & \longrightarrow & \cdots & \longrightarrow & Q_0 & \longrightarrow & M' & \longrightarrow & 0 \end{array}$$

where $P_{\bullet}$, respectively $Q_{\bullet}$ are projective resolutions as in 2.1. Next, the map $\alpha \colon M \to M'$ induces a map $\beta \colon E^{n,n}(M) \to E^{n,n}(M')$ using functoriality. See A.7 Corollary in [L-S] for details. Moreover, since $E^{n+1}(\mathfrak{J}(M)) =$

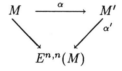

$E^{n+1}(\mathfrak{F}(M')) = 0$ by (2) in Theorem 2.12, it follows from the remarks made above and Proposition 2.5 that we obtain a commutative diagram

$$\begin{array}{ccccccccc} 0 & \longrightarrow & M & \longrightarrow & E^{n,n}(M) & \longrightarrow & E^{n+2}(\mathfrak{F}(M)) & \longrightarrow & 0 \\ & & \alpha\downarrow & & \beta\downarrow & & \downarrow & & \\ 0 & \longrightarrow & M' & \longrightarrow & E^{n,n}(M') & \longrightarrow & E^{n+2}(\mathfrak{F}(M')) & \longrightarrow & 0 \end{array}$$

3.7 **Lemma.** *The map β is an isomorphism.*

Proof. Identify M with a submodule of $E^{n,n}(M)$. Then, the injectivity of α implies that $M \cap \mathrm{Ker}(\beta) = 0$. So the A-module $\mathrm{Ker}(\beta)$ is a submodule of $E^{n,n}(M)$ and is also isomorphic with a submodule of $E^{n,n}(M)/M$. Since $j_A(E^{n,n}(M)/M) \geq n+2$ we get $j_A(\mathrm{Ker}(\beta)) \geq n+2$. Then the purity of $E^{n,n}(M)$ shows that $\mathrm{Ker}(\beta) = 0$ so β is injective. To prove that β is surjective we notice that $\mathrm{Im}(\beta)$ contains $\mathrm{Im}(\alpha)$. Since $j_A(E^{n,n}(M')/M')$ and $j_A(M'/\alpha(M))$ both are $\geq n+2$, it follows from Lemma 2.9 that $j_A(E^{n,n}(M')/\mathrm{Im}(\beta)) \geq n+2$.

Finally, apply Proposition 3.5 to the A-module $E^n(M')$. It follows that $E^{n,n}(M')$ is strongly pure and then $\mathrm{Im}(\beta) = E^{n,n}(M')$ by Proposition 3.4.

Proof of Theorem 3.6. Denote by γ the injective map from M' into $E^{n,n}(M')$. Then we can use the composed map $\beta^{-1}\gamma$ as α' in Theorem 3.6.

Remark. Theorem 3.6 is proved when the ring A is Auslander regular in [L-S] and of course the proof above was inspired by the material in the appendix of [L-S].

3.8 **Corollary.** *Let M be a non-zero and finitely generated A-module with $n = j_A(M)$. Then the following are equivalent:*

(1): *M is strongly pure;* (2): $M = E^{n,n}(M)$; (3): $E^{n+2}(\mathfrak{F}(M)) = E^{n+1}(\mathfrak{F}(M)) = 0$.

4. Filtered Rings

Let A be a ring with a unit element 1_A. The ring A is said to be filtered if we have an increasing sequence of additive subgroups $\cdots A_{v-1} \subset A_v \subset A_{v+1} \subset \cdots$, indexed by integers such that:

$$\cup A_v = A; \quad \cap A_v = 0; \quad 1_A \in A_0; \quad A_v A_k \subset A_{v+k}.$$

Notice that we do not require that $\{A_v\}$ is a strictly increasing sequence. If $A_{-1} = 0$ we say that the filtration is *positive.*

4.1 *The Rees ring.* Let A be a filtered ring. Then we construct a graded ring R_A as follows: Consider first the ring of finite Laurent series in one variable over A. Thus, we take the ring $S = A[T, T^{-1}]$ and define a graded ring structure on S such that T is homogeneous of degree 1 while A is the subring of S-homogeneous elements of degree zero. Hence the vth homogeneous component $S(v)$ is equal to AT^v. Now R_A is the graded subring of S such that $R_A(v) = A_v T^v$ for every integer v.

If v is an integer we denote by j_v the bijective map from A_v into $R_A(v)$, defined by $j_v(x) = xT^v$ for every x in A_v. We notice that if $x \in A_v$ and $y \in A_k$ for some pair of integers, then the product $j_v(x)j_k(y)$ in the ring R_A is equal to $j_{v+k}(xy)$. The ring R_A contains the element T, where $T = j_1(1_A)$. It is obvious that T is a central element in R_A and since the filtration on the ring A is exhaustive, it follows that the localization of R_A with respect to the multiplicative set $\{1, T, T^2, \ldots\}$ is equal to S. In other words, we have

4.2 $R[T^{-1}] = A[T, T^{-1}]$.

Next, denote by $(1 - T)$ the two-sided ideal in the ring R_A generated by the central element $1 - T$. Then we have

4.3 Proposition. $R/(1 - T) = A$.

We refer to [Bj: Proposition 2.13] for the easy proof. Next, let (T) be the two-sided ideal in R_A generated by T. Let us also construct the *associated graded ring* of A, defined by $GA = \oplus A_v/A_{v-1}$. Since $TR_A(v - 1) = A_{v-1}T^v$ holds for every integer v we easily obtain

4.4 Proposition. *The graded rings $R_A/(T)$ and GA are isomorphic.*

4.5 *Filtered A-modules.* Let A be a filtered ring. A filtration on a left A-module M consists of an increasing sequence of additive subgroups $\{\Gamma_v\}$ such that $\cup\Gamma_v = M$ and $A_k\Gamma_v \subset \Gamma_{k+v}$ for all pairs v, k. The filtration $\{\Gamma_v\}$ is called *separated* if $\cap\Gamma_v = 0$. When $\{\Gamma_v\}$ is a filtration on M we refer to the pair (M, Γ) as a *filtered A-module.* Similarly we construct *filtered right* A-modules. Let (M, Γ) be a filtered left A-module. Then we construct a graded R_A-module M_Γ such that the v:th homogeneous component $M_\Gamma(v) = \Gamma_v T^v$. We refer to M_Γ as the *Rees module* associated with (M, Γ).

Concerning the R_A-module M_Γ we notice that the R_A-element T is injective on M_Γ since the map $M_\Gamma(v) \xrightarrow{T} M_\Gamma(v + 1)$ corresponds to the injective inclusion $\Gamma_v \to \Gamma_{v+1}$ in the given A-module M.

Next, denote by GM the *associated graded GA-module* of (M, Γ). Thus, $GM = \oplus \Gamma_v / \Gamma_{v-1}$. Identifying $R_A/(T)$ with GA, it follows easily that the graded GA-modules GM and M_Γ / TM_Γ are the same.

4.7 *The $\mathcal{E}$-functor.* Denote by $G(R_A)$ the category of *graded* left R_A-modules. A graded R_A-module M is *T-torsion free* if T is injective on M. If we instead assume that every element in a graded R-module M is annhilated by some power of T, then we say that M has *T-torsion.* If $M \in G(R_A)$ we see that M contains a unique largest graded submodule with T-torsion. It is denoted by M_t and then M/M_t is T-torsion free.

4.8 **Lemma.** *Let $M \in G(R_A)$. Then $M_t = (1 - T)M_t$ and $1 - T$ is injective on the T-torsion free module M/M_t.*

We leave out the easy proof. Let us now consider a graded R_A-module M. Set $M = \oplus M(v)$. Identifying $R_A/(1 - T)$ with A, it follows that $M/(1-T)M$ is an A-module, denoted by $\mathcal{E}(M)$. We obtain a filtration on the A-module $\mathcal{E}(M)$ such that

$$\mathcal{E}(M)_v = [M(v) + (1 - T)M]/(1 - T)M. \tag{4.9}$$

We see that $M \to \mathcal{E}(M)$ gives a functor from the category of graded left R_A-modules into the category of filtered left A-modules.

4.10 **Proposition.** *Let M be a graded T-torsion free left R_A-module. Then the Rees module of $\mathcal{E}(M)$ is equal to M in the category $G(R_A)$.*

Proof. By 4.9 and *Noether's Isomorphism* we see that $\mathcal{E}(M)_v$ is equal to $M(v)/[M(v) \cap (1-T)M]$. Since M is T-torsion free it is easily seen that $(1 - T)M \cap M(v) = 0$ for every integer v. Then we get Proposition 4.10 by the construction of the Rees module $\oplus \mathcal{E}(M)_v T^v$.

Next, let (M, Γ) be a filtered A-module. Apply the $\mathcal{E}$-functor to the Rees module M_Γ. Using Noether's isomorphism as in the proof of Proposition 4.10 we obtain

$$\mathcal{E}(M_\Gamma)_v = M_\Gamma(v) = \Gamma_v T^v \cong \Gamma_v.$$

We conclude that $\mathcal{E}(M_\Gamma)$ is equal to (M, Γ) in the category of filtered A-modules.

4.11 *Noetherian filtered rings.* Let A be a filtered ring. If R_A is a left and a right noetherian ring then we say that A is a *noetherian filtered ring.* From now on we assume that A is a noetherian filtered ring. Since $A = R_A/(1-T)$ and $GA = R_A/(T)$ it follows that the rings A and GA are noetherian. Denote by $\mathrm{Mod}_f(A)$ the category of finitely generated left A-modules.

4.12 *Definition.* Let $M \in \mathrm{Mod}_f(A)$. A filtration Γ on M such that M_Γ is a finitely generated R_A-module is called a good filtration.

4.13 *Remark.* Let $M \in \mathrm{Mod}_f(A)$. If $m_1, \ldots, m_s$ is a finite set of generators of the A-module M and $k_1, \ldots, k_s$ some s-tuple of integers, then we construct a filtration Γ on M such that

$$\Gamma_v = A_{v-k_1}m_1 + \cdots + A_{v-k_s}m_s.$$

Now the R_A-module M_Γ is generated by $j_{k_1}(m_1), \ldots, j_{k_s}(m_s)$ and hence Γ is a good filtration. Conversely, let Γ be a good filtration on M. We can find a finite set of homogeneous elements $j_{k_1}(m_1), \ldots, j_{k_s}(m_s)$ in M_Γ which generate M_Γ as a left R_A-module. It follows that

$$M_\Gamma(v) = M_v T^v = \Sigma R_A(v-k_i) j_{k_i}(m_i).$$

Then we see that $\Gamma_v = \Sigma A_{v-k_i} m_i$ holds in M, i.e. Γ is of the form above. So in this way we have described the class of good filtrations on M. In particular this description of good filtrations shows that if Γ and Ω are two good filtrations on M, then there exists an integer w such that

$$\Gamma_{v-w} \subset \Omega_v \subset \Gamma_{v+w} \tag{4.14}$$

hold for every v. Thus, up to shifts the two good filtrations increase with the same rate.

4.15 *The use of Rees modules.* Let $M \in \mathrm{Mod}_f(A)$ and choose some good filtration Γ on M. Now M_Γ is a T-torsion free R_A-module and hence M_Γ is a graded R_A-submodule of the localization $M_\Gamma[T^{-1}]$. Recall that $R_A[T^{-1}]$ is equal to $A[T, T^{-1}]$. It follows easily that we have

$$M_\Gamma[T^{-1}] = A[T, T^{-1}] \otimes_A M = M[T, T^{-1}].$$

Hence $M_\Gamma[T^{-1}]$ does not depend on the chosen good filtration. Next, consider $M[T, T^{-1}]$ which is a graded R_A-module. Let $\mathcal{M}$ be a graded R_A- submodule of $M[T, T^{-1}]$ such that the localization $\mathcal{M}[T^{-1}]$ is equal to $M[T, T^{-1}]$.

Applying the $\mathcal{E}$-functor we obtain the filtered A-module $\mathcal{E}(M)$. Moreover, since $\mathcal{M}[T^{-1}]/\mathcal{M}$ has T-torsion, and since the A-module $M[T,T^{-1}]/(1-T)M[T,T^{-1}]$ obviously is equal to M, it follows from the surjectivity of $1-T$ on every T-torsion R_A-module that the A-module $\mathcal{E}(M)$ is equal to M. Thus, $\mathcal{E}(\mathcal{M})$ is a filtered A-module of the form (M,Ω), where Ω is some filtration on M. By Proposition 4.10 we have $M_\Omega = \mathcal{M}$ and hence Ω is a good filtration if and only if $\mathcal{M}$ is a finitely generated R_A-module.

Conversely, let Ω be some good filtration on M which in general is not equal to Γ. Now $M_\Omega[T^{-1}] = M[T,T^{-1}]$ and we conclude that M_Ω is a graded and finitely generated R_A-submodule of $M[T,T^{-1}]$. Summing up, we have proved the following:

4.16 **Proposition.** *Let $M \in \mathrm{Mod}_f(A)$. Then there exists a 1–1 correspondence between the family of good filtrations on M and the family of finitely generated and graded R_A-submodules $\mathcal{M}$ of $M[T,T^{-1}]$ whose localizations $\mathcal{M}[T^{-1}] = M[T,T^{-1}]$.*

4.17 *Remark.* Let $M \in \mathrm{Mod}_f(A)$ and consider a pair of good filtrations Γ and Ω. We say that Ω *increases faster* than Γ if $\Gamma_v \subset \Omega_v$ for every v. Then we write $\Gamma < \Omega$. We notice that $\Gamma < \Omega$ holds if and only if M_Γ is a submodule of M_Ω. Next, if $\Gamma < \Omega$ holds we know that $M_\Gamma[T^{-1}] = M_\Omega[T^{-1}] = M[T,T^{-1}]$. Hence the R_A-module M_Ω/M_Γ has T-torsion. Since it is finitely generated we find a positive integer w such that T^w annihilates M_Ω/M_Γ. This means that $\Omega_{v-w} \subset \Gamma_v$ for every v and this reflects the comparison relation from 4.14.

4.18 *Induced good filtrations.* Let Γ be a good filtration on some finitely generated A-module M. Let N be some A-submodule of M. Put $\Omega_v = N \cap \Gamma_v$. Then Ω is a filtration on N such that N_Ω is a graded R_A-submodule of M_Γ. It follows from the noetherianness of R_A that $N_\Omega \in \mathrm{Mod}_f(R_A)$. Hence Ω is a good filtration on N. So the good filtration on M induces a good filtration on every submodule. Similarly we see that a good filtration on M induces a good filtration on every quotient module.

4.19 *Associated graded modules.* Let Γ be a good filtration on some M in $\mathrm{Mod}_f(A)$. Identifying $R_A/(T)$ with GA, it follows that M_Γ/TM_Γ is a graded GA-module which we denote by $\mathrm{gr}_\Gamma(M)$. So we have $\mathrm{gr}_\Gamma(M) = \oplus\Gamma_v/\Gamma_{v-1}$ and refer to $\mathrm{gr}_\Gamma(M)$ as the *associated graded GA-module* of the filtered A-module (M,Γ). Since $M_\Gamma \in \mathrm{Mod}_f(R_A)$, it follows that $\mathrm{gr}_\Gamma(M) \in \mathrm{Mod}_f(GA)$. Since we have not assumed that filtrations are separated it may occur that $\mathrm{gr}_\Gamma(M)$ is the zero module. Of course, we have $\mathrm{gr}_\Gamma(M) = 0$ if and only if $\Gamma_v = M$ for every integer v. For a given noetherian filtered ring A we obtain a subcategory of $\text{Mod}_f(A)$ whose objects are such that $\text{gr}_\Gamma(M) = 0$ for every good filtration Γ. Denote this subcategory by $\overline{\text{Mod}_f}(A)$. Notice that if $M \in \overline{\text{Mod}_f}(A)$ is such that $\text{gr}_\Gamma(M) = 0$ for some good filtration Γ, then 4.9 implies that $\text{gr}_\Omega(M) = 0$ for any other good filtration Ω and hence M is an object in $\overline{\text{Mod}_f}(A)$.

4.20 *Example.* An important example of a filtered noetherian ring A such that $\overline{\text{Mod}_f}(A)$ contains non-zero modules is the following: Let A be the Weyl algebra $A_1(K)$ in one variable over a commutative field K with its two K-algebra generators t and ∂ satisfying $\partial t - t\partial = 1$. We construct a filtration $\{A_v\}$ such that $t \in A_{-1}$ and $\partial \in A_1$. This is the so-called *V-filtration.* Now $t^2\partial$ belongs to A_{-1} and the A-element $1 - t^2\partial$ is not invertible in the ring A. So the cyclic module $M = A/A(1 - t^2\partial)$ is non-zero. We consider the good filtration on M defined by $\Gamma_v = [A_v + A(1 - t^2\partial)]/A(1 - t^2\partial)$. It is easily seen that $\text{gr}_\Gamma(M) = 0$ and hence M belongs to $\overline{Mod_f}(A)$.

4.21 *Zariskian filtered rings.* A noetherian filtered ring A is called a Zariskian filtered ring if good filtrations on finitely generated left or right A-modules always are separated.

4.22 *Example.* Let A be a positively filtered ring. Then, if GA is a left and right noetherian ring it is well known that R_A is a left and a right noetherian ring. Hence *a positively filtered ring A is filtered noetherian under the sole assumption that GA is noetherian.* It is also obvious that if A is positively filtered with GA noetherian then A is zariskian. In fact, if Γ is a good filtration on some M in $\text{Mod}_f(A)$, then Remark 4.13 and the hypothesis that $A_{-1} = 0$ imply that there exists an integer v_0 such that $v < v_0$ gives $\Gamma_v = 0$. Moreover one has the following

4.23 **Proposition.** *Let A be a positively filtered ring such that GA is noetherian. Then a filtration Γ on some M in* $\text{Mod}_f(A)$ *is good if and only if $\cap\Gamma_v = 0$ and* $\text{gr}_\Gamma(M) \in \text{Mod}_f(GA)$.

We leave out the easy proof. Let us now consider a zariskian filtered ring A where we assume that $A_v \neq 0$ for every $v < 0$. Let $M \in \text{Mod}_f(A)$ and suppose that Γ is a separated filtration on M such that $\text{gr}_\Gamma(M)$ is a finitely generated GA-module. Then we may ask if Proposition 4.23 still holds, i.e. we may ask if the conditions above imply that Γ is a good filtration. The example below shows that Γ is not good in general. Namely, we are going to construct a zariskian filtered ring A and a separated filtration Γ on the free left A-module A such that $\text{gr}_\Gamma(A) \in \text{Mod}_f(GA)$ and yet Γ is not a good filtration. Let us also remark that the example below answers a question raised on page 158 of [Bj].

4.24 *An example.* The constructed example below is due to G.M. Bergman. Let $\mathbf{Z}$ be the ring of integers and consider the ring $A = \{p(t)/q(t) : p(t), q(t) \in \mathbf{Z}[t]$ and $q(0) = 1\}$. We notice that A is a subring of the formal power series ring $\mathbf{Z}[[t]]$ which we denote by $\hat{A}$. Now we have

4.25 **Lemma.** *There exists an element γ in $\hat{A}$ of the form $\gamma = 2 + \varepsilon_1 t + \varepsilon_2 t^2 + \cdots$ such that each ε_i is 0 or 1 and γ does not divide any element of A in A.*

Proof. The ring A has countably many elements and $\hat{A}$ is a unique factorization domain, hence up to associates there are only countably many elements in $\hat{A}$ dividing elements of A. But there are uncountably many γ-elements and hence a sequence $\varepsilon_1, \varepsilon_2, \ldots$ exists so that no non-zero element f in A belongs to the principal ideal in $\hat{A}$ generated by γ.

Having found γ from Lemma 4.25 we consider the localization of the ring A with respect to $1, t, t^2, \ldots$. Denote this ring by B. Since t belongs to the *Jacobson radical* of the ring A we see that B has a *zariskian filtration* given by $B_v = t^v A$ for every integer v. Then, if M is some A-submodule of B such that $A \subset M$ and $\cap t^v M = 0$, we get a *separated filtration* on the free B-module B by $\Gamma_v = t^v M$. We notice that $\mathrm{gr}_\Gamma(B)$ is a finitely generated GB-module if M/tM is a finitely generated $\mathbf{Z}$-module. Moreover, Γ is a *good filtration if and only if M is a finitely generated* A-module. Now we construct M so that $\cap t^v M = 0$ and M/tM is a finitely generated $\mathbf{Z}$-module, but the A-module M is *not* finitely generated. To obtain M we use the localization of $\hat{A}$ with respect to $1, t, t^2, \ldots$. So then $\hat{A}[t^{-1}]$ is the ring $\mathbf{Z}((t))$. Put

$$M = B \cap (\gamma \hat{A}[t^{-1}] + \hat{A}).$$

We see that $A \subset M \subset B$ and M is an A-module. If $n \geq 1$ we see that M contains $2t^{-n} + \varepsilon_1 t^{-n+1} + \cdots + \varepsilon_1 t^{-1}$. It follows that the A-module M is not finitely generated. Next, if $m \in M$ we write $m = \varphi(t)\gamma + \delta(t)$ with $\varphi \in \hat{A}[t^{-1}]$ and $\delta \in \hat{A}$. Now $\delta(0) \in \mathbf{Z}$ and we have $m - \delta(0) = t(t^{-1}\varphi(t)\gamma + t\delta_1(t))$, so $m - \delta(0) \in tM$ and hence $M/tM = \mathbf{Z}$. There remains only to show that $\cap t^v M = 0$. Suppose this intersection contains a non-zero element f. Multiplying f with a t-power we may assume that $f \in A$. If $n \geq 1$ we get $t^{-n} f$ in $\gamma \hat{A}[t^{-1}] + \hat{A}$ and write $t^{-n} f = \varphi_n(t)\gamma + \delta_n(t)$. We see that $t^n \varphi_n(t)$ belongs to $\hat{A}$ and get $f = t^n \varphi_n(t) + t^n \delta_n(t)$. Now $t^n \varphi_n(t)$ converges in $\hat{A}$ to some element w and we obtain $f = w\gamma$. This contradicts the choice of γ. So f must be zero.

5. Filtered Auslander–Gorenstein Rings

First we recall the following result which was proved in [Ek: Theorem 5.3].

5.1 **Theorem.** *Let A be a filtered noetherian ring such that A and GA are Auslander–Gorenstein rings. Then R_A is an Auslander–Gorenstein ring.*

5.2 *Remark.* In the special case when A is a zariskian ring, then Theorem 3.9 in [Bj] shows that if GA is Auslander–Gorenstein so is A and hence also R_A by Theorem 5.1. We also refer to [H-O] where it was proved that if A is a *zariskian* ring such that GA is *Auslander regular* then A and R_A are so.

From now on we consider a filtered noetherian ring A such that A, GA and R_A are Auslander–Gorenstein. We shall begin to study graded R_A-modules. Keeping the ring A fixed we denote the Rees ring by R to simplify the notations.

Graded free resolutions. If k is an integer, then $R[k]$ denotes the free R-module of rank one graded by $R[k](v) = R(v+k)$. If M is a graded and finitely generated R-module we can construct a resolution $\to F_1 \to F_0 \to M \to 0$, where $F_0, F_1, \ldots$ are graded free R-modules of finite rank and the differentials are R-linear maps which are homogeneous of degree zero. We refer to $F_\bullet$ as a graded free resolution of M. Now $\mathrm{Hom}_R(F_\bullet, R)$ is a complex of graded right R-modules whose cohomology groups are given by $E^v_R(M)$. Thus, every $E^v_R(M)$ is a graded and finitely generated right R-module.

Let us now apply the $\mathcal{E}$-functor to $F_\bullet$. Since the R-element $1-T$ is injective on every graded R-module, it follows that $\mathcal{E}(F_\bullet)$ is a free resolution of the A-module $\mathcal{E}(M)$. Moreover, for every graded free R-module F of finite rank it is obvious that $\mathcal{E}(\mathrm{Hom}_R(F,R)) = \mathrm{Hom}_A(\mathcal{E}(F), A)$. Since the cohomology groups of $\mathrm{Hom}_A(\mathcal{E}(F), A)$ are given by $E^v_A(\mathcal{E}(M))$ we conclude

5.2 **Proposition.** *Let M be a finitely generated and graded left R-module. Then $\mathcal{E}(E^v_R(M)) = E^v_A(\mathcal{E}(M))$ for every $v \geq 0$.*

Above we assumed that M is a left R-module. Of course, we get a similar result when M is a graded right R-module. Let us now consider a pair (M, Γ), where Γ is a good filtration on a finitely generated left A-module M. Now M_Γ is a graded and finitely generated R-module and we apply Proposition 5.2. Since $\mathcal{E}(M_\Gamma) = M$, it follows that $\mathcal{E}(E^v_R(M_\Gamma)) =$

$E_A^v(M)$ holds for every integer v. Denote by W_v the T-torsion part of $E_R^v(M_\Gamma)$. Then the material about the $\mathcal{E}$-functor in Section 4 gives

5.3 **Proposition.** *Denote by $\Gamma(v)$ the good filtration on the A-module $\mathcal{E}(E_R^v(M_\Gamma))$. Then the Rees module $E_A^v(M)_{\Gamma(v)}$ is equal to $E_R^v(M_\Gamma)/W_v$.*

5.4 *Remark.* Starting from a good filtration Γ on the left A-module M we notice that the good filtration $\Gamma(v)$ on the right A-module $E_A^v(M)$ is obtained in a unique way since $E_R^v(M_\Gamma)$ is unique in the category of graded R-modules. We refer to $\Gamma(v)$ as the *Γ-induced good filtration* on $E_A^v(M)$.

5.5. *Filt-free resolutions.* If k is an integer then $A[k]$ denotes the free A-module of rank one, filtered by $A[k]_v = A_{k+v}$. If Γ is a good filtration on some finitely generated left A-module M we construct a resolution $\to F_1 \to F_0 \to M \to 0$, where every F_v is a filt-free A-module, i.e. a finite direct sum of filtered A-modules of the form $A[k]$. Moreover, the differentials in the complex are *strictly filter preserving,* i.e. we have $d((F_i)_v) = d(F_i) \cap (F_{i-1})_v$ for $i \geq 1$ and $d((F_0)_v) = \Gamma_v$ for every integer v. If $F_\bullet$ is a graded free resolution of M_Γ we notice that $\mathcal{E}(F_\bullet)$ is a filt-free resolution of (M, Γ).

Let us then consider a filt-free resolution $F_\bullet$ of (M, Γ) such that $F_\bullet = \mathcal{E}(P_\bullet)$, where $P_\bullet$ is a graded free resolution of M_Γ. Now $\mathrm{Hom}_A(F_\bullet, A)$ is a complex of filtered A-modules and its cohomology groups are the filtered right A-modules given by $(E_A^v(M), \Gamma(v))$.

Next, the filtered complex $\mathrm{Hom}_A(F_\bullet, A)$ gives a *spectral sequence* whose E_0-term is the associated graded complex given by a complex of graded right GA-modules. If F is a filt-free A-module we notice that $G(\mathrm{Hom}_A(F, A)) = \mathrm{Hom}_{GA}(GF, GA)$. We conclude that the E_0-term is the complex $\mathrm{Hom}_{GA}(GF_\bullet, GA)$. Next, since the differentials in the filt-free resolution are strictly filter preserving, it follows that $GF_\bullet$ is a free resolution of the GA-module $\mathrm{gr}_\Gamma(M)$. So the cohomology groups of the E_0-term of the spectral sequence are given by $E_{GA}^v(\mathrm{gr}_\Gamma(M))$. Using the *comparison condition* from 4.14, applied to finitely generated right A-modules one can easily show that the *spectral sequence converges.* The convergence implies that for every integer v, it follows that $\mathrm{gr}_{\Gamma(v)}(E_A^v(M))$ is a *subquotient* of the GA-module $E_{GA}^v(\mathrm{gr}_\Gamma(M))$.

In the case when M belongs to $\overline{\mathrm{Mod}_f}(A)$ we conclude that $E_A^v(M)$ belongs to $\overline{\mathrm{Mod}_f}(A)$ for every v. We shall now study the case when M does not belong to $\overline{\mathrm{Mod}_f}(A)$. So then $\mathrm{gr}_\Gamma(M)$ is non-zero and this GA-module has some grade number k. If $v < k$ we have $E_{GA}^v(\mathrm{gr}_\Gamma(M)) = 0$ and since this is the term E_1^v of the spectral sequence, it follows that no coboundaries occur in degree k during the passage to higher terms in the spectral sequence. This gives the following

5.6 **Proposition.** *Put* $k = j_{GA}(\mathrm{gr}_\Gamma(M))$. *Then there exists an exact sequence*

$$0 \to \mathrm{gr}_{\Gamma(k)}(E^k(M)) \to E^k_{GA}(\mathrm{gr}_\Gamma(M)) \to W_k \to 0$$

where $j_{GA}(W_k) \geq j_{GA}(E^{k+1}_{GA}(\mathrm{gr}_\Gamma(M))) \geq k+1$, *unless* $W_k = 0$.

5.7 **Proposition.** *Let M be a finitely generated A-module which does not belong to* $\overline{\mathrm{Mod}}_f(A)$. *Then, for every good filtration we have* $j_{GA}(\mathrm{gr}_\Gamma(M)) \geq j_A(M)$.

Proof. If $k = j_{GA}(\mathrm{gr}_\Gamma(M)) < j_A(M)$ then the exact sequence in Proposition 5.6 gives $j_{GA}(E^k_{GA}(\mathrm{gr}_\Gamma(M))) \geq k+1$. This contradicts Lemma 2.9 applied to the Auslander–Gorenstein ring GA.

5.8 **Corollary.** *Let M be a pure A-module which does not belong to* $\overline{\mathrm{Mod}}_f(A)$. *Then* $j_{GA}(\mathrm{gr}_\Gamma(M)) = j_A(M)$ *for every good filtration* Γ.

Proof. Suppose that $k = j_{GA}(\mathrm{gr}_\Gamma(M)) > j_A(M)$. The purity of M gives $j_A(E^k(M)) > k$ and then $j_{GA}(\mathrm{gr}_{\Gamma(k)}(E^k(M)) > k$ by Proposition 5.7. Now Proposition 5.6 and Lemma 2.8 imply that $j_{GA}(E^k_{GA}(\mathrm{gr}_\Gamma(M)) > k$. But this contradicts Lemma 2.9 and hence $k \leq j_A(M)$. Now $k = j_A(M)$ follows from Proposition 5.7.

Remark. If A is a zariskian filtered ring then the class $\mathrm{Mod}_f(A)$ is empty. So in the zariskian case we see that the equality $j_A(M) = j_{GA}(\mathrm{gr}_\Gamma(M))$ holds for every finitely generated A-module M.

5.9 *A base change formula.* Recall that $GA = R/(T)$. So every R-module M such that $TM = 0$ has a GA-module structure and we denote the resulting GA-module by $g(M)$. Since T is a central non-zero divisor in the ring R, a well known result in homological algebra gives:

5.10 **Proposition.** $g(E^{v+1}_R(M)) = E^v_{GA}(g(M))$ *holds for every finitely generated R-module M annihilated by T and every v.*

Let us again consider a pair (M, Γ), where Γ is a good filtration on some finitely generated left or right A-module M.

5.11 **Lemma.** *Put* $k = j_R(M_\Gamma)$. *Then the R-module* $E^k_R(M_\Gamma)$ *is T-torsion free.*

Proof. Consider the exact sequence

$$0 \longrightarrow M_\Gamma \xrightarrow{T} M_\Gamma \longrightarrow M_\Gamma/TM_\Gamma \longrightarrow 0.$$

Then we conclude that the T-kernel of $E^k_R(M_\Gamma)$ is a quotient of $E^k_R(M_\Gamma/TM_\Gamma)$. Next, by Proposition 5.10 we have $g(E^k_R(M_\Gamma/TM_\Gamma)) = E^{k-1}_{GA}(\mathrm{gr}_\Gamma(M))$. So Lemma 5.11 follows if $E^{k-1}_{GA}(\mathrm{gr}_\Gamma(M)) = 0$. To prove this we use that $j_{GA}(\mathrm{gr}_\Gamma(M)) \geq j_A(M)$ by Proposition 5.7. Next, Proposition 5.2 implies that $j_A(M) \geq j_R(M_\Gamma) = k$. So $j_{GA}(\mathrm{gr}_\Gamma(M)) \geq k$ and hence $E^{k-1}_{GA}(\mathrm{gr}_\Gamma(M)) = 0$.

5.12 **Proposition.** *$j_A(M) = j_R(M_\Gamma)$ holds for every good filtration Γ.*

Proof. Proposition 5.2 gives $j_A(M) \geq j_R(M_\Gamma)$. Next, if $k = j_r(M_\Gamma)$, then $E^k_R(M_\Gamma)$ is T-torsion free and therefore $\mathcal{E}(E^k_R(M_\Gamma))$ is non-zero. Since this A-module is $E^k_A(M)$ we conclude that $k \geq j_A(M)$ and Proposition 5.12 follows.

Let us now consider a Rees module M_Γ and put $n = j_R(M_\Gamma)$. We can construct a Fossum module $\mathcal{F}(M_\Gamma)$ using a graded free resolution $F_{n+1} \to \cdots \to F_0 \to M_\Gamma$. Then we notice that $\mathcal{E}(F_{n+1}) \to \cdots \to \mathcal{E}(F_0) \to \mathcal{E}(M_\Gamma) \to 0$ is of the form in 2.1. Since $\mathcal{E}(M_\Gamma) = M$ we conclude that a similar reasoning as in Proposition 5.2 gives

5.13 **Proposition.** *The A-module M has a Fossum module of the form $\mathcal{E}(\mathcal{F}(M_\Gamma))$.*

5.14 **Corollary.** *The A-module M is pure if and only if M_Γ is a pure R-module for every good filtration Γ.*

Proof. Put $n = j_A(M) = j_R(M_\Gamma)$. Proposition 2.5 shows that $E^{n+1}_R(\mathcal{F}(M_\Gamma))$ is a submodule of M_Γ and hence T-torsion free. Since the $\mathcal{E}$-functor is faithful on T-torsion free graded R-modules, it follows that $E^{n+1}_R(\mathcal{F}(M_\Gamma)) = 0$ if and only if its $\mathcal{E}$-image is zero. Next, using Proposition 5.13 and 5.2, we see that the $\mathcal{E}$-image is equal to $E^{n+1}_A(\mathcal{F}(M))$. Then the purity criterion in (2) of Theorem 2.12 gives Corollary 5.14.

We shall now discuss good filtrations on a *pure* A-module M. Keeping M fixed we put $n = j_A(M)$ and let Γ be some good filtration on M. Then the equality $j_A(M) = j_R(M_\Gamma)$ and Lemma 5.11 imply that $E^n_R(M_\Gamma)$ is T-torsion free. Hence Proposition 5.3 gives

$$E^n_R(M_\Gamma) = E^n_A(M)_{\Gamma(n)}. \tag{5.15}$$

Next, starting from the filtered A-module $(E^n_A(M), \Gamma(n))$ whose grade number is n, we construct the induced good $\Gamma(n)$-filtration on $E^{n,n}_A(M)$ which we denote by $\Gamma(n,n)$. Repeating the previous arguments we obtain

$$(5.16) \qquad E^{n,n}_R(M_\Gamma) = E^{n,n}_A(M)_{\Gamma(n,n)}.$$

Now we consider the pure R-module M_Γ. The purity gives $E^{n+1}_R(\mathcal{F}(M_\Gamma)) = 0$ so Proposition 2.5 gives the exact sequence

$$(5.17) \qquad 0 \longrightarrow M_\Gamma \longrightarrow E^{n,n}_R(M_\Gamma) \xrightarrow{\varphi} E^{n+2}_R(\mathcal{F}(M_\Gamma)) \longrightarrow 0.$$

Let W be the T-torsion submodule of $E^{n+2}_R(\mathcal{F}(M_\Gamma))$ and put $\mathcal{M} = \varphi^{-1}(W)$. Then $\mathcal{M}$ is a graded R-submodule of $E^{n,n}_R(M_\Gamma)$. Repeated use of Lemma 5.11 shows that $E^{n,n}_R(M_\Gamma)$ is T-torsion free and hence $\mathcal{M}$ is. So then $\mathcal{M}$ is a Rees module, say N_Ω. Next, by construction $\mathcal{M}/M_\Gamma$ has T-torsion and hence $\mathcal{E}(\mathcal{M})$ is equal to $\mathcal{E}(M_\Gamma)$. We conclude that $N = M$, i.e. $\mathcal{M} = M_\Omega$ where Ω is a good filtration on the A-module M. Since M_Γ is a submodule of M_Ω, it follows that Ω increases faster than Γ. The good filtration Ω has some special properties.

5.18 **Lemma.** *$\mathrm{gr}_\Omega(M)$ is a GA-submodule of $\mathrm{gr}_{\Gamma(n,n)}(E^{n,n}_A(M))$.*

Proof. The equality in 5.16 and the construction of M_Ω shows that the R-module $E^{n,n}_A(M)_{\Gamma(n,n)}/M_\Omega$ is T-torsion free. Then we divide by T and Lemma 5.18 follows.

5.19 **Lemma.** *$j_R(M_\Omega/M_\Gamma) \geq n+2$.*

Proof. The R-module M_Ω/M_Γ is a submodule of $E^{n+2}_R(\mathcal{F}(M_\Gamma))$ whose grade number is $\geq n+2$.

Now we analyze when GA-modules are pure and begin with

5.20 **Lemma.** *Let Γ be a good filtration on some A-module M. Then, if the R-module M_Γ is strongly pure, it follows that the GA-module $\mathrm{gr}_\Gamma(M)$ is pure.*

Proof. Put $K = M_\Gamma/TM_\Gamma$ so that $g(K) = \mathrm{gr}_\Gamma(M)$. Repeated use of Proposition 5.10 shows that if $v \geq 0$, then $j_{GA}(E^v_{GA}(\mathrm{gr}_\Gamma(M)))$ is equal to $j_R(E^{v+1}_R(K)) - 1$. Next, by a long exact sequence and Lemma 2.8, it follows that

$$j_R(E^{v+1}_R(K)) \geq \inf\{j_R(E^v_R(M_\Gamma)), j_R(E^{v+1}_R(M_\Gamma))\}.$$

Put $n = j_R(M_\Gamma)$ which by 5.12 and 5.8 is equal to $j_{GA}(\mathrm{gr}_\Gamma(M)$. Then, if $v > n$ the inequality above and Definition 3.3 show that $j_R(E_R^{v+1}(K)) \geq v+2$ and hence $j_{GA}(E_{GA}^v(\mathrm{gr}_\Gamma(M))) \geq v+1$. Then $\mathrm{gr}_\Gamma(M)$ is pure by (3) in Theorem 2.12.

5.21 **Corollary.** *Let Ω be the good filtration on M constructed after 5.17. Then $\mathrm{gr}_\Omega(M)$ is a pure GA-module.*

Proof. By Lemma 5.18 it suffices to show that the GA-module $\mathrm{gr}_{\Gamma(n,n)}(E_A^{n,n}(M))$ is pure. But this follows since its Rees module is the strongly pure R-module $E^{n,n}(M_\Gamma)$.

Let us again consider a pure A-module M which does not belong to $\overline{\mathrm{Mod}_f(A)}$. Denote by Λ the family of all good filtrations on M. If Γ and Ω belong to Λ, then $\Gamma + \Omega$ is the good filtration whose Rees module is $M_\Gamma + M_\Omega$, while $\Gamma \cap \Omega$ is the good filtration whose Rees module is $M_\Gamma \cap M_\Omega$.

5.22 *Definition.* Let M be a pure A-module and set $n = j_A(M)$. A pair Γ and Ω in Λ are tamely close if $j_R(M_{\Gamma+\Omega}/M_{\Gamma\cap\Omega}) \geq n+2$.

It is easily seen that Definition 5.22 gives an equivalence relation on Λ.

5.23 **Theorem.** *Every equivalence class in Λ contains a unique good filtration Ω such that $\mathrm{gr}_\Omega(M)$ is a pure GA-module. Moreover, Ω is the unique fastest filtration in the equivalence class, i.e. $\Omega > \Gamma$ for any Γ which is tamely close with Ω.*

Proof. Let Λ_0 denote an equivalence class. Take some Γ in Λ_0. Corollary 5.21 and Lemma 5.19 gives some Ω in Λ_0 such that $\mathrm{gr}_\Omega(M)$ is pure and $\Omega > \Gamma$. There remains to show that Ω is unique and that $\Omega > \Gamma'$ for any Γ' in Λ_0. This follows from the following

Sublemma. *Let Ω be a good filtration on M such that $\mathrm{gr}_\Omega(M)$ is a pure GA-module. Then $\Omega > \Gamma$ for any good filtration Γ which is tamely close to Ω.*

Proof. Set $\mathcal{M} = M_{\Gamma+\Omega}$. It suffices to show that $\mathcal{M} = M_\Omega$. To prove this we put $W = \mathcal{M}/M_\Omega$. Now we have $j_R(W) \geq n+2$, where $n = j_A(M) = j_{GA}(\mathrm{gr}_\Omega(M))$. We notice that the T-kernel of W is a GA-submodule of $\mathrm{gr}_\Omega(M)$. By the base change we have $j_{GA}(g(\mathrm{Ker}_T(W))) = j_R(\mathrm{Ker}_T(W)) - 1 \geq j_R(W) - 1 \geq n+1$. So the purity of $\mathrm{gr}_\Omega(M)$ gives $\mathrm{Ker}_T(W) = 0$. Since $\mathcal{E}(\mathcal{M}) = \mathcal{E}(M_\Omega)$ it follows also that W has T-torsion.

Hence $W = 0$ and the Sublemma is proved.

5.24 *Remark.* The unique fastest good filtration in an equivalence class of Λ is called a *Gabber filtration* on the pure A-module M.

5.25 *The case when M is strongly pure.* Let M be strongly pure and set $n = j_A(M)$. The constructions made after 5.17 show that the good filtration Ω in Corollary 5.21 gives $M_\Omega = E^{n,n}(M_\Omega)$ and hence M_Ω is a strongly pure R-module. Then Theorem 5.23 shows that M_Ω is *strongly pure for every Gabber filtration Ω.* The question arises if $\mathrm{gr}_\Omega(M)$ is a strongly pure GA-module for every Gabber filtration. This is not true as the example below shows.

5.26 *Example.* Suppose that there exists a graded and pure GA-module M which is not strongly pure and has $j_{GA}(M) = 0$. Let us remark that M exists if A is the Weyl algebra $A_1(\mathbf{C})$ endowed with the positive *Bernstein filtration* for then GA is a polynomial ring in two variables and we can choose a torsion free GA-module which is not reflexive. With M as above we regard M as a graded R-module annihilated by T. Let $\alpha: F \to M$ be a surjective map where F is a graded free R-module and α homogeneous of degree zero. Let K be the kernel of α. We notice that $E_R^v(K) = E_R^{v+1}(M)$ holds for every $v \geq 1$. Using Proposition 5.10 and the GA-purity of M we see that the R-module K is strongly pure. Next, since K is T-torsion free we have $K = N_\Omega$ for some A-module N with a good filtration Ω. The strong purity of K and Lemma 5.20 imply that $\mathrm{gr}_\Omega(N)$ is pure and hence Ω is a Gabber filtration. But $\mathrm{gr}_\Omega(N)$ is not strongly pure. Namely, since M is pure but not strongly pure there exists some $v \geq 1$ such that $j_{GA}(E^v(M)) = v+1$. We also have an exact sequence $0 \to M \to \mathrm{gr}_\Omega(N) \to W \to 0$ and conclude that $j_{GA}(E^v(\mathrm{gr}_\Omega(N))) = v+1$ so then $\mathrm{gr}_\Omega(N)$ is not strongly pure.

5.27 *An open problem.* Let M be a strongly pure module over a filtered Auslander–Gorenstein ring. Does there exist some Gabber filtration Ω on M such that $\mathrm{gr}_\Omega(M)$ is a strongly pure GA-module.

A final remark. The existence of Gabber filtrations is non-trivial even in the case when A is commutative. For example, let A be a local regular commutative ring with its maximal ideal $\mathfrak{m}$. The $\mathfrak{m}$-adic filtration on A is zariskian, i.e. we use the filtration defined by $A_{-v} = \mathfrak{m}^v$: $v \geq 1$, while $A = A_0 = A_1 = \ldots$. If $M \in \mathrm{Mod}_f(A)$ we get the good filtration defined by $\{\mathfrak{m}^v M\}$. Now the associated graded module $\oplus \mathfrak{m}^v M/\mathfrak{m}^{v+1} M$ is not pure in general even if M is a pure A-module from the start. In this case we construct the Gabber filtration Ω which is tamely close to the $\mathfrak{m}$-adic filtration. Thus, we have $M = \Omega_0 \supset \Omega_{-1} \supset \ldots$ and $\Omega_{-v} \supset \mathfrak{m}^v M$ for every $v \geq 1$. Moreover, $\cap \Omega_v = 0$ and $\mathrm{gr}_\Omega(M)$ is a pure $\mathrm{gr}_\mathfrak{m}(A)$-module. It would be interesting to see concrete non-trivial examples and if possible find algorithms to compute Gabber's filtration associated with the $\mathfrak{m}$-adic filtration on suitable pure A-modules such as $A/\mathfrak{p}$ where $\mathfrak{p}$ is a prime ideal in A.

REFERENCES

[A-B] M. Auslander and M. Bridger, *Stable module theory,* Mem. Amer. Math. Soc. **94** (1969).

[Ba] H. Bass, *On the ubiquity of Gorenstein rings,* Math. Zeit. **82** (1963), 8–28.

[Bj] J.-E. Björk, *The Auslander condition on noetherian rings,* Sem. Dubreil-Malliavin. Lecture Notes in Math. **1404**, Springer-Verlag (1989), 137–173.

[Ek] E.K. Ekström, *The Auslander condition on graded and filtered noetherian rings,* Ibid. 220–245.

[Fo] R. Fossum, *Duality over Gorenstein rings,* Math. Scand. **26** (1970), 165–176.

[F-G-R] R. Fossum, P. Griffith, and I. Reiten, *Trivial extensions of Abelian categories,* Lecture Notes in Math. **456**, Springer-Verlag (1975).

[H] Li Huishi, *On pure module theory over zariskian filtered rings,* University of Antwerpen, preprint (1989).

[H-O] Li Huishi and F. Oystaeyen, *Global dimension and Auslander regularity of Rees rings.*

[Le 1] T. Levasseur, *Grade des modules sur certains anneaux filtrés,* Comm. in Algebra **9** (1981), 1519–1532.

[Le 2] T. Levasseur, *Complexe bidualisant en algèbre non commutative,* Sem. Dubreil-Malliavin. Lecture Notes in Math. **1146**, Springer-Verlag (1983-84), 270–287.

[L-S] T. Levasseur and J.T. Stafford, *Rings of differential operators on classical rings of invariants,* Mem. Amer. Math. Soc. **81**, no. 412 (1989).

[Re] I. Reiten, *Trivial extensions and Gorenstein rings,* Thesis, Chicago Univ. (1971).
[Z] A. Zaks, *Injective dimensions of semi-primary rings,* J. of Algebra **13** (1969), 73–86.

Received February 21, 1990

Department of Mathematics
University of Stockholm
Box 6701
113 85 Stockholm
Sweden

Relèvements d'opérateurs différentiels sur les anneaux d'invariants

THIERRY LEVASSEUR

A J. Dixmier pour son 65ème anniversaire

O. Introduction et Notations

Le corps de base est celui des nombres complexes noté $\mathbf{C}$.

0.1 Si X est une variété algébrique irréductible affine on note : $\mathcal{O}(X)$ l'anneau des fonctions régulières sur X, $\mathcal{D}(X)$ l'anneau des opérateurs différentiels (o.d. en abrégé) sur $X, \mathcal{D}er^m(X)$ le $\mathcal{O}(X)$-module des o.d. d'ordre $\leq m$ sans terme constant, $\mathcal{D}^m(X) = \mathcal{O}(X) \oplus \mathcal{D}er^m(X)$, cf. 1.1 et 1.2.

Soit $G \to GL(V)$ une représentation linéaire de dimension finie d'un groupe réductif G, on désigne par $V//G$ la variété affine telle que $\mathcal{O}(V//G) = \mathcal{O}(V)^G$. Le groupe G opère naturellement sur $\mathcal{D}(V)$ en laissant stable chaque $\mathcal{D}er^m(V)$. On dit que (V, G) a la propriété de relèvement à l'ordre m si le morphisme de restriction $\varphi : \mathcal{D}er^m(V)^G \to \mathcal{D}er^m(V//G)$ est surjectif; si cette propriété est vérifiée pour tout m on dira que (V, G) a la propriété de relèvement, ce qui implique alors la surjectivité du morphisme d'algèbres $\varphi : \mathcal{D}(V)^G \to \mathcal{D}(V//G)$.

0.2 Si $m = 1\, \mathcal{D}er^1(V//G)$, resp. $\mathcal{D}er^1(V)^G$, n'est autre que l'espace des champs de vecteurs sur $V//G$, resp. des champs de vecteurs invariants sur V. Le relèvement à l'ordre 1 a été étudié par G. Schwarz dans [S]. Rappelons que $V//G$ possède une stratification finie, paramétrée par les classes de conjugaison des groupes d'isotropie des points d'orbite fermée; il existe une strate ouverte que nous appellerons strate principale et noterons $\Omega(V//G)$. On désigne par $\mathcal{A}(V//G)$ l'idéal de $\mathcal{O}(V)^G$ définissant le complémentaire de $\Omega(V//G)$. Il est en particulier démontré dans [S] que si $G \to GL(V)$ est orthogonalisable et si la hauteur de $\mathcal{A}(V//G)$ est ≥ 2 alors (V, G) a la propriété de relèvement à l'ordre 1. Si m est quelconque des conditions nécessaires et suffisantes pour qu'il y ait relèvement à l'ordre m

ont été données, lorsque G est fini dans [K], lorsque G est un tore dans [Mu].

0.3 Nous proposons au §3 une condition suffisante pour que (V, G) possède la propriété de relèvement. Soit $gr\varphi : gr\mathcal{D}(V)^G \to gr\mathcal{D}(V//G)$ le morphisme gradué associé au morphisme filtré φ, posons $K = \mathrm{Ker}(gr\varphi)$, $R = \frac{gr\mathcal{D}(V)^G}{K}$; notons que R est une $\mathbb{C}$-algèbre commutative intègre, de type fini, graduée, contenant $\mathcal{O}(V)^G$ en degré 0. On montre alors, (3.6, théorème) :

Théorème. *Supposons que R vérifie la condition (S_2) et que $\mathcal{A}(V//G)$ engendre un idéal de hauteur ≥ 2 dans R, alors (V, G) a la propriété de relèvement.*

Ce résultat, de démonstration élémentaire, a l'inconvénient de nécessiter une bonne connaissance de R. Néanmoins il permet de traiter les cas G fini et $G = \mathbb{C}^*$ (l'anneau R étant connu dans le cas des tores grâce à [Mu]).

0.4 Une autre application de 0.3 est donnée au §5, elle concerne les o.d. sur les anneaux classiques d'invariants. On étudie les trois cas :

$$(V, G) = \begin{cases} (M_{pk}(\mathbb{C}) \times M_{kq}(\mathbb{C}), GL(k, \mathbb{C})) \\ (M_{kn}(\mathbb{C}), O(k, \mathbb{C})) \\ (M_{2kn}(\mathbb{C}), Sp(2k, \mathbb{C})) \end{cases}$$

où GL, O, Sp désignent respectivement les groupes linéaire, orthogonal et symplectique opérant "naturellement" sur les espaces de matrices de taille appropriée, (cf 5.3). On retrouve un des résultats de [Le-St] :

Théorème. *Soit (V, G) comme ci-dessus ; si $V//G$ n'est pas lisse alors (V, G) a la propriété de relèvement.*

Utilisant les résultats de R. Howe sur les paires réductives duales, on sait qu'il existe un morphisme surjectif $\omega : U(\mathfrak{g}') \to \mathcal{D}(V)^G$ où $\mathfrak{g}'$ est une algèbre de Lie semi-simple d'algèbre enveloppante $U(\mathfrak{g}')$, le noyau de $\varphi\omega$ étant un idéal primitif I. La preuve donnée dans [Le-St] repose en partie sur le fait que I est maximal (lorsque $V//G$ n'est pas lisse). Le théorème de 0.3 permet de s'affranchir de cette vérification (qui utilise la classification des idéaux primitifs) et fournit de plus l'information $\sqrt{gr\,I} = gr\,I$, où $gr\,I$ est l'idéal gradué associé à I pour la filtration naturelle de $U(\mathfrak{g}')$. Malheureusement cette nouvelle démonstration ne donne pas la maximalité de I.

0.5 Les notations qui suivent seront librement utilisées. Si A est une $\mathbb{C}$-algèbre commutative de type fini on désigne par Spec A l'ensemble de ses idéaux premiers et on pose

$$\operatorname{Reg} A = \{\mathcal{P} \in \operatorname{Spec} A / A_{\mathcal{P}} \text{ régulier}\}, \operatorname{Sing} A = \operatorname{Spec} A \backslash \operatorname{Reg} A.$$

Si A est intègre on dit que : A est normale lorsque A est intégralement clos et A vérifie la condition (S_2) si $A = \cap_{ht\mathcal{P}=1} A_{\mathcal{P}}$ où ht est la hauteur d'un idéal. Soit V une variété algébrique définie sur $\mathbb{C}$, $\mathcal{O}_V$ son faisceau structural et $\mathcal{O}(V)$ la $\mathbb{C}$-algèbre des fonctions régulières sur V. Si V est affine on identifie $(V, \mathcal{O}_V)$ à Spec $\mathcal{O}(V)$. Soit p un point (fermé) de V on note $\mathcal{M}_p$ l'idéal maximal de l'anneau local $\mathcal{O}_{V,p}$ de V en p. On définit Reg $V = \{p \in V / \mathcal{O}_{V,p}$ régulier$\}$, Sing $V = V \backslash$ Reg V ; si Reg $V = V$ on dit que V est lisse. Supposons de plus la variété V affine ; si $\mathcal{I}$ est un idéal de $\mathcal{O}(V)$ on pose

$$\mathcal{V}(\mathcal{I}) = \{P \in \operatorname{Spec} \mathcal{O}(V) / \mathcal{P} \supseteq \mathcal{I}\}$$

et inversement si F est fermé dans V, $\mathcal{I}(F)$ est l'idéal radiciel définissant F ; V est dite normale si $\mathcal{O}(V)$ l'est.

Si $t_1, \ldots, t_s$ sont des éléments d'un anneau et $\alpha = (\alpha_1, \ldots, \alpha_s) \in \mathbb{N}^s$ on pose $t^\alpha = t_1^{\alpha_1} t_2^{\alpha_2} \cdots t_s^{\alpha_s}$, $|\alpha| = \alpha_1 + \alpha_2 + \cdots + \alpha_s$.

1. Généralités

1.1 Etant donnée une $\mathbb{C}$-algèbre commutative intègre A, nous désignerons par $\mathcal{D}(A)$ l'anneau des opérateurs différentiels sur A. Rappelons, cf [E.G.A.] ou [M.R.], que $\mathcal{D}(A) = \cup_{m \geq 0} \mathcal{D}^m(A)$ où $\mathcal{D}^m(A)$ est le A-module des opérateurs différentiels (o.d. en abrégé) d'ordre $\leq m$, défini par :

$$\mathcal{D}^m(A) = \{D \in \operatorname{End}_{\mathbb{C}} A / \forall\, a_0, \ldots, a_m \in A \quad [a_0[a_1[\cdots[a_m D]]\cdots] = 0\}.$$

On pose $\mathcal{D}^m(A) = (0)$ si $m < 0$; notons que $\mathcal{D}^0(A) = A$. La sous-algèbre $\mathcal{D}(A)$ de $\operatorname{End}_{\mathbb{C}} A$ est filtrée par les $\mathcal{D}^m(A)$ d'anneau gradué associé $gr\mathcal{D}(A) := \bigoplus_m \frac{\mathcal{D}^m(A)}{\mathcal{D}^{m-1}(A)}$, qui est une $\mathbb{C}$-algèbre commutative intègre. On note $gr_m : \mathcal{D}^m(A) \to gr_m\mathcal{D}(A) := \frac{\mathcal{D}^m(A)}{\mathcal{D}^{m-1}(A)}$ la projection canonique. Si $P \in \mathcal{D}^m(A) \backslash \mathcal{D}^{m-1}(A)$ on dit que P est d'ordre m et on pose $\operatorname{ord} P = m$, le symbole principal de P étant alors $gr_m P$. L'action de $\mathcal{D}(A)$ sur A, provenant de celle de $\operatorname{End}_{\mathbb{C}} A$, sera notée $a \to P * a$, $a \in A$, $P \in \mathcal{D}(A)$.

Le sous-A-module de $\mathcal{D}^m(A)$ formé des o.d. sans terme constant d'ordre $\leq m$ est donc

$$\mathcal{D}er^m(A) = \{D \in \mathcal{D}^m(A)/D * 1 = 0\}.$$

Ainsi $\mathcal{D}^m(A) = A \oplus \mathcal{D}er^m(A)$ et $\mathcal{D}er^1(A)$ est le A-module des dérivations $\mathbb{C}$-linéaires de A. Rappelons que $\mathcal{D}er^1(A) \cong \mathrm{Hom}_A(\Omega^1_A, A)$ où Ω^1_A est le A-module des différentielles d'ordre 1. On pose $\mathcal{D}er(A) = \cup_{m\geq 0}\mathcal{D}er^m(A)$.

Supposons désormais que A soit de plus une $\mathbb{C}$-algèbre de type fini. Alors, les $\mathcal{D}er^m(A)$ sont des A-modules de type fini et si $S \subset A\backslash(0)$ est multiplicativement stable, S est une partie de Ore dans $\mathcal{D}(A)$ et on a les identifications :

$$\mathcal{D}(S^{-1}A) = S^{-1}\mathcal{D}(A) = \mathcal{D}(A)S^{-1},\ \mathcal{D}er^m(S^{-1}A) = S^{-1}A \otimes_A \mathcal{D}er^m(A).$$

En outre, puisque $gr_0\mathcal{D}(A) = A$, les éléments de S s'identifient à leurs images par gr_0 et l'algèbre localisée $S^{-1}gr\,\mathcal{D}(A)$ s'identifie canoniquement à $gr\,\mathcal{D}(S^{-1}A)$.

Lorsque $S = A\backslash\mathcal{P}$, $\mathcal{P} \in \mathrm{Spec}A$, on notera $\mathcal{D}^{\cdot}(A_{\mathcal{P}}), \mathcal{D}er^{\cdot}(A_{\mathcal{P}})$etc... à la place de $S^{-1}\mathcal{D}^{\cdot}(A), S^{-1}\mathcal{D}er^{\cdot}(A)$ etc.... Signalons enfin que si $S^{-1}A$ est un anneau régulier, l'algèbre $\mathcal{D}(S^{-1}A)$ est engendrée sur $\mathbb{C}$ par $S^{-1}A$ et $\mathcal{D}er^1(S^{-1}(A))$, qui est un $S^{-1}A$-module projectif de type fini. L'anneau $gr\,\mathcal{D}(S^{-1}A)$ s'identifie alors à l'algèbre symétrique de ce module et $\mathcal{D}^m(S^{-1}A)$ est égal à l'ensemble des combinaisons linéaires sur $S^{-1}A$ des produits d'au plus m dérivations.

1.2 Soit V une variété irréductible affine ; ce qui a été rappelé en 1.1 s'applique à $A = \mathcal{O}(V)$. Dans ce cas nous substituerons les notations $\mathcal{D}^{\cdot}(V), \mathcal{D}er^{\cdot}(V)$ à $\mathcal{D}^{\cdot}(A), \mathcal{D}er^{\cdot}(A)$ etc... Si p est un point fermé de $V, \mathcal{D}(V)_p$ désignera $\mathcal{D}(\mathcal{O}_{V,p})$. Nous utiliserons parfois la notation Θ_V, à la place de $\mathcal{D}er^1(V)$, pour désigner le module des dérivations.

Il découle de 1.1 que si V est de plus supposée lisse, $gr\mathcal{D}(V)$ est une $\mathbb{C}$-algèbre de type fini régulière, et pour $p \in V$ l'espace tangent à V en p, noté T_pV, s'identifie à l'espace vectoriel $\frac{\mathcal{D}er^1(V)_p}{\mathcal{M}_p\mathcal{D}er^1(V)_p}$ des dérivations ponctuelles sur $\mathcal{O}_{V,p}$.

2. Opérateurs différentiels sur les espaces d'orbites

2.1 On fixe une variété algébrique affine irréductible V et un groupe algébrique réductif G opérant rationnellement sur V. On note $V//G$ la variété affine telle que $\mathcal{O}(V//G) = \mathcal{O}(V)^G$ (anneau des invariants). Le comorphisme de l'inclusion $\mathcal{O}(V)^G \hookrightarrow \mathcal{O}(V)$ sera désigné par π_V (ou π), donc $\pi_V : V \to V//G$, c'est un morphisme surjectif.

Le groupe G opère naturellement sur $\mathcal{D}^m(V)$ pour tout m par la règle:

$$g \cdot D * f = g \cdot (D * g^{-1} \cdot f),\ g \in G,\ D \in \mathcal{D}^m(V) f \in \mathcal{O}(V).$$

Il n'est pas difficile de vérifier que $\mathcal{D}^m(V)$ est un G-module rationnel. On peut donc considérer la $\mathbb{C}$-algèbre $\mathcal{D}(V)^G$ des o.d. invariants ; elle est filtrée par les $\mathcal{D}^m(V)^G$ qui sont des $\mathcal{O}(V)^G$-modules de type fini, cf ([B-H], Theorem 3.1). On dispose alors de morphismes naturels de $\mathcal{O}(V)^G$-modules

$$\varphi : \mathcal{D}^m(V)^G \to \mathcal{D}^m(V//G)$$

φ étant l'application de restriction : $\varphi(P) * f = P * f$, $P \in \mathcal{D}^m(V)^G$, $f \in \mathcal{O}(V)^G$. Ceci donne un morphisme de $\mathbb{C}$-algèbres filtrées $\varphi : \mathcal{D}(V)^G \to \mathcal{D}(V//G)$. On notera J le noyau de φ et U l'algèbre quotient $\frac{\mathcal{D}(V)^G}{J}$; elle est filtrée par les $U^m = \frac{\mathcal{D}^m(V)^G + J}{J} \cong \frac{\mathcal{D}^m(V)^G}{J \cap \mathcal{D}^m(V)^G}$. On pose $gr\, U = \bigoplus_m \frac{U^m}{U^{m-1}}$.

Remarquons que G opère dans $gr_m \mathcal{D}(V)$ pour tout m et que, G étant réductif, on peut identifier les $\mathcal{O}(V)^G$-modules $[gr_m \mathcal{D}(V)]^G$ et $gr_m[\mathcal{D}(V)^G] = \frac{\mathcal{D}^m(V)^G}{\mathcal{D}^{m-1}(V)^G}$, et les algèbres graduées $[gr\mathcal{D}(V)]^G$ et $gr[\mathcal{D}(V)^G]$.

Le morphisme φ induit un morphisme (injectif) filtré que nous noterons encore φ, de U dans $\mathcal{D}(V//G)$, et des morphismes gradués $gr\varphi : gr\mathcal{D}(V)^G \to gr\mathcal{D}(V//G)$ et $gr\varphi : gr\, U \to gr\mathcal{D}(V//G)$. Notons que $gr\, U = \frac{gr\, \mathcal{D}(V)^G}{gr\, J}$, où $gr\, J$ est l'idéal gradué associé, et que $K = \mathrm{Ker}(gr\varphi) \supseteq gr\, J$. On fera l'abus de notation consistant à encore noter K à la place de $\frac{K}{gr\, J}$. Insistons sur la définition de K : $gr_m P \in \mathrm{Ker}(gr\varphi) = K$ équivaut à $\varphi(P) \in \mathcal{D}^{m-1}(V//G)$. Signalons aussi que : $J \cap \mathcal{O}(V)^G = K \cap gr_0 \mathcal{D}(V)^G = (0)$, φ et $gr\varphi$ sont l'identité sur $\mathcal{O}(V)^G$, U^m et $J \cap \mathcal{D}^m(V)^G$ sont des $\mathcal{O}(V)^G$-modules.

2.2 On suppose ici que V est une variété algébrique affine lisse et que G est un groupe réductif opérant sur V (rationnellement). Dans ce cas $gr\, \mathcal{D}(V)^G$ est une $\mathbb{C}$-algèbre de type fini, cf (1.2. et 2.1.), donc $\mathcal{D}(V)^G$ et U le sont également.

Définition. On dit que (V, G) *a la propriété de relèvement à l'ordre* m si $\varphi : \mathcal{D}^m(V)^G \to \mathcal{D}^m(V//G)$ est surjective. Si (V, G) satisfait la propriété de relèvement à tous les ordres on dira que (V, G) a la *propriété de relèvement.*

Notons, qu'avec les notations de 2.1., on peut évidemment remplacer $\mathcal{D}^m(V)^G$ par U^m dans cette définition et que la propriété de relèvement équivaut à $\varphi : U \to \mathcal{D}(V//G)$ est un isomorphisme filtré.

Remarques. (a) Si $Q \in \mathcal{D}(V//G)$ et s'il existe $P \in \mathcal{D}^m(V)$ tel que $P = Q$ sur $\mathcal{O}(V)^G$ alors il existe $D \in \mathcal{D}^m(V)^G$ tel que $\varphi(P) = Q$ (on écrit $\mathcal{D}^m(V) = \mathcal{D}^m(V)^G \oplus \mathcal{D}^m(V)_G$ où $\mathcal{D}^m(V)_G$ est un supplémentaire G-stable).

(b) Soit $H = \{g \in G/g \cdot v = v$ pour tout $v \in V\}$; alors, si $\overline{G} = G/H, (V, G)$ a la propriété de relèvement (à l'ordre m) si $(V, \overline{G})$ l'a. On peut donc se ramener au cas où l'action est fidèle, i.e. $H = \{1\}$.

(c) Supposons que $\rho : G \to GL(V)$ soit une représentation linéaire de G. On peut écrire $V = V^G \oplus V_G$ où V^G est l'espace des points fixes et V_G un supplémentaire G-stable. Comme $\mathcal{O}(V)^G$ est un anneau de polynômes sur $\mathcal{O}(V_G)^G$ on voit que (V, G) a la propriété de relèvement (à l'ordre m) si (V_G, G) l'a. Le cas échéant on pourra supposer $V^G = (0)$.

2.3 On conserve les hypothèses de 2.2.

On note G_x le groupe d'isotropie d'un $x \in V$ et V_{ss} l'ensemble $\{v \in V/G \cdot v$ fermée dans $V\}$ (points semi-simples). Rappelons que pour tout $x \in V$ il existe une unique orbite fermée dans $\overline{G \cdot x}$. Si $\xi \in V//G$ on note $O(\xi)$ l'unique orbite fermée dans $\pi_V^{-1}(\xi)$. Rappelons quelques résultats tirés de [Lu]. Si L est un sous-groupe (réductif) de G on pose

$$(V//G)_{(L)} = \{\xi \in V//G/ \ \forall\, x \in O(\xi) G_x \text{ conjugué à } L\}$$

$$V^{(L)} = \pi^{-1}(V//G)_{(L)} = \{v \in V/\overline{G \cdot v} \supset O(\xi),\ \xi \in (V//G)_{(L)}\}.$$

Les $(V//G)_{(L)}$ forment une stratification finie de $V//G$ et il existe une unique strate, que nous appellerons *strate principale*, $(V//G)_{(C)}$ vérifiant: $(V//G)_{(C)}$ ouvert non vide et $(V//G)_{(L)} \neq \emptyset$ implique C conjugué à un sous-groupe de L.

On notera $\Omega(V//G)$ la strate principale et on pose :

$$\Omega(V, G) = \pi_V^{-1}\Omega(V//G),\ F(V//G) = V//G - \Omega(V//G),$$
$$F(V, G) = V - \Omega(V, G),\ \mathcal{A}(V//G) = \mathcal{I}(F(V//G)),$$
$$\mathcal{A}(V, G) = \mathcal{I}(F(V, G)) = \sqrt{\mathcal{A}(V//G)\mathcal{O}(V)}.$$

Rappelons le résultat, ([Lu], III.3. Corollaire 6) :

Proposition. *Soit $\xi \in V//G$ alors $\xi \in \Omega(V//G)$ si et seulement si $\pi : V \to V//G$ est lisse aux points de $O(\xi)$.*

Donc s'il l'on pose $V^{\text{lisse}} = \{v \in V/\pi \text{ lisse en } v\}$ on a :

$$\Omega(V//G) = \pi(V_{ss} \cap V^{\text{lisse}}), \Omega(V, G) = \{v \in V/\overline{Gv} \supseteq Gx,\ x \in V_{ss} \cap V^{\text{lisse}}\}.$$

Remarque. Supposons que $\rho : G \to GL(V)$ soit une représentation linéaire de G, $V = V^G \oplus V_G$ comme en (2.2 Remarque (c)). On vérifie facilement que : $\mathcal{A}(V,G) = \sqrt{\mathcal{A}(V_G,G)\mathcal{O}(V)}$, si $\dim V_G//G = 0$ alors $\Omega(V//G) = V^G//G \cong V^G$, si $\dim V_G//G > 0$ alors $ht\,\mathcal{A}(V,G) \leq \dim V_G//G$.

2.4 Fixons une représentation linéaire $\rho : G \to GL(V)$ du groupe réductif G. Le problème du relèvement à l'ordre 1, i.e. celui des champs de vecteurs sur $V//G$, a déjà été étudié, spécialement par G. Schwarz dans [S]. Il est résolu par le théorème suivant dans le cas orthogonalisable :

Théorème [S]. *Soit $\rho : G \to GL(V)$ une représentation orthogonalisable de G. L'image de $\varphi : \Theta_V^G \to \Theta_{V//G}$ est égale à l'ensemble des dérivations laissant stables les idéaux de $\mathcal{O}(V)^G$ constitués des fonctions nulles sur les strates de $V//G$.*

L'exemple de ([S], p. 65), où $G = \mathbb{C}^*$, illustre que ce théorème est en défaut si ρ n'est plus orthogonalisable. Le problème du relèvement dans le cas où G est un tore a été résolu par I. Musson dans [Mu]. Dans le cas où G est fini il a été traité par J.M. Kantor dans [K] (voir aussi [I] et [Le]). Nous reviendrons sur les cas G fini et $G = \mathbb{C}^*$ au §4.

La propriété de relèvement se caractérise facilement lorsque $V//G$ est lisse :

Lemme. *Soit $\rho : G \to GL(V)$ une représentation linéaire avec $V//G$ lisse. Alors (V,G) a la propriété de relèvement à l'ordre* 1 *si et seulement si $V_{ss} = V^G$.*

Preuve. Compte tenu de (2.2, remarque (c)) on suppose $V^G = (0)$. Dans ce cas $V_{ss} = (0)$ équivaut à $\mathcal{O}(V)^G = \mathbb{C}$. Supposons $\mathcal{O}(V)^G = \mathbb{C}[q_1,\ldots,q_\ell]$ avec $\ell \geq 1$, $q_1,\ldots,q_\ell$, polynômes sur V algébriquement indépendants, que l'on peut prendre de degré ≥ 2 sans terme de degré ≤ 1. Si $D \in \Theta_V$ alors $D * q_j(0) = 0$ donc il ne peut exister de relèvement de $\frac{\partial}{\partial q_j} \in \Theta_{V//G}$. La réciproque est triviale.

Donnons un exemple montrant que le relèvement à l'ordre 1 ne suffit pas à assurer le relèvement aux ordres supérieurs.

Exemple. Soit $(V,G) = (\mathbb{C}^3, \mathbb{C}^*)$, G opérant linéairement avec les poids $1, 1, -2$. Si $\mathcal{O}(V) = \mathbb{C}[T_1,T_2,T_3]$ on a $\mathcal{O}(V)^G = \mathbb{C}[T_3T_2^2, T_3T_1^2, T_3T_1T_2]$. Le $\mathcal{O}(V)^G$-module $\Theta_{V//G}$ est engendré par les restrictions des $T_i\frac{\partial}{\partial T_j}$, $1 \leq i,j \leq 2$. Soit $P = \frac{1}{2T_3}\frac{\partial^2}{\partial T_2^2} \in \mathcal{D}er^2(V//G)$, il n'existe pas d' élément de

$\mathcal{D}er^2(V)$ envoyant $T_3T_2^2$ sur 1, donc il y a relèvement à l'ordre 1 mais pas à l'ordre 2.

2.5 Soit $\rho : G \to GL(V)$ comme en 2.4. Si $v \in V_{ss}$ de stabilisateur G_v, notons N un supplémentaire G_v-stable de l'espace tangent $T_v(G \cdot v)$ dans $V \cong T_v V$. La représentation de G_v dans N est appelée représentation slice. Le résultat qui suit étend celui démontré pour l'ordre 1 dans [S] ; comme sa preuve est tout à fait semblable et que nous n'aurons pas à l'utiliser nous le donnerons sans démonstration.

Lemme. *Soient (V, G) et $v \in V_{ss}$ comme ci-dessus. Alors si (V, G) a la propriété de relèvement à l'ordre m il en est de même pour la représentation slice (N, G_v).*

Remarque. Si $\xi \in \Omega(V//G)$ et $v \in O(\xi)$ la représentation slice (N, G_v) vérifie $N_{ss} = N^{G_v}$. Il résulte alors de (2.4, lemme) et du théorème du slice étale de Luna qu'il existe un ouvert affine W contenant v, G-stable, tel que (W, G) ait la propriété de relèvement, (cf. 3.5 preuve de la proposition).

3. Condition suffisante pour le relèvement

3.1 On fixe une $\mathbb{C}$-algèbre de type fini intègre A et l'on note $\mathcal{D}$ l'anneau $\mathcal{D}(A)$ de 1.1. On suppose donnée une $\mathbb{C}$-algèbre filtrée $U = \cup_{m \in \mathbb{Z}} U^m$ telle que :

- $U^m = (0)$ si $m < 0$, $U^0 = A$, $U^m \subset U^{m+1}$.
- U^m est un A-module de type fini.
- $gr\, U = \bigoplus\limits_m \frac{U^m}{U^{m-1}}$ est une $\mathbb{C}$-algèbre commutative de type fini.
- Il existe un homomorphisme d'anneaux filtrés $\varphi : U \hookrightarrow \mathcal{D}$, injectif, tel que φ soit un isomorphisme de $U^0 = A$ sur $\mathcal{D}^0 = A$.

Un exemple d'une telle situation est donné par $U = \frac{\mathcal{D}(V)^G}{J}$ comme en 2.1.

Posons, comme d'habitude, $gr_m : U^m \to gr_m U := \frac{U^m}{U^{m-1}}$ et $gr\varphi : grU \to gr\mathcal{D}$ défini par $gr\varphi(gr_m u) = gr_m \varphi(u)$.

Le noyau $K = \operatorname{Ker}(gr\varphi)$ est un idéal gradué égal à $\bigoplus\limits_m K_m$ où $K_m = \{gr_m u \in gr_m U / \varphi(u) \in \mathcal{D}^{m-1}\}$. On a $K_0 = K \cap A = (0)$. On pose $\overline{gr_m U} = \frac{gr_m U}{K_m}$ et $\overline{grU} = \frac{grU}{K} = \bigoplus\limits_m \overline{gr_m U}$, de sorte que $gr\varphi$ induit un morphisme gradué injectif de $\overline{gr\, U}$ dans $gr\mathcal{D}$. Notons que K est un idéal premier et que les éléments de A sont ad-localement nilpotents dans U puisqu'ils le sont dans $\mathcal{D}$ et que φ est injective. On a les lemmes élémentaires :

Lemme 1. *Le morphisme* $gr\varphi$ *est surjectif si et seulement si* $\varphi : U \hookrightarrow \mathcal{D}$ *est un isomorphisme d'anneaux filtrés. Dans ce cas* $K = (0)$.

Lemme 2. *Soit* $S \subset A\backslash(0)$ *une partie multiplicative. Alors :*

(i) S *est une partie de Ore dans* U. *Dans* $S^{-1}U$ *on a* $S^{-1}U^m = U^m S^{-1}$.

(ii) *L'anneau* $S^{-1}U$ *est filtré par les* $S^{-1}A$*-modules* $(S^{-1}U)^m := S^{-1}U^m$. *Si* $T = \varphi(S) \subset A$ *il existe un morphisme injectif d'anneaux filtrés* $S^{-1}\varphi : S^{-1}U \to T^{-1}\mathcal{D}$ *défini par*

$$(S^{-1}\varphi)(s^{-1}u) = \varphi(s)^{-1}\varphi(u).$$

(iii) *Le morphisme gradué associé à* $S^{-1}\varphi$ *s'identifie au morphisme localisé de* $gr\varphi$*:*

$$S^{-1}gr\varphi : S^{-1}grU \to T^{-1}gr\mathcal{D} = gr\, T^{-1}\mathcal{D}.$$

De plus $S^{-1}K = Ker(S^{-1}gr\varphi)$.

3.2 On prend $\varphi : U \hookrightarrow \mathcal{D}$ vérifiant les hypothèses de 3.1. Soit $S \subset A\backslash(0)$ multiplicative telle que $S^{-1}A$ soit régulier et $T^{-1}\varphi(U^1) = T^{-1}\mathcal{D}^1$ où $T = \varphi(S)$.

Lemme 1. *Le morphisme* $S^{-1}\varphi : S^{-1}U \to T^{-1}\mathcal{D}$ *est un isomorphisme d'anneaux filtrés. En particulier* $S^{-1}K = (0)$.

Preuve. Il suffit de prouver la première assertion, cf. (3.1, lemme 2). Comme $T^{-1}A$ est régulier on a $T^{-1}\mathcal{D}^m = (T^{-1}\mathcal{D}^1)(T^{-1}\mathcal{D}^1)\cdots(T^{-1}\mathcal{D}^1)(m$ fois) et de $T^{-1}\varphi(U^1) = T^{-1}\mathcal{D}^1$ on tire $T^{-1}\mathcal{D}^m = (S^{-1}\varphi)(S^{-1}U^m)$.

Dans tout ce qui suit nous considérerons $\mathcal{D}$, resp. $gr\,\mathcal{D}$, comme un U, resp. $gr\,U$ ou $\overline{gr\,U}$, module via φ, resp. $gr\varphi$. On pose $R = gr\varphi(gr\,U)$.

Lemme 2. *Mêmes hypothèses. Soit* $x \in gr\,\mathcal{D}$ *et* $M = \frac{R+Rx}{R} \subset \frac{gr\,\mathcal{D}}{R}$. *Alors le* $gr\,U$*-module* M *vérifie* $S^{-1} \cdot M = (0)$.

Preuve. Par (3.1, lemme 2) et le lemme 1, on a : $T^{-1}R = (S^{-1}gr\varphi)(grU) = gr(T^{-1}\mathcal{D}) = T^{-1}gr\,\mathcal{D}$. Donc $T^{-1}M \subset T^{-1}(\frac{gr\mathcal{D}}{R}) = (0)$ et l'assertion.

3.3 Soit $\varphi : U \hookrightarrow \mathcal{D}$ comme en 3.1. Conformément à 3.2 nous allons considérer $\frac{\mathcal{D}^1}{\varphi(U^1)}$ comme un A-module via φ : $a \cdot [D + \varphi(U^1)] =$

$[\varphi(a)D + \varphi(U^1)]$, $a \in A$, $D \in \mathcal{D}^1$. Définissons un fermé de $Z = \operatorname{Spec} A$ par $F = \operatorname{Sing} Z \cup \operatorname{Supp}_Z \frac{\mathcal{D}^1}{\varphi(U^1)}$ (où Supp_Z est le support). On note $\Omega = Z \backslash F$, $\mathcal{A} = \mathcal{I}(F)$.

Remarquons que $\Omega = \{\mathcal{P} \in Z / \mathcal{P} \in \operatorname{Reg} Z$ et $T^{-1}\varphi(U^1) = T^{-1}\mathcal{D}^1$ où $T = \varphi(A \backslash \mathcal{P})\}$.

Proposition 1. *Soit $x \in gr\,\mathcal{D}$ et $M = \frac{R+Rx}{R}$. Il existe $\nu \in \mathbb{N}$ tel que $\mathcal{A}^\nu \cdot M = (0)$.*

Preuve. Soit $\bar{x} = [x + R]$, on a $A \cdot \bar{x} \subset R \cdot \bar{x} = M$ avec $ann_A A \cdot \bar{x} = ann_A M (ann_A$ étant l'annulateur dans $A)$. Par (3.2, lemme 2) on a $\operatorname{Supp}_Z A \cdot \bar{x} = \mathcal{V}(ann_A M) \subset F = \mathcal{V}(\mathcal{A})$, d'où le résultat.

Pour toute $\mathbb{C}$-algèbre B on note $GK\dim B$ sa dimension de Gelfand-Kirillov, cf [Kr-L]; rappelons que si B est une $\mathbb{C}$-algèbre de type fini ce n'est autre que la dimension de Krull de B.

Proposition 2. *Si $\Omega \neq \emptyset$ alors $dim\; gr\,U = GK dim U = 2 dim A$ et K est un idéal premier minimal de $gr\,U$.*

Preuve. Comme $\Omega \neq \emptyset$ il existe $S \subset A \backslash (0)$ telle que $S^{-1}\overline{gr\,U} = S^{-1}\left(\frac{gr\,U}{K}\right) = S^{-1} gr\,U \xrightarrow{\sim} S^{-1} gr\,\mathcal{D}$, cf (3.2 lemme 1).

Rappelons que $GK\dim \mathcal{D} = GK\dim gr\,\mathcal{D} = 2\dim A$, donc $GK\dim \overline{gr\,U} = 2\dim A \leq GK\dim gr\,U$. Mais $GK\dim gr\,U \leq GK\dim U$, cf ([Kr-L], p. 75), et $GK\dim U \leq GK\dim \mathcal{D}$ implique les assertions.

3.4 On conserve les notations des sections précédentes.

Proposition. *Soit $\varphi : U \hookrightarrow \mathcal{D}$ vérifiant les hypothèses de 3.1. On suppose de plus :*

(i) $\overline{gr\,U}$ vérifie la condition (S_2)

(ii) $dim \frac{\overline{gr\,U}}{\mathcal{A}\overline{gr\,U}} \leq dim \overline{gr\,U} - 2$.

Alors $\varphi : U \hookrightarrow \mathcal{D}$ est un isomorphisme d'anneaux filtrés (et $K = (0)$).

Preuve. Notons que $\Omega \neq \emptyset$, sinon $\mathcal{A} = (0)$ contredisant (ii).

Soit $x \in gr\,\mathcal{D}$ et $M = \frac{R+Rx}{R}$, $R = gr\varphi(grU)$ comme en 3.2. Par (3.3, proposition 1) on a $\mathcal{A}^\nu \overline{gr\,U} \subset ann_{\overline{gr\,U}} M$, donc $\dim_{\overline{gr\,U}} M \leq \dim\left(\frac{\overline{gr\,U}}{\mathcal{A}\overline{gr\,U}}\right) \leq \dim \overline{gr\,U} - 2$.

Remarquons que par (3.2, lemme 1) R et $gr\,\mathcal{D}$ ont même corps de fractions et par ce qui précède $\dim_R \frac{(R+Rx)}{R} \leq \dim R - 2$. La condition

(S_2) sur $R \cong \overline{gr\,U}$ assure alors $x \in R$. Donc $gr\varphi$ est surjective et on conclut par (3.1, lemme 1).

Remarques. (a) On a toujours $\mathcal{A} \subset \mathcal{I}(\mathrm{Sing}\, Z)$. Si φ est un isomorphisme d'anneaux filtrés il y a égalité.

(b) Il n'est pas difficile de voir que les conditions (i) et (ii) de la proposition impliquent $\overline{gr\,U}$ normal.

3.5 Nous allons appliquer ce qui précède à $U = \frac{\mathcal{D}(V)^G}{J}$ où (V, G) est comme en 2.1. Rappelons que $\varphi : \mathcal{D}(V)^G \to \mathcal{D}(V//G)$ est le morphisme de restriction et que $J = \mathrm{Ker}\,\varphi$. On note encore $\varphi : U \hookrightarrow \mathcal{D} := \mathcal{D}(V//G)$ l'application induite. Ici $A = \mathcal{O}(V)^G$ et $Z = \mathrm{Spec}\, A$, cf 3.3, n'est autre que $V//G$. On dispose de deux ouverts dans $V//G$: $\Omega(V//G)$, la strate principale, et l'ouvert Ω défini en 3.3.

Proposition. *On a $\Omega = \Omega(V//G)$, en particulier $\Omega \neq \emptyset$.*

Preuve. Si p est un point fermé de $V//G$ on désigne par U_p^1 et $\mathcal{D}_p^1$ les localisés correspondants des $\mathcal{O}(V)^G$-modules U^1 et $\mathcal{D}^1$. On peut réécrire la définition de Ω sous la forme :

$$\begin{aligned}\Omega &= \{p \in \mathrm{Reg}\, V//G / \varphi(U_p^1) = \mathcal{D}_p^1\} \\ &= \{p \in \mathrm{Reg}\, V//G / \varphi(\Theta_V^G)_p = (\Theta_{V//G})_p\}.\end{aligned}$$

Puisque π est surjective il s'agit de prouver que $\pi^{-1}\Omega = \Omega(V, G) = \{v \in V / \pi$ lisse aux points de $O(\pi(v))\}$, cf (2.3., proposition). Rappelons que $x \in V^{\text{lisse}}$ équivaut à $\xi = \pi(x) \in \mathrm{Reg}\, V//G$ et $d\pi_x : T_x V \to T_\xi V//G$ surjective ; V^{lisse} est un ouvert G-stable de V. Pour simplifier posons $A = \mathcal{O}(V)^G$, $B = \mathcal{O}(V)$.

Soit $v \in \Omega(V, G)$, $\xi = \pi(v) = \pi(x)$ avec $x \in V_{ss} \cap V^{\text{lisse}}$. Puisque Gx et $V \backslash V^{\text{lisse}}$ sont deux fermés G-stables disjoints il existe $f \in A$ valant 0 sur $V \backslash V^{\text{lisse}}$ et 1 sur Gx. Donc $V_f = \mathrm{Spec}\, B_f$ est un ouvert affine G-stable contenu dans V^{lisse} et contenant x. Le comorphisme $\pi_f : V_f \to V_f//G$ de l'inclusion $A_f = B_f^G \hookrightarrow B_f$ est alors lisse ; on en déduit aisément la surjectivité de l'application de restriction $\mathcal{D}er^1(B_f)^G \to \mathcal{D}er^1(A_f)$. Comme $\xi = \pi(x)$ est dans $V_f//G$ il en découle $U_\xi^1 = \frac{\mathcal{D}^1(B)_\xi^G}{J_\xi} \xrightarrow{\sim} \mathcal{D}^1(A_\xi)$, prouvant $\xi \in \Omega$ et $v \in \pi^{-1}\Omega$.

Réciproquement soit $v \in \pi^{-1}\Omega$ et $x \in V_{ss}$ tel que $\pi(v) = \pi(x) = \xi$. Par hypothèse $\xi \in \mathrm{Reg}\, V//G$ il s'agit donc de prouver que $d\pi_x : T_x V \to$

$T_\xi V//G$ est surjective. Cette application s'identifie à l'application naturelle entre dérivations ponctuelles :

$$\frac{\mathcal{D}er^1(B_x)}{\mathcal{M}_x\mathcal{D}er^1(B_x)} \longrightarrow \frac{\mathcal{D}er^1(A_\xi)}{\mathcal{M}_\xi\mathcal{D}er^1(A_\xi)}.$$

Puisque $\xi \in \Omega$ et $A\backslash\mathcal{M}_\xi \subset B\backslash\mathcal{M}_x$, l'application $\mathcal{D}er^1(B)_\xi^G \to \mathcal{D}er^1(A_\xi)$ est surjective et $\mathcal{D}er^1(B)_\xi^G \subset \mathcal{D}er^1(B_x)$. Par conséquent $d\pi_x$ est surjective.

Remarque. Par (3.4, remarque (a)), pour que (V, G) ait la propriété de relèvement il est nécessaire que $\Omega(V//G) = \operatorname{Reg} V//G$. L'exemple de 2.4 montre que cela ne suffit pas.

Signalons le corollaire qui suit, découlant facilement de la proposition et de (3.3, proposition 2).

Corollaire. *Il y a équivalence entre :*

(i) (V, G) *a la propriété de relèvement.*

(ii) $\varphi : \mathcal{D}(V)^G \to \mathcal{D}(V//G)$ *est surjective et* $gr\, J$ *est un idéal premier de* $gr\,\mathcal{D}(V)^G$.

3.6 Nous pouvons donner la condition suffisante pour le relèvement :

Théorème. *Soient* $(V, G), U = \frac{\mathcal{D}(V)^G}{J}, K \subset gr\, U$ *comme en* 2.1. *On suppose :*

(i) $\overline{gr\, U} = \frac{gr\, U}{K}$ *vérifie la condition* (S_2).

(ii) $\dim \frac{\overline{gr\, U}}{\mathcal{A}(V//G)\overline{gr\, U}} \leq \dim \overline{gr\, U} - 2$.

Alors (V, G) *a la propriété de relèvement.*

Preuve. Compte tenu de (3.5, Proposition) et de la définition de $\mathcal{A}(V//G)$ il suffit d'appliquer (3.4, Proposition).

Remarque. Rappelons que si (i) et (ii) sont satisfaits alors $gr\, U$ est un anneau normal et $\Omega(V//G) = \operatorname{Reg} V//G$, cf. (3.4, Remarques) et (3.5, corollaire).

Problème. L'inclusion $gr\, J \subset K$ est évidente, a-t-on toujours $gr\, J = K$? De manière équivalente : si $D \in \mathcal{D}^m(V)^G$ est tel que D restreint à $\mathcal{O}(V)^G$ est d'ordre $\leq m-1$, a-t-on $D \in \mathcal{D}^{m-1}(V)^G + J$?

Le corollaire de 3.5 montre que la réponse est positive si (V, G) a la propriété de relèvement.

4. Exemples : G fini et G = $\mathbb{C}^*$

4.1 Soit (V, G) comme en 2.1. Si (V, G) a la propriété de relèvement alors $\mathcal{A}(V//G) = \mathcal{I}(\mathrm{Sing}\, V//G)$ est de hauteur ≥ 2, car $V//G$ est normale, mais cela n'implique en général pas $ht\,\mathcal{A}(V, G) \geq 2$, cf (2.4, exemple). Nous allons voir que si G est fini ou $\mathbb{C}^*$ alors cette dernière condition assure la propriété de relèvement.

4.2 On suppose ici G fini opérant sur $B = \mathcal{O}(V)$, $\mathbb{C}$-algèbre de type fini régulière. On suppose, cf (2.2, remarque (b)), que $G \subset \mathrm{Aut}_{\mathbb{C}} B$. Soient $A = B^G$ et $\pi : V \to V//G$; rappelons que π est fini et que $V_{ss} = V$. L'ensemble $V_{ss} \cap V^{\mathrm{lisse}}$ coïncide avec V^{et} l'ouvert des points où π est étale.

On retrouve le résultat connu, cf. 2.4. :

Théorème. *Sous les hypothèses précédentes, il y a équivalence entre* (i) (V, G) *a la propriété de relèvement et* (ii) $codim(V \backslash V^{\mathrm{et}}) \geq 2$. *Si* $G \hookrightarrow GL(V)$ *est une représentation linéaire fidèle ces conditions sont équivalentes à* (iii) *G ne contient pas de pseudo réflexion* ($\neq 1$).

Preuve. On remarque $J = \mathrm{Ker}\,\varphi = (0)$, cf ([Le], théorème 5, p. 171). Alors $U = \mathcal{D}(V)^G$, $gr\, U = gr\, \mathcal{D}(V)^G$ est normal.

Si $\mathcal{A} = \mathcal{A}(V//G)$ on a : $\mathcal{A}(V, G) = \sqrt{\mathcal{A}B}$, $\mathcal{V}(\mathcal{A}(V, G)) = V \backslash V^{\mathrm{et}}$, $ht\mathcal{A} = ht\mathcal{A}(V, G) = ht\mathcal{A}\, gr\, \mathcal{D}(V) = ht\mathcal{A}\, gr\, U$. Ceci montre que (i) implique (ii), cf 4.1. Inversement (ii) signifie $ht\mathcal{A} gr\, U \geq 2$ d'où (i) par (3.6, théorème). L'équivalence de (ii) et (iii) est classique.

4.3 On suppose ici que $G = \mathbb{C}^*$ et que $G \hookrightarrow GL(V)$ est une représentation linéaire fidèle. On choisit une base de V de sorte que $\mathcal{O}(V) = \mathbb{C}[T] = \mathbb{C}[T_1, \ldots, T_s]$ et $g \cdot T_j = g^{a_j} T_j$ avec $a_j \in \mathbb{Z}$. Soit $\partial_j = \partial / \partial T_j$, τ_j le symbole de ∂_j de sorte que $\mathcal{D}(V) = \mathbb{C}[T, \partial] = \mathbb{C}[T_1, \ldots, T_s, \partial_1, \ldots, \partial_s]$, $gr\, \mathcal{D}(V) = \mathbb{C}[T, \tau] = \mathbb{C}[T_1, \ldots, T_s, \tau_1, \ldots, \tau_s]$. L'image de l'algèbre de Lie de G dans Θ_V est $\mathbb{C}\theta$ où $\theta = \sum_j a_j T_j \partial_j$, donc θ est central dans $\mathcal{D}(V)^G$ et $\sigma = \sum_j a_j T_j \tau_j$ est le symbole de θ.

Lemme. *Supposons* $dim\, V//G = dim\, V - 1$. *Alors* $J = Ker\varphi = \mathcal{D}(V)^G \theta$, $K = Ker(gr\varphi) = gr\, J = \sigma \cdot gr\, \mathcal{D}(V)^G$, *et* $gr\, U = \overline{gr\, U}$ *vérifie la condition* (S_2).

Preuve. La première assertion résulte de ([Mu], Theorem 4.4). La seconde vient de la première et du fait que $gr\mathcal{D}(V)^G = \mathbb{C}[T,\tau]^G$ est de Cohen-Macaulay.

Théorème. *Soit* $G \hookrightarrow GL(V)$ *une représentation linéaire fidèle de* $G = C^*$. *Si* $ht\mathcal{A}(V,G) \geq 2$ *alors* (V,G) *a la propriété de relèvement.*

Preuve. On voit facilement que l'on peut supposer $V^G = (0)$, $\dim V \geq 2$, et $\dim V//G = \dim V - 1$. Donc en particulier $a_j \neq 0$ pour tout j, si l'on adopte les notations précédentes. Le fermé $F(V,G) = V - \Omega(V,G)$ est stable sous l'action du groupe des matrices diagonales de $GL(s,\mathbb{C})$, ce qui implique que les idéaux premiers minimaux sur $\mathcal{A}(V,G)$ sont de la forme $\mathcal{P}_E = (T_i; i \in E)\mathbb{C}[T]$ où $E \underset{\neq}{\subset} \{1,\ldots,s\}$ (l'inclusion est stricte car $ht\mathcal{A}(V,G) \leq \dim V//G$, cf. 2.3). Comme $\sigma \notin \mathcal{P}_E$, pour E comme ci-dessus, on en déduit que si $\mathcal{A} = \mathcal{A}(V//G)$:

$$\begin{aligned} ht(\mathcal{A},\sigma)\mathbb{C}[T,\tau]^G \geq ht(\mathcal{A},\sigma)\mathbb{C}[T,\tau] &= ht(\mathcal{A}(V,G),\sigma)\mathbb{C}[T,\tau] \\ &= ht\mathcal{A}(V,G) + 1 \geq 3. \end{aligned}$$

Le lemme précédent permet alors d'appliquer (3.6, Théorème) pour avoir le résultat.

Remarques. (a) Dans les hypothèses du théorème on peut montrer que la condition $ht\mathcal{A}(V,G) \geq 2$ est nécessaire.

(b) Rappelons que [Mu] donne des conditions nécessaires et suffisantes pour que (V,G) ait la propriété de relèvement, ceci lorsque G est un tore quelconque.

5. Opérateurs différentiels sur les anneaux classiques d'invariants

5.1 On fixe une $\mathbb{C}$-algèbre U munie d'une filtration $\{U^m\}_{m\in\mathbb{Z}}$ croissante, exhaustive et discrète :

$$U^m = 0 \text{ si } m < 0; \mathbb{C} \subset U^0 \subset U^1 \subset \cdots \subset U^m \subset \cdots; U = \cup_{m\geq 0} U^m$$

On suppose également que U est graduée par $\{U_\ell\}_{\ell\in\mathbb{Z}}$: $U = \bigoplus_\ell U_\ell$, $\mathbb{C} \subset U_0$.

On fait dans ce qui suit l'hypothèse que la filtration et la graduation sont compatibles, c'est à dire :

$$\forall m \geq 0, U^m = \bigoplus_{\ell \in \mathbb{Z}} (U^m \cap U_\ell).$$

On pose alors : $U_\ell^m = U^m \cap U_\ell$ et pour tout $d \in \mathbb{Z}$, $U(d) = \bigoplus_{\ell \in \mathbb{Z}} U_\ell^{d-\ell}$. Notons que $U_k^m U_\ell^n \subset U_{k+\ell}^{m+n}$ et que $U(d) = \sum_{\ell+m \leq d} U_\ell^m$. Il est facile de vérifier que $\{U(d)\}d \in \mathbb{Z}$ est une filtration croissante, exhaustive et séparée sur U.

Aux deux filtrations $\{U^m\}_m$ et $\{U(d)\}_d$ sont associés deux anneaux gradués notés respectivement $gr\, U$ et $\sigma(U)$. On écrit

$$gr\, U = \bigoplus_m gr_m U \qquad gr_m : U^m \to gr_m U = \frac{U^m}{U^{m-1}}$$

$$\sigma(U) = \bigoplus_d \sigma_d(U), \qquad \sigma_d : U(d) \to \sigma_d(U) = \frac{U(d)}{U(d-1)}.$$

Dans les exemples que nous rencontrerons on a $U(d) = 0$ si $d < 0$, cf. 5.2. Si $a \in U^m \backslash U^{m-1}$ et $a \in U(d) \backslash U(d-1)$ on pose $\operatorname{ord} a = m$ et $\operatorname{ord}_\sigma a = d$. Signalons que si $a = \sum_\ell a_\ell$, $a_\ell \in U_\ell$, alors $\operatorname{ord}_\sigma a = d$ équivaut à $\operatorname{ord} a_\ell = d - \ell$ pour un ℓ.

Proposition. *Il existe un isomorphisme d'algèbres, $\phi : \sigma(U) \to gr\, U$, défini sur $\sigma_p(U)$ par :*

$$\forall\; a = \sum_\ell a_\ell \in U(p),\, a_\ell \in U_\ell^{p-\ell},\, \phi(\sigma_p(u)) = \sum_{ord\, a_\ell = p-\ell} gr_{p-\ell}(a_\ell)$$

et prolongé par linéarité à $\sigma(U)$.

Preuve. On définit ϕ sur $\sigma_p(U)$ de la manière suivante. Soit $a = \sum_\ell a_\ell \in U(p)$ avec $a_\ell \in U_\ell^{p-\ell}$, on pose $\phi(\sigma_p(a)) = \sum_{\operatorname{ord} a_\ell = p-\ell} gr_{p-\ell}(a_\ell)$ alors ϕ est bien définie sur $\sigma_p(U)$ et on étend ϕ par linéarité sur $\sigma(U)$. On a clairement $\phi(1) = 1$, il s'agit de vérifier que ϕ est multiplicative et bijective. La multiplicativité se vérifie sur les produits d'éléments homogènes de $\sigma(U)$ ce qui est suffisant. La surjectivité résulte de ce que si $a \in U^m$, $a = \sum_\ell a_\ell$

avec $a_\ell \in U_\ell^m$, alors $gr_m a = \phi(\sum_\ell \sigma_{m+\ell}(a_\ell))$. Pour prouver l'injectivité on considère $x = \sum_{i=1}^{t} \sigma_{p_i}(a^{(i)}) \in \sigma(U)$ où $p_1, \dots, p_t$ sont 2 à 2 distincts et $a^{(i)} = \sum_{\ell_i} a_{\ell_i}^{(i)} \in U(p_i)$, $a_{\ell_i}^{(i)} \in U_{\ell_i}^{pi-\ell_i}$. La condition $\phi(x) = 0$ entraîne $\sum_{p_i - \ell_i = m} a_{\ell_i}^{(i)} \in U^{m-1}$ pour tout m, puis $a_{\ell_i}^{(i)} \in U^{p_i - \ell_i - 1}$ pour tous i et ℓ_i, donc $a^{(i)} \in U(p_i - 1)$ c'est à dire $x = 0$.

Remarque. L'isomorphisme ϕ n'est pas gradué.

5.2 Soient $\mathbb{C}[T] = \mathbb{C}[T_1, \dots, T_s]$ un anneau de polynômes et $d_j \in \mathbb{N}^*$, $1 \le j \le s$. Graduons $\mathbb{C}[T]$ par l'action de $\mathbb{C}^*$: $\lambda \cdot T_j = \lambda^{d_j} T_j$, $\lambda \in \mathbb{C}^*$, $1 \le j \le s$.

Soit $\mathcal{I}$ un idéal premier homogène de $\mathbb{C}[T]$ et $A = \frac{\mathbb{C}[T]}{\mathcal{I}} = \bigoplus_{i \ge 0} A_i$. L'action de $\mathbb{C}^*$ sur A s'étend en une action localement finie sur $\mathcal{D} := \mathcal{D}(A)$; on note $\mathcal{D}_\ell, \ell \in \mathbb{Z}$ les composantes isotypiques correspondantes. On a : $\mathcal{D}_\ell = \{D \in \mathcal{D} / D_* A_i \subset A_{i+\ell} \text{ pour tout } i\}$. La filtration par l'ordre $\{\mathcal{D}^m\}_m$ est compatible avec cette graduation (car $\mathcal{D}^m$ est $\mathbb{C}^*$-stable).

Lorsque $\mathcal{I} = (0)$, $\mathcal{D}$ est égal à l'algèbre de Weyl $\mathbb{C}[T, \partial] = \mathbb{C}[Tj, \partial_j = \partial / \partial T_j ; 1 \le j \le s]$ et alors :

$$\mathbb{C}[T, \partial]_\ell^m = \{\sum_{\alpha\beta} a_{\alpha\beta} T^\alpha \partial^\beta / a_{\alpha\beta} \in \mathbb{C}, \sum_i d_i \alpha_i - \sum_i d_i \beta_i = \ell,\ |\beta| \le m\}.$$

Si $\mathcal{I}$ est quelconque $\mathcal{D}$ se représente sous la forme $\frac{II(\mathcal{I}\mathbb{C}[T,\partial])}{\mathcal{I}\mathbb{C}[T,\partial]}$ où $II(\mathcal{I}\mathbb{C}[T, \partial]) = \{a \in \mathbb{C}[T, \partial] / a\mathcal{I} \subset \mathbb{C}[T, \partial]\}$. Tout élément de $\mathcal{D}_\ell^m$ est alors la classe modulo $\mathcal{I}\mathbb{C}[T, \partial]$ d'un élément de $II(\mathcal{I}\mathbb{C}[T, \partial]) \cap \mathbb{C}[T, \partial]_\ell^m$.

Remarquons que si $d = \operatorname{Max}_{1 \le i \le s} d_i$ alors $\mathbb{C}[T, \partial]_\ell^m = (0)$ pour $\ell + dm < 0$, donc aussi $\mathcal{D}_\ell^m = (0)$.

Prenons $d_j = 1$ pour tout j, i.e. $d = 1$, il vient donc $\mathcal{D}(p) = \bigoplus_\ell \mathcal{D}_\ell^{p-\ell} = (0)$ pour $p < 0$.

Considérons la filtration (par des $\mathbb{C}$-espaces vectoriels de dimension finie) sur $\mathbb{C}[T, \partial]$ donnée par $F^n \mathbb{C}[T, \partial] = \{\sum a_{\alpha\beta} T^\alpha \partial^\beta / |\alpha| + |\beta| \le n\}$ et notons $F^n \mathcal{D}$ la filtration induite sur $\mathcal{D}$ grâce à la description précédente. Il est facile de voir que $\mathcal{D}_\ell^m \subset F^{\ell + 2m} \mathcal{D}$.

5.3 Nous allons appliquer ce qui précède dans la situation provenant des anneaux classiques d'invariants telle qu'elle est exposée dans [Le-St] d'après [H]. Nous commencerons par résumer les notations de [Le-St] (à quelques exceptions près).

On se donne une paire réductive duale classique (G, G') formée de deux groupes classiques contenus dans un groupe symplectique $Sp(\tilde{V})$ où $\tilde{V}$ est un $\mathbb{C}$-espace vectoriel symplectique de polarisation $\tilde{V} = V \oplus V^*$ de sorte que $G' \subset GL(V^*)$. On pose $\mathfrak{sp} = Lie(Sp(\tilde{V}))$, $\mathfrak{g} = Lie(G)$. La représentation dite "métaplectique" consiste (pour nous) en un homomorphisme $\omega : \mathfrak{sp} \rightarrow \mathcal{D}(V)$ provenant de l'identification de $\mathfrak{sp}$ avec la sous-algèbre de Lie de $\mathcal{D}(V)$ formée des anti-commutateurs $ab + ba$ d'éléments a, b de $V \oplus V^*$, où V^* et V sont identifiés à des fonctions et champs de vecteurs sur V (respectivement). Comme G est le commutant de G' dans $Sp(\tilde{V})$ on a un homomorphisme, encore noté ω, de $U(\mathfrak{g})$ vers $\mathcal{D}(V)^{G'}$; par [Ho] il est surjectif. En composant avec $\varphi : \mathcal{D}(V)^{G'} \rightarrow \mathcal{D}(V//G')$ on dispose de $\psi : U(\mathfrak{g}) \rightarrow \mathcal{D}(V//G')$ qui est surjectif si et seulement si φ l'est. Afin d'accorder les notations de [Le.St] avec celles des paragraphes précédents on pose : $J = \mathrm{Ker}\varphi, U = \frac{\mathcal{D}(V)^{G'}}{J}, \overline{\mathcal{X}_k} = V//G'$, $J_k = \mathrm{Ker}\,\psi$, (k étant un entier dont la présence sera justifiée ci-après). On a ainsi un diagramme commutatif :

$$\begin{array}{ccccc}
U(\mathfrak{g}) & \xrightarrow[\omega]{} & \mathcal{D}(V)^{G'} & & \\
\downarrow & & \downarrow & \searrow^{\varphi} & \\
\frac{U(\mathfrak{g})}{J_k} & \xrightarrow[\omega]{\sim} & U := \frac{\mathcal{D}(V)^{G'}}{J} & \hookrightarrow & \mathcal{D}(V//G')
\end{array}$$

Donnons la description de $\overline{\mathcal{X}_k}$:

Cas A. Soient $p \geq q \geq 1$, $k \geq 1$, $G = GL(p+q, \mathbb{C})$, $G' = GL(k, \mathbb{C})$, $V = M_{p,k}(\mathbb{C}) \times M_{k,q}(\mathbb{C}); \overline{\mathcal{X}_k} \cong \{M \in M_{p,q}(\mathbb{C}) / rgM \leq k\}$.

Cas B. Soient $n \geq 1$, $k \geq 1$, $G = Sp(2n, \mathbb{C})$, $G' = O(k, \mathbb{C})$, $V = M_{k,n}(\mathbb{C})$ $\overline{\mathcal{X}_k} \cong \{M \in M_n(\mathbb{C}) / M$ symétrique, $rgM \leq k\}$.

Cas C. Soient $n \geq 1$, $k \geq 1$, $G = O(2n, \mathbb{C})$, $G' = Sp(2k, \mathbb{C}), V = M_{2k,n}(\mathbb{C})$; $\overline{\mathcal{X}_k} \cong \{M \in M_n(\mathbb{C}) / M$ antisymétrique, $rgM \leq 2k\}$.

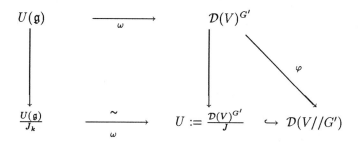

Dans chaque cas l'action de G' sur V est "naturelle", i.e. par translation à gauche (ou à droite, cas A). On a $\mathcal{O}(\overline{\mathcal{X}_k}) = \mathcal{O}(V)^{G'}$. Nous supposerons désormais $\overline{\mathcal{X}_k}$ non lisse, ce qui équivaut à : $\{1 \leq k \leq q-1$, cas A ; $1 \leq k \leq n-1$, cas B ; $1 \leq k \leq \frac{n-2}{2}$, cas C$\}$ c'est à dire k "suffisamment petit" au sens de [Le-St]. L'anneau $\mathcal{O}(\overline{\mathcal{X}_k})$ est isomorphe à $A = \frac{\mathbb{C}[Z]}{I}$ où $Z = \{Z_1, \ldots, Z_n\}$ sont les fonctions coordonnées canoniques sur $M_{p,q}(\mathbb{C})$, $M_n(\mathbb{C})$ (selon le cas) et où I est un idéal homogène de $\mathbb{C}[Z]$ lorsque chaque Z_i est affecté du degré 1. D'autre part $\mathcal{O}(\overline{\mathcal{X}_k}) = \mathcal{O}(V)^{G'}$ est engendré comme sous-anneau de $\mathcal{O}(V)$ par $f_1, \ldots, f_N$ avec $f_i \in S^2(V^*)^{G'}$ (cf [Ho]). Si l'on convient de poser $\mathcal{D} = \mathcal{D}(A) \cong \mathcal{D}(\overline{\mathcal{X}_k})$, l'application $\varphi : \mathcal{D}(V)^{G'} \to \mathcal{D}$ restreinte à $\mathcal{O}(V)^{G'}$ est l'isomorphisme envoyant f_i sur la classe de Z_i modulo I. L'algèbre de Lie $\mathfrak{g}$ se décompose en $\mathfrak{r}^- \oplus \mathfrak{m} \oplus \mathfrak{r}^+$ où $\mathfrak{p} = \mathfrak{m} \oplus \mathfrak{r}^+$ est une sous-algèbre parabolique à radical $\mathfrak{r}^+$ abélien. Si l'on pose $J_k \cap U(\mathfrak{r}^-) = I_k$ alors ω induit un isomorphisme $\frac{U(\mathfrak{r}^-)}{I_k} \to \mathcal{O}(V)^{G'}$ qui composé avec φ fournit un isomorphisme $\psi : \frac{U(\mathfrak{r}^-)}{I_k} \to A = \frac{\mathbb{C}[Z]}{I}$, cf ([Le-St], chap. II). Rappelons que (comme pour toute algèbre de Lie $\mathfrak{g}$) $U(\mathfrak{g})$ est munie d'une filtration canonique : $F^pU(\mathfrak{g}) = \mathbb{C} + \mathfrak{g} + \cdots + \mathfrak{g}^p$. L'anneau gradué associé $gr^F U(\mathfrak{g})$ s'identifiant à $S(\mathfrak{g})$ (l'algèbre symétrique), donc à $\mathcal{O}(\mathfrak{g}^*)$. A l'idéal $J_k = \operatorname{Ker}\psi$ on associe une variété $\mathcal{V}(J_k) := \mathcal{V}(\sqrt{gr^F J_k})$ dans $\mathfrak{g}^*$. On peut voir que $\mathcal{V}(J_k)$ est normale, cf ([Le-St], chap. II).

Lemme-clé ([Le-St], chap. III, 1.6). *Sous les hypothèses et notations précédentes : l'isomorphisme $\psi : \frac{U(\mathfrak{r}^-)}{I_k} \to A$ envoie $\frac{I_{k-1}}{I_k}$ sur $\mathcal{I}(\operatorname{Sing}\overline{\mathcal{X}_k})$ et on a*

$$\dim \frac{S(\mathfrak{g})}{\sqrt{gr^F J_k} + I_{k-1}S(\mathfrak{g})} \leq \dim \frac{S(\mathfrak{g})}{\sqrt{gr^F J_k}} - 2$$

Signalons que I_{k-1} étant un idéal homogène de $U(\mathfrak{r}^-) = S(\mathfrak{r}^-)$, on l'identifie à $gr^F I_{k-1}$ dans $S(\mathfrak{g}) = gr^F U(\mathfrak{g})$.

5.4 Un des buts de [Le-St] est de prouver la surjectivité de ψ (i.e. de φ). La preuve comporte trois étapes :

(a) le lemme clé de 5.3

(b) un lemme (attribué à O. Gabber), cf ([Le-St], chap. III, 1.8) qui grâce à (a) implique que $\mathcal{D}$ est un U-module de type fini.

(c) la maximalité de J_k, ([Le-St], chap. IV), ce qui utilise la théorie des idéaux primitifs.

Nous allons montrer que (3.6, théorème) permet de retrouver cette surjectivité sans avoir à utiliser (b) et (c) ; le (a) nous restant nécessaire.

5.5 Sur $\mathcal{D}(V)$ on dispose de la filtration par l'ordre $\{\mathcal{D}^m(V)\}_m$ et de la graduation $\{\mathcal{D}_\ell(V)\}_\ell$ provenant de celle de $\mathcal{O}(V) = S(V^*) = \underset{i}{\oplus} S^i(V^*)$ avec $S^i(V^*) = \mathcal{O}(V)_i$. Sur l'algèbre $\mathcal{D} = \mathcal{D}(A)$ on a de même $\{\mathcal{D}^m\}_m$ et $\{\mathcal{D}_\ell\}_\ell$ provenant de $A = \underset{\ell}{\oplus} A_\ell$ (rappelons que chaque Z_i a le degré 1). Nous utiliserons les notations de 5.1 et 5.2.

Signalons que $\mathcal{O}(V)^{G'} = S(V^*)^{G'} = \mathbf{C}[f_1, \dots, f_N]$, cf 5.3, est un sous anneau gradué de $\mathcal{O}(V) : \mathcal{O}(V)^{G'} = \underset{i}{\oplus} \mathcal{O}(V)^{G'}_{2i}$ (les f_i sont dans $S^2(V^*)$).

Lemme. $\forall i, \forall m, \varphi(\mathcal{D}^m_{2i}(V)^{G'}) \subset \mathcal{D}^m_i$.

Preuve. Notons que puisque $G' \subset GL(V^*)$ on peut parler de $\mathcal{D}^m_{2i}(V)^{G'}$. La vérification est immédiate lorsqu'on a remarqué que φ envoie $\mathcal{O}(V)^{G'}_{2i}$ sur A_i (car $\varphi(f_i) = Z_i (\operatorname{mod} I)$).

Corollaire. *L'homomorphisme* $\psi : U(\mathfrak{g}) \to \mathcal{D}$ *vérifie :* $\psi(\mathfrak{r}^-) \subset \mathcal{D}^0_1$, $\psi(\mathfrak{r}^+) \subset \mathcal{D}^2_{-1}$, $\psi(\mathfrak{m}) \subset \mathcal{D}^1_0$.

Preuve. Les formules définissant ω, cf ([Le-St], chap IV, 1.9), prouvent que $\omega(\mathfrak{r}^-) \subset \mathcal{D}^0_2(V)^{G'}$, $\omega(\mathfrak{r}^+) \subset \mathcal{D}^2_{-2}(V)^{G'}$, $\omega(\mathfrak{m}) \subset \mathcal{D}^1_0(V)^{G'}$. Il suffit donc d'appliquer le lemme.

5.6 Remarquons que, par (5.5, corollaire), ψ n'est pas un isomorphisme filtré lorsque $U(\mathfrak{g})$, resp. $\mathcal{D}$, est muni de la filtration canonique, resp. $\{\mathcal{D}^m\}_m$. On va montrer que c'est par contre le cas pour $\{\mathcal{D}(p)\}_p$, $\mathcal{D}(p)$ comme en 5.1. Pour $\ell, m \in \mathbf{Z}$ on pose :

$$U^m_\ell(\mathfrak{g}) = \sum_{\substack{\alpha-\beta=\ell \\ \gamma+2\beta\leq m}} F^\alpha U(\mathfrak{r}^-) F^\gamma U(\mathfrak{m}) F^\beta U(\mathfrak{r}^+)$$

(où $F^{\cdot}$ est la filtration canonique).

Lemme 1 (a) $\quad \forall p, F^p U(\mathfrak{g}) = \sum_{\ell+m\leq p} U^m_\ell(\mathfrak{g}) = \sum_{\ell+m=p} U^m_\ell(\mathfrak{g})$.

(b) $\omega(U^m_\ell(\mathfrak{g})) \subset \mathcal{D}^m_{2\ell}(V)^{G'}$ *et* $\varphi(\mathcal{D}^m_{2\ell}(V)^{G'}) \subset \mathcal{D}^m_\ell$.

Preuve. (a) est aisé et (b) résulte des propriétés de ω énoncées dans la preuve de (5.5, corollaire).

Lemme 2. $\forall d \in \mathbf{N}$, $[F^{2d}\mathcal{D}(V)]^{G'} = [F^{2d+1}\mathcal{D}(V)]^{G'} = \omega(F^d U(\mathfrak{g}))$.

Preuve. Rappelons que la filtration $F^{\cdot}\mathcal{D}(V)$ a été introduite en 5.2. Les égalités sont conséquences directes de ([Ho], Th 1.A et Th. 1.B).

Proposition. On a :

$$\omega(U_\ell^m(\mathfrak{g})) = \mathcal{D}_{2\ell}^m(V)^{G'}, \ \omega(F^p U(\mathfrak{g})) = \underset{\ell}{\oplus}\mathcal{D}_{2\ell}^{p-\ell}(V)^{G'}$$

et

$$\psi(F^p U(\mathfrak{g})) \subset \mathcal{D}(p) = \underset{\ell}{\oplus}\mathcal{D}_\ell^{p-\ell}.$$

Preuve. En utilisant 5.2 et le lemme 2 on voit que :

$$\mathcal{D}_{2\ell}^m(V)^{G'} \subset [F^{2\ell+2m}\mathcal{D}(V)]^{G'} = \omega(F^{\ell+m}U(\mathfrak{g}))$$

grâce au lemme 1 on obtient alors

$$\omega(F^p U(\mathfrak{g})) = \underset{\ell+m=p}{\oplus} \omega(U_\ell^m(\mathfrak{g})) = \underset{\ell+m=p}{\oplus} \mathcal{D}_{2\ell}^m(V)^{G'}.$$

Les autres assertions en découlent.

5.7 Rappelons que $\varphi : \mathcal{D}(V)^G \to \mathcal{D} := \mathcal{D}(A)$, $J = \operatorname{Ker}\varphi$ et $U = \frac{\mathcal{D}(V)^G}{J}$. On note $p : \mathcal{D}(V)^G \to U$ la projection et on pose : $U^m = p(\mathcal{D}^m(V)^{G'}), U_\ell = p(\mathcal{D}_{2\ell}(V)^{G'})$.

Lemme. *L'anneau U est gradué par $\{U_\ell\}_\ell$ et la filtration $\{U^m\}_m$ est compatible avec cette graduation. Si $U(d) = \sum_{\ell+m=d} U_\ell^m$, où $U_\ell^m = U_\ell \cap U^m$, alors : $U(d) = p\omega(F^d U(\mathfrak{g}))$, $U_\ell^m = p\omega(U_\ell^m(\mathfrak{g}))$.*

Preuve. Notons que $\mathcal{D}(V)^{G'} = \omega(U(\mathfrak{g})) = \underset{\ell}{\oplus}\mathcal{D}_{2\ell}(V)^{G'}$, cf (5.6, proposition). Par (5.6, lemme 1) on a $\varphi(U_\ell) \subset \mathcal{D}_\ell$; comme $\mathcal{D} = \underset{\ell}{\oplus}\mathcal{D}_\ell$ et φ injective sur U il en résulte $U = \underset{\ell}{\oplus} U_\ell$. La compatibilité résulte de l'égalité $\mathcal{D}^m(V)^{G'} = \underset{\ell}{\oplus}\mathcal{D}_{2\ell}^m(V)^{G'}$. La dernière assertion est conséquence de (5.6, proposition) et des définitions.

Proposition. *L'anneau $gr\,U = \underset{m}{\oplus} \frac{U^m}{U^{m-1}}$ est isomorphe à $\frac{S(\mathfrak{g})}{gr^F J_k}$; il possède un unique idéal premier minimal, égal au noyau K de $gr\varphi : gr\,U \to gr\mathcal{D}$. De plus $\overline{gr\,U} = \frac{gr\,U}{K}$ est normal.*

Preuve. Notons, comme en 5.1, $\sigma(U) = \bigoplus_d \frac{U(d)}{U(d-1)}$. Par le lemme précédent $\omega : \frac{U(\mathfrak{g})}{J_k} \to U$ est un isomorphisme filtré, U étant muni de $\{U(d)\}_d$ et $\frac{U(\mathfrak{g})}{J_k}$ de la filtration canonique. Grâce à (5.1, proposition) il vient : $\frac{S(\mathfrak{g})}{gr^F J_k} \cong \sigma(U) \cong gr\,U$. Puisque $\mathcal{V}(J_k) := \mathcal{V}(\sqrt{gr^F J_k})$ est normale, cf 5.3, $\frac{S(\mathfrak{g})}{gr^F J_k}$ n'a qu'un seul idéal premier minimal, $\frac{\sqrt{gr^F J_k}}{gr^F J_k}$, et $\frac{S(\mathfrak{g})}{\sqrt{gr^F J_k}}$ est normal. Il suffit donc d'utiliser (3.3, proposition 2) et l'isomorphisme précédent.

Remarque. Insistons sur le fait que $\frac{U(\mathfrak{g})}{J_k}$ est filtré par des espaces de dimenstion finie contrairement aux U^m.

5.8 On retrouve maintenant le résultat (k suffisamment petit) :

Théorème. *Le couple (V, G') a la propriété de relèvement.*

Preuve. On va appliquer (3.6, Théorème), vérifions les hypothèses. Remarquons que $\mathcal{A}(V//G') = \mathcal{I}(\mathrm{Sing}\,V//G')$; on peut pour cela : soit déterminer la strate principale de $V//G'$, soit raisonner comme dans ([Le-St], chap III, 1.3 et 1.4).

Par (5.7, proposition) on a : $\overline{gr\,U} \cong \frac{S(\mathfrak{g})}{\sqrt{gr^F J_k}}$. Le lemme clé de 5.1 et la remarque précédente impliquent :

$$\begin{aligned}\dim \frac{\overline{gr\,U}}{\mathcal{A}(V//G')\overline{gr\,U}} &= \dim \frac{S(\mathfrak{g})}{\sqrt{gr^F J_k} + I_{k-1}S(\mathfrak{g})} \leq \dim \frac{S(\mathfrak{g})}{\sqrt{gr^F J_k}} - 2 \\ &= \dim \overline{gr\,U} - 2.\end{aligned}$$

Donc (3.6, théorème) fournit le résultat.

On a la conséquence immédiate, cf (3.5, corollaire) :

Corollaire. *On a un isomorphisme $\psi : \frac{U(\mathfrak{g})}{J_k} \xrightarrow{\sim} \mathcal{D}(\overline{\mathcal{X}_k})$ et $gr^F J_k$ est premier.*

Remarque. (a) La surjectivité de $\psi : U(\mathfrak{g}) \to \mathcal{D}(\overline{\mathcal{X}_k})$ se déduit également d'un résultat plus général dû à A. Joseph , cf [J].

(b) Cette approche ne donne pas la simplicité de $\mathcal{D}(\overline{\mathcal{X}_k})$, i.e. J_k maximal.

(c) Il peut arriver, cf [Le-Sm], qu'un quotient primitif $\frac{U(\mathfrak{g})}{J}$ (en l'occurence pour $\mathfrak{g}$ de type G_2) soit isomorphe à l'anneau des o.d. sur une variété (non lisse) avec $gr^F J$ non premier dans $S(\mathfrak{g})$. Ce phénomène est en relation avec la non normalité de $\frac{S(\mathfrak{g})}{\sqrt{gr^F J}}$.

REFERENCES

[B-H] H. Bass and W. Haboush, *Linearizing certain reductive group actions*, Trans. Amer. Math. Soc. **292** (1985), 463–482.

[E.G.A] A Grothendieck, *Eléments de Géométrie Algébrique. Chap. IV*, Publ. I.H.E.S., **32** (1967).

[H] R. Howe, *Remarks on classical invariant theory*, preprint.

[I] Y. Ishibashi,*Nakai's conjecture for invariant subrings*, Hiroshima Math. J. **15** (1985), 429–436.

[J] A. Joseph, *A surjectivity theorem for rigid highest weight modules*, Inv. Math. **92** (1988), 567–596.

[K] J.M. Kantor, *Formes et opérateurs différentiels sur les espaces analytiques complexes*, Bull. Soc. Math. France **53** (1977), 5–80.

[Kr-L] G. Krause and T. Lenagan, *Growth of Algebras and Gelfand-Kirillov dimension*, Research notes in Math. **116**, Pitman, 1985.

[Le] T. Levasseur, *Anneaux d'opérateurs différentiels*, Dans Séminaire d'Algèbre P. Dubreil et M.P. Malliavin. Lecture Notes in Math. **867**(1981), Springer Verlag, 157–173.

[Le-Sm] T. Levasseur and S.P. Smith, *Primitive ideals and nilpotent orbits in type G_2*, J. Algebra **114** (1988), 81–105.

[Le-St] T. Levasseur and J.T. Stafford, *Rings of differential operators on classical rings of invariants*, Memoirs of the Amer. Math. Soc. **81** 412, (1989).

[Lu] D. Luna, *Slices étales*, Bull. Soc. Math. France. Mémoire **33** (1973), 81–105.

[M.R] J.C. Mc Connell and J.C. Robson, *Non commutative Noetherian Rings*, J. Wiley, New York, 1987.

[Mu] I. Musson*Rings of differential operators on invariant rings of tori*, Trans. Amer. Math. Soc. **303** (1987), 805–827.

[S] G. Schwarz, *Lifting smooth homotopies of orbit spaces*, Publ. I.H.E.S. **51** (1980), 37–135.

Received July 19, 1989

Département de Mathématiques et Informatique
Université de Bretagne Occidentale
29287 Brest Cedex - France

Representations of Quantum Groups at Roots of 1

CORRADO DE CONCINI and VICTOR G. KAC

Dedicated to Jacques Dixmier on his 65th birthday

Introduction.

Quantum group $\mathcal{U}_q = \mathcal{U}_q(\mathfrak{a})$ is a certain (Hopf algebra) deformation of the universal enveloping algebra $\mathcal{U}(\mathfrak{a})$ of a complex simple finite-dimensional Lie algebra $\mathfrak{a}$, introduced by Drinfeld [3], [4] and Jimbo [5] in their study of the quantum Yang–Baxter equation.

An important problem is to describe finite–dimensional representations of the algebra $\mathcal{U}_q$. When q is *generic*, i.e. it is not a root of 1, then the (finite–dimensional) representation theory of $\mathcal{U}_q$ is essentially the same as that of $\mathcal{U}(\mathfrak{a})$, namely representations of $\mathcal{U}_q$ are deformations of representations of $\mathcal{U}(\mathfrak{a})$, so that the latter are obtained as $q \longrightarrow 1$ [12], [18]. Our contribution to the generic case is Proposition 1.9 giving a formula for the determinant of the contravariant form on a Verma module over $\mathcal{U}_q$, and Proposition 2.2 giving an explicit description of the center of $\mathcal{U}_q$. Proposition 2.2 is derived by a method developed in [7]. Proposition 1.9 implies in a usual way the description of irreducible subquotients of Verma modules over $\mathcal{U}_q$; this description was obtained by a different method in [1]. A result, somewhat weaker than Proposition 2.2b was previously obtained by a different method in [19].

When $q = \varepsilon$, a primitive ℓ–th root of 1, the situation changes dramatically. We study this case in §3. Our first key observation is that $\mathcal{U}_\varepsilon$ contains a large "standard" central subalgebra Z_0, so that $\mathcal{U}_\varepsilon$ is finite-dimensional over its center Z_ε. Since $\mathcal{U}_\varepsilon$ has no zero divisors, the algebra $Q(\mathcal{U}_\varepsilon) := Q(Z_\varepsilon) \otimes_{Z_\varepsilon} \mathcal{U}_\varepsilon$ is a division algebra of dimension m^2 over the field of fractions $Q(Z_\varepsilon)$ of Z_ε. Since, moreover, $\mathcal{U}_\varepsilon$ is integrally closed (i.e. is a maximal order in $Q(\mathcal{U}_\varepsilon)$)(Theorem 1.8), Z_ε is integrally closed as well and we may use the theory of finite–dimensional associative algebras. Denoting by Rep $\mathcal{U}_\varepsilon$ the set of equivalence classes of finite-dimensional irreducible representations of $\mathcal{U}_\varepsilon$, we obtain a sequence of

---

Supported in part by NSF grant DMS-8802489 and Sloan grant 88-10-1.

canonical surjective maps: Rep $\mathcal{U}_\varepsilon \xrightarrow{X}$ Spec $Z_\varepsilon \xrightarrow{\tau}$ Spec Z_0. Outside the discriminant $\mathcal{D} \subset$ Spec Z_ε the map X is bijective and $X^{-1}(\chi)$ for $\chi \in$ Spec $Z_\varepsilon \backslash \mathcal{D}$ is a representation of dimension m. Dimensions of all other irreducible representations are less than m. The map τ is finite. In the case when ℓ is odd we find an explicit formula for m and for the degree of τ:

$$m = \ell^N, \ \deg \tau = \ell^n, \tag{0.1}$$

where $N = (\dim \mathfrak{a} - \text{rank } \mathfrak{a})/2$ and $n = \text{rank } \mathfrak{a}$.

The proof of the above formulas consists of two components, both of which, we think, are of independent interest. First, it is the study of a family of "diagonal" modules, which for generic values of parameters are irreducible of dimension ℓ^N (for odd ℓ). Secondly, we introduce a (infinite–dimensional) group G of automorphisms of the algebra $\mathcal{U}_\varepsilon$ (or rather of its analytic completion $\hat{\mathcal{U}}_\varepsilon$). We show that the group G acts on Spec Z_0 by analytic transformations and that the G–orbit of the set $H_0 \subset$ Spec Z_0 corresponding to irreducible diagonal modules contains an open (in metric topology) subset of Spec Z_0. This proves (0.1).

We call the action of G on Spec Z_0 the quantized coadjoint action for the following reasons. As an algebraic variety, Spec Z_0 is isomorphic to $\mathbb{C}^{\dim \mathfrak{a}}$ with n subspaces of codimension 1 removed. The group G has 2^n fixed points in Spec Z_0. In a subsequent paper [2] (written jointly with Procesi)we show that the action of G in the tangent space to a fixed point is precisely the coadjoint action on $\mathfrak{a}$. We hope that this will lead us to the proof of conjectures on the quantized coadjoint action stated in §5. We find it quite remarkable that the quantized coadjoint action is independent of ℓ.

In §4 we study in detail the case of $\mathcal{U}_q(s\ell_2)$.

The most interesting representations of the algebra $\mathcal{U}_\varepsilon$ are the "restricted" representations, i.e. those which correspond to the fixed points of G in Spec Z_0. A remarkable conjecture on the structure of these representations has been proposed recently by Lusztig [14]. We hope that our work will contribute to progress in the solution of this conjecture by some deformation arguments.

Our work was greatly inspired and motivated by the works [24], [21], [20], [23] and [9] on representation theory of simple classical type Lie algebras in characteristic p. We conjecture that the reduction mod p of an irreducible representation of $\mathcal{U}_\varepsilon$, where ε is a primitive p–th root of 1, remains irreducible (for "restricted" representations this has been conjectured by Lusztig [14]). Moreover, the whole "quantum" picture described above descends nicely to the Lie algebra picture in characteristic p. Both theories should undoubtedly benefit from the interaction between these two pictures.

This paper was written in the fall of 1989 when the first author visited MIT. The results of the paper were reported at the International conference on algebraic groups in Hyderabad, India, in December 1989.

We would like to thank G. Lusztig and C. Procesi for stimulating discussions.

§1. Algebras $\mathcal{U}$ and $\mathcal{U}_\varepsilon$ and Verma modules.

1.1. Let q be an indeterminate and let $\mathcal{A} = \mathbb{C}[q, q^{-1}]$, with the quotient field $\mathbb{C}(q)$. For $n \in \mathbb{Z}$ and $d \in \mathbb{N}$, let $[n]_d = (q^{dn} - q^{-dn})/(q^d - q^{-d}) \in \mathcal{A}$. As usual, we define

$$[n]_d! = [n]_d[n-1]_d \cdots [1]_d,$$

and the Gaussian binomial coefficients

$$\begin{bmatrix} n \\ j \end{bmatrix}_d = [n]_d[n-1]_d \dots [n-j+1]_d/[j]_d! \text{ for } j \in \mathbb{N}, \begin{bmatrix} n \\ 0 \end{bmatrix}_d = 1.$$

We shall omit the subscript d when $d = 1$. Here and further we let $\mathbb{N} = \{1, 2, \dots\}$, $\mathbb{Z}_+ = \mathbb{N} \cup \{0\}$.

One knows that $\begin{bmatrix} n \\ j \end{bmatrix}_d \in \mathcal{A}$; this follows from the Gauss binomial formula

$$\prod_{j=0}^{m-1} (1 + q^{2j}x) = \sum_{j=0}^{m} \begin{bmatrix} m \\ j \end{bmatrix} q^{j(m-1)} x^j. \tag{1.1.1}$$

1.2. Fix an $n \times n$ indecomposable matrix (a_{ij}) with integer entries such that $a_{ii} = 2$ and $a_{ij} \leq 0$ for $i \neq j$, and a vector $(d_1, \dots, d_n)$ with relatively prime entries d_i such that the matrix $(d_i a_{ij})$ is symmetric and positive definite. Note that (a_{ij}) is a Cartan matrix of a simple finite-dimensional Lie algebra. Let $q_i = q^{d_i}$.

Following Drinfeld [3], [4] and Jimbo [6], we consider the $\mathbb{C}(q)$-algebra $\mathcal{U}$ defined by the generators E_i, F_i, K_i, K_i^{-1} $(1 \leq i \leq n)$ and the relations (1.2.1)–(1.2.5):

$$K_i K_j = K_j K_i, \; K_i K_i^{-1} = K_i^{-1} K_i = 1, \tag{1.2.1}$$

(1.2.2) $$K_iE_jK_i^{-1} = q_i^{a_{ij}}E_j,\ K_iF_jK_i^{-1} = q_i^{-a_{ij}}F_j,$$

(1.2.3) $$E_iF_j - F_jE_i = \delta_{ij}(K_i - K_i^{-1})/(q_i - q_i^{-1}),$$

(1.2.4) $$\sum_{s=0}^{1-a_{ij}} (-1)^s \begin{bmatrix} 1-a_{ij} \\ s \end{bmatrix}_{d_i} E_i^{1-a_{ij}-s}E_jE_i^s = 0 \text{ if } i \neq j,$$

(1.2.5) $$\sum_{s=0}^{1-a_{ij}} (-1)^s \begin{bmatrix} 1-a_{ij} \\ s \end{bmatrix}_{d_i} F_i^{1-a_{ij}-s}F_jF^si = 0 \text{ if } i \neq j.$$

$\mathcal{U}$ is a Hopf algebra, called the *quantum group* associated to the matrix (a_{ij}), with comultiplication Δ, antipode S and counit ε defined by

(1.2.6) $$\Delta E_i = E_i\otimes 1 + K_i\otimes E_i, \Delta F_i = F_i\otimes K_i^{-1} + 1\otimes F_i, \Delta K_i = K_i\otimes K_i$$

(1.2.7) $$SE_i = -K_i^{-1}E_i, SF_i = -F_iK_i, SK_i = K_i^{-1},$$

(1.2.8) $$\varepsilon E_i = 0, \varepsilon F_i = 0, \varepsilon K_i = 1.$$

Introduce the $\mathbb{C}$–algebra anti–automorphism ω and the $\mathbb{C}$–algebra automorphism φ of $\mathcal{U}$ by

(1.2.9) $$\omega E_i = F_i, \omega F_i = E_i, \omega K_i = K_i^{-1}, \omega q = q^{-1},$$

(1.2.10) $$\varphi E_i = F_i, \varphi F_i = E_i, \varphi K_i = K_i, \varphi q = q^{-1}.$$

Finally, introduce the following elements of $\mathcal{U}$:

$$[K_i; n] = (K_iq^n - K_i^{-1}q^{-n})/(q_i - q_i^{-1}).$$

1.3. One has the following useful relation (due to Kac, see [12]):
(1.3.1)

$$[E_i^m, F_i^s] = \sum_{j=1}^{\min(m,s)} \begin{bmatrix} m \\ j \end{bmatrix}_{d_i} \begin{bmatrix} s \\ j \end{bmatrix}_{d_i} [j]_{d_i}! F_i^{s-j} \prod_{r=j-m-s+1}^{2j-m-s} [K_i; d_ir] E_i^{m-j}$$

This formula is proved in two steps. First, one shows by induction on s that [6]

$$[E_i, F_i^s] = [s]_{d_i} F_i^{s-1}[K_i, d_i(1-s)], \tag{1.3.2}$$

and then proves (1.3.1) by induction on m for arbitrary s. Other useful special cases of (1.3.1) are:

$$[E_i^s, F_i] = [s]_{d_i}[K_i; d_i(1-s)]E_i^{s-1}, \tag{1.3.3}$$

$$[E_i^s/[s]_{d_i}!, F_i^s] = \prod_{j=-s+1}^{0} [K_i; d_i j] + [s]A_i(q), \tag{1.3.4}$$

where $A_i(q) \in \mathcal{U}$ has no poles except for $q = 0$ and $q = \varepsilon$, where $\varepsilon^{2j} = 1$ for $j \in \mathbf{Z}$, $|j| < s$.

Let $\mathcal{U}^+$, $\mathcal{U}^-$ and $\mathcal{U}^0$ be the subalgebras of $\mathcal{U}$ generated by the E_i, the F_i, and the K_i, K_i^{-1} $(i = 1, \dots, n)$ respectively. It follows from (1.2.1)–(1.2.5) and (1.3.1) that $\mathcal{U} = \mathcal{U}^-\mathcal{U}^0\mathcal{U}^+$. Using the comultiplication [18], it is easy to deduce that, moreover, multiplication defines a $\mathbf{C}(q)$–vector space isomorphism

$$\mathcal{U} = \mathcal{U}^- \otimes \mathcal{U}^0 \otimes \mathcal{U}^+. \tag{1.3.5}$$

1.4. Let P be a free abelian group with basis ω_i, $i = 1, \dots, n$. Define the following elements:

$$\rho = \sum_{i=1}^{n} \omega_i, \quad \alpha_j = \sum_{i=1}^{n} a_{ij}\omega_i \quad (j = 1, \dots, n).$$

Let $Q = \sum_i \mathbf{Z}\alpha_i, Q_+ = \sum_i \mathbf{Z}_+\alpha_i$. For $\beta = \sum_i k_i\alpha_i \in Q$ let $\text{ht}\beta = \sum_i k_i$. Introduce a partial ordering on P by $\lambda \geq \mu$ if $\lambda - \mu \in Q_+$.

Define a bilinear pairing $P \times Q \to \mathbf{Z}$ by

$$(\omega_i|\alpha_j) = \delta_{ij} d_i. \tag{1.4.1}$$

Then $(\alpha_i|\alpha_j) = d_i a_{ij}$, so that we get a symmetric $\mathbf{Z}$–valued bilinear form on Q such that $(\alpha|\alpha) \in 2\mathbf{Z}$.

Define automorphisms r_i of P by $r_i\omega_j = \omega_j - \delta_{ij}\alpha_i$ $(i,j = 1,\dots,n)$. Then $r_i\alpha_j = \alpha_j - a_{ij}\alpha_i$. Let W be the (finite) subgroup of $GL(P)$ generated by $r_1,\dots,r_n$. Then Q is W–invariant and the pairing $P\times Q \to \mathbf{Z}$ is W–invariant. Let

$$\Pi = \{\alpha_1,\dots,\alpha_n\},\ R = W\Pi,\ R^+ = R\cap Q.$$

Then, of course, R is a root system corresponding to the Cartan matrix (a_{ij}), W is its Weyl group, R^+ a set of positive roots, etc.

For $\beta = \sum_i n_i\alpha_i \in Q$ we let $K_\beta = \prod_i K_i^{n_i}$. If $\beta = w\alpha_i \in R$ for some $w \in W$, $\alpha_i \in \Pi$, we let:

$$d_\beta = d_i,\ q_\beta = q_i \text{ and } [K_\beta;n] = (K_\beta q^n - K_\beta^{-1}q^{-n})/(q_\beta - q_\beta^{-1}).$$

1.5. Let $U_{\mathcal{A}}$ be the $\mathcal{A}$–subalgebra of U generated by the elements E_i, F_i, K_i, K_i^{-1}, $[K_i;0]$ $(i = 1,\dots,n)$. Let $\mathcal{U}_{\mathcal{A}}^+$ (resp. $\mathcal{U}_{\mathcal{A}}^-$) be the $\mathcal{A}$–subalgebra of $\mathcal{U}_{\mathcal{A}}$ generated by the E_i (resp. F_i) and $\mathcal{U}_{\mathcal{A}}^0$ that generated by the K_i and $[K_i;0]$. Note that relations (1.2.1), (1.2.2), (1.2.4), (1.2.5) together with

$$E_iF_j - F_jE_i = \delta_{ij}[K_i;0], \tag{1.5.1}$$

$$(q_i - q_i^{-1})[K_i;0] = K_i - K_i^{-1}, \tag{1.5.2}$$

are defining relations of the algebra $\mathcal{U}_{\mathcal{A}}$. Note also that $\mathcal{U}_{\mathcal{A}}$ is a Hopf algebra with Δ, S and ε defined by (1.2.6)–(1.2.8) and

(1.5.3)

$$\Delta[K_i;0] = [K_i;0]\otimes K + K^{-1}\otimes[K_i;0],\ S[K_i;0] = -[K_i;0],\ \varepsilon[K_i;0] = 0.$$

Finally, note that the elements $[K_i;d_in]$ lie in $\mathcal{U}_{\mathcal{A}}$ since

$$[K_i;d_in] = [K_i;0]q_i^{-n} + K_i[n]_{d_i}. \tag{1.5.4}$$

Note also the following useful formulas:

(1.5.5)

$$[K_i;d_in]E_j = E_j[K_i;d_i(n+a_{ij})],\ [K_i;d_in]F_j = F_j[K_i;d_i(n-a_{ij})].$$

Given $\varepsilon \in \mathbf{C}^\times$, we may consider the "specialization" $\mathcal{U}_\varepsilon = \mathcal{U}_{\mathcal{A}}/(q-\varepsilon)\mathcal{U}_{\mathcal{A}}$. If $\varepsilon^{2d_i} \neq 1$ for all $i = 1,\dots,n$, then $\mathcal{U}_\varepsilon$ is an algebra over $\mathbf{C}$ on generators E_i, F_i, K_i, K_i^{-1} and defining relations (1.2.1)–(1.2.5), in which $q = \varepsilon$. We denote by $\mathcal{U}_\varepsilon^+$, $\mathcal{U}_\varepsilon^-$, $\mathcal{U}_\varepsilon^0$ the images of $\mathcal{U}_{\mathcal{A}}^+$, $\mathcal{U}_{\mathcal{A}}^-$ and $\mathcal{U}_{\mathcal{A}}^0$ in $\mathcal{U}_\varepsilon$.

Especially important is the "limiting" specialization $\mathcal{U}_1$.

PROPOSITION 1.5. *$\mathcal{U}_1$ is an associative algebra over $\mathbb{C}$ on generators E_i, F_i, K_i and $H_i(=[K_i;0])(i=1,\dots,n)$ and the following defining relations:*

(1.5.6) $$[E_i,F_j]=\delta_{ij}H_i,\ [H_i,E_j]=a_{ij}K_iE_j,\ [H_i,F_j]=-a_{ij}K_iF_j$$

(1.5.7) $$K_i \text{ are central elements and } K_i^2=1,$$

(1.5.8) $$\underbrace{[E_i,\dots[E_i}_{1-a_{ij} times},E_j]=0=\underbrace{[F_i,\dots[F_i}_{1-a_{ij}\text{ times}},F_j]] \text{ if } i\neq j.$$

In particular, $\mathcal{U}_1/(K_i-1;i=1,\dots,n)$ is isomorphic to $U(\mathfrak{g}(a_{ij}))$, the universal enveloping algebra of the simple Lie algebra $\mathfrak{g}(a_{ij})$ over $\mathbb{C}$ associated to the Cartan matrix (a_{ij}).

PROOF: (1.5.6) follows from (1.5.1), (1.5.4) and (1.5.5); (1.5.7) follows from (1.2.2) and (1.5.2); (1.5.8) follows from (1.2.4) and (1.2.5). □

REMARK 1.5. Note that $\mathcal{U}_1^+ := \mathcal{U}_\mathcal{A}^+/(q-1)\mathcal{U}_\mathcal{A}^+$ is an algebra on generators E_i $(i=1,\dots,n)$ and defining relations (1.5.8) for E_i. Hence, being an enveloping algebra of a Lie algebra, $\mathcal{U}_1^+$ has no zero divisors. Hence $\mathcal{U}_\mathcal{A}^+$ and $\mathcal{U}^+$ (and similarly $\mathcal{U}_\mathcal{A}^-$ and $\mathcal{U}^-$) have no zero divisors. It follows from (1.3.3) that $\mathcal{U}_\mathcal{A}$ and $\mathcal{U}$ have no zero divisors. This proof works for an arbitrary symmetrizable generalized Cartan matrix (a_{ij}).

1.6. Given $s\in\mathbb{N}$, we shall often write $E_i^{(s)}$ and $F_i^{(s)}$ for $E_i^s/[s]_{d_i}!$ and $F_i^s/[s]_{d_i}!$, respectively. Following Lusztig [13], introduce the following automorphisms T_i $(i=1,\dots,n)$ of the algebra $\mathcal{U}$:

(1.6.1)
$$T_iE_i=-F_iK_i,\ T_iE_j=\sum_{s=0}^{-a_{ij}}(-1)^{s-a_{ij}}q_i^{-s}E_i^{(-a_{ij}-s)}E_jE_i^{(s)} \text{ if } i\neq j,$$

(1.6.2)
$$T_iF_i=-K_i^{-1}E_i,\ T_iF_j=\sum_{s=0}^{-a_{ij}}(-1)^{s-a_{ij}}q_i^{s}F_i^{(s)}F_jF_i^{(-a_{ij}-s)} \text{ if } i\neq j,$$

(1.6.3) $$T_iK_j=K_jK_i^{-a_{ij}}.$$

Note that

$$T_i\omega = \omega T_i, \tag{1.6.4}$$

$$T_i^{-1} = \varphi T_i \varphi^{-1}. \tag{1.6.5}$$

PROPOSITION 1.6. *[13], [14].*

(a) *Let $w \in W$ and let $r_{i_1} \dots r_{i_k}$ be a reduced expression of w. Then the automorphism $T_w = T_{i_1} \dots T_{i_k}$ of U is independent of the choice of the reduced expression of w.*

(b) *Suppose that $\beta = w\alpha_i \in R^+$, $\alpha_i \in \Pi$. Then $T_w E_i \in U^+$.* □

Proposition 1.6(a) shows that the T_i define a representation of the braid group of W in the group $\mathrm{Aut}_{\mathbf{C}(q)}\mathcal{U}$. Note that when restricted to $\mathcal{U}^0$ this descends to an action of W (given by (1.6.3)), so that $T_w K_\beta = K_{w\beta}$.

REMARK 1.6. Recall that for a Hopf algebra $\mathcal{U}$ one defines the adjoint representation $x \longmapsto \mathrm{ad}x$ in $\mathcal{U}$ by $(\mathrm{ad}x)u = \sum_i a_i u S(b_i)$, where $\Delta x = \sum_i a_i \otimes b_i$. One checks that

$$T_i E_j = (\mathrm{ad} - E_i^{(-a_{ij})})E_j \text{ if } i \neq j.$$

1.7. Fix a reduced expression $r_{i_1} r_{i_2} \dots r_{i_N}$ of the longest element of W. This gives us an ordering of the set of positive roots R^+:

$$\beta_1 = \alpha_{i_1},\ \beta_2 = r_{i_1}\alpha_{i_2},\ \beta_N = r_{i_1} \dots r_{i_{N-1}}\alpha_{i_N}. \tag{1.7.1}$$

Introduce *root vectors* [13], [1]:

$$E_{\beta_s} = T_{i_1} \dots T_{i_{s-1}} E_{i_s},\ F_{\beta_s} = T_{i_1} \dots T_{i_{s-1}} F_{i_s}\ (= \omega E_{\beta_s}).$$

For $k = (k_1, \dots, k_N) \in \mathbf{Z}_+^N$ let $E^k = E_{\beta_1}^{k_1} \dots E_{\beta_N}^{k_N}$, $F^k = \omega E^k$.

For $k, r \in \mathbf{Z}_+^N$ and $u \in U^0$ define the monomial $M_{k,r;u} = F^k u E^r$. Define the height of this monomial by $\mathrm{ht}(M_{k,r;u}) = \prod_i (k_i + r_i)\mathrm{ht}\beta_i \in \mathbf{Z}_+$, and its degree by

$$d(M_{k,r;u}) = (k_N, k_{N-1}, \dots, k_1, r_1, \dots, r_N, \mathrm{ht}(M_{k,r;u})) \in \mathbf{Z}_+^{2N+1}.$$

We shall view $\mathbf{Z}_+^{2N+1}$ as a totally ordered semigroup with the lexicographical order $<$ such that $u_1 < u_2 < \dots < u_{2N+1}$, where $u_i = (\delta_{i,1}, \dots, \delta_{i,2N+1})$ is the standard basis.

The following important lemma was first stated in [11]. Soibelman kindly sent us a letter with its proof. Also Reshetikhin informed us that he and Kirillov independently discovered this result. Finally, Lusztig pointed out that the proof of the lemma at least for some particular choices of the reduced expression can be easily derived from the quiver approach [17] to quantum groups.

LEMMA 1.7 [11]. *For $i < j$ one has:*

$$E_{\beta_j}E_{\beta_i} - q^{(\beta_i|\beta_j)}E_{\beta_i}E_{\beta_j} = \sum_{k\in\mathbb{Z}_+^N} \rho_k E^k,$$

where $\rho_k \in \mathcal{A}$ and $\rho_k = 0$ unless $d(E^k) < d(E_{\beta_i}E_{\beta_j})$. □

Recall that, given a commutative totally ordered semigroup S, an S–filtration of an algebra A is a collection of subspaces $A^{(s)}$, $s \in S$, such that $\bigcup_s A^{(s)} = A$, $A^{(s)} \subset A^{(s')}$ if $s < s'$, and $A^{(s)}A^{(s')} \subset A^{(s+s')}$. The associated graded algebra is $\mathrm{Gr}A = \oplus_{s\in S}\mathrm{Gr}_s A$, where $\mathrm{Gr}_s A = A^{(s)}/\sum_{s'<s} A^{(s')}$, with the usual multiplication. We shall always assume that every subset of S has a minimal element. Then we may define a degree $d(a) \in S$ of $a \in A$ as the minimal s such that $a \in A^{(s)}$. We have a linear isomorphism $A \xrightarrow{\sim} \mathrm{Gr}A$ denoted by $a \longmapsto \overline{a}$, where $\overline{a} = a \bmod \sum_{s<d(a)} A^{(s)} \in \mathrm{Gr}_{d(a)}A$.

Given $s \in \mathbb{Z}_+^{2N+1}$, denote by $\mathcal{U}^{(s)}$ the linear span over $\mathbb{C}(q)$ of the monomials $M_{k,r;u}$ such that $d(M_{k,r;u}) \leq s$. We define $\mathcal{U}_\varepsilon^{(s)} \subset \mathcal{U}_\varepsilon$ similarly.

PROPOSITION 1.7.

(a) *The $\mathcal{U}^{(s)}$, $s \in \mathbb{Z}_+^{2N+1}$, form a filtration of $\mathcal{U}$ (similarly for $\mathcal{U}_\varepsilon$).*

(b) *[13] Elements E^k, $k \in \mathbb{Z}_+^N$, form a basis of $\mathcal{U}^+$ (resp. $\mathcal{U}_\varepsilon^+$) over $\mathbb{C}(q)$ (resp. $\mathbb{C}$).*

(c) *Elements $F^kK_1^{m_1}\ldots K_n^{m_n}E^r$, where $k, r \in \mathbb{Z}_+^N$, $(m_1, \ldots, m_n) \in \mathbb{Z}^n$, form a basis of $\mathcal{U}$ over $\mathbb{C}(q)$ (resp. a basis of $\mathcal{U}_\varepsilon$ over $\mathbb{C}$, provided that $\varepsilon^{2d_i} \neq 1$ for $i = 1, \ldots, n$).*

(d) *The associated graded algebra $\mathrm{Gr}\mathcal{U}$ (resp. $\mathrm{Gr}\mathcal{U}_\varepsilon$ provided that $\varepsilon^{2d_i} \neq 1$, $i = 1, \ldots, n$) is an associative algebra over $\mathbb{C}(q)$ (resp. $\mathbb{C}$) on generators E_α, F_α ($\alpha \in R_+$), $K_i^{\pm 1}$ ($i = 1, \ldots, n$) subject to the following relations:*

$$K_iK_j = K_jK_i,\ K_iK_i^{-1} = 1,\ E_\alpha F_\beta = F_\beta E_\alpha,$$
$$K_iE_\alpha = q^{(\alpha|\alpha_i)}E_\alpha K_i,\ K_iF_\alpha = q^{-(\alpha|\alpha_i)}F_\alpha K_i,$$
$$E_\alpha E_\beta = q^{(\alpha|\beta)}E_\beta E_\alpha, F_\alpha F_\beta = q^{(\alpha|\beta)}F_\beta F_\alpha \text{ if } \alpha > \beta,$$

(resp. same relations with $q = \varepsilon$).

PROOF: It follows from (1.2.3) that $E_\alpha F_\beta = F_\beta E_\alpha$ + (linear combination of the $M_{k,r;u}$ of degree less than $d(E_\alpha F_\beta)$. This together with Lemma 1.7 imply (a) and the relations in (d). The fact that the E^k span

$\mathcal{U}^+$ (resp. $\mathcal{U}_\varepsilon^+$) follows also from Lemma 1.7. Their linear independence is proved as in [13]. This proves (b) and (c). (d) now follows from (c). □

REMARK 1.7. **(a)** Let $\beta = w\alpha_i$. Applying T_w to both sides of (1.3.1) and using (1.5.4), we see that (1.3.1) holds if we replace E_i, F_i and K_i by E_β, F_β and K_β.

(b) For different presentations $\beta = w\alpha_i$ the E_β may be not proportional. For example we have:

(1.7.2) $$E_{ij} := T_j E_i = -E_j E_i + q_j^{-1} E_i E_j \text{ if } a_{ji} = -1.$$

1.8. An algebra $\mathcal{P}$ over $\mathbb{C}$ on generators $x_1, \dots, x_k$ and defining relations $x_i x_j = \lambda_{ij} x_j x_i$ for $i > j$ where $\lambda_{ij} \in \mathbb{C}^\times$, $i, j = 1, \dots, k$, is called a quasipolynomial algebra. One may introduce a $\mathbb{N}^k$-gradation in $\mathcal{P}$ by letting $\deg x_i = (\delta_{i1}, \dots, \delta_{ik})$. It is clear that $\mathcal{P}$ is spanned over $\mathbb{C}$ by monomials $x_1^{r_1} \dots x_k^{r_k}$. Moreover, they form a basis of $\mathcal{P}$. This follows by looking at the representation π of $\mathcal{P}$ in the polynomial algebra $\mathbb{C}[t_1, \dots, t_k]$ defined by

$$\pi(x_i) t_1^{m_1} \dots t_k^{m_k} = \Big(\prod_{j=1}^{i-1} \lambda_{ij}^{m_j}\Big) t_1^{m_1} \dots t_{i-1}^{m_{i-1}} t_i^{m_i+1} t_{i+1}^{m_{i+1}} \dots t_k^{m_k}.$$

As usual, this implies that $\mathcal{P}$ has no zero divisors. Furthermore, let A be a commutative algebra on generators $K_1, \dots, K_s$ with no zero divisors. Let $\mathcal{P}_A = A \otimes_{\mathbb{C}} \mathcal{P}$ with multiplication given by $K_i x_j = \mu_{ij} x_j K_i$, $\mu_{ij} \in \mathbb{C}^\times$. We may extend the $\mathbb{N}^k$-gradation from $\mathcal{P}$ to $\mathcal{P}_A$ by letting $\deg K_i = 0$. This gives $\mathcal{P}_A$ a structure of a free left A-module with basis $\underline{x}^m := x_1^{m_1} \dots x_k^{m_k}$, $m \in \mathbb{Z}_+^k$. It follows that $\mathcal{P}_A$ has no zero divisors as well. Hence $\mathrm{Gr}\mathcal{U}$ and $\mathrm{Gr}\mathcal{U}_\varepsilon$ have no zero divisors by Proposition 1.7d, and we deduce the following corollary.

COROLLARY 1.8. *The algebras $\mathcal{U}$ and $\mathcal{U}_\varepsilon$ have no zero divisors.* □

Let A be an algebra with no zero divisors, let Z be the center of A and $Q(Z)$ the quotient field of Z, and let $Q(A) = Q(Z) \otimes_Z A$. We shall call the algebra A *integrally closed* if for any subring B of $Q(A)$ such that $A \subset B \subset z^{-1}A$ for some $z \in Z$, $z \neq 0$, we have $B = A$. It is clear that for a commutative algebra A this definition of an integrally closed algebra implies the usual one.

PROPOSITION 1.8. *Let A be an integrally closed algebra with no zero divisors. Let C be an S-filtered algebra such that $C_0 = A$ and $\mathrm{Gr}\, C = \mathcal{P}_A$. Assume that each generator x_i of $\mathcal{P} \subset \mathcal{P}_A$ has a preimage $\tilde{x}_i$ in C such that $\tilde{x}_i^\ell$ lies in the center Z of C. Then the algebra C is integrally closed.*

PROOF: Let $z \in Z$, $z \neq 0$, and let B be a subalgebra of $z^{-1}C$ containing C. Let $\varphi \in B$, so that $y := z\varphi \in C$. We can write:

$$y = a\underline{\tilde{x}}^h + \text{ lower degree terms, where } a \in A,\ h \in \mathbf{Z}_+^k,$$

$$z = b\underline{\tilde{x}}^r + \text{ lower degree terms, where } b \in A,\ r \in \mathbf{Z}_+^k.$$

Pick $N \in \mathbf{Z}$ such that $N\ell \geq h_i$ for each i and set

$$t = (N\ell - h_1, \dots, N\ell - h_k) \in \mathbf{Z}_+^k.$$

Since $\underline{x}^h \underline{x}^t = \lambda \underline{x}^{h+t}$, $\lambda \in \mathbf{C}^\times$, we get:

$$Y := \lambda^{-1} y \underline{\tilde{x}}^t = a\underline{\tilde{x}}^{h+t} + \text{ lower degree terms } = \lambda^{-1} z\varphi\underline{\tilde{x}}^t.$$

Letting $\nu = \lambda^{-1}\varphi\underline{\tilde{x}}^t \in B$, we get $Y = z\nu$. For each $m \in \mathbf{N}$ we have:

(1.8.1) $$Y^m = z^{m-1}(z\nu^m) \in z^{m-1}C.$$

Since, by the assumption, $\underline{\tilde{x}}^{h+t} \in Z$, we get:

(1.8.2) $$Y^m = a^m \underline{\tilde{x}}^{m(h+t)} + \text{ lower degree terms.}$$

But $z^j = z^{j-1}(b\underline{\tilde{x}}^r + \text{ lower degree terms}) = bz^{j-1}\underline{\tilde{x}}^r +$ lower degree terms. Hence we have by induction:

(1.8.3) $z^{m-1} = \lambda_m b^{m-1} \underline{\tilde{x}}^{(m-1)r} +$ lower degree terms, where $\lambda_m \in \mathbf{C}^\times$.

Since $\mathrm{Gr}_s C$ is a free A-module of rank 1 generated by $\underline{x}^s$, $s \in \mathbf{Z}_+^k$, we deduce from (1.8.1), (1.8.2) and (1.8.3) that b^{m-1} divides a^m for all $m \in \mathbf{N}$. It follows that the subalgebra of $Q(A)$ generated by A and a/b is contained in $b^{-1}A$. Since A is integrally closed, we deduce that b divides a.

Note now that $y^m = z^{m-1}(z\varphi^m)$, where $z\varphi^m \in C$. It follows that $mh_i \geq (m-1)r_i$ for all $m \in \mathbf{N}$ and all $i = 1, \dots, k$. Hence $h_i \geq r_i$ for all i.

Suppose now that $B \neq C$ and choose $\varphi \in B \backslash C$ such that $y = z\varphi \in C$ has minimal possible degree. Let $c = a/b$, $\underline{s} = \underline{h} - \underline{r}$ and consider the element

$$\varphi' = \varphi - \mu^{-1} c \underline{\tilde{x}}^{\underline{s}},$$

where $\mu \in \mathbf{C}^\times$ is taken from $\underline{x}^h = \mu \underline{x}^r \underline{x}^s$. But $\varphi' \in B \backslash C$ and $d(z\varphi') = d(y - \mu^{-1} c z \underline{\tilde{x}}^s) < \deg y$, a contradiction. □

An immediate corollary of Propositions 1.8 and 1.7d is

THEOREM 1.8. *The algebra $\mathcal{U}_\varepsilon$ is integrally closed.* □

1.9. Let Q_2^* be the group of all homomorphisms of Q to the group $\{\pm 1\}$. Given $\lambda \in P$ and $\delta \in Q_2^*$, define the Verma module of type δ, $M^\delta(\lambda)$ over $\mathcal{U}$ (which is a vector space over $\mathbf{C}(q)$) as the (unique) $\mathcal{U}$-module having a vector v_λ such that

$$\mathcal{U}^+ v_\lambda = 0,\ K_i v_\lambda = \delta(\alpha_i) q^{(\lambda|\alpha_i)} v_\lambda,$$

and the vectors

$$F^k v_\lambda\ (k \in \mathbf{Z}_+^N)$$

form a basis of $M^\delta(\lambda)$.

Such a module exists due to Proposition 1.7c. Any quotient V of $M^\delta(\lambda)$, called a highest weight $\mathcal{U}$-module, admits a weight space decomposition.

$$V = \oplus_{\eta \in Q_+} V_\eta,\ \text{where } V_\eta = \{v \in V | K_i v = \delta(\alpha_i) q^{(\lambda-\eta|\alpha_i)} v\}.$$

We denote the (unique) irreducible quotient of $M^\delta(\lambda)$ by $L^\delta(\lambda) = M^\delta(\lambda)/J^\delta(\lambda)$.

Given a highest weight module V over the algebra $\mathcal{U}$, let v_λ be its highest weight vector and consider its $\mathcal{U}_\mathcal{A}$-submodule (which is an $\mathcal{A}$-module) $V_\mathcal{A} = \mathcal{U}_\mathcal{A} v_\lambda$. This gives rise to a $\mathcal{U}_1$-module $V_1 = V_\mathcal{A}/(q-1)V_\mathcal{A}$. The K_i act on V_1 as $\delta(\alpha_i)$, hence by Proposition 1.5 the module V_1 becomes a highest weight module with highest weight vector v_λ over $U(\mathfrak{g}(a_{ij}))$, the enveloping algebra of the simple finite-dimensional Lie algebra $\mathfrak{g}(a_{ij})$ over $\mathbf{C}$ associated to the Cartan matrix (a_{ij}).

LEMMA 1.9. *[12]. Let V be a Verma module of type δ or an irreducible highest weight module with highest weight λ over $\mathcal{U}$ and let $\eta \in Q_+$. Then*

$$\dim_{\mathbf{C}(q)} V_\eta = \dim_{\mathbf{C}} (V_1)_\eta. \quad \square$$

Given $a = a(q) \in \mathbf{C}(q)$, let $\bar{a} = a(q^{-1})$. A $\mathbf{C}$-bilinear form H on a vector space V over $\mathbf{C}(q)$ with values in $\mathbf{C}(q)$ is called Hermitian if

$$(1.9.1)\qquad \begin{aligned} &H(au, v) = \bar{a} H(u,v),\ H(u, av) = aH(u,v), \text{and} \\ &H(u,v) = \overline{H(v,u)},\ a \in \mathbf{C}(q),\ u, v \in V. \end{aligned}$$

The $\mathcal{U}$-module $M^\delta(\lambda)$ carries a unique Hermitian form H, called the contravariant Hermitian form, such that

(1.9.2)

$$H(v_\lambda, v_\lambda) = 1 \text{ and } H(gu, v) = H(u, \omega(g) v) \text{ for } g \in \mathcal{U},\ u, v \in M^\delta(\lambda).$$

We have: $H(M^\delta(\lambda)_\mu, M^\delta(\lambda)_\nu) = 0$ if $\mu \neq \nu$, and Ker $H = J^\delta(\lambda)$. Denote by H_η the restriction of H to $M^\delta(\lambda)_\eta$, $\eta \in Q_+$, and let $\det_\eta^\delta(\lambda)$ denote the determinant of the matrix of H_η in the basis consisting of elements $F^k v_\lambda$, $k \in \mathrm{Par}\,(\eta)$. Here for $\eta \in Q$ we denote by Par (η) the set of all $k \in \mathbb{Z}_+^N$ such that $\sum_i k_i\beta_i = \eta$. Also, for given $\delta \in Q_2^*$, we view K_β as a function on P defined by $K_\beta(\lambda) = \delta(\beta)q^{(\lambda|\beta)}$; any $\varphi \in \mathcal{U}^0$ thereby becomes a function on P with values in $\mathbb{C}(q)$, which we denote by $\varphi^\delta(\lambda)$.

Given $\beta \in R^+$ and $r \in \mathbb{N}$, let $T_{r\beta} = \{\lambda \in P | 2(\lambda+\rho|\beta) = r(\beta|\beta)\}$, and let $T^0_{r\beta} = \{\lambda \in T_{r\beta} | 2(\lambda+\rho|\gamma) \neq m(\gamma|\gamma)$ for all $\gamma \in R^+\backslash\{\beta\}$ and $m \in \mathbb{N}$ such that $m\gamma < r\beta\}$. Note that if $\varphi \in \mathcal{U}^0$ and φ vanishes on $T^0_{r\beta}$ then φ vanishes on $T_{r\beta}$.

PROPOSITION 1.9.
(a) $\det_\eta = \prod_{\beta\in R^+}\prod_{m\in\mathbb{N}}([m]_{d_\beta}[K_\beta;(\rho|\beta) - \frac{m}{2}(\beta|\beta)])^{|\mathrm{Par}(\eta-m\beta)|}$.
(b) *If* $\lambda \in T^0_{r\beta}$, $r \in \mathbb{N}, \beta \in R^+$, *then* $M^\delta(\lambda)$ *contains a submodule isomorphic to* $M^\delta(\lambda - r\beta)$.

PROOF: Let $\lambda \in T_{r\beta}$, $r \in \mathbb{N}$, $\beta \in R^+$. By Shapovalov's formula [22] for the determinant of the contravariant form of the $U(\mathfrak{g}(A))$–module $M(\lambda)$ it follows that $\dim_{\mathbb{C}}(L^\delta(\lambda)_1)_{r\beta} < \dim_{\mathbb{C}}(M^\delta(\lambda)_1)_{r\beta}$. Hence by Lemma 1.9 we have: $\dim_{\mathbb{C}(q)} L^\delta(\lambda)_{r\beta} < \dim_{\mathbb{C}(q)} M^\delta(\lambda)_{r\beta}$. It follows that $\det_{r\beta}$ is divisible by $[K_\beta;(\rho|\beta)-\frac{r}{2}(\beta|\beta)]$ and that there exists a non–zero vector $v \in M^\delta(\lambda)_{r\beta} \cap J^\delta(\lambda)$, such that $\mathcal{U}^+v = 0$, provided that $\lambda \in T^0_{r\beta}$. Since $\mathcal{U}^-$ has no zero divisors it follows that v generates a submodule of $M^\delta(\lambda)$ isomorphic to $M^\delta(\lambda - r\beta)$ which lies in $J^\delta(\lambda)$, proving (b). Hence $\det_\eta$ is divisible by $[K_\beta;(\rho|\beta) - \frac{r}{2}(\beta|\beta)]^{|\mathrm{Par}(\eta-r\beta)|}$. Thus, $\det_\eta$ is divisble by the right–hand side of the formula in question.

To complete the proof of (a) we calculate the leading term of $\det_\eta$. Using formula (1.3.1) it is easy to see (as in [22]) that it equals to

$$\prod_{\beta\in R^+}\prod_{m\in\mathbb{N}}([m]_{d_\beta}!K_\beta^m)^{|\mathrm{Par}(\eta-m\beta)|-|\mathrm{Par}(\eta-(m+1)\beta)|}$$

which proves (a). □

Let $\varepsilon \in \mathbb{C}^\times$. Given a homomorphism $\lambda : \mathcal{U}^0_\varepsilon \to \mathbb{C}$, we define the Verma module $M_\varepsilon(\lambda)$ over $\mathcal{U}_\varepsilon$ as a vector space over $\mathbb{C}$ with an action of $\mathcal{U}_\varepsilon$ having a vector v_λ such that $\mathcal{U}^+_\varepsilon v_\lambda = 0$, $uv_\lambda = \lambda(u)v_\lambda$ for $u \in \mathcal{U}^0_\varepsilon$ and the vectors $F^k v_\lambda$ ($k \in \mathbb{Z}^N_+$) form a basis of $M_\varepsilon(\lambda)$. Such a module exists due to Proposition 1.7c. The module $M_\varepsilon(\lambda)$ admits a Q_+–gradation $M_\varepsilon(\lambda) =$

$\oplus_{\eta \in Q_+} M_\varepsilon(\lambda)_\eta$, where $M_\varepsilon(\lambda)_\eta$ is the linear span over $\mathbf{C}$ of the $F^k v_\lambda$ with $k \in \mathrm{Par}(\eta)$. Note that $M_\varepsilon(\lambda)_\eta \subset \{v \in M_\varepsilon(\lambda) | K_i v = \lambda(K_i)\varepsilon^{-(\eta|\alpha_i)} v\}$, and that this inclusion is an equality if ε is not a root of 1. Note also that $M^1(\lambda)/(q-\varepsilon)M^1(\lambda) = M_\varepsilon(\lambda_\varepsilon)$, where $\lambda_\varepsilon(K_i) = \varepsilon^{(\lambda|\alpha_i)}$.

Let now $|\varepsilon| = 1$. The contravariant Hermitian form H on $M_\varepsilon(\lambda)$ is defined as above (see (1.9.1) and (1.9.2)) (H is then a usual Hermitian form). We define similarly $\det_{\eta,\varepsilon}(\lambda) (\eta \in Q_+)$ as the determinant of H on $M_\varepsilon(\lambda)_\eta$ in the basis $F^k v_\lambda$, $k \in \mathrm{Par}(\eta)$. Proposition 1.9a gives us immediately the following formula (we assume that $\varepsilon^{2d_i} \neq 1$ for all i):
(1.9.3)

$$\det_{\eta,\varepsilon}(\lambda) = \prod_{\beta \in R^+} \prod_{m \in \mathbf{N}} \left(\frac{\varepsilon^{md_\beta} - \varepsilon^{-md_\beta}}{(\varepsilon^{d_\beta} - \varepsilon^{-d_\beta})^2} \right)^{|\mathrm{Par}(\eta - m\beta)|}$$
$$\times \left(\lambda(K_\beta)\varepsilon^{(\rho|\beta) - \frac{m}{2}(\beta|\beta)} - \lambda(K_\beta)^{-1}\varepsilon^{-(\rho|\beta) + \frac{m}{2}(\beta|\beta)}\right)^{|\mathrm{Par}(\eta - m\beta)|}.$$

We define the $\mathcal{U}_\varepsilon$-module $L_\varepsilon(\lambda)$ as the (unique) quotient of $M_\varepsilon(\lambda)$ by the maximal Q_+-graded submodule $J_\varepsilon(\lambda)$; this module is irreducible.

Formula (1.9.3) implies the following

COROLLARY 1.9. *Let $\lambda(K_i) = \varepsilon^{(\hat{\lambda}|\alpha_i)}$, $i = 1, \ldots, n$, where $\hat{\lambda} \in \mathbf{C} \otimes_{\mathbf{Z}} P$ and ε is not a root of 1. Then $M_\varepsilon(\lambda)$ is an irreducible $\mathcal{U}_\varepsilon$-module if and only if $2(\hat{\lambda} + \rho|\beta) \neq m(\beta|\beta)$ for all $m \in \mathbf{N}$, $\beta \in R^+$.* □

REMARK 1.9. Using the usual argument (see [5] or [8]) one can show that Proposition 1.9b holds for all $\lambda \in T_{r\beta}$ and derive the usual conditions for inclusions of Verma modules over $\mathcal{U}$ and occurence of $L^\delta(\mu)$ in $M^\delta(\lambda)$ (cf. [1]). One can also prove similar results for the $\mathcal{U}_\varepsilon$-modules $M_\varepsilon(\lambda)$ provided that ε is not a root of 1.

1.10. Let $a_{ij} = -a$. Introduce the following notation:

$$E_{j\underbrace{i\ldots i}_{a \text{ times}}} = T_i E_j \text{ if } a = 1, 2 \text{ or } 3,$$
$$E_{ji} = (T_i T_j)^{a-1} E_i \text{ if } a = 2 \text{ or } 3,$$
$$E_{iij} = T_j T_i E_j \text{ if } a = 2, \; E_{iij} = T_j T_i T_j E_i \text{ if } a = 3,$$
$$E_{jii} = T_i T_j E_i \text{ and } E_{iiij} = (T_j T_i)^2 E_j \text{ if } a = 3.$$

Then one has the following important identities:

$$E_i^{(s)} E_j = \sum_{k=0}^{a} q_i^{(a-k)s+k} E_{\underbrace{i\ldots i}_{k \text{ times}} j} E_i^{(s-k)}, \tag{1.10.1}$$

(1.10.2) $$E_j E_i^{(s)} = \sum_{k=0}^{a} q_i^{(a-k)s+k} E_i^{(s-k)} E_{j\underbrace{i\dots i}_{k \text{ times}}},$$

(1.10.3) $$E_i^s E_j^{(s)} = q_j^{s^2} E_j^{(s)} E_i^s + q_j^s E_{ij}^s + [s] A_{ij}(q), \text{ if } a_{ji} = -1,$$

where $A_{ij}(q) \in \mathcal{U}$ has no poles except for $q = 0$ and $q = \varepsilon$, where $\varepsilon^{2j} = 1$ for $j \in \mathbf{Z}$, $|j| < s$.

All these formulas can be deduced directly from [13] or proved in a similar way (by induction).

§2. The center of $\mathcal{U}$.

2.1. Let Z be the center of $\mathcal{U}$ (i.e. the set of elements commuting with all elements of $\mathcal{U}$). Since Z, in particular, commutes with $\mathcal{U}^0$, it is clear (in view of Proposition 1.7c), that any element of Z is of the form

(2.1.1) $$z = \sum_{\eta \in Q_+} \sum_{k,r \in \mathrm{Par}(\eta)} F^k \varphi_{k,r} E^r, \text{ where } \varphi_{k,r} \in \mathcal{U}^0.$$

As usual, the map $z \longmapsto \varphi_{0,0}$ is a homomorphism $h : Z \to \mathcal{U}^0$, called the Harish-Chandra homomorphism.

Fix an element $z \in Z$ of the form (2.1.1). It is clear that $z \in Z$ if and only if z acts as a scalar equal to $\varphi_{0,0}(\lambda)$ on each Verma module $M^\delta(\lambda)$, $\lambda \in P$, $\delta \in Q_2^*$. Denote by ϕ_η the matrix $(\varphi_{k,m})_{k,m \in \mathrm{Par}(\eta)}$. We will compute the matrix ϕ_η by induction on $\eta \in Q_+$. Denote by $G_\gamma^\delta(\lambda)$ the matrix of the operator $\sum_{k,r \in \mathrm{Par}(\gamma)} F^k \varphi_{k,r} E^r$ on $M^\delta(\lambda)_\eta$ in the basis $F^s v_\lambda$ ($s \in \mathrm{Par}(\eta)$). By the inductive assumption, we know G_γ for $\gamma < \eta$. We obviously have

(2.1.2) $$G_\eta = \phi_\eta H_\eta.$$

Hence the fact that z acts as a scalar $\varphi_{00}^\delta(\lambda)$ on $M^\delta(\lambda)_\eta$ can be written as follows:

(2.1.3) $$\phi_\eta H_\eta + \sum_{\gamma < \eta} G_\gamma = \varphi_{0,0} I.$$

This gives us an effective way for calculating the coefficients $\varphi_{k,m}$ of the series (2.1.1), given $\varphi_{0,0}$. By construction, the series z commutes with the F_i and K_i. Since z acts as a scalar also on the dual to a Verma module, we see that z commutes also with the E_i, and hence with $\mathcal{U}$. Denote by S the set of all products of elements $[K_\beta; (\rho|\beta) - \frac{m}{2}(\beta|\beta)], \beta \in R^+$, $m \in \mathbb{N}$. In view of Proposition 1.9a, we have:

$$\varphi_{k,m} \in S^{-1}\mathcal{U}^0. \tag{2.1.4}$$

Note also that we have proved that h is an injective homomorphism.

2.2. Define the homomorphism $\gamma : \mathcal{U}^0 \to \mathcal{U}^0$ by $\gamma K_i = q_i K_i$. Then we have:

$$(\gamma\varphi)(\lambda) = \varphi(\lambda + \rho), \ \lambda \in P. \tag{2.2.1}$$

Given $\varphi \in \mathcal{U}^0$, denote by z_φ the series (2.1.1) with $\varphi_{00} = \gamma(\varphi)$ constructed above.

Note that the group Q_2^* acts on $\mathcal{U}^0$ by

$$\delta K_\beta = \delta(\beta) K_\beta, \ \delta \in Q_2^*, \ \beta \in Q.$$

We thus have an action of the group $W \ltimes Q_2^*$ on $\mathcal{U}^0$. Denote by $\tilde{W}$ the subgroup of this group generated by all subgroups $\sigma W \sigma^{-1}$, $\sigma \in Q_2^*$. Note that in the simply laced case, Q_2^* is canonically isomorphic to $P/2P$, and denoting by $\bar{Q}$ the image of Q, we have: $\tilde{W} = W \ltimes \bar{Q}$.

Proposition 2.2. **(a)** *Given $\varphi_{0,0} \in \mathcal{U}^0$, all coefficients $\varphi_{k,m}$ of the corresponding series z lie in $\mathcal{U}^0$ if and only if*

$$\gamma^{-1}(\varphi_{0,0}) \in \mathcal{U}^{0\tilde{W}} := \{\psi \in \mathcal{U}^0 | w\psi = \psi, w \in \tilde{W}\}. \tag{2.2.2}$$

Moreover, we then have:

$$\deg \varphi_{k,m} + \min(|k|, |m|) \leq \deg \varphi_{0,0}. \tag{2.2.3}$$

(b) *We have:*

$$\gamma^{-1} \circ h : Z \xrightarrow{\sim} \mathcal{U}^{0\tilde{W}},$$

and the map $\varphi \longmapsto z_\varphi$ establishes the inverse isomorphism.

Proof: It is easy to see that condition (2.2.2) is equivalent to

$$\lambda \in T_{r\beta} \ \text{implies} \ \varphi_{00}^\delta(\lambda - r\beta) = \varphi_{00}^\delta(\lambda) \tag{2.2.4}$$

for every $\delta \in Q_2^*$, $r \in \mathbf{N}$ and $\beta \in R^+$.

Suppose now that all coefficients $\varphi_{k,m}$ of the series z lie in $\mathcal{U}^0$. Then z is defined on every Verma module $M^\delta(\lambda)$ and acts on it as the scalar $\varphi_{00}^\delta(\lambda)$. But by Proposition 1.9b, $M^\delta(\lambda) \supset M^\delta(\lambda - r\beta)$ provided that $\lambda \in T_{r\beta}^0$, hence $\varphi_{0,0}^\delta(\lambda - r\beta) = \varphi_{0,0}^\delta(\lambda)$ if $\lambda \in T_{r\beta}^0$. It follows that $\varphi_{0,0}^\delta(\lambda - r\beta) = \varphi_{0,0}^\delta(\lambda)$ for all $\lambda \in T_{r\beta}$, hence (2.2.2) is necessary.

Conversely, suppose that (2.2.2) holds. As has been mentioned above, (2.2.4) holds for $\beta \in R^+$ and $r \in \mathbf{N}$. We shall prove by induction on η that the denominator of $\varphi_{k,m}$ with $k, m \in \mathrm{Par}(\eta)$ does not contain factors $[K_\beta; (\rho|\beta) - \frac{r}{2}(\beta|\beta)]$. In view of (2.1.4), this will imply the sufficiency of (2.2.2) for the $\varphi_{k,m}$ to lie in $\mathcal{U}^0$, and also will complete the proof of (b).

Let $T_{r\beta,\eta}^0 = \{\lambda \in T_{r\beta}^0 | 2(\lambda + \rho|\gamma) \neq m(\gamma|\gamma)$ for all $\gamma \in R^+\backslash\{\beta\}$ and $m \in \mathbf{N}$ such that $m\gamma \leq \eta$. Let $\lambda \in T_{r\beta,\eta}^0$; then, by Proposition 1.9, $M^\delta(\lambda) \supset M^\delta(\lambda - r\beta)$ and, moreover,

$$M^\delta(\lambda)_\eta \cap J^\delta(\lambda) = M^\delta(\lambda - r\beta)_{\eta - r\beta}. \tag{2.2.5}$$

It is clear that z coincides on $V := M^\delta(\lambda - r\beta)_{\eta - r\beta}$ with the operator $\sum_{\gamma<\eta} \sum_{k,m \in \mathrm{Par}\gamma} F^k \varphi_{k,m} E^m$ (and is defined on this subspace by the inductive assumption). On the other hand, z acts on V as the scalar $\varphi^\delta(\lambda - r\beta)$, which is $\varphi_{0,0}^\delta(\lambda)$ by assumption (2.2.4). Hence the matrix $B^\delta(\lambda) := \varphi_{0,0}^\delta(\lambda) I - \sum_{\gamma<\eta} G_\eta^\delta(\lambda)$ is zero on V. Thus, by (2.13) and Proposition 1.7a) we have an equality of matrix-valued functions on V:

$$\phi_\eta^\delta(\lambda) H_\eta^\delta(\lambda) = B^\delta(\lambda),\ \lambda \in P,$$

such that $\mathrm{Ker} H_\eta^\delta(\lambda) \subseteq \mathrm{Ker} B^\delta(\lambda)$ (by (2.2.5)) for $\lambda \in T_{r\beta,\eta}^0$, hence for all $\lambda \in T_{r\beta}$, and $\dim \mathrm{Ker} H_\eta^\delta(\lambda) =$ multiplicity of zero of $\det H_\eta^\delta(\lambda)$ for $\lambda \in T_{r\beta,\eta}^0$. Hence, using [7, Lemma 2], we see that the matrix ϕ_η has a removable singularity in the hyperplane $T_{r\beta}$, as desired.

Inequality (2.2.3) is clear by looking at degrees in (2.1.3), where we let $\deg K_i = 1$, $i = 1, \ldots, n$. □

2.3. Let now $\varepsilon \in \mathbf{C}^\times$ be not a root of 1. Then for every $\varphi \in \mathcal{U}_\varepsilon^{0\tilde{W}}$ we denote by $z_{\varphi,\varepsilon}$ the element z_φ in which q is replaced by ε. This is an element of the center Z_ε of $\mathcal{U}_\varepsilon$. Defining the Harish-Chandra homomorphism $h_\varepsilon : Z_\varepsilon \to \mathcal{U}_\varepsilon^0$ as in 2.1 and $\gamma : \mathcal{U}_\varepsilon^0 \to \mathcal{U}_\varepsilon^0$ by $\gamma_\varepsilon K_i = \varepsilon^{d_i} K_i$ we obtain from Proposition 2.2 that $\gamma_\varepsilon^{-1} \circ h_\varepsilon : Z_\varepsilon \xrightarrow{\sim} \mathcal{U}_\varepsilon^{0\tilde{W}}$ is an isomorphism, the map $\varphi \longmapsto z_{\varphi,\varepsilon}$ being the inverse homomorphism.

§3. The center of $\mathcal{U}_\varepsilon$ and the group G, when ε is a root of 1.

3.1. Let ε be a primitive ℓ-th root of 1. Let $\ell' = \ell$ if ℓ is odd and $\ell' = \frac{1}{2}\ell$ if ℓ is even. We shall assume that

$$\ell' > d := \max_i\{d_i\} \tag{3.1.1}$$

Then $\mathcal{U}_\varepsilon$ is the algebra over $\mathbb{C}$ on generators E_i, F_i, K_i and K_i^{-1} and defining relations (1.2.1)–(1.2.5), where $q = \varepsilon$.

LEMMA 3.1. *The following relations hold in* $\mathcal{U}_\varepsilon$*:*

$$E_\alpha^{\ell'} E_\beta = \varepsilon^{-(\alpha|\beta)\ell'} E_\beta E_\alpha^{\ell'}, \ F_\alpha^{\ell'} F_\beta = \varepsilon^{(\alpha|\beta)\ell'} F_\beta F_\alpha^{\ell'}, \tag{3.1.2}$$

$$E_\alpha^{\ell'} F_\beta = F_\beta E_\alpha^{\ell'}, \ F_\alpha^{\ell'} E_\beta = E_\beta F_\alpha^{\ell'}, \tag{3.1.3}$$

$$K_\beta^{\ell'} E_\alpha = \varepsilon^{(\alpha|\beta)\ell'} E_\alpha K_\beta^{\ell'}, \ K_\beta^{\ell'} F_\alpha = \varepsilon^{-(\alpha|\beta)\ell'} F_\alpha K_\beta^{\ell'}. \tag{3.1.4}$$

PROOF: It suffices to check the relations (3.1.2) and (3.1.3) for $\alpha = \alpha_i$ and $\beta = \alpha_j$ since these relations for arbitrary $\alpha, \beta \in R^+$ follow by using automorphisms T_k; (3.1.4) follows immediately from (1.2.2). If $\alpha = \alpha_i$ and $\beta = \alpha_j$, formula (3.1.3) follows from (1.3.2), formula (3.1.2) for the E's follows from (1.10.1) and for the F's it follows by applying the automorphism φ. □

Denote by Z_ε the center of $\mathcal{U}_\varepsilon$.

COROLLARY 3.1. **(a)** *All elements* E_α^ℓ, F_α^ℓ $(\alpha \in R^+)$ *and* K_β^ℓ $(\beta \in Q)$ *lie in* Z_ε.

(b) *If ℓ is even, then all elements* $E_\alpha^{\ell'}$, $F_\alpha^{\ell'}$ *and* K_β^ℓ *lie in* Z_ε *iff A is of type* B_n $(n \geq 1)$. □

3.2. Consider the Verma module $M_\varepsilon(\lambda)$ over the algebra $\mathcal{U}_\varepsilon$. By formula (3.1.3), we have:

$$\mathcal{U}_\varepsilon^+ F_\alpha^{\ell'} v_\lambda = 0, \ \alpha \in R^+.$$

It follows that $F_\alpha^{\ell'} v_\lambda \in J_\varepsilon(\lambda)$. Let $\overline{M}_\varepsilon(\lambda)$ denote the quotient of $M_\varepsilon(\lambda)$ by the $\mathcal{U}_\varepsilon$-submodule generators by all the vectors $F_\alpha^{\ell'} v_\lambda$, $\alpha \in R^+$. We call $\overline{M}_\varepsilon(\lambda)$ a *diagonal $\mathcal{U}_\varepsilon$-module.* Proposition 1.7b and formulas (3.1.2)–(3.1.4) imply

PROPOSITION 3.2. *Vectors*

$$F_{\beta_1}^{m_1}\dots F_{\beta_N}^{m_N}v_\lambda,\ \textit{where}\ m_i\in\mathbf{Z}_+,\ m_i<\ell', \tag{3.2.1}$$

form a basis over $\mathbf{C}$ *of the space* $\overline{M}_\varepsilon(\lambda)$. □

It is clear that $\overline{M}_\varepsilon(\lambda)$ is irreducible if and only if $\overline{M}_\varepsilon(\lambda)_\eta$, $\eta\neq 0$, contains no singular vectors, i.e. no non–zero vectors v such that $E_iv=0$, $i=1,\dots,n$. Let $\hat\lambda\in\mathbf{C}\otimes_{\mathbf{Z}}P$ be such that $\lambda(K_\beta)=\varepsilon^{(\hat\lambda|\beta)}$. From formula (1.9.3) and Proposition 3.2 we obtain

THEOREM 3.2. *The* $\mathcal{U}_\varepsilon$*–module* $\overline{M}_\varepsilon(\lambda)$ *is irreducible if and only if*
(3.2.2)
$\varepsilon^{2(\hat\lambda+\rho|\beta)-m(\beta|\beta)}\neq 1$ *for all* $\beta\in R^+$ *and* $m\in\mathbf{Z}_+$ *such that* $m<\ell'$. □

COROLLARY 3.2. **(a)** *If* $\lambda(K_\beta^\ell)^2\neq 1$ *for all* $\beta\in R^+$, *then the* $\mathcal{U}_\varepsilon$*–module* $\overline{M}_\varepsilon(\lambda)$ *is irreducible.*

(b) *The* $\mathcal{U}_\varepsilon$*–module* $\overline{M}_\varepsilon(-\rho)$, *where* $(-\rho)(K_\beta)=\varepsilon^{-(\rho|\beta)}$, *is irreducible provided that* $\varepsilon^{2md_i}\neq 1$ *for* $m\in\mathbf{Z}_+$ *such that* $1\le m<\ell'$ *and* $i=1,\dots,n$ *(this condition always holds in the simply laced case or if* ℓ *is odd).* □

3.3. For each $\alpha\in R^+$, let $x_\alpha=E_\alpha^\ell$ and $y_\alpha=F_\alpha^\ell$; for each $\beta\in Q$, let $z_\beta=K_\beta^\ell$. We shall often write x_i and y_i for x_{α_i} and y_{α_i}. Denote by Z_0^+ (resp. Z_0^-) the subalgebra of Z_ε generated by the x_α (resp. y_α), $\alpha\in R^+$, and denote by Z_0^0 the subalgebra generated by the z_β, $\beta\in Q$. Let Z_0 be the subalgebra of Z_ε generated by the subalgebras Z_0^+, Z_0^- and Z_0^0. Then, by Proposition 1.6b, $Z_0^\pm\subset\mathcal{U}_\varepsilon^\pm$ and hence

$$Z_0=Z_0^+\otimes Z_0^0\otimes Z_0^-.$$

It is clear that Z_0^0 is the algebra of Laurent polynomials in z_i and z_i^{-1} ($i=1,\dots,n$), where $z_i=z_{\alpha_i}$.

Denote by Par_ℓ the set of all $k\in\mathbf{Z}_+^N$ such that $0\le k_i<\ell$ ($i=1,\dots,N$), and for $\eta\in Q$ denote by $\mathrm{Par}_\ell(\eta)$ the set of all $k\in\mathrm{Par}_\ell$ such that $\sum_i k_i\beta_i=\eta$.

We obtain from Proposition 1.7c the following

COROLLARY 3.3. **(a)** *The algebra Z_0^+ (resp. Z_0^-) is a polynomial algebra in the x_α (resp. y_α), $\alpha \in R^+$.*

(b) *The algebra $\mathcal{U}_\varepsilon$ (resp. $\mathcal{U}_\varepsilon^-$, or $\mathcal{U}_\varepsilon^+$) is a free module over Z_0 (resp. Z_0^-) with basis $F^k K_1^{m_1} \dots K_n^{m_n} E^r$ (resp. F^k, or E^k), where $k, r \in \mathrm{Par}_\ell$, $0 \le m_i < \ell$.* □

Given a homomorphism $\lambda : \mathcal{U}_\varepsilon^0 \to \mathbb{C}$ and a homomorphism $\nu : Z_0^- \to \mathbb{C}$ we construct the associated *triangular $\mathcal{U}_\varepsilon$-module* $\overline{M}_\varepsilon(\lambda, \nu)$ as the quotient of $\mathcal{U}_\varepsilon$ by the left ideal generated by the elements E_i, $K_i - \lambda(K_i)$, $K_i^{-1} - \lambda(K_i)^{-1}$, $y_\alpha - \nu(y_\alpha)$ ($i = 1, \dots, n$, $\alpha \in R^+$). Denote by v_λ the image of 1 in $\overline{M}_\varepsilon(\lambda, \nu)$. It is clear from Corollary 3.3b, that vectors

$$F_{\beta_1}^{k_1} \dots F_{\beta_N}^{k_N} v_\lambda, \text{ where } (k_1, \dots, k_N) \in \mathrm{Par}_\ell, \tag{3.3.1}$$

form a basis of $\overline{M}_\varepsilon(\lambda, \nu)$. Denoting by $\overline{M}_\varepsilon(\lambda, \nu)_\eta$ the linear span of vectors (3.3.1) with $(k_1, \dots, k_N) \in \mathrm{Par}_\ell(\eta)$, we obtain a Q_+-gradation $\overline{M}_\varepsilon(\lambda, \nu) = \oplus_\eta \overline{M}_\varepsilon(\lambda, \nu)_\eta$, consistent with the action of $\mathcal{U}_\varepsilon^+$ (but not $\mathcal{U}_\varepsilon^-$). It follows that if the only vector in $\overline{M}_\varepsilon(\lambda, \nu)_\eta$, $\eta \neq 0$, killed by $\mathcal{U}^+$ is zero, then $\overline{M}_\varepsilon(\lambda, \nu)$ is irreducible.

Using formula (1.9.3) we deduce from the above remarks

LEMMA 3.3. *If ℓ is odd and λ satisfies (3.2.2), then the $\mathcal{U}_\varepsilon$-module $\overline{M}_\varepsilon(\lambda, \nu)$ is irreducible.* □

REMARK 3.3. (a) The converse to the last statement is false (in contrast to diagonal modules).

(b) $\overline{M}_\varepsilon(\lambda) = \overline{M}_\varepsilon(\lambda, 0)$ if ℓ is odd.

We use triangular modules to prove the following

PROPOSITION 3.3. **(a)** *Let $\alpha = w'\alpha_{i'} \in R^+$ and let $E'_\alpha = T_{w'} E_i$, $F'_\alpha = \omega E'_\alpha$. Then $x'_\alpha = E'^\ell_\alpha$ lies in Z_0^+ and $y'_\alpha = F'^\ell_\alpha$ lies in Z_0^-.*

(b) $\mathcal{U}_\varepsilon^\pm \cap Z_\varepsilon = Z_0^\pm$.

(c) *The subalgebra Z_0 is invariant with respect to the automorphisms T_i.*

PROOF: It suffices to prove a) for y'_α, as the case of x'_α follows by applying ω. Recall that by Proposition 1.6b, $y'_\alpha \in \mathcal{U}_\varepsilon^-$. Hence by Corollary 3.3b, we have:

$$y'_\alpha = \sum_{k \in \mathrm{Par}_\ell} a_k F^k, \text{ where } a_k \in Z_0^-. \tag{3.3.2}$$

Choose λ such that $\overline{M}_\varepsilon(\lambda,\nu)$ is irreducible for all ν (see Lemma 3.3). Since, by Schur's lemma, all elements of Z_0 act as scalars on $\overline{M}_\varepsilon(\lambda,\nu)$, applying both sides of (3.3.2) to v_λ we obtain:

$$\nu(y'_\alpha)I = \sum_{k\in \mathrm{Par}_\ell} \nu(a_k)F^k.$$

By (3.3.1) it follows that $y'_\alpha = a_0 \in Z_0^-$, proving (a). The proof of (b) is similar.

Finally, we prove (c). We have from the definition (1.6.1)–(1.6.3):

$$T_i x_i = (-1)^\ell z_i y_i,\ T_i y_i = (-1)^\ell z_i^{-1} x_i, \tag{3.3.3}$$

$$T_i z_\beta = z_{r_i\beta}. \tag{3.3.4}$$

It is known that for every i there exists a reduced expression $r_{i_1} r_{i_2} \dots r_{i_N}$ of the longest element of W such that $i_1 = i$. Then we have:

$$R^+ = \{\alpha_{i_1}, r_{i_1}\alpha_{i_2}, r_{i_1}r_{i_2}\alpha_{i_3}, \dots, r_{i_1}\dots, r_{i_{n-1}}\alpha_{i_n}\},$$

and

$$R^+\backslash\{\alpha_i\} = \{\alpha_{i_2}, r_{i_2}\alpha_{i_3}, \dots, r_{i_2}\dots r_{i_{n-1}}\alpha_{i_n}\}.$$

Since $E_{i_2}^\ell, (T_{i_2}E_{i_3})^\ell, \dots, (T_{i_2}\dots T_{i_{n-1}}E_{i_n})^\ell$ lie in Z_0^+, we obtain, by Proposition 1.6b that their images under T_i also lie in Z_0^+. □

3.4. Let from now on ℓ be an odd integer greater than d (see (3.1.1)). For each $\alpha \in R^+$ define derivations e_α, f_α and $k'_{\pm\alpha}$ of the algebra $\mathcal{U}$ as follows:

(3.4.1)

$$e_\alpha(u) = [E_\alpha^\ell/[\ell]_{d_\alpha}!, u],\ f_\alpha(u) = [F_\alpha^\ell/[\ell]_{d_\alpha}!, u],\ k'_{\pm\alpha}(u) = [K_{\pm\alpha}^\ell/[\ell]_{d_\alpha}!, u].$$

Let $e_i = e_{\alpha_i}$, $f_i = f_{\alpha_i}$, $k'_i = k'_{\alpha_i}$. Note that if $\alpha = w\alpha_i$, then

$$e_\alpha = T_w e_i T_w^{-1},\ f_\alpha = T_w f_i T_w^{-1}. \tag{3.4.2}$$

A remarkable fact is that the derivations e_α, f_α and $k'_{\pm\alpha}$ can be pushed down to $\mathcal{U}_\varepsilon$, where ε is a primitive ℓ–th root of 1, ℓ odd (if ℓ is even one should replace ℓ by ℓ' in (3.4.1)). For $k'_{\pm\alpha}$ this is clear. Due to (3.4.2) it suffices to check this for the e_i and f_i. In the latter case this is clear from the formulas (1.3.2), (1.3.3), and (1.10.1) and the formula obtained from

(1.10.1) by applying ω. In fact, using them we obtain explicit formulas for the derivations e_i, f_i and $k_{\pm\alpha} = c_\alpha^{-\ell} k'_{\pm\alpha}$. Here and further we let

$$c_\alpha = \varepsilon^{d_\alpha} - \varepsilon^{-d_\alpha} \text{ for } \alpha \in R^+, \ c_j = c_{\alpha_j}.$$

We have (cf. [13] where analogous derivations are considered in the "restricted" case):

$$(3.4.3) \qquad e_i(F_j) = \frac{\delta_{ij} c_i^{\ell-2}}{\ell} (K_i \varepsilon^{d_i} - K_i^{-1} \varepsilon^{-d_i}) E_i^{\ell-1},$$

$$(3.4.4) \qquad e_i(K_j^{\pm 1}) = \mp \frac{(\alpha_i|\alpha_j) c_i^\ell}{(\alpha_i|\alpha_i)\ell} E_i^\ell K_j^{\pm 1},$$

$$(3.4.5) \qquad e_i(E_j) = \frac{1}{2} \sum_{s=1}^{-a_{ij}} \begin{bmatrix} -a_{ij} \\ s \end{bmatrix} (E_i^{(s)} E_j E_i^{(\ell-s)} - E_i^{(\ell-s)} E_j E_i^{(s)}),$$

$$(3.4.6) \qquad k_\beta(E_\alpha) = \frac{(\alpha|\beta)}{\ell(\alpha|\alpha)} z_\beta E_\alpha.$$

(We have specialized $q = \varepsilon$ in (3.4.5)). Indeed (3.4.3, 4 and 6) are easy using (1.3.2), (1.3.3) and the formula

$$[\ell - 1]! = \ell / (\varepsilon - \varepsilon^{-1})^{\ell-1}.$$

Formula (3.4.5) is obtained by taking the difference of (1.10.1) and (1.10.2) and specializing $q = \varepsilon$. Formulas for the f_i follow immediately from (3.4.3)–(3.4.5) by applying the automorphism φ:

$$\varphi e_i \varphi^{-1} = f_i, \ \varphi E_i = F_i, \ \varphi K_i = K_i, \ \varphi \varepsilon = \varepsilon^{-1}.$$

Denote by $\hat{\mathfrak{g}}$ (resp. $\hat{\mathfrak{g}}^+$, or $\hat{\mathfrak{g}}^-$) the subalgebra over Z_0 of the Lie algebra of all derivations of the algebra $\mathcal{U}_\varepsilon$ generated by all the e_α, $k_{\pm\alpha}$ and f_α (resp. e_α, or f_α), $\alpha \in R^+$.

Lemma 3.4. *The Lie algebra $\hat{\mathfrak{g}}$ is normalized by the T_i.*

Proof: This follows from (3.4.2) and

$$(3.4.7) \qquad T_i e_i T_i^{-1} = -z_i f_i - y_i k'_i. \quad \square$$

PROPOSITION 3.4. *The subalgebra Z_0 is invariant with respect to $\hat{\mathfrak{g}}$.*

PROOF: Trivially, $\hat{\mathfrak{g}} Z_\varepsilon \subset Z_\varepsilon$ and $\hat{\mathfrak{g}}^\pm Z_0^\pm \subset \mathcal{U}_\varepsilon^\pm$. Hence, by Proposition 3.3b:

$$\hat{\mathfrak{g}}^\pm Z_0^\pm \subset Z_0^\pm. \tag{3.4.8}$$

Furthermore, we have by (3.4.4) and (3.4.6):

$$e_i(z_j^{\pm 1}) = \mp \frac{d_j a_{ji} c_i^\ell}{2d_i} z_j^{\pm 1} x_i. \tag{3.4.9}$$

$$k_\beta(x_\alpha) = \frac{(\alpha|\beta)}{(\beta|\beta)} x_\alpha z_\beta. \tag{3.4.10}$$

Now we show that

$$e_\alpha(y_\beta) \in Z_0. \tag{3.4.11}$$

Indeed, if $\alpha = \beta$, (3.4.11) follows from

$$e_\alpha(y_\alpha) = (z_\alpha - z_\alpha^{-1}) c_\alpha^{-\ell}. \tag{3.4.12}$$

The latter formula is clear if $\alpha = \alpha_i$, by (1.3.4); for general $\alpha \in R^+$ it is deduced by applying T_w. Finally, if $\alpha \neq \beta$, then, using (3.4.2), we may assume that $\alpha = \alpha_i$. We have:

$$T_i e_i(y_\beta) = T_i e_i T_i^{-1} T_i(y_\beta) = -(z_i f_i + y_i k_i') T_i(y_\beta)$$

by (3.4.7). Since $T_i(y_\beta) \in Z_0^-$ and, by (3.4.8), $f_i Z_0^- \subset Z_0^-$, the proof of (3.4.11) is completed. The inclusion $f_\alpha(x_\beta) \in Z_0$ follows from (3.4.11) by applying ω. □

For $\gamma \in R^+$ we shall often let $e_{-\gamma} = f_\gamma$, $x_{-\gamma} = y_\gamma$. Proposition 3.4 implies the following

COROLLARY 3.4. *We have for $\alpha, \beta \in R$:*

$$e_\alpha(x_\beta)|_{x_\gamma = 0, \gamma \in R} = 0 \ \textit{if}\ \alpha + \beta \neq 0, \ \ e_\alpha(z_\beta)|_{x_\gamma = 0, \gamma \in R} = 0. \quad \square$$

The following useful formula follows from (3.4.9) by applying the T_i:

$$e_\alpha(z_\beta) = -\frac{(\alpha|\beta)c_\alpha^\ell}{(\alpha|\alpha)} z_\beta x_\alpha, \ \alpha \in R, \beta \in Q. \tag{3.4.13}$$

We deduce immediately from (1.10.3) one more formula (in which we put $x_{ij} = E_{ij}^\ell$):

$$\frac{d_i}{d_j}\left(\frac{c_j}{c_i}\right)^\ell e_i(x_j) = -e_j(x_i) = \frac{1}{2}c_j^\ell x_i x_j + x_{ij}, \text{ if } a_{ji} = -1. \tag{3.4.14}$$

REMARK 3.4. Note that in coordinates $\overline{x}_\alpha = c_\alpha^\ell x_\alpha$ and z_i all our formulas for the derivations e_α, f_α and $k_{\pm\alpha}$ are independent of ℓ. We can show that indeed the action of the $e_\alpha (\alpha \in R)$ on Z_0 is independent of ℓ in these coordinates [2] (for the k_α this is clear by (3.4.10)).

3.5. Denote by $\hat{Z}_0$ the algebra of all formal power series in the x_α ($\alpha \in R$), z_i and z_i^{-1} ($i = 1, \dots, n$) which converge to a holomorphic function for all complex values of the x_α and all non-zero complex values of the z_i. Let $\hat{\mathcal{U}}_\varepsilon = \hat{Z}_0 \otimes_{Z_0} \mathcal{U}_\varepsilon$, $\hat{Z}_\varepsilon = \hat{Z}_0 \otimes_{Z_0} Z_\varepsilon$. It is clear by (3.4.10) that the series $\exp tk_\alpha$, $t \in \mathbb{C}$, converges to an automorphism of $\hat{\mathcal{U}}_\varepsilon$.

PROPOSITION 3.5. *The series* $\exp te_\alpha$, $t \in \mathbb{C}$, $\alpha \in R$, *converges to an automorphism (over* $\mathbb{C}$*) of the algebra* $\hat{\mathcal{U}}_\varepsilon$.

PROOF: Due to the automorphisms T_i it suffices to show that the series $\exp te_i$ when applied to the K_β, F_j and E_j converges to an element of the algebra $\hat{\mathcal{U}}_\varepsilon$. We obtain from (3.4.4):

$$(\exp te_i)K_\beta = e^{-((\alpha_i|\beta)c_i^\ell/(\alpha_i|\alpha_i)\ell)tx_i} K_\beta. \tag{3.5.1}$$

Furthermore, using (3.4.3), we obtain:

$$e_i^n(F_i) = (-1)^{n-1}c_i^{-2}(c_i^\ell/\ell)^n x_i^{n-1}(\varepsilon^{d_i}K_i + (-1)^n\varepsilon^{-d_i}K_i^{-1})E_i^{\ell-1},$$

It follows that:
(3.5.2)

$$(\exp te_i)F_j = F_j - \delta_{ij}\left(\frac{e^{-t(c_i^\ell/\ell)x_i} - 1}{c_i^2 x_i}\varepsilon^{d_i}K_i + \frac{e^{t(c_i^\ell/\ell)x_i} - 1}{c_i^2 x_i}\varepsilon^{-d_i}K_i^{-1}\right)E_i^{\ell-1}.$$

Finally, define endomorphisms ψ_s of $\mathcal{U}_\varepsilon$ for $s = 1, \dots, \ell - 1$ by: $\psi_s(u) = E_i^s u E_i^{\ell-s}$. Then (3.4.5) can be rewritten as follows:

$$e_i(E_j) = \frac{1}{2} \sum_{s=1}^{-a_{ij}} \begin{bmatrix} -a_{ij} \\ s \end{bmatrix} ([s]_{d_i}![\ell - s]_{d_i}!)^{-1} (\psi_s - \psi_{\ell-s}) E_j$$

Since ψ_i and ψ_j commute, it suffices to show that $(\exp t\psi_s)u$ converges to an element of $\hat{\mathcal{U}}_\varepsilon$ if $u \in \hat{\mathcal{U}}_\varepsilon$. We have:

$$(\exp t\psi_s)u = \sum_{r=0}^{\ell-1} t^r E_i^{rs} u E_i^{r(\ell-s)} \sum_{m=0}^{\infty} \frac{(tx_i)^{m\ell}}{(m\ell + r)!}. \quad \square$$

One deduces from (3.4.12 and 13) the following formulas for the action on $\hat{Z}_0$:

$$(3.5.3) \qquad (\exp te_i) y_j = y_j - \delta_{ij} c_i^{-2\ell} \left(z_i \frac{e^{-c_i^\ell t x_i} - 1}{x_i} + z_i^{-1} \frac{e^{c_i^\ell t x_i} - 1}{x_i} \right),$$

$$(3.5.4) \qquad (\exp te_i) z_\beta = e^{-((\alpha_i | \beta) c_i^\ell / (\alpha_i | \alpha_i)) t x_i} z_\beta .$$

3.6. Recall that we have: $\mathcal{U}_\varepsilon \supset Z_\varepsilon \supset Z_0$, where Z_ε is the center of the algebra $\mathcal{U}_\varepsilon$, $Z_0 = \mathbb{C}[x_\alpha (\alpha \in R); z_i, z_i^{-1} \ (i = 1, \dots, n)]$, and that $\mathcal{U}_\varepsilon$ is a free module over Z_0 of rank ℓ^{2N+n} (Corollary 3.3b).

Since $\mathcal{U}_\varepsilon$ is a finitely generated module over Z_0, it follows [see e.g. [15, (3.5)]) that Z_ε is as well, hence Z_ε is integral over Z_0 and it is a finitely generated algebra.

As usual, for a finitely generated algebra R over $\mathbb{C}$, we denote by Spec R the affine algebraic variety of algebra homomorphisms $R \longrightarrow \mathbb{C}$. For example, Spec $Z_0 \simeq \mathbb{C}^{2N} \times \mathbb{C}^{\times n}$; denote by H the (n-dimensional) subvariety of Spec Z_0 consisting of the points such that $x_\alpha = 0$ for all $\alpha \in R$ and by H_0 the subset of H consisting of the points such that $z_\alpha^2 \neq 1$ for all $\alpha \in R$.

Denote by G the subgroup of the automorphism group of the algebra $\hat{\mathcal{U}}_\varepsilon$ generated by all automorphisms $\exp te_\alpha$ and $\exp tk_\alpha$, $\alpha \in R$. This group leaves invariant $\hat{Z}_\varepsilon$ and $\hat{Z}_0$ (by Proposition 3.4), hence acts by holomorphic transformations on the algebraic varieties Spec Z_ε and Spec Z_0.

REMARK 3.6. The action of G on Spec Z_0 may be viewed as a "quantization" of the coadjoint action, H being a "quantization" of a Cartan subalgebra and H_0 being the set of regular elements. Note also, that by Proposition 3.3c, the T_i act on Spec Z_0. This action leaves H invariant and induces the action of W on H. Note finally that, by Remark 3.4, the action of G on Spec Z_0 in coordinates $\overline{x}_\alpha$, z_i is independent of ℓ.

LEMMA 3.6. *The set GH_0 contains a non-empty open (in metric topology) subset of Spec Z_0.*

PROOF: Consider the (holomorphic) map $\Psi : \mathbb{C}^{2N} \times H_0 \to \text{Spec } Z_0$ given by $\Psi((t_\alpha)_{\alpha\in R}, u) = (\prod_{\alpha\in R} \exp t_\alpha e_\alpha)u$. It follows from formula (3.4.10) and Corollary 3.4 that the differential $d\Psi$ is non-degenerate at all points $(0, u)$, $u \in H_0$. The lemma follows. □

3.7. Denote by Rep $\mathcal{U}_\varepsilon$ the set of all (non-zero) irreducible finite-dimensional representations (up to equivalence) of the algebra $\mathcal{U}_\varepsilon$ over $\mathbb{C}$. Then we have a sequence of canonical maps:

$$\text{Rep } \mathcal{U}_\varepsilon \xrightarrow{X} \text{Spec } Z_\varepsilon \xrightarrow{\tau} \text{Spec } Z_0. \tag{3.7.1}$$

The map τ is induced by the inclusion $Z_0 \subset Z_\varepsilon$. Since Z_ε is integral over Z_0, τ is a finite and hence surjective morphism. The map X is defined as follows. By Schur's lemma, for $\pi \in \text{Rep } \mathcal{U}_\varepsilon$ we have

$$\pi(u) = \chi_\pi(u)I, \quad u \in Z_\varepsilon, \text{ where } \chi_\pi(u) \in \mathbb{C},$$

and we let $X(\pi) = \chi_\pi$.

Given $\chi \in \text{Spec } Z_\varepsilon$, let I^χ denote the ideal of $\mathcal{U}_\varepsilon$ generated by the kernel of the homomorphism $\chi : Z_\varepsilon \to \mathbb{C}$. Let $\mathcal{U}_\varepsilon^\chi = \mathcal{U}_\varepsilon / I^\chi$. Since $\mathcal{U}_\varepsilon$ has no zero divisors (Corollary 1.8b), it is a torsion free (finitely generated) module over Z_ε, hence (see e.g. [15, (3.10)]) $\mathcal{U}_\varepsilon^\chi \neq 0$. Hence the map X is surjective.

Since $\mathcal{U}_\varepsilon$ has no zero divisors, the same is true for Z_ε and we can consider its quotient field $Q(Z_\varepsilon)$. Let $Q(\mathcal{U}_\varepsilon) = Q(Z_\varepsilon) \otimes_{Z_\varepsilon} \mathcal{U}_\varepsilon$; this algebra is finite-dimensional over the field $Q(Z_\varepsilon)$ and has no zero divisors. Hence $Q(\mathcal{U}_\varepsilon)$ is a division algebra over its center $Q(Z_\varepsilon)$, hence it is the quotient algebra of $\mathcal{U}_\varepsilon$. It follows that $\mathcal{U}_\varepsilon$ is a $\mathbb{C}$-subalgebra of a full matrix algebra $\text{Mat}_m(\mathsf{F})$ over some finite extension F of $Q(Z_\varepsilon)$, such that $\mathsf{F} \otimes_{Q(Z_\varepsilon)} \mathcal{U}_\varepsilon = \text{Mat}_m(\mathsf{F})$, for some $m \in \mathbb{N}$ (see e.g. [16]). Then m is called the *degree* of $\mathcal{U}_\varepsilon$. Hence we may consider the characteristic polynomial of $x \in \mathcal{U}_\varepsilon$:

$$\det(\lambda - x) = \lambda^m - (\text{tr } x)\lambda^{m-1} + \ldots + (-1)^m \det x,$$

its norm $\mathcal{N}(x) = \det x$ and the bilinear form $(x, y) = \operatorname{tr} xy$.

Being the center of an integrally closed algebra $\mathcal{U}_\varepsilon$ (see Theorem 1.8), Z_ε is integrally closed. Hence all coefficients of the characteristic polynomial of $x \in \mathcal{U}_\varepsilon$ lie in Z_ε. The determinants $\det((u_i, u_j)_{i,j=1}^{m^2}$, where $u_1, \dots, u_{m^2} \in \mathcal{U}_\varepsilon$, generate a non-zero ideal of Z_ε called the discriminant ideal of $\mathcal{U}_\varepsilon$. The set $\mathcal{D} \subset \operatorname{Spec} Z_\varepsilon$ of zeros of the discriminant ideal is a well-defined closed (proper) subvariety of Spec Z_ε, called the *discriminant subvariety.* The following facts are special cases of well-known general results:

LEMMA 3.7. (a) *$\mathcal{U}_\varepsilon^\chi$ is isomorphic to $\mathrm{Mat}_m(\mathbb{C})$ if and only if $\chi \notin \mathcal{D}$.*

(b) *If $\chi \in \mathcal{D}$, then $\dim_{\mathbb{C}} \mathcal{U}_\varepsilon^\chi \geq m^2$, but the dimension of every irreducible representation of $\mathcal{U}_\varepsilon^\chi$ is less than m.*

PROOF: Let $\varphi : Z_\varepsilon \to \mathbb{C}$ be a point of Spec Z_ε outside $\mathcal{D}$. Then there exist elements $u_1, \dots, u_{m^2}$ in $\mathcal{U}_\varepsilon$ such that $\varphi(\det(u_i, u_j)) \neq 0$. We have: $D := \det((u_i, u_j)) \in Z_\varepsilon$; let $d = \varphi(D)$. Consider the algebra $A := \mathcal{U}_\varepsilon \otimes_\varphi \mathbb{C}$. We claim that the elements $\overline{u}_i := u_i \otimes 1$ form a basis of A over $\mathbb{C}$. Indeed, the u_i being linearly independent over $Q(Z_\varepsilon)$, form a basis of $Q(\mathcal{U}_\varepsilon)$ over $Q(Z_\varepsilon)$. Let $\{u^i\}$ be the dual basis with respect to the trace form. Then $Du^i \in \mathcal{U}_\varepsilon$ for each i. For $u \in \mathcal{U}_\varepsilon$ we have:

$$Du = \sum_i (Du, u^i) u_i = \sum_i (u, Du^i) u_i.$$

Passing to A we get:

$$d\overline{u} = \sum_i \varphi((u, Du^i)) \overline{u}_i.$$

Thus, the $\overline{u}_i$ span A over $\mathbb{C}$ and $\det((\overline{u}_i, \overline{u}_j)) \neq 0$. Hence A is a m^2-dimensional semisimple algebra over $\mathbb{C}$. On the other hand, each element of $\mathcal{U}_\varepsilon$ satisfies a polynomial of degree m with coefficients in Z_ε, hence each element of A satisfies a polynomial of degree m with complex ccefficients. It follows that $A = \mathrm{Mat}_m(\mathbb{C})$ (since $m^2 = n_1^2 + n_2^2 + \dots$, $n_i \geq 0$ and $n_1 + n_2 + \dots \leq m$ imply that $m = n_i$ for some i).

Let now $\rho : \mathcal{U}_\varepsilon \to \mathrm{Mat}_m(\mathbb{C})$ be an irreducible representation and let $\varphi = X(\rho) : Z_\varepsilon \to \mathbb{C}$ be the corresponding element of Spec Z_ε. Let $u_1, \dots, u_{m^2}$ be elements of $\mathcal{U}_\varepsilon$ such that $\rho(u_i)$ form a basis of $\mathrm{Mat}_m(\mathbb{C})$ over $\mathbb{C}$, and let $D = \det((u_i, u_j))$. Then $\varphi(D) = \det(\operatorname{tr} \rho(u_i)\rho(u_j)) \neq 0$, hence $\varphi \notin \mathcal{D}$.

In order to prove (b) notice that since each element of $\mathcal{U}_\varepsilon$ satisfies a polynomial of degree m with coefficients in Z_ε, every irreducible representation of $\mathcal{U}_\varepsilon$ has dimension $\leq m$. This and (a) give the second part of (b), while the first part follows from the fact that $\mathcal{U}_\varepsilon$ is a torsion free rank m^2 module over Z_ε. □

3.8. Consider a finite-dimensional irreducible representation π of $\mathcal{U}_\varepsilon$ in a complex vector space V. Since Z_0 acts by scalar operators on V, π extends in an obvious way to $\hat{\mathcal{U}}_\varepsilon$. Given $g \in G$, denote by π^g the "twisted" (irreducible) representation of $\mathcal{U}_\varepsilon$ in V defined by

$$\pi^g(u) = \pi(gu), \ u \in \mathcal{U}_\varepsilon.$$

Note that

$$\chi_{\pi^g} = \chi_\pi \circ g, \ g \in G.$$

We call the representation π *diagonalizable* (resp. *triangulizable*) if there exists $g \in G$ such that π^g is a diagonal (resp. triangular) representation.

Denote by Ω the set of all $\lambda \in \text{Spec } Z_0$ such that $(\tau \cdot X)^{-1}\lambda$ consists of ℓ^n irreducible diagonalizable representations of (maximal) dimension ℓ^N. Note that by Corollary 3.2a it follows $H_0 \subset \Omega$. An immediate corollary of Lemma 3.6 and Corollary 3.2a is

LEMMA 3.8. *Ω contains a non-empty open (in metric topology) subset of Spec Z_0.* □

Since, due to Corollary 3.3b:

$$\dim_{Q(Z_0)} Q(\mathcal{U}_\varepsilon) = \ell^{2N+n}, \tag{3.8.1}$$

Lemma 3.8 gives us immediately

$$\dim_{Q(Z_0)} Q(Z_\varepsilon) = \ell^n, \tag{3.8.2}$$

$$m^2 := \dim_{Q(Z_\varepsilon)} Q(\mathcal{U}_\varepsilon) = \ell^{2N}. \tag{3.8.3}$$

The following theorem summarizes the obtained results.

THEOREM 3.8. *Let ℓ be an odd integer, $\ell > \max_i\{d_i\}$ and let ε be a primitive ℓ-th root of 1.*

(a) *Spec Z_ε is a normal affine algebraic variety and the map τ : Spec $Z_\varepsilon \to$ Spec Z_0 is finite (surjective) of degree ℓ^n.*

(b) *If $\chi \in$ Spec $Z_\varepsilon \backslash \mathcal{D}$ (where the discriminant $\mathcal{D}$ is a closed proper subvariety), then $X^{-1}(\chi)$ consists of a single irreducible representation of $\mathcal{U}_\varepsilon$ of dimension ℓ^N. If $\chi \in \mathcal{D}$, then $X^{-1}(\chi)$ consists of a finite number of irreducible representations of $\mathcal{U}_\varepsilon$ of dimension less than ℓ^N.* □

REMARK 3.8. Let the Cartan matrix A be of type B_n, $n \geq 1$. Let ℓ be a positive even integer, and let, as before, $\ell' = \ell/2$. Then elements $E_\alpha^{\ell'}, F_\alpha^{\ell'}$ ($\alpha \in R^+$) and $K_\beta^{\ell'}$ ($\beta \in Q$) lie in the center Z_ε of the algebra $\mathcal{U}_\varepsilon$ (by Lemma 3.1). Denote by Z_0 the subalgebra generated by all these elements. Then all the results proved above still hold if ℓ is replaced by ℓ'. In particular:

$$\dim_{Q(Z_\varepsilon)} Q(\mathcal{U}_\varepsilon) = \ell'^{2N}, \ \dim_{Q(Z_0)} Q(Z_\varepsilon) = \ell'^n. \tag{3.8.4}$$

3.9. Given $\varphi \in \mathcal{U}_\varepsilon^{0\tilde{W}}$, let $\varphi_{0,0} = \varphi(\varepsilon^{d_1}K_1, \dots, \varepsilon^{d_n}K_n)$. Furthermore, for each pair, $k, r \in \text{Par}_\ell$ construct elements $\varphi_{k,r;\varepsilon}$ by formula (2.1.3), in which q is replaced by ε. By formula (1.9.3) and the argument proving Proposition 2.2 (in which Verma modules are replaced by diagonal modules), we see that $\varphi_{k,r;\varepsilon} \in \mathcal{U}_\varepsilon^0$. We let

$$z_{\varphi,\varepsilon} = \sum_{k,r\in\text{Par}_\ell} F^k \varphi_{k,r;\varepsilon} E^r \in \mathcal{U}_\varepsilon.$$

LEMMA 3.9. $z_{\varphi,\varepsilon} \in Z_\varepsilon$.

PROOF: Note that the intersection of kernels of all triangular modules is $M := \sum_{\alpha\in R^+} x_\alpha \mathcal{U}_\varepsilon$. Since, by the construction, $z_{\varphi,\varepsilon}$ acts on each triangular module as a scalar, we obtain: $[z_{\varphi,\varepsilon}, F_i] \in M$. It follows, using Corollary 3.3b, and commutation relations (1.2.1)–(1.2.3), that $[z_{\varphi,\varepsilon}, F_i] = 0$. Since $\omega(z_{\varphi,\varepsilon}) = z_{\varphi,\varepsilon}$ (see §2.1), we obtain that also $[z_{\varphi,\varepsilon}, E_i] = 0$. □

Thus, we have an injective homomorphism $\mathcal{U}_\varepsilon^{0\tilde{W}} \to Z_\varepsilon$ defined by $\varphi \longmapsto z_{\varphi,\varepsilon}$.

§4. The example of $\mathcal{U}$ and $\mathcal{U}_\varepsilon$ of type A_1.

4.1. We consider here the simplest example, that of the quantum algebra $\mathcal{U}$, associated to the matrix (2), which first appeared in the work of Kulish-Reshetikhin and Sklyanin. This is a $\mathbb{Q}(q)$–algebra on generators E, F, K and K^{-1} and defining relations

$$KK^{-1} = 1, \ KEK^{-1} = q^2E, \ KFK^{-1} = q^{-2}F, \tag{4.1.1}$$

(4.1.2) $$EF - FE = (K - K^{-1})/(q - q^{-1}).$$

The center Z of $\mathcal{U}$ contains the well-known element

(4.1.3) $$c = \frac{Kq + K^{-1}q^{-1}}{(q - q^{-1})^2} + FE.$$

Note that $c = z_\varphi$, where $\varphi = (K + K^{-1})/(q - q^{-1})^2$ (see § 2.2). Since φ generates $\mathcal{U}^{0\tilde{W}}$, we deduce that c generates Z.

It is well-known that all finite-dimensional irreducible representations over $\mathbb{Q}(q)$ of $\mathcal{U}$ are equivalent to representations $\pi_n^\pm$, $n \in \mathbb{Z}_+$, of dimension $n+1$, which in some basis $v_0, \dots, v_n$ are given as follows (we let $v_{-1} = 0 = v_{n+1}$):

(4.1.4)
$$\pi_n^\pm(K)v_j = \pm q^{n-2j}v_j,\ \pi_n^\pm(E)v_j = \pm[n-j+1]v_{j-1},\ \pi_n^\pm(F)v_j = [j+1]v_{j+1}.$$

These facts still hold for the specialization of q to any complex number different from 0 and a root of 1.

4.2. Let now $\ell > 2$ be an integer, and let, as before $\ell' = \ell$ if ℓ is odd and $\ell' = \ell/2$ if ℓ is even. Let ε be a primitive ℓ'th root of 1, and let $\varepsilon' = \varepsilon$ if ℓ is odd and $= \varepsilon^2$ if ℓ is even. Denote by $\mathcal{U}_\varepsilon$ the algebra over $\mathbb{C}$ on generators E, F, K and K^{-1} and defining relations (4.1.4), (4.1.2) where q is replaced by ε. Let Z_0 be the subalgebra of the center Z_ε generated by $x = E^{\ell'}$, $y = F^{\ell'}$, $z = K^{\ell'}$ and z^{-1}. We have by (3.8.2), (3.8.3) and (3.8.4):

(4.2.1) $$\dim_{Q(Z_\varepsilon)} Q(\mathcal{U}_\varepsilon) = \ell'^2,\ \dim_{Q(Z_0)} Q(Z_\varepsilon) = \ell'.$$

Note that the norm of an element $f(K) \in \mathcal{U}^0$ over $Q(Z_\varepsilon)$ can be calculated as follows:

(4.2.3) $$\mathcal{N}(f(K)) = \prod_{j=0}^{\ell'-1} f(\varepsilon'^j K).$$

We also have:

(4.2.4) $$\mathcal{N}(E) = (-1)^{\ell'+1}x,\ \mathcal{N}(F) = (-1)^{\ell'+1}y.$$

Denote again by c the element of Z_ε given by formula (4.1.3) where q is replaced by ε. Taking norm of both sides of the equality

$$c - \frac{K\varepsilon + K^{-1}\varepsilon^{-1}}{(\varepsilon - \varepsilon^{-1})^2} = FE,$$

we obtain:

$$\prod_{j=0}^{\ell'-1}\left(c-\frac{K\varepsilon\varepsilon'^j+K^{-1}\varepsilon^{-1}\varepsilon'^{-j}}{(\varepsilon-\varepsilon^{-1})^2}\right)=xy. \tag{4.2.5}$$

The left-hand side of (4.2.5) is of the form:

$$c^{\ell'}+a_1c^{\ell'-1}+\ldots+a_{\ell'-1}c+a,$$

where, clearly, $a_i\in\mathbb{C}$ for $i=1,\ldots,\ell'-1$, and

$$a=(-1)^{\ell}(\varepsilon-\varepsilon^{-1})^{-2\ell'}(z+z^{-1}).$$

Hence we can rewrite (4.2.5) in either of the following two forms, where $c_j^{\pm}\in\mathbb{C}$:

$$\prod_{j=0}^{\ell'-1}(c-c_j^{\pm})=xy+(-1)^{\ell+1}\frac{z+z^{-1}\pm 2}{(\varepsilon-\varepsilon^{-1})^{2\ell'}}. \tag{4.2.6$_\pm$}$$

In order to calculate the constants $c_j^{\pm}(\varepsilon)$, consider the representations $\pi_{n,\varepsilon}^{\pm}$ of $\mathcal{U}_\varepsilon$ over $\mathbb{C}$ defined as follows. If ℓ is odd (resp. even), $\pi_{n,\varepsilon}^{\pm}$ (resp. $\pi_{n,\varepsilon}^{(-)^n}$) is defined for each $n=0,1,\ldots,\ell-1$ by formulas (4.1.4) for $\pi_n^{\pm}$ (resp. π_n^{+}), where q is replaced by ε and $0\le j\le n$ (resp. $0\le j\le n$ if $n<\ell'$ and $0\le j\le n-\ell'$ if $n\ge\ell'$).

The proof of the following lemma is standard.

LEMMA 4.2. **(a)** *Representations $\pi_{n,\varepsilon}^{\pm}$ form a complete non-redundant list of all finite-dimensional irreducible representations of $\mathcal{U}_\varepsilon$ over $\mathbb{C}$ such that $x=y=0$, $z^2=1$.*

(b) *$\pi_{n,\varepsilon}^{\pm}(c)=(\pm1)^{\ell}c_n(\varepsilon)$, where*

$$c_n(\varepsilon)=\frac{\varepsilon^{n+1}+\varepsilon^{-n-1}}{(\varepsilon-\varepsilon^{-1})^2}. \quad \square \tag{4.2.7}$$

Comparing Lemma 4.2 with (4.2.6$_\pm$), we obtain an explicit form of (4.2.6$_\pm$):

$$\prod_{j=0}^{\ell-1}(c\pm c_j(\varepsilon))=xy+\frac{z+z^{-1}\pm 2}{(\varepsilon-\varepsilon^{-1})^{2\ell}} \text{ if } \ell \text{ is odd,}$$

(4.2.8$_\pm$)

$$\prod_{\substack{j=0\\ j \text{ even}\\ \text{(odd)}}}^{\ell-1}(c-c_j(\varepsilon))=xy-\frac{z+z^{-1}\mp 2}{(\varepsilon-\varepsilon^{-1})^{\ell}} \text{ if } \ell \text{ is even.}$$

Finally, Z_ε is generated by Z_0 and Z_1. Indeed, if $z \in Z_\varepsilon$, then $z = \sum_{n \geq 0} z_n^{\pm}$, where $z_n^+ = \sum_{k \geq 0} F^k \varphi_k E^{k+n}, z_n^- = \sum_{k \geq 0} F^{k+n} \psi_k E^k$, $\varphi_k, \psi_k \in \mathcal{U}_\varepsilon^0$ and $z_n^{\pm} \in Z_\varepsilon$. Since $z_n^{\pm}$ commute with K, $z_n^{\pm} = 0$ unless n is a multiple of ℓ'. Hence Z_ε is generated by x, y and central elements of the form $\sum_{k \geq 0} F^k \varphi_k E^k = \sum_{k \geq \ell'} F^k \varphi_k E^k +$(element from Z_1). Noting that Z_1 is generated by c, we are done. (An alternative proof: due to Section 5.3 it suffices to show that Spec Z'_ε is normal, which is the case since this is a hypersurface, non-singular in codimension 1.)

We can state the results obtained in the A_1 case in the following form:

THEOREM 4.2. *Let $\mathcal{U}_\varepsilon$ be the quantum group of type A_1 at a primitive ℓ'th root of unity ε, $\ell > 2$. Let $\ell' = \ell$ (resp. $\ell/2$) if ℓ is odd (resp. even). Then:*

(a) *The center Z_ε of $\mathcal{U}_\varepsilon$ is generated by the elements x, y, z, z^{-1} and c with defining relations $zz^{-1} = 1$ and (4.2.8$_+$) (or (4.2.8$_-$)).*

(b) *Z_ε is integral over the subalgebra $Z_0 = \mathbb{C}[x, y, z][z^{-1}]$.*

(c) $\dim_{Q(Z_\varepsilon)} Q(\mathcal{U}_\varepsilon) = \ell'^2$, $\dim_{Q(Z_0)} Q(Z_\varepsilon) = \ell'$.

(d) *Spec Z_ε is a 3-dimensional normal affine algebraic variety with the following singular points given in coordinates (x, y, z, c):*

$$a_j^{\pm} = (0, 0, \pm 1, \pm c_j(\varepsilon)) \text{ for } j = 0, 1, \ldots, \ell - 2 \text{ if } \ell \text{ is odd,}$$
$$a_j = (0, 0, (-1)^j, c_j(\varepsilon)) \text{ for } j = 0, 1, \ldots, \ell - 2,\ j \neq \ell' - 1 \text{ if } \ell \text{ is even.}$$

(e) *The map X : Rep $\mathcal{U}_\varepsilon \to$ Spec Z_ε is surjective and $X^{-1}(a)$ is a single representation of dimension ℓ' if and only if a is a non-singular point of Spec Z_ε. Furthermore $X^{-1}(a_j^{\pm})$ consists of two representations $\pi_{j,\varepsilon}^{\pm}$ and $\pi_{\ell-j-2,\varepsilon}^{\pm}$ of dimensions $j+1$ and $\ell-j-1$ if ℓ is odd, and $X^{-1}(a_j)$ consists of two representations $\pi_{j,\varepsilon}^{(-)^j}$ and $\pi_{\ell-j-2,\varepsilon}^{(-)^j}$ of dimensions $j+1$ and $\ell' - j - 1$ (resp. $j - \ell' + 1$ and $\ell - j - 1$) if $j < \ell'$ (resp. $j \geq \ell'$) if ℓ is even.* □

REMARK 4.2. **(a)** Since $e(c) = 0 = f(c)$, it follows from (4.2.8) that the polynomial

$$P = (-1)^{\ell+1}(\varepsilon - \varepsilon^{-1})^{2\ell'} xy + z + z^{-1}$$

is fixed by G. Moreover, Z_0^G is generated by P. Denote by $\mathcal{O}_a$ the hypersurface $P = a$, $a \in \mathbb{C}$, in Spec Z_0. It is easy to show that $\mathcal{O}_a$ is a G-orbit if $a \neq \pm 2$, that points $(0, 0, \pm 1)$ are fixed and the complements to them in $\mathcal{O}_{\pm 2}$ are G-orbits. Hence every irreducible representations of $\mathcal{U}_\varepsilon$ of type A_1 is triangulizable, and up to the action of G, the only non–diagonalizable representations are those corresponding to the points $(0, 1, \pm 1)$ of Spec Z_0.

(b) Every ℓ'–dimensional irreducible representation of $\mathcal{U}_\varepsilon$ of type A_1 can be written in some basis $v_0, v_1, \dots, v_{\ell'-1}$ in the following form, for some $\lambda \in \mathbb{C}^\times$, $a, b \in \mathbb{C}$:

$$Kv_j = \lambda\varepsilon^{-2j}v_j,\ Fv_j = v_{j+1}\ (j = 0, \dots, \ell' - 2),\ Fv_{\ell'-1} = bv_0,$$

$$Ev_j = \Big(\frac{(\lambda\varepsilon^{1-j} - \lambda^{-1}\varepsilon^{j-1})(\varepsilon^j - \varepsilon^{-j})}{(\varepsilon - \varepsilon^{-1})^2} + ab\Big)v_{j-1}\ (j = 1, \dots \ell' - 1),$$

$$Ev_0 = av_{\ell'-1}.$$

Indeed, we let v_0 be an eigenvector of K, and let $v_j = F^j v_0$ for $j = 1, \dots, \ell' - 1$.

§5. Open problems.

5.1. The case of even ℓ seems to be more difficult than the case of odd ℓ. First, Z_0 probably should be replaced by Z_0', the intersection of Z_ε with the subalgebra generated by the $K_\beta^{\ell'}$ $(\beta \in Q)$, $E_\alpha^{\ell'}$ and $F_\alpha^{\ell'}$ $(\alpha \in R^+)$, which is a more complicated algebra. Second, the set of diagonalizable modules does not contain an open subset in general.

The only case when these difficulties can be easily resolved is the case of the matrix (a_{ij}) of type B_n, $n \geq 1$, as explained by Remark 3.8.

Let s denote the cardinality of the set $\{\beta \in R^+ | (\beta|Q) \subset 2\mathbb{Z}\}$.

Conjecture 5.1. $m = \ell^N/2^s$ if ℓ is even.

5.2.*Conjecture 5.2.* **(a)** $\Omega = \text{Spec } Z_0 \backslash \tau(\mathcal{D})$.

(b) $\tau(\mathcal{D}) = \overline{G(H\backslash H_0)}$.

(c) $\mathcal{D}$ is the set of singular points of Spec Z_ε.

5.3. Denote by Z_1 the image of the homomorphism $\mathcal{U}_\varepsilon^{0\tilde{W}} \to Z_\varepsilon$ constructed in Section 3.9 and by Z_ε' the subalgebra of Z_ε generated by Z_0 and Z_1. It is clear that $\dim_{Q(Z_0)} Q(Z_\varepsilon') \geq \ell^n$. It follows from (3.7.3) that $Q(Z_\varepsilon') = Q(Z_\varepsilon)$. Since Z_ε' is integral over Z_0, hence over Z_ε, we conclude that the embedding $Z_\varepsilon' \subset Z_\varepsilon$ induces the normalization map:

Spec $Z_\varepsilon \to$ Spec Z'_ε. Note also that using diagonal modules, it is easy to show that

$$Z_\varepsilon \subset Z'_\varepsilon + \sum_{\alpha\in R} x_\alpha \mathcal{U}_\varepsilon .$$

Conjecture 5.3. $Z'_\varepsilon = Z_\varepsilon$.
This is checked in the A_1 case in § 4.

5.4. Let $\lambda \in H_0$, so that the diagonal representation π_λ of $\mathcal{U}_\varepsilon$ in $\overline{M}(\lambda)$ is irreducible. Then

$$\pi_\lambda^{T_i} = \pi_{r_i.\lambda}, \tag{5.4.1}$$

where $(r_i.\lambda)(K_\beta) = \varepsilon^{-(\beta|\alpha_i)}\lambda(r_i K_\beta)$. This follows from two easy facts: v_λ is a unique up to a constant factor singular vector of π_λ, and $F_i^{\ell-1}v_\lambda$ is a singular vector of $\pi_\lambda^{T_i}$. It follows from (3.8.1), that $\pi_\lambda(T_i z)v_\lambda = \pi_\lambda^{T_i}(z)v_\lambda = \pi_{r_i.\lambda}(z)v_\lambda = \pi_\lambda(z)v_\lambda$ for $z \in Z_1$, and any $\lambda \in H$. Thus,

$$T_i z = z \bmod \sum_{\alpha\in R} x_\alpha Z_\varepsilon \text{ if } z \in Z_1. \tag{5.4.2}$$

Conjecture 5.4 (a) All elements of Z_1 are fixed by the T_i.
(b) $Z_1 = \mathcal{U}_\varepsilon^G$.

5.5. We propose below the following hypothetical picture for the action of G on Spec Z_0, similar to the coadjoint action of a simple algebraic group (see [10] in characteristic 0 and [9] in characteristic p).

Denote by $N(H)$ (resp. $C(H)$) the subgroup of those elements of G that leave H invariant (resp. pointwise fixed, and let $W_0 = N(H)/C(H)$.

Conjecture 5.5.1 (a) Closed G–orbits are precisely those which intersect H.
(b) The intersection of a closed G–orbit with H is a W_0–orbit.
(c) W_0 is a finite group.
(d) The restriction homomorphism from Spec Z_0 to H induces an isomorphism of algebras of invariants:

$$Z_0^G \simeq \mathbb{C}[z_1, z_1^{-1}, \dots, z_n, z_n^{-1}]^{W_0}.$$

It is easy to show that the fixed points of G are $\chi \in$ Spec Z_0 such that $\chi(x_\alpha) = 0$ ($\alpha \in R$) and $\chi(z_i)^2 = 1$ ($i = 1, \dots, n$). We call a G–orbit *nilpotent* if its closure contains a fixed point of G.

Conjecture 5.5.2. There are finitely many nilpotent orbits.

Conjecture 5.5.3. Any element of Spec Z_0 can be transformed by G to an element χ such that $\chi(y_\alpha) = 0$, and $\chi(z_\alpha)^2 \neq 1 \Longrightarrow \chi(x_\alpha) = 0$ ($\alpha \in R^+$). In particular every irreducible finite dimensional representation of $\mathcal{U}_\varepsilon$ is triangulizable.

Given χ as in Conjecture 5.5.3, we define χ_s, the *semisimple part* of χ by $\chi_s(z_\alpha) = \chi(z_\alpha)$, $\chi_s(x_\alpha) = 0$ ($\alpha \in R$).

Conjecture 5.5.4. (a) The semisimple part χ_s of an element $\chi \in$ Spec Z_0 is well-defined up to a G-conjugacy.

(b) The closure of the G-orbit of χ contains χ_s.

REFERENCES

[1] Andersen, H.H., Polo, P., Wen K., *Representations of quantum algebras*, preprint.

[2] De Concini, C., Kac, V.G., Procesi, C., *The quantum coadjoint space*, in preparation.

[3] Drinfeld, V.G., *Hopf algebras and quantum Yang-Baxter equation*, Soviet Math. Dokl. **32** (1985), 254–258.

[4] Drinfeld, V.G., *Quantum groups*, Proc. ICM, Berkeley (1986), 798–820.

[5] Jantzen, J.C., *Moduln mit einem höchsten Gewicht*, Lecture Notes in Math 750, Springer Verlag 1979.

[6] Jimbo, M., *A q-difference analogue of $\mathcal{U}(\mathfrak{g})$ and the Yang-Baxter equation*, Lett. Math. Phys. **10** (1985), 63–69.

[7] Kac, V.G., *Laplace operators of infinite-dimensional Lie algebras and theta function*, Proc. Nat'l. Acad. Sci. USA **81** (1984), 645-647.

[8] Kac, V.G., Kazhdan, D.A., *Structure of representations with highest weight of infinite-dimensional Lie algebras*, Adv. Math **34** (1979), 97–108.

[9] Kac, V.G., Weisfeiler, B.Yu., *Coadjoint action of a semi-simple algebraic group and the center of the enveloping algebra in characteristic p*, Indag. Math **38** (1986), 136–151.

[10] Kostant, B., *Lie group representation on polynomial rings*, Amer. J. Math **86** (1963), 327–402.

[11] Levendorskii, S.Z., Soibelman, Ya.S., *Some applications of quantum Weyl group I*, preprint.

[12] Lusztig, G., *Quantum deformations of certain simple modules over enveloping algebras*, Adv. in Math. **70** (1988), 237–249.

[13] Lusztig, G., *Quantum groups at roots of* 1, Geom. Ded. (1990).

[14] Lusztig, G., *Finite-dimensional Hopf algebras arising from quantum groups*, J. Amer. Math. Soc. **3** (1990), 257–296.

[15] Nagata, M., *Local rings*, Interscience publ., 1962.

[16] Pierce, R.S., *Associative algebras*, Springer Verlag, 1982.

[17] Ringel, C.M., *Hall algebras and quantum groups*, 1989, preprint.

[18] Rosso, M., *Finite dimensional representations of the quantum analogue of the enveloping algebra of a complex simple Lie algebra*, Comm. Math. Phys. **117** (1988), 581–593.

[19] Rosso, M., *Analogues de la forme de Killing et du théorème d'Harish-Chandra pour les groupes quantiques*, 1989, preprint.

[20] Rudakov, A.N., *On representations of classical Lie algebras in characteristic p*, Izv. AN USSR Ser. matem. **34** (1970), 735–743.

[21] Rudakov, A.N., Shafarevich, I.R., *Irreducible representations of the simple three-dimensional Lie algebra over a field of finite characteristic*, Mat. Zametki **2** (1967), 439–454.

[22] Shapovalov, N.N., *On a bilinear form on the universal enveloping algebra of a complex semisimple Lie algebra*, Funct. Anal. Appl. **6** (1972), 307–312.

[23] Weisfeiler, B. Yu, Kac, V.G., *On irreducible representations of Lie p-algebras*, Funct. Anal. Appl. **5:2** (1971), 28–36.

[24] Zassenhaus, H., *The representations of Lie algebras in prime characteristic*, Proc. Glasgow Math. Assoc. **2** (1954), 1–36.

Received May 23, 1990

C. De Concini
Scuola Normale Superiore
Pisa, Italy

V. G. Kac
Department of Mathematics
MIT
Cambridge, MA 02139, USA

III. INVARIANT THEORY

Action d'un tore dans une variété projective

MICHEL BRION et CLAUDIO PROCESI

A J. Dixmier, pour son soixante–cinquième anniversaire

Introduction

Lorsqu'un groupe algébrique réductif G opère dans une variété algébrique projective X, le tout sur un corps algébriquement clos, la donnée d'un fibré en droites ample G–linéarisé L sur X permet de définir l'ouvert des points stables de X. Le quotient (au sens habituel d'espace d'orbites) de cet ouvert $X^s = X^s(L)$ par G existe ; c'est une variété quasiprojective notée X^s/G. De plus X^s est inclus dans l'ouvert $X^{ss} = X^{ss}(L)$ des points semi–stables, et on peut encore définir un "quotient" Y de X^{ss} par G (c'est l'espace des orbites fermées de G dans X^{ss}). La variété Y est projective, et contient X^s/G comme ouvert. Pour la définir, on introduit l'algèbre $A := \oplus_{n=0}^{\infty} \Gamma(X, L^n)$ et sa sous–algèbre A^G formée des invariants de G ; alors Y est le **Proj** de l'algèbre graduée A^G. Donc Y est munie de faisceaux $\mathcal{O}(n)$ pour tout entier n, et l'un d'eux est inversible. Ainsi on peut définir la classe $1/n[\mathcal{O}(n)]$ dans le groupe de Picard de Y, tensorisé avec $\mathbf{Q}$. Cette classe est ample, et ne dépend que de X et de L.

Notre but est d'étudier ces objets, introduits par Mumford dans [MF], lorsque G est un tore (noté T). L'idée est de considérer simultanément les quotients associés aux fibrés T–linéarisés $L^n \otimes \mathcal{O}(\chi)$ où n est un entier positif, et $\mathcal{O}(\chi)$ est le fibré en droites trivial sur X, le tore opérant dans chaque fibre de $\mathcal{O}(\chi)$ par multiplication par le caractère χ. La notion de point stable ou semi–stable pour $L^n \otimes \mathcal{O}(\chi)$ ne dépend en fait que de χ/n.

A tout quotient p d'un caractère de T par un entier, on peut ainsi associer $X^s(p)$, $X^{ss}(p)$, $Y(p)$ et une classe ample L_p dans $\mathrm{Pic}_{\mathbf{Q}}(Y(p))$; lorsque $p = 0$, on retrouve les notions précédentes.

Nous nous plaçons pour simplifier dans la situation où L est très ample, c'est-à-dire où X est une sous-variété d'un espace projectif $\mathbf{P}(V)$, le tore T opérant linéairement dans V, et où L est la restriction de $\mathcal{O}(1)$ à X. Nous montrons que $X^s(p)$, $X^{ss}(p)$, $Y(p)$ ne dépendent que de la position de p par rapport à un certain ensemble fini Π de caractères de T (l'ensemble des poids de T dans V). Plus précisément, nous définissons (1.1, 1.2) une partition de l'enveloppe convexe $\mathcal{C}$ de Π en "faces". Chaque face est l'intérieur d'un polyèdre convexe, et $X^s(p)$, $X^{ss}(p)$, $Y(p)$ ne dépendent que de la face de p ; pour toute face F, on peut définir $X^s(F)$, $X^{ss}(F)$, $Y(F)$. Nous montrons en outre (1.3) que l'application $p \in F \to L_p \in \mathrm{Pic}_{\mathbf{Q}}(Y(F))$ est affine.

Soit F une face ouverte de $\mathcal{C}$. Nous montrons que tout point semi-stable pour F est stable ; donc si X est lisse, alors $Y(F)$ n'a que des singularités quotients par des groupes abéliens finis. Les quotients associés aux faces ouvertes sont donc relativement simples. Lorsque F est une face quelconque, soit F' une face ouverte dont l'adhérence contient F. Nous définissons en 1.4 un morphisme $\pi_{F,F'} : Y(F') \to Y(F)$, presque toujours birationnel. En 1.5, nous étudions le cas où F est de codimension un, contenue dans l'adhérence de deux faces ouvertes F_- et F_+. Les morphismes π_{F,F_-} et π_{F,F_+} sont en général assez compliqués ; par contre, si $\hat{Y}$ désigne le produit de $Y(F_-)$ et $Y(F_+)$ au-dessus de $Y(F)$, les morphismes de $\hat{Y}$ vers $Y(F_-)$, $Y(F)$ et $Y(F_+)$ sont des éclatements de sous-variétés, avec le même diviseur exceptionnel (voir 2.3 pour un énoncé précis).

Prolongeons à $\mathcal{C}$ l'application affine $p \to L_p$ de F_+ vers $\mathrm{Pic}_{\mathbf{Q}}(Y_+)$. Relevons chaque L_p à $\hat{Y}$, en une classe L_p^+ ; définissons de même L_p^-. Nous montrons en 2.3 que la différence $L_p^+ - L_p^-$ est un multiple du diviseur exceptionnel de l'éclatement $\hat{Y} \to Y(F)$, et est nulle pour tout $p \in F$. Par suite, lorsqu'on considère le produit $\tilde{Y}$ des quotients associés à toutes les faces, au-dessus de tous les morphismes $\pi_{F,F'}$, les applications $p \to L_p$ se recollent en une application affine par morceaux, et continue, de $\mathcal{C}$ vers $\mathrm{Pic}_{\mathbf{Q}}(\tilde{Y})$.

Dans la troisième partie de ce travail, nous appliquons ces résultats à l'étude asymptotique des modules multigradués ; nous considérons diverses extensions à ces modules des notions de multiplicité et de fonction d'Hilbert-Samuel pour un module gradué. Plus précisément, nous considérons une algèbre de polynômes A, graduée par $\mathbf{N} \times \mathbf{Z}^l$ (la graduation par $\mathbf{N}$ étant définie par le degré des polynômes). Notons T le tore ayant $\mathbf{Z}^l$ comme groupe des caractères. Tout A-module gradué par $\mathbf{N} \times \mathbf{Z}^l$ définit un faisceau T–linéarisé $\mathcal{M}$ sur l'espace projectif $\mathbf{P}(V)$ associé à A. Soit Π l'ensemble des poids de T dans V, c'est-à-dire des opposés des $\mathbf{Z}^l$–degrés des générateurs de A. Pour toute face F associée à Π, soit $\mathcal{M}^{(F)}$ le faisceau formé des invariants de T dans $(\pi_F)_*(\mathcal{M})$, où $\pi_F : X^{ss}(F) \to Y(F)$ est le "quotient". Lorsque M est un A–module de type fini, nous montrons que $\mathcal{M}^{(F)}$ est un faisceau cohérent sur $Y(F)$. Pour tout $p \in F$, notons $\mu_M(p)$ le degré de $\mathcal{M}^{(F)}$ relativement à la classe ample L_p (voir [Kle; I.3]). Nous montrons que la fonction μ_M est polynomiale sur chaque face. En outre, elle intervient comme densité dans diverses expressions intégrales d'invariants asymptotiques de M : des "polynômes d'Hilbert–Samuel généralisés" (3.5), les "polynômes de Joseph" ([Jos]; voir 3.6). Lorsque M est l'anneau des coordonnées homogènes d'une sous-variété projective lisse X de $\mathbf{P}(V)$, la fonction μ_M est continue, et sa transformée de Fourier ne dépend que des points fixes de T dans X, et de leur fibré normal (3.4).

Enfin, dans un appendice, nous exposons brièvement les liens entre nos résultats et des théorèmes dus à Atiyah, Duistermaat–Heckman, Guillemin–Sternberg en géométrie symplectique (voir [Ati], [DH1&2], [GS1 &2]). Il y a en effet d'étroites relations, étudiées dans [Kir], entre l'action algébrique d'un tore complexe dans une sous-variété X d'un espace projectif, et l'action hamiltonienne de son sous-tore compact maximal T_c dans X vue comme variété symplectique (la forme symplectique étant la partie imaginaire d'une forme kählerienne invariante par T_c sur $\mathbf{P}(V)$. Les objets de notre étude s'interprètent en termes de l'application moment J : son image s'identifie au polyèdre convexe $\mathcal{C}$; les faces ouvertes sont formées de valeurs régulières de J; le quotient $Y(p)$ est "l'espace des phases réduit" $J^{-1}(p)/T_c$; la classe de L_p dans $H^2(Y(p), \mathbf{Q})$ provient de la classe de cohomologie de la forme symplectique. De plus, si M est l'anneau des coordonnées homogènes de X, la fonction μ_M est la densité de la mesure image de l'application moment.

Après avoir obtenu les résultats de ce travail, nous avons pris connaissance (en mai 1989) de l'article [GS3] de V. Guillemin et S. Sternberg, qui démontrent des résultats voisins de ceux de notre deuxième partie, dans le cadre des actions hamiltoniennes de tores compacts dans les variétés symplectiques.

1. Quotients par un tore.

1.1. Notations.

Soit T un tore opérant linéairement dans un espace vectoriel V (le corps de base k étant algébriquement clos). Soit X une sous-variété fermée irréductible de l'espace projectif $\mathbf{P}(V)$, stable par T. On suppose pour simplifier que T opère fidèlement dans V et ne contient pas les homothéties, et que de plus X n'est contenue dans aucun sous–espace linéaire de $\mathbf{P}(V)$.

On note $\mathfrak{X}(T)$ le groupe des caractères de T, et E (resp. $E_{\mathbf{Q}}$) l'espace vectoriel $\mathfrak{X}(T)\otimes_{\mathbf{Z}}\mathbf{R}$ (resp. $\mathfrak{X}(T)\otimes_{\mathbf{Z}}\mathbf{Q}$). Soit $V = \oplus_{\chi\in\mathfrak{X}(T)} V_\chi$ la décomposition de V en sous–espaces propres de T. Notons Π l'ensemble des χ tels que $V_\chi \neq \{0\}$, c'est–à–dire l'ensemble des poids de T dans V. Pour tout $x \in \mathbf{P}(V)$, on note $\tilde{x}$ un représentant de x dans V, et $\tilde{x} = \sum v_\chi$ sa décomposition en vecteurs propres de T. Notons $\Pi(x)$ l'ensemble des $\chi \in \Pi$ tels que $v_\chi \neq 0$; on dit que c'est l'ensemble des poids de x.

Pour tout $p \in E$, soit $\overline{F}(p)$ l'intersection des simplexes dont tous les sommets sont dans Π, et qui contiennent p. C'est un polyèdre convexe dans E, à sommets dans $E_{\mathbf{Q}}$. Notons $F(p)$ son intérieur relatif, c'est–à–dire l'intérieur de $\overline{F}(p)$ dans l'espace affine qu'il engendre ; appelons $F(p)$ la *face* de p. D'après le théorème de Carathéodory [Val; Theorem 1.21], tout point de l'enveloppe convexe $\mathcal{C}$ de Π appartient à un simplexe ayant tous ses sommets dans Π. Par suite, les $F(p)_{p\in E}$ forment une partition de $\mathcal{C}$. De plus, toute face est contenue dans l'adhérence d'une face ouverte. On en déduit que toute face de codimension un est contenue dans l'adhérence d'au plus deux faces ouvertes. Tout hyperplan affine engendré par une face de codimension un, est aussi engendré par l points de Π, où l est la dimension de E.

1.2. **Faces et quotients.**

Pour tout $p \in E$, on note $X^{ss}(p)$ (resp. $X^s(p)$) l'ensemble des $x \in X$ tels que l'enveloppe convexe de $\Pi(x)$ contient p (resp. contient p dans son intérieur). Il est clair que ces deux ensembles ne dépendent que de la face de p ; pour toute face F, on peut donc définir $X^{ss}(F)$ et $X^s(F)$. L'énoncé suivant est évident.

Proposition.
(i) $X^s(F)$, $X^{ss}(F)$ sont des ouverts de X, stables par T.
(ii) $X^{ss}(F)$ est non vide $\Leftrightarrow$ F est non vide.
(iii) $X^s(F)$ est non vide $\Leftrightarrow$ F est dans l'intérieur de $\mathcal{C}$.
(iv) $X^{ss}(F) = X^s(F)$ si F est ouverte dans E. Réciproquement, si $X = \mathbf{P}(V)$ et $X^{ss}(F) = X^s(F)$, alors F est ouverte dans E.

Un point de $X^s(F)$ (resp. $X^{ss}(F)$) est dit *stable* (resp. *semistable*) pour F. Le théorème ci–dessous, et sa démonstration, justifient cette terminologie.

Théorème. *Il existe une variété projective $Y(F)$, avec action triviale* *de T, et un morphisme affine T–équivariant $\pi_F : X^{ss}(F) \to Y(F)$, tels que $\mathcal{O}_{Y(F)}$ soit formé des invariants de T dans $(\pi_F)_*(\mathcal{O}_{X^{ss}(F)})$. La restriction de π_F à $X^s(F)$ a pour fibres les orbites de T.*

Démonstration. Remarquons d'abord que si $0 \in \mathcal{C}$, alors $X^{ss}(0)$ (resp. $X^s(0)$) est l'ensemble des points semi–stables (resp. stables) de $X \subset \mathbf{P}(V)$; cela résulte en effet de [MF;Theorem 2.1].

Soit p un point de $F \cap E_{\mathbf{Q}}$ (un tel point existe car $\overline{F}$ est un polyèdre convexe à sommets dans $E_{\mathbf{Q}}$). Ecrivons $p = \chi/n$ où n est le plus petit entier positif tel que $np \in \mathfrak{X}(T)$. Définissons une action linéaire de T dans la n–ième puissance symétrique S^nV par: $t \cdot m^n = \chi(t)^{-1}(t \cdot m)^n$. Plongeons X dans $\mathbf{P}(S^nV)$ par le plongement de Veronese. On vérifie immédiatement que $X^{ss}(F) = X^{ss}(p)$ est l'ensemble des points semi–stables de $X \subset \mathbf{P}(S^nV)$, et que $X^s(F)$ est l'ensemble des points stables. L'énoncé résulte alors de [MF; 1.4].

Remarque. Dans le langage de [MF; Chapitre 1], que nous adopterons, π_F est un quotient catégorique universel, et sa restriction à $X^s(F)$ un quotient géométrique universel.

1.3. **Une classe de faisceaux sur les quotients.**

Le fibré en droites $\mathcal{O}(1)$ sur $\mathbf{P}(V)$ se restreint à X en un fibré T–linéarisé, noté encore $\mathcal{O}(1)$. Pour tout caractère χ de T, notons $\mathcal{O}(\chi)$ le fibré en droites T–linéarisé suivant : c'est le fibré en droites trivial sur X, et T opère dans chaque fibre via le caractère $-\chi$. L'application $\chi \to \mathcal{O}(\chi)$ est un homomorphisme de groupes, de $\mathfrak{X}(T)$ vers $\mathrm{Pic}^T(X)$ (le groupe des classes d'isomorphisme des fibrés en droites T–linéarisés sur X). On étend cet homomorphisme de $E_{\mathbf{Q}}$ vers $\mathrm{Pic}^T(X) \otimes_{\mathbf{Z}} \mathbf{Q}$, en $p \to \mathcal{O}(p)$.

Reformulons la démonstration du théorème 1.2 dans le langage des fibrés en droites. Soit $A = \oplus_{n=0}^{\infty}\Gamma(X, \mathcal{O}(n))$; c'est une algèbre graduée dans laquelle T opère. D'après [MF; 1.11], on a $Y(0) = \mathbf{Proj}(A^T)$ où A^T est l'algèbre des invariants de T dans A. Donc $Y(0)$ est muni canoniquement de faisceaux notés $\mathcal{O}(n)$ pour chaque entier n. D'après [EGA II; 8.14.4], il existe un entier $n > 0$ tel que $\mathcal{O}(n)$ soit inversible et très ample. Soit $[\mathcal{O}(n)]$ sa classe dans $\mathrm{Pic}(Y(0))$. Posons $[\mathcal{O}(1)] := 1/n[\mathcal{O}(n)]$; c'est un élément de $\mathrm{Pic}_{\mathbf{Q}}(Y(0)) := \mathrm{Pic}(Y(0)) \otimes_{\mathbf{Z}} \mathbf{Q}$. D'après [EGA II; 8.14.12], cette classe ne dépend pas du choix de n (voir [Dem] pour plus de détails sur cette construction). Par définition, $[\mathcal{O}(1)]$ est la classe de $(\pi_0)_*^T\mathcal{O}(1)$ dans $\mathrm{Pic}_{\mathbf{Q}}(Y(0))$ (image directe invariante de $\mathcal{O}(1)$).

Plus généralement, soit $p = \chi/n$ un point rationnel de $\mathcal{C}$. Remarquons que $X^{ss}(p)$ est l'ensemble des points semistables de X associés au fibré

T–linéarisé $\mathcal{O}(n) \otimes \mathcal{O}(\chi)$. Par suite, $Y(p)$ est le **Proj** de l'algèbre graduée

$$\oplus_{m=0}^{\infty}\Gamma(X, \mathcal{O}(mn) \otimes \mathcal{O}(m\chi))^T = \oplus_{m=0}^{\infty}\Gamma(X, \mathcal{O}(mn))_{m\chi}$$

(on rappelle que pour tout T–module rationnel W, et tout caractère ψ de T, on note W_ψ le sous–espace des vecteurs propres de T dans W, de poids ψ). On a donc sur $Y(p)$ un faisceau canonique $\mathcal{O}(1)$, qui dépend de χ et de n. Posons $[\mathcal{O}_p(1)] = 1/n[\mathcal{O}(1)]$ dans $\mathrm{Pic}_{\mathbf{Q}}(Y(F))$; cette classe ne dépend que de p.

Proposition. *Soit p un point rationnel d'une face F. Alors $[\mathcal{O}_p(1)]$ est la classe de $(\pi_F)_*^T(\mathcal{O}(1) \otimes \mathcal{O}(p))$ dans $\mathrm{Pic}_{\mathbf{Q}}(Y(F))$.*

Démonstration. Par définition, $\mathcal{O}_{\chi,n}(1)$ est l'image directe invariante du faisceau canonique sur le **Proj** de $\oplus_{m=0}^{\infty}\Gamma(X, \mathcal{O}(mn) \otimes \mathcal{O}(m\chi))$, c'est–à–dire $\mathcal{O}_{\chi,n}(1) = (\pi_p)_*^T(\mathcal{O}(n) \otimes \mathcal{O}(n\chi))$. L'énoncé s'en déduit aussitôt.

Corollaire. *Pour toute face F, l'application $p \to [\mathcal{O}_p(1)]$ est la restriction d'une application affine.*

Démonstration. Soit χ/n un point rationnel de l'espace affine engendré par F. On vérifie immédiatement que la restriction de $\mathcal{O}(n) \otimes \mathcal{O}(\chi)$ à toute orbite de T dans $X^{ss}(F)$ est le T–fibré en droites trivial. D'après [Kra2; Proposition 3], il existe un fibré en droites $\mathcal{L}_{n,\chi}$ sur $Y(F)$ tel que $\mathcal{O}(n) \otimes \mathcal{O}(\chi) = \pi_F^*\mathcal{L}_{n,\chi}$. Par la formule de projection, on a $\mathcal{L}_{n,\chi} = (\pi_F)_*^T(\mathcal{O}(n) \otimes \mathcal{O}(\chi))$, donc $\mathcal{L}_{n,\chi}$ est unique. Puisque l'application $\chi/n \to \mathcal{O}(1) \otimes \mathcal{O}(\chi/n)$ est affine, il en est de même de $\chi/n \to (1/n)\mathcal{L}_{n,\chi}$. On conclut grâce au fait que $(1/n)\mathcal{L}_{n,\chi} = \mathcal{O}_{\chi/n}(1)$ pour tout $\chi/n \in F$.

1.4. **Relations entre les quotients.**

Soient F, F' deux faces distinctes telles que $F \subset \overline{F'}$.

Théorème.
(i) $X^s(F) \subset X^s(F') \subset X^{ss}(F') \subset X^{ss}(F)$.
(ii) Il existe un morphisme $\pi_{F,F'}$ faisant commuter le diagramme

$$\begin{array}{ccc} X^{ss}(F') & \longrightarrow & X^{ss}(F) \\ \pi_{F'}\downarrow & & \pi_F\downarrow \\ Y(F') & \xrightarrow{\pi_{F,F'}} & Y(F) \end{array}$$

(iii) $\pi_{F,F'}$ est birationnel si F n'est pas dans le bord de $\mathcal{C}$.

Démonstration.

(i) est une vérification directe.

(ii) D'après [MF; 1.11]et la démonstration du théorème 1.2, $\pi_F : X^{ss}(F) \to Y(F)$ est le quotient par la relation d'équivalence: $x \sim y \Leftrightarrow$ les adhérences des T–orbites de x et y se rencontrent. Par suite, l'inclusion de $X^{ss}(F')$ dans $X^{ss}(F)$ passe au quotient en $\pi_{F,F'} : Y(F') \to Y(F)$.

(iii) Si F n'est pas contenue dans le bord de $\mathcal{C}$, alors $X^s(F)$ n'est pas vide d'après 1.2 ; de plus les restrictions de π_F et $\pi_{F'}$ à $X^s(F)$ coïncident, car $X^s(F) \subset X^s(F')$. Donc $\pi_{F,F'}$ induit un isomorphisme de $\pi_{F'}(X^s(F))$sur son image.

Remarque. Supposons X lisse. Pour toute face ouverte F', la variété $Y(F')$ n'a que des singularités quotients par des groupes finis cycliques. Par suite, si F est une face contenue dans $\overline{F'}$ et non dans le bord de $\mathcal{C}$, le morphisme $\pi_{F,F'}$ est une désingularisation partielle de $Y(F)$. A l'aide du théorème du slice étale [Lun; III.1], on peut montrer que $Y(F)$ n'a que des singularités quotients par des groupes diagonalisables, de dimension au plus la codimension de F dans V.

Etudions maintenant le quotient associé à une face F incluse dans le bord de $\mathcal{C}$. Soit $\langle F\rangle$ l'espace affine engendré par F. On peut choisir un sous–groupe à un paramètre λ de $\mathcal{C}$ (c'est-à-dire un élément du réseau dual de $\mathfrak{X}(T)$) tel que λ est constant sur F, égal à $a \in \mathbf{Z}$, et $\langle \lambda, p\rangle > a$ pour tout $p \in \mathcal{C} \setminus \langle F\rangle$. Soit X^F l'ensemble des points fixes de λ dans $X^{ss}(F)$, c'est-à-dire l'ensemble des $x \in X$ tels que l'enveloppe convexe de $\Pi(x)$ contienne F, et soit contenue dans $\langle F\rangle$. C'est un ouvert de l'ensemble X^λ des points fixes de λ dans X.

Pour tout $x \in X$, le morphisme $t \longrightarrow \lambda(t)x$ de k^* vers la variété complète X, se prolonge en un morphisme de k vers X. On note $\lim_{t\to 0} \lambda(t)x$ l'image de 0 par ce morphisme.

Proposition.

(i) $X^{ss}(F)$ *est l'ensemble des* $x \in X$ *tels que* $\lim_{t\to 0} \lambda(t)x \in X^F$.

(ii) La restriction r *de* π_F *à* X^F *est un quotient géométrique universel, et* π_F *se factorise en* $q : \begin{array}{ccc} X^{ss}(F) & \longrightarrow & X^F \\ x & \longrightarrow & \lim_{t\to 0} \lambda(t)x \end{array}$ *suivi de* r.

(iii) Lorsque X *est lisse,* X^F *est aussi lisse, et* q *est une fibration localement triviale en espaces affines.*

Démonstration.

(i) Soient $x \in X$ et $y = \lim_{t\to 0} \lambda(t)x$. Par définition de λ, on a:

$F \subset$ enveloppe convexe de $\Pi(x) \Leftrightarrow F \subset$ enveloppe convexe de $\Pi(y)$.
(ii) X^λ est une sous-variété fermée T-stable de X, et $\Pi(X^\lambda) \subset \langle F\rangle$, donc F est ouvert dans l'enveloppe convexe de $\Pi(X^\lambda)$. De plus $X^F = X^\lambda(F)$. La première assertion suit donc de 1.2. La deuxième est évidente.
(iii) résulte de [BB; Theorem 4.4].

Corollaire. *On suppose X lisse. Soit $F' \neq F$ une face dont l'adhérence contient F. Alors $\pi_{F,F'}$ n'est pas birationnel.*

Démonstration. En considérant l'intersection de $X^{ss}(F)$ et d'une fibre de π_F, on se ramène au cas où F et X^F sont des points, et où $X^{ss}(F)$ est un T-module. Alors $Y(F)$ est un point, mais on vérifie sans peine que $Y(F')$ a une dimension positive.

1.5. **Cas des faces de codimension un.**

Soit F une face de codimension 1. Supposons d'abord F incluse dans le bord de $\mathcal{C}$. Alors F est contenue dans l'adhérence d'une unique face ouverte F_+. Choisissons comme précédemment un sous-groupe à un paramètre λ tel que l'équation de $\langle F\rangle$ soit $\langle \lambda, p\rangle = a$, et que $\langle \lambda, p\rangle > a$ pour tout $p \in F_+$; notons X^F l'ensemble des points fixes de λ dans $X^{ss}(F)$. L'énoncé suivant est immédiat.

Proposition.
(i) $X^{ss}(F_+) = X^{ss}(F)\backslash X^F$.
(ii) π_{F,F_+} est une fibration localement triviale pour la topologie étale, avec pour fibres les quotients par λ des $\{x \in X^{ss}(F)\backslash X^F \mid \lim_{t\to 0} \lambda(t)x = y\}$ ($y \in X^F$). Lorsque X est lisse, les fibres de π_{F,F_+} sont des espaces projectifs avec poids.

(Rappelons qu'un *espace projectif avec poids* est une variété de la forme $(V\backslash\{O\})/k^*$, où k^* opère linéairement dans l'espace vectoriel V, avec tous ses poids positifs).

Considérons maintenant une face de codimension 1, rencontrant l'intérieur de $\mathcal{C}$. Soient F_-, F_+ les faces ouvertes dont l'adhérence contient F. Soit λ un sous-groupe à un paramètre de T tel que $\langle \lambda, p\rangle = a$ sur F, et $\langle \lambda, p\rangle > a$ sur F_+. Soit X^F comme précédemment. Nous posons $X_- = \{x \in X \mid \lim_{t\to\infty} \lambda(t)x \in X^F\}$; $X_+ = \{x \in X \mid \lim_{t\to 0} \lambda(t)x \in X^F\}$; $\dot{X}_+ = X_+\backslash X^F$; $\dot{X}_- = X_-\backslash X^F$. Remarquons que $\dot{X}_-$ est inclus dans $X^{ss}(F_-)$; notons Y_- son image dans $Y(F_-)$. Notons enfin Y^F l'image de X^F dans $Y(F)$.

Théorème.
(i) $X^s(F_-)\backslash X^s(F) = \dot{X}_-$ *et* $X^{ss}(F)\backslash X^{ss}(F_-) = X_+$.
(ii) $\pi_{F_-,F} : Y(F_-) \to Y(F)$ *induit un isomorphisme de* $Y(F_-)\backslash Y_-$ *sur* $Y(F)\backslash Y^F$.
(iii) La restriction de $\pi_{F_-,F}$ *à* Y_- *fait commuter le diagramme :*

$$\begin{array}{ccc} x & \longrightarrow & \lim_{t\to\infty} \lambda(t)x \\ \dot{X}_- & \longrightarrow & X^F \\ \pi_{F_-}\big\downarrow & & \pi_F\big\downarrow \\ Y_- & \xrightarrow{\pi_{F_-,F}} & Y^F \end{array}$$

C'est une fibration localement triviale pour la topologie étale, avec pour fibres les quotients par λ *de* $\{x \in \dot{X}_- \mid \lim_{t\to\infty} \lambda(t)x = y\}$ $(y \in X^F)$.

Démonstration.
(i) se vérifie immédiatement.
(ii) D'après (i), on a : $X^s(F_-)\backslash \dot{X}_- = X^s(F) \subset X^s(F_-)$, donc π_{F_-} et π_F coïncident sur $X^s(F_-)\backslash \dot{X}_-$.
(iii) Remarquons que X_- est un fermé T-stable de $X^{ss}(F)$, et que $X_- \cap X^{ss}(F_-) = \dot{X}_-$. En outre, la restriction de π_{F_-} à X_- se factorise en l'application $x \to \lim_{t\to\infty} \lambda(t)x$, suivie du quotient de X^F par T. L'assertion en résulte aussitôt.

On conclut que $Y(F_+)$ s'obtient à partir de $Y(F_-)$ en remplaçant Y_- par Y_+. L'objectif de la seconde partie est de préciser ces résultats.

2. Structure des morphismes entre certains quotients.

2.1. Action linéaire d'un tore de dimension un.

Considérons une action linéaire de $T = k^*$ dans un k-espace vectoriel N de dimension finie. Décomposons N en $N_+ \oplus N_0 \oplus N_-$ où N_+ (resp. N_0, N_-) est la somme des sous-espaces propres associés aux caractères positifs (resp. nuls, négatifs) de T. On suppose pour simplifier que $N_0 = \{0\}$, c'est-à-dire que T opère dans N avec pour seul point fixe l'origine. Soit $Z = N//T$ le quotient (au sens de Mumford) de N par T : si $k[N]$ est l'algèbre des fonctions polynomiales sur N, et $k[N]^T$ la sous-algèbre formée des fonctions invariantes par T, alors $Z =$**Spec** $k[N]^T$. C'est une variété affine, en général singulière en 0 (point associé à l'idéal maximal homogène de $k[N]^T$). En-dehors de 0, elle n'a que des singularités quotients par des groupes finis cycliques.

Notons $\dot{N}_+ = N_+ \backslash \{0\}$, et $\mathbf{P}(N_+)$ le quotient de $\dot{N}_+$ par T ; c'est un espace projectif avec poids. Soit Z_+ le quotient de $\dot{N}_+ \times N_-$ par l'action de T. La première projection induit un morphisme $p_+ : Z_+ \to \mathbf{P}(N_+)$ qui fait de Z_+ le quotient d'un fibré vectoriel sur $\mathbf{P}(N_+)$, par un groupe fini cyclique.

L'inclusion de $\dot{N}_+ \times N_-$ dans N induit un morphisme $f_+ : Z_+ \to Z$. On voit sans mal que f_+ contracte la section nulle $(\dot{N}_+ \times \{0\})/T$ sur 0, et se restreint en un isomorphisme sur le complémentaire de la section nulle. Définissons de même $\dot{N}_-$, $\mathbf{P}(N_-)$, Z_-, p_- et f_-. Soit $\hat{Z}$ le produit fibré $Z_+ \times_Z Z_-$. On a le diagramme

$$\begin{array}{ccccc}
\hat{Z} & \xrightarrow{\phi_-} & Z_- & \xrightarrow{p_-} & \mathbf{P}(N_-) \\
{\scriptstyle \phi_+}\downarrow & \overset{\psi}{\searrow} & {\scriptstyle f_-}\downarrow & & \\
Z_+ & \xrightarrow{f_+} & Z & & \\
{\scriptstyle p_+}\downarrow & & & & \\
\mathbf{P}(N_+) & & & &
\end{array}$$

L'algèbre $k[N_+]$ est graduée (par le poids par rapport à T), et $\mathbf{P}(N_+)$ est le **Proj** de $k[N_+]$. Par suite, $\mathbf{P}(N_+)$ est muni canoniquement de faisceaux qu'on notera $\mathcal{O}_+(n)$.

Proposition. *On a*

$$\hat{Z} = \mathbf{Spec} \bigoplus_{n=0}^{\infty} \mathcal{O}_+(n) \otimes \mathcal{O}_-(n)$$

(faisceau d'algèbres graduées sur $\mathbf{P}(N_+) \times \mathbf{P}(N_-)$), et

$$Z_+ = \mathbf{Spec} \bigoplus_{n=0}^{\infty} \mathcal{O}_+(n) \otimes k[N_-]_n$$

où $k[N_-]_n$ désigne les fonctions polynomiales sur N_-, de poids $-n$ par rapport à T. Le diagramme

$$\begin{array}{ccc}
\hat{Z} & \xrightarrow{\phi_-} & Z_- \\
{\scriptstyle \phi_+}\downarrow & \overset{\psi}{\searrow} & {\scriptstyle f_-}\downarrow \\
Z_+ & \xrightarrow{f_+} & Z
\end{array}$$

est dual de

$$\begin{array}{ccc} \bigoplus_{n=0}^{\infty} \mathcal{O}_+(n) \otimes \mathcal{O}_-(n) & \longleftarrow & \bigoplus_{n=0}^{\infty} k[N_+]_n \otimes \mathcal{O}_-(n) \\ \uparrow & & \uparrow \\ \bigoplus_{n=0}^{\infty} \mathcal{O}_+(n) \otimes k[N_-]_n & \longleftarrow & \bigoplus_{n=0}^{\infty} k[N_+]_n \otimes k[N_-]_n \end{array}$$

Démonstration. Il suffit de prouver que $Z_+ = \mathbf{Spec} \bigoplus_{n=0^\infty} \mathcal{O}_+(n) \otimes k[N_-]_n$; l'assertion sur $\mathcal{O}_{\hat{Z}}$ en résulte, car $\mathcal{O}_{\hat{Z}} = \mathcal{O}_{Z_+} \otimes_{\mathcal{O}_Z} \mathcal{O}_{Z_-}$. Soit f une fonction polynomiale de poids p sur N_+. Posons $Z_{+f} = p_+^{-1}(\mathbf{P}(N_+)_f)$; c'est un ouvert affine de Z_+. On a

$$k[Z_{+f}] = k[N_{+f} \times N_-]^T = \bigoplus_{n=0}^{\infty} (k[N_+][1/f])_n \otimes k[N_-]_n$$

D'autre part,

$$\Gamma(\mathbf{P}(N_+)_f, \bigoplus_{n=0}^{\infty} \mathcal{O}_+(n) \otimes k[N_-]_n) = \bigoplus_{n=0}^{\infty} \Gamma(\mathbf{P}(N_+)_f, \mathcal{O}_+(n)) \otimes k[N_-]_n$$

$$= \bigoplus_{n=0}^{\infty} (k[N_+][1/f]_n \otimes k[N_-]_n$$

Définition. Soit X une variété avec action de k^*, telle que la graduation associée de $\mathcal{O}_X$ soit en degrés positifs ou nuls: $\mathcal{O}_X = \oplus_{n=0}^{\infty} \mathcal{O}_{X,n}$. Soit Y la sous-variété de X telle que $\mathcal{O}_Y = \mathcal{O}_{X,0}$. *L'éclatement de* Y *dans* X est le **Proj** de l'Algèbre graduée $\oplus_{n=0}^{\infty} \mathcal{A}_n$ où $\mathcal{A}_n = \oplus_{m \geq n} \mathcal{O}_{X,m}$. Son diviseur exceptionnel (considéré ici comme variété réduite) est le **Proj** du faisceau d'algèbres graduées $\mathcal{O}_X$.

Corollaire.
(i) f_+ *et* f_- *sont propres.*
(ii) ϕ_+, ϕ_-, ψ *sont des éclatements (resp. des sections nulles de* p_+ *et* p_-, *et de* 0*), avec pour diviseur exceptionnel* $\mathbf{P}(N_+) \times \mathbf{P}(N_-)$.

Démonstration. Montrons que ψ est l'éclatement de 0 dans X, où la graduation de $k[X]$ est définie par $k[X]_n = k[N_+]_n \otimes k[N_-]_n$ (les assertions sur ϕ_+ et ϕ_- se démontrent de façon analogue). Cela résulte du lemme suivant [Fle; 2.5]:

Lemme. *Soit* $B = \oplus_{n=0}^{\infty} B_n$ *une algèbre graduée de type fini, avec* $B_0 = k$. *L'éclatement de* 0 *dans* B *est le spectre du faisceau d'algèbres* $\oplus_{n=0}^{\infty} \mathcal{O}_{\mathbf{P}(B)}(n)$ *où* $\mathbf{P}(B)$ *est le* **Proj** *de* B.

2.2. Comparaison de faisceaux sur les quotients.

On conserve les notations précédentes. Soit $\pi_+ : \dot{N}_+ \times N_- \to Z_+$ le quotient par T. Comme en 1.2, notons $\mathcal{O}(\chi)$ le fibré en droites T-linéarisé sur N associé au caractère $-\chi$ de T. Posons $\mathcal{O}_\chi^+(1) = (\pi_+)_*^T(\mathcal{O}(\chi))$; définissons de même $\mathcal{O}_\chi^-(1)$. On a d'autre part un faisceau $\mathcal{O}(1)$ sur $\hat{Z}$, associé au $\mathcal{O}_{\hat{Z}}$-Module $\bigoplus_{n=0}^\infty \mathcal{O}_+(n) \otimes \mathcal{O}_-(n)$. On peut encore voir $\mathcal{O}(1)$ comme le faisceau associé au diviseur (de Weil) $\psi^{-1}(0)$ sur $\hat{Z}$, où $\psi^{-1}(0)$ est la fibre ensembliste de ψ en 0. Pour tout $\chi \in \mathbf{Z}$, on a donc la classe de $\chi.\mathcal{O}(1)$ dans $\mathrm{Pic}_{\mathbf{Q}}(\hat{Z})$.

Théorème. *Pour tout $\chi \in \mathbf{Z}$, on a*

$$\phi_+^* \mathcal{O}_\chi^+(1) - \phi_-^* \mathcal{O}_\chi^-(1) = \chi.\mathcal{O}(1) = \chi.\psi^{-1}(0)$$

dans $\mathrm{Pic}_{\mathbf{Q}}(\hat{Z})$.

Démonstration. Soit f une fonction polynomiale, vecteur propre de T de poids p, sur N_-. Avec les notations de la preuve du théorème 2.1, on a:

$$\Gamma(Z_{+f}, \mathcal{O}_\chi^+(1)) = k[N_{+f} \times N_-]_\chi = \bigoplus_{n=0}^{\infty} (k[N_+][1/f]_{n+\chi} \otimes k[N_-]_n$$

Autrement dit, $\mathcal{O}_\chi^+(1)$ est le Module $\bigoplus_{n=0}^\infty \mathcal{O}_+(n+\chi) \otimes k[N_-]_n$ sur $\mathcal{O}_{Z_+}$. De même, $\mathcal{O}_\chi^-(1)$ est le Module $\bigoplus_{n=0}^\infty k[N_+]_n \otimes \mathcal{O}_-(n-\chi)$ sur $\mathcal{O}_{Z_-}$. Par suite, $\phi_+^* \mathcal{O}_\chi^+(1) - \phi_-^* \mathcal{O}_\chi^-(1)$ est le Module $\bigoplus_{n=0}^\infty \mathcal{O}_+(n+\chi) \otimes \mathcal{O}_-(n+\chi)$ sur $\bigoplus_{n=0}^\infty \mathcal{O}_+(n) \otimes \mathcal{O}_-(n)$.

Remarque. Ces résultats sont encore valables dans un cadre un peu plus général. Soient en effet deux k-algèbres de type fini A^+ et A^-, graduées en degrés positifs, telles que $A_0^+ = A_0^- = k$. Le *produit de Segre* [GW; Chapter 4] de A^+ et A^- est la k-algèbre graduée $A = \bigoplus_{n=0}^\infty A_n^+ \otimes A_n^-$. Soient X_+ (resp. $\dot{X}_+$) le spectre de A^+ (resp. le spectre de A^+, privé de l'idéal maximal homogène); le groupe multiplicatif T opère dans X_+ et $\dot{X}_+$. Notons $\pi_+ : \dot{X}_+ \times X_- \to Z_+$ le quotient par T, et $\mathcal{O}_\chi^+(1) = (\pi_+)_*^T(\mathcal{O}(\chi)$ comme ci-dessus. On a un morphisme $f_+ : Z_+ \to Z := \mathbf{Spec}(A)$. Posons $\hat{Z} = Z_+ \times_Z Z_-$. Les énoncés qui précèdent se généralisent sans difficulté en la

Proposition.
(i) Dans le diagramme

$$\begin{array}{ccc} \hat{Z} & \xrightarrow{\phi_-} & Z_- \\ {\scriptstyle \phi_+}\Big\downarrow & \overset{\psi}{\searrow} & {\scriptstyle f_-}\Big\downarrow \\ Z_+ & \xrightarrow{f_+} & Z \end{array}$$

f_+ et f_- sont birationnelles propres ; ψ (resp. ϕ_+, ϕ_-) est l'éclatement de 0 dans Z (resp. de $\mathbf{Proj}(A_+) = \dot{X}_+/k^*$ *dans Z_+; de* $\mathbf{Proj}(A_-) = \dot{X}_-/k^*$ *dans Z_-).*
(ii) Pour tout $\chi \in \mathbf{Z}$, on a

$$\phi_+^* \mathcal{O}_\chi^+(1) - \phi_-^* \mathcal{O}_\chi^-(1) = \chi.\psi^{-1}(0)$$

dans $Pic_{\mathbf{Q}}(\hat{Z})$.

Dans le cas étudié précédemment, on prend $A_\pm = k[N_\pm]$, avec la graduation induite par l'action de T dans N, et non la graduation comme algèbre de fonctions polynomiales.

2.3. **Quotients associés à deux faces ouvertes contigües.**

Revenons à une situation déjà abordée en 1.5. Soient F_+, F_- deux faces ouvertes contenant dans leur adhérence une même face F de codimension un. Soient λ un sous-groupe à un paramètre de T, et $a \in \mathbf{Z}$, tels que $\lambda > a$ sur F_+, et $\lambda = a$ sur F; on suppose de plus que λ est indivisible dans le groupe dual de $\chi(T)$. Soit $\mathrm{Im}(\lambda)$ son image dans T. Notons pour simplifier $f_+ : Y(F_+) \to Y(F)$ et $f_- : Y(F_-) \to Y(F)$ les morphismes définis en 1.4. Posons $\hat{Y} = Y(F_+) \times_{Y(F)} Y(F_-)$. Soient X^F, $\dot{X}_+$, $\dot{X}_-$ comme en 1.5, et Y^F, Y_+, Y_- leurs quotients respectifs par T.

Pour tout point rationnel p de F_+, on a la classe de $\mathcal{O}_p(1)$ dans $\mathrm{Pic}_{\mathbf{Q}}(Y(F_+))$, qui dépend de façon affine de p. On peut définir cette classe pour tout $p \in E_{\mathbf{Q}}$ par $\mathcal{O}_p^+(1) = (\pi_{F_+})_*^T(\mathcal{O}(1) \otimes \mathcal{O}(p))$; si $p \in F$ on retrouve la définition précédente d'après 1.3. Introduisons de même $\mathcal{O}_p^-(1)$ dans $\mathrm{Pic}_{\mathbf{Q}}(Y(F_-))$, et $\mathcal{O}_p(1)$ dans $\mathrm{Pic}_{\mathbf{Q}}(Y(F))$. Le résultat qui suit permet de comparer ces classes, relevées dans $\mathrm{Pic}_{\mathbf{Q}}(\hat{Y})$.

Théorème.
(i) Dans le diagramme

$$\begin{array}{ccc} \hat{Y} & \xrightarrow{\phi_-} & Y(F_-) \\ \phi_+ \downarrow & \overset{\psi}{\searrow} & f_- \downarrow \\ Y(F_+) & \xrightarrow{f_+} & Y(F) \end{array}$$

ψ, ϕ_+, ϕ_- sont des éclatement (de Y^F, Y_+, Y_- respectivement) avec pour diviseur exceptionnel $Y_+ \times_{Y^F} Y_-$.
(ii) Le diagramme

$$\begin{array}{ccc} \psi^{-1}(Y^F) & \xrightarrow{\phi_-} & Y_- \\ \phi_+ \downarrow & & f_- \downarrow \\ Y_+ & \xrightarrow{f_+} & Y^F \end{array}$$

provient de

$$\begin{array}{cccc} \dot{X}_+ \times_{X^F} \dot{X}_- & \longrightarrow & \dot{X}_- & x \\ \downarrow & & \downarrow & \downarrow \\ \dot{X}_+ & \longrightarrow & X^F & \lim\limits_{t\to\infty} \lambda(t)x \\ x & \longrightarrow & \lim\limits_{t\to 0} \lambda(t)x & \end{array}$$

par passage au quotient géométrique universel.
(iii) Dans $\mathrm{Pic}_{\mathbf{Q}}(\hat{Y})$, on a

$$\phi_+^* \mathcal{O}_p^+(1) - \phi_-^* \mathcal{O}_p^-(1) = (\langle \lambda, p\rangle - a)\psi^{-1}(Y^F)$$

pour tout $p \in E_{\mathbf{Q}}$.
(iv) Les classes $\phi_+^* \mathcal{O}_p^+(1)$, $\phi_-^* \mathcal{O}_p^-(1)$ et $\psi^* \mathcal{O}_p(1)$ coïncident pour tout $p \in E$ tel que $\langle \lambda, p\rangle = a$.

Démonstration. Remarquons d'abord qu'il suffit de prouver le théorème pour $X = \mathbf{P}(V)$. Décomposons V en $N_0 \oplus N_+ \oplus N_-$ où N_0 (resp. N_+, N_-) est la somme des espaces propres V_χ avec $\langle \chi, \lambda\rangle = a$ (resp. $> a$, $< a$). L'ensemble $\mathcal{N} = \{[n_0 + n_+ + n_-] \in \mathbf{P}(V) | \, n_0 \neq 0, [n_0] \in \mathbf{P}(V)^F\}$ est un voisinage ouvert de $\mathbf{P}(V)^F$ dans $\mathbf{P}(V)$, stable par T. Les assertions du théorème étant locales en $\mathbf{P}(V)^F = \mathcal{N}^F$, on peut remplacer X par $\mathcal{N}$.

Posons $\mathcal{N}_+ = \mathcal{N} \cap \mathbf{P}(N_0 \oplus N_+)$, et $\dot{\mathcal{N}}_+ = \mathcal{N}_+ \setminus \mathcal{N}^F$; définissons de même $\mathcal{N}_-$ et $\dot{\mathcal{N}}_-$. Soit $\mathcal{N}//\lambda$ le quotient (au sens de Mumford) de $\mathcal{N}$ par l'action de k^* via λ. Le groupe $T/\mathrm{Im}(\lambda)$ opère dans $\mathcal{N}//\lambda$, et on a un

3. Comportement asymptotique de modules multigradués.

3.1. Modules multigradués et action de tores.

Soit A la k–algèbre des polynômes en les indéterminées $X_0, \ldots, X_r$. Soit $\chi_0, \ldots, \chi_r$ une suite d'éléments de $\mathbf{Z}^l$. On munit A d'une graduation par le monoïde $\mathbf{N} \times \mathbf{Z}^l$ en attribuant à chaque X_i le multidegré $(1, \chi_i)$. Le monôme $X_0^{a_0} \cdots X_r^{a_r}$ sera dit de degré $\sum_{i=0}^r a_i$, et de poids $\sum_{i=0}^r a_i \chi_i$.

Soit $M = \oplus M_{m,\chi}$ un A–module multigradué. Pour tout élément (n, δ) de $\mathbf{N} \times \mathbf{Z}^l$, on note $M(n, \delta)$ le A–module multigradué défini par $M(n, \delta)_{m,\chi} = M_{m+n,\chi+\delta}$. On pose $M(n) := M(n, 0)$.

La proposition suivante est une généralisation immédiate d'un résultat classique sur les modules de type fini sur les anneaux noethériens.

Proposition. *Soit M un A–module multigradué de type fini. Il existe des idéaux premiers multihomogènes $(I_i)_{1 \leq i \leq s}$ de A, et des éléments $(n_i, \delta_i)_{1 \leq i \leq s}$ de $\mathbf{N} \times \mathbf{Z}^l$ tels que M admette une suite de sous–modules multigradués avec sous–quotients isomorphes à $(A/I_i)(n_i, \delta_i)$.*

Soit T le tore ayant $\mathbf{Z}^l$ comme groupe des caractères. Soit V le T–module dont les poids sont $-\chi_0, \ldots, -\chi_r$ (avec leur multiplicité). Le tore T opère dans A, qui s'identifie aux fonctions polynomiales sur V. Un A–module multigradué n'est autre qu'un A–module gradué avec action de T.

Pour toute variété X dans laquelle T opère, on a la notion de faisceau T–linéarisé sur X [MF], qu'on abrégera en T–faisceau. Soit $\mathcal{M}$ un T–faisceau; l'espace de ses sections est alors un T–module rationnel. Tout A–module multigradué M définit un faisceau $\mathcal{M}$ sur $\mathbf{Proj}(A) = \mathbf{P}(V)$, muni d'une T–linéarisation canonique. Réciproquement, tout T–faisceau quasi–cohérent $\mathcal{M}$ définit un A–module multigradué $\Gamma_*(\mathcal{M}) := \oplus_{n=-\infty}^{\infty} \Gamma(\mathbf{P}(V), \mathcal{M} \otimes \mathcal{O}(n))$, dont le T–faisceau associé est naturellement isomorphe à $\mathcal{M}$.

Voici la traduction de la proposition ci–dessus, dans le langage des T–faisceaux.

Proposition (bis). *Soit $\mathcal{M}$ un T–faisceau cohérent sur $\mathbf{P}(V)$. Il existe des sous–variétés irréductibles $X_1, \ldots, X_s$ de $\mathbf{P}(V)$, stables par T, et des éléments $(n_i, \delta_i)_{1 \leq i \leq s}$ de $\mathbf{N} \times \mathbf{Z}^l$ tels que $\mathcal{M}$ admette une suite de T–faisceaux avec sous–quotients isomorphes à $\mathcal{O}_{X_i} \otimes \mathcal{O}(n_i) \otimes \mathcal{O}(\delta_i)$.*

3.2. Décomposition des modules multigradués.

On conserve les notations de 3.1. Pour tout point rationnel p de $E := \mathbf{Z}^l \otimes_{\mathbf{Z}} \mathbf{R}$, posons $A^{(p)} = \oplus_{n=0}^{\infty} A_{n,np}$. C'est une sous–algèbre de A, graduée par $A_n^{(p)} = A_{n,np}$. Pour tout A-module multigradué M, posons $M^{(p)} = \oplus_{n=0}^{\infty} M_{n,np}$; c'est un $A^{(p)}$–module gradué, et M est somme directe des $M^{(p)}$, $p \in E$.

Proposition. *La k–algèbre $A^{(p)}$ est de type fini. De plus, pour tout A–module M, multigradué de type fini, le $A^{(p)}$–module $M^{(p)}$ est de type fini.*

Démonstration. Soit n le dénominateur de p. Notons $k[-p]$ la k–algèbre multigraduée engendrée par une indéterminée de degré $-n$ et de poids $-np$. Alors $M^{(p)}$ s'identifie à l'espace des éléments de multidegré 0 dans $M \otimes k[-p]$. L'énoncé résulte donc de [Kra1;II.3.2].

Le sous–ensemble $\Pi := \{-\chi_0, \dots, -\chi_r\}$ de E détermine une décomposition en faces de son enveloppe convexe. Soient F une de ces faces, et p un point rationnel de F. D'après 1.3, le **Proj** de $A^{(p)}$ s'identifie au quotient $Y(F)$ de $\mathbf{P}(V)^{ss}(F)$ par T ; de plus le faisceau $\mathcal{O}(n)$ sur $\mathbf{Proj}(A^{(p)})$ définit la classe $n[\mathcal{O}_p(1)]$ dans $\mathrm{Pic}_{\mathbf{Q}}(Y(F))$.

Notons $K(Y(F))$ le groupe de Grothendieck des faisceaux cohérents sur $Y(F)$. Le groupe $\mathrm{Pic}(Y(F))$ opère dans $K(Y(F))$ par produit tensoriel, d'où un module $K_{\mathbf{Q}}(Y(F))$ sur $\mathrm{Pic}_{\mathbf{Q}}(Y(F))$. Puisque $\pi_F : \mathbf{P}(V)^{ss}(F) \to Y(F)$ est affine, et que $(\pi_F)_*^T$ transforme tout T–faisceau cohérent en un faisceau cohérent [Kra1; loc.cit.], on a une application $(\pi_F)_*^T$ du groupe de Grothendieck $K^T(\mathbf{P}(V)^{ss}(F))$ des T–faisceaux cohérents sur $\mathbf{P}(V)^{ss}(F)$, vers $K(Y(F))$.

Théorème. *Soit M un A–module multigradué, définissant le T–faisceau $\mathcal{M}$ sur $\mathbf{P}(V)$. Pour tout point rationnel p de la face F, soit $\mathcal{M}^{(p)}$ le faisceau sur $Y(F)$ associé au $A^{(p)}$–module $M^{(p)}$. Alors pour tout $n \in \mathbf{Z}$, on a*

$$\mathcal{M}^{(p)}(n) = (\pi_F)_*^T(\mathcal{M} \otimes \mathcal{O}(n) \otimes \mathcal{O}(np)) = (\pi_F)_*^T(\mathcal{M}) \otimes \mathcal{O}_p(1)^{\otimes n}$$

dans $K_{\mathbf{Q}}(Y(F))$.

Démonstration. Soit m un entier tel que $\mathcal{O}_p(1)^{\otimes m}$ soit un fibré en droites très ample sur $Y(F)$; en particulier mp est entier, et $\mathcal{O}_p(1)^{\otimes m}$ est le faisceau $\mathcal{O}(m)$ sur $Y(F) = \mathbf{Proj}(A^{(p)})$. Soit $f \in A_{m,mp} = \Gamma(Y(F), \mathcal{O}(m))$. Notons $Y(F)_f$ l'ouvert de $Y(F)$ où f ne s'annule pas. On a

$$\Gamma(Y(F)_f, (\pi_F)_*^T(\mathcal{M} \otimes \mathcal{O}(nm) \otimes \mathcal{O}(nmp))$$

$$= \Gamma(\mathbf{P}(V)_f, \mathcal{M} \otimes \mathcal{O}(nm) \otimes \mathcal{O}(nmp)) = (M(nm, nmp)[1/f])_{0,0}$$

$$= \cup_{q\geq 0} f^{-q} M_{(n+q)m,(n+q)mp} = \Gamma(Y(F)_f, \mathcal{M}^{(p)}(nm))$$

D'où $\mathcal{M}^{(p)}(n) = (\pi_F)_*^T(\mathcal{M} \otimes \mathcal{O}(n) \otimes \mathcal{O}(np))$; en particulier $\mathcal{M}^{(p)} = (\pi_F)_*^T(\mathcal{M})$. Comme $\mathcal{O}(n) = \mathcal{O}_p(1)^{\otimes n}$, la seconde égalité en résulte.

Corollaire. *Le faisceau $\mathcal{M}^{(p)}$ ne dépend que de la face de p.*

3.3. Fonction multiplicité d'un module multigradué.

Rappelons d'abord comment on définit la multiplicité d'un module gradué (voir [Smo]). Soit B une k-algèbre de type fini, graduée en degrés positifs et telle que $B_0 = k$ (mais non nécessairement engendrée par B_1). Soit N un B-module gradué de type fini; notons d sa dimension. La limite quand $n \longrightarrow \infty$ de $d!n^{-d}\sum_{m=0}^{n}\dim(N_m)$ existe, et est un nombre rationnel, appelé *multiplicité de* N ; notons-la μ_N. Elle se lit aussi sur la série de Poincaré F_N de N, définie par $F_N(z) = \sum_{m=0}^{\infty} z^m \dim(N_m)$. En effet, cette série converge pour $|z| < 1$, vers une fraction rationnelle, et μ_N est la limite quand $z \longrightarrow 1$ de $(1-z)^d F_N(z)$.

Donnons maintenant une interprétation géométrique de la multiplicité, à l'aide du théorème de Riemann-Roch (voir [Ful]). Soit X le **Proj** de B; soit $\mathcal{N}$ le faisceau cohérent sur X associé au B-module N. Notons $\mathcal{O}(1)$ l'élément de $\mathrm{Pic}_{\mathbf{Q}}(X)$ défini de façon analogue à 1.3. Pour simplifier, notons encore $\mathcal{O}(1)$ sa classe de Chern (à coefficients rationnels). Soient $\tau_X(\mathcal{N})$ la "classe de Todd" de $\mathcal{N}$ (cf. [Ful; Theorem 18.3]), et $\tau_{X,d}$ sa composante de degré d. Alors la multiplicité de N est donnée par

$$\mu_N = \int_X \mathcal{O}(1)^d \cap \tau_{X,d}(\mathcal{N})$$

(cf.[Ful;18.3.6] et aussi [Kle; Chapter I]).

Revenons à la situation de 3.1 et 3.2. Pour tout A-module M, multigradué de type fini, soit $\mathcal{C}(M)$ l'ensemble des p tels que la dimension de $M^{(p)}$ soit maximale, et non nulle ; notons d cette dimension. L'ensemble $\mathcal{C}(M)$ est une réunion de faces d'après le corollaire 3.2. Pour tout $p \in \mathcal{C}(M)$, soit $\mu_M(p)$ la multiplicité du $A^{(p)}$-module $M^{(p)}$. Ce nombre mesure la croissance de M dans la direction de p ; il s'interprète aussi comme le degré du faisceau $\mathcal{M}^{(p)}$ sur $Y(p)$, relativement à la classe ample $\mathcal{O}_p(1)$ (voir [Kle; I.3]).

Proposition. *La fonction $p \longrightarrow \mu_M(p)$ est polynomiale, de degré au plus $d-1$, sur toute face contenue dans $\mathcal{C}(M)$.*

Démonstration. Soit F une telle face. Pour tout point rationnel p

de F, on a

$$\mu_M(p) = \int_{Y(F)} \mathcal{O}_p(1)^{d-1} \cap \tau_{Y(F),d-1}(\mathcal{M}^{(p)})$$

D'après 1.3, la classe $\mathcal{O}_p(1)$ dépend de façon affine de p ; d'autre part, $\mathcal{M}^{(p)}$ est indépendant de p d'après 3.2.

Soit I un idéal premier multihomogène de A. On va étudier la fonction μ_M (notée simplement μ) lorsque $M = A/I$. Notons X le **Proj** de A/I, c'est-à-dire l'ensemble des zéros de I dans $\mathbf{P}(V)$. Le faisceau $\mathcal{M}$ associé à M, n'est autre que $\mathcal{O}_X$. On suppose que les hypothèses simplificatrices de 1.1 sont vérifiées. Soit F une face de codimension 1, contenue dans l'adhérence de deux faces ouvertes F_- et F_+. Soient λ le sous-groupe à un paramètre indivisible de T, et a l'entier tels que $\lambda = a$ sur F, et $\lambda > a$ sur F_+. Définissons comme en 1.2 et 2.3 les variétés $Y(F)$, Y^F, $Y(F_-)$, $Y(F_+)$ et le diagramme

$$\begin{array}{ccc} \hat{Y} & \xrightarrow{\phi_-} & Y(F_-) \\ \phi_+\big\downarrow & \overset{\psi}{\searrow} & f_-\big\downarrow \\ Y(F_+) & \xrightarrow{f_+} & Y(F) \end{array}$$

Lorsque Y^F est vide, les ouverts $X^{ss}(F_-)$, $X^{ss}(F)$ et $X^{ss}(F_+)$ coïncident d'après 1.5. On peut donc identifier $Y(F_-)$, $Y(F)$ et $Y(F_+)$, et l'application $p \to \mathcal{O}_p(1)$ est affine sur $F_- \cup F_+$; de plus, $\mathcal{O}_X^{(p)}$ ne dépend pas de $p \in F_- \cup F_+$. Par suite, μ est polynomiale sur cet ensemble.

On suppose désormais que Y^F est non vide. Notons μ_+ (resp. μ_-) la fonction polynomiale sur E dont la restriction à F_+ (resp. F_-) est μ.

Théorème. *Le polynôme $\mu_+(p) - \mu_-(p)$ est divisible par $(\lambda.p - a)^{c_F}$, où c_F est la codimension de Y^F dans $Y(F)$.*

Démonstration. Choisissons une projection u de E sur $\langle F \rangle$; alors $u(p) \in F$ pour tout p assez voisin de F. D'après le théorème 2.3, la classe de $\phi_+^*(\mathcal{O}_p^+(1)) - \psi^*(\mathcal{O}_{u(p)}(1))$ dans $\mathrm{Pic}_{\mathbf{Q}}(\hat{Y})$ est une fonction affine de p, nulle lorsque $\lambda.p = a$. Il existe donc une classe D_+ telle que $\phi_+^*(\mathcal{O}_p^+(1)) = \psi^*(\mathcal{O}_{u(p)}(1)) + (\lambda.p - a)D_+$. On a de même $\phi_-^*(\mathcal{O}_p^-(1)) = \psi^*(\mathcal{O}_{u(p)}(1)) + (\lambda.p - a)D_-$. Par suite $\mu_+(p) - \mu_-(p)$ est égal à

$$\int_{Y(F_+)} \mathcal{O}_p^+(1)^{d-1} - \int_{Y(F_-)} \mathcal{O}_p^-(1)^{d-1} = \int_{\hat{Y}} \phi_+^*(\mathcal{O}_p^+(1))^{d-1} - \phi_-^*(\mathcal{O}_p^-(1))^{d-1}$$

$$= \int_{\hat{Y}} (\psi^*(\mathcal{O}_{u(p)}(1)) + (\lambda.p - a)D_+)^{d-1} - (\psi^*(\mathcal{O}_{u(p)}(1) + (\lambda.p - a)D_-)^{d-1}$$

C'est donc une combinaison linéaire de termes

$$\int_{\hat{Y}} \psi^*(\mathcal{O}_{u(p)}(1)^\alpha(\lambda.p-a)^\beta(D_+^\beta - D_-^\beta)$$

avec $\beta > 0$ et $\alpha+\beta = d-1$. Or $D_+ - D_- = (\lambda.p-a)^{-1}(\phi_+^*(\mathcal{O}_p^+(1) - \phi_-^*(\mathcal{O}_p^-(1))$ est égal à $\psi^{-1}(Y^F)$ d'après le théorème 2.3. Par suite, $\mu_+(p) - \mu_-(p)$ est une combinaison linéaire de termes

$$(\lambda.p-a)^\beta \int_{\psi^{-1}(Y^F)} \psi^*(\mathcal{O}_{u(p)}(1)|_{Y^F})^\alpha\, M$$

où M est un monôme en D_+ et D_-. Mais $(\mathcal{O}_{u(p)}(1)|_{Y^F})^\alpha$ est nul pour tout $\alpha > \dim(Y^F)$, c'est-à-dire pour $\beta < c_F$.

Corollaire. *La fonction μ est continue sur $\mathcal{C}$.*

3.4. **Transformée de Fourier de la fonction multiplicité.**

Soit W un T-module rationnel de dimension finie. On définit son caractère par $\mathrm{car}(W) = \sum_{\chi\in\mathfrak{X}(T)} \dim(W_\chi)e^\chi$ où les e^χ forment la base canonique de l'algèbre du groupe $\mathfrak{X}(T)$ sur $\mathbf{Z}$. Pour toute forme linéaire λ sur E, on pose $\langle \mathrm{car}(W), \lambda\rangle = \sum_{\chi\in\mathfrak{X}(T)} \dim(W_\chi)\exp\langle\chi,\lambda\rangle$ où exp désigne l'exponentielle ordinaire.

Soit M un A-module multigradué de type fini, de dimension D. Soient d, $\mathcal{C}(M)$ et μ_M comme en 3.3. Notons $\mathcal{F}(\mu_M)$ la *transformée de Fourier* de μ_M, définie par

$$\mathcal{F}(\mu_M)(\lambda) = \int_{\mathcal{C}(M)} \mu_M(p)\exp\langle p,\lambda\rangle dp$$

pour tout $\lambda \in E^*$. Ici dp est la mesure de Lebesgue sur E, normalisée de façon que $E/\mathfrak{X}(T)$ soit de mesure 1.

Proposition. *On a: $D = d + \dim\mathcal{C}(M)$, et*

$$\mathcal{F}(\mu_M)(\lambda) = \lim_{n\to\infty} d!n^{-D} \sum_{m=0}^{n} \langle \mathrm{car}(M_m), \lambda/m\rangle$$

pour tout $\lambda \in E^*$.

Démonstration.

$$\sum_{m=0}^{n} \langle \mathrm{car}(M_m), \lambda/m\rangle = \sum_{0\le m\le n;\chi\in\mathfrak{X}(T)} \dim(M_{m,\chi})\exp\langle\chi,\lambda/m\rangle$$

$$= \sum_{0 \le m \le n; p \in E} \dim(M_{m,mp}) \exp\langle p, \lambda \rangle$$

Or pour p fixé,

$$\sum_{m=0}^{n} \dim(M_{m,mp}) \sim_{n \to \infty} d!^{-1} n^d \mu_M(p)$$

par définition de μ_M. Par un raisonnement analogue à [Bri; 3.2], on en déduit que

$$\sum_{m=0}^{n} \langle \mathrm{car}(M_m), \lambda/m \rangle \sim_{n \to \infty} d!^{-1} n^{d + \dim \mathcal{C}(M)} \int_{\mathcal{C}(M)} \mu_M(p) \exp\langle p, \lambda \rangle dp$$

En particulier, lorsque $\lambda = 0$, on a:

$$\sum_{m=0}^{n} \dim(M_m) \sim_{n \to \infty} d!^{-1} n^{d + \dim \mathcal{C}(M)} \int_{\mathcal{C}(M)} \mu_M(p) dp$$

Par suite, la dimension de M est égale à $d + \dim \mathcal{C}(M)$.

Corollaire. *La multiplicité de M est égale à*

$$d!^{-1} D! \int_{\mathcal{C}(M)} \mu_M(p)\, dp$$

On va déterminer $\mathcal{F}(\mu_M)$ lorsque $M = A/I$ pour un idéal premier multihomogène I, et que $X := \mathbf{Proj}(A/I)$ est lisse. Alors le sous–schéma X^T des points fixes de T dans X est aussi lisse. Pour toute composante connexe Z de X^T, on note d_Z sa dimension; χ_Z le poids de tout élément de Z; α_Z la classe de Chern de la restriction de $\mathcal{O}(1)$ à Z. Soit N_Z le fibré conormal à Z dans X. Toute fibre de N_Z est un T–module rationnel, donc on peut décomposer N_Z en composantes isotypiques $N_{Z,\chi}$. Soient $\alpha_{i,Z}$ les racines de Chern de N_Z, chacune provenant de la composante isotypique de type $\chi_{i,Z}$. Définissons formellement un "fibré conormal divisé" $\mathcal{N}_Z$ par ses racines de Chern $\alpha_{i,Z}/\chi_{i,Z}$. Pour tout entier $j \ge 0$, on note $S^j\mathcal{N}_Z$ sa j–ième puissance symétrique, et $\mathrm{ch}(S^j\mathcal{N}_Z)$ son caractère de Chern ; c'est la j–ième fonction symétrique élémentaire des $\alpha_{i,Z}/\chi_{i,Z}$.

Théorème. *La transformée de Fourier de μ_M est égale à*

$$d! \sum_{Z \subset X^T} \exp\langle \chi_Z, \lambda \rangle \prod_{i=1}^{D-d_Z-1} \langle \chi_{i,Z}, \lambda \rangle^{-1}$$

$$\sum_{j=0}^{d_Z} (-1)^{D-d_Z+j-1} \int_Z (d_Z - j)!^{-1} \alpha_Z^{d_Z-j} \langle \mathrm{ch}(S^j\mathcal{N}_Z), \lambda \rangle$$

Démonstration. Pour tout entier n assez grand, on a: $\mathrm{car}(M_n) = \mathrm{car}\ \Gamma(X, \mathcal{O}(n)) = \mathrm{car}\ \chi(X, \mathcal{O}(n))$ où χ désigne la caractéristique d'Euler. Notons $i : X^T \to X$ l'inclusion. D'après la formule de Lefschetz–Riemann–Roch [Nie], le caractère de $\chi(X, \mathcal{O}(n))$ est égal à

$$\mathrm{car}\ \chi(X^T, i^*\mathcal{O}(n) \otimes (\lambda_{-1} N_{X^T,X})^{-1}) = \sum_{Z \subset X^T} \mathrm{car}\ \chi(Z, \mathcal{O}(n)|_Z \otimes (\lambda_{-1} N_Z)^{-1})$$

où $N_{X^T,X}$ est le fibré conormal à X^T dans X, et $\lambda_{-1}E$ désigne la somme alternée des puissances extérieures du fibré vectoriel E. Par le théorème de Riemann–Roch, car $\chi(Z, \mathcal{O}(n)|_Z \otimes (\lambda_{-1}N_Z)^{-1}$ est égal à

$$\int_Z e^{n\alpha_Z} e^{n\chi_Z} \prod_{i=1}^{D-d_Z-1} (1 - e^{\alpha_{i,Z}} e^{\chi_{i,Z}})^{-1}\, \mathrm{Td}(T_Z)$$

où T_Z est le fibré tangent à Z, et $\mathrm{Td}(T_Z)$ sa classe de Todd. Donc $\langle \mathrm{car}(M_n), \lambda/n \rangle$ est équivalent quand $n \to \infty$ à

$$\sum_{Z \subset X^T} \int_Z \exp\langle \chi_Z, \lambda \rangle\, e^{n\alpha_Z} \prod_{i=1}^{D-d_Z-1} (1 - e^{\alpha_{i,Z}} \exp\langle \chi_{i,Z}, \lambda/n \rangle)^{-1}\, \mathrm{Td}(T_Z)$$

Remarquons que $(1 - e^{\alpha_{i,Z}} \exp\langle \chi_{i,Z}, \lambda/n \rangle)^{-1}$ est le produit de $-(\alpha_{i,Z} + \langle \chi_{i,Z}, \lambda/n \rangle)^{-1}$ par une série entière E_i en $\alpha_{i,Z} + \langle \chi_{i,Z}, \lambda/n \rangle$, dont le premier terme est 1. De plus $\prod_{i=1}^{D-d_Z-1} -(\alpha_{i,Z} + \langle \chi_{i,Z}, \lambda/n \rangle)^{-1}$ est égal à

$$\prod_{i=1}^{D-d_Z-1} \langle \chi_{i,Z}, \lambda \rangle^{-1} (-n)^{D-d_Z-1} \sum_{j=0}^{d_Z} (-n)^j \langle \mathrm{ch}(S^j \mathcal{N}_Z), \lambda \rangle$$

Chaque intégrale sur Z est donc égale à

$$\exp\langle \chi_Z, \lambda \rangle \prod_{i=1}^{D-d_Z-1} \langle \chi_{i,Z}, \lambda \rangle^{-1} (-n)^{D-d_Z-1}$$
$$\sum_{j=0}^{d_Z} (-n)^j \int_Z e^{n\alpha_Z} \langle \mathrm{ch}(S^j \mathcal{N}_Z), \lambda \rangle \left(\prod E_i\right) \mathrm{Td}(T_X)$$

La plus grande puissance de n qui apparaît dedans est donc n^{D-1}, et son coefficient est

$$\exp\langle \chi_Z, \lambda \rangle \prod_{i=1}^{D-d_Z-1} \langle \chi_{i,Z}, \lambda \rangle^{-1}$$
$$\sum_{j=0}^{d_Z} (-1)^{D-d_Z-1+j} \int_Z (d_Z - j)!^{-1} \alpha_Z^{d_Z-j} \langle \mathrm{ch}(S^j \mathcal{N}_Z), \lambda \rangle$$

On conclut grâce à la proposition précédente.

Corollaire. *Lorsque T n'a qu'un nombre fini de points fixes dans X, la transformée de Fourier de μ_M est*

$$d! \sum_{z \in X^T} \exp\langle \chi_z, \lambda \rangle \prod_{i=1}^{D-1} \langle \delta_{i,z}, \lambda \rangle^{-1}$$

où χ_z est le poids de z, et les $\delta_{i,z}$ sont les poids de T dans l'espace tangent en z à X.

3.5. Une généralisation des polynômes d'Hilbert–Samuel

Soit M un A–module multigradué de type fini, de dimension D. Soient φ une fonction polynomiale sur E, et s son degré. Posons

$$H_{M,\varphi}(n) := \sum_{\chi \in \mathfrak{X}(T)} \dim(M_{n,\chi})\, \varphi(\chi)$$

En particulier, $H_{M,1}$ est la fonction de Hilbert–Samuel de M.

Théorème. *La fonction $n \to H_{M,\varphi}(n)$ est polynomiale pour n assez grand ; le degré de ce polynôme est au plus $D+s-1$, et son terme de degré $D+s-1$ est $d!^{-1} \int_{C(M)} \mu_M(p)\, \varphi_s(p)\, dp$ où φ_s est la composante homogène de degré s de φ.*

Démonstration. Lorsque M est égal à A, la dimension de $M_{n,\chi}$ est le nombre de solutions du système $x_0 + \cdots + x_r = n$; $x_0\chi_0 + \cdots + x_r\chi_r = \chi$ où les x_i sont des entiers non négatifs. Par suite

$$H_{A,\varphi}(n) = \sum_{x_0 + \cdots + x_r = n} \varphi(x_0\chi_0 + \cdots + x_r\chi_r)$$

En prenant pour φ un monôme, on voit sans mal que la série génératrice $\sum_{n=0}^{\infty} z^n H_{A,\varphi}(n)$ est une fonction rationnelle, ayant pour dénominateur $(1-z)^{r+s+1}$. On en déduit que $H_{A,\varphi}$ est un polynôme en n pour n assez grand.

Considérons maintenant le cas où $M = A(n, \delta)$ pour un $(n, \delta) \in \mathbf{N} \times \mathfrak{X}(T)$. Alors

$$H_{M,\varphi}(m) = \sum_{\chi \in \mathfrak{X}(T)} \dim(A_{m+n,\chi+\delta})\, \varphi(\chi) = H_{A,\varphi_\delta}(m+n)$$

où $\varphi_\delta(\chi) := \varphi(\chi - \delta)$. Par suite $H_{M,\varphi}(n)$ est encore polynomiale pour n assez grand.

Tout A–module multigradué de type fini admet une résolution libre multigraduée par des sommes directes de modules de la forme $A(n, \delta)$. La première assertion du théorème s'en déduit aussitôt. Pour montrer les deux autres, il suffit de prouver que $n^{-D-s} \sum_{m=0}^{n} H_{M,\varphi}(m)$ a pour limite $d!^{-1} \int_{C(M)} \mu_M(p) \varphi_s(p) dp$ quand $n \to \infty$. Or

$$\sum_{n=0}^{\infty} H_{M,\varphi}(m) = \sum_{0 \leq m \leq n; \chi \in \mathfrak{X}(T)} \dim(M_{m,\chi})\ \varphi(\chi)$$

$$= \sum_{0 \leq m \leq n; p \in E} \dim(M_{m,mp})\ \varphi(mp)$$

est équivalent quand $n \to \infty$ à $d!^{-1} n^{d+s+\dim C(M)} \int_{C(M)} \mu_M(p)\ \varphi_s(p)\ dp$.

Le corollaire suivant précise la proposition 3.4 ; il semble délicat de le prouver directement.

Corollaire. *Pour tout A–module multigradué de type fini M, de dimension D, la transformée de Fourier de μ_M est la limite quand $n \to \infty$ de $d!\ n^{-D+1} \langle \mathrm{car}(M_n), \lambda/n \rangle$.*

Démonstration.

$$\langle \mathrm{car}(M_n), \lambda/n \rangle = \sum_{\chi \in \mathfrak{X}(T)} \dim(M_{n,\chi}) \exp\langle \chi, \lambda/n \rangle$$

$$= \sum_{r=0}^{\infty} \sum_{\chi \in \mathfrak{X}(T)} \dim(M_{n,\chi})\ r!^{-1} n^{-r} \langle \chi, \lambda \rangle^r = \sum_{r=0}^{\infty} r!^{-1} n^{-r}\ H_{M,\varphi_{r,\lambda}}(n)$$

où $\varphi_{r,\lambda}(p) := \langle p, \lambda \rangle^r$. Le r–ième terme de cette série est majoré en valeur absolue par

$$r!^{-1} n^{-r}\ \dim(M_n) \sup |\langle \chi, \lambda \rangle^r|$$

où le sup est pris sur tous les χ tels que $M_{n,\chi} \neq 0$. Soit $(m_1, \ldots, m_p)$ un système générateur de M, formé d'éléments hommogènes de multidegré (d_i, ω_i). Si $M_{n,\chi} \neq 0$, alors il existe i tel que $\chi + \omega_i \in (n + d_i)\mathcal{C}$. Par suite, $\sup |\langle \chi, \lambda \rangle^r| = O(n^r)$; donc la série

$$\sum_{r=0}^{\infty} r!^{-1} n^{-r-D+1}\ H_{M,\varphi_{r,\lambda}}(n) = n^{-D+1} \langle \mathrm{car}(M_n), \lambda/n \rangle$$

converge uniformément par rapport à n.

3.6. Polynômes de Joseph.

Soit B une k–algèbre de type fini, graduée par $\mathbf{Z}^l$, avec $B_0 = k$. On suppose que tous les poids de B sont dans un même demi–espace ouvert. Notons σ l'ensemble des formes linéaires λ sur $\mathbf{R}^l$ telles que $\langle\lambda, \chi\rangle > 0$ pour tout $\chi \in \mathbf{Z}^l$ tel que $B_\chi \neq 0$. C'est l'intérieur d'un cône convexe polyédral rationnel, c'est–à–dire de l'enveloppe convexe d'un nombre fini de demi–droites de $(\mathbf{R}^l)^*$, engendrées par des formes à valeurs entières sur $\mathbf{Z}^l$.

Soit M un B–module gradué de type fini. Pour tout $\lambda \in \sigma$, à valeurs entières sur $\mathbf{Z}^l$, définissons une $\mathbf{Z}$–graduation sur $M = M_{(\lambda)}$ par $(M_{(\lambda)})_n = \oplus_{\langle\chi,\lambda\rangle=n} M_\chi$. Il est clair que $(M_{(\lambda)})_n = 0$ pour tout n assez petit. Notons $f_M(\lambda)$ la multiplicité de $M_{(\lambda)}$. Un résultat dû à Joseph ([Jos]; voir aussi [BBM]) affirme que f_M est la restriction à σ d'une fonction rationnelle sur $\mathbf{R}^l$. De plus, si S désigne la suite des degrés d'un système générateur de B formé d'éléments homogènes, alors $\prod_{\alpha\in S} \alpha$ est un dénominateur de cette fonction rationnelle.

Nous allons lier cette fonction de Joseph aux notions précédentes. Remarquons d'abord qu'on peut prendre pour B une k–algèbre de polynômes. La filtration de B par le degré total induit une filtration sur tout B–module M. Lorsque M est $\mathbf{Z}^l$–gradué de type fini, il est clair que les fonctions de Joseph de M et de son gradué associé sont les mêmes. On peut donc supposer que $B = A$ et que M est gradué par $(\mathbf{N} \times \mathbf{Z}^l)$. Pour tout $(q, \lambda) \in \mathbf{N} \times (\mathbf{Z}^l)^*$, définissons alors une $\mathbf{Z}$–graduation sur $M = M_{(q,\lambda)}$ par $(M_{(q,\lambda)})_m = \oplus_{nq+\langle\chi,\lambda\rangle=m} M_{n,\chi}$. Lorsque $\lambda \in \sigma$, on a $(M_{(q,\lambda)})_n = 0$ pour tout n assez petit ; notons $f_M(q, \lambda)$ la multiplicité de $M_{(q,\lambda)}$. On a $f_M(0, \lambda) = f_M(\lambda)$.

Remarquons que $f_M(q, \lambda)$ est la fonction de Joseph du module $\mathbf{Z}^{l+1}$–gradué M (dont les degrés sont dans le demi–espace ouvert où la première coordonnée est positive). On peut donc supprimer l'hypothèse que les poids de A sont dans un même demi-espace ouvert.

Proposition. *Soit M un A–module multigradué de type fini, de dimension D. Soient d, $\mathcal{C}(M)$ et μ_M comme en 3.3. Alors*

$$f_M(q, \lambda) = D!d!^{-1} \int_{\mathcal{C}(M)} \mu_M(p)\, (q + \langle p, \lambda\rangle)^{-d}\, dp$$

pour tout (q, λ) tel que $\lambda > -q$ sur $\mathcal{C}(M)$.

Démonstration. Le module $M_{(q,\lambda)}$ est de dimension D, donc

$$\sum_{m=0}^{n} \dim(M_{(q,\lambda)})_n = \sum_{0\leq mq+\langle\chi,\lambda\rangle\leq n} \dim(M_{m,\chi})$$

est équivalent quand $n \to \infty$ à $D!^{-1}n^D f_M(q,\lambda)$. D'autre part

$$\sum_{0\leq mq+\langle\chi,\lambda\rangle\leq n} \dim(M_{m,\chi}) = \sum_{0\leq m(q+\langle p,\lambda\rangle)\leq n} \dim(M_{m,mp})$$

est équivalent à $d!^{-1}n^{d+\dim C(M)} \int_{C(M)} \mu_M(p)\,(q+\langle p,\lambda\rangle)^{-d}\,dp$.

Comme en 3.4, déterminons f_M lorsque $M = A/I$ pour un idéal premier multihomogène I, et que $X := \mathbf{Proj}(A/I)$ est lisse. On reprend les notations X^T, Z, α_Z, χ_Z, $\alpha_{i,Z}$, $\chi_{i,Z}$, $\mathcal{N}_Z$ de 3.4.

Théorème. *La fonction rationnelle f_M est égale à*

$$(-1)^D \sum_{Z\subset X^T} (q+\langle\chi_Z,\lambda\rangle)^{-1} \prod_{i=1}^{D-d_Z-1} \langle\chi_{i,Z},\lambda\rangle^{-1}$$
$$\int_Z \langle \operatorname{ch} S^{d_Z}((q+\chi_Z)^{-1}\mathcal{O}(1)|_Z + \mathcal{N}_Z),\lambda\rangle$$

Démonstration. Considérons la série $F(z) := \sum z^n \mathrm{car}(M_n)$. D'après la démonstration du théorème 3.4, on a

$$F(z) = P(z) + \sum_{Z\subset X^T} \int_Z (1-ze^{\alpha_Z}e^{\chi_Z})^{-1} \prod_{i=1}^{D-d_Z-1} (1-e^{\alpha_{i,Z}}e^{\chi_{i,Z}})^{-1}\,\mathrm{Td}(T_Z)$$

où P est un polynôme en z et les e^χ. Notons $F_{q,\lambda}$ la série de Poincaré de $M_{(q,\lambda)}$. Alors $F_{q,\lambda}(z) = \sum_{m,\chi} \dim(M_{m,\chi})\, z^{mq+\langle\chi,\lambda\rangle}$ donc $F_{q,\lambda}(\exp t) = \langle F(\exp tq), t\lambda\rangle$. Par suite $F_{q,\lambda}(\exp t)$ est équivalente quand $t\to 0$ à la sommme des

$$\int_Z (1-e^{\alpha_Z}\exp t(q+\langle\chi_Z,\lambda\rangle))^{-1} \prod_{i=1}^{D-d_Z-1} (1-e^{\alpha_{i,Z}}\exp t\langle\chi_{i,Z},\lambda\rangle)^{-1}\,\mathrm{Td}(T_Z)$$

Le terme sous l'intégrale est le produit de

$$(-1)^{D-d_Z}(\alpha_Z + t(q+\langle\chi_Z,\lambda\rangle))^{-1} \prod_{i=1}^{D-d_Z-1} (\alpha_{i,Z}+t\langle\chi_{i,Z},\lambda\rangle)^{-1}$$

par une série entière en t et les α, dont le premier terme est 1. Comme dans la démonstration du théorème 3.4, on voit donc que $t^D F_{q,\lambda}(\exp t)$ est équivalente quand $t \to 0$ à l'expression cherchée. On conclut grâce au fait rappelé en 3.3 que $\lim_{z\to 1}(1-z)^D F_{q,\lambda}(z)$ est la multiplicité de $M_{(q,\lambda)}$.

Corollaire. *Lorsque T n'a qu'un nombre fini de points fixes dans X, on a*

$$f_M(q,\lambda) = -\sum_{z\in X^T} (q + \langle \chi_z, \lambda\rangle)^{-1} \prod_{i=1}^{D-1} \langle \delta_{i,z}, \lambda\rangle^{-1}$$

où χ_z est le poids de z, et les $\delta_{i,Z}$ sont les poids de T dans l'espace tangent en z à X.

Appendice: liens avec la géométrie symplectique.

A 1. Action hamiltonienne d'un tore compact dans une variété symplectique.

Le but de cette section est de fixer des notations ; nous renvoyons à [DH1&2], [GS1&2], [Kir] pour plus de détails.

Soit X une variété C^∞ compacte connexe. On suppose que X est *symplectique*, c'est-à-dire munie d'une 2-forme différentielle alternée σ, fermée et non dégénérée. Soit T_c un tore compact qui opère fidèlement dans X en préservant σ. L'action de T_c est dite *hamiltonienne* s'il existe une application différentiable $J : X \to \mathfrak{t}_c^*$ (où $\mathfrak{t}_c$ est l'algèbre de Lie de T_c, et $\mathfrak{t}_c^*$ son dual) telle que

(i) Pour tout $x \in X$, la différentielle de J en x est la composée de $\sigma_x : T_xX \to T_x^*X$, et de $T_x^*X \to \mathfrak{t}_c^*$ duale de l'application canonique de $\mathfrak{t}_c$ vers T_xX.

(ii) J est invariante par T_c.

Sous ces hypothèses, J est appelée *application moment* ; elle est unique à translation près par un élément de $\mathfrak{t}_c^*$.

Soit p une valeur régulière de J. Alors T_c opère dans la variété $J^{-1}(p)$. Notons Y_p le quotient de $J^{-1}(p)$ par T_c ; on l'appelle l'*espace des phases réduit*. Il est muni d'une structure de "V-variété" (modelée sur le quotient d'une variété par un groupe fini). La restriction de σ à $J^{-1}(p)$ dégénère exactement le long des orbites de T_c, donc se descend en une forme symplectique σ_p sur Y_p. La variété symplectique X est munie d'une mesure canonique (la *mesure de Liouville*) $d\beta := e^\sigma$. Notons $d\beta_p$ la mesure de Liouville de Y_p, et $\mu(p)$ son volume. D'après [DH1; Proposition 3.2], la fonction μ est la densité de la mesure image $J_*(d\beta)$, par rapport à la mesure de Lebesgue dp sur $\mathfrak{t}_c^*$ (normalisée pour le réseau $\mathfrak{X}(T_c)$ des caractères de T_c).

Résumons maintenant quelques résultats remarquables (voir [Ati], [GS1&2], [DH1&2]).
a) L'ensemble X^{T_c} des points fixes de T_c dans X, est une sous-variété symplectique non vide de X. De plus $J(X^{T_c})$ est fini, et $J(X)$ est l'enveloppe convexe de $J(X^{T_c})$.
b) Soit U un ouvert convexe de valeurs régulières de J. Identifions tous les Y_p ($p \in U$) à une fibre fixée Y_q. Alors la classe de σ_p dans $H^2(Y_p, \mathbf{R})$ est égale à $\sigma_q + \partial(p-q)$ où $\partial : \mathfrak{X}(T_c) \otimes_{\mathbf{Z}} \mathbf{R} = H^1(T_c, \mathbf{R}) \to H^2(Y_q, \mathbf{R})$ est l'homomorphisme de connection associé à $J^{-1}(q) \to J^{-1}(q)/T_c = Y_q$.
c) La fonction μ est polynomiale sur tout ouvert connexe de valeurs régulières de J.
d) La transformée de Fourier de μ, définie par

$$\lambda \longrightarrow \int_{J(X)} \mu(p) \exp\langle ip, \lambda\rangle \, dp = \int_X \exp\langle iJ(x), \lambda\rangle \, d\beta(x)$$

s'exprime uniquement en fonction de la restriction de J à X^{T_c}, et du fibré normal à X^{T_c} dans X (pour des formules explicites, voir [DH1&2]).

A 2. Le cas d'une variété projective.

Soit V un espace vectoriel complexe dans lequel le tore compact T_c opère de façon linéaire et continue. Choisissons un produit scalaire hermitien invariant par T_c sur V, noté $(x, y) \to x \cdot y$. Il définit une structure kählerienne sur l'espace projectif $\mathbf{P}(V)$, invariante par T_c. Ainsi $\mathbf{P}(V)$ est une variété symplectique, et la classe de cohomologie de la forme symplectique σ est la classe de Chern de $\mathcal{O}(1)$. En outre, l'action de T_c dans $\mathbf{P}(V)$ est hamiltonienne, l'application moment étant donnée par $J(x)(\xi) = (\tilde{x} \cdot \xi\tilde{x})(\tilde{x} \cdot \tilde{x})^{-1}$ pour tout $x \in \mathbf{P}(V)$, avec pour représentant $\tilde{x}$ dans V, et pour tout $\xi \in \mathfrak{t}_c$. Ces considérations s'étendent à toute sous-variété algébrique fermée X de $\mathbf{P}(V)$, lisse et stable par T_c.

Soit T le tore complexifié de T_c ; il opère aussi dans V et $\mathbf{P}(V)$. Pour tout $x \in \mathbf{P}(V)^T = \mathbf{P}(V)^{T_c}$, on peut identifier le poids de x avec $J(x)$. On en déduit facilement que l'enveloppe convexe des poids de x est $J(\overline{T.x})$, où $\overline{T.x}$ désigne l'adhérence de la T-orbite de x. L'énoncé a) ci-dessus est donc immédiat pour une variété algébrique projective ; l'image de l'application moment coïncide avec le polyèdre $\mathcal{C}$ défini en 1.1. Nous allons rapprocher les énoncés b), c), d) des résultats de ce travail, grâce aux liens entre la théorie géométrique des invariants et les actions de groupes compacts introduits par Kempf et Ness, et développés par F. Kirwan et L. Ness [KN], [Kir], [Nes].

Soit X une sous-variété fermée de $\mathbf{P}(V)$, lisse et stable par T. Avec les notations de 1.2, la fibre de J en 0 est contenue dans $X^{ss}(0)$, et toute orbite fermée de T dans $X^{ss}(0)$ rencontre $J^{-1}(0)$ suivant une unique orbite de T_c. Par suite, l'inclusion $J^{-1}(0) \to X^{ss}(0)$ induit un homéomorphisme de $Y_0 = J^{-1}(0)/T_c$ sur $X^{ss}(0)//T = Y(0)$. De plus, la classe de la forme symplectique σ_0 est la classe de Chern de $\mathcal{O}_0(1)$ (défini en 1.3) dans $H^2(Y(0), \mathbf{Q})$.

On en déduit comme en 1.3 que pour tout point rationnel p de $\mathcal{C}$, la fibre de J en p est contenue dans $X^{ss}(p)$, et que l'application induite de Y_p vers $Y(p)$ est un homéomorphisme. De plus $\sigma_p = \mathcal{O}_p(1)$ dans $H^2(Y(p), \mathbf{Q})$. L'énoncé b) ci-dessus résulte donc (dans le cas d'une variété algébrique projective) du corollaire 1.3. Le volume $\mu(p)$ de Y_p n'étant autre que le degré de $\mathcal{O}_p(1)$, la densité de la mesure image de J s'interprète comme la fonction μ_M définie en 3.3, où M est l'algèbre des fonctions polynomiales sur le cône sur X ; on peut aussi prendre $M = \oplus_{n=0}^{\infty} \Gamma(X, \mathcal{O}(n))$. L'énoncé c) correspond à la proposition 3.3, et d) au théorème 3.4.

REFERENCES

[Ati] M. F. Atiyah: *Convexity and commuting Hamiltonians, Bull. London Math. Soc.* **14** (1982), 1-15.

[BB] A. Bialynicki-Birula: *Some theorems on actions of algebraic groups, Ann. Math.* **98** (1973), 480-497.

[BBM] W. Borho, J. L. Brylinski, R. MacPherson: *Equivariant K-theory approach to nilpotent orbits*, preprint I.H.E.S. (mars 1986).

[Bri] M. Brion: *Points entiers dans les polyèdres convexes, Ann. sci. Ecole Norm. Sup.* **21** (1988), 653-663.

[Dem] M. Demazure: *Anneaux gradués normaux.* Séminaire sur les singularités des surfaces, Ecole polytechnique 1978-79.

[DH1&2] J. J. Duistermaat, G. J. Heckman: *On the variation in the cohomology of the symplectic form of the reduced phase space* **1**, *Invent. math.* **69** (1982), 259-268 et **2**, *Invent. math.* **72** (1983), 153-158.

[EGA II] A. Grothendieck: *Elements de géométrie algébrique II, Publications mathématiques de l'I.H.E.S.* **8** (1961).

[Fle] H. Flenner: *Rationale quasihomogene Singularitäten, Archiv der Math.* **36** (1981), 35-44.

[Ful] W. Fulton: *Intersection theory*, Ergebnisse der Math. **2** (Springer-Verlag 1984).

[GS1&2] V. Guillemin, S. Sternberg: *Convexity properties of of the moment mapping* **1**, *Invent. math.* **67** (1982), 491-513 et **2**, *Invent. math.* **77** (1984), 533-546.

[GS3] V. Guillemin et S. Sternberg: *Birational equivalence in the symplectic category, Invent. math.* **97** (1989), 485-522.

[GW] S. Goto, K. i. Watanabe: *On graded rings I, J. Math. Soc. Japan* **30** (1978), 179-213.

[Jos] A. Joseph: *On the variety of a highest weight module, J. of Alg.* **88** (1984), 238-278.

[KN] G. Kempf, L. Ness: *The length of vectors in representation spaces*, dans Algebraic Geometry, Springer Lecture Notes in Math. **732** (1979).

[Kir] F. Kirwan: *Cohomology of quotients in symplectic and algebraic geometry*, Princeton University Press, Mathematical Notes **31** (1984).

[Kle] S. L. Kleiman: *Towards a numerical theory of ampleness, Ann. Math.* **84** (1966), 293-344.

[Kol] J. Kollar: *The structure of algebraic threefolds: An introduction to Mori's program, Bull. Amer. Math. Soc.* **17** (1987), 211-273.

[Kra1] H. Kraft: *Geometrische Methoden in der Invariantentheorie*, Aspekte der Mathematik, Vieweg 1985.

[Kra2] H. Kraft: *G-vector bundles and the linearization problem*, à paraître dans un volume de Contemporary Mathematics.

[Lun] D. Luna: *Slices étales*, Mémoire de la Société mathématique de France **33** (1973), 81-105.

[MF] D. Mumford, J. Fogarty: *Geometric invariant theory*, second enlarged edition, Springer Verlag (1982).

[Nes] L. Ness: *A stratification of the nilcone via the moment map, Amer. J. Math.* **106** (1984), 1281-1330.

[Nie] H. Nielsen: *Diagonalizably linearized coherent sheaves, Bull. Soc. Math. France* **102** (1974), 85-97.

[Smo] W. Smoke: *Dimension and multiplicity for graded algebras, J. of Alg.* **21** (1972), 149-173.

[Val] F. A. Valentine, *Convex sets*, McGraw Hill (1964).

Received October 23, 1989

Michel Brion
Institut Fourier
Université de Grenoble-I
F-38402 Saint-Martin d'Hères

Claudio Procesi
Istituto Matematico
Città Universitaria
I-00185 Roma

When are the Stabilizers of all Nonzero Semisimple Points Finite?

VLADIMIR L. POPOV

Dedicated to J. Dixmier on his 65th birthday

1. Introduction

Let G be a reductive algebraic group and V a finite dimensional algebraic G-module over $\mathbf{C}$. It is well known that closed G-orbits in V play a special role in Invariant theory. Particularly, the conjugacy classes of stabilizers G_v of semisimple points $v \in V$ (i.e. such points v that the orbit Gv is closed in V) define the Luna stratification of the varieties V and $V/G := \operatorname{Spec} \mathbf{C}[V]^G$, and the local structure of V/G in the neighbourhood of the point $\pi_{V,G}(v)$, where $\pi_{V,G} : V \to V/G$ is the canonical morphism, is described by the structure of the neighbourhood of zero in the (categorical) quotient of the slice-module of the group G_v [1]. This is the key point of the solution of such problems as the classification of G-modules with free algebra of invariants, [2], or G-modules which have the property that V/G is a complete intersection, [3], as well as of some other problems, [8]. The usual arguments in these cases are as follows: one takes v to be such a point that G_v is a nontrivial group which has a sufficiently simple structure; then one can control the slice-module of this point, and if this module appears to be "bad" (i.e. it does not have the property which is under investigation) then V itself is also "bad"; this makes it possible to exclude from the consideration a lot of G-modules as *a priory* being "bad". This scheme does not work if G_v is trivial for each semisimple point $v \in V, v \neq 0$. This leads to the problem of classification of G-modules with this property. Clearly, when solving this problem one can assume that G^0 is simply connected, the ineffective kernel of the action of G on V (i.e. the group $\{g \in G | gv = v \text{ for all } v \in V\}$ is finite and that $V^G = 0$. This problem was explicitly formulated in a discussion with N.L. Gordeev whose interest in these G-modules was inspired by his approach to the problem of estimation of the homological dimension of algebras of invariants of reductive linear groups, [17], see also [8]. For finite groups G the solution to this problem is known: clearly, it is identical with the problem of classification of finite linear groups acting freely on spheres, and these groups were classified by J.A.Wolf, [18]. Here I shall give the solution to this problem for *connected simple* groups G. As a matter of fact, this solution will be derived from the following classification of a more general class of G-modules:

Theorem. *Let G be a connected, simply connected simple algebraic group and V a finite dimensional algebraic G-module with a finite ineffective kernel and such that $V^G = 0$. Then the stabilizer of each nonzero semisimple point of V is finite if and only if V or its dual V^* is contained in the following list:*

$G = SL_2$, $V = R(p_1\varpi_1) + \ldots + R(p_s\varpi_s)$, $s \geq 1$, *and all* p_i *are odd;*

$G = SL_{l+1}$, $l \geq 2$, $V = mR(\varpi_1)$, $m \geq 1$;

$G = SL_{l+1}$, $l \geq 2$, l *is even,* $V = R(\varpi_2)$, $V = 2R(\varpi_2)$ *and* $V = R(\varpi_2) + R(\varpi_l)$.

$G = \mathrm{Spin}_{10}$, $V = R(\varpi_4)$

$G = \mathrm{Sp}_l$, $l \geq 2$, $V = R(\varpi_1)$.

Here and further $R(\lambda)$ is a simple G-module with highest weight λ with respect to a fixed root decomposition defined by a maximal torus S and a Borel subgroup containing S. We use the standard notations and assume that the numeration of simple roots $\alpha_1, \ldots, \alpha_l$ and fundamental weights $\varpi_1, \ldots, \varpi_l$, as well as their expressions via ε_i, are taken from [9], [10].

Modern developments in Invariant theory are based on the philosophy that the classes of G-modules V distinguished by various "nice"properties (such as freeness of the algebra of invariants, freeness of the module of covariants, equidimensionality of the morphism $\pi_{V,G}$, nontriviality of the generic stabilizer, polarity etc.) are of paramount importance for Invariant theory while the others are in some way "wild", see [19]. It seems to me that one can find new interesting sources of "nice"properties considering the singularities of the variety V/G and the Luna stratification of V and V/G, [1]. One has to consider the theorem above and its corollary below along the lines of this program. Namely, the theorem gives the classification of such G-modules (G is connected and simple) that all the singularities of the variety V/G, except the "zero-point"$\pi_{V,G}(0)$, come from "finite"Invariant theory, i.e. are the quotient singularities by finite linear groups. All the G-modules which are considered in the corollary below have exactly two Luna's strata, and all of the fibers of the principal stratum of these G-modules are G-isomorphic to the group G itself (with the action by left translations). More generally, it seems to me to be interesting the following

Problem. *Classify G-modules with exactly two Luna's strata (2-strata G-modules).*

As it is already marked above, this problem was solved by J.A. Wolf for finite groups [18], so it would be interesting to solve this problem for connected groups G, and one has to consider the corollary below along the lines of this problem. One can systematically produce examples of 2-strata modules using the fact that there are canonically associated with any G-module V some 2-strata modules (of other groups), namely, the slice-modules of subprincipal strata of V. If V is a 2-strata G-module, then $P := V \setminus \pi_{V,G}^{-1}(\pi_{V,G}(0))$ is the principal stratum. Therefore $B := V/G \setminus \pi_{V,G}(0)$ is a smooth variety and $\pi_{V/G} : P \to B$ is a fiber space (with respect to complex topology), [1]. Hence, considering the loops around $\pi_{V/G}(0)$ in V/G, one obtains some monodromy effect.

Problem. *Calculate this monodromy.*

It would be also interesting to describe the topological invariants of the manifold B for 2-strata G-modules V.

Remarks. (a) All the G-modules which are listed in the theorem are locally transitive (i.e. have a dense orbit) except the SL_2-modules $R(p_1\varpi_1)+\ldots+R(p_s\varpi_1)$, where all p_i are odd and $p_1 \neq 1$ if $s = 1$, and the SL_{l+1}-modules $mR(\varpi_1)$, where $m \geq l+1 \geq 3$.

(b) It would be interesting to extend the classification to the case when G is a *connected semisimple* group. The following example shows that for such groups G the class of G-modules which have the desired property is definitely more wide, see also [8]:

Claim. *Let $G = \mathrm{SL}_2 \times SL_p$, $V = R(q\varpi_1) \otimes R(\varpi_1)$ and p and q are odd. Then the stabilizer of each nonzero semisimple point of V is finite.*

Proof of the claim. Assume that there exists such a nonzero semisimple point $v \in V$ that $\dim G_v \geq 1$. Since Gv is closed, it follows that G_v is reductive, cf., e.g., [1]. Hence G_v contains a one-dimensional torus T. Let $\phi : k^* \to T$ be a fixed isomorphism. Replacing v by gv for a suitable $g \in G$, we may assume that $\phi(t) = \mathrm{diag}(t^a, t^{-a}) \times \mathrm{diag}(t^{b_1}, \ldots, t^{b_p}) \in G = SL_2 \times SL_p$ for all $t \in k^*$ and some $a, b_1, \ldots, b_p \in \mathbf{Z}$. Clearly, $b_1+\ldots+b_p = 0$. We can find such a weight basis $v_0, v_1, \ldots, v_q$, resp., $u_1, \ldots, u_p$, relative to the diagonal torus in SL_2-module $R(q\varpi_1)$, resp., SL_p-module $R(\varpi_1)$, that v_i, resp., u_j, is an eigenvector of $\mathrm{diag}(t^a, t^{-a})$, resp., $\mathrm{diag}(t^{b_1}, \ldots, t^{b_p})$, with the eigenvalue $t^{a(q-2i)}$, resp., t^{b_j}. Then the vectors $v_i \otimes u_j \in V$ form a weight basis relative to T in V, and $\phi(t)(v_i \otimes u_j) = t^{a(q-2i)+b_j} v_i \otimes u_j$. Therefore

V^T is the linear span of all $v_i \otimes u_j$ for which $a(q-2i)+b_j = 0$. We will prove that for some j the vector $v_i \otimes u_j$ does not lie in V^T for any t. Assume the contrary; since q is odd, we then have $b_j = af_j$ for any $j = 0, \ldots, p$, where f_j is some odd number. Thus, $0 = b_1 + \ldots + b_p = a(f_1 + \ldots + f_p)$, and, therefore, since p is odd, $a = 0$. Consequently, V^T is the linear span of those $v_i \otimes u_j$ for which $b_j = 0$. Clearly, we do not have $b_j = 0$ for all j (however there exists a j for which $b_j = 0$, since $v \in V^T$ and therefore $V^T \neq 0$). Suppose, for definiteness, $b_0 \neq 0$. Consider in G the one-dimensional torus $\{\delta(t) = \text{diag}(1,1) \times \text{diag}(t^{-p+1}, t, \ldots, t) \in SL_2 \times SL_p | t \in k^*\}$. If $j > 0$, then $\delta(t)(v_i \otimes u_j) = t(v_i \otimes u_j)$. Consequently, $\delta(t)$ acts on V^T as a homothety with coefficient t. Since $v \in V$, this contradicts the fact that the orbit Gv is closed.

Thus, for some j the vector $v_i \otimes u_j$ does not lie in V^T for any i; suppose, for definiteness, this is true for $j = 0$. Then, as above, $\delta(t)$ acts on V^T as a homothety with coefficient t, which contradicts the fact that Gv is closed. Q.E.D.

Corollary. *The connected simply connected non locally transitive simple linear algebraic groups $G \subseteq \text{GL}(V)$ which have the property that the stabilizer of each nonzero semisimple point of V is trivial are exactly the groups $G = SL_d, d \geq 2, V = mR(\varpi_1)$, where $m \geq d$.*

Proof of the Corollary. If $G = SL_2$, $V = R(p\varpi_1)$, p is odd, $p \geq 3$, then the property does not hold (see Lemma 2.1 in [8]). Therefore it follows from Remark (a) above that one has only to check the property for the groups which are pointed out in the formulation of the Corollary. In this case the point v is the vector $(v_1, \ldots, v_m), v_i \in R(\varpi_1)$. It follows from here that G_v is trivial if $\text{rk}\{v_1, \ldots, v_m\} = d$, and that the condition $\text{rk}\{v_1, \ldots, v_m\} \neq d$ implies for each $t \in \mathbf{C}$ the existence of an element $g \in G$ which acts on the linear span of the set $\{v_1, \ldots, v_m\}$ as a homothety with the coefficient t. Hence $0 \in \overline{Gv}$ if $rk\{v_1, \ldots, v_m\} \neq d$. Therefore it is sufficient to check that $rk\{v_1, \ldots, v_m\} = d$ implies $\overline{Gv} = Gv$. Assume that this is not the case and let Gu, $u = (u_1, \ldots, u_m)$, be the (unique) closed orbit in $\overline{Gv}$. It follows from $\dim Gu < \dim Gv$ that $\dim G_u > 0$. Therefore $\text{rk}\{u_1, \ldots, u_m\} < d$, and hence $u = 0$. It follows from here that all the homogeneous nonconstant invariants of G-module V vanish at v. In particular, if one identifies V with the coordinate space (of columns) $\mathbf{C}^d$ then the determinant of any d vectors taken from $v_1, \ldots, v_m$ equals to zero. Therefore $\text{rk}\{v_1, \ldots, v_m\} < d$ which is a contradiction. Q.E.D.

The claim of the theorem follows immediately from the propositions 6, 7, 8, 9 and 10 which are proved in the next paragraphs. We consider separately each of the types of simple groups; the case $G = SL_d$ appears to be the most difficult one.

Notations

V^H is the fixed point set of the group H which acts on V.

$Z_G(H)$ is the centralizer of a subgroup H of G.

$\Delta(V)$ and Σ are resp. the weight system of V and the root system of G with respect to S.

X_{++} is the monoid of dominant weights of S.

W is the Weyl group of G. The numbers $a_1, \ldots, a_l$ are called *numerical labels* of the element $a_1\varpi_1 + \ldots + a_l\varpi_l$.

We denote by 1 the trivial one-dimensional module and by V^* the dual of V. We abbreviate s.g.p. for the stabilizer of general position.

2. Some preliminary results. The cases B_l, C_l, D_{2s}, E_7, E_8, F_4 and G_2.

We shall say briefly that V is a *G-module of type* (F) if the G-stabilizer of each nonzero semisimple point of V is *finite*.

Proposition 1. *V is a G-module of type (F) iff all of the fibers of the canonical morphism $\pi_{V,G}$, except the "zero-fiber" $N = \pi_{V,G}^{-1}(\pi_{V,G}(0))$, are closed orbits with finite stabilizers, or, equivalently, iff each point of the set $V \setminus N$ is semisimple and its stabilizer is finite.*

Proof. The "if" part is evident, so assume that $v \in V \setminus N$ and Gu is the (unique) closed orbit in $X = \pi_{V,G}^{-1}(\pi_{V,G}(v)) \subset V \setminus N$. Then G_u is finite and therefore $\dim Gu = \dim G$. On the other hand, one has $Gu \subseteq \overline{Gu}$, and hence $\dim G \geq \dim Gv \geq \dim Gu = \dim G$. Therefore $\dim Gv = \dim Gu$, hence $Gu = Gv = X$. Q.E.D.

Proposition 2. *If a direct sum of G-modules is of type (F) then each summand is also of type (F).*

Proof. This is evident. Q.E.D.

Recall [13], that a module is called *stable* if its generic orbit is closed. If the group is semisimple then the stability of a module is equivalent to the reductivity of its s.g.p.,[13]; we shall use this systematically below.

Proposition 3. *Assume that there exists a reductive subgroup H of G such that* $\dim H > 0$ *and that the $Z_G(H)/H$-module V^H has a nonzero stable submodule. Then V is not of type (F).*

Proof. It follows from the assumption that $Z_G(H)$-orbit of some nonzero point of V^H is closed. Therefore G-orbit of this point is also closed, [4], and the assertion follows from the inequality $\dim H > 0$. Q.E.D.

The group $Z_G(H)/H$ in the proposition 3 is always reductive. In what follows we shall systematically use this proposition together with the following assertion:

Proposition 4. *Let H be a reductive subgroup of $G = G^0$ and $D = Z_G(H)/H$. Let $R(\lambda_i)$, $1 \leq i \leq s$, be a set of simple G-modules and assume that D^0-module $R(\lambda_i)^H$ contains a simple submodule M_i with the multiplicity $n_i \geq 1$. Then the Cartan product of D^0-modules $M_1, \ldots, M_s$ is contained in $R(\lambda_1 + \ldots + \lambda_s)^H$ with a multiplicity* $\geq \max\{n_1, \ldots, n_s\}$.

Proof. This is a special case of theorem 4.1 in [8]. Q.E.D.

Proposition 5. *Assume that there exist such $\lambda_1, \ldots, \lambda_d \in \Delta(V)$ that*

(1) $\operatorname{rk}\{\lambda_1, \ldots, \lambda_d\} < \dim S$;

(2) *zero is an inner point of the convex envelope of the set $\{\lambda_1, \ldots, \lambda_d\}$;*

(3) *$\lambda_i - \lambda_j \notin \Sigma$ for each $i \neq j$.*

Then V is not of type (F).

Proof. Let v_i is a nonzero weight vector of the weight λ_i and let $v = v_1 + \ldots + v_d$. Then it follows from [5] and (2) and (3) that Gv is closed. It follows from (1) that the intersection of all the subgroups $\operatorname{Ker}\lambda_i$ of S, $1 \leq i \leq d$, is a group which has positive dimension. Now the claim follows from the fact that this group is contained in G_v. Q.E.D.

Corollary 1. *Let* $\operatorname{rk} G > 1$ *and V be a module of type (F) such that $k\lambda \in \Delta(V)$ for some $\lambda \in \Delta(V)$ and $k \in \mathbf{Q}, k < 0$. Then $\alpha/(1-k) \in X_{++}$ for some $\alpha \in \Sigma \cap X_{++}$. (Therefore if $\alpha/(1-k)$ is not an element of X_{++} for each $\alpha \in \Sigma \cap X_{++}$ then each G-module is not of type (F)).*

Proof. The set $\Delta(V)$ is W-invariant and each W-orbit in this set intersects the Weyl chamber. Therefore we can assume that $\lambda \in X_{++}$. It follows from the inequalities $\operatorname{rk}\{\lambda, k\lambda\} \leq 1 < \operatorname{rk} G$ and $k < 0$ that the convex envelope of the set $\{\lambda, k\lambda\}$ contains zero, therefore we derive from the proposition 5 that $\alpha = \lambda - k\lambda = (1-k)\lambda \in \Sigma$. Since $1 - k > 0$, we have $\alpha \in X_{++}$. Q.E.D.

In the same way one obtains that if $0 \in \Delta(V)$ and $\operatorname{rk} G$ is arbitrary then V is not of type (F) (one has to take $\lambda = 0$).

Corollary 2. *Let* $\operatorname{rk} G > 1$ *and* $-1 \in W$. *Then the highest weight of each simple G-module V of type (F) is of the form* $\alpha/2$, *where* $\alpha \in \Sigma \cap X_{++}$.

Proof. Applying the transformation $-1 \in W$ to $\lambda \in \Delta(V)$ one obtains that $k\lambda \in \Delta(V)$ for $k = -1$ and then can argue as in the corollary 1. Q.E.D.

We consider now the connected simple groups which satisfy the assumptions of the corollary 2, i.e. the groups of the types B_l, C_l, D_{2s}, G_2, F_4, E_7 and E_8. In these cases one obtains the desired result just straightforwardly from this corollary:

Proposition 6. *If G is the group of type* $B_l (l \geq 3), D_{2s}, G_2, F_4, E_7$ *or* E_8 *then each G-module is not of type (F). If G is the group of type* C_l *then the unique simple G-module of type (F) is* $R(\varpi_1)$, *and this module is locally transitive.*

Proof. Let G be the group of type B_l $(l \geq 3)$. Then it is easy to see, [9], that $\Sigma \cap X_{++}$ consists only of ϵ_1 and $\epsilon_1 + \epsilon_2$ which are not divisible by 2 in X_{++}. The claim now follows from corollary 2 of proposition 5 and from proposition 3.

Let G be the group of type C_l $(l > 2)$. Then $\Sigma \cap X_{++}$ consists only of $2\epsilon_1$ and $\epsilon_1 + \epsilon_2$.Therefore $\alpha/2 \in X_{++}$ for $\alpha \in \Sigma \cap X_{++}$ if and only if $\alpha = 2\epsilon_1$. The group G acts on $R(\varpi_1) \setminus 0$ transitively because of Witt's theorem.

If G is the group of type D_{2s}, G_2, F_4, E_7, E_8 then $\Sigma \cap X_{++}$ consists, resp., only of ϖ_2, ϖ_1 and ϖ_2, ϖ_1 and ϖ_4, ϖ_1, ϖ_8 and all these elements are not divisible by 2 in X_{++}. Q.E.D.

3. The cases E_6 and D_{2s+1}

In these cases the group W does not contain -1 and we have to use another arguments. These arguments are based on propositions 3 and 4.

Proposition 7. *If G is the group of type* E_6 *then each G-module in not of type (F).*

Proof. Let T be the one-dimensional torus in G defined by the highest root. Then $Z_G(T)/T$ is a group of type A_5 and one has the following table (see [8]):

Table 1

G-module V	structure of $Z_G(T)^0/T$-module V^T
$R(\varpi_1)$	$R(\varpi_4)$
$R(\varpi_2)$	$R(\varpi_1+\varpi_5)+1$
$R(\varpi_3)$	contains $R(\varpi_3+\varpi_5)$
$R(\varpi_4)$	contains $R(2\varpi_1+\varpi_4)+R(\varpi_2+2\varpi_5)+R(\varpi_1+\varpi_5)$
$R(\varpi_5)$	contains $R(\varpi_1+\varpi_3)$
$R(\varpi_6)$	$R(\varpi_2)$

As it follows from [13] and [10], $R(\varpi_1)$ and $R(\varpi_5)$ are the unique simple nonstable modules of the group of type A_5. Therefore it follows from proposition 4 and table 4 that for each simple G-module V there exists a nonzero stable submodule of $Z_G(T)^0/T$-module V^T. Now the claim follows from propositions 3 and 2. Q.E.D.

Proposition 8. *If G is the group of type D_l, l is odd and ≥ 5, then the unique simple G-modules of type (F) are $R(\varpi_4)$ and $R(\varpi_5)$ for $l=5$, and these modules are locally transitive.*

Proof. Let T be the one-dimensional torus in G defined by the highest root. Then $Z_G(T)$ is a group of type A_1+D_{l-2} (where, by definition, $D_3=A_3$) and one has the following table (see [8]):

Table 2

G-module V	structure of $Z_G(T)^0/T$-module V^T
$R(\varpi_p)$, $1\leq p\leq l-2$	$(1\otimes R_p)+(1\otimes R(\varpi_{p-2}))+$ $(1\otimes R(\varpi_{p-4}))+(R(2\varpi_1)\otimes R(\varpi_{p-2}))$, where, by definition, $R(\varpi_i)=1$ if $i=0, R(\varpi_i)=0$ if $i<0$ and $R_p=\begin{cases} R(\varpi_p) & \text{if } 1\leq p\leq l-4,\\ R(\varpi_{l-3}+\varpi_{l-2}) & \text{if } p=l-3,\\ R(2\varpi_{l-3})+R(2\varpi_{l-2}) & \text{if } p=l-2\end{cases}$
$R(\varpi_{l-1})$	$R(\varpi_1)\otimes R(\varpi_{l-2})$
$R(\varpi_l)$	$R(\varpi_1)\otimes R(\varpi_{l-3})$

It follows from table 2 and proposition 4 that if V is a simple G-module and $l\geq 7$ then $Z_G(T)^0/T$-module V^T contains a submodule of type $1\otimes R(\lambda)$, $l\neq \varpi_{l-3}$, ϖ_{l-2}, if the $(l-1)$-th and l-th numerical labels a_{l-1} and a_l of the highest weight of G-module V are equal to 0, and of type $R(\mu)\otimes R(\nu)$, $\mu\neq 0$, $\nu\neq 0$ otherwise. It is not difficult to see, using [12],

[10] and [11], that such a submodule is stable. Let now $l = 5$. The similar arguments show that if $a_{l-1} = a_l = 0$ then $Z_G(T)^0/T$-module V^T contains a submodule of type $1 \otimes R(\lambda)$, $\lambda \neq \varpi_1, \varpi_3$, and that such a submodule is stable. If $a_{l-1} + a_l \neq 0$ and $V \neq R(\varpi_4)$ or $R(\varpi_5)$ then V^T contains $R(\mu) \otimes R(\nu)$, $\mu \neq 0$, $\nu \neq \varpi_1$, and this submodule is also stable. Therefore the claim follows from proposition 3 and the fact that $R(\varpi_4)$ and $R(\varpi_5)$ are locally transitive ($l = 5$), see [11]. Q.E.D.

4. The case $\mathbf{A_l}$

We assume here that $G = SL_{l+1}$ and V is a simple G-module with highest weight $a_1\varpi_1 + \ldots + a_l\varpi_l$. As the dual modules define the same linear group, we need to consider only one of them.

Let T be the one-dimensional torus in G defined by the root α_1. Then $Z_G(T)/T \simeq Z \times SL_{l-1}$, where Z is a one-dimensional torus, and, if χ is a fixed generator of the group of characters of Z, one has the following table (see [8]):

Table 3

G-module V	structure of $Z_G(T)/T$-module V^T
$R(\varpi_p), 1 \leq p \leq l$	$(\chi^{p-1-l} \otimes R(\varpi_{p-2})) + (\chi^p \otimes R(\varpi_p))$
$Ad = R(\varpi_1 + \varpi_l)$	$(\chi^0 \otimes Ad) + 2(\chi^0 \otimes 1)$
$R(p\varpi_1)$	$\oplus_{r=0}^{[p/2]}(\chi^{p-r(l+1)} \otimes R((p-2r)\varpi_1))$

The group $Z_G(T)/T$ is not semisimple and therefore we cannot apply the stability criterion from [13] to $Z_G(T)/T$-modules U. The criterion which we need is proved in [8] and (in a slightly modified form) is as follows: $Z \times SL_{l-1}$-module U is stable iff SL_{l-1}-module U is stable and there exist in its algebra of invariants $\mathbf{C}[U]^{SL_{l-1}}$ two Z-semi-invariants with the weights χ^a and χ^b, where $ab < 0$. It follows from [8] that the second condition is fullfiled if there exist in U two such non locally transitive SL_{l-1}-submodules that the weights of the action of Z on these submodules have different signs. We shall systematically use these results below, so when we claim that a $Z_G(T)/T$-module is stable we have in mind that this follows from these results. Also, before we start to consider the different concrete cases, we have to notice that a simple SL_{l-1}-module is not stable iff it is locally transitive and that the complete list of such modules is: $R(\varpi_1)$, $R(\varpi_{l-2})$ and, if l is even, also $R(\varpi_2)$ and $R(\varpi_{l-3})$.

(1) The case $\mathbf{a_1 = a_l = 0}$, $\mathbf{l \geq 3}$.

(1a) It follows from table 3 and proposition 4 that $Z_G/(T)$-module V^T contains a submodule $(\chi^a \otimes 1) + (\chi^b \otimes 1), ab < 0$, if $l = 3$, and a submodule $(\chi^a \otimes R(\lambda)) + (\chi^b \otimes R(\mu)), ab < 0, \lambda \neq 0, \mu \neq 0$, if $l \geq 4$, and also that: (a) if l is odd then $R(\lambda)$ and, resp., $R(\mu)$ can be locally transitive only if $(a_3, a_4, \ldots, a_{l-1}) = (1, 0, \ldots, 0)$ and, resp., $(a_2, a_3, \ldots, a_{l-2}) = (0, 0, \ldots, 0, 1)$; (b) if l is even then $R(\lambda)$ and, resp., $R(\mu)$ can be locally transitive only if $(a_3, a_4, \ldots, a_{l-1}) = (1, 0, \ldots, 0), (0, 1, \ldots, 0)$ or $(0, \ldots, 0, 1)$ and, resp., $(a_2, a_3, \ldots, a_{l-2}) = (1, 0, \ldots, 0), (0, \ldots, 0, 1)$ or $(0, \ldots, 0, 1, 0)$. Therefore, by proposition 3, we see that if $a_1 = a_l = 0$ and $l \geq 3$ then all G-modules of type (F) are contained (to within taking of dual module) only among G-modules of type $R(a_2\varpi_2 + \varpi_3)$, $l \geq 4$, and if l is even, of types $R(a_2\varpi_2 + \varpi_4)$, $l \geq 6$, and $R(a_2\varpi_2 + \varpi_{l-1})$, $l \geq 4$.

(1b) *Let* $V = R(a_2\varpi_2 + \varpi_3)$, $l \geq 4$, $a_2 > 0$. It follows from table 3 that $Z_G(T)/T$-module V^T contains a submodule $(\chi^{(2-l)+2+(1-l)(a_2-1)} \otimes R(\varpi_1 + \varpi_2)) + (\chi^{2a_2+3} \otimes R(a_2\varpi_2 + a_3))$. The exponent of χ is negative in the first summand if $l \geq 5$, and it is positive in the second. Therefore V is not of type (F) if $l \geq 5$. If $l = 4$ then V^T contains a submodule $(\chi^{-3a_2+3} \otimes 1) + (\chi^{2a_2-2} \otimes R(\varpi_1 + a_2\varpi_2))$. Therefore V is not of type (F) also if $l = 4$.

(1c) *Let* $V = R(a_2\varpi_2 + \varpi_4), l$ *be even,* $l \geq 6$ *and* $a_2 > 0$. It follows from table 3 that $Z_G(T)/T$-module V^T contains a submodule $(\chi^a \otimes R(2\varpi_2)) + (\chi^b \otimes R(a_2\varpi_2 + \varpi_4))$, where $a < 0, b > 0$. Therefore V is not of type (F).

(1d) *Let* $V = R(a_2\varpi_2 + \varpi_{l-1}), l$ *be even,* $l \geq 4$ *and* $a_2 > 0$. It follows from table 3 that $Z_G(T)/T$-module V^T contains a submodule $(\chi^{(1-l)(a_2-1)} \otimes R(\varpi_2 + \varpi_{l-3})) + (\chi^{2a_2+l-1} \otimes R(a_2\varpi_2))$. This shows that V is not of type (F) if $a_2 \geq 2$. If $a_2 = 1$ then V^T contains $\chi^0 \otimes 1$ and therefore again V is not of type (F). Finally, l being even, V is locally transitive if $a_2 = 0$ and therefore V is of type (F).

(1e) *Let* $V = R(\varpi_3), l \geq 5$. If $l = 5$ or 6 then, as it follows from [11] and [12], V is stable and s.g.p. of V has positive dimension; therefore V is not of type (F) in this case and one can assume that $l \geq 7$. Take in $G = SL_{l+1}$ two-dimensional torus $T_1 = \{\text{diag}(t_1, t_2, t_3, 1, \ldots, 1) | t_1t_2t_3 = 1\}$. Then $Z_G(T_1) = T_1 \times Z_1 \times Z'_G(T_1)$, where $Z'_G(T_1) = \text{diag}(1, 1, 1, SL_{l-2})$ and $Z_1 = \{\text{diag}(t^{-l+2}, 1, 1, t, \ldots, t)\}$. Let ψ be the character of Z_1 defined by the formula $\psi(\text{diag}(t^{-l+2}, 1, 1, t, \ldots, t)) = t$. Now the same arguments as in [8] show that if one identifies $Z_G(T_1)/T_1$ with $Z_1 \times SL_{l-2}$ then $Z_G(T_1)/T_1$-module V^{T_1} is of the form $(\psi^{-l+2} \otimes 1) + (\psi^3 \otimes R(\varpi_3))$. It follows from the inequality $l \geq 7$ that this module is stable and therefore V is not of type (F).

(1f) *Let* $V = R(\varpi_4), l$ *be even and* $l \geq 8$. Take in $G = SL_{l+1}$

a three-dimensional torus $T_2 = \{\text{diag}(t_1, t_2, t_3, t_4, 1, \ldots, 1) | t_1t_2t_3t_4 = 1\}$. Then $Z_G(T_2) = T_2 \times Z_2 \times Z'_G(T_2)$, where $Z'_G(T_2) = \{\text{diag}(1,1,1,1, SL_{l-3})\}$, $Z_2 = \{\text{diag}(t^{3-l}, 1, 1, 1, t, \ldots, t)\}$. Let φ be the character of Z_2, defined by the formula $\varphi(\text{diag}(t^{3-l}, 1, 1, 1, t, \ldots, t)) = t$. Now, as in [8], one can show that if one identifies $Z_G(T_2)/T_2$ with $Z_2 \times SL_{l-3}$ then $Z_G(T_2)/T_2$-module V^{T_2} is of the form $(\varphi^{3-l} \otimes 1) + (\varphi^4 \otimes R(\varpi_4))$. It follows from the restrictions on l that SL_{l-3}-module $R(\varpi_4)$ is locally transitive only for $l = 8$. Therefore, if l is even and $l > 8$ then V is not of type (F).

Consider now the case $l = 8$ separately. We return back to the torus T. Then $Z_G(T)/T$ can be identified with $Z \times SL_7$. It follows from table 3 that $Z_G(T)/T$-module V^T is of the form $(\chi^{-5} \otimes R(\varpi_2)) + (\chi^4 \otimes R(\varpi_4))$. We obtain from [11] and [13] that SL_7-module $R(\varpi_2) + R(\varpi_4)$ is stable. Use now that the algebra of invariants of SL_7-module $R(\varpi_4)$ is not trivial (because this module is not locally transitive and the group SL_7 is semisimple). It follows from this fact that there exist in this algebra a Z-semi-invariant of weight $\chi^a, a < 0$. Let us show that one can find in this algebra also a Z-semi-invariant of weight $\chi^b, b > 0$. Indeed, it is clear that the algebra of polynomial functions on the space of SL_7-module $R(\varpi_2) + R(\varpi_4)$, considered as SL_7-module, contains a submodule of the form $S^3(R(\varpi_2)^*) \otimes R(\varpi_4)^*$ an which Z acts scalarly with the weight χ^{11}. But one has the following equality of SL_7-modules, [6]: $S^3R(\varpi_2) = R(3\varpi_2) + R(\varpi_1 + \varpi_2) + R(\varpi_3)$ and hence $S^3(R(\varpi_2)^*) = R(3\varpi_5) + R(\varpi_5 + \varpi_6) + R(\varpi_4)$. Therefore $S^3(R(\varpi_2)^*) \otimes R(\varpi_4)^* \supset R(\varpi_4) \otimes R(\varpi_4)^* \supset 1$. This means that the algebra of invariants of SL_7-module $R(\varpi_2) + R(\varpi_4)$ contains a homogeneous invariant of degree $(3,1)$, which is a Z-semi-invariant of the weight χ^{11}. Therefore, $Z_G(T)/T$-module V^T is stable and hence V is not of type (F).

So we proved in this section that if $a_1 = a_l = 0, l \geq 3$ then only $R(\varpi_2)$ and $R(\varpi_{l-1})$ for even l are unique simple G-modules of type (F).

(2) The case l = 2.

Let $V = R(p\varpi_1)$. It follows from table 3 that if we identify $Z_G(T)/T$ with a one-dimensional torus Z then $Z_G(T)/T$-module V^T is of the form $\chi^p + \chi^{p-3} + \chi^{p-6} + \ldots + \chi^{p-[p/2]3}$, and hence contains a submodule $\chi^p + \chi^{p-[p/2]3}$. We have $p - [p/2]3 \leq 0$ for $p \geq 2$ (and the equality takes place only for $p = 3$). Therefore V is not of type (F) if $p \geq 2$. If $p = 1$ then V is locally transitive and hence of type (F). It follows from the equality $R(p\varpi_1)^* = R(p\varpi_2)$ that $Z_G(T)/T$-module V^T for $V = R(q\varpi_2)$ is of the form $\chi^{-q} + \chi^{-q+3} + \chi^{-q+6} + \ldots + \chi^{-q+[q/2]3}$. We can conclude from here and from proposition 4 that $Z_G(T)/T$-module V^T for $V = R(p\varpi_1 + q\varpi_2), pq \neq 0$, contains a submodule of type $\chi^a + \chi^b, ab < 0$, if either both of p and q are not equal to 1 and 3, or $p = 1, q \geq 2$, or $p = 3, q \geq 2$, and it contains a submodule of the type χ^0 if $p = q = 1$. This submodule is stable in each of the cases and therefore V is not of type (F).

So we conclude that if $l = 2$ then $R(\varpi_1)$ and $R(\varpi_2)$ are unique simple G-modules of type (F).

(3) The case $\mathbf{a_1 \neq 0}$, $\mathbf{a_l = 0}$, $\mathbf{l \geq 3}$. .

(3a) *Let at first be $a_1 = p \geq 2$.* It follows from table 3 that $Z_G(T)/T$-module V^T contains a submodule of type $(\chi^{p-([p/2]-1)(l+1)+a} \otimes R((p - 2([p/2]-1))\varpi_1 + \alpha)) + (\chi^{p+b} \otimes R(p\varpi_1 + \beta)) + (\chi^{p-[p/2](l+1)+c} \otimes R((p - 2[p/2]\varpi_1 + \gamma))$ for some $\alpha, \beta, \gamma \in X_{++}$ and $a, b, c \in \mathbf{Z}, a < 0, b > 0, c < 0$. If we consider now the dependence on p of the signs of the exponents of χ and of the coefficients of ϖ_1 in this sum, then we obtain, after some calculations, that this submodule contains a stable submodule when either $l = 3, a_1 \geq 6$ or $l \geq 4, a_1 \geq 4$ and $a_1 \neq 5$ for $l = 4$. Therefore V is not of type (F).

Let $l = 3$ and $a_1 = p = 5$. If $a_2 \neq 0$, then it follows from table 3 that $Z_G(T)/T$-module V^T contains a stable submodule of type $(\chi^a \otimes R(5\varpi_1)) + (\chi^b \otimes R(3\varpi_1))$. Therefore V is not of type (F) in this case. If $a_2 = 0$, then V^T contains a submodule L of type $(\chi^1 \otimes R(3\varpi_1)) + (\chi^{-3} \otimes R(\varpi_1))$. In this case one can identify $(Z_G(T)/T)'$ with SL_2. It follows from [13] and [11] that SL_2-module $R(3\varpi_1) + R(\varpi_1)$ is stable. It is known, [14], that (1,3), (2,2), (3,3), (4,0) are the degrees of elements of a minimal system of homogeneous generators of algebra of invariants of this SL_2-module. Therefore these elements, as Z-semi-invariants, have, resp., the weights $\chi^{-8}, \chi^{-4}, \chi^{-6}, \chi^4$. This shows that L is stable and hence V is not of type (F).

Let $l = 4, a_1 = p = 5$. It follows from table 3 that if $a_2 + a_3 \neq 0$ then $Z_G(T)/T$-module V^T contains a submodule of type $(\chi^a \otimes R(5\varpi_1 + \alpha)) + (\chi^b \otimes R(3\varpi_1 + \beta))$, $ab < 0$, $\alpha, \beta \in X_{++}$. This submodule is stable and therefore V is not of type (F). Otherwise, if $a_2 = a_3 = 0$, then V^T contains a stable submodule of type $\chi^0 \otimes R(3\varpi_1)$, and hence V is again is not of type (F).

Let $l = 3, a_1 = 4$. Then $Z_G(T)/T$-module V^T contains a ubmosule of type $(\chi^a \otimes R(4\varpi_1)) + (\chi^b \otimes 1), ab < 0$, which is stable. Therefore V is not of type (F).

Let $l \geq 3, a_1 = p = 3$. It follows from table 3 that $Z_G(T)/T$-module V^T contains a submodule of type $(\chi^a \otimes R(3\varpi_1 + \alpha)) + (\chi^b \otimes R(\varpi_1 + \beta))$, where $ab < 0, \alpha, \beta \in X_{++}$, and $\beta \neq 0$ if $a_3 + \ldots + a_{l-1} \neq 0$. In the latter case this submodule is stable and hence V is not of type (F). Assume now that $a_3 = \ldots = a_{l-1} = 0$. Consider at first the case $a_2 = d > 0$. It follows from table 3 that $Z_G(T)/T$-module V^T contains a submodule of type $(\chi^{3+2d} \otimes R(3\varpi_1 + 2\varpi_2)) + (\chi^{3+d(1-l)} \otimes R(\varpi_1 + \varpi_2))$, which is stable when $(l, d) \neq (3, 1)$ or $(4, 1)$. If $(l, d) = (4, 1)$ then the second summand of this submodule is stable, and if $(l, d) = (3, 1)$ then it follows from table 3 that V^T contains a submodule of type $(\chi^5 \otimes R(3\varpi_1)) + (\chi^{-3} \otimes R(\varpi_1))$. The same arguments which we use when considered the case $l = 3, a_1 = p = 5$ show that this latter submodule is stable. Finally, assume that $a_2 = 0$ and consider the same torus T_1 as in (1e). Then it is not difficult to see that $Z_G(T_1)/T_1$-module V^{T_1} is of the type $(\chi^{-l+2} \otimes 1) + (\chi^3 \otimes R(3\varpi_1))$ and hence is stable. Therefore we see that V is not of type (F) if $l \geq 3$, $a_1 = p = 3$.

Let now $a_2 = 2$. It follows from table 3 that $Z_G(T)/T$-module V^T contains a submodule of type $(\chi^a \otimes R(2\varpi_1 + \alpha)) + (\chi^b \otimes R(\beta))$; $a > 0$, $b < 0$; $\alpha, \beta \in X_{++}$, where $R(\beta)$ is locally transitive only when either $(a_3, \ldots, a_{l-1}) = (1, 0, \ldots, 0)$ or l is even and $(a_3, \ldots, a_{l-1}) = (0, 1, 0, \ldots, 0)$, $(0, \ldots, 0, 1)$. In all of the other cases this submodule is stable and therefore V is not of type (F). We shall consider these special cases separately. *Assume that* $(a_3, \ldots, a_{l-1}) = (1, 0, \ldots, 0)$ (*and hence* $l \geq 4$). Denote $a_2 = d$. If $d > 0$ then it follows from table 3 that $Z_G(T)/T$-module V^T contains a submodule of type $(\chi^a \otimes R(2\varpi_1 + \alpha)) + (\chi^b \otimes R(\varpi_1 + \varpi_2)), \alpha \in X_{++}, a > 0, b < 0$, which is stable. If $d = 0$ then V^T contains a submodule of type $(\chi^{4-l} \otimes R(3\varpi_1)) + (\chi^5 \otimes R(2\varpi_1 + \varpi_3))$ which is stable when $l > 4$. If $l = 4$ then the first summand of this submodule is stable. Therefore V is not of type (F) in the case under consideration.

Assume that l *is even an* $(a_3, \ldots, a_{l-1}) = (0, 1, 0, \ldots, 0)$ (*and hence* $l \geq 6$). It follows from table 3 that if $d > 0$ and, resp., $d = 0$ then $Z_G(T)/T$-module V^T contains a submodule of type $(\chi^{2+2d} \otimes R(2\varpi_1 + d\varpi_2)) + (\chi^{5-l+d(1-l)} \otimes R(2\varpi_2))$ and, resp., of type $(\chi^6 \otimes R(2\varpi_1 + \varpi_4)) + (\chi^{5-l} \otimes R(2\varpi_1 + \varpi_2))$. Both of these submodules are stable and hence V is not of type (F).

Finally, assume that l *is even and* $(a_3, \ldots, a_{l-1}) = (0, \ldots, 0, 1)$, $l \geq 6$. We obtain from table 3 that if $d > 0$ and, resp., $d = 0$ then $Z_G(T)/T$-module V^T contains a submodule of the type $(\chi^{1+2d+l} \otimes R(2\varpi_1 + d\varpi_2)) + (\chi^{d(1-l)} \otimes R(\varpi_2 + \varpi_{l-3}))$ and, resp., of the type $\chi^0 \otimes 1$. Both of them are stable and hence V is not of type (F).

So we conclude that V *is not of type* (F) *if* $a_1 \geq 2$, $a_l = 0$, $l \geq 3$.

(3b) *Let consider the case* $a_1 = 1$, $l \geq 3$. Assume at first that $l \geq 5$ and $a_3 + \ldots + a_{l-2} \neq 0$. Then it follows from table 3 that $Z_G(T)/T$-module V^T contains a submodule of type $(\chi^a \otimes R(\varpi_1 + \alpha)) + (\chi^b \otimes R(\varpi_1 + \beta))$, $a < 0$; $b > 0$; $\alpha, \beta \neq 0$. This submodule is stable. Hence V is not of type (F).

Now assume that $l \geq 4$, $a_2 = s$, $a_{l-1} = t$ *and* $a_i = 0$ *if* $i \neq 1, 2, l-1$.

It follows from table 3 that $Z_G(T)/T$-module V^T contains a submodule of type $(\chi^{2s+(l-1)t+1}\otimes R(\varpi_1+s\varpi_2))+(\chi^{(1-l)s-2t+1}\otimes R(\varpi_1+t\varpi_{l-3}))$ if $st>0$, of type $(\chi^{(l-1)(t-1)-1}\otimes R(\varpi_1+\varpi_{l-3}))+(\chi^{1-2t}\otimes R(\varpi_1+t\varpi_{l-3}))$ if $s=0, t\geq 2$, and of type $(\chi^{(1-l)(s-1)+3}\otimes R(\varpi_1+\varpi_2))+(\chi^{2s+1}\otimes R(\varpi_1+s\varpi_2))$ if $s\geq 2$, $t=0$. These submodules are stable if $(l,s)\neq(4,2)$. On the other hand, the first summand of the first of these submodules is stable if $(l,s)=(4,2)$. Therefore V is not of type (F) in the cases under consideration.

Assume at last that $l=3$, $a_2=s>0$. If $s=1$ then, as it follows from [15], V is not of type (F). Let $s\geq 2$. Then, using the equalities of SL_4-modules $R(\varpi_2)=\wedge^2(\varpi_1)$ and $S^2R(\varpi_2)=R(2\varpi_1)+1$, [16], one can easily show that $Z_G(T)/T$-module $R(2\varpi_2)^T$ is of the form $(\chi^0\otimes 1)+(\chi^4\otimes 1)+(\chi^{-4}\otimes 1)+(\chi^0\otimes R(2\varpi_1))$. It follows from here and from table 3 that V^T contains a submodule of type $(\chi^{1-2s}\otimes R(\varpi_1))+(\chi^{2s-3}\otimes R(3\varpi_1))$. Now the same arguments as in the case (3a), $l=3, a_1=5$, show that this submodule is stable. Hence V is not of type (F).

So we conclude that if $a_1\neq 0$, $a_l=0$, $l\geq 3$, $(a_1,\dots,a_l)\neq(1,0,\dots,0)$, $(1,1,0,\dots,0)$ *and* $(1,0,\dots,0,1,0)$ *then* V *is not of type* (F). Notice also that G-module $R(\varpi_1)$ is locally transitive and hence of type (F).

(4) The case $a_1a_l\neq 0$, $l\geq 3$.

(4a) *Let* $a_1=a_l=1$. If all the other numerical labels are equal to zero then V is adjoint G-module and therefore, as it is well known, is not of type (F). If $a_2+\ldots+a_{l-1}\neq 0$ then it follows from table 3 that $Z_G(T)/T$-module V^T contains a submodule of type $(\chi^a\otimes R(\varpi_1+\varpi_{l-2}+\alpha))+(\chi^b\otimes R(\varpi_1+\varpi_{l-2}+\beta))$, $a>0$; $b<0$; $\alpha,\beta\in X_{++}$. This submodule is stable and therefore V is not of type (F).

(4b) *Let* $a_1\geq 2, a_l\geq 2$. It follows from table 3 that $Z_G(T)/T$-module V^T contains a submodule of the type $(\chi^a\otimes R(p\varpi_1+\alpha))+(\chi^b\otimes R(q\varpi_{l-2}+\beta))$, $a>0$; $b<0$; $p\geq 2$; $q\geq 2$; $\alpha,\beta\in X_{++}$, which is stable. Therefore V is not of type (F).

(4c) *Let* $a_1=p\geq 3$, $a_l=1$. It follows from table 3 that $Z_G(T)/T$-module V^T contains a submodule of type $(\chi^a\otimes R(p\varpi_1+\varpi_{l-2}+\alpha))+(\chi^b\otimes R((p-2[p/2])\varpi_1+\varpi_{l-2}+\beta))+(\chi^c\otimes R(p-2([p/2]-1)\varpi_1+\varpi_{l-2}+\gamma))$, $a>0$; $b<0$; $\alpha,\beta\in X_{++}$, and, if p is even, also $c<0$. Considered as a SL_{l-1}-module, the first summand is always not locally transitive, the second is not locally transitive if p is odd and the third if p is even. Therefore V is not of type (F).

(4d) *Let* $a_1=2, a_l=1$. If $a_3+\ldots+a_{l-1}\neq 0$ (hence $l\geq 4$) then $Z_G(T)/T$-module V^T contains a submodule of type $(\chi^a\otimes R(\varpi_{l-2}+\alpha))+(\chi^b\otimes R(2\varpi_1+\varpi_{l-2}+\beta))$, $a<0, b>0$, $\alpha,\beta\in X_{++}$, $\alpha\neq 0$. This submodule is stable and therefore V is not of type (F) in this case. If $a_3=\ldots=a_{l-1}=$

0 (hence $l \geq 4$) and $a_2 = d$ then $Z_G(T)/T$-module V^T contains a submodule of type $(\chi^{1+2d} \otimes R(2\varpi_1 + \varpi_{l-2} + d\varpi_2)) + (\chi^{-l+(1-l)(d-1)+2} \otimes R(\varpi_2 + \varpi_{l-1}))$ when $d \geq 2$, and of type $(\chi^3 \otimes R(2\varpi_1 + \varpi_2 + \varpi_{l-2})) + (\chi^{-l+2} \otimes R(\varpi_2 + \varpi_{l-2}))$ when $d = 1$. The exponents of χ in the first summands of these submodules are positive and in the second summands are negative, and these summands, as SL_{l-1}-modules, are not locally transitive. Therefore V is not of type (F) in this case also. If $l = 3$ and $a_2 = d \geq 1$ then it follows from table 3 that $Z_G(T)/T$-module V^T contains a submodule of type $(\chi^{2d+1} \otimes R(3\varpi_1)) + (\chi^{-2d+1} \otimes 2R(\varpi_1))$. This submodule is stable because SL_2-modules $R(3\varpi_1), 2R(\varpi_1)$ are not locally transitive, and $2d + 1 > 0, -2d + 1 < 0$. Hence V is not of type (F) again.

So we conclude that if $a_1 a_l \neq 0, l \geq 3$ *and* $(a_1, \ldots, a_l) \neq (2, 0, \ldots, 0, 1)$ *then* V *is not of type* (F).

(5) The case $(\mathbf{a_1}, \ldots, \mathbf{a_l}) = (\mathbf{2}, \mathbf{0}, \ldots, \mathbf{0}, \mathbf{1})$, $\mathbf{l} \geq \mathbf{3}$.

One has the following equalities of G-modules (e.g., [16]): $R(2\varpi_1) \otimes R(\varpi_l) = R(\varpi_1) + R(2\varpi_1 + \varpi_l)$, $R(\varpi_1)^* = R(\varpi_l)$ and $R(2\varpi_1) = S^2 R(\varpi_1)$. It is well known, [10], that the weights of G-module $R(\varpi_1)$ are $\varepsilon_1, \ldots, \varepsilon_{l+1}$ and all of them have multiplicity 1. It easily follows from these facts that the weights of G-module $R(2\varpi_1 + \varpi_l)$ are $\varepsilon_i + \varepsilon_j - \varepsilon_k$, where $1 \leq i, j, k \leq l + 1$, and the multiplicity of the weight equals to 1 if $k \neq i, j$ and equals to l otherwise. Now consider the following l weights of this G-module: $\lambda_1 = \varepsilon_1 + \varepsilon_{l+1} - \varepsilon_2$, $\lambda_2 = 2\varepsilon_2 - \varepsilon_3$, $\lambda_3 = 2\varepsilon_3 - \varepsilon_4, \ldots, \lambda_{l-1} = 2\varepsilon_{l-1} - \varepsilon_l$, $\lambda_l = 2\varepsilon_l$. They are linearly dependent with positive coefficients: $\lambda_1 + \lambda_2 + \ldots + \lambda_l = 0$. Therefore zero is the internal point of the convex envelope of the set $\{\lambda_1, \ldots, \lambda_l\}$, and the rank of this set is less than $l = \operatorname{rk} G$. Moreover, $\lambda_i - \lambda_j \notin \Sigma$ for each $i \neq j$. It follows now from proposition 5 that V is not of type (F).

(6) The case $(\mathbf{a_1}, \ldots, \mathbf{a_l}) = (\mathbf{1}, \mathbf{1}, \mathbf{0}, \ldots, \mathbf{0})$, $\mathbf{l} \geq \mathbf{3}$. Using the equalities of G-modules $R(\varpi_1) \otimes R(\varpi_2) = R(\varpi_1 + \varpi_2) + R(\varpi_3)$, $R(\varpi_2) = \wedge^2 R(\varpi_1)$, $R(\varpi_3) = \wedge^3 R(\varpi_1)$, see [16], one can show that the weigts of G-module $R(\varpi_1 + \varpi_2)$ are $\varepsilon_i + \varepsilon_j - \varepsilon_k$, where at least two of the indices i, j, k are different, and also that the multiplicity of the weight is equal to 2 if all the indices are different and is equal to 1 otherwise.

Let $l + 1 = 3k$, $k \in \mathbf{Z}$. Then $\lambda_1 = \varepsilon_1 + \varepsilon_2 + \varepsilon_3$, $\lambda_2 = \varepsilon_4 + \varepsilon_5 + \varepsilon_6$, $\ldots, \lambda_k = \varepsilon_{3k-2} + \varepsilon_{3k-1} + \varepsilon_{3k}$ are the weights of G-module V which have the properties $\lambda_i - \lambda_j \notin \Sigma$ for each $i \neq j$, and $\lambda_1 + \ldots \lambda_k = 0$. As in n.(5), it follows from here that V is not of type (F).

Let $l + 1 = 3k + 1$, $k \in \mathbf{Z}$. Then we consider the following weights of G-module V: $\lambda_1 = \varepsilon_1 + \varepsilon_2 + \varepsilon_3$, $\lambda_2 = \varepsilon_4 + \varepsilon_5 + \varepsilon_6$, $\ldots$, $\lambda_{k-1} = \varepsilon_{3k-5} +$

$\varepsilon_{3k-4}+\varepsilon_{3k-3}$, $\lambda_k = 2\varepsilon_{3k-2}+\varepsilon_{3k-1}$, $\lambda_{k+1} = 2\varepsilon_{3k-1}+\varepsilon_{3k}$, $\lambda_{k+2} = 2\varepsilon_{3k}+\varepsilon_{3k+1}$, $\lambda_{k+3} = 2\varepsilon_{3k+1}+\varepsilon_{3k-2}$. We have $\lambda_i - \lambda_j \notin \Sigma$ for each $i \neq j$, and $3\lambda_1+3\lambda_2+\ldots+3\lambda_{k-1}+\lambda_k+\lambda_{k+1}+\lambda_{k+2}+\lambda_{k+3} = 0$. Therefore, as above, we conclude that V is not of type (F).

Let $l+1 = 3k+2$, $k \in \mathbf{Z}$. Then we consider the following weights of G-module V: $\lambda_1 = \varepsilon_1+\varepsilon_2+\varepsilon_3$, $\lambda_2 = \varepsilon_4+\varepsilon_5+\varepsilon_6, \ldots, \lambda_k = \varepsilon_{3k-2}+\varepsilon_{3k-1}+\varepsilon_{3k}$, $\lambda_{k+1} = \varepsilon_{3k}+\varepsilon_{3k+1}+\varepsilon_{3k+2}$, $\lambda_{k+2} = 2\varepsilon_{3k+1}+\varepsilon_{3k-2}$, $\lambda_{k+3} = 2\varepsilon_{3k+2}+\varepsilon_{3k-1}$. We have $\lambda_i - \lambda_j \notin \Sigma$ for each $i \neq j$, and $3\lambda_1 + 3\lambda_2 + \ldots + 3\lambda_{k-1} + 2\lambda_k + \lambda_{k+1} + \lambda_{k+2} + \lambda_{k+3} = 0$. Hence V is not of type (F).

(7) The case $(\mathbf{a}_1, \ldots, \mathbf{a}_l) = (1, 0, \ldots, 0, 1, 0)$, $l \geq 4$.

Here we use the following equalities of G-modules: $R(\varpi_1 \otimes R(\varpi_{l-1}) = R(\varpi_1+\varpi_{l-1})+R(\varpi_l)$, $R(\varpi_l) = R(\varpi_1)^*$, $R(\varpi_{l-1}) = (\wedge^2 R(\varpi_1))^*$ (see [16]). Using these equalities, one can easily see that the weights of G-module $R(\varpi_1 + \varpi_{l-1})$ are $\varepsilon_k - \varepsilon_i - \varepsilon_j$, $1 \leq i, j, k \leq l+1$, $i \neq j$, and also that the multiplicity of the weight is equal to 1 if $i, j \neq k$ and equal to $l-1$ otherwise.

Let $l+1 = 2k$, $k \in \mathbf{Z}$. We consider the following weights of G-module V: $\lambda_1 = \varepsilon_{2k} - \varepsilon_1 - \varepsilon_2$, $\lambda_2 = \varepsilon_{2k} - \varepsilon_3 - \varepsilon_4$, $\lambda_3 = \varepsilon_{2k} - \varepsilon_5 - \varepsilon_6, \ldots,$ $\lambda_{k-1} = \varepsilon_{2k} - \varepsilon_{2k-3} - \varepsilon_{2k-2}$, $\lambda_k = \varepsilon_1 - \varepsilon_2 - \varepsilon_{2k-1}$, $\lambda_{k+1} = \varepsilon_2 - \varepsilon_1 - \varepsilon_{2k-1}$, $\lambda_{k+2} = -\varepsilon_{2k}$. Then we have $\lambda_i - \lambda_j \notin \Sigma$ for each $i \neq j$, and $2\lambda_1 + 2\lambda_2 + \ldots + 2\lambda_{k-1} + \lambda_k + \lambda_{k+1} + (2k-1)\lambda_{k+2} = 0$. Hence V is not of type (F).

Let $l+1 = 2k+1$, $k \in \mathbf{Z}$. Now we consider the following weights of G-module V: $\lambda_1 = \varepsilon_{2k+1} - \varepsilon_1 - \varepsilon_2, \lambda_2 = \varepsilon_{2k+1} - \varepsilon_3 - \varepsilon_4, \lambda_3 = \varepsilon_{2k+1} - \varepsilon_5 - \varepsilon_6, \ldots, \lambda_k = \varepsilon_{2k+1} - \varepsilon_{2k-1} - \varepsilon_{2k}, \lambda_{k+1} = -\varepsilon_{2k+1}$. Then $\lambda_i - \lambda_j \notin \Sigma$ if $i \neq j$, and $\lambda_1 + \lambda_2 + \ldots + \lambda_k + (k+1)\lambda_{k+1} = 0$. Hence V is not of type (F).

(8) The case $\mathbf{l} = \mathbf{1}$.

If a_1 is even then zero is the weight of $V = R(a_1\varpi_1)$ and therefore V is not of type (F) (see section 2 above). If a_1 is odd then zero is not the weight of V. It is well known that the stabilizer of any semisimple point is reductive (e.g., see [1]). But it is also well known that the identity component of any reductive subgroup of SL_2 is either identity or a maximal torus of SL_2. This shows that V is G-module of type (F) if a_1 is odd.

The results of this section give, all together, the proof of the following assertion which is the main result of the section:

Proposition 9. *If $G = SL_{l+1}$ then, to within of taking the duals, the unique simple G-modules of type (F) are $R(\varpi_1)$; $R(\varpi_2)$ if l is even; $R(p\varpi_1)$ if $l = 1$ and p is odd.*

5. Nonsimple G-modules of type (F)

Our starting point here will be proposition 2 and the description of simple G-modules of type (F) given in propositions 6, 7, 8, 9.

(1) *Let G be the group of type D_5.* It follows from [11] and [13] that G-modules $2R(\varpi_4)$, $2R(\varpi_5)$ and $R(\varpi_4)+R(\varpi_5)$ are stable and that their s.g.p. have positive dimension. Therefore these G-modules are not of type (F). This shows that there are no nonsimple G-modules of type (F).

(2) *Let G be the group of type C_l.* Denote by H the s.g.p. of G-module $V = mR(\varpi_1)$, $m \geq 2$. Then it follows from [11] that $\dim H > 0$ iff either $m = 2k, k < l$ (and H is reductive in this case), or $m = 2k-1$, $k \leq l$ (and H is not reductive in this case). In the first case V is stable because of [13], and hence V is not of type (F). The group G being semisimple, we have $\dim V/G = \dim V - \dim G + \dim H$. It follows from here and from the information on $\dim H$, see [11], that one has $\dim V/G > 0$ in the second case. Therefore, as proposition 1 shows, in this case V is also not of type (F).

Assume now that neither of these cases takes place. Let T be the one-dimensional torus defined by the root $\alpha = \varpi_1$. As it is proved in [8], $Z_G(T)/T$ is the group of type C_{l-1}, and if L is G-module of type $R(\varpi_1)$ then L^T is a $Z_G(T)^0/T$-module of type $R(\varpi_1)$. Therefore V^T is $Z_G(T)^0/T$-module of type $mR(\varpi_1)$. It follows from the previous discussion that the s.g.p. of this module is finite and hence this module is stable. So in this case V is also not of type (F).

This shows that there are no nonsimple G-modules of type (F).

(3) *Let $G = SL_2$.* Then the same arguments as we used in n.(8) of section 4 show that the sum of any simple G-modules of type (F) also is a G-module of type (F).

(4) *Let $G = SL_{l+1}, l \geq 2$.* It follows from [11], [13] that G-modules $R(\varpi_1)+R(\varpi_l)$ and, if l is even, $l \geq 4$, $R(\varpi_1)+R(\varpi_2)$, $R(\varpi_2)+R(\varpi_{l-1})$ are stable and have infinite s.g.p. Hence these G-modules are not of type (F).

It follows from [11] that $2R(\varpi_2)$ for l even, $l \geq 4$, is locally transitive, and hence of type (F).

If l is even, $l \geq 4$, and $V = 3V(\varpi_2)$, then it follows from table 3 that V^T is $Z_G(T)/T$-module of type $(\chi^{1-l} \otimes (3 \cdot 1)) + (\chi^2 \otimes (3R(\varpi_2))$. Using that SL_{l-1}-module $3R(\varpi_2)$ is not locally transitive, [11], we obtain from here that V^T is stable, and hence V is not of type (F).

It follows from [11] that $mR(\varpi_1)$ is locally transitive (and hence of type (F)) if $m < l+1$. Otherwise it is not locally transitive but still of type

(F) (see the proof of corollary to theorem in section 1).

Finally, let l is even, $l \geq 4$, and $V = R(\varpi_2) + mR(\varpi_l)$. It follows from the tables in [11] that V is locally transitive (and hence of type (F)) if $m = 1$, and that s.g.p is infinite and $\dim V/G > 0$ (and hence V is not of type (F), see n.(2) of this section) if $2 \leq m < l$. If $m \geq l$ then s.g.p. is finite and we need another arguments. We consider the two-dimensional torus T_1 of G which we intriduced in n.(1e) of section 4. Then it is not difficult to check that V^{T_1} is $Z_G(T_1)/T_1$-module of type $(\psi^2 \otimes R(\varpi_2)) + (\psi^{-1} \otimes mR(\varpi_{l-3})$. Using that SL_{l-2}-modules $R(\varpi_2)$ and $mR(\varpi_{l-3})$ for $m \geq l > l-3$ are not locally transitive, we obtain that V^{T_1} is stable and hence V is not of type (F).

Therefore we proved the following

Proposition 10. *Nonsimple G-modules V of type (F) of connected simple groups G are exhausted, to within of taking the duals, by the following list:*

$G = SL_2$, $V = R(p_1\varpi_1) + \ldots + R(p_s\varpi_s)$, $s \geq 2$, *all p_i are odd,*
$G = SL_{l+1}$, $l \geq 2$, $V = mR(\varpi_1)$, $m \geq 2$,
$G = \mathrm{SL}_{l+1}$, *l is even* ≥ 4, $V = 2R(\varpi_2)$ *or* $R(\varpi_2) + R(\varpi_l)$.

Acknowledgement: I wish to thank the Mathematical Institute of Basel University, where this paper was written up, for its hospitality. Especially I am grateful to H. Kraft and also to D. Wehlau for the help on preparing the paper.

REFERENCES

[1] Luna, D.: *Slices étales.* Bull. Soc. Math. France, Mem. **33** (1973), 81–105

[2] Kac, V.G.; Popov, V.L. Vinberg, È.B.: *Sur les groupes linéaires algébriques dont l'algèbre des invariants est libre.* C. R. Acad. Sci. Paris **283 A** (1976), 865–878

[3] Nakajima, H.: *Representations of a reductive algebraic group whose algebras of invariants are complete intersections.* J. reine angew. Math. **367** (1986), 115–138

[4] Luna, D.: *Adherences d'orbite et invariants.* Inv. math. **29** (1975), 231–238

[5] Dadok, J.; Kac, V.G.: *Polar representations.* J. Algebra **92** (1985), 504–524

[6] Schwarz, G.W.: *Representations of simple Lie groups with regular rings of invariants.* Inv. math. **49** (1978), 167–191

[7] Gordeev, N.L.: *Finite linear groups whose algebra of invariants are complete intersections.* Math. USSR-Izv. **28** (1987), 335- 379

[8] Popov, V.L.: *Syzygies in the theory of invariants.* Math. USSR-Izv. **22** (1984), 507–585

[9] Bourbaki, N.: *Groupes et algèbres de Lie. Ch. 4-6.* Hermann, Paris (1968)

[10] Bourbaki, N.: *Groupes et algèbres de Lie. Ch. 7-8.* Hermann, Paris (1975)

[11] Èlashvili, A.G.: *A canonical form and stationary subalgebras of points in general position for simple linear groups.* Funct. Anal. Appl. **6** (1972), 44–53

[12] Èlashvili, A.G.: *Stationary subalgebras of points in general position for irreducible linear Lie groups..* Funct. Anal. Appl. **6** (1972), 139–148

[13] Popov, V.L.: *Stability criteria for the action of a semisimple group on a factorial manifold.* Math. USSR-Izv. **4** (1970), 527–535

[14] Springer, T.A.: *Invariant Theory.* Lect. Notes Math., Springer Verlag **585** (1977)

[15] Panyushev, D.I.: *Classification of four-dimensional anti-commutative algebras with a nontrivial group of automorphisms.* In: Questions of theory of groups and of homological algebra. Yaroslavl. (In Russian) (1978), 98–106

[16] Onishchik, A.L.; Vinberg, È.B.: *Seminar on algebraic groups and Lie groups (1967/68).* Izdat. Moskov. Gos. Univ. (In Russian) (1969)

[17] Gordeev, N.L.: *Coranks of elements of linear groups and complexity of algebras of invariuants.* To appear (In Russian) (1989)

[18] Wolf, J.A.: *Spaces of Constant Curvature.* McGraw-Hill, New York (1967)

[19] Popov, V.L.: *Modern developments in Invariant theory.* Proceedings of the International Congress of Mathematicians. Berkeley,U.S.A. (1986), 394–405

Received February 9, 1990

Moscow Institute of Electronical Constructions (MIEM)
per B. Vuzovskii 3/12
Moscow 109028, USSR

$\mathbf{C}^*$-Actions On Affine Space

HANSPETER KRAFT*

Dedicated to Jacques Dixmier on his 65th birthday

Introduction

Over the last few years algebraic group actions on affine space have been intensively studied. One of the central questions in this area is the following. (We refer to [K2] for a survey and related problems).

Linearization Problem. *Is every action of a reductive algebraic group G on affine space $\mathbf{C}^n$ linearizable, i.e., G-equivariantly isomorphic to a linear action?*

For example, every reductive group action on $\mathbf{C}^2$ is linearizable. This is a consequence of the structure of the group of algebraic automorphisms of $\mathbf{C}^2$ as an *amalgamated product*, a result which can be traced back to VAN DER KULK. Such a structure theorem does not hold in higher dimension, due to an example of BASS. Nevertheless, the linearization problem has got a positive answer for semisimple groups in dimension $n \leq 4$, by the work of KRAFT, POPOV and PANYUSHEV ([KP], [Pa]).

All these results depend on some "smallness" assumptions. E.g., linearization always holds whenever the only G-invariant functions are constant. The more general situation where the ring of invariants has (Krull-) dimension 1 has been studied by LUNA, SCHWARZ and KRAFT (see [KS1], [KS2]). It turns out that in many cases one can still prove linearization, but in general there exist infinite families of nonequivalent actions on affine space. The first examples of such non-linearizable actions were found by SCHWARZ [Sch], e.g., actions of O_2 on $\mathbf{C}^4$ and of SL_2 on $\mathbf{C}^7$.

Using these examples and techniques KNOP showed that for every *semisimple* group there exist non-linearizable actions on some $\mathbf{C}^n$, and

* Partially supported by Schweizerischer Nationalfonds

addition, the quotient $\mathbf{C}^3/\!/\mathbf{C}^*$ is smooth then $\alpha = 1$, and vice versa. In this case the quotient map $\pi : \mathbf{C}^3 \to \mathbf{C}^2$ is given by $(x, y, z) \mapsto (x^\beta y, x^\gamma z)$.

1.3. Theorem. *Let X be a smooth affine variety of dimension 3. Assume that $\mathbf{C}^*$ acts semifreely on X with a unique fixed point $x_0 \in X$ and that $X/\!/\mathbf{C}^* \simeq \mathbf{C}^2$. Then the action is linearizable, i.e., X is $\mathbf{C}^*$-equivariantly isomorphic to the tangent representation $T_{x_0}(X)$.*

For the proof we need some preparation. First, we remark that the representation on $V := T_{x_0}(X)$ is semifree and has a smooth quotient: $V/\!/\mathbf{C}^* \simeq \mathbf{C}^2$; this follows from the assumptions and the slice theorem. As a consequence, the weights on V are $(-1, \beta, \gamma)$ where $\beta, \gamma > 0$ and the quotient map $\pi_V : V = \mathbf{C}^3 \to \mathbf{C}^2$ is given by $\pi_V(x, y, z) = (x^\beta y, x^\gamma z)$ with respect to suitable coordinates (see Example 1.2). We fix an identification $X/\!/\mathbf{C}^* = \mathbf{C}^2$ such that $\pi_X(x_0) = 0$. In order to obtain a $\mathbf{C}^*$-equivariant isomorphism $\varphi : X \xrightarrow{\sim} V$ we proceed in three steps:

(I) We construct an isomorphism $\dot\varphi : \dot X \xrightarrow{\sim} \dot V$ of the complements of the zero fibers $\dot X := X \setminus \pi_X^{-1}(0)$ and $\dot V := V \setminus \pi_V^{-1}(0)$ such that $\pi_V \circ \dot\varphi = \pi_X$.

(II) From the slice theorem we obtain an isomorphism $\hat\varphi : \hat X \xrightarrow{\sim} \hat V$ of étale neighborhoods $\hat X \supset \pi_X^{-1}(0)$ and $\hat V \supset \pi_V^{-1}(0)$ of the zero fibers.

(III) Finally, we modify the isomorphisms $\dot\varphi$ and $\hat\varphi$ in such a way that they coincide on the intersection $\dot X \cap \hat X$. This will define, by gluing, an isomorphism $\varphi : X \xrightarrow{\sim} V$.

The same method applies to higher dimensional varieties X and yields the following result. Only step III needs some modification; we leave the details to the reader.

1.4. Theorem. *Consider a semifree action of $\mathbf{C}^*$ on a smooth affine variety X with a unique fixed point $x_0 \in X$. Assume that $X/\!/\mathbf{C}^* \simeq \mathbf{C}^k$ for some k. Then X is $\mathbf{C}^*$-equivariantly isomorphic to $T_{x_0}X$.*

1.5. We now give the details of the construction outlined above.

Step I: Define $\dot{\mathbf{C}}^2 := \mathbf{C}^2 \setminus \{0\}$. Since the actions on X and V are both semifree (with a unique fixed point) it follows from the slice theorem that the induced maps

$$\pi_V : \dot V \to \dot{\mathbf{C}}^2 \quad \text{and} \quad \pi_X : \dot X \to \dot{\mathbf{C}}^2$$

are principal $\mathbf{C}^*$-fibrations. They are trivial, because $\mathbf{C}^2$ is factorial and so every line bundle on $\mathbf{C}^2$ or $\dot{\mathbf{C}}^2$ is trivial. Thus we get a $\mathbf{C}^*$-equivariant isomorphism $\dot{\varphi} : \dot{X} \xrightarrow{\sim} \dot{V}$ with a commutative diagram

$$\begin{array}{ccc} \dot{X} & \xrightarrow{\dot{\varphi}} & \dot{V} \\ {\scriptstyle \pi_X}\downarrow & & \downarrow{\scriptstyle \pi_V} \\ \mathbf{C}^2 & = & \mathbf{C}^2 \end{array} \tag{$*$}$$

This finishes the first step.

Step II: We put $\widehat{\mathbf{C}^2} := \operatorname{Spec} \mathbf{C}[\![u,v]\!]$. There is a canonical morphism $\widehat{\mathbf{C}^2} \to \mathbf{C}^2$ from which we obtain the following fiber products (which can be seen as limits over all "saturated" étale neighborhoods[1] of the zero fiber):

$$\begin{array}{ccc} V \times_{\mathbf{C}^2} \widehat{\mathbf{C}^2} =: \hat{V} & \longrightarrow & V \\ \downarrow & & \downarrow \\ \widehat{\mathbf{C}^2} & \longrightarrow & \mathbf{C}^2 \end{array} \qquad \begin{array}{ccc} X \times_{\mathbf{C}^2} \widehat{\mathbf{C}^2} =: \hat{X} & \longrightarrow & X \\ \downarrow & & \downarrow \\ \widehat{\mathbf{C}^2} & \longrightarrow & \mathbf{C}^2 \end{array}$$

The coordinate ring of $\hat{V}$ is given by

$$\mathcal{O}(\hat{V}) = \mathcal{O}(V) \otimes_{\mathbf{C}[u,v]} \mathbf{C}[\![u,v]\!] \quad \text{where} \quad u := x^{\beta} y, \; v := x^{\gamma} z,$$

i.e., it is the $\mathbf{C}^*$-finite part of the completion of $\mathcal{O}(V)$ in the ideal generated by u and v (cf. [Ma]), and similarly for $\hat{X}$, where we use the identification $\mathcal{O}(X)^{\mathbf{C}^*} = \mathbf{C}[u,v]$. It is again a consequence of the slice theorem that there is a $\mathbf{C}^*$-equivariant isomorphism $\hat{\varphi} : \hat{X} \xrightarrow{\sim} \hat{V}$. This finishes the second step.

Step III: The two isomorphisms $\hat{\varphi}$ and $\dot{\varphi}$ both induce $\mathbf{C}^*$-equivariant isomorphisms $\dot{\hat{X}} \xrightarrow{\sim} \dot{\hat{V}}$ of the "intersections"

$$\begin{aligned} \dot{\hat{X}} &:= \dot{X} \cap \hat{X} := \hat{X} \setminus \pi_X^{-1}(0) = \dot{X} \times_{\mathbf{C}^2} \widehat{\mathbf{C}^2} \qquad \text{and} \\ \dot{\hat{V}} &:= \dot{V} \cap \hat{V} := \hat{V} \setminus \pi_V^{-1}(0) = \dot{V} \times_{\mathbf{C}^2} \widehat{\mathbf{C}^2}. \end{aligned}$$

This is clear for $\hat{\varphi}$ since $\hat{\varphi}$ maps $\pi_X^{-1}(0)$ onto $\pi_V^{-1}(0)$. For $\dot{\varphi}$ it follows from the diagram $(*)$ by base change. The next lemma is easy.

1.6. Lemma. *Assume that $\hat{\varphi}$ and $\dot{\varphi}$ induce the same isomorphism $\dot{\hat{X}} \xrightarrow{\sim} \dot{\hat{V}}$. Then there is a $\mathbf{C}^*$-equivariant isomorphism $\varphi : X \xrightarrow{\sim} V$ such that $\varphi|_{\dot{X}} = \dot{\varphi}$ and $\varphi|_{\hat{X}} = \hat{\varphi}$.*

Proof. Clearly, we have $\mathcal{O}(\dot{V}), \mathcal{O}(\hat{V}) \subset \mathcal{O}(\dot{\hat{V}})$ in a canonical way, and $\mathcal{O}(V) = \mathcal{O}(\dot{V}) \cap \mathcal{O}(\hat{V})$ since every function $f \in \mathcal{O}(\dot{V}) \setminus \mathcal{O}(V)$ has poles along

1 saturated means inverse image of a subset of the quotient

the plane $\{x = 0\} \subset \pi_V^{-1}(0)$ and does therefore not belong to $\mathcal{O}(\hat{V})$. Similarly, $\mathcal{O}(\dot{X}), \mathcal{O}(\hat{X}) \subset \mathcal{O}(\hat{\dot{X}})$, and $\mathcal{O}(X) = \mathcal{O}(\dot{X}) \cap \mathcal{O}(\hat{X})$. By assumption, the two isomorphisms $\dot{\varphi}^* : \mathcal{O}(\dot{V}) \xrightarrow{\sim} \mathcal{O}(\dot{X})$ and $\hat{\varphi}^* : \mathcal{O}(\hat{V}) \xrightarrow{\sim} \mathcal{O}(\hat{X})$ are both induced from the same isomorphism $\psi : \mathcal{O}(\hat{\dot{V}}) \xrightarrow{\sim} \mathcal{O}(\hat{\dot{X}})$. Hence, ψ determines an isomorphism $\varphi^* : \mathcal{O}(V) = \mathcal{O}(\dot{V}) \cap \mathcal{O}(\hat{V}) \xrightarrow{\sim} \mathcal{O}(\dot{X}) \cap \mathcal{O}(\hat{X}) = \mathcal{O}(X)$. Q.E.D.

We cannot assume a priori that $\dot{\varphi}$ and $\hat{\varphi}$ induce the same isomorphism $\hat{\dot{X}} \xrightarrow{\sim} \hat{\dot{V}}$. Therefore, we consider the automorphism $\rho := \dot{\varphi} \circ \hat{\varphi}^{-1}$ of $\hat{\dot{V}}$. By the following proposition there is a $\mathbf{C}^*$-equivariant automorphism $\dot{\psi}$ of $\dot{V}$ such that the composition $\dot{\psi} \circ \rho$ extends to an automorphism $\hat{\psi}$ of $\hat{V}$. Putting $\dot{\varphi}_1 := \dot{\psi} \circ \dot{\varphi} : \dot{X} \xrightarrow{\sim} \dot{V}$ and $\hat{\varphi}_1 := \hat{\psi} \circ \hat{\varphi} : \hat{X} \xrightarrow{\sim} \hat{V}$ we obtain

$$\hat{\varphi}_1|_{\hat{\dot{X}}} = \dot{\psi} \circ \rho \circ \hat{\varphi} = \dot{\psi}|_{\hat{\dot{V}}} \circ \dot{\varphi}_{\hat{\dot{X}}} = \dot{\varphi}_1|_{\hat{\dot{X}}}.$$

By Lemma 1.6 this finishes the last step of the proof of Theorem 1.3. Q.E.D.

1.7. Proposition. *Let ρ be a $\mathbf{C}^*$-equivariant automorphism of $\hat{\dot{V}}$. Then there is a $\mathbf{C}^*$-equivariant automorphism $\dot{\psi}$ of $\dot{V}$ such that the composition $\dot{\psi} \circ \rho$ extends to an automorphism $\hat{\psi}$ of $\hat{V}$.*

This is a special case of Proposition 2.3 of the next paragraph whose proof will be given in 3.3.

§ 2. A generalization

For our main results we need a more general version of Theorem 1.3.

2.1. Consider the linear action of $\mathbf{C}^*$ on $W := \mathbf{C}^3$ with weights $(-1, \beta, \gamma)$ where $\beta, \gamma > 0$ (see Example 1.2). Let α be a positive integer. Assume that the cyclic group $\mu_\alpha := \{s \in \mathbf{C}^* \mid s^\alpha = 1\}$ acts on W by $\mathbf{C}^*$-equivariant automorphisms. The zero fiber $W^0 := \pi_W^{-1}(0)$ is the union of the (y,z)-plane and the x-axis, and both components are stable under the action of μ_α. In particular, there is a $\lambda \in \mathbf{Z}$ such that $s \cdot (x, 0, 0) = (s^\lambda x, 0, 0)$ for all $s \in \mu_\alpha, x \in \mathbf{C}$.

2.2. Theorem. *The $\mathbf{C}^* \times \mu_\alpha$-action on W is linearizable.*

Proof. We follow the same steps as in the proof of Theorem 1.3. Let $V := T_0 W$ be the tangent representation of $\mathbf{C}^* \times \mu_\alpha$ in the origin. By assumption, $V /\!\!/ \mathbf{C}^*$ and $W /\!\!/ \mathbf{C}^*$ are both isomorphic to $\mathbf{C}^2$. Using the slice theorem we see that the μ_α-actions on the quotients are locally isomorphic (i.e., induce equivalent tangent representation at $\pi_W(0)$ and $\pi_V(0)$). Since every action of a reductive group on $\mathbf{C}^2$ is linearizable (see [K2, 1.3 Corollary 2]) we can identify $W /\!\!/ \mathbf{C}^*$ and $V /\!\!/ \mathbf{C}^*$ with $\mathbf{C}^2$ where μ_α acts linearly on $\mathbf{C}^2$.

There is a $\mathbf{C}^*$-equivariant isomorphism $\mathbf{C}^* \times W /\!\!/ \mathbf{C}^* \xrightarrow{\sim} W \setminus \{x = 0\}$ given by $(t, (u, v)) \mapsto (t, t^\beta u, t^\gamma v)$, and similarly for V. It is easy to see that every lift of the μ_α-action on $W /\!\!/ \mathbf{C}^*$ to an action on $\mathbf{C}^* \times W /\!\!/ \mathbf{C}^*$ commuting with the $\mathbf{C}^*$-action is of the form $s \cdot (t, \bar{w}) = (s^\lambda t, s \cdot \bar{w})$ with some $\lambda \in \mathbf{Z}$. But this λ is the same for W and V: It is given by the action of μ_α on the x-axis (see 2.1). Hence there is a $\mathbf{C}^* \times \mu_\alpha$-equivariant isomorphism $\dot{\varphi} : W \setminus \{x = 0\} \xrightarrow{\sim} V \setminus \{x = 0\}$ with a commutative diagram

$$\begin{array}{ccc} W \setminus \{x = 0\} & \longrightarrow & V \setminus \{x = 0\} \\ {\scriptstyle \pi_W}\big\downarrow & & \big\downarrow{\scriptstyle \pi_V} \\ \mathbf{C}^2 & = & \mathbf{C}^2 \end{array}$$

This finishes the first step.

The second step is again a consequence of the slice theorem: There is a $\mathbf{C}^* \times \mu_\alpha$-equivariant isomorphism $\hat{W} \xrightarrow{\sim} \hat{V}$.

For the last step we want to apply again Lemma 1.6. It is clear that we can follow the same lines as in the proof of Theorem 1.3 once we have established the following generalization of Proposition 1.7. This will finish the proof of Theorem 2.2. Q.E.D.

2.3. Proposition. *Let ρ be a $\mathbf{C}^* \times \mu_\alpha$-equivariant automorphism of $\dot{\hat{V}}$. Then there is a $\mathbf{C}^* \times \mu_\alpha$-automorphism $\dot{\psi}$ of $\dot{V}$ such that the composition $\dot{\psi} \circ \rho$ extends to an automorphism $\hat{\psi}$ of $\hat{V}$.*

(The proof will be given in the next paragraph.)

Combining 1.3 and 2.2 we obtain the following result.

2.4. Theorem. *Let X be a smooth affine variety of dimension 3. Assume that $\mathbf{C}^* \times \mu_\alpha$ acts on X satisfying the following conditions:*

(a) *The $\mathbf{C}^*$-action is semifree with a unique fixed point $x_0 \in X$.*

(b) $X /\!\!/ \mathbf{C}^* \simeq \mathbf{C}^2$.

Then X is $\mathbf{C}^* \times \mu_\alpha$-equivariantly isomorphic to $T_{x_0}X$.

§ 3. Automorphisms of $\hat{V}$

Let $\mathrm{Aut}_0\,\mathbf{C}[u,v]$ denote the group of those automorphisms of the $\mathbf{C}$-algebra $\mathbf{C}[u,v]$ which stabilize the homogeneous maximal ideal. We have a canonical inclusion $\mathrm{Aut}_0\,\mathbf{C}[u,v] \hookrightarrow \mathrm{Aut}\,\mathbf{C}[\![u,v]\!]$.

3.1. Lemma. *Given an automorphism $\sigma \in \mathrm{Aut}\,\mathbf{C}[\![u,v]\!]$ and a natural number $k > 0$, there is an $\eta \in \mathrm{Aut}_0\,\mathbf{C}[u,v]$ such that the composition $\tilde{\sigma} := \eta\circ\sigma$ satisfies the following condition:*

(C_k) *$\tilde{\sigma}(v) = vf + u^k h$ for suitable $h \in \mathbf{C}[\![u]\!]$, $f \in \mathbf{C}[\![u,v]\!]$, $f(0,0) = 1$.*

(Condition C_k essentially means that the power series $\tilde{\sigma}(v)|_{v=0}$ has a zero of order $\geq k$ in $u = 0$.)

Proof. First we find an $\eta_1 \in \mathrm{GL}_2(\mathbf{C}) \subset \mathrm{Aut}_0\,\mathbf{C}[u,v]$ such that $\sigma_1 := \eta_1\circ\sigma$ has the form

$$\sigma_1(u) = u + a, \qquad \sigma_1(v) = v + b$$

where $a, b \in (u,v)^2 \subset \mathbf{C}[\![u,v]\!]$. In particular, σ_1 satisfies condition C_2. Next, by induction, we may assume that σ_1 satisfies condition C_i. Define $\eta_2 \in \mathrm{Aut}_0\,\mathbf{C}[u,v]$ by $\eta_2(u) = u$, $\eta_2(v) = v - h(0)u^i$. An easy calculation then shows that $\tilde{\sigma} := \eta_2\circ\sigma_1$ satisfies condition C_{i+1}. Q.E.D.

3.2. Remark. In the notations of the lemma above assume in addition that the cyclic group μ_α acts linearly on $\mathbf{C}[\![u,v]\!]$ with weights β, γ, i.e., $s \cdot u = s^\beta u$ and $s \cdot v = s^\gamma v$. Then an equivariant version of the lemma holds as well:

Given a μ_α-equivariant automorphism σ of $\mathbf{C}[\![u,v]\!]$ and a natural number $k > 0$ there is a μ_α-equivariant automorphism $\eta \in \mathrm{Aut}_0\,\mathbf{C}[u,v]$ such that the composition $\eta\circ\sigma$ satisfies condition C_k of Lemma 3.1.

(In fact, in the proof above the linear automorphism η_1 is clearly μ_α-equivariant. Hence, we can write $\sigma_1(v)$ in the form $\sigma_1(v) = vf + u^i h$ where $h \in \mathbf{C}[\![u]\!]$, $h(0) \neq 0$ and $f(0,0) = 1$. Since $\sigma_1(v)$ is a semi-invariant of weight γ (with respect to μ_α), u^i is also a semi-invariant of weight γ, i.e., $\beta i \equiv \gamma \bmod \alpha$. Hence, the automorphism η_2 defined by $\eta_2(u) = u$, $\eta_2(v) = v - h(0)u^i$ is μ_α-equivariant.)

3.3. We come back to the situation of § 1: $V = \mathbf{C}^3$ is the $\mathbf{C}^*$-module with weights $(-1, \beta, \gamma)$ and quotient map $\pi_V : V \to \mathbf{C}^2$, $(x,y,z) \mapsto (x^\beta y, x^\gamma z)$. The zero fiber $\pi^{-1}(0)$ has two components, the (y,z)-plane $\{x = 0\}$ and the x-axis $\{y = z = 0\}$. We have a $\mathbf{C}^*$-equivariant isomorphism

$$\mathbf{C}^* \times \mathbf{C}^2 \xrightarrow{\sim} V \setminus \{x = 0\}, \quad (t,(u,v)) \mapsto (t, t^\beta u, t^\gamma v)$$

In particular, $\mathcal{O}(V \setminus \{x=0\}) = \mathbf{C}[u,v,x,x^{-1}]$ where $u = x^\beta y$, $v = x^\gamma z$, and $\mathcal{O}(V \setminus \{x=0\})^{\mathbf{C}^*} = \mathbf{C}[u,v]$. Define $\tilde{V} := \hat{V} \setminus \{x=0\} \simeq \mathbf{C}^* \times \widehat{\mathbf{C}^2}$. Then $\hat{\dot{V}} = \tilde{V} \setminus \{x\text{-axis}\}$, hence

$$\mathcal{O}(\hat{\dot{V}}) = \mathcal{O}(\tilde{V}) = \mathcal{O}(\hat{V})_x = \mathbf{C}[\![u,v]\!][x,x^{-1}] \subset \mathbf{C}[\![x,y,z]\!][x^{-1}].$$

Clearly, $\mathcal{O}(\hat{\dot{V}})^{\mathbf{C}^*} = \mathcal{O}(\tilde{V})^{\mathbf{C}^*} = \mathbf{C}[\![u,v]\!]$. On the other hand, $\mathcal{O}(\hat{V})$ is the completion of $\mathcal{O}(V)$ in the ideal generated by u and v (see 1.5, Step II), and so $\mathcal{O}(\hat{V}) \subset \mathbf{C}[\![x,y,z]\!]$ in a canonical way. It follows that

$$\mathcal{O}(\hat{V}) = \mathcal{O}(\hat{\dot{V}}) \cap \mathbf{C}[\![x,y,z]\!].$$

Proof of Proposition 2.3: The $\mathbf{C}^* \times \mu_\alpha$-automorphism ρ of $\hat{\dot{V}}$ induces an automorphism ρ^* of $\mathcal{O}(\tilde{V})$ which stabilizes the $\mathbf{C}^*$-invariants $\mathbf{C}[\![u,v]\!]$. We have to find a $\mathbf{C}^* \times \mu_\alpha$-automorphism $\dot{\psi}$ of $\dot{V}$ such that the composition $\dot{\psi}\circ\rho$ extends to an automorphism of $\hat{V}$. The automorphism ρ defines a μ_α-automorphism σ of $\mathbf{C}[\![u,v]\!]$ and satisfies $\rho^*(x) = xe$, where $e \in \mathbf{C}[\![u,v]\!]^*$ is a unit which is fixed under μ_α. (This follows from the fact that the units of $\mathcal{O}(\tilde{V})$ are all of the form $x^i e$ where $i \in \mathbf{Z}$ and $e \in \mathbf{C}[\![u,v]\!]^*$.) Clearly, there is a $\mathbf{C}^* \times \mu_\alpha$-automorphism of $\hat{V}$ which fixes the invariants and sends x to ex: It is given by $x \mapsto ex, y \mapsto e^{-\beta}y, z \mapsto e^{-\gamma}z$. Hence, we may assume that ρ^* fixes x,

$$\rho^*(x) = x, \quad \rho^*(u) = p, \quad \rho^*(v) = q,$$

where $p, q \in \mathbf{C}[\![u,v]\!]$. Such an automorphism ρ extends to a morphism $\hat{\rho} : \hat{V} \to \hat{V}$ if (and only if) the two functions $\rho^*(y)$ and $\rho^*(z)$ belong to $\mathcal{O}(\hat{V})$, i.e.,

$$x^{-\beta}p(x^\beta y, x^\gamma z) \in \mathbf{C}[\![x,y,z]\!], \quad x^{-\gamma}q(x^\beta y, x^\gamma z) \in \mathbf{C}[\![x,y,z]\!]. \qquad (**)$$

(We use the fact mentioned above that $\mathcal{O}(\hat{V}) = \mathcal{O}(\hat{\dot{V}}) \cap \mathbf{C}[\![x,y,z]\!]$). We claim that this condition can always be satisfied by a suitable change of coordinates. In addition, we will see that this extension $\hat{\rho}$ is automatically an isomorphism.

Case 1: $\beta = \gamma$. Here condition (**) is automatically satisfied because p and q belong to the maximal ideal $(u,v) \subset \mathbf{C}[\![u,v]\!]$. In addition, the Jacobian of $\hat{\rho}$ in the origin (with respect to the variables x, y, z) is given by

$$\operatorname{Jac}\hat{\rho}\,(0) = \det\begin{pmatrix} 1 & 0 & 0 \\ \star & \frac{\partial p}{\partial u}|_0 & \frac{\partial p}{\partial v}|_0 \\ \star & \frac{\partial q}{\partial u}|_0 & \frac{\partial q}{\partial v}|_0 \end{pmatrix} = \operatorname{Jac}\rho\,(0) \neq 0.$$

Hence, ρ extends to an automorphism $\hat{\rho}$ of $\hat{V}$.

Case 2: $\beta < \gamma$. Here the first condition in $(**)$ is clearly satisfied, and the second is equivalent to the fact that the power series $q(u,v)$ has the form

$$q(u,v) = vf + u^k h$$

where $f \in \mathbf{C}[u,v]$, $h \in \mathbf{C}[\![u]\!]$ and $k\beta \geq \gamma$. It follows from 3.1 and 3.2 that there is a μ_α-equivariant automorphism η of $\mathbf{C}[u,v]$, stabilizing the homogeneous maximal ideal (u,v), such that the second component $q := \tilde{\rho}(v)$ of the composition $\tilde{\rho} = \eta \circ \rho$ has the required form. This automorphism η extends to a $\mathbf{C}^* \times \mu_\alpha$-automorphism $\tilde{\eta}$ of $\mathbf{C}[u,v,x,x^{-1}]$ by putting $\tilde{\eta}(x) = x$. It is clear now that the corresponding automorphism of $V \setminus \{x=0\}$ leaves the x-axis invariant (since $\tilde{\eta}$ stabilizes the ideal $(y,z) \subset \mathcal{O}(\dot{V})$) and defines an $\mathbf{C}^* \times \mu_\alpha$-equivariant automorphism $\dot{\psi}$ of $\dot{V}$ which has the property that the composition $\dot{\psi} \circ \rho$ satisfies the condition $(**)$, i.e., extends to a morphism $\hat{\rho} : \hat{V} \to \hat{V}$. It remains to see that $\hat{\rho}$ is an isomorphism. Again, one easily calculates the Jacobian and finds

$$\operatorname{Jac} \hat{\rho}(0) = \det \begin{pmatrix} 1 & 0 & 0 \\ \star & \frac{\partial p}{\partial u}|_0 & \star \\ \star & 0 & \frac{\partial q}{\partial v}|_0 \end{pmatrix} = \operatorname{Jac} \rho(0) \neq 0$$

(We use the fact that $q(u,v) = vf + u^k h$ where $k \geq 2$). Hence, $\hat{\rho}$ is an isomorphism. This finishes the proof of Proposition 2.3. Q.E.D.

§ 4. The hyperplane F

We begin with an example which will serve as a guideline for the further considerations in this paragraph.

4.1. Example. Consider the linear action of $\mathbf{C}^*$ on $V = \mathbf{C}^3$ with weights $(-\alpha, \beta, \gamma)$, $\alpha, \beta, \gamma > 0$, $\gcd(\alpha, \beta, \gamma) = 1$. The plane

$$E := \{(1, y, z) \mid y, z \in \mathbf{C}\} \simeq \mathbf{C}^2$$

is stable under the subgroup μ_α of $\mathbf{C}^*$, and the quotient map $\pi_V : V \to V/\!\!/\mathbf{C}^*$ induces an isomorphism $\bar{\pi} : E/\mu_\alpha \xrightarrow{\sim} V/\!\!/\mathbf{C}^*$. Consider the (reduced) fiber product

$$\begin{array}{ccc} V \times_{V/\!\!/\mathbf{C}^*} E & \xrightarrow{\widetilde{\pi_E}} & V \\ \downarrow & & \downarrow \pi_V \\ E & \xrightarrow{\pi_E} & V/\!\!/\mathbf{C}^* \end{array}$$

Claim. *Then the normalization of $V \times_{V/\!\!/\mathbf{C}^*} E$ is $\mathbf{C}^*$-equivariantly isomorphic to the representation $\tilde{V}$ with weights $(-1, \beta, \gamma)$.*

Proof. Consider the finite $\mathbf{C}^*$-equivariant morphism $\zeta : \tilde{V} \to V$ given by $(x, y, z) \mapsto (x^\alpha, y, z)$, and the quotient morphism $\pi_{\tilde{V}} : \tilde{V} \to E$, $(x, y, z) \mapsto (1, x^\beta y, x^\gamma z)$. It is easy to see that the compositions $\pi_V \circ \zeta$ and $\pi_E \circ \pi_{\tilde{V}}$ are equal. Hence, we obtain a finite $\mathbf{C}^*$-morphism $\eta : \tilde{V} \to V \times_{V/\!\!/\mathbf{C}^*} E$ which induces the identity on the quotient:

$$\begin{array}{ccccc}
\tilde{V} & \xrightarrow{\eta} & V \times_{V/\!\!/\mathbf{C}^*} E & \xrightarrow{\widetilde{\pi_E}} & V \\
\downarrow{\scriptstyle \pi_{\tilde{V}}} & & \downarrow & & \downarrow{\scriptstyle \pi_V} \\
E & = & E & \xrightarrow{\pi_E} & V/\!\!/\mathbf{C}^*
\end{array}$$

Since ζ and $\widetilde{\pi_E}$ have both the same degree α the morphism η is birational, and the claim follows. Q.E.D.

We remark that $\mathbf{C}^* \times \mu_\alpha$ acts on $\tilde{V}$, where the action of μ_α is given by $s \cdot (x, y, z) = (s^\alpha x, y, z)$, and that $\zeta = \widetilde{\pi_E} \circ \eta$ is the quotient under μ_α.

4.2. For the rest of the paragraph we consider an affine variety X with a faithful $\mathbf{C}^*$-action which satisfies the following assumptions:

(i) X is irreducible of dimension 3;

(ii) X is smooth and factorial;

(iii) $x_0 \in X$ is an isolated fixed point.

The tangent representation $V := T_{x_0}X$ has weights $(-\alpha, \beta, \gamma)$, $\alpha, \beta, \gamma > 0$, $\gcd(\alpha, \beta, \gamma) = 1$. As before, we denote by $\pi_X : X \to X/\!\!/\mathbf{C}^*$ and $\pi_V : V \to V/\!\!/\mathbf{C}^*$ the quotient maps. As a consequence of the slice theorem we know that the zero fibers $X^0 := \pi_X^{-1}(\pi_X(x_0))$ and $V^0 := \pi_V^{-1}(\pi_V(0))$ are $\mathbf{C}^*$-isomorphic. Clearly, V^0 is the union of the (y, z)-plane and the x-axis. Since X is factorial the two-dimensional component $F_0 (\simeq \mathbf{C}^2)$ of X^0 is defined by an irreducible function $f \in \mathcal{O}(X)$, and this function is a semi-invariant of weight $-\alpha$, i.e., $f(t \cdot w) = t^{-\alpha} f(w)$ for $w \in X$. Define $F := f^{-1}(1) \subset X$.

4.3. Lemma. *F is stable under μ_α and $(\mathbf{C}^* \times F)/\mu_\alpha \xrightarrow{\sim} X \setminus f^{-1}(0)$ where μ_α acts by $s \cdot (t, w) := (ts^{-1}, s \cdot w)$ for $s \in \mu_\alpha$, $t \in \mathbf{C}^*$, $w \in F$. In particular, F is smooth and π_X induces an isomorphism $\bar{\pi} : F/\mu_\alpha \xrightarrow{\sim} X/\!\!/\mathbf{C}^*$.*

Proof. Since $f : X \to \mathbf{C}$ is a semi-invariant of weight $-\alpha$, we see that $X = \mathbf{C}^* F \cup f^{-1}(0)$. Hence, we have a surjection $\mathbf{C}^* \times F \twoheadrightarrow X \setminus f^{-1}(0)$ given by $(t, w) \mapsto t \cdot w$, which induces an isomorphism $(\mathbf{C}^* \times F)/\mu_\alpha \xrightarrow{\sim} X \setminus f^{-1}(0)$. Now the lemma follows easily. Q.E.D.

4.4. Theorem. *Assume that $F \simeq \mathbf{C}^2$. Then the $\mathbf{C}^*$-action on X is linearizable, i.e., X is $\mathbf{C}^*$-equivariantly isomorphic to $T_{x_0}X$.*

Proof. (a) Since every reductive group action on $\mathbf{C}^2$ is linearizable (see [K2, 1.3 Corollary 2]) we can assume that $F = \mathbf{C}^2$ with a linear action of μ_α with weights (β, γ). Consider the diagram

$$\begin{array}{ccccc} \tilde{X} & \xrightarrow{\eta} & X \times_{X/\!\!/\mathbf{C}^*} F & \xrightarrow{\widetilde{\pi_F}} & X \\ \downarrow{\scriptstyle \pi_{\tilde{X}}} & & \downarrow{\scriptstyle \overline{\pi_X}} & & \downarrow{\scriptstyle \pi_X} \\ F & = & F & \xrightarrow{\pi_F} & X/\!\!/\mathbf{C}^* \end{array}$$

where $\pi_F := \pi_X|_F$ is the quotient by μ_α, $X \times_{X/\!\!/\mathbf{C}^*} F$ is the (reduced) fiber product, and $\eta : \tilde{X} \to X \times_{X/\!\!/\mathbf{C}^*} F$ the normalization (cf. Example 4.1). Clearly, $\mathbf{C}^* \times \mu_\alpha$ acts on $\tilde{X}$ such that η is equivariant. The composition $\zeta := \widetilde{\pi_F} \circ \eta$ is the quotient by μ_α, and $\pi_{\tilde{X}}$ is the quotient by $\mathbf{C}^*$. Since $\mathbf{C}^* \times F$ is smooth and $X \setminus f^{-1}(0) \simeq (\mathbf{C}^* \times F)/\mu_\alpha$ we obtain

$$\widetilde{\pi_F}^{-1}(X \setminus f^{-1}(0)) \simeq \mathbf{C}^* \times F \simeq \zeta^{-1}(X \setminus f^{-1}(0)).$$

Hence, $\tilde{X}$ is smooth outside the zero fiber $\tilde{X}^0 := \pi_{\tilde{X}}^{-1}(0)$.

(b) We now claim that $\tilde{X}$ is everywhere smooth. If the $\mathbf{C}^*$-action on X is linearizable this follows from Example 4.1. In general, $\tilde{X}$ is smooth in a neighborhood of the zero fiber $\tilde{X}^0$ by the slice theorem. In fact, setting $\hat{Q} := \operatorname{Spec} \hat{\mathcal{O}}(X/\!\!/\mathbf{C}^*)_{\pi(p)}$ and $\hat{F} := \operatorname{Spec} \hat{\mathcal{O}}(F)_0$ where the hat ^ denotes the completion of the local ring in the maximal ideal, we obtain by base change the following commutative diagram:

$$\begin{array}{ccccc} \hat{\tilde{X}} & \xrightarrow{\hat{\eta}} & \hat{X} \times_{\hat{Q}} \hat{F} & \longrightarrow & \hat{X} \\ \downarrow & & \downarrow & & \downarrow \\ \hat{F} & = & \hat{F} & \longrightarrow & \hat{Q} \end{array}$$

Again, $\hat{\eta}$ is the normalization of the (reduced) fiber product $\hat{X} \times_{\hat{Q}} \hat{F}$. This diagram is isomorphic to the corresponding diagram in the linear situation, hence $\hat{\tilde{X}}$ is isomorphic to $\hat{\tilde{V}}$ with a $\mathbf{C}^* \times \mu_\alpha$-equivariant isomorphism. On the other hand, we have seen in (a) that $\tilde{X} \setminus \tilde{X}^0$ is smooth, hence the claim.

(c) The considerations above also show that the $\mathbf{C}^*$-action on $\tilde{X}$ is semifree with a unique fixed point $\tilde{x_0} \in \pi_{\tilde{X}}^{-1}(0)$. Thus we can apply Theorem 2.4 to obtain a $\mathbf{C}^* \times \mu_\alpha$-equivariant isomorphism $\tilde{\varphi} : \tilde{X} \xrightarrow{\sim} \tilde{V} := T_{\tilde{x_0}}\tilde{X}$. By construction, the composition $\tilde{X} \to X \times_{X/\!\!/\mathbf{C}^*} F \to X$ is the quotient by

μ_α and therefore $X \simeq \tilde{X}/\mu_\alpha \simeq \tilde{V}/\mu_\alpha \simeq V$, where the isomorphisms are all $\mathbf{C}^*$-equivariant. Q.E.D.

4.5. Remark. Assume that X is isomorphic to $\mathbf{C}^3$. Then the semi-invariant $f \in \mathbf{C}[x, y, z]$ defining the two-dimensional component F_0 of the zero fiber in $\mathbf{C}^3$ (see 4.2) has the following properties:

(i) $F_0 = f^{-1}(0) \simeq \mathbf{C}^2$;

(ii) $f : \mathbf{C}^3 \setminus F_0 \to \dot{\mathbf{C}}$ is an (étale) fibration with fiber $F := f^{-1}(1)$.

Question. *Do these conditions imply that F is isomorphic to $\mathbf{C}^2$?*

As we have seen above, a positive answer would imply that any $\mathbf{C}^*$-action on $\mathbf{C}^3$ is linearizable.

§ 5. Reduction to the semifree case

Let X, V, α, β, γ, etc. be as in 4.2. In addition to the conditions (i), (ii) and (iii) of 4.2 we assume that

(iv) X is acyclic.

Here *acyclic* means that X has the $\mathbf{Z}$-homology of a point. Similarly we define *$\mathbf{Z}/p$-acyclic.*

Let d be one of the numbers $\gcd(\alpha, \beta)$, $\gcd(\alpha, \gamma)$, $\gcd(\beta, \gamma)$, and consider the subgroup $\mu_d \subset \mathbf{C}^*$ acting on X.

5.1. Proposition. *If the $\mathbf{C}^*$-action on X/μ_d is linearizable, then the $\mathbf{C}^*$-action on X is linearizable as well.*

This result allows to reduce the linearization problem to the case where $\gcd(\alpha, \beta) = \gcd(\alpha, \gamma) = \gcd(\beta, \gamma) = 1$. This means that we can assume that the action on X is semifree. In fact, if X is acyclic and the action on the tangent space $T_{x_0}X$ semifree, then the action on X is semifree, too. (The fixed point set in X of every cyclic p-group is connected ([Bo, III.4.6]); thus, it is contained in the zero fiber). A first consequence is the following result:

5.2. Corollary. *If $X/\!\!/\mathbf{C}^*$ is isomorphic to $\mathbf{C}^2$ then the $\mathbf{C}^*$-action on X is linearizable.*

Proof. By 5.1 we can assume that the action is semifree. Since the quotient is smooth we have $\alpha = 1$ (see Example 1.2), and so $F \xrightarrow{\sim} F/\mu_\alpha \xrightarrow{\sim} \mathbf{C}^2$. Now the claim follows from Theorem 4.4. Q.E.D.

From this we obtain the theorem mentioned in the introduction.

5.3. Theorem. *Consider a* $\mathbf{C}^*$*-action on* $\mathbf{C}^3$ *and assume that the quotient* $\mathbf{C}^3/\!\!/\mathbf{C}^*$ *is smooth. Then the action is linearizable.*

Proof. If the quotient is zero-dimensional this is an immediate consequence of the slice theorem [Lu1]. For a one-dimensional quotient we can apply the methods developed by KRAFT and SCHWARZ ([KS1, Theorem 0.2(3)], cf. [KS2]).

Finally, the characterization of $\mathbf{C}^2$ due to FUJITA, MIYANISHI and SUGIE shows that a smooth two-dimensional quotient of the form $\mathbf{C}^3/\!\!/\mathbf{C}^*$ is isomorphic to $\mathbf{C}^2$ (see [K2, Theorem 2.2]). If the fixed point set is zero-dimensional then the claim follows from Proposition 5.1. Otherwise we can apply Corollary 6.7. Q.E.D.

5.4. Remark. In [KR2] one can find more general conditions which imply linearization. They are formulated in terms of the type of the singularity of $\pi_X(x_0) \in (\mathbf{C}^3)/\!\!/\mathbf{C}^*$.

For the proof of Proposition 5.1 we need the following two results.

5.5. Proposition. *Let* Z *be an affine smooth factorial* $\mathbf{C}^*$*-variety, and let* $\mu_e \subseteq \mathbf{C}^*$ *be the cyclic subgroup of order* e. *Assume that there is a* $\mathbf{C}^*$*-equivariant isomorphism* $\varphi : Z/\mu_e \xrightarrow{\sim} W$ *where* W *is a* $\mathbf{C}^*$*-module, such that the image of the fixed point set* Z^{μ_e} *in* W *is a linear subspace of codimension 1. Then the* $\mathbf{C}^*$*-action on* Z *is linearizable.*

Proof. By assumption, we have $\mathcal{O}(Z)^{\mu_e} = \mathbf{C}[x_1, x_2, \ldots, x_n]$ where the x_i are algebraically independent semi-invariants. Let $f = 0$ be the equation defining the hypersurface Z^{μ_e}. Then $f^d = 0$ defines the image of Z^{μ_e} in Z/μ_e, and we can arrange that $f^d = x_n$. Hence, $\mathbf{C}[x_1, x_2, \ldots, x_{n-1}, f] = \mathcal{O}(Z)$ and the claim follows. Q.E.D.

5.6. Lemma. *Let* $S \subset V$ *be a closed smooth* $\mathbf{C}^*$*-stable surface. Assume that* $0 \in S$ *and that* $S/\!\!/\mathbf{C}^* \xrightarrow{\sim} \mathbf{C}$. *Choose a* $\mathbf{C}^*$*-equivariant linear projection* $V = T_0V \to T_0S$. *This projection induces an isomorphism* $\varphi : S \xrightarrow{\sim} T_0S$. *In particular, there is a* $\mathbf{C}^*$*-equivariant automorphism of* V *which maps* S *onto a coordinate plane.*

Proof. It follows from the assumptions that $W := T_0S$ has a one-dimensional quotient $W/\!\!/\mathbf{C}^* = \mathbf{C}$, and we get the following diagram:

$$\begin{array}{ccc} S & \xrightarrow{\varphi} & W \\ {\scriptstyle\pi_S}\downarrow & & \downarrow{\scriptstyle\pi_W} \\ S/\!\!/\mathbf{C}^* & \xrightarrow{\bar\varphi} & \mathbf{C} \end{array}$$

Since $S/\!\!/\mathbf{C}^* \simeq \mathbf{C}$ we see that $\bar\varphi$ is a finite (surjective) morphism. On the other hand, φ is étale at 0, hence $\bar\varphi$ is étale at $\pi_S(0)$ ([Lu2, 1.3 Lemme fondamental]; cf. [Sl, § 4]). Furthermore, $\varphi^{-1}(0) = \{0\}$ because φ is $\mathbf{C}^*$-equivariant, and so $\bar\varphi$ is an isomorphism. But this implies that φ is bijective, hence an isomorphism, too.

The second claim is clear: S is the graph of the function $f := \mathrm{pr} \circ \varphi^{-1} : W \to \mathbf{C}$. Q.E.D.

Proof of Proposition 5.1: If d divides $\gcd(\beta, \gamma)$ then X^{μ_d} is equal to F_0, the two-dimensional component of the zero fiber of X (4.2). It follows that the image of X^{μ_d} in the $\mathbf{C}^*$-module X/μ_d is the coordinate plane spanned by the positive weight vectors. Hence, the claim follows from Proposition 5.5.

Now consider the case where d divides $\gcd(\alpha, \beta)$, the remaining one is similar. Here, the image of X^{μ_d} in the $\mathbf{C}^*$-module X/μ_d is a smooth surface S containing the origin. Let p be a prime divisor of d. Then $X^{\mu_d} = X^{\mu_p}$, hence X^{μ_d} is $\mathbf{Z}/p$-acyclic ([Bo, III.4.6]). This implies that $S/\!\!/\mathbf{C}^* \simeq X^{\mu_d}/\!\!/\mathbf{C}^* \simeq \mathbf{C}$ ([KPR, Theorem A]). Hence, we can apply Lemma 5.6, and the claim follows again from Proposition 5.5. Q.E.D.

§ 6. Actions with many fixed points

6.1. Let X be a smooth affine variety with an action of a torus T. Assume that the fixed point set X^T is connected and non-empty. The aim of this paragraph is to compare X with the normal bundle $N := \mathcal{N}(X^T)$ of the fixed point set X^T and to find conditions under which they are isomorphic. Recall that the normal bundle is a T-vector bundle over X^T and that N is trivial, i.e., $N \xrightarrow{\sim} X^T \times W$ where W is a T-module, in case every vector bundle on X^T is trivial ([K2, Corollary 2.1]).

We denote by $p : N \to X^T$ the canonical projection. The zero section of N induces an isomorphism $\sigma_0 : X^T \xrightarrow{\sim} N^T$ and the quotient map $\pi_X : X \to X/\!\!/T$ an isomorphism $X^T \xrightarrow{\sim} \pi_X(X^T)$. We use these maps to identify X^T with N^T and with $\pi_X(X^T)$.

6.2. Proposition. *Assume that there is a retraction $\rho : X \to X^T$ which is T-equivariant. Then there exists a T-equivariant morphism $\varphi : X \to N$ such that $\varphi|_{X^T} = \sigma_0$, and an open neighborhood U of $\pi_X(X^T)$ in $X/\!\!/T$ such that the diagram*

$$\begin{array}{ccccc} X & \supset & \pi^{-1}(U) & \xrightarrow{\varphi} & N \\ \downarrow{\scriptstyle \pi_X} & & \downarrow{\scriptstyle \pi_X} & & \downarrow{\scriptstyle \pi_N} \\ X/\!\!/T & \supset & U & \xrightarrow{\bar{\varphi}} & N/\!\!/T \end{array}$$

is cartesian with étale morphisms φ and $\bar{\varphi}$.

Outline of Proof: Let $\boldsymbol{a} \subset \mathcal{O}(X)$ be the ideal of X^T and put $A := \mathcal{O}(X^T) = \mathcal{O}(X)/\boldsymbol{a}$. Then $\boldsymbol{a}/\boldsymbol{a}^2$ is a projective A-module and $\mathcal{O}(N) \simeq S_A(\boldsymbol{a}/\boldsymbol{a}^2)$, the symmetric A-algebra of $\boldsymbol{a}/\boldsymbol{a}^2$. The retraction ρ defines an inclusion $\rho^* : A \hookrightarrow \mathcal{O}(X)$ such that $\mathcal{O}(X) = A \oplus \boldsymbol{a}$. It is easy to see that there exists a T-stable A-submodule $P \subset \boldsymbol{a}$ such that $P \oplus \boldsymbol{a}^2 = \boldsymbol{a}$. Hence, P is projective and $S_A(P) \simeq \mathcal{O}(N)$. It follows that the canonical homomorphism $S_A(P) \to \mathcal{O}(X)$ gives rise to a T-equivariant morphism $\varphi : X \to N$ extending σ_0. By construction, φ is étale in a neighborhood of X^T. Now the claim is a consequence of the slice theorem (see [Sl, § 4]). Q.E.D.

An action on X is called *fix-pointed* if every closed orbit is a fixed point, or equivalently, if the quotient map π_X induces an isomorphism $X^T \xrightarrow{\sim} X/\!\!/T$. The next two corollaries follow immediately from the proposition above (cf. [BH, 10.3, 10.5] or [K2, 5.5]).

6.3. Corollary. *Assume that the action on X is fix-pointed. Then X is T-equivariantly isomorphic to the normal bundle N.*

6.4. Corollary. *Consider a fix-pointed action of T on $\mathbf{C}^n$. Then $\mathbf{C}^n$ is T-equivariantly isomorphic to $(\mathbf{C}^n)^T \times W$ where W is a T-module. In particular, if $\dim(\mathbf{C}^n)^T \leq 2$ then the action is linearizable.*

Proof. Since every vector bundle on $\mathbf{C}^n$ is trivial by Quillen and Suslin, the same holds for $(\mathbf{C}^n)^T$, because there is a retraction $\rho : \mathbf{C}^n \to (\mathbf{C}^n)^T$. It follows that the normal bundle N is a trivial T-vector bundle (6.1), hence the first claim. The second part is a consequence of the characterization of $\mathbf{C}^2$ due to Fujita, Miyanishi and Sugie which implies that $\mathbf{C}^n/\!\!/T \simeq \mathbf{C}^2$ (cf. 5.3). Q.E.D.

6.5. Remark. It is clear that Proposition 6.2 and its corollaries hold for any reductive group G.

Now assume that the image $\pi(X^T)$ of the fixed point set is of codimension

1 in $X/\!\!/T$. Then $N/\!\!/T$ is a line bundle over $\pi(X^T)$. Hence, if X^T is factorial then $N/\!\!/T \simeq X^T \times \mathbf{C}$. This suggests to make the following assumptions:

6.6. Theorem. *Assume that the following conditions hold:*

(i) *The action on X is semifree;*

(ii) *There is an isomorphism $\mu : X^T \times \mathbf{C} \xrightarrow{\sim} X/\!\!/T$ which sends $(x, 0)$ to $\pi(x)$;*

(iii) *X^T (or X) is factorial and $\mathcal{O}(X^T)^* = \mathbf{C}^*$.*

Then X is T-equivariantly isomorphic to the normal bundle N.

Proof. It follows from (iii) that the line bundle $N/\!\!/T \to \pi(X^T)$ is trivial. Hence, the quotient map can be written as $\pi_N : N \to X^T \times \mathbf{C}$, $x \mapsto (p(x), z)$ where $\pi_N(\lambda x) = (p(x), \lambda^d z)$ for a suitable d and all $\lambda \in \mathbf{C}$. Similarly, we can assume that $\pi_X : X \to X^T \times \mathbf{C}$ satisfies $\pi_X(x) = (x, 0)$ for $x \in X^T$. Let $\hat{X}$ and $\hat{N}$ denote the completions with respect to the ideal defining the "zero fiber" $X^0 := \pi_X^{-1}(X^T)$ and $N^0 := \pi_N^{-1}(X^T)$, respectively. Then, by Proposition 6.2, there is a T-equivariant isomorphism $\varphi : \hat{X} \xrightarrow{\sim} \hat{N}$ which induces an automorphism $\bar{\varphi} : X^T \times \hat{\mathbf{C}} \xrightarrow{\sim} X^T \times \hat{\mathbf{C}}$ of the quotient.

We claim that we can arrange that $\bar{\varphi} = id$. In fact, the automorphism $\bar{\varphi}$ is given by a morphism $\tau : X^T \to \mathbf{C}[\![t]\!]^*$. Since $\mathcal{O}(X^T)^* = \mathbf{C}^*$ there exists a $\tilde{\tau} : X^T \to \mathbf{C}[\![t]\!]^*$ such that $\tilde{\tau}(x)^d = \tau(x)$. Now define $\hat{\varphi}_1 : \hat{X} \to \hat{N}$ by $\hat{\varphi}_1(x) := \tilde{\tau}(p(x))^{-1} \cdot \hat{\varphi}(x)$. Then $\hat{\varphi}_1$ induces the identity on the quotient.

Define $\dot{X} := X \setminus X^0$ and $\dot{N} := N \setminus N^0$. By assumption (i), $\dot{X}$ and $\dot{N}$ are principal T-fiber bundles over $X^T \times \dot{\mathbf{C}}$, hence trivial by assumption (iii). Thus, we get an isomorphism $\dot{\varphi} : \dot{X} \xrightarrow{\sim} \dot{N}$ which induces the identity on the quotient $X^T \times \dot{\mathbf{C}}$.

Consider the automorphism $\rho := \hat{\varphi} \circ \dot{\varphi}^{-1}$ of $\hat{\dot{N}} := \hat{N} \setminus N^0$. It remains to see that every such ρ can be written as a product $\rho = \hat{\rho} \circ \dot{\rho}$ where $\hat{\rho}$ extends to an automorphism of $\hat{N}$ and $\dot{\rho}$ to an automorphism of $\dot{N}$ (cf. 1.6). But this is clear: ρ is given by a morphism $\tilde{\rho} : X^T \times \hat{\dot{\mathbf{C}}} \to T$ (where $\hat{\dot{\mathbf{C}}} := \dot{\mathbf{C}} \cap \hat{\mathbf{C}} = \operatorname{Spec} \mathbf{C}(\!(t)\!)$) which does not depend of X^T, and since $\mathbf{C}(\!(t)\!)^* = \mathbf{C}[\![t]\!] \cdot \mathbf{C}[t, t^{-1}]$ we obtain the required decomposition. Q.E.D.

6.7. Corollary. *Let T act on $\mathbf{C}^n$. Assume that there exists an orbit of codimension ≤ 2 and that* $\dim(\mathbf{C}^n)^T \geq 1$. *Then the action is linearizable.*

Proof. By assumption, we have $\dim \mathbf{C}^n/\!\!/T \leq 2$. Since the fixed point set has dimension ≥ 1 if follows from the slice theorem that the quotient is smooth. Hence, either the action is fix-pointed and therefore linearizable by 6.4, or else $\mathbf{C}^n/\!\!/T \simeq \mathbf{C}^2$ and $(\mathbf{C}^n)^T \simeq \mathbf{C}$ (see 5.3). But this implies that

$\mathbf{C}^n /\!\!/ T \simeq \pi(\mathbf{C}^n) \times \mathbf{C}$ by the theorem of ABHYANKAR-MOH [AM]. It is clear now that all conditions of Theorem 6.6 are satisfied, and the claim follows since the normal bundle N is trivial (see 6.1). Q.E.D.

REFERENCES

[AM] Abhyankar, S.S; Moh, T.-T.: *Embeddings of the line in the plane.* J. Reine Angew. Math. **276** (1975), 149–166

[BH] Bass, H.; Haboush, W.: *Linearizing certain reductive group actions.* Trans. Amer. Math. Soc. **292** (1985), 463–482

[BB1] Białynicki-Birula, A.: *Remarks on the action of an algebraic torus on k^n, I.* Bull. Acad. Polon. Sci. Sér. Sci. Math. **14** (1966), 177–181

[BB2] Białynicki-Birula, A.: *Remarks on the action of an algebraic torus on k^n, II.* Bull. Acad. Polon. Sci. Sér. Sci. Math. **15** (1967), 123–125

[Bo] Borel, A.: *Seminar on transformation groups.* Ann. of Mathematical Studies **46**, Princeton University Press, 1960

[KR1] Koras, M.; Russell, P.: *G_m-actions on $\mathbf{A}^3$.* Canad. Math. Soc. Confer. Proc. **6** (1986), 269–276

[KR2] Koras, M.; Russell, P.: *On linearizing "good" $\mathbf{C}^*$-actions on $\mathbf{C}^3$.* Proceedings of the Conference on "Group Actions and Invariant Theory", Montreal 1988. Canad. Math. Soc. Confer. Proc. **10** (1989), 93–102

[KR3] Koras, M.; Russell, P.: *Codimension 2 torus actions on affine n-space.* Proceedings of the Conference on "Group Actions and Invariant Theory", Montreal 1988. Canad. Math. Soc. Confer. Proc. **10** (1989), 103–110

[K1] Kraft, H.: *Geometrische Methoden in der Invariantentheorie.* Aspekte der Mathematik **D1**, Vieweg-Verlag, Braunschweig 1984

[K2] Kraft, H.: *Algebraic automorphisms of affine space.* In: "Topological methods in algebraic transformation groups," edited by H. Kraft et al. Progress in Math. **80** (1989), 81–105, Birkhäuser Verlag

[K3] Kraft, H.: *G-vector bundles and the linearization problem.* Proceedings of the Conference on "Group Actions and Invariant Theory", Montreal 1988. Canad. Math. Soc. Confer. Proc. **10** (1989), 111–123

[KPR] Kraft, H.; Petrie, T.; Randall, J.: *Quotient varieties.* Adv. in Math. **74** (1989), 145–162

[KP] Kraft, H.; Popov, V. L.: *Semisimple group actions on the three dimensional affine space are linear.* Comment. Math. Helv. **60** (1985), 466–479

[KS1] Kraft, H.; Schwarz, G.: *Linearizing reductive group actions on affine space with one-dimensional quotient.* Proceedings of the Conference on "Group Actions and Invariant Theory", Montreal 1988. Canad. Math. Soc. Confer. Proc. **10** (1989), 125–132

[KS2] Kraft, H.; Schwarz, G.: *Reductive group actions on affine space with one-dimensional quotient.* Preprint 1990 , to appear

[Lu1] Luna, D.: *Slices étales.* Bull. Soc. Math. France, Mémoire **33** (1973), 81–105

[Lu2] Luna, D.: *Adérences d'orbite et invariants.* Invent. Math. **29** (1975), 231–238

[Ma] Magid, A. R.: *Equivariant completions and tensor products.* Proceedings of the Conference on "Group Actions and Invariant Theory", Montreal 1988. Canad. Math. Soc. Confer. Proc. **10** (1989), 133–136

[MP] Masuda, M.; Petrie, T.: *Equivariant Serre conjecture and algebraic actions on* $\mathbf{C}^n$. Preprint (1990)

[Pa] Panyushev, D. I.: *Semisimple automorphism groups of four-dimensional affine space.* Math. USSR-Izv. **23** (1984), 171–183

[Sch] Schwarz, G. W.: *Exotic algebraic group actions.* C. R. Acad. Sci. Paris **309** (1989), 89–94

[Sl] Slodowy, P.: *Der Scheibensatz für algebraische Transformationsgruppen.* In: Algebraic Tranformation Groups and Invariant Theory, DMV Seminar **13** (1989), 89–113, Birkhäuser Verlag

Received April 2, 1990

Hanspeter Kraft
Mathematisches Institut
Universität Basel
Rheinsprung 21
CH-4051 Basel
Switzerland